D01455888

PROGRAMMING THE 80286, 80386, 80486, AND PENTIUM-BASED PERSONAL COMPUTER

BARRY B. BREY
DeVry Institute of Technology

Prentice Hall

Englewood Cliffs, New Jersey   Columbus, Ohio

To my brother, Jack A. Brey

Library of Congress Cataloging-in-Publication Data

Brey, Barry B.
Programming the 80286, 80386, 80486, and pentium-based personal computer / Barry B. Brey
p. cm.
Includes index.
ISBN 0-02-314263-4
1. Intel 80xxx series microprocessors--Programming. 2. Pentium (Microprocessor)--Programming. I. Title
QA76.8.I2674B77 1996
005.265--dc20 95-32795
CIP

Cover photo: Tom Pantages photo/courtesy Intel Corp.
Editor: Charles E. Stewart, Jr.
Production Editor: Mary Irvin
Production Coordination: Custom Editorial Productions, Inc.
Cover Designer: Dennis M. Stredney
Production Manager: Laura Messerly

This book was set in Times Roman and was printed and bound by R.R. Donnelley & Sons Company/Virginia. The cover was printed by Phoenix Color Corp.

Printed in the United States of America

10 9 8 7 6 5 4 3 2 1

ISBN 0-02-314263-4

Prentice-Hall International (UK) Limited, *London*
Prentice-Hall of Australia Pty. Limited, *Sydney*
Prentice-Hall Canada Inc., *Toronto*
Prentice-Hall Hispanoamericana, S. A., *Mexico*
Prentice-Hall of India Private Limited, *New Delhi*
Prentice-Hall of Japan, Inc., *Tokyo*
Simon & Schuster Asia Pte. Ltd., *Singapore*
Editora Prentice-Hall do Brasil, Ltda., *Rio de Janeiro*

PREFACE

This text is written for students in any course of study who require a thorough knowledge of the configuration and programming of the Intel-based personal computer. These courses of study may include computer science, engineering, or engineering technology, to name a few. It is a very practical reference text for anyone interested in the programming, and other aspects, of this important microprocessor family and the personal computer system.

Today, anyone in a field of study that uses computers must understand assembly language programming and the personal computer. This is especially true when high-level languages such as C/C++ are used for software development. Even though C/C++ can perform most machine functions, assembly language software is used to improve the performance of a program and to control hardware aspects of a system. Because machine control is addressed with assembly language, the intimate details of assembly language must be understood. Assembly language is the only choice for video gaming applications and virtual reality, because of the efficiency it provides.

It is helpful if the student has a foundation in number systems. If not, Section 1–3 provides a primer on number systems and may be included as part of this course. There are no other requirements for a course based on this text, except access to a Microsoft MASM assembler package version 5.10 or, even better, version 6.X.

Intel microprocessors have gained wide application and acceptance in many areas of electronics, communications, control systems, and desktop computer systems. The personal computer is often a launching point for system development, and is even available in integrated form as the 80386EX—a PC on a chip.

ORGANIZATION AND COVERAGE

Each chapter of the text begins with a set of objectives that briefly define the contents of the chapter. This is followed by the body of the chapter, which includes many programming applications that illustrate the main topics of the chapter. At the end of each chapter, a summary, which doubles as a study guide, reviews the information presented in the chapter. Finally, questions and problems are provided to promote practice and mental exercise with the concepts presented in the chapter.

This text contains hundreds of complete programming examples using the Microsoft Macro Assembler program, thus providing an opportunity to learn how to program the Intel family of microprocessors and the personal computer. Operation of the programming environment includes the linker, library, macros, DOS function, and BIOS functions. Batch programs and the configuration of the personal computer are also covered.

APPROACH

Because the Intel family of microprocessors is quite diverse, this text initially concentrates on real-mode assembly language programming, which is compatible with all versions of the Intel family of microprocessors and all personal computers based on them. The 80286, 80386, 80486, and Pentium microprocessors are compared and contrasted to the 8086/8088 microprocessor. This entire series of microprocessors is very similar; more advanced versions are easily learned once the basics are understood.

The text also develops an understanding of the CONFIG.SYS file and the AUTOEXEC.BAT file as they apply to DOS and the configuration of the personal computer. An understanding of configuration is fostered through a study of utility programs and batch programs that illustrate the customization of DOS. The configuration of Windows and the structure of its WIN.INI and SYSTEM.INI files are developed and discussed.

In addition to explaining the programming and operation of the microprocessor, this text explains the programming and operation of the numeric coprocessor (8087/80287/80387/80486/7 and Pentium). The numeric coprocessor functions in a system to provide access to floating-point calculations, which are important in applications such as control systems, video graphics, and computer-aided design (CAD). The numeric coprocessor allows a program to perform complex arithmetic operations that would be difficult with normal microprocessor programming.

Advanced programming techniques include learning to access the expanded memory system using VCPI and the EMS functions provided by EMM386.EXE. Access to the extended memory system programming is provided by DPMI and the XMS functions provided by HIMEM.SYS. Protected-mode programming and applications are provided by DPMI. The structure of the device driver is studied and several device driver programs are illustrated. Other advanced programming techniques include the TSR, hot key, and mouse driver.

SELECTED OBJECTIVES

1. Develop control software to control an application interfaced to the 8086/8088, 80286, 80386, 80486, and Pentium microprocessor. Generally, the software developed will function on all versions of the microprocessor. This software includes DOS-based applications.
2. Program using DOS function calls to control the keyboard, video display system, and disk memory in assembly language.
3. Use DOS internal and external commands to control the personal computer system and write batch programs.

4. Use the BIOS functions to control the keyboard, display, and other components of the computer system.
5. Use INT 33H to access the mouse through the mouse driver.
6. Develop software that uses interrupt hooks and hot keys to gain access to terminate-and-stay-resident software.
7. Program the numeric coprocessor to solve complex equations.
8. Explain the differences between the family members and highlight the features of each member.
9. Describe and use real- and protected-mode operation of the 80286, 80386, 80486, and Pentium microprocessors.
10. Use the EMS, XMS, VCPI, and DPMI drivers to access protected-mode operation, expanded memory, and extended memory.
11. Write device drivers using the IOCTL functions provided by DOS.
12. Execute another program using the EXEC function provided by DOS.
13. Invoke assembly language procedures from high-level languages like C/C++, PASCAL, BASIC, and FORTRAN.

CONTENT OVERVIEW

Chapters 1 through 3 introduce the Intel family of microprocessors, with an emphasis on the microprocessor-based personal computer system. These early chapters introduce the microprocessor, its history, its operation, and the methods used to store data in a microprocessor-based system. The structure of the personal computer, its memory map, its I/O map, disk system, and DOS internal and external commands are introduced. Also explored are the DOS configuration files—CONFIG.SYS and AUTOEXEC.BAT—as well as the Windows configuration files—WIN.INI and SYSTEM.INI. The construction and development of batch programs for controlling the personal computer system are explored. Once an understanding of the basic machine is grasped, Chapters 4–8 explain the basic programming model of the microprocessor, real- and protected-mode operation, and how each instruction functions with the Intel family of microprocessors. As instructions are explained, simple applications are presented to illustrate the operation of the instructions and develop basic programming concepts.

Once the basis for programming is developed, Chapters 9, 10, and 11 provide applications using the assembler program. These applications include programs that use DOS and BIOS function calls. Disk files are explained, as well as keyboard and video operation of a personal computer system. The mouse and video systems are explained as well as their use with application software. These chapters provide the tools required to develop virtually any program on a personal computer system.

Chapter 12 presents the operation and construction of TSR (terminate-and-stay-resident) software. This includes the use of the timer and hot keys to access the TSR. A simple TSR that beeps the speaker once per half hour and twice on the hour is developed to illustrate these techniques. A screen saver program and other applications are also developed.

Chapter 13 details the operation and programming of the 8087–Pentium family of arithmetic coprocessors. Few applications today function efficiently without the power of the arithmetic coprocessor. Quadratic equations and a program that displays sinusoidal harmonic content are developed as examples.

Chapter 14 presents advanced programming concepts, including the EXEC function, EMS and XMS control, access to the protected-mode microprocessor, device drivers, and interface to high-level languages.

Appendices enhance the application of the text. These include:

Appendix A: A complete listing of the DOS INT 21H function calls. This appendix details the use of the assembler program and many of the BIOS function calls, including BIOS function call INT 10H. Also included in this appendix are the INT 33H mouse functions, INT 67H EMS functions, XMS functions, and VCPI and DPMI functions.

Appendix B: A complete listing of all 8086 through Pentium instructions, including many example instructions and machine coding in hexadecimal, as well as clock timing information.

Appendix C: A compact listing of all the instructions that change the flag bits.

Appendix D: Answers for the even-numbered questions and problems.

A glossary of the many important terms presented throughout the book is also appended to the text.

APPENDUM

A disk containing many of the more advanced example programs is included with the text. This example disk contains the unassembled source files of many of the example programs. These files can be assembled and examined. Each program is keyed by the example number in the text, for example EXA10-2.ASM. The disk also contains a macro include file that eases the task of assembly language programming by providing macros that perform most of the I/O tasks associated with assembly language programming.

ACKNOWLEGMENTS

Thanks are extended to the following individuals who reviewed the manuscript of this text: Charles F. Solomon, Texas State Technical College; Professor Chandra R. Sekhar, Purdue University of Calumet; and Professor Omer Farook, Purdue University of Calumet.

CONTENTS

8 PROGRAM CONTROL INSTRUCTIONS 261

9 KEYBOARD/DISPLAY DOS AND BIOS FUNCTIONS 293

10 CONDITIONAL ASSEMBLY AND VIDEO DISPLAY 330

11 DISK MEMORY FUNCTIONS AND THE MOUSE 368

12 INTERRUPT HOOKS AND THE TSR 395

CHAPTER 1

Introduction to the Microprocessor and Personal Computer

INTRODUCTION

This chapter provides an overview of the microprocessor. Included are a discussion of the history of computers as well as the function of the microprocessor in the microprocessor-based computer system or personal computer. Also introduced are the terms and jargon of the computer field so that *computerese* may be used and understood when discussing microprocessors and computers.

This chapter describes the block diagram which, along with a description of the function of each block, details the operation of a computer system. The chapter also details the memory and input/output system of the personal computer.

Finally, the way various data are stored in the memory is described, so each data type can be used with software development. Numeric data are stored as integers—either floating-point or binary-coded decimal (BCD)—while alphanumeric data is stored in ASCII (American Standard Code for Information Interchange) code. The steps needed to convert between decimal and any other number base, such as binary, octal, and hexadecimal are also provided.

OBJECTIVES

Upon completion of this chapter, you will be able to:

1. Converse using appropriate computer terminology including: bit, byte, data, real memory system, expanded memory system (EMS), extended memory system (XMS), DOS, BIOS, I/O, and so forth.
2. Briefly describe the history of the computer and list applications performed by the computer.
3. Name the pioneering software developers and describe the programs that they created.
4. Draw a block diagram of a computer system and explain the purpose of each block.
5. Describe the function of the microprocessor and detail its basic operation.

6. Define the contents of the memory system in the personal computer.
7. Convert between decimal, binary, and hexadecimal data.
8. Differentiate between integers, floating-point data, binary-coded decimal data (BCD), and ASCII data, and represent numeric and alphanumeric information as appropriate.

1–1 HISTORICAL BACKGROUND

This first section outlines the historical events leading to the development of the microprocessor, specifically the extremely powerful 80386, 80486, and Pentium microprocessors currently in use. Although a study of history is not essential to understanding the microprocessor, it provides a historical perspective on the fast-paced evolution of the computer.

The Mechanical Age

The idea of a computing system is not new—it has been around since long before modern electrical and electronic devices. The idea of calculating with a machine dates to before 500 B.C. when the Babylonians invented the *abacus*, the first mechanical calculator. The abacus used strings of beads to perform calculations. It was used by the ancient Babylonian priesthood to keep track of their vast storehouses of grain. The abacus, which was used extensively and is still in use today, was not improved until 1642, when Blaise Pascal, the mathematician, invented a calculator constructed of gears and wheels. Each gear contained ten teeth that, when moved one complete revolution, advanced a second gear one place. This is the same principal employed in automobile odometers and is the basis of all mechanical calculators. Incidentally, the PASCAL programming language is named in honor of Blaise Pascal for his pioneering work in mathematics and with the mechanical calculator.

The arrival of the first practical geared, mechanical machines used to automatically compute information dates to the early 1800s. This was before the invention of the light bulb, before much was known about electricity. In this dawn of the computer age, humans dreamed of mechanical machines that could compute numerical facts with a program—not merely calculate facts like a calculator.

One early pioneer of mechanical computing machinery was Charles Babbage, who was aided by Augusta Ada Byron, the Countess of Lovelace. Babbage was commissioned in 1823 by the Royal Astronomical Society of Great Britain to produce a programmable calculating machine. This machine was to generate navigational tables for the Royal Navy. He accepted the challenge and began to create what he called his *analytical engine.* This engine was a mechanical computer that stored 1,000 twenty-digit decimal numbers and a variable program that could modify the machine to perform various calculating tasks. Input to his engine was through punched cards, similar to those used by computers in the 1950s and 1960s. It is assumed that he obtained the idea of using punched cards from Joseph Jacquard, a Frenchman who in 1801 used punched cards as input to a weaving machine that is today called Jacquard's loom. Jacquard's loom used punched cards to select intricate weaving patterns in the cloth that it produced. The punched cards programmed the loom.

After many years of work, Babbage's dream began to fade when he realized that the machinists of his day were unable to create the mechanical parts needed to complete his work. The analytical engine required more than 50,000 machined parts, which could not be made with enough precision to allow his engine to function reliably.

The Electrical Age

The 1800s saw the advent of the electric motor (conceived by Michael Faraday) and with it, a multitude of motor-driven adding machines based on Pascal's mechanical calculator. These electrically-driven mechanical calculators were common pieces of office equipment until well into the early 1970s, when the small hand-held electronic calculator, first introduced by Bomar, appeared. Monroe was also a leading pioneer of electronic calculators, but their desktop, four-function machines were the size of cash registers.

In 1889 Herman Hollerith developed the punched card for storing data. Like Babbage he apparently borrowed the idea of a punched card from Jacquard. He developed a mechanical machine—driven by one of the new electric motors—that counted, sorted, and collated information stored on punched cards. The idea of calculating by machinery intrigued the United States government so much that Hollerith was commissioned to use his punched-card system to store and tabulate information for the 1890 census.

In 1896, Hollerith formed a company called the Tabulating Machine Company. This company developed a line of machines that used punched cards for tabulation. After a number of mergers, the Tabulating Machine Company became the International Business Machines Corporation, now referred to as IBM. We often refer to the punched cards used in computer systems as *Hollerith cards* in honor of Herman Hollerith. The 12-bit code used on a punched card is called the *Hollerith code*.

Mechanical machines driven by electric motors continued to dominate the information processing world until the advent of the first electronic calculating machine in 1942. Konrad Zuse, a German inventor, developed an electronic calculating computer, the Z3, that was used by the Germans for aircraft and missile design during World War II. Had Zuse been given adequate funding by the German government, he most likely would have developed a much more powerful computer system. Today, Zuse is finally receiving some belated honor for his pioneering work in the area of digital electronics in the 1930s.

It has been recently discovered (through the declassification of British military documents) that the first truly electronic computer was placed into operation in 1943 to break secret German military codes. This first electronic computer system, which used vacuum tubes, was invented by Alan Turing. Turing called his machine *Colossus*, most likely because of its size. A problem with Colossus was that its design allowed it to break secret German military codes generated by the mechanical *Enigma* machine, but it could not solve other problems. Colossus was not programmable; it was a fixed-program computer system, which today is often called a **special purpose computer.**

The first general purpose, programmable electronic computer system was developed in 1946 at the University of Pennsylvania. This first modern computer was called the **ENIAC** (*Electronics Numerical Integrator and Calculator*). The ENIAC was a huge machine containing more than 17,000 vacuum tubes and over 500 miles of wires. This massive machine weighed more than 30 tons, yet performed only about 100,000 operations per second. The ENIAC thrust the world into the age of electronic computers. The ENIAC was programmed by rewiring its circuits—a process that took many days to accomplish. Workers changed electrical connections on plug-boards that looked much like early telephone switchboards. Another problem with the ENIAC was the life of the vacuum tube components, which required frequent maintenance.

Breakthroughs that followed were the development of the transistor in 1948 at Bell Labs, followed by the invention of the integrated circuit in 1958 by Jack Kilby of Texas Instruments. The integrated circuit led to the development of digital integrated circuits (RTL or resistor-to-transistor logic) in the 1960s and the first microprocessor in 1971 at Intel Corporation. At this time Intel, and one of its engineers, Marcian T. "Ted" Hoff, developed the 4004 microprocessor—the device that started the microprocessor revolution that continues today at an ever-accelerating pace.

Programming Advances

Once programmable machines had been developed, programs and programming languages began to appear. The first truly programmable electronic computer system was programmed by rewiring its circuits. This proved too cumbersome for practical application, so early in the evolution of computer systems, computer languages began to appear to control the computer. The first such language was *machine language,* which used binary code to store ones and zeros in the memory system in groups called programs. This was more efficient than rewiring a machine to program it, but it was still extremely time-consuming. John von Neumann, a mathematician, was the first to develop a system that accepted instructions and stored them in memory. Computers are often called *von Neumann machines* in honor of John von Neumann.

Once computer systems such as the UNIVAC I and II became available in the early 1950s, **assembly language** was used to simplify the chore of entering binary code into a computer as its instructions. This allowed the programmer to use mnemonic codes, such as ADD for addition, in place of a cryptic binary number, such as 01000111. Assembly language was an aid to programming, but it wasn't until Grace Hopper developed the first high-level programming language called *FLOW-MATIC,* in 1957 that computers became easier to program. Later the same year, IBM developed FORTRAN (*FORmula TRANslator*) for its computer systems. The FORTRAN language allowed programmers to develop programs that used formulas to solve mathematical problems. FORTRAN is still in use by some scientists. A similar language, introduced about a year after FORTRAN, was ALGOL (*ALGOrithmic Language*).

The first truly successful, widespread programming language for business applications was COBAL (*COmputer Business-oriented Algorithmic Language*). Although COBAL use has diminished somewhat in recent years, it is still a major player in many large business systems. Another fairly popular business language is RPG (*Report Program Generator*), which allows programming by specifying the form of the input, output, and calculations.

Since these early days of programming, additional languages have appeared. Some of the more common are BASIC, C/C++, PASCAL, and ADA. The BASIC and PASCAL languages were both designed as teaching languages, but have escaped the classroom and are used in many computer systems. The BASIC language is probably the easiest of all to learn. Some estimates indicate that the BASIC language is used for 80% of the programs written by users for use in personal computers. Recently, a new version of BASIC, called VISUAL BASIC, has appeared to make programming in the WINDOWS environment easier. The VISUAL BASIC language may eventually supplant C/C++ and PASCAL.

In the scientific community, C/C++ and PASCAL are used as control programs. Both languages, especially C/C++, allow the programmer almost complete control over the programming environment and computer system. In many cases, C/C++ is replacing assembly language. Even so, assembly language still plays an important role in programming. Most video games written for the personal computer are written almost exclusively in assembly language. Assembly language is also interspersed with C/C++ and PASCAL to perform machine control functions efficiently.

The ADA language is used heavily by the U.S. Department of Defense. The ADA language was named after Augusta Ada Byron, Countess of Lovelace, who worked with Charles Babbage in the early 1800s.

The Microprocessor Age

The world's first microprocessor, the Intel 4004, was a 4-bit microprocessor—a programmable controller on a chip that is meager by today's standards. It addressed only 4,096 four-bit wide memory locations. (A **bit** is a binary digit with a value of one or zero. A 4-bit wide memory location is often called a **nibble** or *nybble*.) The 4004 instruction set contained only 45 instructions. It was fabricated with then-current P–channel MOSFET technology that only allowed it to execute instructions at a rate of 50 **KIPs** (*kilo-instructions per second*). This was slow when compared to the 100,000 instructions executed per second by the 30-ton ENIAC computer in 1946. But the 4004 weighed much less than an *ounce*.

At first, applications abounded for this device. The 4-bit microprocessor debuted in early video game systems and small microprocessor-based control systems. One such early video game, a shuffleboard game, was produced by Balley. The main problems with this early microprocessor were its speed, word width, and memory size. The evolution of the 4-bit microprocessor ended when Intel released the 4040, an updated version of the earlier 4004. The 4040 operated at a higher speed, although it lacked improvements in word width and memory size. Other companies, in particular Texas Instruments (TMS–1000), also produced 4-bit microprocessors. The 4-bit microprocessor still survives in low-end applications such as microwave ovens and small control systems, and is still available from some microprocessor manufacturers. Most calculators are also still based on 4-bit microprocessors that process 4-bit **BCD** (*binary-coded decimal*) codes.

In 1971, realizing that the microprocessor was a commercially viable product, Intel Corporation released the 8008—an extended 8-bit version of the 4004 microprocessor. The 8008 addressed an expanded memory size (16K bytes) and contained additional instructions (a total of 48) that provided an opportunity for its application in more advanced systems. (A **byte** is generally an 8-bit wide binary number and a **K** is 1,024 bytes. Often, memory size is specified in K bytes.)

As engineers developed more demanding uses for the 8008 microprocessor, they discovered that its somewhat small memory size, slow speed, and instruction set limited its usefulness. Intel recognized these limitations and, in 1973, introduced the 8080 microprocessor—the first of the modern 8-bit microprocessors. About 6 months after Intel released the 8080 microprocessor, Motorola introduced its MC6800 microprocessor. The flood gates opened: the 8080 and, to a lesser degree, the MC6800 ushered in the age of the microprocessor. Soon, other companies began to introduce their own versions of the 8-bit microprocessor. Table 1–1 lists several of these early microprocessors and their manufacturers. Of these early microprocessor producers, only Intel and Motorola continue to successfully manufacture newer and improved versions of the microprocessor. Zilog still manufactures microprocessors, but has decided, fairly successfully, to concentrate on microcontrollers and embedded controllers rather than general-purpose microprocessors. Rockwell has all but abandoned microprocessor development in favor of modem circuitry.

What Was Special About the 8080? Not only could the 8080 address more memory and execute additional instructions, but it executed them ten times faster than the 8008. An addition that took 20 μs (50,000 instructions per second) on an 8008-based system, required only 2.0 μs (500,000 instructions per second) on an 8080-based system. Also, the 8080 was directly compatible with TTL (transistor-transistor logic); the 8008 was not. This made interfacing much easier and less expensive. The 8080 addressed four times more memory than the 8008 (64K bytes vs. 16K bytes). These improvements are responsible for ushering in the era of the 8080, and the continuing saga of the microprocessor.

The first personal computer, the MITS Altair 8800, was released in 1974. (The model name 8800 was probably chosen to avoid copyright violations with Intel). The BASIC language interpreter written for the Altair 8800 computer was developed by Bill Gates, who was later to establish the Microsoft Corporation. The assembler program for the Altair 8800 was written by Digital Research Corporation, which now produces DR–DOS for the personal computer.

The 8085 Microprocessor. In 1977, Intel Corporation introduced an updated version of the 8080—the 8085. This was to be the last 8-bit general-purpose microprocessor developed by Intel. Although only slightly more advanced than an 8080, the 8085 executes software at an even higher speed. An addition that took 2.0 μs (500,000 instructions per second) on the 8080, requires only 1.3 μs (769,230 instructions per second) on the 8085. The main advantages of the 8085 are its internal clock generator, internal system controller, and higher clock frequency. This higher level of component integration reduced the cost and increased the usefulness of the 8085 microprocessor. Intel has sold more than 100 million 8085 microprocessors, its most successful 8-bit microprocessor. Because the 8085 is also manufactured by many other companies (*second-sourced*), there are well over 200

TABLE 1–1 Early 8-bit microprocessors

Manufacturer	*Part Number*
Fairchild	F-8
Intel	8080
MOS Technology	6502
Motorola	MC6800
National Semiconductor	IMP-8
Rockwell International	PPS-8
Zilog	Z-8

million of these microprocessors in existence. Applications that contain the 8085 are still being used and designed and will likely continue to be popular well into the future. Another company, Zilog Corporation, sold more than 500 million 8-bit microprocessors. Their Z-80 uses a language code compatible with the 8085, which means that there are well over 700 million microprocessors that execute 8085/Z-80 compatible code!

The Modern Microprocessor

Intel released the 8086 microprocessor in 1978 and the 8088 a year or so later. Both devices are 16-bit microprocessors, which execute instructions in as little as 400 ns (2.5 MIPs or 2.5 millions of instructions per second). This represented a major improvement over the execution speed of the 8085. In addition, the 8086 and 8088 address 1M byte of memory, 16 times more memory than the 8085. (A **1M byte memory** contains 1,024K byte-sized memory locations, or 1,048,576 bytes.) This higher execution speed and larger memory size allowed the 8086 and 8088 to replace smaller minicomputers in many applications. One other feature found in the 8086/8088 is an instruction cache or queue that prefetches a few instructions before they are executed. The queue speeds the operation of many sequences of instructions and proved to be the basis for the much larger instruction caches found in modern microprocessors.

The increase in memory size and additional instruction capacity of the 8086 and 8088 have led to many sophisticated applications for microprocessors. Improvements to the instruction set included a multiply and divide instruction, which were missing on earlier microprocessors. Also, the number of instructions increased from 45 on the 4004, to 246 on the 8085, to well over 20,000 on the 8086 and 8088 microprocessors. Note that these microprocessors are called **CISC** (*complex instruction set computers*) because of the number and complexity of instructions. The additional instructions ease the task of developing efficient and sophisticated applications, even though their number is at first overwhelming and time-consuming to learn. The 16-bit microprocessor also provides more internal register storage space than the 8-bit microprocessor. The additional registers allow software to be written more efficiently.

The 16-bit microprocessor evolved mainly because of the need for larger memory systems. The popularity of the Intel family was ensured when, in 1981, IBM decided to use the 8088 microprocessor in its personal computer. Applications such as spreadsheets, word processors, spelling checkers, and computer-based thesauruses are memory intensive and require more than the 64K bytes of memory found in 8-bit microprocessors. The 16-bit 8086 and 8088 microprocessors provide 1M byte of memory for these applications. Soon, even 1M byte of memory proved limiting for large spread sheets and other applications. This led to the introduction of the 80286 microprocessor, an updated 8086, by Intel in 1983.

The 80286 Microprocessor. The 80286 microprocessor (also a 16-bit microprocessor) is almost identical to the 8086 and 8088, except it addresses a 16M byte memory system instead of 1M byte. The instruction set of the 80286 is also almost identical to the 8086 and 8088, except for a few additional instructions that manage the extra 15M bytes of memory. The clock speed of the 80286 is increased so it executes some instructions in as little as 250 ns (4.0 MIPs) with the original release 8.0 MHz version. Some changes also occurred in the internal execution of the instructions that led to an eight-fold increase in speed for many instructions when compared to 8086/8088 instructions.

The 32-bit Microprocessor. Applications began to demand faster microprocessor speeds, more memory, and wider data paths. This led to Intel's introduction, in 1986, of the 80386. The 80386 represented a major overhaul of the 16-bit 8086–80286 microprocessor's architecture. The 80386 is Intel's first microprocessor to contain a 32-bit data bus and a 32-bit memory address. (Note that Intel produced an earlier, unsuccessful 32-bit microprocessor called the iapx-432). Through these 32-bit buses, the 80386 addresses up to 4G bytes of memory. (*1 G* of memory contains 1,024M or 1,073,741,824 locations.) 4G bytes of memory can store an astounding 1,000,000 typewritten, double-spaced pages of data. The 80386 is also available in a few modified versions such as the 80386SX, which addresses 16M bytes of memory through a 16-bit data and 24-bit address bus, and the 80386SL/80386SLC, which address 32M byte of memory through a 16-bit data and 25-bit address bus. An 80386SLC version contains an internal cache memory that allows it to process data at even higher rates. Recently Intel has released the 80386EX microprocessor. The 80386EX is called an *embedded PC* because it contains all the components of the AT class personal computer on a single integrated circuit. The 80386EX also contains 24 lines for input/output data, a 26-bit address bus, a 16-bit data bus, a DRAM refresh controller, and programmable chip selection logic.

Applications that require higher microprocessor speeds and large memory systems include software systems that use a GUI or **graphical user interface.** Modern graphical displays often contain 256,000 or more picture elements (**pixels** or **pels**). The least-sophisticated **VGA** (*variable graphics array*) video display has a resolution of 640 pixels per scanning line with 480 scanning lines. In order to display one screen of information, each picture element must be changed. This requires a high-speed microprocessor. Many new software packages use this type of video interface. These GUI-based packages require high microprocessor speeds and, often, accelerated video adapters for quick and efficient manipulation of video text and graphical data. The most striking system, which requires high-speed computing for its graphical display interface, is Microsoft Corporation's Windows.[1] We often call a GUI a **WYSIWYG** (what you see is what you get) display.

The 32-bit microprocessor is needed because of the size of its data bus, which transfers real (single-precision floating-point) numbers that require 32-bit wide memory. In order to efficiently process 32-bit real numbers, the microprocessor must efficiently pass them between itself and memory. It takes four read-or-write cycles for them to pass throught an 8-bit data bus, but only one read-or-write cycle for them to pass through a 32-bit data bus. This significantly increases the speed of any program that manipulates real numbers. Most high-level languages, spread sheets, and data base management systems use real numbers for data storage. Real-numbers are also used in graphical design packages that use vectors to plot images on the video screen. These include such **CAD** (*computer aided drafting/design*) systems as AUTOCAD, ORCAD, and so forth.

Besides providing higher clocking speeds, the 80386 includes a memory management unit that allows memory resources to be allocated and managed by the operating system. Earlier microprocessors left memory management completely to the software. The 80386 includes hardware circuitry for memory management and memory assignment, which improves its efficiency and reduces software overhead.

---

[1]Windows is a registered trademark of Microsoft Corporation.

The instruction set of the 80386 microprocessor is compatible with the earlier 8086, 8088, and 80286 microprocessors. Additional instructions reference the 32-bit registers and manage the memory system. Note that memory management instructions and techniques used by the 80286 are also compatible with the 80386 microprocessor. These features allow older, 16-bit software to operate on the 80386 microprocessor.

The 80486 Microprocessor. In 1989, Intel released the 80486 microprocessor, which incorporates an 80386-like microprocessor, an 80387-like numeric coprocessor, and an 8K byte cache memory system into one integrated package. Although the 80486 is not radically different from the 80386, it does include one substantial change: the internal structure of the 80486 is modified so that about half of its instructions execute in one clock instead of two clocks. Because the 80486 is available in a 50 MHz version, about half the instructions execute in 25ns (50 MIPs). The average speed improvement for a typical mix of instructions is about 50 percent over an 80386 operated at the same clock speed. Newer versions of the 80486 execute instructions at even higher speeds with a 66 MHz, double-clocked version. The double-clocked 66 MHz version executes instructions at the rate of 66 MHz with memory transfers executed at the rate of 33 MHz. A triple-clocked version from Intel, the 80486DX4, improves the internal execution speed to 100 MHz, with memory transfers at 33 MHz. Note that the 80486DX4 performs integer operations at nearly the same speed as the 60 MHz Pentium. It also contains an expanded 16K byte cache in place of the 8K byte cache that is standard on 80486 microprocessors. The future promises to bring microprocessors that internally execute instructions at rates of up to 250 MHz or higher.

Other versions of the 80486 are called overdrive[2] processors. The overdrive processor is actually a double-clocked version of an 80486DX that replaces an 80486SX or slower speed 80486DX. When the overdrive processor is plugged into its socket, it disables or replaces the 80486SX or 80486DX and functions as a doubled-clocked version of the microprocessor. For example, if an 80486SX operating at 25 MHz is replaced with an overdrive microprocessor, it functions as a 80486DX2 operating at 50MHz, using a memory transfer rate of 25 MHz.

Table 1–2 lists many microprocessors produced by Intel and Motorola with information about their word and memory sizes. Other companies produce microprocessors, but none have attained the success of Intel or Motorola.

The Pentium Microprocessor. The Pentium, in 1993, is similar to the 80386 and 80486 microprocessor. This microprocessor was originally labeled the P5 or 80586, but Intel decided not to use a number because it appears to be impossible to copyright a number. The two introductory versions of the Pentium operate with a clocking frequency of 60 MHz and 66 MHz and a speed of 110 MIPs, with a higher frequency (100 MHz) version, operating at 150 MIPs. A Double-clocked Pentium, operating at 120 MHz will soon be available. The Pentium's cache size is increased to 16K bytes from the 8K cache found in the basic version of the 80486, (an 8K byte instruction cache and an 8K byte data cache). This allows a program that transfers a large amount of memory data to benefit from a cache. The memory system contains up to 4G bytes, with the data bus width increased from the 32 bits

[2]Overdrive is a registered trademark of Intel Corporation.

TABLE 1–2 Many modern microprocessors

Manufacturer	*Part*	*Data Bus Width*	*Memory Size*
Intel	8048	8	2K internal
	8051	8	8K internal
	8085A	8	64K
	8086	16	1M
	8088	8	1M
	8096	16	8K internal
	80186	16	1M
	80188	8	1M
	80286	16	16M
	80386EX	16	64M
	80386DX	32	4G
	80386SL	16	32M
	80386SLC	16	32M + 8K cache
	80386SX	16	16M
	80486DX/DX2	32	4G + 8K cache
	80486SX	32	4G + 8K cache
	80486DX4	32	4G + 16K cache
	Pentium	64	4G + 16K cache
Motorola	6800	8	64K
	6805	8	2K
	6809	8	64K
	68000	16	16M
	68008Q	8	1M
	68008D	8	4M
	68010	16	16M
	68020	32	4G
	68030	32	4G + 256 cache
	68040	32	4G + 8K cache
	68050	32	proposed, but never released
	68060	64	4G + 16K cache
	PowerPC	64	4G + 32K cache

found in the 80386 and 80486 to a full 64 bits. The data bus transfer speed is either 60 MHz or 66 MHz, depending on the version of the Pentium. This wider data bus width accommodates double-precision floating-point numbers used for high-speed vector-generated graphical displays. It also transfers data between the memory system and microprocessor at a higher rate. This should allow *virtual reality*—software that operates at realistic rates—on current and future Pentium-based platforms. The widened data bus and higher execution speed of the Pentium should also allow full-frame video displays that operate at scan rates of 30 Hz or higher—comparable to commercial television.

Probably the most ingenious feature of the Pentium is its dual integer processors. The Pentium executes two instructions not dependent on each other simultaneously, because it contains two independent internal integer processors called *superscaler tech-*

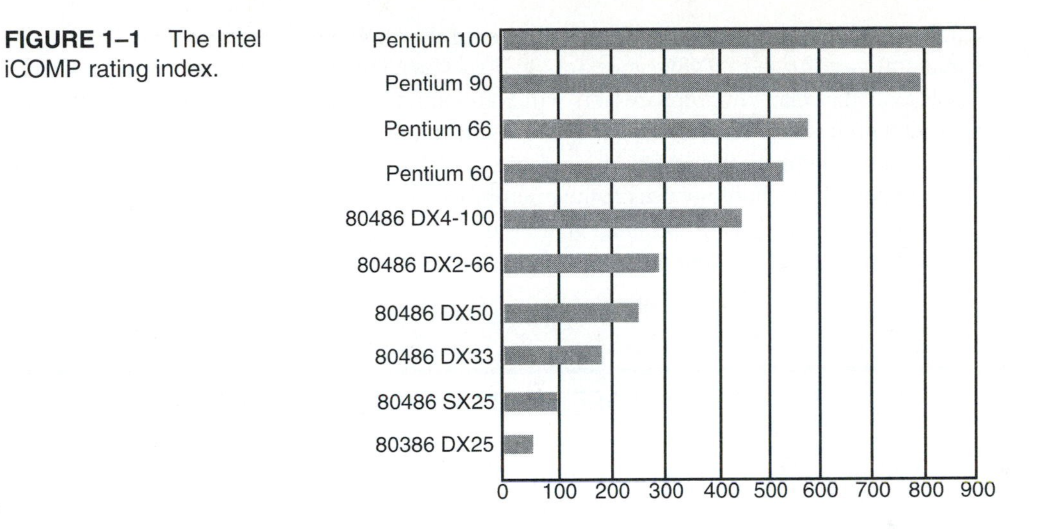

FIGURE 1–1 The Intel iCOMP rating index.

nology. This allows the Pentium to execute two instructions per clocking period. Another feature that enhances performance is a jump prediction technology that speeds the execution of programs that include loops. Like the 80486, the Pentium employs an internal floating-point coprocessor to handle floating-point data, albeit at about a five times speed improvement. These features portend continued success for the Intel family of microprocessors. They also may allow the Pentium to replace some of the **RISC** (*reduced instruction set computer*) machines that currently execute one instruction per clock. Note that some newer RISC processors execute more than one instruction per clock through the introduction of superscaler technology. Motorola, Apple, and IBM have recently produced the PowerPC, a RISC microprocessor that has two integer units and one floating-point unit. The PowerPC certainly boosts the performance of the Apple Macintosh, but is slow at emulating the Intel family of microprocessors. Tests indicate that the PowerPC executes DOS and WINDOWS applications at about the speed of the slower 80486SX 25 MHz microprocessor. Note that currently 4 million personal computers use Apple Macintosh[3] systems, compared to more than 160 million personal computers using Intel microprocessors.

In order to compare the speeds of various microprocessors, Intel devised the iCOMP rating index. This index is a composite of SPEC92, ZD Bench, and Power Meter. Figure 1–1 shows the relative speeds of the more commonly used microprocessors.

No one can really make accurate predictions, but the success of the Intel family should continue for quite a few years. It is rumored that Intel will soon bring out a P6 microprocessor containing more than six million transistors, four integer units, and a floating-point unit. The basic clock frequency will be 150 MHz. In addition to the internal 16K level one (L1) cache, the P6 will contain a 256K level two (L2) cache. This will obviously increase the performance of most software.

While there may be a change to RISC technology, it is more likely that a new technology being jointly developed by Intel and Hewlett-Packard, called hypercube tech-

[3]Macintosh is a trademark of Apple Computer Corporation.

nology, will embody the CISC instruction set of the 80X86 family of microprocessors, so that software for the 80X86 processors will survive. The basic premise behind this technology is that many microprocessors will communicate directly with each other, allowing parallel processing without any change to the instruction set or program. Currently, the superscaler technology uses many microprocessors, but they all share the same register set. The hypercube technology uses many microprocessors, each containing their own register sets, that are linked with the other microprocessors. Hypercube technology should offer parallel processing without the need for any special program.

1–2 THE MICROPROCESSOR-BASED PERSONAL COMPUTER SYSTEM

Computer systems have undergone many changes in recent history. Machines which once filled large areas have been reduced to small desk-top computer systems because of the microprocessor. These desk-top computers possess power that was only dreamed of a few years ago. Million-dollar mainframe computer systems developed in the early 1980s are not as powerful as the 80386, 80486, or Pentium microprocessor-based computers of today. In fact, many smaller companies are replacing their mainframe computers with microprocessor-based systems. Companies like DEC (Digital Equipment Corporation), have stopped producing mainframe computer systems in order to concentrate on microprocessor-based computer systems.

This section examines the structure of microprocessor-based personal computers, including information about the memory and operating systems used in many microprocessor-based computer systems.

Refer to Figure 1–2 for a block diagram of the personal computer. This diagram applies to any computer system from the early mainframe computers to the latest micro-

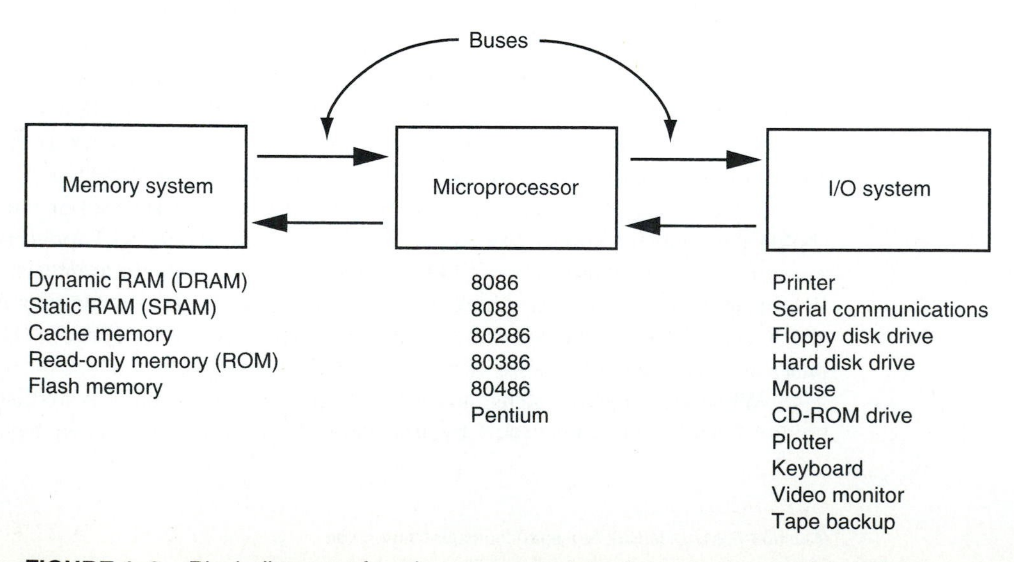

FIGURE 1–2 Block diagram of a microprocessor-based personal computer system.

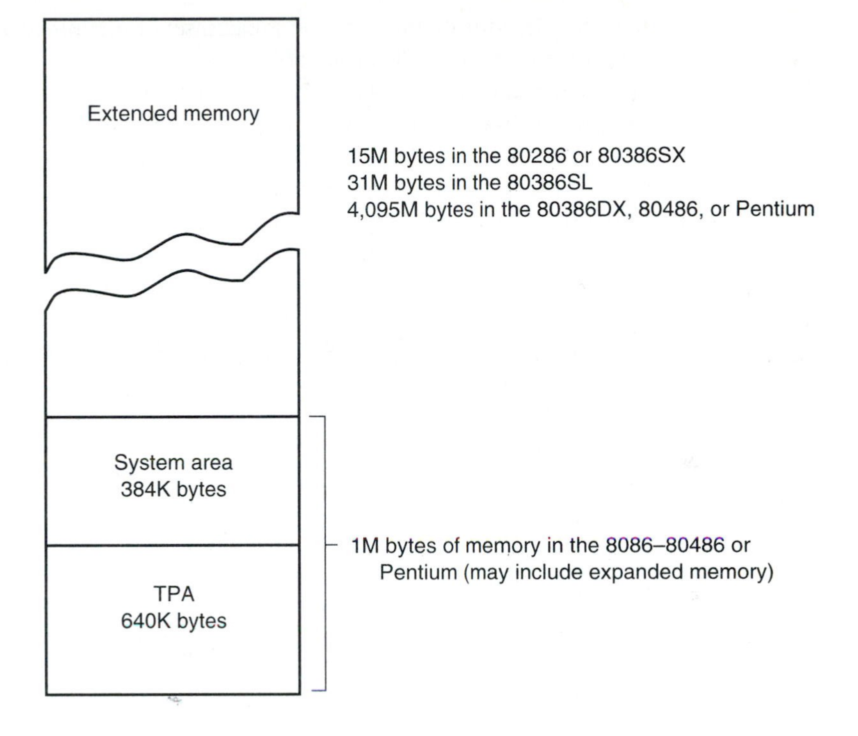

FIGURE 1–3 Memory map of a personal computer system.

processor-based systems. The block diagram is composed of three blocks that are interconnected by buses. (A **bus** is a set of common connections that carry the same type of information. For example, the address bus, which contains 20 to 32 connections, is a bus that conveys the memory address to the memory.) These blocks and their function in a personal computer are outlined in this section of the text.

The Memory and I/O System

The memory structures of all Intel 8086–80486 and Pentium-based personal computer systems are similar. This includes everything from the first personal computers introduced by IBM in 1981 to the most powerful, high-speed versions available today, which are based on the Pentium microprocessor. Figure 1–3 illustrates the memory map of a personal computer system. This map applies to any IBM personal computer or any of the many IBM compatible clones in existence.

The memory system is divided into three main parts: **TPA** (*transient program area*), **system area,** and **XMS** (*extended memory system*). The type of microprocessor in your computer determines whether an extended memory system exists. If the computer is based upon an older 8086 or 8088 (a PC[4] or XT[5]), the TPA and system areas exist, but there is no extended memory area. The PC and XT contain 640K bytes of TPA and 384K bytes of system memory, for a total memory size of 1M bytes. We often call the first 1M byte of

[4]PC is a trademark of IBM Corporation for a personal computer.

[5]XT is a trademark of IBM Corporation for the extended technology personal computer.

memory the **real memory** because each Intel microprocessor is designed to function in this area in its real mode of operation.

Computer systems based on the 80286 through the Pentium not only contain the TPA (640K bytes) and system area (384K bytes), they may also contain extended memory. These machines are often called AT[6] class machines. The PS/1 and PS/2, produced by IBM, are other versions of the same basic memory design. Sometimes these machines are also referred to as **ISA** (*industry standard architecture*) or **EISA** (*extended ISA*) machines. The PS/2 is referred to as a *micro-channel*[7] *architecture* system or ISA system, dependent on the model number. Recently, a new bus, the **PCI** (*peripheral control interconnect*) bus, has begun to appear in Pentium-based systems. Extended memory contains up to 15M bytes in the 80286- and 80386SX-based computer and up to 4,095M bytes in the 80386DX-, 80486-, and Pentium-based computer. This is in addition to the first 1M byte of real memory. The ISA machine contains an 8-bit peripheral bus used to interface 8-bit devices to the computer in the 8086/8088-based PC or XT computer system. The AT class machine, also an ISA machine, uses a 16-bit peripheral bus for interface and may contain an 80286–80486 or Pentium microprocessor. The EISA bus is a 32-bit peripheral interface bus found in 80386DX, 80486, or Pentium-based systems. Note that each of these buses are compatible with the earlier versions. That is, an 8-bit interface card functions in the 16-bit ISA or 32-bit EISA bus standards; likewise, a 16-bit interface card functions in the 16-bit ISA or 32-bit EISA standard. Another recent bus type is called the **VESA**[8] **local bus** or *VL bus.* The local bus interfaces disk and video to the microprocessor at the local bus level. This allows 32-bit interfaces to function at the same clock speed as the microprocessor. A recent modification to the VESA local bus supports the 64-bit data bus of the Pentium microprocessor and competes directly with the PCI bus. The ISA and EISA standards function at only 8 MHz, which reduces the performance of disk and video interfaces using these standards. The PCI bus is either a 32- or 64-bit bus that is specifically designed to function with the Pentium microprocessor at 60 or 66MHz.

The TPA. The transient program area (TPA) holds the operating system and other programs that control the computer system. The TPA also stores any currently active or inactive application program. The length of the TPA is 640K bytes. As mentioned, this area of memory holds the operating system, which requires a portion of the TPA. In practice, the amount of memory remaining for application software is about 628K bytes if MSDOS[9] version 5.0 or 6.X is used as an operating system. Earlier versions of MSDOS required more of the TPA and often left 530K bytes or less for application programs. Another operating system found in personal computers is PCDOS.[10] PCDOS and MSDOS are compatible, so they function in the same manner with application programs. Windows and OS/2[11] are other operating systems that are compatible with DOS and allow DOS programs to

---

[6]AT is a trademark of IBM Corporation used to denote an advanced class computer system.

[7]Micro-channel is a registered trademark of IBM Corporation.

[8]VESA is the Video Electronics Standards Association.

[9]MSDOS (Microsoft Disk Operating System) is a trademark of Microsoft Corporation.

[10]PCDOS (Personal Computer Disk Operating System) is a trademark of IBM Corporation.

[11]OS/2 (operating system version 2) is a trademark of IBM Corporation.

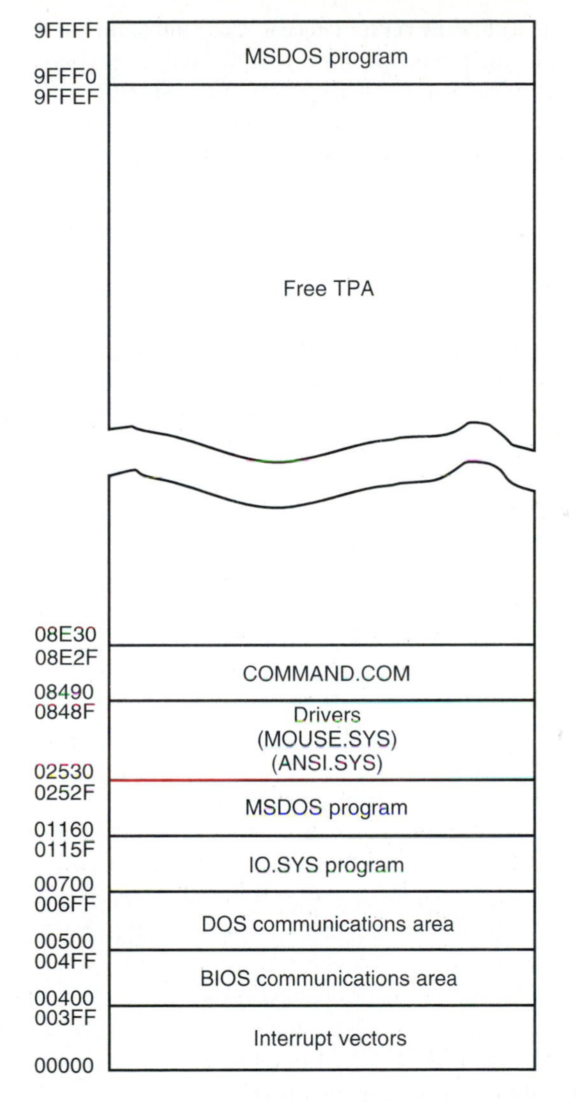

FIGURE 1–4 Memory map of the DOS TPA. This map varies from one version of DOS to another, and also changes with different drivers and configurations.

execute. The **DOS** (*disk operating system*) controls both the way the disk memory is organized and the way some of the I/O devices connected to the system are controlled. Figure 1–4 shows the organization of the TPA in a computer system.

The memory map shows how the many areas of the TPA are used for system programs, data, and drivers. It also shows a large area of memory available for application programs. To the left of each area is a hexadecimal number that represents the memory addresses that begin and end each data area. Hexadecimal **memory addresses** or *memory locations* are used to number each byte of the memory system. (A **hexadecimal** number is a number represented in radix 16 or base 16 with each digit representing a value from 0–9 and A–F. We often write a hexadecimal number with an H at the end to indicate that it is a hexadecimal value. For example, 1234H is 1234 hexadecimal.)

The **interrupt vectors** access various features of the DOS, BIOS (basic I/O system), and applications. The **BIOS** is a collection of programs stored in a read-only memory (ROM) or flash memory that operate many of the I/O devices connected to your computer system. Note that a flash memory is an EEPROM (electrically erasable read-only memory) that is erased in the system electrically, while a ROM must be programmed at the factory for an EPROM (erasable/programmable read-only memory). These programs are stored in the system area, defined later in this section of the chapter.

The BIOS and DOS communications areas contain transient data used by programs to access I/O devices and the internal features of the computer system. (Refer to Appendix A for a complete listing of the BIOS and DOS communications area.) These are stored in the TPA, so they can be changed as the system operates. The TPA contains read/write memory (called **RAM** or *random access memory*), so it can change as a program executes.

The IO.SYS is a program that loads into the TPA from the disk whenever a MSDOS or PCDOS system is started. The IO.SYS contains programs that allow DOS to use the keyboard, video display, printer, and other I/O devices often found in the computer system. The IO.SYS program links DOS to the programs stored on the BIOS ROM.

The MSDOS (PCDOS) program occupies two areas of memory. One area is 16 bytes in length and is located at the top of the TPA. The other is much larger and is located near the bottom of the TPA. The MSDOS program controls the operation of the computer system. The size of the MSDOS area depends on the version of MSDOS installed in the computer memory and how the MSDOS program is installed. If MSDOS is installed in **high memory** with the HIMEM.SYS driver, most of the TPA is free to hold application programs. Note that high memory is described later in this text and only applies to 80286 or newer microprocessors.

The size of the **driver** area and number of drivers change from one computer to another. Drivers are programs that control installable I/O devices such as a mouse, disk cache, hand scanner, CDROM memory (Compact Disk Read-Only Memory), or other installable devices and programs. *Installable drivers* are programs that control or drive devices or programs that are added to the computer system. Drivers are normally files that have an extension of .SYS (such as MOUSE.SYS), or, in DOS version 3.2 and later, the extension .EXE (such as EMM386.EXE). Because few computer systems are identical, the driver area varies in size and contains different numbers and types of drivers.

The COMMAND.COM program—command processor—controls the operation of the computer from the keyboard. The COMMAND.COM program processes the DOS commands as they are typed on the keyboard. For example, if DIR is typed, the COMMAND.COM program displays a directory of the disk files in the current disk directory. If the COMMAND.COM program is erased, the computer cannot be used from the keyboard. Never erase COMMAND.COM, IO.SYS or MSDOS.SYS to make room for other software or your computer will not function. Note that these programs can be reloaded to the disk if erased with the SYS.COM program located in the DOS directory.

The free TPA area holds application programs as they are executed. These application programs include word processors, spreadsheet programs, CAD programs, and so forth. The TPA also holds *TSR* (terminate and stay resident) programs that remain in memory in an inactive state until activated by a hot-key sequence or other event such as an interrupt. A calculator program is an example TSR that activates whenever an ALT C key (hot-key) is typed. A *hot-key* is a combination of keys on the keyboard that activate a TSR program. A TSR program is often also called a **pop-up program**, because, when activated, it pops up inside another program.

The System Area. The system area, although smaller than the TPA, is just as important. The system area contains programs on a read-only memory (ROM) or flash memory and also areas of read/write (RAM) memory for data storage. Figure 1–5 shows the system area of a typical computer system. As with the map of the TPA, this map also includes the hexadecimal memory addresses of the various areas.

The first area of the system space contains video display RAM and video control programs on ROM or flash memory. This area generally starts at location A0000H and extends to location C7FFFH. The size and amount of memory used depends on the type of video display adapter attached to the system. Display adapters that are often attached to a computer include the earlier **CGA** (*color graphics adaptor*) and **EGA** (*enhanced graphics adapter*) or one of the many newer forms of **VGA** (*variable graphics array*). Generally,

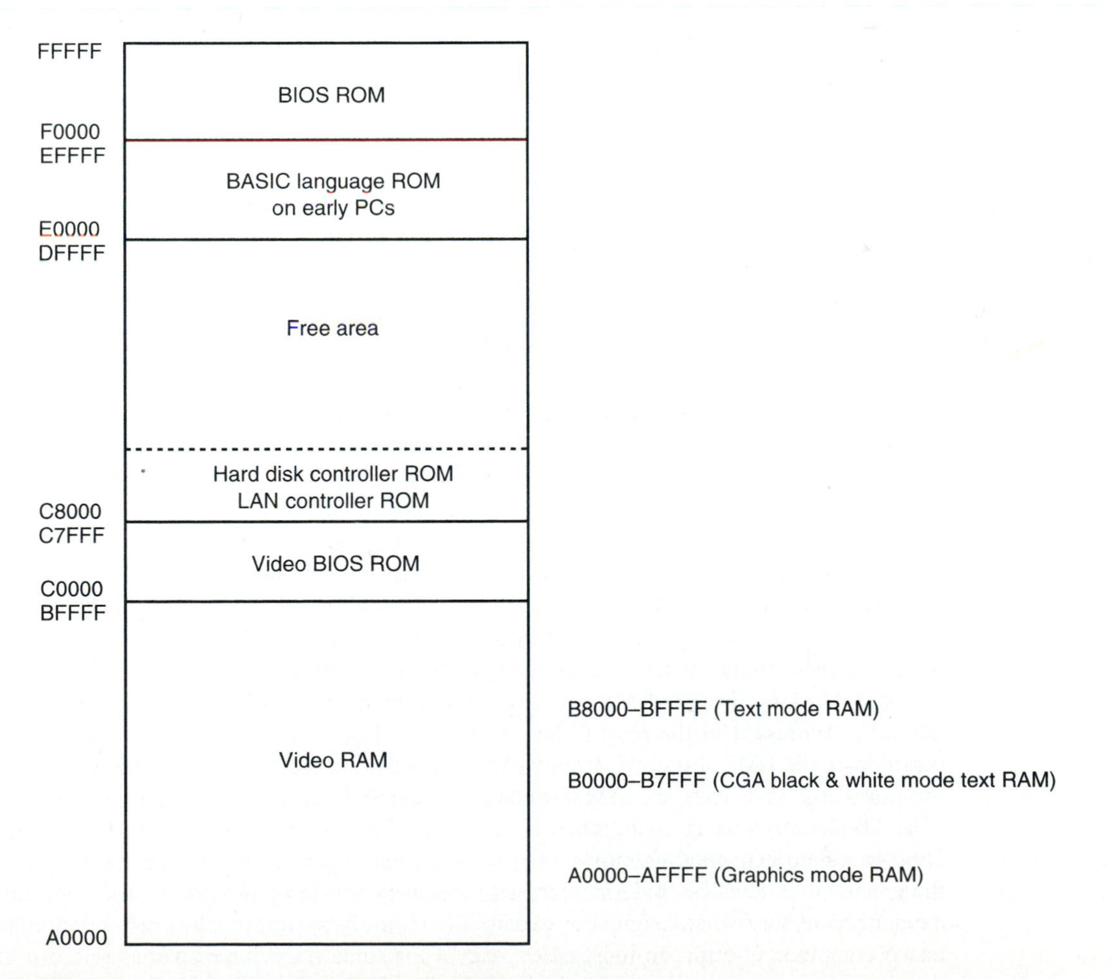

FIGURE 1–5 System area of a typical personal computer system.

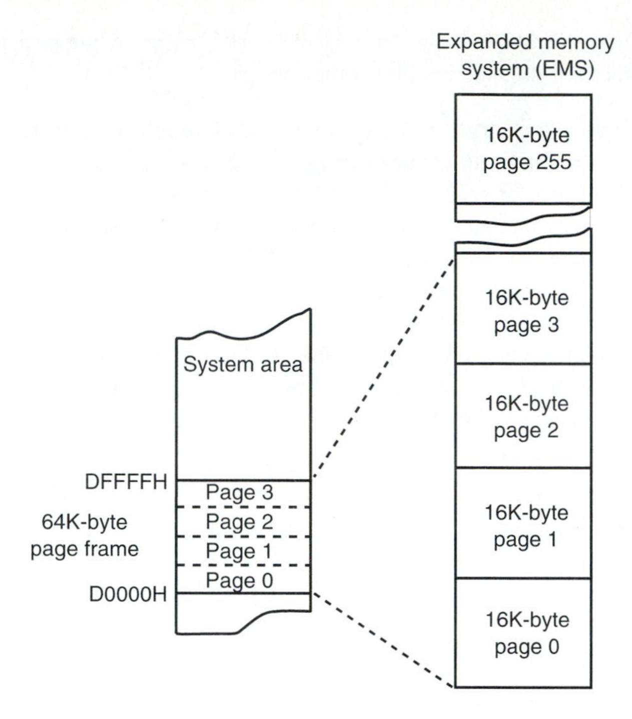

FIGURE 1–6 The expanded memory system (EMS) showing the page frame that holds four 16K-byte pages from the expanded memory system. In this illustration expanded memory contains 256 pages of 16K bytes each for a total expanded memory system of 4M bytes.

the video RAM located at A0000H–AFFFFH stores graphical or bit-mapped data and the memory at B0000H–BFFFFH stores text data. The **video BIOS,** located on a ROM or flash memory, is found at locations C0000H–C7FFFH and contains programs that control the video display.

If a hard disk memory is attached to the computer, the interface card might contain a ROM. The ROM, often found with older MFM or RLL hard disk drives, holds low-level format software, at location C8000H. The size, location, and presence of the ROM depends on the type of hard disk adapter attached to the computer.

The area at locations C8000H–DFFFFH is often open or free. This area is used for the **expanded memory system** (EMS) in a PC or XT or the **upper memory system** in an AT system. Its use depends on the system and its configuration. The expanded memory system allows a 64K byte page frame of memory to be used by application programs. This 64K byte page frame (usually locations D0000H–DFFFFH) is expandable by switching pages of memory from the EMS into this range of memory addresses. Note that the information is addressed in the page frame as 16K byte pages of data that are swapped with pages from the EMS. Figure 1–6 shows the expanded memory system. Most application programs that state they are LIM 4.0 driver compatible can use expanded memory. The LIM 4.0 memory management driver is the result of Lotus, Intel, and Microsoft standardizing access to expanded memory systems. Note that expanded memory is slow because the change to a new 16K byte memory page requires action by the driver. Also note that expanded memory was designed to expand the memory system of the early 8086/8088-based computer systems. In most cases, except for some DOS-based games that use the sound card, expanded memory should be avoided in 80386/80486/Pentium-based systems.

Memory locations E0000H–EFFFFH contain the cassette BASIC language on ROM found in early IBM personal computer systems. This area is often open or free in newer computer systems. In newer systems, we often back-fill this area with extra RAM called **upper memory.**

Finally, the system BIOS ROM is located in the top 64K bytes of the system area (F0000H–FFFFFH). This ROM controls the operation of the basic I/O devices connected to the computer system. It does not control the operation of the video system, which has its own BIOS ROM at location C0000H. The first part of the BIOS (F0000H–F7FFFH) contains programs that set up the computer and the second part contains procedures that control the basic I/O system. Once the system is set up, upper memory blocks at locations F0000H–F7FFFH is available, if EMM386.EXE is installed. Also available for upper memory blocks are locations B0000H–B7FFFH, provided black and white video is not needed in the CGA mode.

I/O Space. The I/O (*input/output*) space in a computer system extends from I/O port 0000H to port FFFFH. (An *I/O port address* is similar to a memory address except that instead of addressing memory, it addresses an I/O device.) The *I/O devices* allow the microprocessor to communicate with the outside world. The I/O space allows the computer to access up to 64K different 8-bit I/O devices. A great number of these locations are available for expansion in most computer systems. Figure 1–7 shows the I/O map found in many personal computer systems.

The I/O area contains two major sections. The area below I/O location 0500H is considered reserved for system devices (see Figure 1–7). The remaining I/O space, from I/O port 0500H through FFFFH, is available for expansion.

Various I/O devices that control the operation of the system are usually not directly addressed. Instead, the BIOS ROM addresses these basic devices, which can vary slightly in location and function from one computer to the next. Access to most I/O devices should always be made through DOS or BIOS function calls to maintain compatibility from one computer system to another. This map is provided as a guide to illustrate the I/O space in the system.

The DOS Operating System

The operating system is the program that operates the computer. This text assumes that the operating system is either **MSDOS** or **PCDOS,** which are the operating systems found in 85 percent of the more than 160 milllion personal computers in use, according to a recent *PC Magazine* article.[12] The Windows operating system is used on 65 million personal computers, according to the same article. The operating system is stored on a disk that is placed in one of the floppy disk drives or found on a hard disk drive that is either resident to the computer or to a *local area network* (LAN). Some dedicated systems store the DOS on a ROM. An example is the Tandy Corporation personal computer. Each time the computer is powered up or reset, the operating system is read from the disk or LAN. We call this operation **booting** the system. Once DOS is installed in the memory by the boot, it controls the operation of the computer system, its I/O devices, and application programs. In addition to

[12]*PC Magazine* is a Ziff-Davis publication.

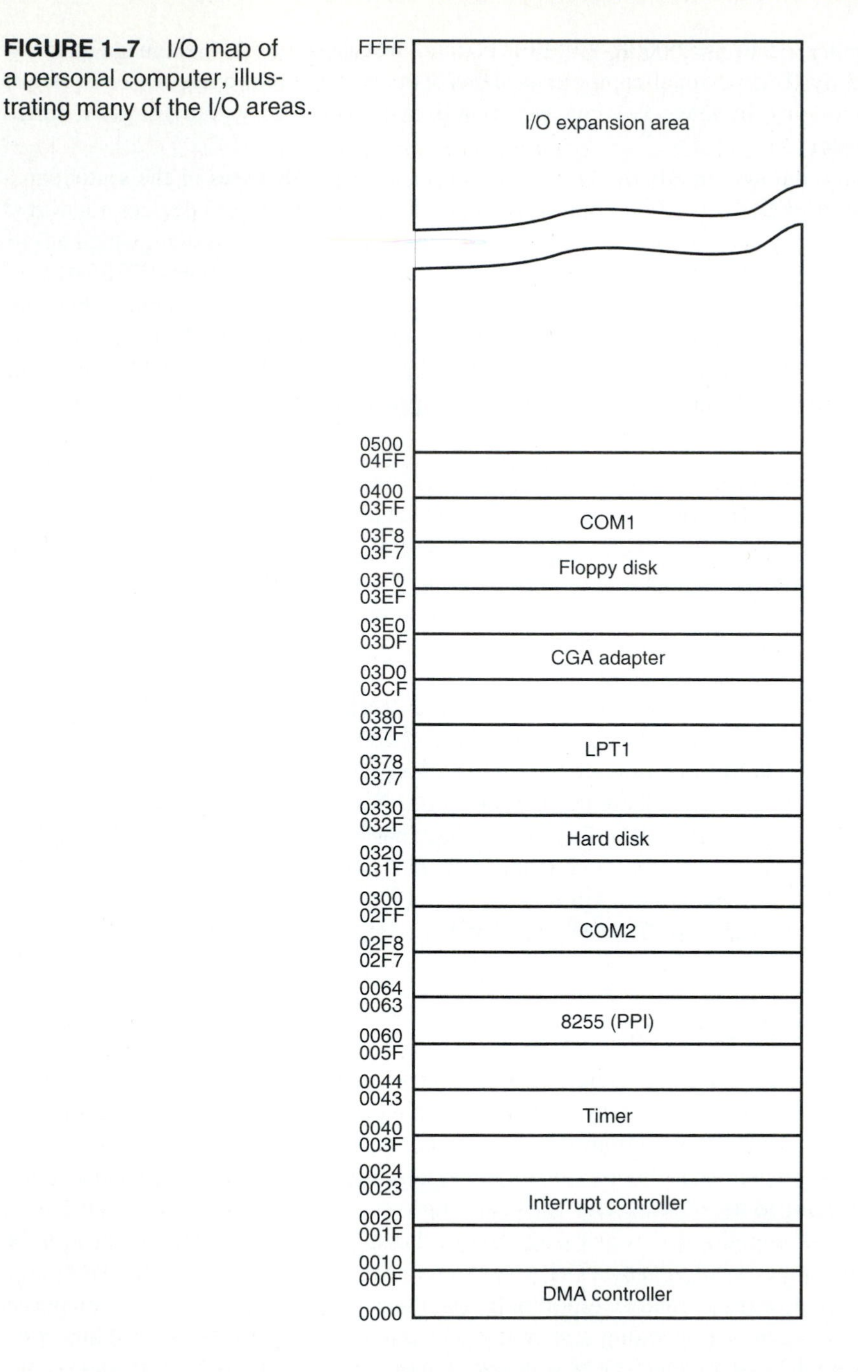

FIGURE 1–7 I/O map of a personal computer, illustrating many of the I/O areas.

the DOS operating system, other operating systems are sometimes used to control or operate the computer. These other operating systems include: Windows from Microsoft (more than 65 million users), OS/2 from IBM (more than 4 million users), Unix from AT&T (more than 2 million users), and many others.

The first task of the DOS operating system, after loading into memory, is to use a file called the CONFIG.SYS file, which should not be erased. This file specifies various drivers that load into the memory, setting up or *configuring* the machine for operation. Example 1–1 lists a **CONFIG.SYS** file for DOS version 5.0. Chapter 2 shows a CONFIG.SYS file for DOS version 6.X and provides greater detail on the structure of this important file. Note that the statements in this file vary from machine to machine; the one illustrated is just an example.

EXAMPLE 1–1

```
REM DOS VERSION 5.0 CONFIG.SYS FILE
REM
FILES=30
BUFFERS=30
STACKS=64,128
FCBS=48
SHELL=C:\DOS\COMMAND.COM C:\DOS\ /E:256 /P
DEVICE=C:\DOS\HIMEM.SYS
DOS=HIGH,UMB
DEVICE=C:\DOS\EMM386.EXE I=C800-EFFF NOEMS
DEVICEHIGH SIZE=1EB0 C:\LASERLIB\SONY_CDU.SYS
DEVICEHIGH SIZE=0190 C:\DOS\SETVER.EXE
DEVICEHIGH SIZE=3150 C:\MOUSE1\MOUSE.SYS
LASTDRIVE = F
```

The first four statements in this CONFIG.SYS file set up the number of files, buffers, stacks, and file control blocks required to execute various programs. These settings should be adequate for just about any program loaded into memory using DOS. In general, if a program requires more buffers and so forth, the documentation indicates that the CONFIG.SYS file must be changed to reflect this increased need. Many modern programs automatically adjust the CONFIG.SYS file when installed by changing these parameters or by adding additional statements.

The SHELL command specifies the command processor used with DOS. In this example the COMMAND.COM file is the command processor (also selected by default) using the E:256 /P switches. The E:256 switch sets the environment size to 256 bytes. The /P switch tells the command processor to make COMMAND.COM permanent. If COMMAND.COM is not permanent, it must be loaded into memory from the disk each time DOS returns to the command prompt. Although this may free a small amount of memory, the constant access to the disk for the COMMAND.COM program increases wear and tear on the disk drive and also lengthens the time required to return to the DOS prompt.

The first DEVICE (driver) loaded into memory in this example is a program called HIMEM.SYS (high memory driver). A **driver** is usually a program that controls an I/O device or a program that must remain in the computer system memory. The HIMEM.SYS program allows a 64K minus 16 byte section of extended memory (100000H–10FFEFH), just above the first 1M byte of memory, to be used for programs in the 80286 through the Pentium-based system. (This extra 64K memory space is supported by most of these computer systems, but not all 80286 systems.) This driver allows a DOS-based system access to 1M plus 64K bytes of memory. This extra 64K byte section of **high memory** holds most of the MSDOS version 5.0 or 6.2 program, freeing additional space in the TPA. The next command (DOS=HIGH,UMB) tells the computer to load DOS into this high part of memory and to allow the use of upper memory blocks.

In order to enable the upper memory blocks, available only in an 80386, 80486, or Pentium-based system, the EMM386.EXE (**extended memory manager**) program is loaded into memory. The extended memory manager is a driver that emulates expanded memory in extended memory and also the extended memory system. This program backfills free areas of memory within the system area, so programs can be loaded into this area and accessed directly by DOS applications. The I=C800–EFFF switch tells EMM386.EXE to use memory area C8000H–EFFFFH for upper memory or upper memory blocks (*UMB*). Drivers and programs are loaded into upper memory, freeing even more area in the TPA for application programs. Before using the I=C800–EFFF switch, make sure that your computer does not contain any system ROM/RAM in this area of the memory. Note that NOEMS tells EMM386.EXE to exclude expanded memory. Expanded memory can also be installed by replacing NOEMS with a number that indicates how much extended memory to allocate to LIM 4.0 expanded memory system (EMS). Today most systems should not use expanded memory. If expanded memory is required, the NOEMS is replaced with RAM 1024 to enable 1,024 bytes of expanded memory. The FRAME=D000 statement places the page frame for expanded memory at location D0000H–DFFFFH if expanded memory is enabled.

The DEVICEHIGH command loads drivers and programs into the upper memory blocks allocated by the EMM386.EXE driver. In the CONFIG.SYS file illustrated, three drivers are loaded into upper memory blocks beginning at location C8000H. The first is a program that operates a SONY CDROM drive, the second loads a program called SETVER, and the third loads the MOUSE driver.

The last statement in the CONFIG.SYS file in Example 1–1 shows the LASTDRIVE statement. This tells DOS which is the last disk drive connected to your computer system. By using the LASTDRIVE statement, more memory can be freed for use in the TPA. Each drive requires a buffer area and if you use the actual last drive with this statement, extra memory is made available. Other drivers may also be loaded into memory using the CONFIG.SYS file, such as a PRINT.SYS driver, ANSI.SYS driver, or any other program that functions as a driver. Driver programs normally contain the DOS extension .SYS used to indicate a system file. Be sure to be very careful when changing the CONFIG.SYS file, because an error locks up the computer system (except for DOS 6.X, which can exit this type of system lockup). Once the computer is locked up by a CONFIG.SYS error, the only way to recover is to boot off a DOS floppy disk that contains the operating system with a functioning CONFIG.SYS file.

Once the operating system completes its configuration as dictated by CONFIG.SYS, the **AUTOEXEC.BAT** (*automatic execution batch*) file is executed by the computer. If none exists, the computer asks for the time and date. Example 1–2 shows a typical AUTOEXEC.BAT file for either DOS version 5.0 or 6.X. This is only an example; variations occur from system to system. The AUTOEXEC.BAT file contains commands that execute when power is first applied to the computer. These are the same commands that could be typed from the keyboard, but the AUTOEXEC.BAT saves us from doing so each time the computer is powered up.

EXAMPLE 1–2

```
PATH C:\DOS;C:\;C:\MASM\BIN;C:\MASM\BINB\;C:\UTILITY
PATH C:\WS;C:\LASERLIB
SET BLASTER=A220 I7 D1 T3
```

```
SET INCLUDE=C:\MASM\INCLUDE\
SET HELPFILES=C:\MASM\HELP\*.HLP
SET INIT=C:\MASM\INIT\
SET ASMEX=C:\MASM\SAMPLES\
SET TMP=C:\MASM\TMP
SET SOUND=C:\SB
LOADHIGH C:\LASERLIB\MSCDEX.EXE /D:SONY_001 /L:F /M:10
LOADHIGH C:\LASERLIB\LLTSR.EXE ALT-Q
LOADHIGH C:\DOS\FASTOPEN C:=256
LOADHIGH C:\DOS\DOSKEY /BUFSIZE=1024
LOADHIGH C:\LASERLIB\PRINTF.COM
DOSKEY GO=DOSSHELL
DOSSHELL
```

The PATH statement specifies the search paths whenever a program name is typed at the command line. The order of the path search is the same as the order of the paths in the path statement. For example, if PROG is typed at the command line, the machine first searches C:\DOS, then the root directory C:\, then C:\MASM\BIN, and so forth until the program named PROG is found. If it isn't found, the command interpreter (COMMAND.COM) informs the user that the program is not found.

The SET statement sets a variable name to a path. This allows names to be associated with paths for batch programs. It's also used to set command strings (environments) for various programs. The first SET command sets the environment for the sound blaster card. The second SET command sets INCLUDE to the path C:\MASM\INCLUDE\. Note that the SET statements are stored in the DOS environment space that was reserved in the CONFIG.SYS file using the SHELL statement. If the environment becomes too large, you must change the SHELL statement to allow more space.

LOADHIGH or LH places programs into upper memory blocks defined by EMM386.EXE program. LOADHIGH is used at any DOS command prompt for loading a program into the high memory area, as long as the computer is an 80386 or above. The last command in this AUTOEXEC.BAT file is DOSSHELL. The DOSSHELL program is a menu program included with MSDOS version 5.0 or 6.X. This last command is replaced by WIN, WIN /S (standard mode), WIN /R (real mode in Windows 3.1), or WIN /3 (enhanced mode) to run Microsoft Windows in place of DOSSHELL.

Once the CONFIG.SYS and AUTOEXEC.BAT files are executed, the program name last shown in the AUTOEXEC.BAT file is executed. In this example, the system operates from the DOSSHELL program.

The Microprocessor

At the heart of the microprocessor-based computer system is the microprocessor integrated circuit. The **microprocessor** is the controlling element in a computer system and is sometimes referred to as the **CPU** (*central processing unit*). The microprocessor controls memory and I/O through a series of connections called **buses.** Buses select an I/O or memory device, transfer data between an I/O device or memory and the microprocessor, and control the I/O and memory system. Memory and I/O are controlled through instructions that are stored in the memory and executed by the microprocessor.

The microprocessor performs three main tasks for the computer system: (1) data transfer between itself and the memory or I/O systems, (2) simple arithmetic and logic operations, and (3) program flow via simple decisions. Although these are simple tasks, it is through them that the microprocessor performs virtually any series of operations or tasks.

TABLE 1–3 Simple arithmetic and logic operations

Operation	*Comment*
Addition	
Subtraction	
Multiplication	
Division	
AND	Logical multiplication
OR	Logical addition
NOT	Logical inversion
NEG	Arithmetic negation
Shift	
Rotate	

The power of the microprocessor is in its ability to execute millions of instructions per second from a **program** or **software** (group of instructions) stored in the memory system. This *stored program concept* has made the microprocessor and computer system a very powerful device. Recall that Babbage wanted to use the stored program concept in his analytical engine.

Table 1–3 shows the arithmetic and logic operations that the Intel family of microprocessors execute. These operations are very basic, but through them, very complex problems are solved. Data is operated upon from the memory system or internal registers. Data widths are variable and include a **byte** (8 bits), **word** (16 bits), and **doubleword** (32 bits). Note that only the 80386 through the Pentium directly manipulate 8-, 16-, and 32-bit numbers. The earlier 8086–80286 directly manipulate 8- and 16-bit numbers, but not 32-bit numbers. The 80486DX and Pentium also contain a numeric coprocessor that allows them to perform complex arithmetic using floating-point arithmetic. The numeric coprocessor was an additional component in the 8086 through the 80386-based personal computer.

Another feature that makes the microprocessor powerful is its ability to make simple decisions. Decisions are based upon numerical facts. For example, a microprocessor can decide if a number is zero, if it is positive, and so forth. These simple decisions allow the microprocessor to modify the program flow so programs appear to think. Table 1–4 lists the decision-making abilities of the Intel family of microprocessors.

TABLE 1–4 Decisions found in an 8086–80486 and Pentium microprocessor

Decision	*Comment*
Zero	Test a number for zero or not-zero
Sign	Test a number for positive or negative
Carry	Test for a carry after an addition or a borrow after a subtraction
Parity	Test a number for the count of the number of ones expressed as even or odd
Overflow	Test for an invalid signed result

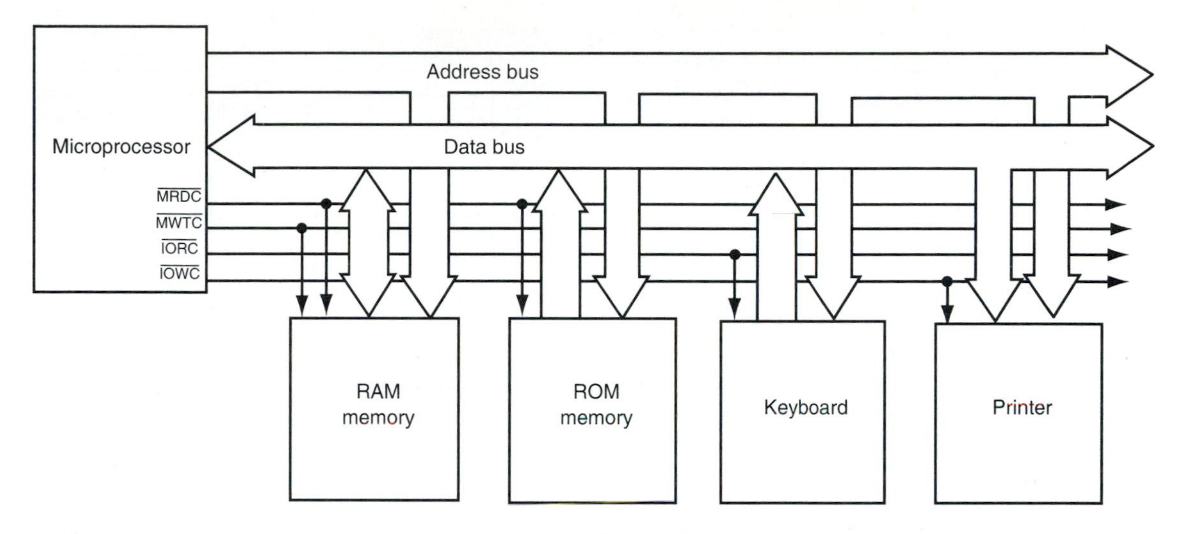

FIGURE 1–8 Block diagram of a computer system showing the address, data, and control buses. Decoders are not shown in the illustration. $\overline{\text{MRDC}}$ (memory read), $\overline{\text{MWTC}}$ (memory write), $\overline{\text{IORC}}$ (I/O read), and $\overline{\text{IOWC}}$ (I/O write) comprise the control bus.

Buses. A *bus* is a common group of wires that interconnect components in a computer system. The buses that interconnect the sections of a computer system transfer address, data, and control information between the microprocessor and its memory and I/O systems. In the microprocessor-based computer system, three buses exist for this transfer of information: address, data, and control. Figure 1–8 shows how these buses interconnect various system components such as the microprocessor, read/write memory (RAM), read-only memory (ROM), and a few I/O devices.

The address bus requests a memory location from the memory or an I/O location from the I/O devices. If I/O is addressed, the address bus contains a 16-bit I/O address of 0000H–FFFFH. The 16-bit I/O address or port number selects one of 64K different I/O devices. If memory is addressed, the address bus contains a memory address. The memory address varies in width with the different versions of the microprocessor. The 8086 and 8088 address 1M byte of memory using a 20-bit address that selects locations 00000H–FFFFFH. The 80286 and 80386SX address 16M bytes of memory using a 24-bit address that selects locations 000000H–FFFFFFH. The 80386SL/80386SLC/80386EX address 32M bytes of memory using a 25-bit address that selects locations 0000000H–1FFFFFFH. The 80386DX, 80486SX, and 80486DX address 4G bytes of memory using a 32-bit address that selects locations 00000000H–FFFFFFFFH. The Pentium also addresses 4G bytes of memory, but it uses a 64-bit data bus to access up to 8 bytes of memory at a time.

The data bus transfers information between the microprocessor and its memory and I/O address space. Data transfers vary in size from 8-bits wide to 64-bits wide in various members of the Intel microprocessor family. The 8088 contains an 8-bit data bus that transfers 8-bits of data at a time. The 8086, 80286, 80386SL, 80386SX, and 80386EX transfer 16-bit data through their data buses; the 80386DX, 80486SX, and 80486DX

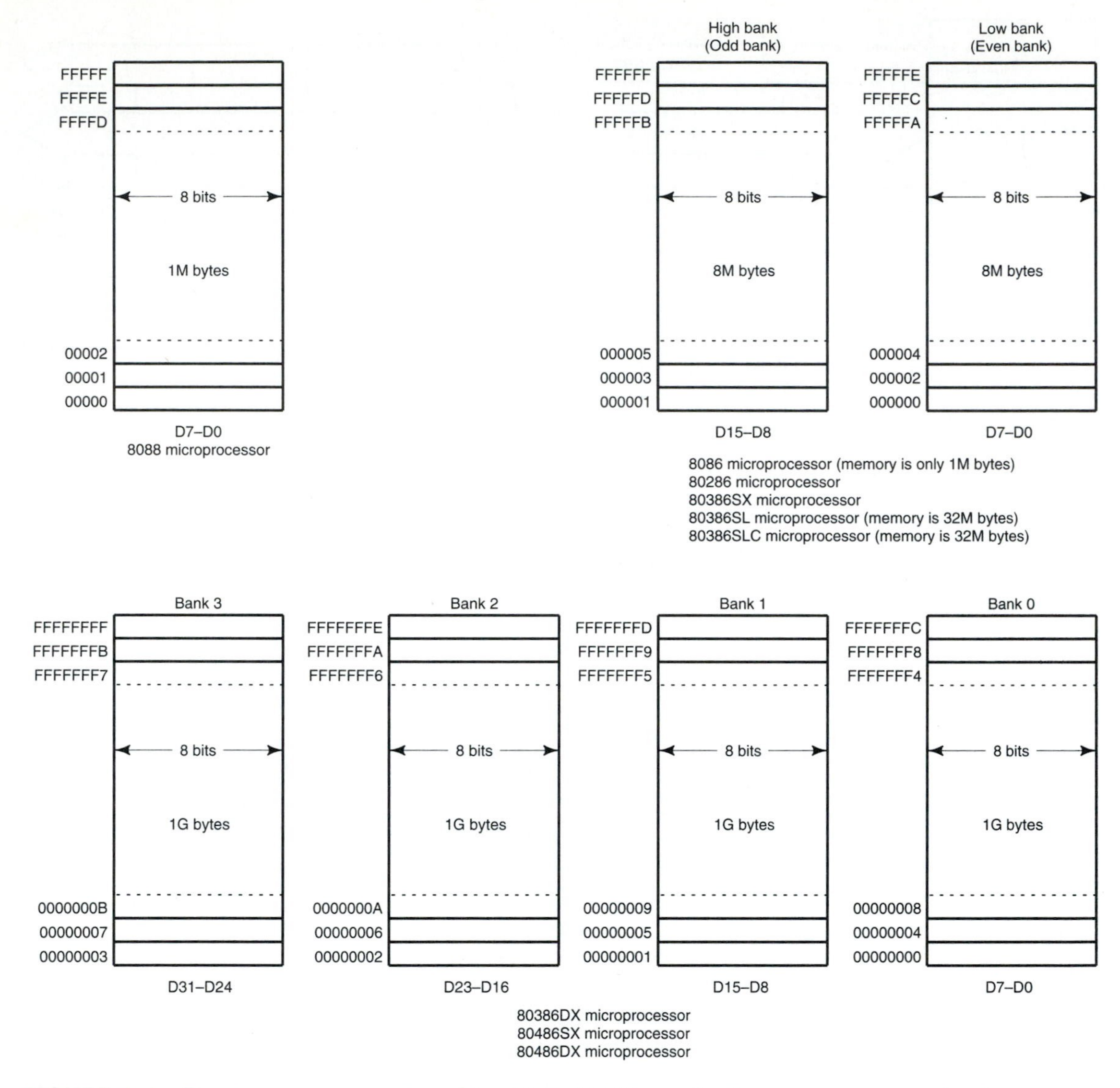

FIGURE 1–9 Physical memory system of the 8086–80486 and Pentium microprocessor family.

transfer 32-bit data; and the Pentium transfers 64-bit data. The advantage of a wider data bus is speed in applications that use wide data. For example, if a 32-bit number is stored in memory, it takes the 8088 microprocessor four transfer operations to complete, because its data bus is only 8-bits wide. The 80486 accomplishes the same task with one transfer, because its data bus is 32-bits wide. Figure 1–9 shows the memory widths and sizes of the 8086–80486 and Pentium microprocessors. Notice how the memory sizes and organizations differ between various members of the Intel microprocessor family. In all family members, the memory is numbered by byte. Notice that the Pentium contains a 64-bit wide data bus.

FIGURE 1–9 *(continued)*

The control bus contains lines that select the memory or I/O and cause them to perform a read or a write operation. In most computer systems, there are four control bus connections: $\overline{\text{MRDC}}$ (*memory read control*), $\overline{\text{MWTC}}$ (*memory write control*), $\overline{\text{IORC}}$ (*I/O read control*), and $\overline{\text{IOWC}}$ (*I/O write control*). The overbar indicates that the control signal is active-low; that is, it is active when a logic zero appears on the control line. For example, if $\overline{\text{IOWC}} = 0$, the microprocessor is writing data from the data bus to an I/O device whose address appears on the address bus.

The microprocessor reads the contents of a memory location by sending the memory an address through the address bus. Next it sends the memory read control signal ($\overline{\text{MRDC}}$) to cause memory to read data. Finally the data read from the memory is passed to the microprocessor through the data bus. Whenever a memory write, I/O write, or I/O read occurs, the same sequence ensues, except different control signals are issued and the data flow out of the microprocessor through its data bus for a write operation.

1–3 NUMBER SYSTEMS

The use of the microprocessor requires a working knowledge of binary, decimal, and hexadecimal number systems. This section of the text provides a background on these systems for those unfamiliar with number systems. Conversions between decimal and binary, decimal and hexadecimal, and binary and hexadecimal.

Digits

Before numbers are converted from one number base to another, the digits of a number system must be understood. Early in our education, it was learned that a decimal, or base 10, number was constructed with 10 digits: 0 through 9. The first digit in any numbering system is always a zero. The same rule applies to any other number system. For example, a base 8 (*octal*) number contains 8 digits: 0 through 7, while a base 2 (*binary*) number contains 2 digits: 0 and 1. If the base of a number exceeds 10, the additional digits use the letters of the alphabet beginning with an A. For example, a base 12 number contains 12 digits: 0 through 9, followed by A for 10 and B for 11. Note that a base 10 number does not contain a 10 digit, just as a base 8 number does not contain an 8 digit. The most common number systems in widespread use with computers are decimal, binary, octal, and hexadecimal (base 16). Each system is described and used in this section of the chapter.

Positional Notation

Once the digits of a number system are understood, larger numbers can be constructed using positional notation. In grade school, we all learned that the position to the left of the units position was the tens position, the position to the left of the tens position was the hundreds position, and so forth. What we probably did not learn was that the units position has a weight of 10^0 or 1, the tens position has weight of 10^1 or 10, and the hundreds position has a weight of 10^2 or 100. The powers of the positions are critical in understanding numbers in other numbering systems. The position to the left of the radix (*number base*) point, called a decimal point only in the decimal system, is always the units position in any number system. For example, the position to the left of the binary point is 2^0 or 1, while the position to the left of the octal point is 8^0 or 1. In any case, any number raised to its zero power is always 1, or the units position.

The position to the left of the units position is always the number base raised to the first power. In decimal, this is 10^1 or 10; in binary, it is 2^1 or 2, and in octal, 8^1 or 8. Therefore, an 11 decimal has a different value than a 11 binary. The 11 decimal signifies 1 ten plus 1 unit, and has a value of 11 units. The 11 binary signifies 1 two plus 1 unit for a value of 3 units. The 11 octal has a value of 9 units.

Positions to the right of the radix point have negative powers. In the decimal system, the first digit to the right of the decimal point has a value of 10^{-1} or 0.1. In the binary system, the first digit to the right of the binary point has a value of 2^{-1} or 0.5. In general, the principles that apply to decimal numbers apply to numbers in any other number system.

Conversion to Decimal

To convert from any number base to decimal, multiply the weight by the value (or number) of each digit, then sum the products to form the decimal aquivalants. Example 1–3 shows the power and weight of each digit position for the number 110.101 in binary. To convert a binary number to decimal, multiply the weight of each (1) digit by the value of that digit to find that digit's numeric value, then sum all the numeric values. The 110.101 binary is equivalent to 6.625 in decimal (4 + 2 + 0.5 + 0.125).

EXAMPLE 1–3

Power	2^2		2^1		2^0		2^{-1}		2^{-2}		2^{-3}		
Weight	4		2		1		0.5		0.25		.125		
Number	1		1		0	.	1		0		1		
Numeric Value	4	+	2	+	0	+	0.5	+	0	+	0.125	=	6.625_{10}

Let's apply the conversion technique to a base 6 number, 25.2. Example 1–4 shows the powers and weights of each digit of this number. In this example, there are 2 units under 6^1, which have a value of 12 (2×6), and 5 units under 6^0, which have a value of 5 (5×1). Note how the weight is multiplied by the digit to form the numeric value. So, the whole number portion has a decimal value of 12 + 5 or 17 units. The number to the right of the hex point is 2 units under 6^{-1}, which represents a value of .333 ($2 \times .167$). The number 25.2 in base 6 therefore has a decimal value of 17.333.

EXAMPLE 1–4

Power	6^1		6^0		6^{-1}		
Weight	6		1		.167		
Number	2		5	.	2		
Numeric Value	12	+	5	+	0.333	=	17.333_{10}

EXAMPLE 1–5

Power	8^2		8^1		8^0		8^{-1}		
Weight	64		8		1		.125		
Number	1		2		3	.	7		
Numeric Value	64	+	16	+	5	+	.875	=	85.875_{10}

Suppose 125.7 octal is to be converted to decimal. To accomplish this conversion, first write down the weight of each digit position of the number, then multiply the weight by the value of the digit. As shown in Example 1–5, 125.7 octal is 85.875 decimal, or 1×64 plus 2×8 plus 5×1 plus $7 \times .125$. The weight of the position to the left of the units position is 16 (8×2). The weight of the next position is 64 (8×8). If another position existed, its weight would be 512 (64×8). To find the weight of the next higher-order position, multiply the weight of the current position by the number base, or 8 in this example. To calculate the weights of positions to the right of the radix point, divide by the number base. In the octal system, the position immediately to the right of the octal point is $^1/_8$ or .125. Thus, the value of .7 octal is .875 decimal ($.125 \times 7$). The next position is .125/8 or .015625, which can also be written as $^1/_{64}$. Also note that the number in Example 1–5 can also be written as the decimal number $85^7/_8$.

Example 1–6 shows the binary number 11011.0111 written with the weights and powers of each position. If these (1) digit weights are summed, the value of the binary number converted to decimal is 27.4375.

EXAMPLE 1–6

```
Power           2^4  2^3  2^2  2^1  2^0  2^-1 2^-2   2^-3    2^-4
Weight          16   8    4    2    1    0.5  0.25   .125    .0625
Number          1    1    0    1    1  . 0    1      1       1
Numeric Value   16 + 8 +  4 +  2 +  1 +  0 +  .25 +  .125 +  .0625 = 27.4375(10)
```

It is interesting to note that 2^{-1} is also $^1/_2$, 2^{-2} is also $^1/_4$, 2^{-4} is $^1/_{16}$, and so forth. The fractional part of this number is $^7/_{16}$, or .4375 decimal. In binary code, 0111 is a 7 for the numerator, and the rightmost one is in the $^1/_{16}$ position for the denominator. The binary fraction of .101 is $^5/_8$ and the binary fraction of .001101 is $^{13}/_{64}$.

Hexadecimal numbers are often used with computers. 6A.CH (H for hexadecimal) is illustrated with its weights in Example 1–7. The sum of its digits are 106.75, or $106^3/_4$. The whole number part is represented with 6×16 plus 10 (A) $\times$ 1. The fraction part is 12 (C) as a numerator and 16 (16^{-1}) as the denominator or $^{12}/_{16}$, which is $^3/_4$.

EXAMPLE 1–7

```
Power     16^1  16^0    16^-1
Weight    16    1       .0625
Number    6     A   .   C
```

Conversion from Decimal

Conversions *from* decimal to other number systems are more difficult to accomplish than conversion *to* decimal. To convert the whole number portion of a number to decimal, divide by the radix. To convert the fractional portion multiply by the radix.

Whole Number Conversion from Decimal. To convert a decimal whole number to another number system, divide by the radix and save the remainders as significant digits of the result. An algorithm for this conversion follows:

1. Divide the decimal number by the radix (number base).
2. Save the remainder (the first remainder is the least significant digit).
3. Repeat Steps 1 and 2 until the quotient is zero.

For example, to convert a 10 decimal to binary divide it by 2. The result is 5, with a remainder of 0. The first remainder is the units position of the result; in this example, 0. Next divide the 5 by 2. The result is 2 with a remainder of 1. The 1 is the value of the twos (2^1) position. Continue the division until the quotient is a zero. Example 1–8 shows this conversion process. The result is written as 1010 binary (read from the bottom to the top).

EXAMPLE 1–8

```
2) 10   remainder = 0
 2) 5   remainder = 1
  2) 2  remainder = 0
   2) 1 remainder = 1
      0
```

To convert a 10 decimal into base 8, divide by 8 as shown in Example 1–9. 10 decimal is a 12 octal.

EXAMPLE 1–9

```
8) 10   remainder = 2
 2) 1   remainder = 1
    0
```

Conversion from decimal to hexadecimal is accomplished by dividing by 16. The remainders will range in value from 0 through 15. Any remainder of 10 though 15 is then converted to the letters A through F for the hexadecimal number. Example 1–10 shows the decimal number 109 converted to a 6DH.

EXAMPLE 1–10

```
16) 109   remainder = 13 (D)
 16)  6   remainder = 6
      0
```

Converting from a Decimal Fraction. Conversion from a decimal fraction to another number base is accomplished with multiplication by the radix. For example, to convert a decimal fraction into binary, multiply by 2. After the multiplication, the whole number portion of the result is saved as a significant digit of the result and the fractional remainder is again multiplied by the radix. When the fraction remainder is zero, multiplication ends. Some numbers are never-ending; that is, a zero is never a remainder. An algorithm for conversion from a decimal fraction follows:

1. Multiply the decimal fraction by the radix (number base).
2. Save the whole number portion of the result (even if zero) as a digit. Note that the first result is written immediately to the right of the radix point.
3. Repeat Steps 1 and 2 using the fractional part of Step 2 until the fractional part of Step 2 is zero.

Suppose .125 decimal is converted to binary. This is accomplished with multiplications by 2 as illustrated in Example 1–11. Notice that the multiplication continues until the fractional remainder is zero. The whole number portions are written as the binary fraction (0.001) in this example.

EXAMPLE 1–11

```
   .125
x     2
  0.25 digit is 0

   .25
x    2
  0.5   digit is 0

   .5
x   2
 1.0    digit is 1. The result is written as 0.001 binary
```

This same technique is used to convert a decimal fraction into any number base. Example 1–12 shows the same decimal fraction, .125, converted to octal, by multiplying with an 8.

EXAMPLE 1–12

```
  .125
x____8
   1.0 digit is 1. The result is written as 0.1 octal
```

Conversion to a hexadecimal fraction appears in Example 1–13. Here decimal .046875 is converted to hexadecimal by multiplying by 16. Note that .046875 decimal is a 0.0CH.

EXAMPLE 1–13

```
  .046875
x ______16
     0.75 digit is 0

  .75
x__16
 12.0 digit is 12 (C). The result is written as 0.0C hexadecimal
```

Binary-Coded Hexadecimal

Binary-coded hexadecimal (BCH) is used to represent hexadecimal data in binary code. A binary-coded hexadecimal number is a hexadecimal number written so each digit is represented by a 4-bit binary number. The values for the BCH digits appear in Table 1–5.

TABLE 1–5 Binary-coded hexadecimal (BHC) code

Hexadecimal Digit	*BCH Code*
0	0000
1	0001
2	0010
3	0011
4	0100
5	0101
6	0110
7	0111
8	1000
9	1001
A	1010
B	1011
C	1100
D	1101
E	1110
F	1111

Hexadecimal numbers are represented in BCH code by converting each digit to BCH code, with a space between each coded digit. Example 1–14 shows 2AC converted to BCH code. Note that each BCH digit is separated by a space.

EXAMPLE 1–14

```
2AC = 0010 1010 1100
```

The purpose of BCH code is to allow a binary version of a hexadecimal number to be written in a form that can easily be converted between BCH and hexadecimal. Example 1–15 shows a BCH-coded number converted to hexadecimal code.

EXAMPLE 1–15

```
1000 0011 1101 . 1110 = 83D.E
```

Complements

At times, data are stored in complement form to represent negative numbers. There are two systems that are used to represent negative data: radix and radix –1 complements. To represent a negative number with the radix –1 complement, each digit of the number is subtracted from the radix.

EXAMPLE 1–16

```
  2 2 2 2  2 2 2 2
- 0 1 0 0  1 1 0 0
  1 0 1 1  0 0 1 1
```

Example 1–16 shows how the 8-bit binary number 01001100 is one's (radix – 1) complemented to represent it as a negative value. Notice that each digit of the number is subtracted from the radix to generate the radix –1 (one's) complement. In this example, the negative of 01001100 is 10110011. The same technique can be applied to any number system, as illustrated in Example 1–17. There, 15's (radix –1) complement of a 5CD hexadecimal is computed by subtracting each digit from fifteen.

EXAMPLE 1–17

```
  15 15 15
-  5  C  D
   A  3  2
```

Today, the radix –1 complement is not used by itself, but it is used as a step for finding the radix complement. The radix complement is the way negative numbers are represented in modern computer systems; radix –1 complement was used in the early days of computer technology. The main problem with the radix –1 complement is that a negative or a positive zero exists, where in the radix complement system, only a positive zero can exist.

To form the radix complement, first find the radix –1 complement and then add a one to the result. Example 1–18 shows how the number 0100 1000 is converted to a negative value by two's (radix) complementing it.

EXAMPLE 1–18

```
  2 2 2 2  2 2 2 2
- 0 1 0 0  1 0 0 0
  1 0 1 1  0 1 1 1   (one's complement)
+ 0 0 0 0  0 0 0 1
  1 0 1 1  1 0 0 0   (two's complement)
```

To prove that 0100 1000 is the inverse (negative) of 1011 0111, add the two together to form an 8-digit result. The ninth digit is dropped and the result is zero, because 0100 1000 is a positive 72 while 1011 0111 is a negative 72. The same technique applies to any number system. Example 1–19 shows how the inverse of 345 hexadecimal is found by first fifteen's complementing the number and then adding one to the result to form the sixteen's complement. As before, if the original 3-digit number 345 is added to the inverse of CBB, the result is a 3-digit 000. The fourth bit (carry) is dropped. This proves that 345 is the inverse of CBB. Additional information about one's and two's complements are presented with signed numbers in the next section of the text.

EXAMPLE 1–19

```
  15 15 15
-  3  4  5
   C  B  A   (fifteen's complement)
+  0  0  1
   C  B  B   (sixteen's complement)
```

1–4 COMPUTER DATA FORMATS

Successful programming requires a precise understanding of data formats. In this section, many computer data formats used with the Intel family of microprocessors are described. Commonly, data appear as ASCII, BCD, signed and unsigned integers, and floating-point numbers (real numbers). Other forms are used, but are not presented here because they are not common.

ASCII Data

ASCII (*American Standard Code for Information Interchange*) data (see Table 1–6) represents alphanumeric characters in the memory of a computer system. The standard *ASCII code* is a 7-bit code with the eighth, and most significant, bit used to hold parity in some systems. If ASCII data are used with a printer, the most significant bit is 0 for alphanumeric printing, and 1 for graphics printing. In the personal computer, an extended ASCII character set is selected by placing a logic 1 in the left-most bit. Table 1–7 shows the extended ASCII character set using code 80H–FFH. The extended ASCII characters store some foreign letters and punctuation, Greek characters, mathematical characters, box drawing characters, and other special characters. Note that extended characters can vary from one printer to another. The list provided is designed to be used with the IBM ProPrinter;[13] it matches the special character set found with some word processors.

[13]ProPrinter is a trademark of International Business Machine Corporation.

TABLE 1–6 ASCII code

	Second															
	X0	X1	X2	X3	X4	X5	X6	X7	X8	X9	XA	XB	XC	XD	XE	XF
First																
0X	NUL	SOH	STX	ETX	EOT	ENQ	ACK	BEL	BS	HT	LF	VT	FF	CR	SO	SI
1X	DLE	DC1	DC2	DC3	DC4	NAK	SYN	ETB	CAN	EM	SUB	ESC	FS	GS	RS	US
2X	SP	!	“	#	$	%	&	’	(	)	*	+	,	-	.	/
3X	0	1	2	3	4	5	6	7	8	9	:	;	<	=	>	?
4X	@	A	B	C	D	E	F	G	H	I	J	K	L	M	N	O
5X	P	Q	R	S	T	U	V	W	X	Y	Z	[	\	]	^	_
6X	‘	a	b	c	d	e	f	g	h	i	j	k	l	m	n	o
7X	p	q	r	s	t	u	v	w	x	y	z	{	|	}	~	:::

The ASCII control characters, also listed in Table 1–6, perform control functions in a computer system. These include: clear screen, backspace, line feed, etc. To enter the control codes through the computer keyboard, the control key is held down while typing a letter. To obtain the control code 01H, type control A. Control code 02H is obtained by typing control B, etc. Note that the control codes appear on the screen as ^A for control A, ^B for control B, and so forth. Also note that the carriage return code (CR) is the **enter key** on most modern keyboards. The purpose of CR is to return the cursor or print-head to the left margin. Another code that appears in many programs is the line feed code (LF) that moves the cursor down one line.

To use Table 1–6 or 1–7 for converting alphanumeric or control characters into ASCII characters, first locate the alphanumeric code for conversion. Next, find the first digit of the hexadecimal ASCII code. Then, find the second digit. For example, the letter *A* is ASCII code 41H, the letter *a* is ASCII code 61H.

TABLE 1–7 Extended ASCII code as printed by the IBM ProPrinter

First	Second															
	X0	X1	X2	X3	X4	X5	X6	X7	X8	X9	XA	XB	XC	XD	XE	XF
0X		☺	☻	♥	♦	♣	♠	•	◘	○	◙	♂	♀	♪	♫	☼
1X	►	◄	↕	‼	¶	§	▬	↨	↑	↓	→	←	∟	↔	▲	▼
8X	Ç	ü	é	â	ä	à	å	ç	ê	ë	è	ï	î	ì	Ä	Å
9X	É	æ	Æ	ô	ö	ò	û	ù	ÿ	Ö	Ü	¢	£	¥	₧	ƒ
AX	á	í	ó	ú	ñ	Ñ	ª	º	¿	⌐	¬	½	¼	¡	«	»
BX	░	▒	▓	│	┤	╡	╢	╖	╕	╣	║	╗	╝	╜	╛	┐
CX	└	┴	┬	├	─	┼	╞	╟	╚	╔	╩	╦	╠	═	╬	╧
DX	╨	╤	╥	╙	╘	╒	╓	╫	╪	┘	┌	█	▄	▌	▐	▀
EX	α	β	Γ	π	Σ	σ	μ	γ	Φ	Θ	Ω	δ	∞	∮	∈	∩
FX	≡	±	≥	≤	⌠	⌡	÷	≈	°	·	·	√	n	2	■	

ASCII data are most often stored in memory using a special directive to the assembler program called *define byte(s)* or **DB.** (The *assembler* is a program that is used to program a computer in its native binary machine language.) The DB directive, along with several examples of its usage with ASCII-coded character strings, is listed in Example 1–20. Notice how each character string is surrounded by apostrophes (')—never use the quote ('). Also notice that the assembler lists the ASCII-coded value for each character to the left of the character string. To the far left is the hexadecimal memory location where the character string is stored in the memory system. For example, the character string WHAT is stored beginning at memory address 001D and the first letter is stored as 57 (W) followed by 68 (h), and so forth.

EXAMPLE 1–20

```
0000  42 61 72 72 79   NAMES  DB   'Barry B. Brey'
      20 42 2E 20 42
      72 65 79
000D  57 68 65 72 65   MESS   DB   'Where can it be?'
      20 63 61 6E 20
      69 74 20 62 65
      3F
001D  57 68 61 74 20   WHAT   DB   'What is on first.'
      69 73 20 6F 6E
      20 66 69 72 73
      74 2E
```

BCD (Binary-Coded Decimal) Data

Binary-coded decimal (BCD) information is stored in either packed or unpacked forms. Packed BCD data are stored as two digits per byte and unpacked BCD data are stored as one digit per byte. The range of a BCD digit extends from 0000_2 to 1001_2, or 0–9 decimal. Unpacked BCD data is often returned from a keypad or keyboard, while packed BCD data is used for some of the instructions included for BCD addition and subtraction in the instruction set of the microprocessor.

Table 1–8 shows some decimal numbers converted to both packed and unpacked BCD forms. Applications that require BCD data are point-of-sale terminals, or other devices that perform a minimal amount of simple arithmetic. If a system requires complex arithmetic, BCD data is seldom used, because there is no simple and efficient method of performing complex BCD arithmetic.

Example 1–21 shows how to use the assembler to define both packed and unpacked BCD data. In all cases, the convention of storing the least-significant data first is followed. This means that to store a 83 into memory the 3 is stored first, followed by the 8. With

TABLE 1–8 Packed and unpacked BCD data

Decimal	*Packed*		*Unpacked*		
12	0001 0010		0000 0001	0000 0010	
623	0000 0110	0010 0011	0000 0110	0000 0010	0000 0011
910	0000 1001	0001 0000	0000 1001	0000 0001	0000 0000

packed BCD data, the letter H (hexadecimal) follows the number to ensure that the assembler stores the BCD value rather than a decimal value for packed BCD data. Notice how the numbers are stored in memory as unpacked (one digit per byte) or packed (2 digits per byte).

EXAMPLE 1–21

```
                              ;Unpacked BCD data (least significant first)
                              ;
0000  03 04 05                NUMB1  DB    3,4,5        ;defines the number 543
0003  07 08                   NUMB2  DB    7,8          ;defines the number 87
                              ;
                              ;Packed BCD data (least significant first)
                              ;
0005  37 34                   NUMB3  DB    34H,37H      ;defines the number 3437
0007  03 45                   NUMB4  DB    3,45H        ;defines the number 4503
```

Byte-Sized Data

Byte-sized data are stored as unsigned and signed integers. Figure 1–10 illustrates both the unsigned and signed forms of the byte-sized integer. The difference in these forms is the weight of the left-most bit position. Its value is 128 for the unsigned integer and –128 for the signed integer. In the signed integer format, the left-most bit represents the sign bit of the number as well as a weight of –128. For example, 80H represents a value of 128 as an unsigned number and –128 as signed number. Unsigned integers range in value from 00H–FFH (0–255). Signed integers range in value from –128 to 0 to +127.

Although negative signed numbers are represented in this way, they are stored in the **two's complement form.** The method of evaluating a signed number using the weights of each bit position is much easier than two's complementing a number to find its value. This is especially true in the world of calculators designed for programmers.

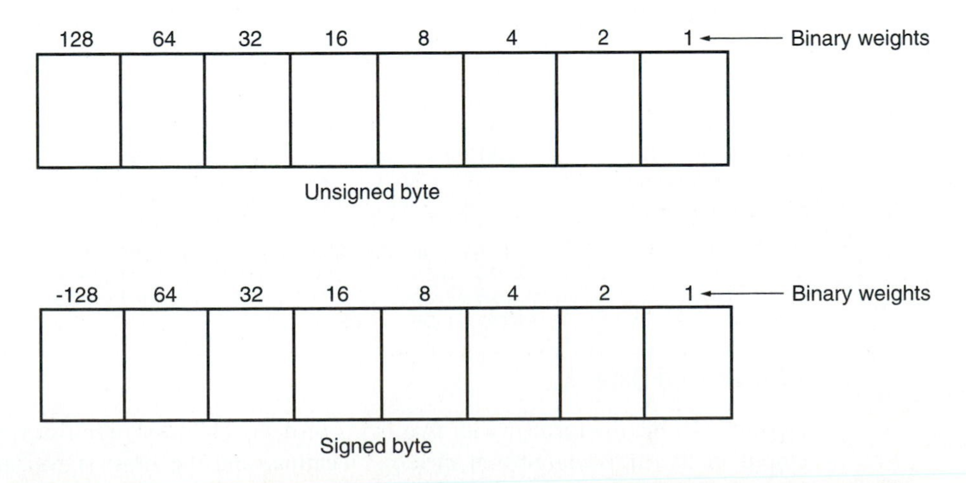

FIGURE 1–10 Unsigned and signed bytes illustrating the weights of each binary-bit position.

Whenever a number is two's complemented, its sign changes from negative to positive or positive to negative. For example, the number 00001000 is a +8. Its negative value (–8) is found by two's complementing the +8. To form a two's complement, first one's complement the number, then add a 1. To one's complement a number, invert each bit of a number from zero to one or from one to zero. Example 1–22 shows how numbers are two's complemented using this technique.

EXAMPLE 1–22

```
+8 = 00001000
     11110111 (one's complement)
+           1
-8 = 11111000 (two's complement)
```

Another, and probably simpler, technique for two's complementing a number starts with the right-most digit. Start writing down the number from right to left. Write the number exactly as it appears until the first one. Write down the first one, and then invert of complement all remaining ones to its left. Example 1–23 shows this technique with the same number as in Example 1–22.

EXAMPLE 1–23

```
+8 = 00001000
         1000 (write number to first 1)
     1111     (invert the remaining bits)
-8 = 11111000
```

To store 8-bit data in memory using the assembler program, use the DB directive as in prior examples. Example 1–24 lists many forms of 8-bit numbers stored in memory using the assembler program. Notice in this example that a hexadecimal number is defined by the letter H following the number, and a decimal number is written as is.

EXAMPLE 1–24

```
                    ;Unsigned byte-sized data
                    ;
0000  FE            DATA1   DB   254         ;define 254 decimal
0001  87            DATA2   DB   87H         ;define 87 hexadecimal
0002  47            DATA3   DB   71          ;define 71 decimal
                    ;
                    ;Signed byte-sized data
                    ;
0003  9C            DATA4   DB   -100        ;define a -100 decimal
0004  64            DATA5   DB   +100        ;define a +100 decimal
0005  FF            DATA6   DB   -1          ;define a -1 decimal
0006  38            DATA7   DB   56          ;define a 56 decimal
```

Word-Sized Data

A *word* (16 bits) is formed with two bytes of data. The least significant byte is always stored in the lowest-numbered memory location, and the most significant byte in the highest. This method of storing a number is called the **little endian** format. An alternate method, not used with the Intel family of microprocessors, is called big endian format.

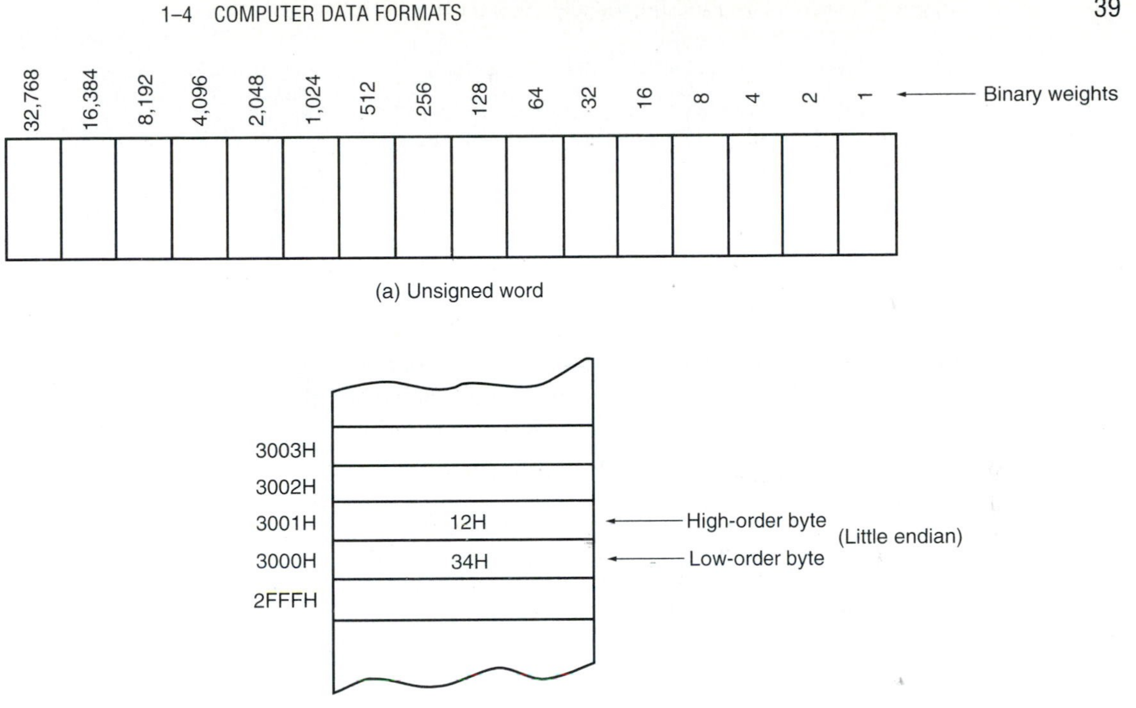

FIGURE 1–11 Storage format for a 16-bit word in (a) a register and (b) in two bytes of memory.

With the **big endian** format, numbers are stored with the lowest location containing the most-significant data. The big endian format is used with the Motorola family of microprocessors. Figure 1–11(a) shows the weights of each bit position in a word of data, and Figure 1–11(b) shows how the number 1234H appears when stored in the memory locations 3000H and 3001H. The only difference between a signed and an unsigned word is the left-most bit position. In the unsigned form, the left-most bit is unsigned and in the signed form its weight is a –32,768. As with byte-sized signed data, the signed word is in two's complement form when representing a negative number. Also notice that the low-order byte is stored in the lowest-numbered memory location (3000H) and the high-order byte is stored in the highest-numbered location (3001H).

Example 1–25 shows several signed and unsigned word-sized data stored in memory using the assembler program. Notice that the *define word(s)* directive, or **DW,** causes the assembler to store words in the memory instead of bytes as in prior examples. Also notice that the word data is displayed by the assembler in the same form as entered. For example, 1000H is displayed by the assembler as 1000. This is for our convenience, because the number is actually stored in the memory as a 00 10 in two consecutive memory bytes.

EXAMPLE 1–25

```
                        ;Unsigned word-sized data
                        ;
0000   09F0             DATA1    DW    2544           ;define 2544 decimal
```

```
0002    87AC          DATA2    DW     87ACH          ;define 87AC hexadecimal
0004    02C6          DATA3    DW     710            ;define 710 decimal
                      ;
                      ;Signed word-sized data
                      ;
0006    CBA8          DATA4    DW     -13400         ;define a -13400 decimal
0008    00C6          DATA5    DW     +198           ;define a +198 decimal
000A    FFFF          DATA6    DW     -1             ;define a -1 decimal
```

Doubleword-Sized Data

Doubleword-sized data requires four bytes of memory because it is a 32-bit number. Doubleword data appears as a product after a multiplication and also as a dividend before a division. In the 80386 through the Pentium, memory and registers are also 32 bits in width. Figure 1–12 shows the form used to store doublewords in the memory and the binary weights of each bit position.

When a doubleword is stored in memory, its least significant byte is stored in the lowest-numbered memory location, and its most significant byte is stored in the highest-numbered memory location using the little endian format. This is also true for word-sized data. For example, a 12345678H stored in memory location 00100H–00103H is stored with the 78H in memory location 00100H, the 56H in location 00101H, the 34H in location 00102H, and the 12H in location 00103H.

To define doubleword-sized data, use the assembler directive *define doubleword(s)*, **DD.** Example 1–26 shows both signed and unsigned numbers stored in memory using the DD directive.

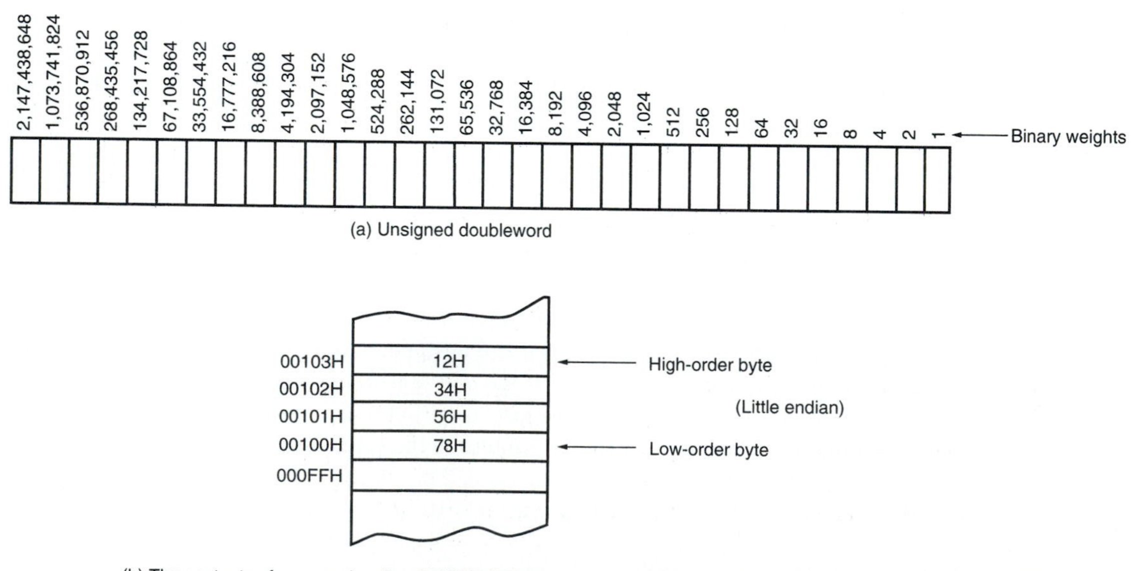

FIGURE 1–12 Storage format for a 32-bit doubleword in (a) a register and (b) in four bytes of memory.

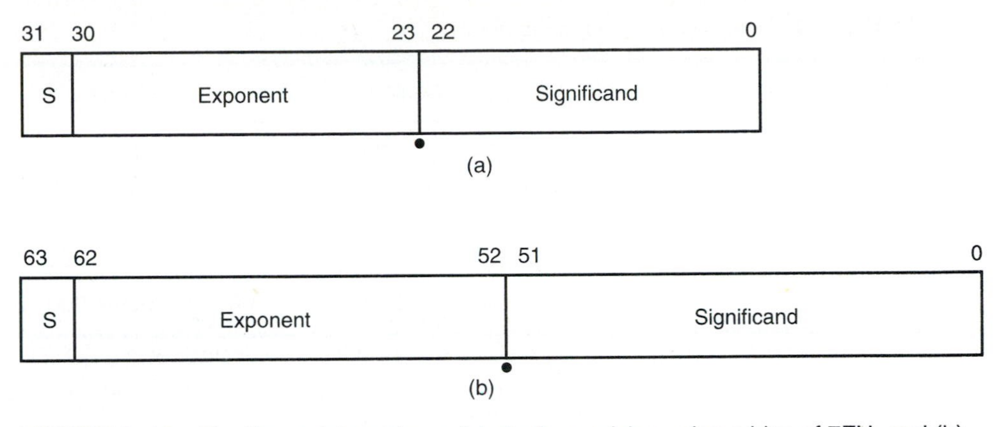

FIGURE 1–13 Floating-point numbers: (a) single-precision using a bias of 7FH, and (b) double-precision using a bias of 3FFH.

EXAMPLE 1–26

```
                          ;Unsigned doubleword-sized data
                          ;
0000    0003E1C0          DATA1   DD    254400       ;define 254400 decimal
0004    87AC1234          DATA2   DD    87AC1234H    ;define 87AC1234 hexadecimal
0008    00000046          DATA3   DD    70           ;define 70 decimal
                          ;
                          ;Signed doubleword-sized data
                          ;
000C    FFEB8058          DATA4   DD    -1343400     ;define a -1343400 decimal
0010    000000C6          DATA5   DD    +198         ;define a +198 decimal
0014    FFFFFFFF          DATA6   DD    -1           ;define a -1 decimal
```

Integers of any width may also be stored in memory. The forms listed here are standard forms, but that doesn't mean that a 128-byte wide integer can't be stored in the memory. The microprocessor is flexible enough to allow any size data. When nonstandard width numbers are stored in memory, the DB directive is normally used to store them. For example, the 24-bit number 123456H is stored using a DB 56H,34H,12H directive. Note that this conforms to the little endian format.

Real Numbers

Because many high-level languages use the Intel family of microprocessors, real numbers are often encountered. A real number, or as it is often called, a **floating-point** number, contains two parts: a **mantissa,** (*significand* or *fraction*) and an **exponent.** Figure 1–13 depicts both the 4- and 8-byte forms of real numbers as they are stored in any Intel system. Note that the 4-byte real number is called **single-precision** and the 8-byte form is called **double-precision.** The form presented here is the same form specified by IEEE[14] standard IEEE–754, version 10.0. This standard has been adopted as the standard form of

[14]IEEE is the Institute of Electrical Engineers.

TABLE 1–9 Single-precision real numbers

Decimal	*Binary*	*Normalized*	*Sign*	*Biased Exponent*	*Mantissa*
+12	1100	1.1×2^3	0	10000010	1000000 00000000 00000000
−12	1100	-1.1×2^3	1	10000010	1000000 00000000 00000000
+100	1100100	1.1001×2^6	0	10000101	1001000 00000000 00000000
−1.75	1.11	-1.11×2^0	1	01111111	1100000 00000000 00000000
+0.25	.01	1.0×2^{-2}	0	01111101	0000000 00000000 00000000
+0.0	0	0	0	00000000	0000000 00000000 00000000

real numbers with virtually all high-level programming languages and many applications packages. The standard also applies to data manipulated by the numeric coprocessor in the personal computer. Figure 1–13(a) shows the single-precision form that contains a sign bit, an 8-bit exponent, and a 24-bit fraction (mantissa). Note that because applications often require double-precision floating-point numbers (see Figure 1–13(b)), the Pentium, with its 64-bit data bus, performs memory transfers at twice the speed of the 80386/80486 microprocessors.

Simple arithmetic indicates that it should take 33 bits to store all three pieces of data. Not true—the 24-bit mantissa contains an implied (hidden) one-bit that allows the mantissa to represent 24-bits while being stored in only 23-bits. The *hidden bit* is the first bit of the normalized real number. When normalizing a number, it is adjusted so its value is at least 1, but less than 2. For example, if a 12 is converted to binary (1100) it is normalized and the result is 1.1×2^3. The 1. is not stored in the 23-bit mantissa portion of the number. The 1. is the hidden one-bit. Table 1–9 shows the single-precision form of this number and others.

The exponent is stored as a biased exponent. With the single-precision form of the real number, the bias is 127 (7FH); with the double-precision form, it is 1023 (3FFH). The *bias* is added to the exponent before it is stored in the exponent portion of the floating-point number. In the previous example, there is an exponent of 2^3, represented as a biased exponent of 127 + 3 or 130 (82H) in the short form or 1026 (402H) in the long form.

There are two exceptions to the rules for floating-point numbers. The number 0.0 is stored as all zeros. The number infinity is stored as all ones in the exponent and all zeros in the mantissa. The sign-bit indicates a ± 0.0 or a ± ∞.

As with other data types, the assembler can be used to define real numbers in both single- and double-precision forms. Because single-precision numbers are 32-bit numbers, use the DD directive or the **DQ** directive (*define quadwords(s)*) to define 64-bit double-precision real numbers. Optional directives for real numbers are REAL4, REAL8, and REAL10 for defining single-, double-, and extended-precision real numbers. Example 1–27 shows numbers defined in real number format.

EXAMPLE 1–27

```
                    ;Single-precision real numbers
                    ;
0000 3F9DF3B6       NUMB1  DD       1.234         ;define 1.234
0004 C1BB3333       NUMB2  DD       -23.4         ;define -23.4
```

```
0008    43D20000    NUMB3  REAL4    4.2E2        ;define 420
000C    3F9DF3B6    NUMB4  REAL4    1.234        ;define a 4-byte real number
                    ;
                    ;Double-precision real numbers
                    ;
0010                NUMB5  DQ       123.4        ;define 123.4
     405ED9999999999A
0018                NUMB6   REAL8   -23.4        ;define -23.4
     C1BB333333333333
0028                NUMB7   REAL8   123.4        ;define an 8-byte real number
     405ED9999999999A
                    ;
                    ;Extended-precision real numbers
                    ;
0030                NUMB8  REAL10   123.4        ;define a 10-byte real number
     4005F6CCCCCCCCCCCCCD
```

1–5 SUMMARY

1. The first mechanical computer was the abacus in 500 B.C. It remained unchanged until 1642, when Blaise Pascal improved it. An early mechanical computer system was the analytical engine developed by Charles Babbage in 1823. This machine never functioned because of the impossibility of obtaining properly machined parts.
2. The first electronic calculating machine was developed during World War II by Konrad Zuse, an early pioneer of digital electronics. His computer, the Z3, was used in aircraft and missile design for the German war effort.
3. The first electronic computer, which used vacuum tubes, was placed into operation in 1943 to break secret German military codes. This computer system, the Colossus, was invented by Alan Turing. Its program was fixed and could not be changed.
4. The first general-purpose, programmable, electronic computer system was developed in 1946 at the University of Pennsylvania. This first modern computer was called the ENIAC (Electronics Numerical Integrator and Calculator).
5. The first high-level programming language, called FLOW-MATIC, was developed for the UNIVAC I computer by Grace Hopper in the early 1950s. This led to FORTRAN and other early programming languages.
6. The world's first microprocessor, the Intel 4004, was a 4-bit microprocessor with a programmable controller on a chip. It was meager by today's standards, addressing a mere 4,096 four-bit memory locations. Its instruction set contained only 45 different instructions.
7. Microprocessors that are common today include the 8086/8088, (the first 16-bit microprocessors), the 80286, 80386, 80486, and Pentium microprocessor. With each newer version, improvements increased speed and performance. This process of speed and performance improvement will continue.
8. Microprocessor-based personal computers contain memory systems that include three main areas: TPA (transient program area), system area, and extended memory. The TPA holds application programs, the operating system, and drivers. The system area contains memory used for video display cards, disk drives, and the BIOS ROM. The

extended memory area is only available to the 80286 through the Pentium microprocessor in an AT-style personal computer system.

9. The 8086/8088 address 1M bytes of memory from location 00000H through location FFFFFH. The 80286 and 80386SX address 16M bytes of memory from location 000000H through FFFFFFH. The 80386SL addresses 32M bytes of memory from location 0000000H through 1FFFFFFH. The 80386DX, 80486, and Pentium address 4G bytes of memory from location 00000000H through FFFFFFFFH.
10. All versions of the 8086-80486 and Pentium microprocessor address 64K bytes of I/O address space. These I/O ports are numbered from 0000H through FFFFH, with I/O ports 0000H–04FFH reserved for use by the personal computer system.
11. The operating system in many personal computers is either MSDOS (Microsoft disk operating system) or PCDOS (personal computer disk operating system from IBM). The operating system performs the task of operating or controlling the computer and its I/O devices.
12. The microprocessor is the controlling element in a computer system. The microprocessor performs data transfers, simple arithmetic, and logic operations, and makes simple decisions. The microprocessor executes programs stored in the memory system to perform complex operations in short periods of time.
13. All computer systems contain three buses to control memory and I/O. The address bus is used to request a memory location or I/O device. The data bus transfers data between the microprocessor and its memory and I/O spaces. The control bus controls the memory and I/O and requests reading or writing of data. Control is accomplished with $\overline{\text{IORC}}$ (I/O read control), $\overline{\text{IOWC}}$ (I/O write control), $\overline{\text{MRDC}}$ (memory read control), and $\overline{\text{MWTC}}$ (memory write control).
14. Numbers are converted from any number base to decimal by noting the weights of each position. The weight of the position to the left of the radix point is always the units position in any number system. The position to the left of the units position is always the radix times one. Succeeding positions are determined by multiplying by the radix. The weight of the position to the right of the radix point is always determined by dividing by the radix.
15. Conversion from a whole decimal number to any other base is accomplished by dividing by the radix. Conversion from a fractional decimal number is accomplished by multiplying by the radix.
16. Hexadecimal data are represented in hexadecimal or, at times, in binary-coded hexadecimal (BCH). A binary-coded hexadecimal number is one that is written with a 4-bit binary number that represents each hexadecimal digit.
17. The ASCII code is used to store alphabetic or numeric data. The ASCII code is a 7-bit code and can use an eighth bit to extend the character set from 128 codes to 256 codes. The carriage return (enter) code returns the print head or cursor to the left margin. The line feed code moves the cursor or print head down a line.
18. Binary-coded decimal (BCD) data are sometimes used in a computer system to store decimal data. This data is stored in either packed (two digits per byte) or unpacked (one digit per byte) form.
19. Binary data are stored as a byte (8 bits), word (16 bits), or doubleword (32 bits) in a computer system. This data may be unsigned or signed. Signed negative data are always stored in the two's complement form. Data that are wider than 8 bits are always stored using the little endian format.

20. Floating-point data are used to store whole, mixed, and fractional numbers. A floating-point number is composed of a sign, a mantissa, and an exponent.
21. We use the assembler directive DB to define bytes, DW to define words, DD to define doublewords, and DQ to define quadwords.
22. Example 1–28 shows the assembly language formats for storing numbers as bytes, words, doublewords, and real numbers. Also shown are ASCII-coded character strings.

EXAMPLE 1–28

```
                              ;ASCII data
                              ;
0000  54 68 69 73 20 69       MES1    DB     'This is a character string in ASCII'
         73 20 61 20 63 68
         61 72 61 63 74 65
         72 20 73 74 72 69
         6E 67 20 69 6E 20
         41 53 43 49 49
0023  53 6F 20 69 73 20       MES2    DB     'So is this'
         74 68 69 73
                              ;BYTE data
                              ;
002D   17                     DATA1   DB     23        ;23 decimal
002E   DE                     DATA2   DB     -34       ;-34 decimal
002F   34                     DATA3   DB     34H       ;34 hexadecimal
                              ;
                              ;WORD data
                              ;
0030   1000                   DATA4   DW     1000H     ;1000 hexadecimal
0032   FF9C                   DATA5   DW     -100      ;-100 decimal
0034   000C                   DATA6   DW     +12       ;=12 decimal
                              ;
                              ;DOUBLEWORD data
                              ;
0036   00001000               DATA7   DD     1000H     ;1000 hexadecimal
003A   FFFFFED4               DATA8   DD     -300      ;-300 decimal
003E   00012345               DATA9   DD     12345H    ;12345 hexadecimal
                              ;
                              ;Real data
                              ;
0042   4015C28F               DATA10  REAL4  2.34     ;2.34 decimal
0046   C00CCCCD               DATA11  REAL4  -2.2     ;-2.2 decimal
004A                          DATA12  REAL8  100.3    ;100.3 decimal
         4059133333333333
```

1–6 QUESTIONS AND PROBLEMS

1. Who developed the Analytical Engine?
2. The 1890 census used a new device called a punched card. Who developed the punch card?
3. Who was the founder of IBM Corporation?
4. Who developed the first electronic calculator?
5. The first truly electronic computer system was developed for what purpose?

6. The first general-purpose programmable computer was called the _______________.
7. The world's first microprocessor was developed in 1971 by _______________.
8. Who was the Countess of Lovelace?
9. Who developed the first high-level programming language called FLOW-MATIC?
10. What is a von Neumann machine?
11. Which 8-bit microprocessor ushered in the age of the microprocessor?
12. The 8085 microprocessor, introduced in 1977, has sold _______________ copies.
13. Which Intel microprocessor was the first to address 1M bytes of memory?
14. The 80386SL addresses _______________ bytes of memory.
15. How much memory is available to the 80486 microprocessor?
16. When did Intel introduce the Pentium microprocessor?
17. What is the acronym MIPs?
18. What is the acronym CISC?
19. A binary bit stores a _______________ or a _______________.
20. A computer K is equal to _______________ bytes.
21. A computer M is equal to _______________ K bytes.
22. A computer G is equal to _______________ M bytes.
23. How many typewritten pages of information are stored in a 4G byte memory system?
24. The first 1M byte of memory in a computer system contains a _______________ and a _______________ area.
25. How much memory is found in the transient program area?
26. How much memory is found in the systems area?
27. The 8086 microprocessor addresses _______________ bytes of memory.
28. The 80286 microprocessor addresses _______________ bytes of memory.
29. Which microprocessors address 4G bytes of memory?
30. Memory above the first 1M bytes is called _______________ memory.
31. What is the BIOS?
32. What is DOS?
33. What is the difference between an XT and an AT computer system?
34. What is the VESA local bus?
35. The ISA bus holds _______________-bit interface cards.
36. What is the XMS?
37. What is the EMS?
38. A driver is stored in the _______________ area.
39. What is a TSR?
40. How is a TSR often accessed?
41. What is the purpose of the CONFIG.SYS file?
42. What is the purpose of the AUTOEXEC.BAT file?
43. The COMMAND.COM program processes what information?
44. The personal computer system addresses _______________ bytes of I/O space.
45. Where is the high memory located in a personal computer?
46. The DEVICE or DEVICEHIGH statement is found in what file?
47. Where are the upper memory blocks used by MSDOS version 5.0 or 6.2?
48. Where is the video BIOS?
49. Draw a block diagram of a computer system.
50. What is the purpose of the microprocessor in a microprocessor-based computer system?

51. List the three buses found in all computer systems.
52. Which bus transfers the memory address to the I/O device or the memory device?
53. Which control signal causes the memory to perform a read operation?
54. What is the purpose of the $\overline{\text{IORC}}$ signal?
55. If the $\overline{\text{MRDC}}$ signal is a logic 0, which operation is performed by the microprocessor?
56. Convert the following binary numbers into decimal:
    (a) 1101.01
    (b) 111001.0011
    (c) 101011.0101
    (d) 111.0001
57. Convert the following octal numbers into decimal:
    (a) 234.5
    (b) 12.3
    (c) 7767.07
    (d) 123.45
    (e) 72.72
58. Convert the following hexadecimal numbers into decimal:
    (a) A3.3
    (b) 129.C
    (c) AC.DC
    (d) FAB.3
    (e) BB8.0D
59. Convert the following decimal integers into binary, octal, and hexadecimal:
    (a) 23
    (b) 107
    (c) 1238
    (d) 92
    (e) 173
60. Convert the following decimal numbers into binary, octal, and hexadecimal:
    (a) .625
    (b) .00390625
    (c) .62890625
    (d) 0.75
    (e) .9375
61. Convert the following hexadecimal numbers into binary-coded hexadecimal code (BCH):
    (a) 23
    (b) AD4
    (c) 34.AD
    (d) BD32
    (e) 234.3
62. Convert the following binary-coded hexadecimal numbers into hexadecimal:
    (a) 1100 0010
    (b) 0001 0000 1111 1101
    (c) 1011 . 1100
    (d) 0001 0000
    (e) 1000 1011 1010

63. Convert the following binary numbers to the one's complement form:
    (a) 1000 1000
    (b) 0101 1010
    (c) 0111 0111
    (d) 1000 0000
64. Convert the following binary numbers to the two's complement form:
    (a) 1000 0001
    (b) 1010 1100
    (c) 1010 1111
    (d) 1000 0000
65. Define *byte, word,* and *doubleword.*
66. Convert the following words into ASCII-coded character strings.
    (a) FROG
    (b) Arc
    (c) Water
    (d) Well
67. What is the ASCII code for the enter key and what is its purpose?
68. Use an assembler directive to store the ASCII-character string 'What time is it?' in memory.
69. Convert the following decimal numbers into 8-bit signed binary numbers.
    (a) +32
    (b) –12
    (c) +100
    (d) –92
70. Convert the following decimal numbers into signed binary words.
    (a) +1,000
    (b) –120
    (c) +800
    (d) –3,212
71. Use an assembler directive to store –34 in the memory.
72. Show how the following 16-bit hexadecimal numbers are stored in the memory system.
    (a) 1234H
    (b) A122H
    (c) B100H
73. What is the difference between the *big endian* and *little endian* formats for storing numbers larger than 8 bits in width.
74. Use an assembler directive to store 123A hexadecimal into the memory.
75. Convert the following decimal numbers into both packed and unpacked BCD forms.
    (a) 102
    (b) 44
    (c) 301
    (d) 1,000
76. Convert the following binary numbers into signed decimal numbers.
    (a) 10000000
    (b) 00110011

(c) 10010010
(d) 10001001

77. Convert the following BCD numbers (assume that these are packed numbers) into decimal numbers.
(a) 10001001
(b) 00001001
(c) 00110010
(d) 00000001

78. Convert the following decimal numbers into single-precision floating-point numbers.
(a) +1.5
(b) –10.625
(c) +100.25
(d) –1,200

79. Convert the following single-precision floating-point numbers into decimal.
(a) 0 10000000 11000000000000000000000
(b) 1 01111111 00000000000000000000000
(c) 0 10000010 10010000000000000000000
(d) 1 01111110 11000000000000000000000

CHAPTER 2

Disks, DOS, and Batch Programs

INTRODUCTION

This chapter continues the introduction to DOS. In order to set up and operate a personal computer, an understanding of DOS is required. This will be accomplished by highlighting the DOS commands and some of the commonly used features of DOS. (Note that this chapter presents the commands and features of the latest DOS version: 6.X).

One of these features is the batch program. Batch programs are used to store DOS commands for later playback. This saves the computer user from typing a long and involved series of DOS commands.

The DOS commands or directives and programs include: FORMAT, CD, RD, MD, DIR, TYPE, ATTRIB, COPY, and many batch directives. These will be presented in a logical way so they can be used to control the personal computer.

The format of the disk memory is also explained in this chapter. This includes a discussion of disks and disk formats and the terms used to describe them. Defined are *track, sector, cylinder, directory,* etc.

OBJECTIVES

Upon completion of this chapter, you will be able to:

1. Explain the format of a file name and describe wild-card applications with DOS commands.
2. Describe the types and organization of information stored on a disk in the root directory, subdirectories, file allocation table, and boot sector.
3. State how many bytes are stored on various sizes of mini- and micro-floppy disks.
4. Define such terms as *sector, track, cluster, head,* and *cylinder.*
5. Explain the operation and features of compressed disks, as determined by the Double-Space (DriveSpace) program.

6. Detail the operation of the more commonly used DOS commands such as DIR, MD, RD, CD, ATTRIB, COPY, FORMAT, and so forth.
7. Write simple batch programs to control the operation of the personal computer.
8. Develop a CONFIG.SYS file that allows DOS versions 6.0 and 6.2 to boot up different system configurations.

2–1 DISKS AND DISK ORGANIZATION

Data are stored on the disk in the form of files that contain either data or programs. Before we examine files, let's look at how information is placed onto the surface of the disk. The data on a disk is organized in four main areas: the boot sector, the file allocation table (FAT), the root directory, and the data storage area. The first sector on the disk is called the boot sector. The **boot sector** contains a program called a **bootstrap loader** that reads the disk operating system (usually DOS) from the disk into the memory system when power is applied to the computer. The **FAT** stores codes that indicate which disk clusters or sectors contain data, which are free, and which are bad. All references to disk files are handled through the FAT and a directory entry. The **root directory** is where directory names, files, and programs are referenced. The disk files are all considered sequential access files, meaning that they are accessed one byte at a time from the beginning of the file to the end.

File Names

Files and programs are stored on a disk and referenced by a file name and an extension to the file name. With the DOS operating system, the file name can be 1 to 8 characters in length. The file name can contain just about any ASCII character except for spaces or the " \ . / [ ] * , : < > | ; ? = characters. In addition to the file name, the file can have an optional 1 to 3 digit extension to the file name. Table 2–1 shows a few file names with and without extensions. Note that the name of a file and its extension are always separated by a period.

Directory and Subdirectory Names. The DOS file management system arranges the data and programs on a disk into directories and subdirectories. The rules that apply to file names also apply to directory and subdirectory names. That is, the name can be up to eight characters in length and an extension can appear that is up to three characters in length. An

TABLE 2–1 DOS file names and extensions

Name.Extension
TEST.TXT
READ-ME.DOC
ALWAY12.COM
RUN_IT~.EXE
CHAPTER.02
NOEXTEN1
1.2

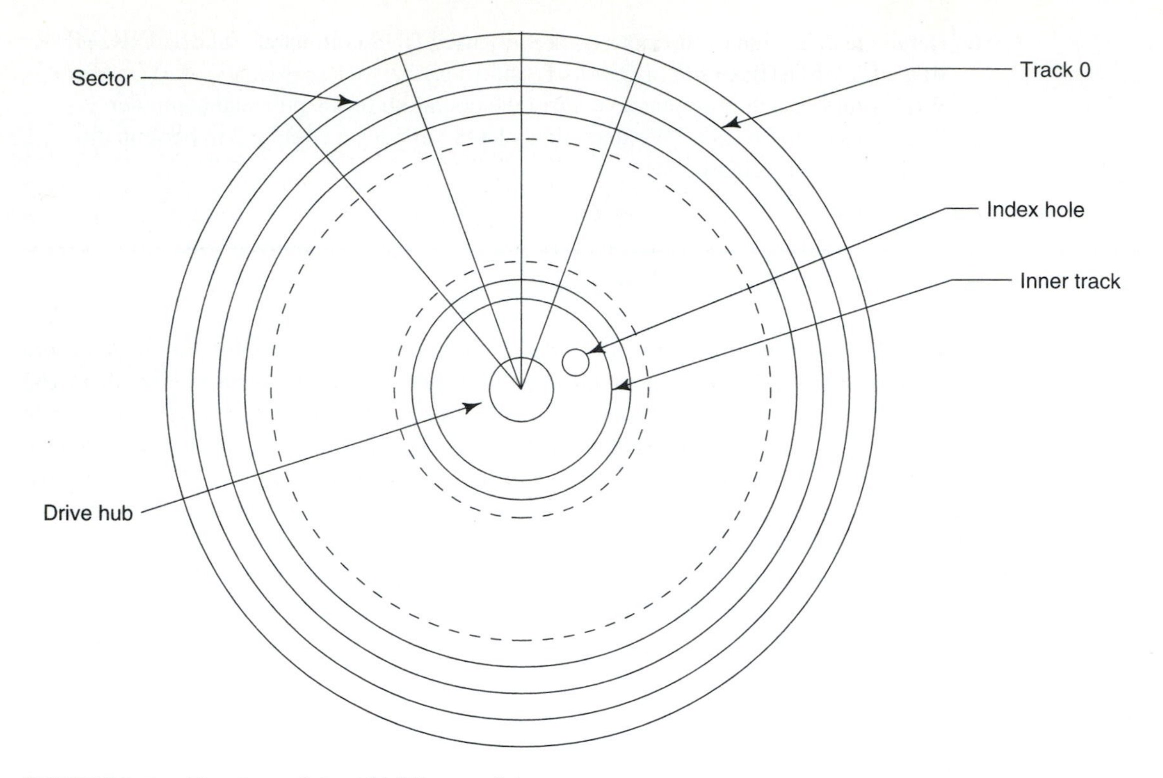

FIGURE 2–1 Structure of the $5^{1}/_{4}''$ floppy disk.

extension is not usually used with a directory name, but can be used if needed. The disk is structured so it contains a root directory when first formatted. The root directory for a hard disk used as drive C is C:\. Any other directory is placed in the root directory. For example, C:\DATA is directory DATA in the root directory. Each directory placed in the root directory can have subdirectories. Examples are the subdirectories C:\DATA\AREA1 and C:\DATA\AREA2 where the directory DATA contains two subdirectories: AREA1 and AREA2. Subdirectories, too, can have subdirectories. For example, C:\DATA\AREA2\LIST depicts directory DATA, subdirectory AREA2, which contains subdirectory LIST. More detail on directories and subdirectories is provided below and in Section 2–2.

Disk Organization

Figure 2–1 illustrates the organization of data stored in sectors and tracks on the surface of a $5^{1}/_{4}''$ floppy disk. $3^{1}/_{2}''$ floppy disks and hard disks are similar, but do not have an index hole. The surface of the disk is divided into concentric, nonspiraling rings of data called **tracks.** The outer track is always track 0 and the inner track is always track 39, on a double-density, $5^{1}/_{4}''$ floppy disk, or 79 on all other floppy disks. The number of tracks on a hard disk is determined by the disk size and could be 1,000 or more for very large hard disks. Note that the current version of WINDOWS can have trouble creating a permanent virtual swap file on hard disks that contain more than 1024 tracks.

Each track is divided into sections, or groups of data areas, called *sectors*. A sector almost always contains 512 bytes of data. When a program, or a file containing data, is stored on the disk it is stored in sectors that may or may not be contiguous to each other. It is not unusual to find that one program or data file occupies sectors scattered about the surface of the disk. This is not desirable, as will be explained later.

Floppy Disks. Modern floppy disks are available in two different sizes, the 5¼″ mini-floppy and the 3½″ **micro-floppy disk.** The 5¼″ **mini-floppy disk,** which is being phased out, is currently available as a double-sided, double-density (DSDD) disk that stores 360K bytes and a high-density (HD) disk that stores 1.2M bytes. Early 5¼″ floppy disks were single-sided and stored either 180K (double-density) or 92K (single-density) bytes. Earlier floppies were a standard 8″ in size and stored 256K bytes of data. The 3½″ micro-floppy is available as a double-sided, double-density (DSDD) disk that stores 720K bytes, a high-density (HD) disk that stores 1.44 M bytes, and an extra-high-density (EHD or ED) disk that stores 2.88M bytes. The 2.88M byte floppy disks are most popular.

Each type of disk memory organizes its surface into *tracks* that contains **sectors.** Each sector on a disk usually contains 512 bytes of data. **Clusters** are groups of sectors used to access information on the disk. The number of sectors and tracks varies from one disk type to another. For example, the 5¼″ double-density, double-sided disk contains 40 tracks with 9 sectors per track ($40 \times 9 \times 512 = 180$K bytes) per side. A pair of tracks (top and bottom) is called a **cylinder.** We often use the term *cylinder* in place of *track* and *upper* or *lower* **head** in place of *side.* Some fixed or hard disk drives have up to 16 heads (tracks) per cylinder because they contain more than one disk platter. Table 2–2 lists the common sizes of floppy disks and summarizes information about the number of tracks, sectors, clusters, root directory size, and disk capacities. These figures are for unformatted disks; actual capacity will be less due to loss of space to the boot sector, FAT, root directory, and operating system.

Figure 2–2 shows the organization of major areas on the disk. The length of the FAT (file allocation table) is determined by the size of the disk. Disk sectors or groups of sectors are represented in the FAT, usually by a 16-bit number. Some extremely large fixed disks use a 32-bit number as an entry in the FAT. Many floppy disks use a 12-bit FAT. The length of the root directory is usually fixed at 3,584 bytes on a 5¼″ DSDD floppy disk. The root directory is larger on a 3½″ floppy or hard disk. The boot sector is always a single *512 bytes* long and located in the outer track (Track 0) at sector 0, the first sector on the disk.

The boot sector contains a bootstrap loader program, which is read into RAM when the system is powered up. After the bootstrap loader is read into RAM, it loads the

TABLE 2–2 5¼″ and 3½″ floppy disks

Size	*Tracks per Side*	*Sectors per Track*	*Sectors per Cluster*	*Bytes in Root*	*Unformatted Capacity per Disk*
5¼″ DSDD	40	9	2	3,584	368,640 (360K) bytes
5¼″ HD	80	15	1	7,168	1,228,880 (1.2M) bytes
3½″ DSDD	80	9	2	3,584	737,280 (720K) bytes
3½″ HD	80	18	1	7,168	1,474,560 (1.44M) bytes
3½″ ED	80	36	1	7,168	2,949,120 (2.88M) bytes

Note: All sectors are 512 bytes in length.

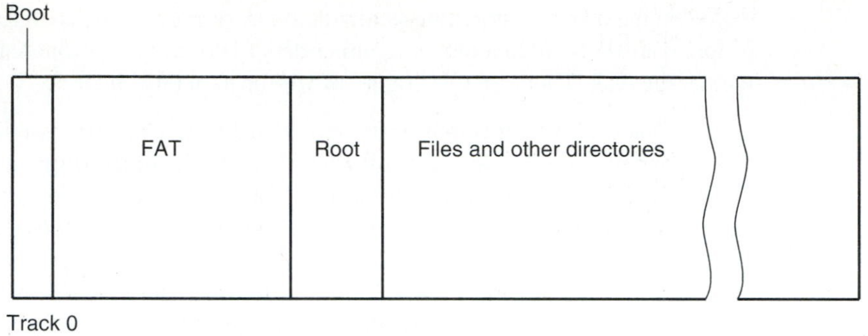

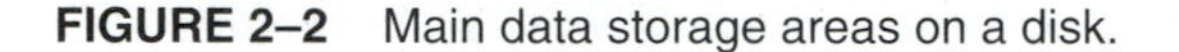

FIGURE 2–2 Main data storage areas on a disk.

IO.SYS[1] and MSDOS.SYS[2] programs into RAM. Next, the bootstrap loader passes control to the MSDOS control program, which uses the CONFIG.SYS file to configure the system. After the CONFIG.SYS file configures the system, the MSDOS.SYS program loads the DOS command processor COMMAND.COM into memory. The AUTOEXEC.BAT file is then executed and control is passed to the user. User control is often provided at the DOS prompt, from either DOSSHELL (available with DOS version 5.0) or WINDOWS.

The File Allocation Table (FAT). The FAT indicates which sectors or clusters of sectors are free, which are corrupted (unusable), and which contain valid data or programs. Data are accessed by cluster number on all hard disk drives. The FAT is referenced each time DOS writes data to the disk, in order to find a free sector. The FAT is also accessed for a read operation, to determine the location of the next cluster in a chain. Each free cluster is indicated by a 0000H in the FAT and each occupied cluster is indicated by the cluster number of the next cluster in the file chain or the last cluster code, FFFFH. A cluster can be any size, from one or two sectors, on many floppy disks up to 16 sectors on a hard disk memory. Many hard disk memory systems use four or more sectors per cluster, which means the smallest file allocation unit is 512 × 4, or 2,048, bytes in length if a cluster contains four sectors. Note that clusters are also referred to as *file allocation units*. The larger the hard disk drive, the more sectors in a cluster. A 16-bit FAT table accesses 64K—17 clusters. A 0000H code indicates a free cluster, FFF7H indicates a bad cluster, FFFFH indicates the end of a file chain, and the remaining cluster numbers 0001H–FFF0H are available. If a cluster is four sectors of 512 bytes, a fixed disk containing a 16-bit FAT accesses a maximum of 65,519 clusters. Because a sector is 512 bytes and a cluster is 2,048 bytes or four sectors, this hard disk drive contains a maximum of 134,182,912 bytes of unformatted data. If a cluster contains eight sectors (4,096 bytes), then the maximum capacity of the hard disk drive is 268,365,824 bytes. With 16 sectors per cluster the hard disk can be a maximum of 536,731,648 bytes. If a larger hard disk exists, either more sectors per clusters are allocated or the FAT is increased to 24 bits. With DOS, the FAT is currently 16

[1]IO.SYS is an I/O control progam provided by Microsoft Corporation with Microsoft DOS.

[2]MSDOS.SYS is the Microsoft Disk Operating System.

bits for most hard disk drives. Note that if the cluster size is 4K bytes, the smallest file allocated by DOS is 4K bytes even if a file contains only 1 byte of data.

Directories and Subdirectories. Figure 2–3 shows the format of each directory entry in the root or any other directory or subdirectory. For example, in the DOS command line C:\GAME\TEST.TXT, the disk drive is C, the directory is GAMES and the file is TEST.TXT. Each entry contains a name, extension, attribute, time, date, location, and length. The length of the file is stored as a 32-bit number. This means that a file can have a maximum length of 4G bytes. The location is the starting cluster number. The root directory on a 5¹/₄″ DSDD floppy disk has room for 112 of these 32-byte directory entries. On other floppy disks or on a hard disk drive the root directory has room for 224 or more entries. The root directory is a fixed area on the disk. When the root directory is full, the disk is full. This is why it is important to use directories and subdirectories, which can contain any number of files. A single directory requires one entry in the root directory. The entry points to another cluster or clusters that contain the directory. A directory can be any length. A subdirectory entry or a file is stored in the directory created, not in the root directory. This allows an almost infinite number of files stored in any subdirectories (see Figure 2–4). The same file name can appear in different directories and subdirectories and be considered different.

The location area in the directory entry holds the first cluster number where the file is stored on the disk. The FAT location that pertains to this cluster address holds either the cluster number where the file continues, or an FFFFH to indicate the end of the file. This way, DOS can store files on the disk in fragments if required. Fragments cause problems after a disk is in use for a long period of time. As programs and files are stored and erased, they always seem to occupy a larger number of smaller and smaller fragments of disk space. Eventually, a file that requires four clusters will be stored at four different places on the surface of the disk. This significantly increases the time required to access and load

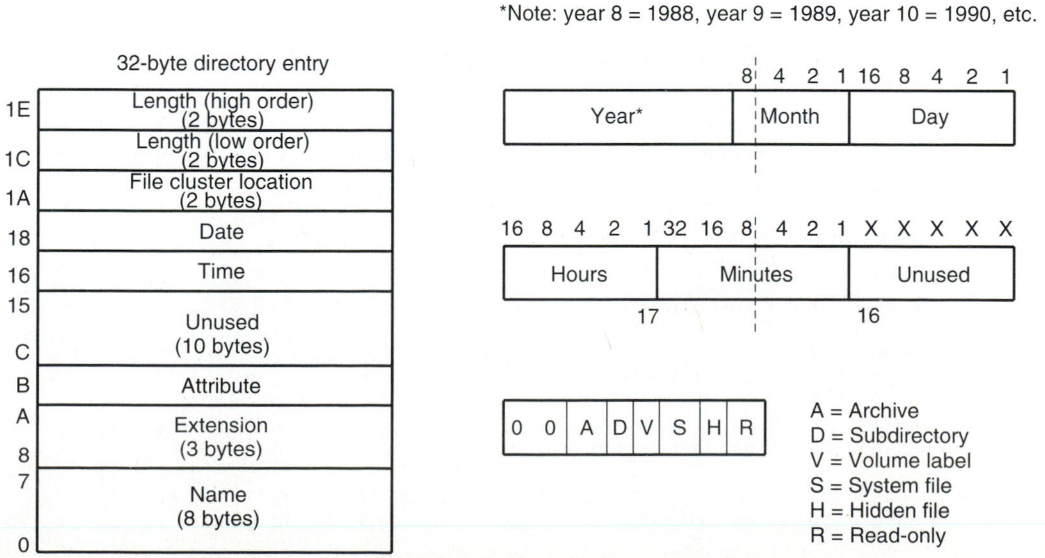

FIGURE 2–3 Format for any directory or subdirectory entry.

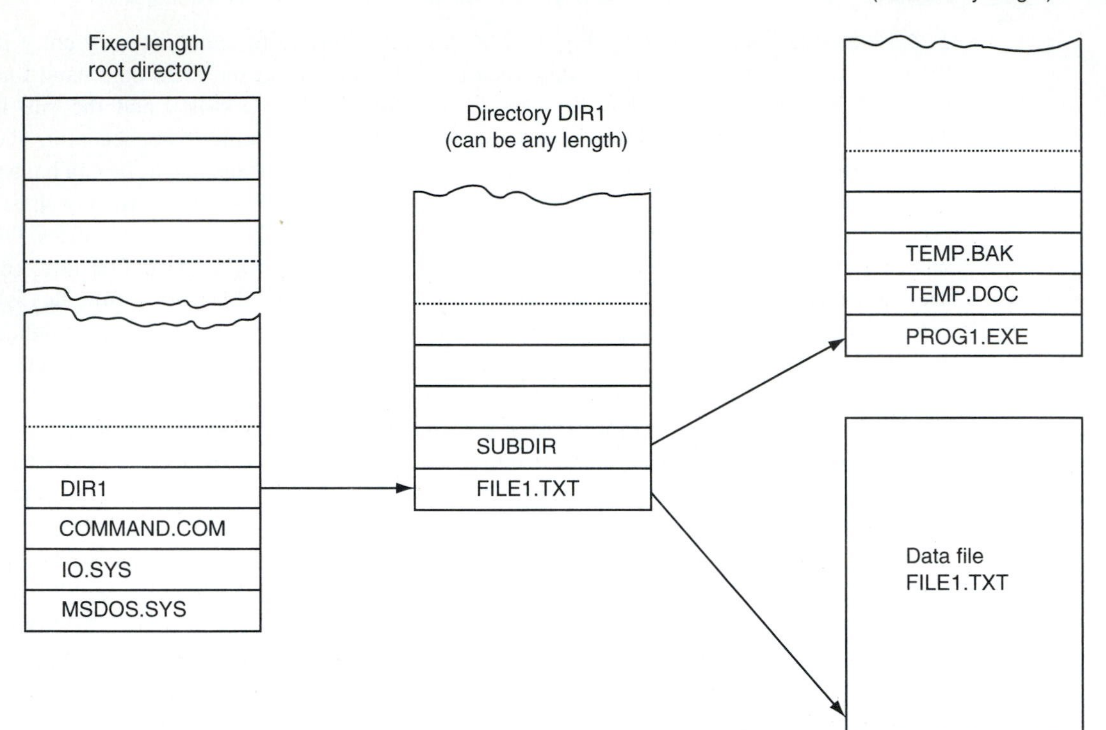

FIGURE 2–4 The root directory, a directory, a subdirectory, and a data file. Notice how the root directory entry addresses the directory, and how the directory addresses the subdirectory and a a data file FILE1.TXT.

large files. Fragmentation is corrected by using a defragmentation program such as PC-TOOLS[3] or the DOS DEFRAG program. It is recommended that the hard disk be defragmented occasionally, especially if load times become a noticeable problem. In practice, it is recommended that the hard disk be defragmented no more than one time per month because of the length of time required to defragment a disk.

EXAMPLE 2–1

```
0000    49 4F 20 20 20 20 20 20 53 59 53 07 00 00 00 00 IO SYS
0010    00 00 00 00 00 00 00 00 93 11 02 00 39 82 00 00

0020    4D 53 44 4F 53 20 20 20 53 59 53 07 00 00 00 00 MSDOS SYS
0030    00 00 00 00 00 00 C0 44 93 12 13 00 92 00 00 00

0040    43 4F 4D 4D 41 4E 44 20 43 4F 4D 00 00 00 00 00 COMMAND COM
0050    00 00 00 00 00 00 00 00 93 11 26 00 B5 92 00 00

0060    42 41 52 52 59 20 42 52 45 59 20 28 00 00 00 00 BARRY BREY
0070    00 00 00 00 00 00 E0 AD 6A 13 00 00 00 00 00 00
```

[3]PCTOOLS is registered to Central Point Software.

```
0080   50 43 54 4F 4F 4C 53 20 20 20 20 10 00 00 00 00 PCTOOLS
0090   00 00 00 00 00 00 80 AE 6A 13 5C 00 00 00 00 00

00A0   44 4F 53 20 20 20 20 20 20 20 20 10 00 00 00 00 DOS
00B0   00 00 00 00 00 00 E0 B0 6A 13 4E 00 00 00 00 00

00C0   52 55 4E 5F 46 57 20 20 42 41 54 00 00 00 00 00 FUN_FW BAT
00D0   00 00 00 00 00 00 40 BD 6A 13 97 0F 4A 00 00 00

00E0   46 4F 4E 54 57 41 52 45 20 20 20 10 00 00 00 00 FONTWARE
00F0   00 00 00 00 00 00 60 BD 6A 13 6E 00 00 00 00 00
```

Example 2–1 shows how part of the root directory appears in a hexadecimal dump. Identify the date, time, location, and length of each entry. Also identify the attribute for each entry. The listing shows hexadecimal data and ASCII data, as is customary in most computer hexadecimal dumps.

Compact Disk Read-Only Memory (CDROM). Another common storage medium is the compact disk read-only memory (CDROM) or optical disk memory. The CDROM is a low cost, read-only, optical disk memory. Access times for a single-speed CDROM are typically 300 ms or longer—about the same as a floppy disk. By comparison, a hard disk magnetic memory can have access times as little as 7 ms. Data transfer rates are 150K bytes per second for the single-speed CDROM, 300K bytes for the double-speed, 450K bytes for the triple-speed, and 600K bytes for the quad-speed version. Data transfer speeds on a hard disk memory are often well over 1M bytes per second and can be much higher—up to 132M bytes per second in a PCI or VESA local bus system. The CDROM software is available for most programs and applications. The CDROM is available with large volume data storage such as the Bible, encyclopedias, clip art, magazine articles, etc. These applications have wide appeal at current prices. A CDROM stores 660M bytes of data, or a combination of data and musical passages. The data stored on a CDROM is in the same form as that stored on a floppy or hard disk drive—a FAT, root directory, and data organized in clusters. Musical passages are always stored using the 150K bytes per second transfer rate; one CDROM can hold up to 70 minutes of music. Because the transfer rate is limited, music is often stored in a compressed form that must be decoded by a digital sound card. CDROM musical passages are stored in the same format as that found on an audio CD. As systems develop and become more visually active, a CDROM drive will become a system requirement.

Figure 2–5 illustrates the operation of the CDROM drive. A solid state laser beam is focused on the surface of the CDROM through a set of lenses. If the laser light strikes a pit (a depression in the surface of the CDROM), little light is reflected to a photodetector and no voltage is generated. If the laser light strikes a land (an unpitted surface), most of it is reflected to the photodetector and a voltage is produced. The lack of voltage at the photodetector represents a logic zero, and the presence of voltage represents a logic one. In this way, binary digital data are read from the surface of the CDROM.

Another type of optical memory is the WORM (write-once/read mostly) drive, which sees some commercial use. Its application is very specialized due to the nature of WORM. Because data may only be written once, its main application is in the banking industry, the insurance industry, and other massive data storing organizations. WORM is normally used to form an audit trail of transactions that are retrieved only during an audit.

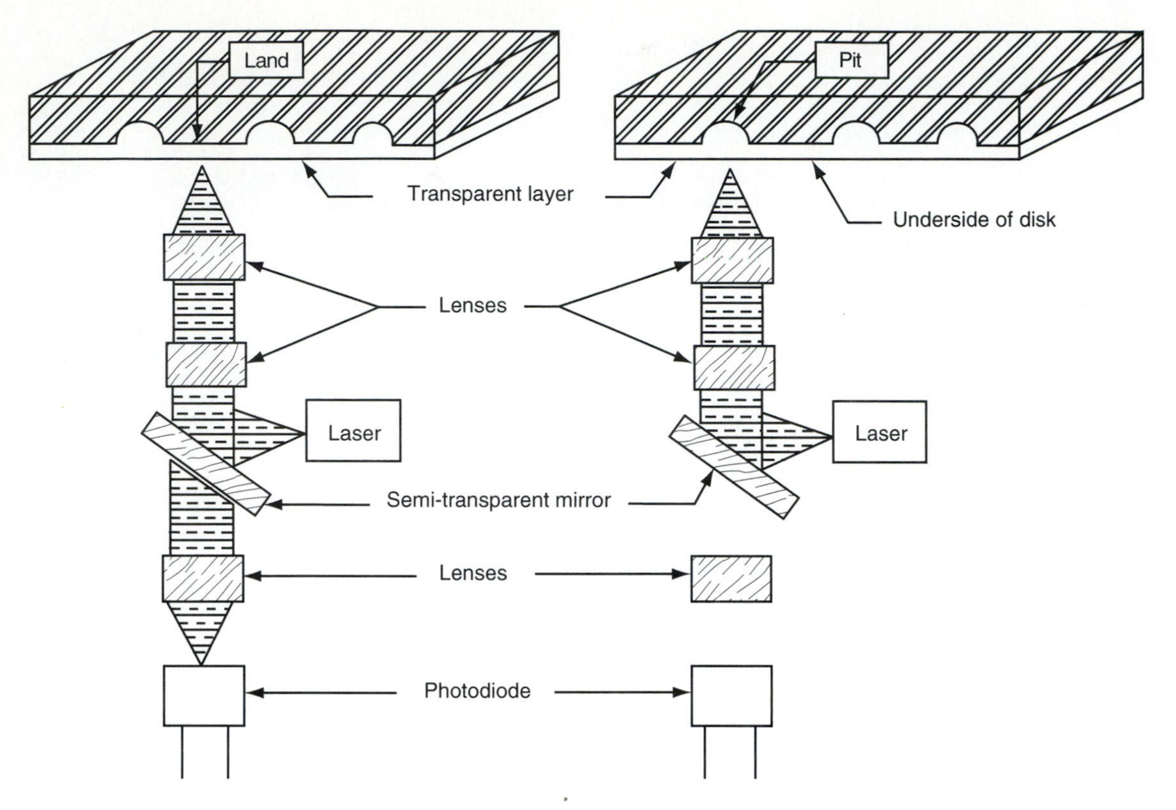

FIGURE 2–5 The mechanism for reading a CDROM.

You might call WORM an archiving device. The WORM drive is not compatible with CDROM.

Read/write CDROMs are beginning to appear. These presently cost 10 to 20 times more than the read-only CDROM and do not have a standard data format. Various manufacturers produce these devices with different capacities and formats. A few are beginning to produce them in the same format as the CDROM, but the price is prohibitive for most users.

The main advantage of the optical disk is its durability. Because a solid state laser beam is used to read the data from the disk, and the focus point is below a protective plastic coating, the surface of the disk may contain small scratches and dirt particles and still be read correctly. This feature means the optical disk requires less care than a comparable floppy disk. The only way to destroy data on an optical disk is to break it or deeply scar it.

Disk Partitions

Another feature of DOS is its ability to organize the surface of the disk into partitions. A **partition** is a section of the surface of the hard disk drive that contains file storage space for an operating system or other logical disk drives. The DOS operating system is only one of many that are available. Recently, partitions have been used to boost the storage capacity of the IDE hard disk from 500M to over 1G byte. The ability to partition the drive into sections that contain different operating systems is advantageous if, for example, you plan to use DOS and OS/2[4] from the same hard disk. You could set up part of the disk as a DOS partition for storing DOS programs and files, and part as an OS/2 partition for storing OS/2 files and programs.

The DOS partitions are organized and managed by a program that comes with DOS, called FDISK. The FDISK program sets up a primary DOS partition, extended DOS partitions (up to 23) where other DOS logical drives can be organized, and non-DOS partitions, which are actually extended partitions. Non-DOS partitions are not controlled or initialized by FDISK, but are controlled by other operating systems. The extended partitions were placed in DOS versions before 4.01. Before DOS 4.01, the largest partition size was 32M bytes, because the system used a 12-bit FAT. Beginning with DOS 4.01, the partition size was increased to 524M bytes, because the system uses a 16-bit FAT. Today DOS rarely uses extended partitions to set up other logical disk drives; the non-DOS partition is used to store files for other operating systems such as OS/2. The primary DOS partition is where the MSDOS, IO.SYS and COMMAND.COM are stored, along with DOS files and programs. The primary partition is usually specified as disk drive C and the extended partitions, if needed, are organized as logical drives E through up to Z. Drive D is usually reserved for a second hard disk that can be added to the system.

Figure 2–6 shows a disk that has been partitioned into three sections. The outer boot section is called the primary DOS partition. The other two partitions have been set up as two additional logical disk drives (drives D and E). These logical disk drives are accessed just as if they were separate, discrete, disk drives.

The FDISK Program. The FDISK program has four options if executed. To execute FDISK for disk drive C type: C:\>FDISK at the DOS prompt. When accessed, FDISK displays the information illustrated in Example 2–2. Notice that four options exist to allow the management of the partitions. In most cases the disk is set up with a single DOS partition that is active and contains the DOS operating system. It is rare today for other logical drives to be used on hard disks, although it may become commonplace for large IDE disk drives.

EXAMPLE 2–2

```
         MS-DOS Version 6
      Fixed Disk Setup Program
  (C)Copyright Microsoft Corp. 1983-1993

            FDISK Options

Current fixed drive: 1

Choose one of the following:

1. Create DOS partition or Logical DOS Drive
2. Set active partition
3. Delete DOS partition or Logical DOS Drive
4. Display partition information

Enter choice: [1]

Press Esc to exit FDISK
```

---

[4]OS/2 is a registered trademark of International Business Machines, Incorporated.

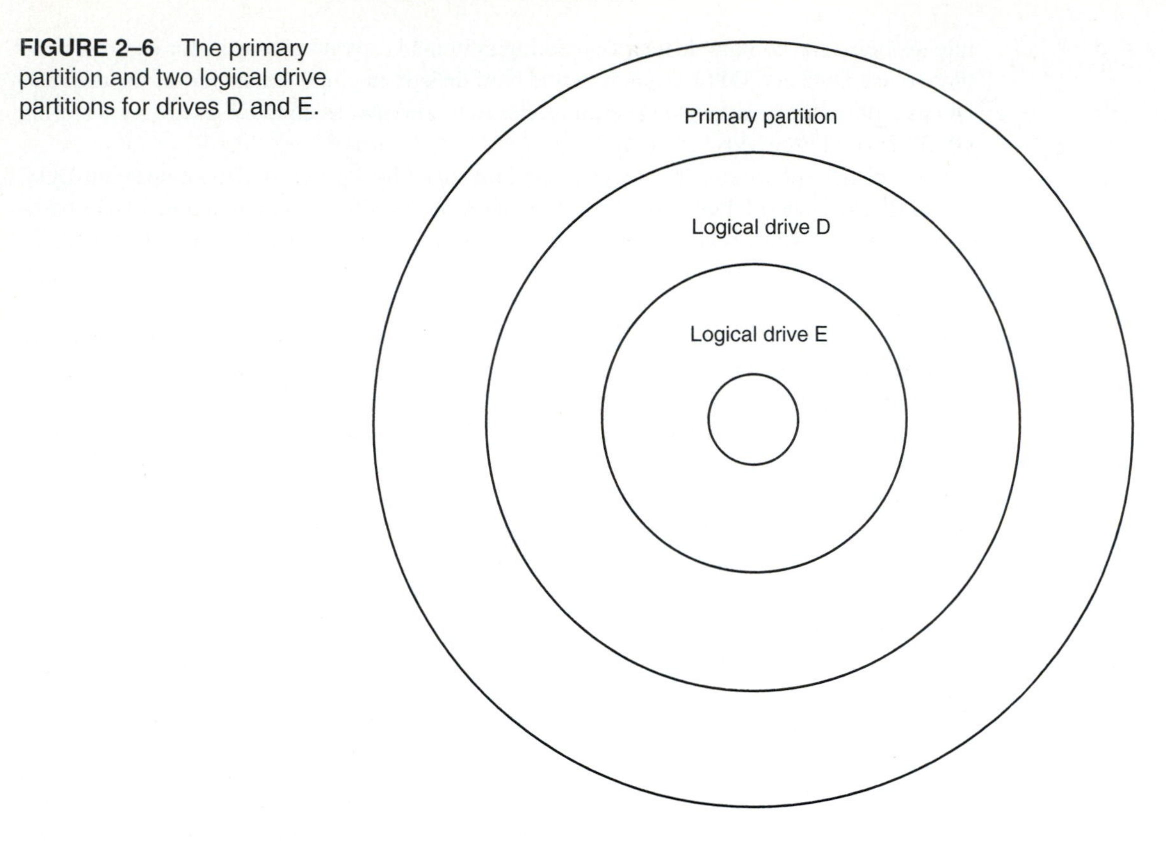

FIGURE 2–6 The primary partition and two logical drive partitions for drives D and E.

Compressed Disk Partitions

A new feature is added to DOS version 6.X called DoubleSpace. (DOS version 6.21 is being released without the DoubleSpace option and DOS version 6.22 is being released with DriveSpace[5], a modified version of DoubleSpace.) The DoubleSpace[6] (DriveSpace) program creates a partition, in the form of a hidden system file, on the disk drive that acts as another compressed disk drive. A compressed disk is one where the data has been reformatted, encoded, and stored at a higher density. With the DoubleSpace program, storage space is increased approximately 1.8 times for most files. Never store compressed files created by other programs on the compressed disk drive. Common examples of compressed files are ARC (archive) and ZIP (compressed) files. Compressed files often become larger when stored on a compressed disk drive. Other programs similar to DoubleSpace (such as Stacker) can increase storage space by an average of 2.8 times, according to Microsoft. In practice, program files usually compress at a much lower ratio, while ASCII text files compress at a higher ratio.

To run the DoubleSpace program type DBLSPACE at the DOS command prompt. This executes the DoubleSpace program. It is suggested that custom installation be chosen,

---

[5]DriveSpace is a trademark of Microsoft Corporation.

[6]DoubleSpace is a trademark of Microsoft Corporation.

to allow DoubleSpace to create a new disk drive without converting or compressing all the files on the hard disk drive. Never compress the Windows files, especially the virtual swap file created with Windows operating in the enhanced mode. Choose the default setting, which leaves 2M bytes of space on the current hard disk drive (most likely drive C). The remaining room will be converted to a new and empty compressed drive (usually drive H or I). For example, if drive C currently has 40M bytes of free space, DoubleSpace will use 38M bytes to create a new compressed disk drive (drive H), with about 70M bytes of free space.

Once the compressed drive is created, copy files to this new compressed drive and begin to free space on disk drive C. Be sure to change any .BAT or other files so they reflect the change of location from drive C to drive H or I. If the amount of room freed by DoubleSpace is considerable, run DoubleSpace again and add more of drive C to the compressed drive H using the change size option. If the compressed drive must be removed for any reason, this is accomplished with the DBLSPACE command and the uncompress feature on the tools menu. (Note that this feature is not available in DOS version 6.0.) Be aware that the compressed drive is stored in the root directory as the hidden system file DBLSPACE.001. Also hidden in the root directory are DBLSPACE.INI and DBLSPACE.BIN. Never remove or erase any of these files or your compressed information will be erased and forever lost.

DoubleSpace works best with program files and data files. A slight decrease in performance is noted, but because DoubleSpace frees up so much additional disk memory space, the slight change in performance can usually be tolerated. If a program runs sluggishly, it is best left as an uncompressed file. Additional information about DoubleSpace is available in the DOS manual and is well worth the consideration. Other programs such as DoubleDisk and Stacker are also worth a look.

2–2 INTERNAL DOS COMMANDS

There are many DOS commands available to control the organization of data and programs on the disk. If the C> prompt (DOS version 5.0) or C:\> prompt (DOS version 6.X) appears on the video screen, the computer is executing the COMMAND.COM program. When COMMAND.COM executes, the computer waits for a command with, in this case, the C disk drive selected as the default or currently logged drive. Once the command prompt is displayed, any DOS command or the name of any program is entered for execution.

To change to (*log to*) another disk drive, type the letter of the drive followed by a colon and then hit enter. For example, to log onto disk drive A, type A: at the DOS prompt followed by enter. Note that the A and B disk drives are often floppy disks and C is often a hard disk drive located inside the computer. If your computer has two hard drives, the second is often the D drive. A CDROM drive is often the E drive. A compressed disk drive appears as either drive H or I. The F drive is often the network logon drive, if a network is present.

The DOS commands and many programs consist of the command or program name followed by a switch or switches. A **switch** is selected with the slash (/) key. Don't confuse the / (switch) key with the \ (directory) key. For example, the DIR /W command typed at the DOS prompt will display the directory (DIR) and switch to wide format because of the /W (*wide*) switch. The /? switch, which functions with most programs, displays abbreviated help information for the program or command. For example, the DIR /? command displays help for DIR command. More extensive help on DIR, or any other command, is available by typing HELP DIR to display complete help for the DIR command.

Table 2–3 lists the internal DOS commands. Internal commands are stored inside the COMMAND.COM program and require no additional disk access to execute. This makes them function faster. Many of these commands are used for normal operation and control of information stored on the disk. There are also other DOS control programs or external commands, but these are programs, not internal commands.

Organizing Disk Data

The way data are stored on a hard disk is important. As mentioned in Section 2–1, the root directory can only store a limited number of file and directory entries. Because of this, the files stored on the hard disk must be organized and managed in directories and subdirecto-

TABLE 2–3 Internal DOS commands

Command	*Function*
*CHCP	Display or change the code page number
CD (CHDIR)	Display or change current directory
CLS	Clear video screen
COPY	Copies one or more files
*CTTY	Change terminal device
DATE	Display or change date
DEL (ERASE)	Erase file or files
DIR	Display directory
EXIT	Quits the COMMAND.COM program
LH (LOADHIGH)	Loads a program into upper memory
MD (MKDIR)	Make directory
PATH	Sets a search path
PROMPT	Changes the DOS command prompt
RD (RMDIR)	Removes an empty directory
RENAME	Renames a file or files
SET	Display, set or remove environment variables
TIME	Display or change the time
TYPE	Display the contents of an ASCII file
VER	Reports the DOS version number
VERIFY	Controls the verification of a disk write operation
VOL	Reports the disk volume label and serial number

Note: * = seldom used command.

ries. This is accomplished with the directory control instructions: **CD** (*change directory*), **MD** (*make directory*), and **RD** (*remove directory*). In early versions of DOS, no directory/subdirectory structure existed, which meant that once the root directory filled, the disk was full.

It is fairly important that the root directory contain just a few files. These files are AUTOEXEC.BAT, COMMAND.COM, and CONFIG.SYS. Also present in the root directory are MSDOS.SYS and IO.SYS files, but these don't appear in the directory listing because they are hidden system files. If DoubleSpace is installed, the root also contains three hidden files used for the control of the compressed drive and other DoubleSpace information. No other program file should normally be placed in the root directory. The remainder of the root directory should be used for directory entries. Recall that directories other than the root may contain any number of files and subdirectory entries.

EXAMPLE 2–3

```
C:\>CD C:\
C:\>MD WORDP
C:\>MD GAMES
C:\>MD SPREAD
```

Suppose that the system has a word processor, some games, and a drawing program. To make the most effective use of the disk drive, it is suggested that one directory be created for the word processor, another for the games, and a third for the drawing program. This is accomplished by using the MD (*make directory*) command as illustrated in Example 2–3. Notice that the root directory of the C drive is first selected by using the CD (*change directory*) command. Next the three directories are created with the MD command. Once these directories are created, a switch to any of them is made with the CD command.

The TREE.COM program is used to view the directory tree of any directory or the root directory. The TREE.COM program displays the directory tree on the video screen. Refer to Figure 2–7 for a typical directory tree as generated by the TREE.COM program. Note that WINDOWS, the last directory entry in this example tree, contains two subdirectories: SYSTEM and TEMP.

The RD (*remove directory*) command removes a directory or subdirectory as long as the directory or subdirectory contains no other files or subdirectories. If an attempt is made to remove a directory or subdirectory that contains files, DOS indicates that the directory cannot be removed because it is not empty. With DOS version 6.X a new utility program called *DELTREE* is available. The DELTREE program erases the entire directory tree including all its directories, subdirectories, and files. The DELTREE program should not be used, because if a mistake is made, it takes a considerable amount of time to reconstruct the erased directories and their contents. Imagine executing the DELTREE command at the root directory, and erasing the entire disk!

The CD (*change directory*) command changes the default directory or displays the current path. If the CD command is typed at the root directory, DOS displays C:\. If the CD command is used at the WINDOWS directory, DOS displays C:\WINDOWS. If CD is followed by a directory name, the default directory is changed. If the CD command is followed by a .. the currently logged (*child*) directory reverts to the prior (*parent*) directory in

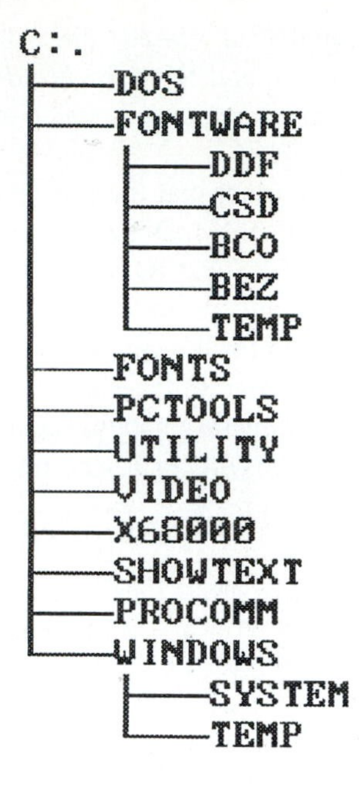

FIGURE 2–7 Typical directory tree as generated by the TREE.COM program.

the directory tree. For example, if the current directory is C:\WINDOWS\SYSTEM and the CD.. command is typed, the new directory is C:\WINDOWS. Here WINDOWS is the parent and SYSTEM is the child.

Using DIR, DEL, COPY, and RENAME

The DIR command displays a list of files and directories. An entire directory is displayed with DIR. If desired, a single file or several files are displayed with the use of wild cards. A **wild card** is a character that acts as any other character. The two wild cards used with DOS commands are the question mark (?) and the asterisk (*). The question mark emulates or substitutes for any single character in a file name and the asterisk acts as the entire file name or entire extension. An example is DIR *.TXT, which displays all the files in the current directory that have the file extension TXT. Another example is DIR C???????.TXT that displays all the files that begin with the letter C and have the TXT file extension.

EXAMPLE 2–4

```
H:\WS\PC>DIR

Volume in drive H is COMPRESSED
Volume Serial Number is 1804-1649
Directory of H:\WS\PC

.         <DIR>          01-10-94 6:09p
..        <DIR>          01-10-94 6:09p
TEMPS     <DIR>          01-10-94 6:09p
-WS25     BAT        245 01-01-92 1:45a
FORMAT               768 01-01-93 1:46p
TEST               1,664 01-04-94 8:53p
TEMP               1,408 01-05-94 5:07p
```

```
TREE     PIX  18,448 01-05-94 11:03p
EXAMPLES      17,536 01-10-94 10:58p
TREE WSG      23,692 01-05-94 11:03p
FRONT         11,648 01-09-94 12:56p
CHAPTER  02   73,600 01-10-94 10:53p
EXAMPLES BAK  17,664 01-10-94 10:54p
CHAPTER  01  101,376 01-10-94 6:19p
CHAPTER  BAK  72,960 01-10-94 10:35p
  14 file(s) 341,009 bytes
          16,416,768 bytes free

H:\WS\PC>
```

The DIR command followed by enter displays the entire contents of the currently logged directory. Example 2–4 shows how the DIR command displays a small directory located on compressed drive H. In this example, the DIR command lists the disk volume name, disk serial number, and the current directory, which is H:\WS\PC. Following this are the back-paths, . (**child** or current **directory**) and . . (**parent directory**) and then the disk files located within this directory. Also displayed are directory names, indicated by <DIR> in the directory listing. Each directory entry contains the file name with its extension, the number of bytes in the file, and the creation date and time. For example, the FORMAT file has no extension, is 768 bytes in length, was created on January 1, 1993 at 1:46 P.M. At the very end of the directory, the number of files in the directory and the number of bytes they occupy on the disk are listed. In this case, there are 14 files and directory back tracks that use 341,009 bytes of disk space. Following this is the total number of free bytes remaining on the disk.

EXAMPLE 2–5

```
H:\WS\PC>DIR /W

Volume in drive H is COMPRESSED
Volume Serial Number is 1804-1649
Directory of H:\WS\PC

[.]            [..]          -WS25.BAT    FORMAT       TEST
TEMP           TREE.PIX      EXAMPLES.BAK TREE.WSG     FRONT
CHAPTER.BAK    CHAPTER.02    EXAMPLES     CHAPTER.01
  14 file(s) 341,393 bytes
          16,416,768 bytes free

H:\WS\PC>
```

Another way to display the directory is by using the /W (*wide*) switch. The /W switch causes the directory command to display the information in wide format, as illustrated in Example 2–5.

The DIR C???????.* command displays all files that begin with the letter C. Refer to example 2–6 for this command and its effect.

Another useful switch is the /P (*pause*) switch, which displays the directory a page at a time. This is particularly useful for displaying long directories. For example, DIR /W /P displays the directory in wide format with a pause after each full screen of information is displayed.

EXAMPLE 2–6

```
C:\WS\PC>DIR C???????.*

Volume in drive H is COMPRESSED
Volume Serial Number is 1804-1649
Directory of H:\WS\PC

CHAPTER BAK    72,960  01-10-94 10:35p
CHAPTER 01    101,376  01-10-94  6:19p
CHAPTER 02     73,600  01-10-94 10:53p
   3 file(s) 247,936  bytes
          16,416,768  bytes free
```

DEL (Erase). The DEL command allows any file to be erased or deleted from the disk. It also allows groups of files, represented by wild cards, to be erased or deleted from the disk. For example, the DEL C:\WINDOWS\TEMP\~???????.* command deletes or erases all files in directory C:\WINDOWS\TEMP that begin with the tilde (~). These are temporary files, generated by the operation of the WINDOWS program, that can be safely erased.

Another use for DEL is to erase all of the files in a given directory. This is accomplished by typing DEL *.* to erase all files in the current directory. (Note that *.* is often pronounced *star-dot-star*.) Whenever all files in a directory are deleted in this manner, DOS asks if all files need to be erased.

COPY. The COPY command allows files to be copied from one part of the disk to another or from one disk to another. As with other DOS commands, COPY can use wild cards and any directory desired. For example, the COPY AB.TXT CD.TXT command copies the contents of the AB.TXT file into a new file called CD.TXT in the current default directory.

Suppose that all the files in one directory must be moved to another directory. Example 2–7 shows how a new directory (NEW) is created and then how all the files from the old directory (OLD) are copied into this new directory. Notice that DOS displays all files copied from C:\OLD. After the copy operation, the old directory is cleared by erasing all files.

EXAMPLE 2–7

```
C:\>MD C:\NEW
C:\>COPY C:\OLD\*.* C:\NEW\*.*
C:\OLD\FILE1.TXT
C:\OLD\FILE2.TXT
C:\OLD\FILE3.TXT
C:\>DEL C:\OLD\*.*
All files in directory will be deleted!
Are you sure (Y/N)?Y
C:\>RD C:\OLD
```

RENAME. The RENAME command allows files to be renamed. If the file TEXT1.TXT is to be renamed TEXT1.DOC, enter the following command at the DOS prompt: RENAME TEXT1.TXT TEXT1.DOC. The old name is entered first, followed by a space and the new name.

Changing the Dos Prompt

The DOS prompt, by default C:\>, may be changed to customize the system. To change the DOS prompt, use the PROMPT command followed by a new prompt. For example, the PROMPT BARRY command will change the prompt to BARRY. Typing PROMPT by itself returns the prompt to its default prompt, C>. This alternate default DOS prompt is generated by typing PROMPT or PROMPT NG (default drive, >) as described next. The normal default prompt is generated by typing PROMPT PG (default path, >).

The DOS prompt command also contains other special codes that allow the addition of system information to the new prompt. These special codes are listed in Table 2–4.

Suppose a customized prompt that contains the time and current path followed by an enter/linefeed is required. This new prompt is created by typing PROMPT TP$_ at the current DOS prompt. By using this as a template, a customized prompt can be created for any computer system.

2–3 EXTERNAL DOS COMMANDS

The external DOS commands are programs that appear in the DOS directory of your system. All external DOS command are suffixed with the extension .COM (*command*), .EXE (*execute*), or .BAT (*batch*). External commands are often referred to as *utility* programs or extensions to the COMMAND.COM program. The difference between the external and internal command is that the external command requires additional time to read from the DOS directory before it executes. The internal command is already in the memory, so DOS responds faster to an internal command. Another difference is that if you

TABLE 2–4 Special codes for the PROMPT command

Code	*Action*
$Q	Equal sign (=)
$$	Dollar sign ($)
$T	Current time
$D	Current date
$P	Current drive and path
$V	MSDOS version number
$N	Current disk drive
$G	Greater than (>)
$L	Less than (<)
$B	Pipe (|)
$_	Enter/linefeed
$E	Escape code (1BH)
$H	Backspace

TABLE 2–5 External DOS commands

Command	*Function*
*APPEND	Adds additional directories to the current directory
**ASSIGN	Redirects a disk drive request from one drive to another
ATTRIB	Displays or changes a file/directory attribute
**BACKUP	Copies files to a backup disk
CHKDSK	Reports the status of a disk drive
COMP	Compares two files
†DBLSPACE	Creates and manages compressed disk drives
*DEBUG	Tests and debugs assembly language programs
†DEFRAG	Defragments a disk
†DELTREE	Deletes an entire directory tree and all its files
DISKCOMP	Compares two disks
DISKCOPY	Copies one disk to another
DOSKEY	Command line macro processor
DOSSHELL	Directory system for DOS
EDIT	DOS ASCII editor program
**EDLIN	DOS ASCII editor program
*EXE2BIN	Converts an .EXE file to a binary file (.COM)
EXPAND	Uncompresses a compressed DOS file
**FASTHELP	Displays the same help available with the /? switch
**FASTOPEN	File directory caching program
FC	Compares two files and displays the difference
*FDISK	Controls disk partitions
FIND	Finds a text string in a file or files
FORMAT	Formats a disk
**GRAFTABL	Enables some older monitors for the display of extended ASCII codes
*GRAPHICS	Allows DOS to print the displayed contents of the screen
HELP	Displays help on DOS commands
†INTERLNK	Allows files from two computers to be exchanged
*JOIN	Joins the contents of one disk drive to another

don't use some of the external commands, they can be removed or erased from the disk drive allowing more room for storing other programs and files.

Although the list of external commands is fairly extensive, only the most commonly used external DOS commands are presented in Table 2–5. For further reading on less commonly used externals, refer to the DOS manual.

FORMAT

One important DOS command required to use a new, unformatted disk is FORMAT. The format command prepares the disk surface by creating tracks and sectors and information about the tracks and sectors required by DOS to read or write the disk. Usually the FORMAT command appears in two forms: FORMAT drive: and FORMAT drive: /S.

TABLE 2–5 (continued)

Command	*Function*
*KEYB	Configures the keyboard for another language
LABEL	Creates or changes a label for a disk
LOADFIX	Fixes the "packed file corrupt" error
MEM	Displays the amount and organization of the memory
†MEMMAKER	Configures the systems for all available memory
MIRROR	Records information about the format of a disk
MODE	Configures some system devices
MORE	Displays one screen output at a time
†MSAVE	Combats viruses
†MSBACKUP	Replaces the backup command
MSD	Microsoft diagnostics program
*NLSFUNC	Loads the country support code
†POWER	Conserves power in a lap-top computer
PRINT	Prints a text file
QBASIC	Starts quick basic
**RECOVER	Recovers readable information from a defective disk
*REPLACE	Replaces files in the destination directory with files from the source directory
RESTORE	Reloads files saved with BACKUP
†SCANDISK	Repairs many maladies on floppy and hard disk drive
SETVER	Sets the version number DOS returns to a program
*SHARE	Allows files to be shared
SORT	Sorts data
*SUBST	Associates a path with a drive
SYS	Installs the system and COMMAND.COM file onto a disk
TREE	Displays a directory tree
UNDELETE	Restores a deleted file
UNFORMAT	Restores a disk erased with FORMAT
VERIFY	Enables double verification on all disk writes
†SAFE	Used to protect against viruses
XCOPY	Copies files and directories

Note: * = seldom used DOS command. ** = do not use this command; it is obsolete. † = new to DOS version 6.X.

The FORMAT drive: command allows a disk placed into any disk drive to be formatted for use by DOS. For example, if the disk in drive A requires formatting, type FORMAT A: followed by the enter key. This prepares a data disk for use with DOS. If the disk in drive A must be a *bootable* disk—one that brings the system up from a cold start—then type FORMAT A: /S. The /S switch informs the FORMAT command to place the system files, IO.SYS and MSDOS.SYS, and the COMMAND.COM program on the disk. Note that the *SYS* program is also used to add the system to the disk to make it bootable at any time after a FORMAT. The SYS program can also repair a disk that has lost its system files. Other switches are available and can be viewed by typing FORMAT /? or by using the HELP FORMAT command. Example 2–8 shows the information provided by the FORMAT /? command.

EXAMPLE 2–8

```
C:\>FORMAT /?

Formats a disk for use with MS-DOS.

FORMAT drive: [/V[:label]] [/Q] [/U] [/F:size] [/B | /S]
FORMAT drive: [/V[:label]] [/Q] [/U] [/T:tracks /N:sectors] [/B | /S]
FORMAT drive: [/V[:label]] [/Q] [/U] [/1] [/4] [/B | /S]
FORMAT drive: [/Q] [/U] [/1] [/4] [/8] [/B | /S]

/V[:label]  Specifies the volume label.
/Q          Performs a quick format.
/U          Performs an unconditional format.
/F:size     Specifies the size of the floppy disk to format (such
            as 160, 180, 320, 360, 720, 1.2, 1.44, 2.88).
/B          Allocates space on the formatted disk for system files.
/S          Copies system files to the formatted disk.
/T:tracks   Specifies the number of tracks per disk side.
/N:sectors  Specifies the number of sectors per track.
/1          Formats a single side of a floppy disk.
/4          Formats a 5.25-inch 360K floppy disk in a high-density drive.
/8          Formats eight sectors per track.
/C          Tests clusters that are currently marked "bad".

C:\>
```

The FORMAT A: /V:FROG command formats a data disk in drive A and labels it as FROG. The FORMAT B: /F:720 command formats the $3^{1}/_{2}''$ disk in drive B as a 720K DSDD disk, even if the drive and disk are high-density. A FORMAT B: command, with the same disk drive, formats a high-density 1.44M disk if the disk drive is high-density.

CHKDSK

The **CHKDSK** (check disk) command allows the status of any disk—floppy, hard, or compressed—to be tested or checked. Note that CHKDSK should never be used from within Windows or from the Windows shell to DOS (MSDOS icon). CHKDSK is used to locate and correct a problem called *lost allocation units* that may occur with a disk. An allocation unit is a file cluster. If some allocation units are lost, it is because of a faulty program or because the computer was reset while a program was writing data to the disk. For example, if the computer locks up because of an error and the control-alternate-delete key sequence is activated, the system reboots and can leave lost allocation units in its wake. The loss of file allocation units can cause problems when the hard disk is defragmented. They also reduce the amount of available room on the disk. To repair lost allocation units, use the CHKDSK command with the /F (*fix*) switch. The fix switch closes the lost file allocation units and places them in the root directory as FILE0000.CHK, FILE0001.CHK, and so forth. These can be either erased or renamed and used, but they probably won't be complete files.

EXAMPLE 2–9

```
H:\>CHKDSK

Volume COMPRESSED created 01-09-1994 4:58p
Volume Serial Number is 1804-1649
```

```
45,400,064 bytes total disk space
 253,952   bytes in 31 directories
28,753,920 bytes in 991 user files
16,392,192 bytes available on disk

   8,192   bytes in each allocation unit
   5,909   total allocation units on disk
   2,368   available allocation units on disk

   655,360 total bytes memory
   632,944 bytes free

Instead of using CHKDSK, try using SCANDISK. SCANDISK can reliably detect
and fix a much wider range of disk problems. For more information,
type HELP SCANDISK from the command prompt.

H:\>
```

The CHKDSK program also reports the size of the disk, amount of free space remaining, number of directories, number of files, and so forth. It reports any bad sectors that may be present and other disk related problems. Bad sectors may occasionally appear on some hard disk drives when the disk is first placed into operation. These are marked bad, and will forever remain bad.

For a much more extensive test of the disk, DOS version 6.X contains the SCANDISK command. The SCANDISK command tests many more aspects of the disk than CHKDSK. Even though SCANDISK is available, CHKDSK is still useful to repair the lost file chains error, a common problem. CHKDSK takes much less time to do its work than SCANDISK. Occasionally, the SCANDISK program should be executed to perform a *surface scan*. The surface scan detects areas on the surface of the disk that may soon fail. Any such area is detected and the cluster is copied elsewhere by the surface scan. Surface scan marks the cluster bad so it is never again used by DOS. If areas continue to fail despite the surface scan, this is often an indication of an imminent hard drive or controller failure.

ATTRIB

Attributes are special features assigned to a file or directory. These features are changed or displayed with the **ATTRIB** external command. Any file or directory can be modified with ATTRIB. These file/directory modifications include: read-only, hidden, system, or archive. A hidden file is not displayed in the file list when the DIR command is executed. The read-only attribute prevents a file from being written or erased. The system attribute marks a file as a system file. The archive attribute is set whenever DOS writes to a file; it is used for backing up only those files that have been written since the last backup, and is cleared when the MSBACKUP command is executed. Table 2–6 lists the permissible attribute switches for the ATTRIB command. MSBACKUP should be executed to back up the hard disk drive at least once per month. If a personal computer is used heavily, a weekly or even daily backup is suggested.

The ATTRIB +H C:\GAMES\SETUP.COM command hides the SETUP.COM file in directory GAMES on disk drive C. Any of the attribute switches can be used in any combination to modify the attributes of a file. To hide the name of a directory and make it read-

TABLE 2–6 Aribute switches used with the DOS external ATTRIB command

Switch	*Function*
+R	Selects read-only
–R	Removes read-only
+A	Sets archive
–A	Removes archive
+S	Marks a file as a system file
–S	Removes the system file mark
+H	Hides a file
–H	Removes the hide-a-file attribute

only, use the ATTRIB command. For example, ATTRIB +H+R C:\GAMES will make the directory GAMES appear as a hidden directory that is not displayed by the DIR command. It also makes it a read-only directory name. Even though the directory is hidden, it can be used by DOS commands. The directory name is not displayed unless you use DIR /A (display all) or DIR /AH (display hidden) to display hidden files and directories.

To view the attributes of a file just type ATTRIB followed by the file name. This displays any attributes assigned to the file. Note that wild cards may also be used with this command. For example, ATTRIB +H C:\GAMES*.* hides all the files in the GAMES directory.

DOSKEY

The DOSKEY program, new beginning with DOS version 5.0, is an improved command line processor that allows for macro commands, the display of prior commands, and a command history. The DOSKEY program is a **TSR** or *terminate and stay resident* program. This means that the program remains in memory (it uses about 3K bytes in its default configuration) until activated by a combination of keystrokes. The DOSKEY program is active only when the system is at the DOS command line.

The most important feature of DOSKEY is that it allows the previous DOS commands to be redisplayed and used. This saves a lot of time when performing repetitive disk maintenance operations. The up and down arrow keys are used to scroll through previously typed commands. The left and right arrow keys allow you to position the cursor and change these commands. By default, the DOSKEY program stores the past 512 bytes of commands, but this can be changed if necessary by using the /bufsize= switch. If 512 bytes is not enough space, try using DOSKEY /bufsize=1024 to change the size of the buffer to 1,024 bytes.

In addition to providing access to previously typed commands, DOSKEY allows the creation of macros to replace long and involved keying sequences. An example is the MEM /C |MORE command. This command displays the contents of memory; the |MORE part of the command causes the machine to fill one screen at a time and wait. |MORE can be used with many DOS commands to pause the display of information. The MEM /C |MORE command is useful, but requires time to type. The MEM /C /P also functions with

TABLE 2–7 Shorthand keying sequences for DOSKEY

Sequence	*Function*
$G	Redirects the output to another file. An example is the macro DOSKEY DD=DIR $GFROG.DIR, which writes the current directory to file FROG.DIR whenever DD is typed.
GG	Appends the output to a file
$L	Redirects input
$B	Sends macro output to a command such as MORE. This is equivalent to using the DOS pipe (|).
$T	Separates commands
$$	Specifies a dollar sign
$1–$9	Represents one piece of command line information
$*	Represents the entire command line

a pause. This keying sequence can be made into a macro by typing the following once: DOSKEY MC=MEM /C $BMORE. (The $B is used to enter the DOS pipe symbol (|), which allows more than one command on a line.) This keying sequence assigns MC to MEM /C |MORE, so the next time the contents of memory must be displayed, just type MC. The MC becomes the MEM /C |MORE command.

Table 2–7 lists common shorthand keying sequences used when creating macros for the DOSKEY program. Suppose that a macro (SW) is needed to create a new directory name in the root directory and then switch to that new directory. This is accomplished by typing DOSKEY SW=CD C\:$TMD 1TCD $1, which enters the CD, MD, and CD commands into macro SW. Notice how $T is used to separate commands and how $1 is used to obtain the file name from the command line. If needed, $2 could be used to obtain the second file name from command line, and so forth. After this macro is entered, the SW BOTTLE command switches to the root directory, makes a new directory called BOTTLE, and then switches to directory BOTTLE.

MEM

The MEM command displays the size and contents of the memory attached to the system. The MEM command uses four different switches: /P (*program*), /D (*debug*), /F (*free*), and /C (*classify*). If just the MEM command is typed on a line, DOS reports the amount of free and occupied memory in the system. Refer to Example 2–10 for the effect of the MEM command.

EXAMPLE 2–10

```
C:\>MEM

Memory Type      Total = Used + Free
______________   _____   ____   ____

Conventional      640K    22K   618K
```

```
Upper                139K    124K     15K
Reserved               0K      0K      0K
Extended (XMS)       245K    245K      0K
--------------     ------  ------  ------
Total memory       1,024K    392K    632K

Total under 1 MB     779K    147K    632K

Largest executable program size 618K (632,576 bytes)
Largest free upper memory block   6K (6,032 bytes)
MS-DOS is resident in the high memory area.

C:\>
```

When the CONFIG.SYS file is installed, the MEM /C command reports how much memory remains to load system programs into high memory. Example 2–11 shows the effect of executing the MEM /C command.

EXAMPLE 2–11

```
C:\>MEM /C

Modules using memory below 1 MB:

 Name      Total       =   Conventional  +  Upper Memory
 -----     ----------      ------------     ------------

 MSDOS      15,037 (15K)  15,037 (15K)           0  (0K)
 HIMEM       1,168  (1K)   1,168  (1K)           0  (0K)
 EMM386      3,120  (3K)   3,120  (3K)           0  (0K)
 COMMAND     3,008  (3K)   3,008  (3K)           0  (0K)
 SETVER        832  (1K)       0  (0K)         832  (1K)
 RAMBIOS    33,056 (32K)       0  (0K)      33,056 (32K)
 IMOUSE     14,464 (14K)       0  (0K)      14,464 (14K)
 DBLSPACE   39,488 (39K)       0  (0K)      39,488 (39K)
 HHSCAND     8,832  (9K)       0  (0K)       8,832  (9K)
 SMARTDRV   26,000 (25K)       0  (0K)      26,000 (25K)
 DOSKEY      4,656  (5K)       0  (0K)       4,656  (5K)
 Free      647,600 632K) 632,672(618K)      14,928 (15K)

Memory Summary:

 Type of Memory         Total =   Used   +  Free
 --------------         -------   -------   -------

 Conventional           655,360    22,688   632,672
 Upper                  142,256   127,328    14,928
 Reserved                     0         0         0
 Extended (XMS)         250,960   250,960         0
 --------------         -------   -------   -------

 Total memory         1,048,576   400,976   647,600

 Total under 1 MB       797,616   150,016   647,600

 Largest executable program size    632,576  (618K)
 Largest free upper memory block      6,032    (6K)
 MS-DOS is resident in the high memory area.

C:\>
```

UNDELETE

Another commonly used external DOS command is the UNDELETE command. The UNDELETE command allows files to be undeleted in many cases. If a file has just been erased, it can almost always be undeleted; if it was erased last week, it may be impossible to undelete. The UNDELETE command is used to correct deletion errors at the time of the error.

EXAMPLE 2–12

```
C:\>UNDELETE

Directory: C:\DOS
File Specifications: *.*

   Deletion-tracking file not found.

   MS-DOS directory contains 2 deleted files.
   Of those, 2 files may be recovered.

Using the MS-DOS directory.

   ?ODE     COM 23537 4-09-91 5:00a .... Undelete (Y/N)?Y
   Please type the first character for?ODE .COM: M

File successfully undeleted.

   ?OSSHELL INI 21452 1-08-94 12:28p .... Undelete (Y/N)?N

C:\>
```

In most cases, the UNDELETE command appears on the command line by itself. If desired, the directory can be specified or wild cards can be used. Example 2–12 shows a session started by the UNDELETE command. Here, DOS discovered two deleted files in the DOS directory. Notice that the first letter of a deleted file does not appear. This is because DOS places an E5H in place of the first letter of the file name whenever a file is deleted or erased. The undelete command looks for this and displays it as a question mark. If you choose to undelete a file, DOS asks for the first letter of the file name.

Other DOS Commands

There are other external DOS commands, but many are either obsolete or so specialized that they are beyond the scope of this text. Some of them are new to DOS version 6.X and are explained using the DOS HELP command or are self-explanatory.

The most useful of the new DOS version 6.X commands are DEFRAG, DBLSPACE (explained earlier), SCANDISK (also explained earlier), and MEMMAKER. The DEFRAG program is important because, as mentioned in Section 2–1, disk drives are fragmented by DOS in the course of normal disk maintenance. Fragmentation reduces the performance of the drive after a month or two of use. For this reason, it is recommended that the disk drive is defragmented monthly.

The MEMMAKER command is used to initially setup the CONFIG.SYS file, to optimize the operation of the memory manager program EMM386.EXE. The MEMMAKER program should be executed whenever more memory is added to the system, or when you change the configuration of the system.

The HELP and EDIT commands are written in QBASIC and depend on QBASIC for operation. Never erase QBASIC from the DOS directory to create extra space on the disk, or the EDIT and HELP programs will not function.

2–4 THE CONFIG.SYS FILE

The CONFIG.SYS file is probably one of the most important files in your personal computer system. This file is used by DOS to setup or configure the computer each time you apply power. Without it, the computer will function, but probably not in a desirable way.

Configuration Commands

The CONFIG.SYS file contains its own set of unique commands that are not DOS commands. Table 2–8 shows a list of the commands used in the CONFIG.SYS file. Note that these commands do not work at the DOS command line; they only work inside the CONFIG.SYS file. The CONFIG.SYS file is edited using the DOS EDIT program located

TABLE 2–8 Commands available to the CONFIG.SYS file

Command	*Function*
BREAK	Causes DOS to test for the control-C or control-break key combinations
BUFFER	Sets the number of buffer areas DOS uses for disk data transfer
COUNTRY	Chooses the country and language conventions for the system (by default American)
DEVICE	Loads an installable device into the memory
DEVICEHIGH	Loads an installable device into high memory
DOS	Specifies how DOS is loaded into memory
DRIVPARM	Chooses the characteristics of the a: disk drive
FCBS	Selects the number of file control blocks (FCB) that DOS can access
FILES	Specifies how many files DOS can open at one time
INSTALL	Executes a terminate and stay resident program
LASTDRIVE	Chooses the last valid disk drive letter
REM	A remark or comment
SHELL	Indicates information about the command interpreter program COMMAND.COM
STACKS	Sets the number of stacks used for processing hardware interrupts
SWITCHES	Specifies the use of the conventional keyboard

in the DOS directory. Access to this program and the CONFIG.SYS file, which must be located in the root directory, is gained by typing C:\DOS\EDIT C:\CONFIG.SYS. What appears on the screen is the edit screen and the contents of the current CONFIG.SYS file.

Using the CONFIG.SYS File. A CONFIG.SYS file appears in Example 2–13. This example contains almost all the dialog that normally appears in most CONFIG.SYS files, except for some drivers unique to specific systems.

The first few lines install the HIMEM.SYS and EMM386.EXE memory management programs. A DEVICE statement is used to install devices (programs and drivers) in the memory. The first DEVICE statement in this example installs the HIMEM.SYS driver. The HIMEM.SYS driver manages an area of memory called the high memory area that exists just above the first 1M byte of memory on 80286 or newer machines. This area begins at location 100000H and extends to location 10FFEFH, providing approximately 64K bytes of additional memory for DOS. In most cases, no switches are required for this program to function properly.

Once HIMEM.SYS is loaded into the system, the DOS= command specifies where DOS is loaded into the memory and whether or not it can use upper memory blocks. The DOS=UMB command states that DOS will use upper memory blocks (UMB). Later in the CONFIG.SYS file, DOS=HIGH loads DOS high, which frees a considerable amount of memory in the TPA for application programs. The DEVICEHIGH and LOADHIGH commands are used to store programs and drivers in the upper memory blocks.

The EMM386.EXE program manages upper memory blocks as well as extended and expanded memory. In this example, the I switch indicates that upper memory blocks are to appear at location C8000H through EFFFFH. (Because of the way that the microprocessor accesses memory, only the first four digits of the address are used with EMM386.EXE.) The values used in this example are correct for most systems, unless a network card, some video adapters, older RLL hard drives, or ROM Basic are in place. In some systems, an additional 32K bytes of upper memory can be obtained with the I=C800–F7FF switch. This allows upper memory to replace that part of the BIOS memory which is used only when the DEL key is typed at boot time. The NOEMS switch tells EMM386.EXE not to use expanded memory, but to use upper memory blocks. Expanded memory is only used with certain game programs, and is wasted if Windows is used in the system. If games need expanded memory, remove the NOEMS switch and replace it with the number of bytes for expanded memory and the word RAM, for access to both expanded memory and upper memory. For example, the 2048 RAM switch allows 2M for expanded memory and also allows for the use of upper memory. When expanded memory is installed, you will lose 64K bytes of upper memory to the expanded memory page frame. Once EMM386.EXE is installed, upper memory can be used to hold devices and programs that are placed there with DEVICEHIGH and LOADHIGH.

The next commands configure the memory for areas that are used for disk file data. The FILES statement specifies the number of files that DOS can access at one time. This is usually set for 20 or 30, and can be adjusted by the installation of most software packages.

EXAMPLE 2–13

```
DEVICE=C:\DOS\HIMEM.SYS
DEVICE=C:\DOS\EMM386.EXE NOEMS i=c800-efff
```

```
DOS=UMB
FILES=30
BUFFERS=10
LASTDRIVE=D
FCBS=4,0
DEVICEHIGH /L:1,12048 =C:\DOS\SETVER.EXE
SHELL=C:\DOS\COMMAND.COM C:\DOS\ /E:256 /p
DOS=HIGH
DEVICEHIGH /L:1,33024 =C:\VIDEO\RAMBIOS.SYS
DEVICEHIGH /L:1,34320 =C:\MOUSE\IMOUSE.SYS /S75
DEVICEHIGH=C:\DOS\DBLSPACE.SYS /MOVE
DEVICEHIGH /L:1,10816 =H:\SCANMAN\HHSCAND.SYS /A=280/I=9/D=3
```

The BUFFERS statement specifies how many 512 byte-long buffers are set up for DOS disk access. In most cases, 10 to 50 buffers should be adequate for DOS. The number of buffers is adjusted for the size of the hard drive in the system. A buffer size of 50 is better for a large hard disk drive; 30 is adequate for a medium-size hard disk drive. If SMARTDRV is used, the buffers statement is set to 10 buffers.

The STACKS statement, which does not appear in this example, allocates memory to internal hardware interrupts. The number of stacks and their size can be adjusted. The minimum values (and default values) should be 9 stacks with 128 bytes for each stack, entered as STACKS 9,128. Below these values, errors and lockups may occur. The DOS system automatically sets up this statement, and it should not be used unless your machine locks up with the message "internal stack overflow".

The FCBS statement sets the number of file control blocks for DOS. The file control block stores information about files that were designed to operate with DOS versions 2.0 and earlier. Most programs today do not use file control blocks; instead, they use file handles.

The LASTDRIVE statement indicates the last disk drive in the system. This system uses A and B for two floppy drives, C and D for two hard disk drives and E for the CDROM drive. Each drive requires a small amount of memory. If five drives are connected to the system, use LASTDRIVE=E. If more drives are added, change the last drive statement. Never use LASTDRIVE=Z, or memory is wasted.

The SHELL command specifies the command line processor for the system. In the example, the COMMAND.COM program is selected as the command processor, just as it would be if no SHELL statement existed. The difference is that the switches applied to the COMMAND.COM program make it permanent, and specify 256 bytes for the DOS environment. The /P switch causes COMMAND.COM to remain in memory, eliminating the need for DOS to go to the disk drive and read the command processor each time DOS is entered. The environment space (/E:256) is used by programs to store information used by the programs.

In the example CONFIG.SYS file, the five DEVICEHIGH commands load five drivers into upper memory blocks. These are the set version program (SETVER) from Microsoft, Video RAMBIOS driver, mouse driver, DoubleSpace driver, and hand scanner driver. The sizes for these drivers are obtained by using the DEVICE command to place them in normal memory. The MEM /C command is then executed to find the size of each driver. This is the size used with the DEVICEHIGH command. Alternatively, the MEMMAKER program determines drivers' sizes and loads the sizes into the CONFIG.SYS file. The /L:1 switch tells the DEVICEHIGH command which region of upper memory to use for storing the device. The /L:1 switch stored the driver into region 1. Because there is no

TABLE 2–9 Codes used with COUNTRY

Country	*Country Code*	*Code Pages*
United States	001	437 or 850
Canadian/French	002	863 or 850
Latin America	003	850 or 437
Netherlands	031	850 or 437
Belgium	032	850 or 437
France	033	850 or 437
Spain	034	850 or 437
Hungary	036	852 or 850
Yugoslavia	038	852 or 850
Italy	039	850 or 437
Switzerland	041	850 or 437
Czechoslovakia	042	852 or 850
United Kingdom	044	437 or 850
Denmark	045	850 or 865
Sweden	046	850 or 437
Norway	047	850 or 865
Poland	048	852 or 850
Germany	049	850 or 437
Brazil	055	850 or 437
English	061	437 or 850
Portugal	351	850 or 860
Finland	358	850 or 437

way to determine the regions, the /L switch should be formed with the MEMMAKER command. The /L:2,128 switch uses region 2 and specifies a minimum size of 128 bytes.

The CONFIG.SYS file illustrated should be adequate for almost all systems. If a problem is encountered, check the DOS manual for additional information.

Additional Notes. Another feature of DOS is its ability to function with different languages, using the COUNTRY command. The COUNTRY command configures DOS to recognize character sets and punctuation conventions of many other languages besides U.S. English. These sets often change the keyboard layout and the information displayed by some ASCII codes. By default, DOS is configured for the United States and does not require the COUNTRY command for this operation. Table 2–9 lists the supported languages and codes required for the COUNTRY command. For example, to use DOS in Poland, the COUNTRY command is COUNTRY 048,852 C:\DOS\COUNTRY.SYS. The country is 048 and the code page is 852 (refer to the DOS manual for the ASCII codes in this code page). The COUNTRY.SYS program contains the country specific information and must be included with the COUNTRY command.

TABLE 2–10 Special DOS batch program commands

Command	*Function*
CALL	Used to call another batch program (.BAT)
*CHOICE	Causes the batch program to wait for a certain key
ECHO	Displays a message or enables/disables the command echoing feature of the COMMAND.COM program
FOR	Executes a command for each file in a set of files
GOTO	Allows branching to lines that are labeled
IF	Allows conditional branching
PAUSE	Displays the "Press any key to continue..." message
REM	Comment or remark statement
SHIFT	Changes the position of replaceable parameters

Note: * = new to DOS version 6.X.

2–5 BATCH PROGRAMS (.BAT)

Batch files, with the extension .BAT, hold the normal DOS commands that are used to control the computer. Batch files are an important feature of DOS because they mimic the keyboard and provide the command processor with DOS commands. This means commands do not have to be keyed on the keyboard every time they are needed. One of the most important batch programs is the AUTOEXEC.BAT file, which is automatically executed each time the system boots.

Special Batch File Commands

The DOS system contains special commands for batch programs that control the flow of the batch program and test certain conditions as signaled by DOS applications. Table 2–10 lists the special batch program commands used in preparing batch programs. These special commands are used with any other DOS command when writing a batch program.

AUTOEXEC.BAT

Batch files can be written to perform any sequence of tasks at the DOS command prompt. One of the most common batch programs is the AUTOEXEC.BAT file. Example 2–14 shows an AUTOEXEC.BAT file executed when the computer is first powered up.

EXAMPLE 2–14

```
PATH C:\DOS;C:\;C:\UTILITY
SET BLASTER=A220 I7 D1 T3
SET SOUND=C:\SB
LOADHIGH C:\DOS\DOSKEY /BUFSIZE=1024
DOSKEY GO=DOSSHELL
LOADHIGH DOSSHELL
```

The first statement in this batch program is the PATH statement. The PATH statement defines the search path whenever a file name is typed for execution at the DOS prompt. In this example, DOS searches first the C:\DOS directory, then the root directory (C:\), and, finally, the C:\UTILITY directory. The last directory searched is the current default directory. The longer the PATH statement, the longer it takes DOS to find the program name. For this reason, the length of the PATH should be limited to only directories that are likely to hold the program name. In most cases, these directories are root, DOS, and WINDOWS.

The SET statement places information that is a portion of the COMMAND.COM program into the environment space. In this sample AUTOEXEC.BAT file, SET BLASTER=A220 I7 D1 T3 sets the environment name BLASTER with the information listed. The A220 is the I/O port for the Sound Blaster[7] card, I7 is the interrupt port number, D1 is the number of DMA channels, and T3 specifies DMA channel number 3. This information is used in a batch program that is used to enable and select Sound Blaster. Other SET statements can appear to communicate information to batch programs. If a batch file is created using the command DIR %SOUND%, it displays a directory of C:\SB, because SOUND is set to C:\SB. Anytime information stored in a SET variable is used in a batch file, surround the variable name with percent symbols.

The LOADHIGH (or LH) command loads programs into upper memory blocks if they are available. In this example, the TSR program and DOSKEY are loaded into upper memory blocks.

The DOSKEY GO=DOSSHELL programs the macro GO with the program name DOSSHELL. This allows the word GO to be used from the DOS prompt whenever the DOSSHELL program must be executed. Finally, the last entry into the AUTOEXEC.BAT file is the LOADHIGH C:\DOS\DOSSHELL command. This transfers control from the AUTOEXEC.BAT program to the DOSSHELL program.

Writing Batch Files

Batch files can be written with the DOS editor program EDIT or any other ASCII editor or word processor that can edit ASCII data. The only requirement for a batch program file is that the extension must be .BAT to show the command processor that the file is a batch program.

A batch program should have a unique name. For example, the COPY.BAT file will not execute if the current directory or path contains a COPY.COM or COPY.EXE file. The command processor always searches for a COM file first, and, if found, it executes the COM file and ignores the EXE or BAT file. The DOS execution order is always COM, EXE, and BAT for files that have the same name.

Suppose a batch program is needed to format both 720K and 1.44M floppy disks in drive B. A batch program can be written to format the 720K disks. The normal format command is then used to format the 1.44M byte disks. Example 2–15 shows a short batch program that formats a 720K DSDD 3½″ disk in drive B. The FORMAT command uses the /F:720 switch to format a 720K DSDD disk. It also uses the /U switch to force a format

[7]Sound Blaster is a trademark of Creative Labs, Incorporated.

for the disk. The REM statement places remarks or comments in the batch program, and the PAUSE statement pauses for any keystroke.

EXAMPLE 2–15

```
@ECHO OFF
REM The MAKE720.BAT file
ECHO.
ECHO This program formats a 720K disk in Drive B.
ECHO Place disk in drive B now . . .
ECHO.
PAUSE
FORMAT B: /F:720 /U
```

The ECHO statement is used in several different ways. If the ECHO statement ends with a period, a blank line is displayed. If ECHO is followed by text, the text displays on the screen. The ECHO OFF command prevents DOS commands from displaying or echoing data to the video display. Note that any line preceded with the @ symbol will not display. The @ symbol prevents the ECHO OFF command from being displayed in the example programs in this section.

To illustrate the power of the batch program and some other commands, Example 2–15 is rewritten in Example 2–16 to show the use of the CHOICE, IF, and GOTO statements. The batch program in Example 2–16 allows the user the choice of formatting the disk in drive B as either a 720K or 1.44M disk.

EXAMPLE 2–16

```
@ECHO OFF
REM A batch file to format a disk in drive B
ECHO.
ECHO A - Format a 720K disk in drive B
ECHO B - Format a 1.44M disk in drive B
ECHO.

CHOICE /C:AB Select choice

IF ERRORLEVEL 2 GOTO HD
IF ERRORLEVEL 1 GOTO DSDD

:HD
ECHO.
FORMAT B: /F:1440 /U
GOTO END

:DSDD
ECHO.
FORMAT B: /F:720 /U

:END
```

Notice that the ECHO commands list two choices, A–720K or B–1.44M. The CHOICE command allows the batch program to ask the user to enter either an A or a B. The /C:AB switch places the choices in a box surrounded by brackets [A,B] followed by a question mark. The “Select choice” part of the CHOICE command is the prompt that appears before the bracketed choices. In this example, choice displays: "Select choice [A,B]?". If the letter A is typed, CHOICE returns with the ERRORLEVEL set to a 1 and if B is typed ERRORLEVEL is set to a 2. If a third choice were given, ERRORLEVEL returns a 3 and so forth. Note that a /C:123 allows the choice of 1, 2, or 3. Any letter or number can be used for choices.

The IF ERRORLEVEL statement checks the value in ERRORLEVEL and performs a GOTO if true. In this example, IF ERRORLEVEL 2 GOTO HD will continue the program at label :HD if the letter B is typed. (Note that labels must begin with a colon.) If the letter A is typed, the program continues at label :DSDD. Always test the largest value in ERRORLEVEL first, because the if statement tests for an equal or greater than condition. The ERRORLEVEL command is also used to test error information returned from DOS programs. In most cases, a value of zero indicates no error.

Using Data from the Command Line. Command line data are used in a batch program with the symbols %1 through %9 and are often called *replaceable variables*. If one entry exists, always use %1. If two entries exist, use %1 for the first, %2 for the second, and so forth. The %0 entry returns the batch program name and is also used in some batch programs.

To illustrate the use of command line data, a batch program called MAKE is written in Example 2–17. The MAKE batch program contains one command line parameter in the form of drive letter A or drive letter B. The MAKE batch program formats a disk in either drive A or B and formats it as either a 720K or 1.44M 3½″ floppy disk. To format the disk in drive A, type MAKE A; to format the disk in drive B, type MAKE B.

EXAMPLE 2–17

```
@ECHO OFF
REM A batch file to format a 3-1/2" disk in any drive
REM Command line syntax is MAKE A or MAKE B
IF "%1"=="A" GOTO CONT
IF "%1"=="B" GOTO CONT
IF "%1"=="a" GOTO CONT
IF "%1"=="b" GOTO CONT
ECHO.
ECHO You must type MAKE A or MAKE B
ECHO.
GOTO END

:CONT
ECHO.
ECHO A - Format a 720K disk
ECHO B - Format a 1.44M disk
ECHO.

CHOICE /C:AB Select choice

IF ERRORLEVEL 2 GOTO HD
IF ERRORLEVEL 1 GOTO DSDD

:HD
ECHO.
FORMAT %1: /F:1440 /U
GOTO END

:DSDD
```

```
ECHO.
FORMAT %1: /F:720 /U

:END
```

The IF "%1"=="A" GOT CONT instruction tests the command line data for the letter A. If an A exists, the program continues at label :CONT. Note that %1 and A are both surrounded by quotes. The quotes are required to prevent a command that erroneously contains no parameter from displaying a syntax error for each IF statement. The == in the IF statement is used to mean "is the same as." Therefore, the statement IF "%1"=="B" GOTO CONT means "if the command line parameter is the same as B, GOTO CONT."

The replaceable parameter %1 is also used in the two FORMAT commands found in this batch program. The FORMAT %1 /F:720 /U statement formats drive A if MAKE A is typed at the command line or drive B if MAKE B is typed.

Using a Batch Program to List Selected Files. The directory command is extremely powerful, as shown in Example 2–18. This batch program does a directory of files with a given extension in all directories and subdirectories. It either displays them or stores them in a file. If the batch program is called DISPALL, typing DISPALL BAT will display a list of all batch files on the disk. If DISPALL BAT TEMP is typed, the batch program stores a listing of the files into TEMP.

EXAMPLE 2–18

```
@ECHO OFF
REM The DISPALL batch program displays or saves all the files on drive C
REM with a given extension.
REM To use DISPALL, type DISPALL BAK to display all backup files.
REM Type DISPALL BAK C:\SAVE.IT to save the selected file names
REM in C:\SAVE.IT

ECHO.
ECHO One moment please ...
ECHO.

IF "%1"=="" GOTO MES
IF "%2"=="" GOTO LIST

REM Saves the selected file names in a file
DIR *.%1 /S /B > %2
GOTO END

:LIST
REM Display files with selected extension
DIR *.%1 /S /B /P
ECHO.
GOTO END

:MES
ECHO You must include the extension after DELALL
ECHO For example, DISPALL BAK displays all backup files
ECHO DISPALL BAK TEMP save the file names in TEMP
ECHO.

:END
```

TABLE 2–11 CONFIG.SYS commands that control multiple configurations

Command	*Function*
[MENU]	Initializes multiple configurations
[COMMON]	Indicates common areas of the multiple configurations
[menuitem]	Programmable menu item added with the MENUITEM command
MENUCOLOR	Sets the colors of the text and background of the menu
MENUITEM	Sets a menu item for display
MENUDEFAULT	Selects a default menu item after a selectable timeout value
SUBMENU	Allows submenus to be added to the main configuration menu

This batch program illustrates how the DIR command is used to search all directories on the disk for a specific file name extension. The /S switch causes all directories and subdirectories to be searched. The /B switch lists only the path and name of the file. Note how the DIR instruction is used with the > symbol to redirect the output of the directory command into file %2. The < symbol is also used, but not in this example, to direct the flow in the opposite direction. An example is SORT < DATA.TXT where the contents of the DATA.TXT file are input to the SORT command. The >> symbol can also be used to direct output data to append or add to the end of a file. An example is DIR >> TEMP.$$$, where the directory listing appends the TEMP.$$$ file.

Using Multiple Configurations

A new feature of DOS version 6.X is its ability to allow multiple boot configurations. In order to control this process some new commands have been added that provide a menu and multiple paths in the CONFIG.SYS file. Table 2–11 lists these additional commands.

EXAMPLE 2–19

```
[MENU]
MENUCOLOR=0,7
MENUITEM=NORM,Normal memory system
MENUITEM=EXPANDED, Expanded memory system enabled
MENUDEFAULT=NORM,30

[COMMON]
DEVICE=C:\DOS\HIMEN.SYS

[NORM]
DEVICE=C:\DOS\EMM386.EXE NOEMS i=c800-efff

[EXPANDED]
DEVICE=C:\DOS\EMM386.EXE i=c800-efff RAM 2048

[COMMON]
DOS=UMB
BUFFERS=10,0
```

```
FILES=30
LASTDRIVE=D
FCBS=4,0
DEVICEHIGH /L:1,12048 =C:\DOS\SETVER.EXE
SHELL=C:\DOS\COMMAND.COM C:\DOS\ /E:256 /p
DOS=HIGH
DEVICEHIGH /L:1,33024 =C:\VIDEO\RAMBIOS.SYS
DEVICEHIGH /L:1,34320 =C:\MOUSE\IMOUSE.SYS /S75
DEVICEHIGH=C:\DOS\DBLSPACE.SYS /MOVE
DEVICEHIGH /L:1,10816 =H:\SCANMAN\HHSCAND.SYS /A=280/I=9/D=3
```

Example 2–19 shows a configure systems file that allows two menu choices. One, NORM, sets up the memory system without any expanded memory. The other, EXPANDED, sets up the system with 2M bytes of expanded memory. Although expanded memory is still needed for some DOS game programs, it is not used by WINDOWS. If games are to be played, choose expanded memory. If Windows is required, choose normal configuration.

The very first line of the CONFIG.SYS file must be [MENU], to enable multiple configurations. Under the [MENU] section are the MENUCOLOR, MENUITEM, and MENUDEFAULT commands. The MENUCOLOR command in this example chooses color number 0 for text and color number 7 for the menu background. These are standard DOS colors as listed in Table 2–12.

The MENUITEM command creates a menu entry and defines a section of the CONFIG.SYS file that is executed only when the menu item is selected. For example, the

TABLE 2–12 DOS color numbers for MENUCOLOR

Number	*Color*
0	Black
1	Blue
2	Green
3	Cyan (blue-green)
4	Red
5	Magenta (violet)
6	Brown
7	White
8	Gray
9	Bright Blue
10	Bright Green
11	Bright Cyan
12	Bright Red
13	Bright Magenta
14	Yellow
15	Intense White

Note: Colors 8 through 15 may blink on some display systems.

MENUITEM=NORM statement, or Normal memory system, displays "Normal memory system" as item number 1 in the menu. The menu section is defined as [NORM]. The MENUDEFAULT=NORM,30 command selects NORM as the default entry after 30 seconds have elapsed. This allows the system to boot if no menu choice is selected. The [COMMON] sections of the CONFIG.SYS file are always executed, no matter what menu item is selected.

2–6 SUMMARY

1. Data are stored in tracks, sectors, clusters, and cylinders on a disk. A track is a concentric ring of data, a sector is a section of a track, clusters are groups of sectors, and a cylinder is a group of tracks.
2. The disk is formatted with four main storage areas: the boot sector, the file allocation table (FAT), the root directory, and the file storage area. The boot sector is 512 bytes in length and it contains a program that boots the system or loads DOS into memory. The FAT is a table that indicates which clusters are free, which are occupied, and which are corrupt. The root directory is the main directory on the disk that stores file names and directory names. The file storage area is where the data and programs are stored.
3. File and directory names can be up to eight characters in length, with an extension of up to three characters. Data are organized into directories and subdirectories and stored as files and programs.
4. The fixed or hard disk drive can be partitioned into several different logical disk drives by using FDISK. Additional storage can be obtained by using DoubleSpace to nearly double the amount of room on the disk.
5. Internal DOS commands are stored within the COMMAND.COM program because these are the most commonly used DOS commands. Some of them include: DIR, DEL, CD, RD, and MD.
6. The DOS prompt is C:\> when DOS is started and can be changed by using the PROMPT command. If just the prompt command is typed, the DOS prompt becomes C>.
7. The directory structure of the disk is maintained by using internal DOS commands MD (make directory), RD (remove directory), and CD (change directory).
8. Disk files are managed with internal DOS commands DEL (delete), COPY, and RENAME.
9. The external DOS commands are programs or utilities that are stored in the DOS directory. They provide additional control over the directory structure, file structure, and overall performance of the computer system.
10. External commands that are new features in DOS version 6.X are DoubleSpace, DEFRAG, SCANDISK, DELTREE, VSAFE, INTERLNK, MEMMAKER, MSAVE, POWER, and MSBACKUP.
11. The FORMAT command is used to prepare a disk for use in the system by placing tracks, sectors, and other information on the surface of the disk. The FORMAT command can format any size disk by using the /F: switch.
12. The CHKDSK and SCANDISK programs are used to search a disk in order to find any problems that may occur and correct some of these problems. The CHKDSK program

also indicates the size and structure of the disk drive. Another command, MEM, is used to check the size and structure of the memory system. Never use CHKDSK from within WINDOWS.

13. The ATTRIB command provides control over the attributes of files and directory entries. ATTRIB displays the current attributes or changes them to any setting. This includes hiding directories and making directories read-only if needed.
14. The DEFRAG program defragments a disk to enhance its performance. The hard disk drive in a personal computer should be defragmented once a month for optimal performance.
15. The DoubleSpace program (DBLSPACE) creates and maintains a compressed disk drive. The compressed disk drive typically stores about 1.8 times more information than an uncompressed drive.
16. The DOSKEY program, which first appeared in DOS version 5.0, is an important DOS feature that allows prior command scrolling and the addition of DOS macro sequences. A macro allows DOS commands to be stored with DOSKEY for activation with a single command line entry.
17. The CONFIG.SYS file sets up or configures the personal computer for operation. The CONFIG.SYS file has its own set of unique commands to direct the configuration process. Some command load drivers control the memory and the way the memory is structured.
18. The AUTOEXEC.BAT file is a batch program executed at boot time. This program enters a series of DOS commands to load programs, set environment variables, define search paths, and so forth.
19. A batch program is a collection of normal DOS commands and special batch commands that direct the operation of the system from the DOS command line. The batch program relieves the user from entering long and arduous series of DOS commands from the keyboard. Instead, they are placed in a batch program and executed by typing the name of the batch program at the DOS prompt.
20. The IF statement is used to test the ERRORLEVEL or compare two values. Associated with the IF statement is the GOTO statement, which causes the program to continue at the label defined by the GOTO if the IF statement is true.
21. Multiple configurations are allowed by DOS version 6.X using the [MENU], MENUITEM, MENUCOLOR, MENUDEFAULT, and SUBMENU commands. Multiple configurations allow the user to set up the system for multiple memory configurations using different drivers if required.

2–7 QUESTIONS AND PROBLEMS

1. A file name may contain up to ______________ characters in a DOS-based system.
2. A file name extension may contain ______________ characters in a DOS-based system.
3. Which of the following file names are valid?
   (a) FRIENDLY1
   (b) 123.FIL

   (c) A*BC.EXT
   (d) WALL-EYE.FIS
   (e) FROG.LEG
4. What is the root directory?
5. In the line C:\DATA1\DATA2, identify the purpose of each of the following parts:
   (a) C:
   (b) DATA1
   (c) DATA2
6. What is another name for a hard disk drive?
7. What is a track?
8. When a track is divided into parts, each part is called a ______________.
9. Define the term *cylinder*.
10. How many bytes of memory are stored on a $3^1/_2''$ floppy formatted as a HD disk?
11. How many bytes of memory are stored on a $3^1/_2''$ floppy formatted as a DSDD disk?
12. What is the purpose of the boot sector on a disk memory?
13. What information is stored in the file allocation table on a disk memory?
14. What is a cluster?
15. If a cluster contains eight sectors, the minimum file allocation unit size is ______________ bytes.
16. How many directory entries fit in the root directory of a $3^1/_2''$ DSDD floppy disk?
17. What is fragmentation?
18. What information is found in a directory entry?
19. What is a disk partition and what program is used to control it?
20. What is a compressed disk partition?
21. What program is used to create and maintain compressed disk partitions?
22. Where is the data for a compressed disk stored?
23. What is an internal DOS command?
24. What program processes DOS commands?
25. What command prompt is used by default with DOS version 6.2?
26. Explain what each of the following DOS commands produce on the video display:
   (a) DIR
   (b) DIR /W
   (c) DIR /P
   (d) DIR /?
27. What is the purpose of DOS wild card ? as it applies to a file name or extension?
28. What is the purpose of DOS wild card * as it applies to a file name or extension?
29. Develop a series of DOS commands that change to the root directory and create a new directory called TEST1.3.
30. Develop a series of DOS commands that remove directory TEST1.3 from the root directory.
31. Which DOS command is used to remove the contents of all directories and directory names from a tree?
32. What is the purpose of the TREE.COM program?
33. What do the symbols . and .. indicate in a directory?
34. Develop DOS commands to accomplish the following:
   (a) erase all files with the extension .TXT from the current directory
   (b) erase file FROG.ANY from directory C:\TEST

(c) erase all files in the current directory

(d) erase any file that begins with the letter D from the current directory

35. What is accomplished by the COPY C:\TEMP*.* C:\NEW\ instruction?
36. What DOS command is used to change the DOS prompt?
37. Select a DOS command that changes the DOS prompt so it appears as:

(a) C> (b) WHAT NOW> (c) C:> (d) C<>

38. What is the difference in execution speed between an external and internal DOS command?
39. Can external DOS commands be removed from the DOS directory if more space is needed?
40. Select a DOS command that formats the disk in drive A as a 720K floppy disk.
41. Select a DOS command that formats the disk in drive B as a bootable 1.44M floppy disk.
42. What is the purpose of the CHKDSK program?
43. What is a lost allocation unit?
44. Develop a DOS command that makes file C:\TEMP\TEST.TXT a hidden, read-only file.
45. Develop a macro for the DOSKEY program that changes the DOS prompt to C> when OLD is typed at the command line.
46. Develop a macro for the DOSKEY program that erases all files with the extension .BAK in the current directory whenever DELL is typed at the command line.
47. How is the buffer size changed for the DOSKEY program?
48. Develop a CONFIG.SYS file that loads HIMEM.SYS and EMM386.EXE into the system. You must load DOS high, and use upper memory block with an expanded memory size of 1024K bytes. No other feature is needed for this CONFIG.SYS file.
49. Create a batch program that loads the DOSKEY program high and programs the following macros:
    (a) display all root directory files that begin with the letters DR
    (b) display all files with the extension .BAK that appear in the current directory in the current directory BAK
    (c) display the contents of all files in the current directory with the extension .TXT
50. Create a batch program that displays the directory in Example 2–20 and performs the tasks listed in the directory.

EXAMPLE 2–20

```
         FORMAT MENU

1 Format Drive A as a 360K DSDD 5 1/4″ floppy disk
2 Format Drive A as a 1.2M HD 5 1/4″ floppy disk

Enter choice(1,2)?
```

51. Create a batch program called MOVE one two. Parameters one and two contain file names. The batch program should copy file name one to file name two and then erase file name one.
52. What is the purpose of the [MENU] command when found in a CONFIG.SYS file?
53. How is MENUCOLOR used to change the colors of the menu in a CONFIG.SYS file?
54. What is the purpose of the number associated with the MENUDEFAULT command in a CONFIG.SYS file?

CHAPTER 3

Windows and Configuring Windows

INTRODUCTION

The Windows program, from the Microsoft Corporation, is becoming the secondary standard operating system for the personal computer. Currently it is found in more than 40 percent of all personal computer systems—about 60 million personal computers. As its popularity continues to increase, it will replace DOS as the primary operating system. For this reason, a chapter discussing the basic operation and configuration of Windows is included in this text.

The **Windows** program is a user-friendly graphical user interface **(GUI)** that allows relatively unskilled computer users to operate the personal computer with more ease and less training than DOS. One of the differences between DOS and Windows is in the way that data are displayed. The DOS system uses a text or character-based display; Windows uses a graphical display. The graphical display allows the integration of text and pictures, while the text display does not easily allow the integration of graphics. Another difference is in the way Windows functions with the printer and other system components. In DOS, the printer and mouse are treated as add-on features, while Windows treats them as essential, integrated components.

OBJECTIVES

Upon completion of this chapter, you will be able to:

1. Describe the Windows operating system and contrast it with DOS.
2. Modify the WIN.INI and SYSTEM.INI files to customize the operation of Windows.
3. Optimize the system for Windows.
4. Develop PIF files for DOS-based applications.

3–1 INTRODUCTION TO WINDOWS

The Windows program provides a unified approach to operating the personal computer not provided by the DOS operating system. A user must undergo considerable training before using a personal computer operated with DOS. The Windows operating system eliminates the need for much of that training, because of the way it presents the user with tasks. In place of the command line prompt, so familiar to DOS users, Windows is accessed via the program manager. The **program manager** provides the user with a display of icons (see Figure 3–1) that select and **launch** (execute) various tasks. There is no need to memorize commands or program names and extensions. This is the key to the ease of use of Windows—once the icons and purpose of each application are learned.

One major problem with DOS is that, as each program is added to the DOS-based personal computer, it must be configured for the hardware in the system. Each program is supplied with a series of driver programs that allow it to function with various printers and displays. Even the mouse requires a special driver for many DOS applications. When a DOS program is installed, these features must be selected and enabled. This often requires considerable time and training and is often fraught with problems, such as incorrect, incompatible, or missing drivers. The reason each DOS application must be individually set up for its hardware interface is that DOS applications do not talk to each other; each requires individual initialization.

The Windows system must also be configured, but usually only once. The Windows program, unlike DOS, allows each application to communicate with a common set of hardware devices connected to the computer through the application programming interface, or API. The configuration of the Windows program is handled when a new device such as a printer is added. Each new program placed in the Windows system uses the same basic configuration. This significantly reduces the amount of time required to add and use a new program. It provides consistency from one program to the next, because the video display interface is nearly identical. Windows is a load and play environment. Once the program is loaded, it usually functions without any other changes. With DOS, each program contains its own unique video display interface, requiring considerable training for each application. Thus, the Windows system provides almost instant access to the personal computer with a minimal amount of training.

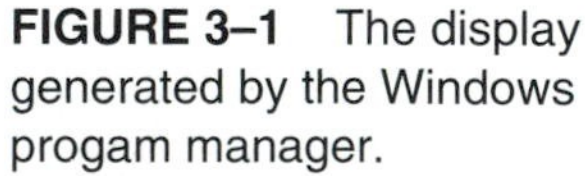
FIGURE 3–1 The display generated by the Windows progam manager.

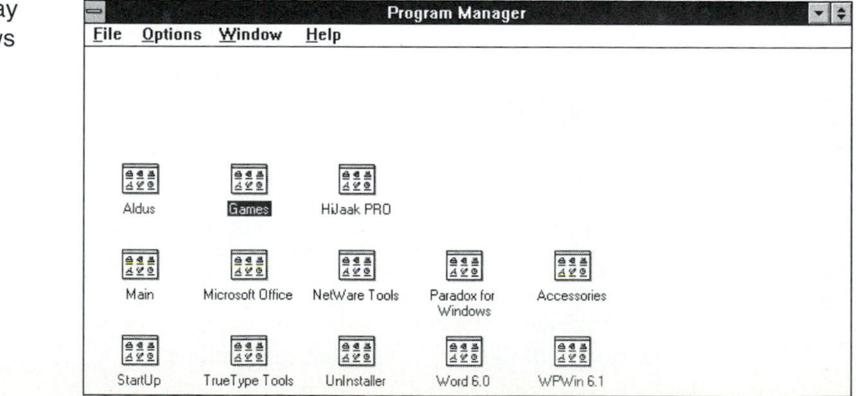

The main disadvantage of the Windows system is the amount of memory required for decent, high-speed performance (2M bytes of memory is minimum, although 4M, 8M, or 16M bytes are better), the need for a high-speed or accelerated video graphics card, and the need for a high-speed microprocessor. These problems have mostly been eliminated by the current generation of high-performance microprocessors, lower-cost memory, and accelerated video graphics cards. There is no problem with the operation and speed of Windows in a modern personal computer.

Probably one of the greatest advantages of Windows is its ability to exchange data between applications through the DDE (dynamic data exchange). For example, enter the encyclopedia program and copy a passage to the ClipBoard. The passage in the ClipBoard is held by Windows even when the encyclopedia is exited. If the word processor is launched, the contents of the ClipBoard can be copied into a word processor document. Pictures can be clipped and copied in the same way. With DOS this task was either impossible, or it required a considerable amount of data conversion and often quite a few programs to accomplish.

Using Windows

The Windows operating system is accessed via the program manager. To enter Windows, DOS is accessed at the DOS prompt or in the AUTOEXEC.BAT file with the command C:\WINDOWS\WIN. Once DOS activates Windows, the program manager is in control of the personal computer. The program manager displays groups of icons or pictures that represent the various programs and features available to the user. The user merely moves the mouse pointer to one of the icons and double-clicks the left mouse button to launch or execute the selected application.

For example, double-clicking on the *main group* icon in the program manager displays a group of icons that control various features of Windows. If the mouse pointer is moved to the **file manager** icon (a filing cabinet) and double-clicked, access is provided to the file manager program. The file manager (see Figure 3–2) allows the user to accomplish

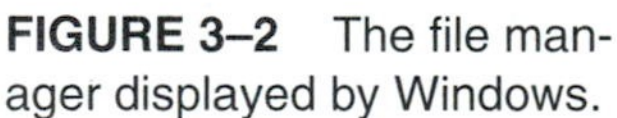

FIGURE 3–2 The file manager displayed by Windows.

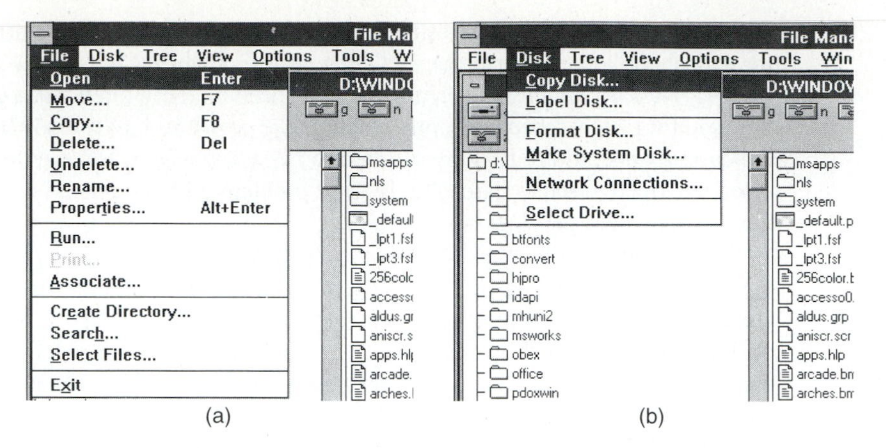

FIGURE 3–3 The file (a) and disk (b) pull-down memus found in Windows file manager.

the same basic tasks that many of the DOS commands perform. These tasks can be performed without memorizing any commands, because the commands are displayed in menu form. Figure 3–3 shows the menus displayed by the file manager for File and Disk functions. Notice that many of the DOS commands discussed in Chapter 2 are embodied in these two simple menus.

The file manager allows any directory to be displayed by double-clicking on the directory name in the directory tree. To change to a new disk drive, double-click on any of the drive icons displayed above the directory tree and directory. To obtain help on any aspect of the file manager, click on the word *help* under the word *file manager* in the menu bar. Operating the file manager requires that the user is proficient with the mouse and knows how to click and double-click on objects. It also requires an understanding of the basic DOS concepts learned in Chapter 2. The on-line help provided with the file manager is substantial and can usually provide a complete understanding of any feature in the file manager.

Using the Mouse. A mouse and a few simple directions are all that is needed to become proficient at managing the file system. For example, to delete or erase a file in the file manager, move the mouse pointer to the file name and click. Next, press the delete button on the keyboard. The file will be deleted after the file manager asks if you really want to delete the file.

Windows is easy to use, but what about configuring Windows and operating it with maximum efficiency? The next section of the text deals with configuring the SYSTEM.INI file to configure Windows.

3–2 THE WIN.INI FILE

The WIN.INI file is used to configure Windows, much the same way the CONFIG.SYS file is used to configure DOS. The **WIN.INI** (Windows initialization file) contains sections that appear in the listing as [Sound], [Desktop], and so forth. These sections configure the operation of various Windows components. Most sections in the WIN.INI file are configured by selecting the MAIN menu of the Windows program manager. If desired, the WIN.INI file can also be modified using the Windows note-pad program or the DOS EDIT program. Table 3–1 lists the Windows sections that may appear in the WIN.INI file, and describes the purpose of each section.

[Windows]

The [Windows] section specifies how the mouse and keyboard function with KeyboardSpeed, CursorBlinkRate, and DoubleClickSpeed. The Programs= statement lists the allowable program name extensions of COM, EXE, BAT, and PIF. Devices that are installed for use by Windows, such as the Canon Bubble Jet color printer, are also listed in this section. In addition to the items shown in the [Windows] section of this example, the Run= statement is added to automatically execute a program when Windows is started. An example is Run=C:\AMIPRO\AMIPRO.EXE to execute the Ami Pro word processor when Windows starts. Other command lines may appear in the [Windows] section, but are not common, so are not explained in this text. Please refer to the Windows resource kit or Windows software development kit for additional detail on [Windows] and other sections of the WIN.INI file.

TABLE 3–1 Basic sections of the WIN.INI file

Section	*Function*
[Windows]	Defines applications that start with Windows, printing, mouse, and other settings
[Desktop]	Controls the appearance of the desktop
[Extensions]	Similar to SET for DOS
[Intl]	Selects international settings
[Ports]	Lists all available I/O ports
[Fonts]	Lists screen fonts
[FontSubstitutes]	Lists interchangeable fonts
[TrueType]	Describes TrueType fonts
[MCI Extensions]	Media control interface devices
[Network]	Network settings
[Embedding]	Lists server objects for OLE (object linking and embedding)
[Windows Help]	Settings for help
[Sound]	Source files assigned to system events
[PrinterPorts]	Lists ports assigned to the printer
[Devices]	Used for compatibility with earlier versions of Windows
[Program]	Search paths for program names
[Colors]	Assigns colors to the Windows display

[Desktop]

The [Desktop] section controls the appearance of the screen, the placement of windows, and position of the icons. This section of the WIN.INI file is easily modified with the *control panel* application in MAIN.

[Extensions]

The [Extensions] section is similar to the SET command for DOS, except it is much more powerful. An extension for any program is defined in this section so that whenever a file with the extension is selected, the program associated with the extension automatically executes. An example is DATUM.WRI. Because the extension .WRI is associated to the program WRITE.EXE, Windows automatically starts to the write program.

Modifying WIN.INI

The WIN.INI file can be modified with any ASCII text editor. The DOS Edit program or Windows Notepad program are examples of ASCII text editors. Whenever the WIN.INI file is changed, change one entry at a time, then test for proper operation by restarting Windows. This makes it easy to modify the WIN.INI file without causing any long-lasting or severe problems. Because only one entry at a time is changed, it is simple to reverse the change if an error is detected. For the changes to become effective, Windows must be closed and restarted.

3–3 THE SYSTEM.INI FILE

The SYSTEM.INI file is similar to the DOS CONFIG.SYS file, but more powerful. The SYSTEM.INI file contains global information that Windows uses when it starts. Table 3–2 shows the sections found in the SYSTEM.INI file. These sections are delineated using brackets, just as they are in the WIN.INI file.

TABLE 3–2 Basic sections of the SYSTEM.INI file

Section	*Function*
[Boot]	Lists drives and modules used by Windows
[Boot.description]	The names of devices that change with Windows setup
[Drivers]	Names assigned to the installable driver files
[Keyboard]	Information about the keyboard
[MCI]	Media control interface drivers
[NonWindowsApp]	Information used by non-Windows applications
[Standard]	Used for standard mode operation
[386enh]	Used for enhanced 80386 mode operation

[Boot]

The [Boot] section contains entries that direct the operation of Windows when it is first activated. Never delete any entries in this section because, a deletion will most likely prevent Windows from starting or operating correctly. Windows setup and installation inserts many of the driver names in this section of the SYSTEM.INI file. Never change any of them unless you are directed to or completely understand the effect of a change on the operation of Windows. This section details the portions of [Boot] that may be modified without causing problems to the Windows system.

The shell=progman.exe line can be changed to cause Windows to run another shell manager in place of the program manager. Unless you need to run another shell in place of the program manager, you should not change this entry. If shell= is left blank, no shell is executed.

Another entry that may be changed is the mouse.drv= line. This selects the default Windows mouse driver. It is best to use the driver provided by Microsoft unless you have a special application, such as trackball, that might require the installation of a different mouse driver.

[Keyboard]

The [Keyboard] section specifies the type of keyboard connected to the computer and the language assigned to the system. In the example, as with most U.S. systems, only the keyboard type will contain a value. The keyboard type number specifies the type of keyboard connected to the system. This should be 4 for a 101- or 102-key keyboard. If an older, 83-key, keyboard is attached, the value is 1.

[386ENH]

This section is the one most often modified in the SYSTEM.INI file if you operate using an 80386—Pentium microprocessor and your machine contains more than 2M bytes of RAM. If your machine does not meet these requirements then you will not change the [Standard] section.

For optimal disk interface speed it is very important to enable 32-bit disk access. This is enabled by using 32BitDiskAcess=on. If the machine balks at this setting—it rarely will—disable it with off. The 32-bit disk access is turned off by the normal Windows setup and should be enabled for maximum disk transfer speed.

Another increase in disk performance is obtained by changing the size of the DMA buffer. By default DMABufferSize=16 (16K bytes). Changing this default to 64 will increase the performance of the system if you have more than 2M bytes of system memory.

The INT28Critical=on is the default setting for handling critical INT 28H interrupts. These are used by some network cards, hence the default setting. This can be turned off in many systems to improve performance.

The swap file is controlled by two entries in the SYSTEM.INI file. The PermSwapDOSDrive=C selects drive C for a permanent swap file and the PermSwapSizeK=15702 selects a permanent swap file size of 15702K bytes. In most cases, the permanent swap file allows faster Windows performance, but should be turned off and a temporary swap file used if little disk space remains. These values are usually changed by Control Panel under MAIN in the enhanced section (the IC chip). The permanent swap file must never be stored on a compressed disk drive, or Windows will lock and crash.

[Standard]

The [Standard] section contains information for operating Windows in a system that contains less than 2M bytes of RAM. It is doubtful that many such systems successfully execute Windows applications, other than a few of the built-in games.

3–4 OTHER .INI FILES

Each application normally contains one or more .INI files. The Windows directory contains at least three of these, which are used to configure the program manager, control panel, and the file manager. The contents of an .INI file can be modified to customize the Windows application.

EXAMPLE 3–1

```
[Settings]
Window=2 16 514 299 1
SaveSettings=0
MinOnRun=0
AutoArrange=0

[Groups]
Group1=C:\WINDOWS\ATIUTIL.GRP
Group2=C:\WINDOWS\ACCESSOR.GRP
Group3=C:\WINDOWS\MAIN.GRP
Group4=C:\WINDOWS\GAMES.GRP

[Restrictions]
NoClose=1
```

The PROGMAN.INI file (see Example 3–1) contains [Settings] and [Group] sections. A third section, called [Restrictions], can be added to disable program manager features. The [Restrictions] section is normally used in a network installation.

To disable the run function from the program manager, insert line NoRun=1 under the [Restrictions] section. (Note that 1 is on and 0 is off.) This prevents the user from running programs that do not appear as an icon. To prevent a user from exiting Windows insert the line: NoClose=1. To remove the file menu from the program manager, insert the line: NoFileMenu=1. The edit level available to Windows can be changed with the line: EditLevel=n. The value of n determines the edit restrictions of the program manager. Refer to Table 3–3 for a list of the values of n and the restrictions they impose. An example is EditLevel=1, which prevents a user from modifying program groups.

TABLE 3–3 Restriction imposed by *n* in the [Restrictions] section of the PROGMAN.INI file

n	*Restrictions*
0	No edit restrictions imposed
1	Prevents creation, deletion, or renaming of program groups and also prevents the function of New, Move, Copy, and Delete in the File Menu
2	Prevents all restrictions of n=1 and prevents the user from changing program items or applications
3	Prevents all restrictions of n=1 and n=2 and prevents the user from changing program command lines
4	Prevents all restrictions of n=1 through n=3 and prevents all areas of the properties box from functioning

3–5 IMPROVING PERFORMANCE

To reduce the amount of hard disk storage space required by Windows, some files may be deleted from the Windows and DOS directories. This section of the text lists the files and their purpose. Also described are tips on improving the performance of the Windows program.

DOS Setup

The best Windows performance can only be achieved by properly configuring DOS. Hopefully you are using DOS version 6.0 or 6.2 to operate Windows. These versions of DOS provide some features that significantly improve the performance of the Windows environment.

Make sure that the HIMEM.SYS and EMM386.EXE drivers are installed in the CONFIG.SYS file for DOS. The EMM386.EXE program should not allocate extended memory. Use EMM386.EXE RAM and any other switches identified by MEMMAKER. The RAM switch is very important for Windows, because it allows Windows to access extended memory and allows DOS applications running under Windows to use expanded memory. This frees all extended memory for extended memory use, unless a DOS application running under Windows requires extended memory. The new version of EMM386.EXE dynamically allocates and releases memory for use as expanded memory when needed.

The SMARTDRV.EXE program is also improved. This program now caches CDROM reads in the cache. This improves the performance of the CDROM drive when accessed via WINDOWS.

Do not use the RAMDRIVE.SYS driver to create a RAM drive. The SMARTDRV.SYS program is much more efficient at accessing and caching common disk data. It is better to use RAM for SMARTDRV than for RAMDRIVE.

Load both DOS HIGH and UMB (upper memory blocks) in the CONFIG.SYS file. The use of upper memory blocks frees more conventional memory for Windows and DOS applications that run under Windows.

Windows Setup

Optimal performance of Windows uses at least 4M bytes of RAM. The more RAM, the better, at least up to 16M bytes. Speed improvement is far greater when memory is upgraded from 4M to 8M bytes than when it is upgraded from 8M to 16M bytes. If only 2M bytes are available, most applications will crawl and make Windows annoying to use, especially for word processing.

As mentioned in the last section, use the 32-bit access in the enhanced mode of operation. This affords the most significant increase in system speed. Make sure that a permanent swap file is used instead of a temporary swap file. This also speeds system performance.

Always use the lowest resolution display driver. This increases performance with most video display adapters. The size of the monitor should dictate the resolution selected for Windows. The 640 × 480 mode is about all a 12″ monitor can comfortably use. The 15″ monitor functions best with 800 × 600 resolution. A 17″ monitor can use the 1024 × 768 mode. These are the author's preferences for most application software; you might experience sufficient comfort with a higher resolution. If possible, use or purchase an accelerated video card using the VESA local bus or PCI bus.

TABLE 3–4 DOS files that can often be deleted

File N	*Comments*
APPEND	Unless this is in the AUTOEXEC.BAT file
ASSIGN	Do not use
DEBUG	Delete unless hexadecimal machine language is used
DOSSHELL	Delete unless DOSSHELL is installed
EDLIN	Do not use
EXE2BIN	Delete unless .COM files must be created with an assembler
FASTHELP	An older version of help that can be deleted
FASTOPEN	Causes problems with SMARTDRV
FDISK	Unless you create additional logical drives
GRAFTABL	Used with CGA monitors
GRAPHICS	Used with CGA monitors
JOIN	Do not use
KEYB	For foreign keyboards
NLSFUNC	For foreign language support
SHARE	Causes problems
SUBST	Do not use
DELTREE	Can cause major problems; better to delete it

Files You May Not Need

Some of the files found in the DOS and Windows directories may not be needed for operation of DOS or Windows. Table 3–4 lists the DOS files that can normally be deleted from the directory to add additional space to the hard disk memory.

Some files that appear in the Windows directory can also be removed from the system to free additional hard disk space. If enhanced mode is used, delete the following files to free 183K bytes on the hard disk drive: DOSX.EXE, DSWAP.EXE KRNL-286.EXE, WINOLDAP.MOD, WSWAP.EXE, and *.2GR. Removal of these files permanently disables standard mode operation.

If standard mode is used, delete the following files to free 781K bytes on the hard disk drive: CGA?????.FON, EGA?????.FON, CPWIN386.CPL, DOSAPP.FON, *.3GR, *.386, WIN386.EXE, WIN386.PS2, and WINOA386.MOD. Removal of these files permanently disables enhanced mode operation.

3–6 DOS APPLICATION PIF FILES

To execute a DOS program from Windows, a *program information file* (PIF) is created to define the DOS application. The creation of the PIF is handled by the PIF editor, located in the Main section of program manager. The PIF editor allows the requirements for a DOS application to be specified.

Figure 3–4 shows the first window of the PIF editor. The program filename is the name of the DOS program to be run from Windows. The window title is the name applied to the icon designated for the program manager or other programming icon group. The startup directory indicates which disk drive and directory hold the DOS application. The optional parameters block holds any switches that are applied to the DOS program such as /W or others. These are the main parameters that are defined for a PIF.

Other parameters define how the DOS application uses the memory and video display. If the DOS application is a text mode application, choose Test. If the DOS application uses graphics, choose Graphics/Multiple text. The memory size required by the DOS

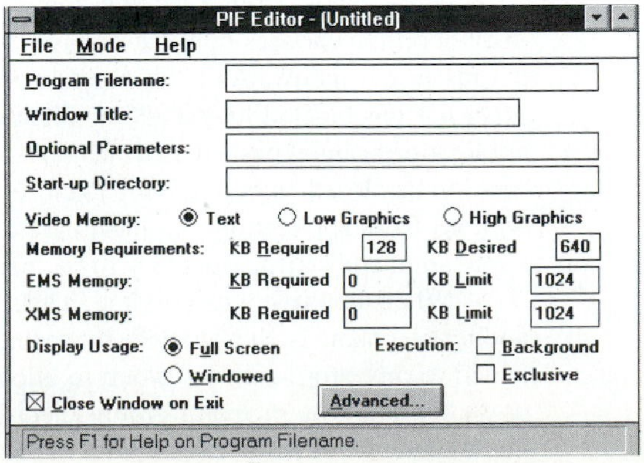

FIGURE 3–4 The first screen of the program information file (PIF) editor.

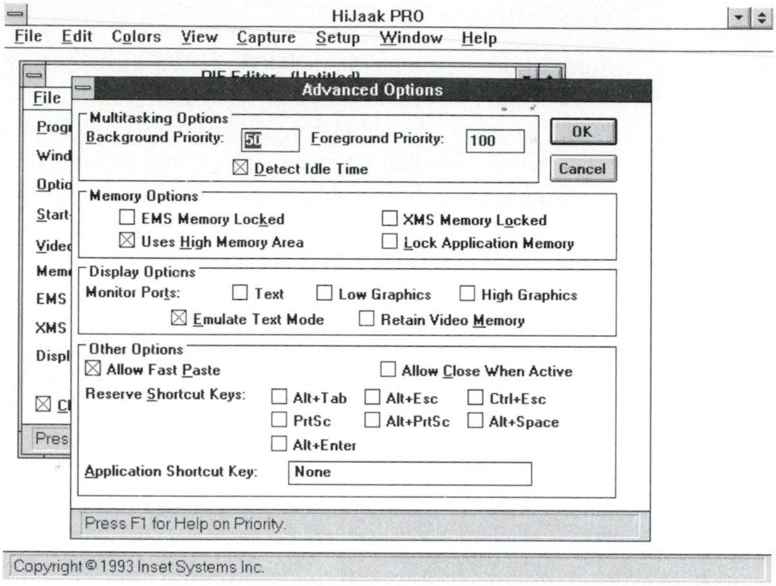

FIGURE 3–5 The advanced options menu from the PIF editor.

application is specified using a –1 to give the application all available memory, or some other value depending on the size of the application. In most cases the –1 is used to specify memory size.

If the DOS application directly modifies the keyboard or a COM port, check the box or boxes that apply. In most cases, choose full screen for a DOS application unless you specifically need to run it in a window. Running a DOS application inside a window may cause problems with the text display or graphics.

The advanced PIF options are available in enhanced 80386 mode and normally do not need to be modified. Figure 3–5 shows the advanced options box. There are four areas that are modified in the advanced options section of the PIF. The multitasking option adjusts the processing time between foreground (active window) and background execution of a DOS application.

The memory option section controls how the DOS application uses memory. The display option chooses how the output from the DOS application appears on the screen if it runs in a window. As mentioned, most DOS applications run better full-screen. The other options block allows DOS applications to use the clipboard for pasting. Most DOS applications cannot paste data to windows. The allows close, when active box allows Windows to be closed without having to close the DOS application. This may or may not work with all DOS applications. In most cases, DOS applications only require the name of the application and startup directory to be specified for proper operation.

When the PIF editor is exited, it asks for the name of the PIF. In most cases, the DOS application name is chosen, with the extension .PIF. Once the PIF is constructed, it can be added to any programming group to allow the DOS application to be executed by Windows. To add to a program group select New from the File menu in the group. Use the name of the PIF file developed with the PIF editor.

3–7 SUMMARY

1. Windows uses a graphical user interface (GUI) to decrease the complexity of the system as viewed by the user. GUI allows information to be displayed in WYSIWYG form.
2. The program manager allows the user to select applications via icons displayed on the video monitor. This eliminates the need to learn program names and type them from the keyboard.
3. The amount of memory required by Windows is 4M bytes for good performance or 8M bytes for better performance.
4. The file manager program replaces many of the DOS commands learned in Chapter 2. The file manager allows files to be moved, deleted, renamed, and so forth, without memorizing or using any of the DOS commands.
5. The mouse is an essential part of the Windows environment, used to select applications and manipulate data. Without a mouse, Windows is extremely difficult to use. Once the mouse is mastered, it becomes second nature.
6. Two files are used to initialize and control Windows: WIN.INI and SYSTEM.INI. These files perform many of the same tasks as the CONFIG.SYS and AUTOEXEC.BAT files used with DOS.
7. Other .INI files are used to define how each application uses Windows. Often these are changed by the user to modify the behavior of a Windows application.
8. The program information file (PIF) specifies how a DOS application performs with Windows.

3–8 QUESTIONS AND PROBLEMS

1. What is a GUI?
2. What is an icon?
3. How does the Windows program reduce the need for training?
4. What does the term *launch* imply?
5. What is the program manager?
6. How is an application launched from the program manager?
7. Why doesn't DOS share printer drivers with all of its applications?
8. How much memory is required for Windows?
9. What is the purpose of the file manager?
10. Describe the difference between DOS commands and commands issued from the file manager.
11. How is the disk drive selected in the file manager?
12. How is the WIN.INI or SYSTEM.INI file modified in Windows?
13. What is the purpose of the run= entry in the WIN.INI file?
14. What is the purpose of 32BitDiskAccess in the SYSTEM.INI file?
15. What is the difference between a permanent and temporary swap file?
16. What is the difference between standard and enhanced mode operation of Windows?

17. In the PROGMAN.INI file, what is the purpose of the NoClose= statement in the [Restrictions] section?
18. Is it important to use the SMARTDRV program with Windows?
19. What purpose does the RAMDRIVE system have with Windows?
20. List five DOS files that can be deleted from the DOS directory when running Windows.
21. What files can be deleted from the Windows directory if standard mode is never used?
22. What files can be deleted from the Windows directory if enhanced mode is never used?
23. What is the purpose of the PIF?
24. How is the PIF modified or created?
25. When a –1 is used in a PIF to specify memory size, what is implied?
26. What mode of operation allows the advanced options menu to be selected in the PIF?

CHAPTER 4

The Microprocessor and Its Architecture

INTRODUCTION

This chapter presents the microprocessor as a programmable device, first looking at its internal programming model and then at how it addresses its memory space. The architecture of the entire family of Intel microprocessors is presented simultaneously, as are the ways family members address the memory system.

The addressing modes for this powerful family of microprocessors are described for both real and protected modes of operation. Real-mode memory exists at locations 00000H–FFFFFH (the first 1M byte of the memory system) and is present on all versions of the microprocessor. Protected-mode memory exists at any location in the entire memory system, but is only available on the 80286–80486 and Pentium microprocessors. for the 80286, protected-mode memory contains 16M bytes; for the 80386 and above, 4G bytes.

OBJECTIVES

Upon completion of this chapter, you will be able to:

1. Describe the function and purpose of each program-visible register in the 8086–80486 and Pentium microprocessors.
2. Detail the flag register and the purpose of each flag bit.
3. Describe how memory is accessed using real-mode-memory-addressing techniques.
4. Describe how memory is accessed using protected-mode memory-addressing techniques.
5. Describe the program-invisible registers found in the 80286, 80386, 80486, and Pentium microprocessors.
6. Detail the operation of the memory paging mechanism.

4–1 80386, 80486, AND PENTIUM INTERNAL ARCHITECTURES

Before a program is written or any instruction investigated, the internal configuration of the microprocessor must be known. This section of the chapter details the program-visible internal architecture of the 8086–80486 and Pentium microprocessors. Also detailed are the function and purpose of each of these internal registers.

The Programming Model

The programming model of the 8086 through Pentium is considered **program visible** because its registers are used during programming and are specified by the instructions. Other models, detailed later in this chapter, are considered **program invisible** because their registers are not addressable directly during applications programming, but may be used indirectly during system programming. Only the 80286 and above contain the program-invisible registers used to control and operate the protected memory system.

Figure 4–1 illustrates the programming model of the 8086 through Pentium microprocessor. The earlier 8086, 8088, and 80286 contain 16–bit internal architectures, a subset of the registers shown in Figure 4–1. The 80386, 80486, and Pentium microprocessors contain full 32-bit internal architectures. The architectures of the earlier 8086 through the 80286 are fully upward compatible with the 80386 through Pentium. The shaded areas in this illustration represent registers that are not found in the 8086, 8088, or 80286 microprocessors—they are enhancements provided on the 80386, 80486, and Pentium microprocessors.

The programming model contains 8-, 16-, and 32-bit registers. The 8-bit registers are AH, AL, BH, BL, CH, CL, DH, and DL, and are referred to when an instruction is formed using these two-letter designations. For example, an ADD AL,AH instruction adds AH to AL. This instruction is interpreted as add to AL, AH. The 16-bit registers are AX, BX, CX, DX, SP, BP, DI, SI, IP, FLAGS, CS, DS, ES, SS, FS, and GS. These registers are also referenced with two-letter designations. For example, an ADD DX,CX instruction adds CX to DX. The 32-bit extended registers are EAX, EBX, ECX, EDX, ESP, EBP, EDI, ESI, EIP, and EFLAGS. These 32-bit extended registers and 16-bit registers FS and GS are only available in the 80386 and above. They are referenced using the register designations FS or GS for the two new 16-bit registers and one of the three-letter designations for the 32-bit registers. For example, an ADD ECX,EBX instruction adds EBX to ECX.

Some registers are multipurpose registers, while some have special purposes. The **multipurpose registers** include EAX, EBX, ECX, EDX, EBP, EDI, and ESI. These registers hold various data sizes (bytes, words, or doublewords) and are used for almost any purpose as dictated by a program.

Multi-Purpose Registers

EAX (accumulator) EAX is referenced as a 32-bit register (EAX), as a 16-bit register (AX), or as either of two 8-bit registers (AH or AL). Note that if an 8- or 16-bit register is addressed, only that portion of the 32-bit register changes without affecting the remaining bits. The accumulator is used for instructions such as multiplication, division, and some of the adjustment

FIGURE 4–1 The internal program-visible register set of the 8086–Pentium microprocessor.

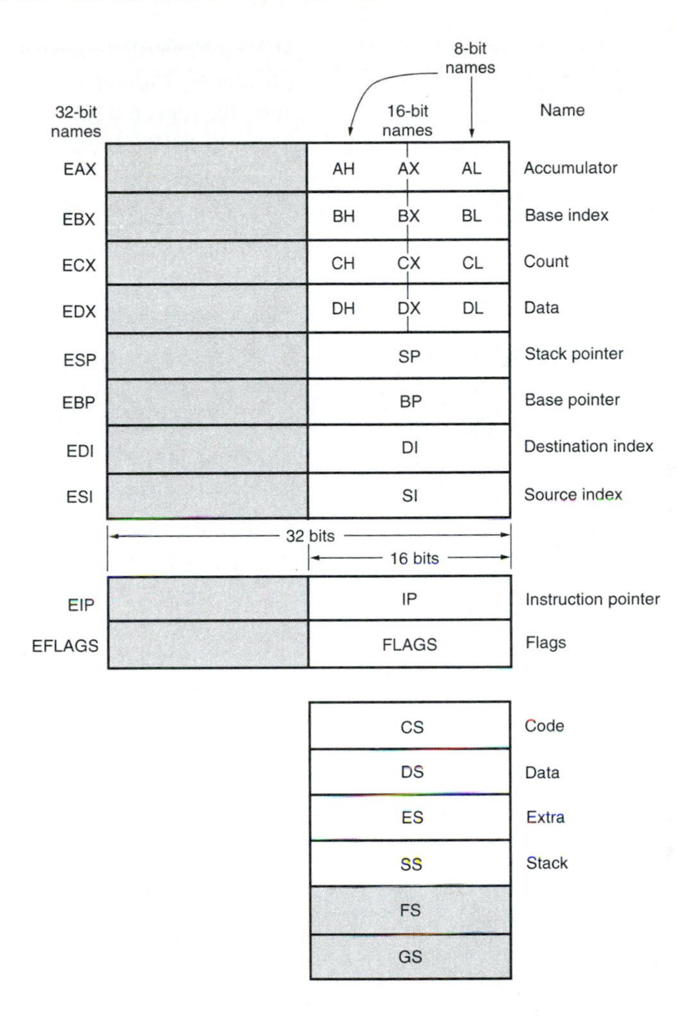

instructions. For these instructions, the accumulator has a special purpose, but it is generally considered a multipurpose register. In the 80386 and above, the EAX register may also hold an address to access a location in the memory system.

EBX (base index) EBX is also addressable as EBX, BX, BH, or BL. The BX register is sometimes used to address memory in all versions of the microprocessor. In the 80386 and above, EBX can also address memory data.

ECX (count) ECX is a multipurpose register that also holds the count for various instructions. In the 80386 and above, the ECX register can also address memory data. Instructions that use a count are the repeated string instructions

(REP/REPE/REPNE), shift, rotate, and LOOP/LOOPD instructions. The shift and rotate instructions use CL as the count, the repeated string instructions use CX as the count, and the LOOP/LOOPD instructions use either CX or ECX as the count.

EDX (data) EDXis a multipurpose register that holds part of the result from a multiplication or part of the dividend before a division. In the 80386 and above, this register can also address memory data.

EBP (base pointer) EBP points to a memory location for memory data transfers. This register is addressed as either BP or EBP.

EDI (destination index) EDI often addresses string destination data for the string instructions. It also functions as either a 32-bit (EDI) or 16-bit (DI) multipurpose register.

ESI (source index) ESI addresses source string data for the string instructions. Like EDI, ESI also functions as a multipurpose register. As a 16-bit register, it is addressed as SI; as a 32-bit register it is addressed as ESI.

Special-Purpose Registers. The special-purpose registers include: EIP, ESP, EFLAGS, discussed below, and the segment registers CS, DS, ES, SS, FS, and GS, discussed in the next section. **Special-purpose registers** perform special tasks in a microprocessor.

EIP (instruction pointer) EIP addresses the next instruction in a section of memory defined as a code segment. This register is IP (16-bits) when the microprocessor operates in the real mode and EIP (32-bits) when the 80386 and above operate in the protected mode. Note that the 8086, 8088, and 80286 contain EIP, but only the 80286 and above operate in the protected mode. The instruction pointer, which points to the next instruction in a program, is used by the microprocessor to find the next sequential instruction in a program located within the code segment. The instruction pointer can be modified with a jump or a call instruction.

ESP (stack pointer) ESP addresses an area of memory called the stack. The stack memory is explained later in the text, with instructions that address stack data. This register is referred to as SP if used as a 16-bit register and ESP if referred to as a 32-bit register.

EFLAGS EFLAGS indicate the condition of the microprocessor and controls its operation. Figure 4–2 shows the flag registers of all versions of the microprocessor. Note that the flags are upward compatible from the 8086/8088 to the Pentium microprocessor. The 8086–80286 contain a FLAG register (16-bits) and the 80386 and above contain an EFLAG register (32-bit extended flag register).

FIGURE 4–2 The flag registers of all family members. Note how they are all upward-compatible.

The five rightmost flag bits and the overflow flag change after many arithmetic and logic instructions execute. Some of the flags are also used to control features found in the microprocessor. Following is a list of each flag bit with a brief description of its function. As instructions are introduced in subsequent chapters, additional detail on the flag bits is provided. The five rightmost flags and the overflow flag are changed by most arithmetic and logic operations, but data transfers do not affect them.

C (carry) The carry flag holds the carry after addition or the borrow after subtraction. The carry flag also indicates error conditions as dictated by some programs and procedures. This is especially true of the DOS function calls detailed in later chapters and Appendix A.

P (parity) The parity flag is a logic 0 for odd parity and a logic 1 for even parity. **Parity** is a count of ones in a number expressed as even or odd. For example, if a number contains three binary one bits, it has odd parity. If a number contains zero one bits has even parity. The parity flag finds little application in modern programming. It was implemented in early Intel microprocessors for checking data in data communications environments. Today, parity checking is often accomplished by data communications equipment instead of by the microprocessor.

A (auxiliary carry) The auxiliary carry flag holds the carry (half-carry) after addition or the borrow after subtraction between bit positions 3 and 4 of the result. This highly specialized flag bit is tested by the DAA and DAS instructions to adjust the value of AL after a BCD addition or subtraction. Otherwise, the A flag bit is not used by the microprocessor or any other instructions.

Z (zero) The zero flag shows that the result of an arithmetic or logic operation is zero. If Z = 1, the result is zero, and if Z = 0, the result is not zero.

S (sign) The sign flag holds the arithmetic sign of the result after an arithmetic or logic instruction executes. If S = 1, the sign bit (leftmost bit of a number) is set, or negative; if S = 0, the sign bit is cleared, or positive.

T (trap) The trap flag enables trapping through an on-chip debugging feature. (A program is *debugged* to find an error, or *bug*.) If the T flag is enabled (1), the microprocessor interrupts the flow of the program as indicated by the debug registers and control registers. If the T flag is a logic 0, the trapping (debugging) feature is disabled. The CodeView program can use the trap feature and debug registers to debug faulty software.

I (interrupt) The interrupt flag controls the operation of the INTR (interrupt request) input pin. If I = 1, the INTR pin is enabled; if I = 0, the INTR pin is disabled. The state of the I flag bit is controlled by the STI (set I flag) and CLI (clear I flag) instructions.

D (direction) The direction flag selects either the increment or decrement mode for the DI and/or SI registers during string instructions. If D = 1, the registers are automatically decremented; if D = 0, the registers are automatically incremented. The D flag is set with the STD (set direction) and cleared with the CLD (clear direction) instructions.

O (overflow) The overflow flag occurs when signed numbers are added or subtracted. An overflow indicates that the result has exceeded the capacity of the machine. For example, if 7FH (+127) is added—using an 8-bit addition—to 01H (+1), the result is 80H (–128). This result represents an overflow condition, indicated by the overflow flag for signed addition. For unsigned operations, the overflow flag is ignored.

IOPL (input/output privilege level) The input/output privilege level flag is used in protected-mode operation to select the privilege level for I/O devices. If the current privilege level is higher, or more trusted, than the IOPL, then I/O executes without hindrance. If the IOPL is lower than the current privilege level, an interrupt occurs, causing execution to suspend. Note that an IOPL of 00 is the highest or most trusted level, and an IOPL of 11 is the lowest or least trusted level.

NT (nested task) The nested task flag indicates that the current task is nested within another task in protected-mode operation. This flag is set when the task is nested by software.

RF (resume) The resume flag is used with debugging to control the resumption of execution after the next instruction.

VM (virtual mode) The virtual-mode flag selects virtual-mode operation in a protected-mode system. A virtual-mode system allows multiple DOS memory partitions, 1M byte in length, to coexist in the memory system. This allows the system program to execute multiple DOS programs.

AC (alignment check) The alignment check flag activates if a word or doubleword is addressed on a nonword or nondoubleword boundary. Only the 80486SX microprocessor contains the alignment check bit, which is primarily used by its companion numeric coprocessor, the 80487SX, for synchronization.

VIF (virtual interrupt) The virtual interrupt flag is a virtual-mode copy of the interrupt flag bit, available to the Pentium microprocessor.

VIP (virtual interrupt pending) The vertical interrupt pending flag provides information about a virtual-mode interrupt for the Pentium microprocessor. This is used in multitasking environments to provide the operating system with virtual interrupt flags and interrupt pending information.

ID (identification) The identification flag indicates that the Pentium microprocessor supports the CPUID instruction. The CPUID instruction provides the system with information about the Pentium microprocessor, such as its version number and manufacturer.

Additional registers, called segment registers, generate memory addresses when combined with other registers in the microprocessor. There are either four or six segment registers in various versions of the microprocessor. A segment register functions differently in the real mode than in protected mode, as detailed later in this chapter. Following is a list of each segment register, along with its function in the system:

CS (code segment) The **code segment** register is a section of memory that holds the code (*programs and procedures*) used by the microprocessor. The code segment register defines the starting address of the section of memory holding code. In real-mode operation, it defines the start of a 64K byte section of memory; in protected mode, it selects a descriptor that describes the starting address and length of a section of memory holding code. In protected mode, the code segment is limited to 64K bytes in the 8088–80286 and 4G bytes in the 80386 and above.

DS (data) The **data segment** register is a section of memory that contains most data used by a program. Data are accessed in the data segment by an offset address or the contents of other registers that hold the offset address. Like the code segment, its length is limited to 64K bytes in the 8086–80286 and 4G bytes in the 80386 and above.

ES (extra) The **extra segment** register is an additional data segment used by some of the string instructions to hold destination data.

SS (stack) The **stack segment** register defines the area of memory used for the stack. The location of the current entry point in the stack segment is determined by the stack pointer register. The BP register also addresses data within the stack segment.

FS and **GS** These supplemental segment registers are available in the 80386, 80486, and Pentium microprocessors to allow two additional memory segments for access by programs.

4–2 REAL-MODE MEMORY ADDRESSING

The 80286 and above operate in either the real or protected mode. Only the 8086 and 8088 operate exclusively in the real mode. This section details the operation of the microprocessor in the real mode. In real-mode operation, the microprocessor can address only the first 1M byte of memory space, even if it is a Pentium microprocessor. Note that the first 1M byte of memory is called the **real memory** or *conventional memory* system. Both MSDOS and PCDOS assume the microprocessor is operated in real mode at all times. Real-mode operation allows application software written for the 8086/8088 (which only contain 1M byte of memory) to function in the 80286 and above without changing the software. This upward compatibility of software is partially responsible for the continuing success of the Intel family of microprocessors. In all cases, these microprocessors begin operation in the real mode by default whenever power is applied or the microprocessor is reset. The DOS environment is a real-mode environment.

Segments and Offsets

All real-mode memory addresses consist of a segment address plus an offset address. The **segment address,** located within one of the segment registers, defines the beginning address of any 64K-byte memory segment. The **offset address** selects any location within the 64K-byte memory segment. Figure 4–3 shows how the segment-plus-offset addressing scheme selects memory location. This illustration shows a memory segment 64K bytes in length that begins at location 10000H and ends at location 1FFFFH. It also shows how an offset (sometimes called a **displacement**) of F000H selects location 1F000H in the memory system. Note that the *offset* or *displacement* is the distance above the start of the segment.

The segment register in Figure 4–3 contains a 1000H, yet it addresses a starting segment at location 10000H. In the real mode, each segment register is internally appended with a *0H* on its rightmost end. This forms a 20-bit memory address, allowing it to access the start a segment at any 16-byte boundary within the first 1M byte of memory. This is required in early versions of the microprocessor to generate a 2-bit memory address. For example, if a segment register contains a 1200H, it addresses a 64K-byte memory segment beginning at location 12000H. Likewise, if a segment register contains a 1201H, it addresses a memory segment beginning at location 12010H. Because of the internally appended 0H, real mode segments can only begin at a 16-byte boundary in the memory system. This 16-byte boundary is often called a **paragraph.**

Because a real-mode segment of memory is 64K in length, once the beginning address is known, the ending address is found by adding FFFFH. For example, if a segment register contains 3000H, the first address of the segment is 30000H and the last address is 30000H + FFFFH, or 3FFFFH. Table 4–1 shows several examples of segment register contents and the starting and ending addresses of the memory segments selected by each segment address.

The offset address is added to the start of the segment to address a memory location in the memory segment. For example, if the segment address is 1000H and the offset address is 2000H, the microprocessor addresses memory location 12000H. The segment-plus-offset address is sometimes written as 1000:2000 for a segment address of 1000H with an offset of 2000H.

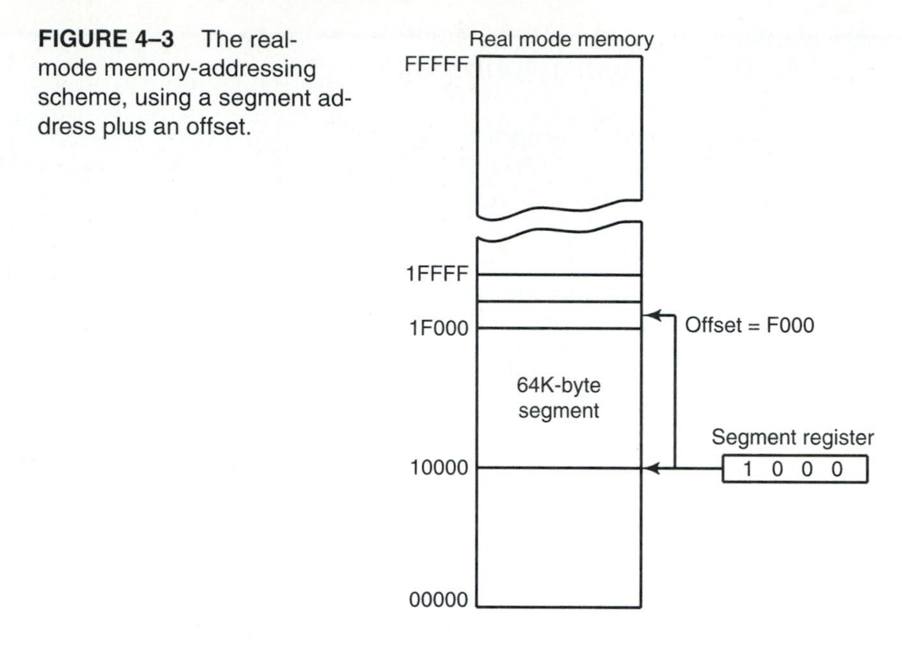

FIGURE 4–3 The real-mode memory-addressing scheme, using a segment address plus an offset.

In the 80286 (with special external circuitry), and the 80386 through Pentium, an extra 64K minus 16 bytes of memory is addressable when the segment address is FFFFH and the HIMEM.SYS driver is installed in the system. This area of memory (0FFFF0H-10FFEFH) is referred to as high memory. When an address is generated using a segment address of FFFFH, the A20 address pin is set (if supported) when an offset is added. For example, if the segment address is FFFFH and the offset address is 4000H, the machine addresses memory location FFFF0H + 4000H, or 103FF0H.

Some addressing modes combine more than one register and an offset value to form an offset address. When this occurs, the sum of these values may exceed FFFFH. For example, the address accessed in a segment whose segment address is 3000H and whose offset address is specified as the sum of F000H plus 3000H will access memory location 32000H, instead of location 42000H. When the F000H and 3000H are added, they form a 16-bit (modulo 16) sum of 2000H used as the offset address, not 12000H, the true sum. Note that the carry of 1 (F000H + 3000H = 12000H) is dropped for this addition to form the offset address of 2000H.

TABLE 4–1 Example segment addresses

Segment Register	*Starting Address*	*Ending Address*
2000H	20000H	2FFFFH
2001H	20010H	3000FH
2100H	21000H	30FFFH
AB00H	AB000H	BAFFFH
1234H	12340H	2233FH

Default Segment and Offset Registers

The microprocessor has a set of rules that apply to segments whenever memory is addressed. These rules, which apply in either the real or protected mode, define the segment-plus-offset address used by certain addressing modes. For example, the code segment register is always used with the instruction pointer to address the next instruction in a program. This combination is either CS:IP or CS:EIP, depending upon the microprocessor and mode of operation. The code segment register defines the start of the code segment and the instruction pointer locates the next instruction within the code segment. This combination (CS:IP or CS:EIP) locates the next instruction executed by the microprocessor. For example, if CS = 1400H and IP/EIP = 1200H, the microprocessor fetches its next instruction from memory location 14000H + 1200H, or 15200H.

Another of the default combinations is the stack. Stack data are referenced through the stack segment at the memory location addressed by either the stack pointer (SP/ESP) or the base pointer (BP/EBP). These combinations are referred to as SS:SP (SS:ESP) or SS:BP (SS:EBP). For example, if SS = 2000H and BP = 3000H, the microprocessor addresses memory location 23000H for a stack segment memory location. Note that in real mode, only the rightmost 16 bits of the extended register address a location within the memory segment. In the 80386–Pentium, never place a number larger than FFFFH into an offset register if the microprocessor is operated in the real mode. This causes the system to halt and indicate an addressing error.

Other defaults for addressing memory using any Intel microprocessor with 16-bit registers are shown in Table 4–2. Table 4–3 shows the defaults assumed in the 80386 and above, using 32-bit registers. The 80386 and above have a far greater selection of segment-plus-offset address combinations than the 8086–80286 microprocessors.

The 8086–80286 allow four memory segments and the 80386 and above allow six memory segments. Figure 4–4 shows a system that contains four memory segments. Note that memory segments can touch or even overlap if 64K bytes of memory are not required for a segment. Think of segments as windows that can be moved over any area of memory to access data or code. A program can have more than four or six segments, but can only access four or six segments at a time.

Suppose an application program requires 1000H bytes of memory for its code, 190H bytes of memory for its data, and 200H bytes of memory for its stack. This application does not require an extra segment. When this program is placed in the memory system by DOS, it is loaded in the TPA at the first available area of memory above the drivers and

TABLE 4–2 8086–80486 and Pentium default 16-bit segment-plus-offset address combinations

Segment	*Offset*	*Special Purpose*
CS	IP	Instruction address
SS	SP or BP	Stack address
DS	BX, DI, SI, or a 16-bit number	Data address
ES	DI for string instructions	String destination address

TABLE 4–3 8086–80486 and Pentium default 32-bit segment-plus-offset address combinations

Segment	*Offset*	*Special Purpose*
CS	EIP	Instruction address
SS	ESP or EBP	Stack address
DS	EAX, EBX, ECX, EDX, EDI, ESI, an 8-bit number, or a 32-bit number	Data address
ES	EDI for string instructions	String destination address
FS	No default	General address
GS	No default	General address

other TPA programs. This area is indicated by a free pointer maintained by DOS. Program loading is handled automatically by the DOS program loader. Figure 4–5 shows how this application is stored in the memory system. The segments show an overlap because the amount of data in them does not require 64K bytes of memory. The side view of the segments clearly shows the overlap and how segments can be slid to any area of memory by changing the segment starting address. Fortunately, DOS calculates and assigns segment starting addresses. This is explained later, in Chapter 9.

Advantages of Segment-Plus-Offset Addressing

The segment-plus-offset addressing scheme may seem unduly complicated. It *is* complicated, but it also affords advantages to the system. This complicated scheme of segment-plus-offset addressing allows programs to be relocated in the memory system and allows programs written to function in the real mode to operate in the protected mode.

A **relocatable program** is one that can be placed into any area of memory and be executed without change. **Relocatable data** is data that can be placed in any area of memory and used without any change to the program. The segment-plus-offset addressing scheme allows both programs and data to be relocated without changing a thing in a program or data. This makes it ideal for use in computer systems because memory structure of personal computers differs from machine to machine.

Because memory is addressed within a segment by an offset address, the memory segment can be moved to any place in the memory system without changing any of the offset addresses. This is accomplished by moving the entire program, as a block, to a new area and then changing only the contents of the segment registers. If an instruction is 4 bytes above the start of the segment, its offset address is 4. If the entire program is moved to a new area of memory, this offset address of 4 still points to 4 bytes above the start of the segment. Only the contents of the segment register must be changed to address the program in the new area of memory. Without this addressing feature, a program would have to be extensively rewritten or altered before it was moved. This would require many versions of a program for the many different configurations of computer systems.

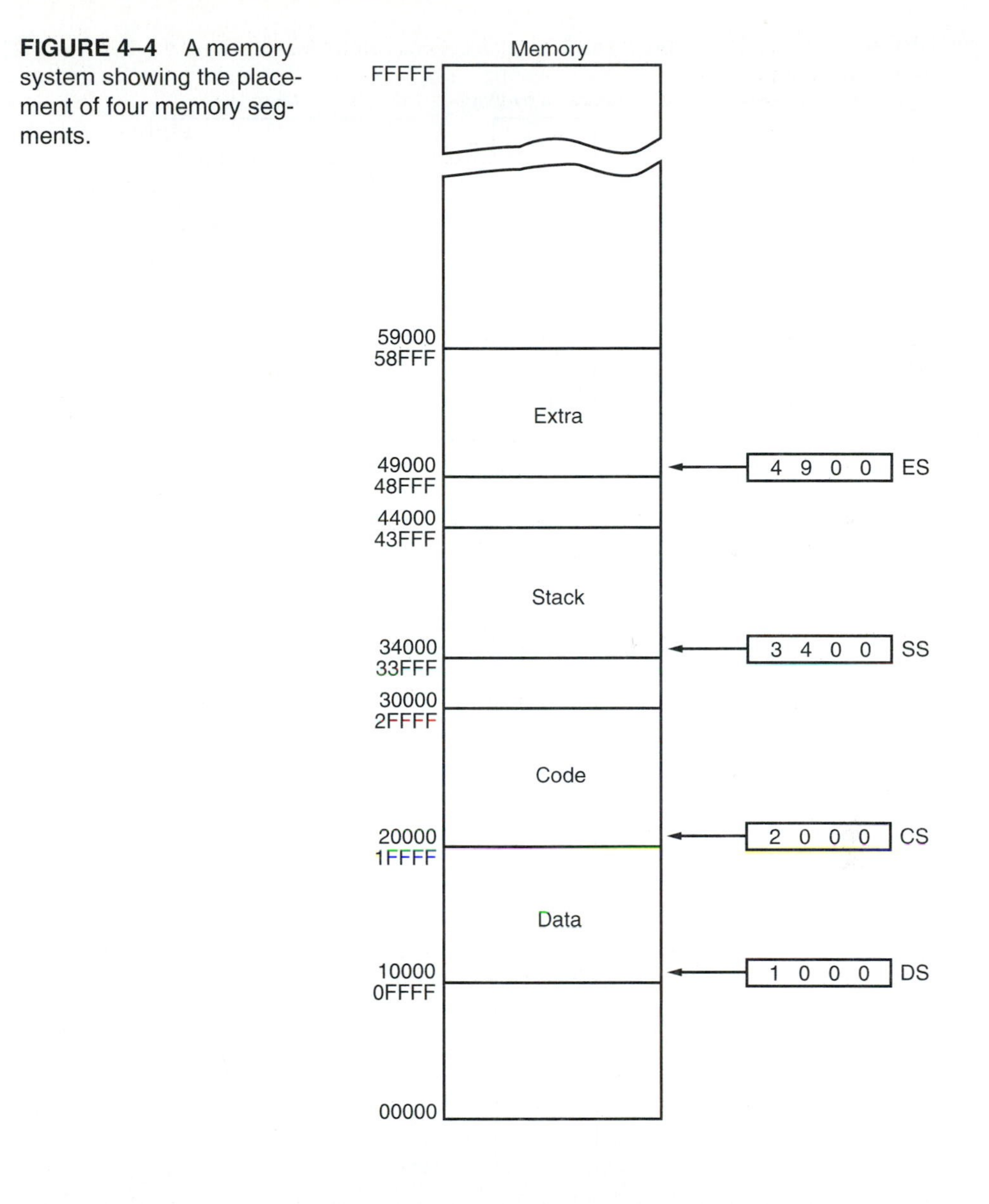

FIGURE 4–4 A memory system showing the placement of four memory segments.

4–3 PROTECTED-MODE MEMORY ADDRESSING

Protected-mode memory addressing (80286 and above) allows access to data and programs located above the first 1M byte of memory, as well as within the first 1M byte of memory. Addressing this extended section of the memory system requires a change to the segment-plus-offset addressing scheme used with real-mode memory addressing. In the protected mode, when data and programs in the extended memory are addressed, the offset address is

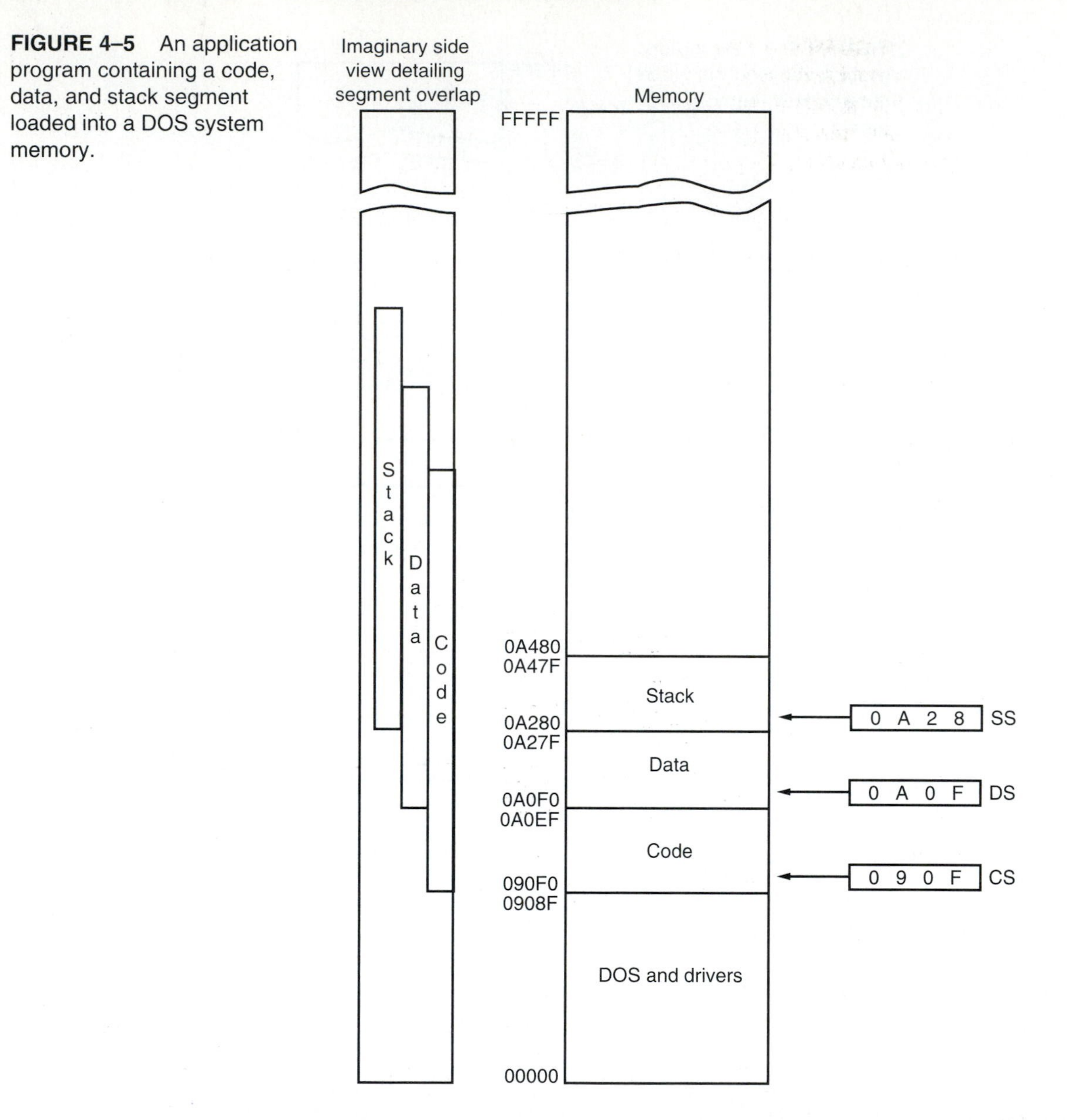

FIGURE 4–5 An application program containing a code, data, and stack segment loaded into a DOS system memory.

used but the segment address is no longer present. In place of the segment address, the segment register contains a **selector** that selects a descriptor from a descriptor table. The **descriptor** describes the memory segment's location, length, and access rights. Because the segment-plus-offset address still accesses data within the first 1M byte of memory, protected-mode instructions are identical to real-mode instructions. In fact, most programs written to function in the real mode will function without change in the protected mode. The difference between modes is in the way the segment register is interpreted by the microprocessor to access the memory segment.

Selectors and Descriptors

The selector, located in the segment register, selects one of 8,192 descriptors from one of two tables of descriptors. The descriptor describes the location, length, and access rights of the segment of memory. Indirectly, the segment register still selects a memory segment, but not in the direct way it does in the real mode. For example, in the real mode, if CS = 0008H, the code segment begins at location 00080H. In the protected mode, this segment number can address any memory location in the entire system for the code segment.

There are two descriptor tables used with the segment registers: one contains global descriptors and the other contains local descriptors. The **global descriptors** contain segment definitions that apply to all programs, while the **local descriptors** are usually unique to an application. Each descriptor table contains 8,192 descriptors, so a total of 16,384 descriptors are available to an application at any time. Because the descriptor describes a memory segment, this allows up to 16,384 memory segments to be described for each application.

Figure 4–6 shows the format of a descriptor for the 80286 through Pentium. Note that each descriptor is 8 bytes in length, so the global and local descriptor tables are each a maximum of 64K bytes in length. Descriptors for the 80286 and the 80386 through the Pentium differ slightly, but the 80286 descriptor is upward compatible.

The base address portion of the descriptor indicates the starting location of the memory segment. For the 80286 microprocessor, the base address is a 24-bit address, so segments begin at any location in its 16M bytes of memory. Note that the paragraph boundary limitation is removed in the protected mode. The 80386 and above use a 32-bit base address that allows segments to begin at any location in its 4G bytes of memory. Notice how the 80286 descriptor's base address is upward compatible to the 80386 through Pentium, because its most significant 8 bits are 00H.

The segment limit contains the last offset address found in a segment. For example, if a segment begins at memory location F00000H and ends at location F000FFH, the base address is F00000H and the limit is FFH. For the 80286 microprocessor, the base address is F00000H and the limit is 00FFH. For the 80386 and above, the base address is 00F00000H and the limit is 000FFH. Notice the limit for the 80286 is a 16 bit and the limit for the 80386 through Pentium is 20 bits. The 80286 accesses memory segments that are between 1 and 64K bytes in length. The 80386 and above access memory segments that are between 1 and 1M byte, or 4K and 4G bytes, in length.

80286 descriptor

7	0 0 0 0 0 0 0 0	0 0 0 0 0 0 0 0	6
5	Access rights	Base (B23–B16)	4
3	Base (B15–B0)		2
1	Limit (L15–L0)		0

80386/80486/Pentium descriptor

7	Base (B31–B24)	G	D	O	AV	Limit (L19–L16)	6
5	Access rights				Base (B23–B16)		4
3	Base (B15–B0)						2
1	Limit (L15–L0)						0

FIGURE 4–6 The descriptor formats for the 80286 and 80386/80486/Pentium microprocessors.

Another feature found in the 80386 through Pentium descriptor that is not found in the 80286 descriptor is the G bit or **granularity** bit. If G = 0, the limit specifies a segment from 1 to 1M bytes in length. If G = 1, the limit is any multiple of 4K bytes, because the value of the limit is multiplied by 4K bytes (appended with 000H). This allows a segment length from 4K to 4G bytes, in steps of 4K bytes. The reason the segment length is 64K bytes in the 80286 is that the offset address is always 16 bits because of the 80286's 16-bit internal architecture. The 80386 and above use a 32-bit architecture which allow an offset address of 32 bits in the protected mode. This 32-bit offset address allows segment lengths of 4G bytes. The 16-bit offset address allows segment lengths of 64K bytes.

In the 80386 and above, the AV bit is used to indicate that the segment is available (AV = 1) or not available (AV = 0). The D bit indicates how instructions access register and memory data in the protected or real mode. If D = 0, the instructions are 16-bit instructions compatible with the 8086-80286 microprocessors. This means that the instructions use 16-bit offset addresses and 16-bit registers by default. This mode is often called the *16-bit instruction mode*. If D = 1, the instructions are 32-bit instructions. The *32-bit instruction mode* assumes all offset addresses and registers are 32 bits by default. Note that default for register size and offset address size can be overridden in both the 16- and 32-bit instruction modes. The MSDOS or PCDOS operating system requires that the instructions are always used in the 16-bit instruction mode. Windows also requires that the 16-bit instruction mode is selected. 32-bit instruction is only accessible in a protected-mode system, such as Windows NT or OS/2. The next version of Windows will support 32-bit instruction mode. More detail on these modes and their application to the instruction set appears in Chapters 5 and 6.

The access rights byte (Figure 4–7) controls access to the memory segment. This byte describes how the segment functions in the system. The access rights byte allows complete control over the segment. If the segment is a data segment, the direction of growth is specified. If the segment grows beyond its limit, the microprocessor's program is interrupted, indicating a general protection fault. You can even specify whether a data segment can be written or is write-protected. The code segment is controlled in a similar fashion and can be reading-inhibited to protect software.

Descriptors are chosen from the descriptor table by the segment register. Figure 4–8 shows how the segment register functions in a protected-mode system. The segment register contains a 13-bit selector field, a table selector bit, and a requested privilege level field. The 13-bit selector chooses one of the 8,192 descriptors from the descriptor table. The TI bit selects either the global descriptor table (TI = 0) or the local descriptor table (TI = 1). The requested privilege level (RPL) requests the access privilege level of a memory segment. The highest privilege level is 00 and the lowest is 11. If the requested privilege level matches or is higher in priority than the privilege level set by the access rights byte, access is granted. For example, if the requested privilege level is 10 and the access rights byte sets the segment privilege level at 11, access is granted, because 10 is higher in priority than privilege level 11. Privilege levels are used in multi-user environments. If the privilege level is violated, the system normally indicates a privilege violation.

Figure 4–9 shows how the segment register, containing a selector, chooses a descriptor from the global descriptor table. The entry in the global descriptor table selects a segment in the memory system. In this illustration, DS contains 0008H, which accesses Descriptor 1 from the global descriptor table, using a requested privilege level of 00. Descriptor 1 contains a descriptor that defines the base address as 00100000H with a segment

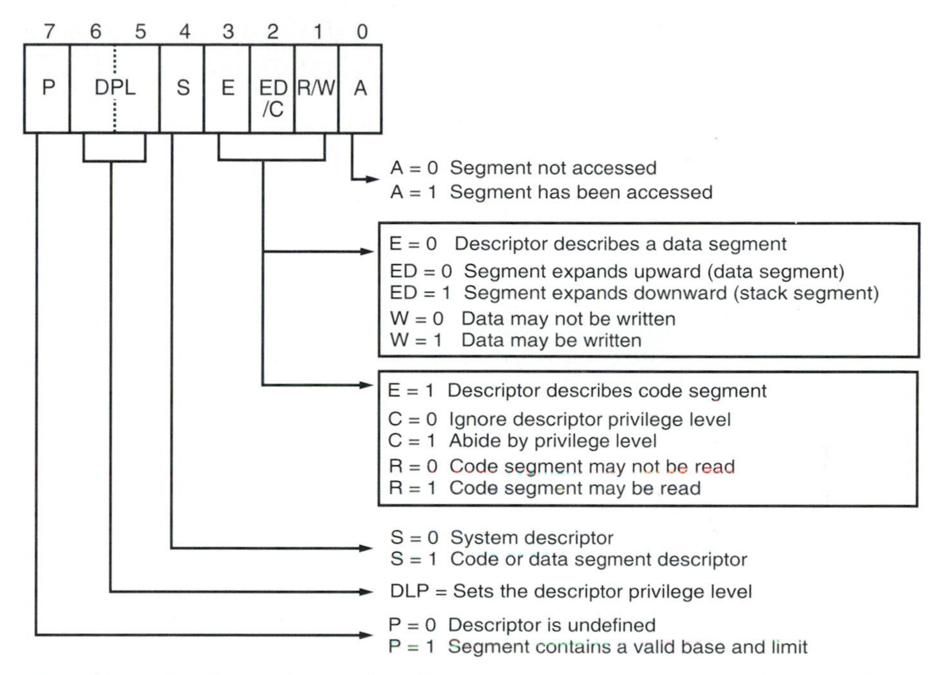

Note: Some of the letters used to describe the bits in the access rights bytes vary in Intel documentation.

FIGURE 4–7 The access rights byte for the 80286, 80486, and Pentium descriptor.

limit of 000FFH. This means that a value of 0008H loaded into DS causes the microprocessor to use memory locations 00100000H–001000FFH for the data segment. Note that Descriptor 0 is called the null descriptor and may not be used for accessing memory.

Program-Invisible Registers

The global and local descriptor tables are found in the memory system. In order to access and specify the address of these tables, the 80286, 80386, 80486, and Pentium contain pro-

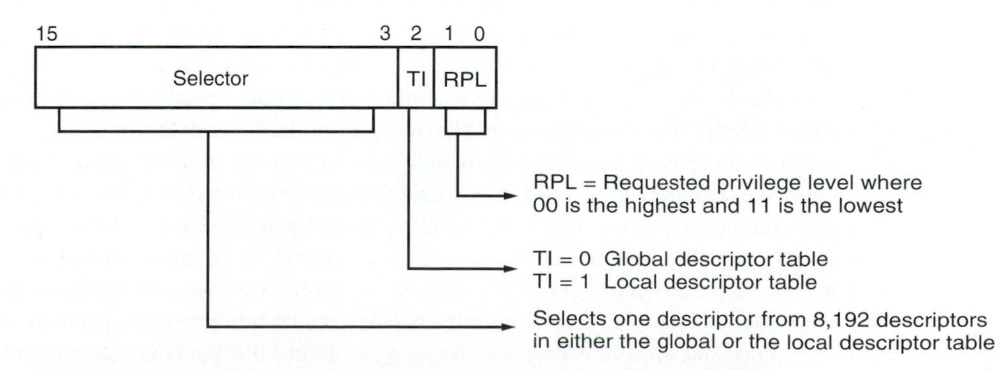

FIGURE 4–8 The contents of a segment register during protected-mode operation of the 80286, 80386, 80486, or Pentium microprocessor.

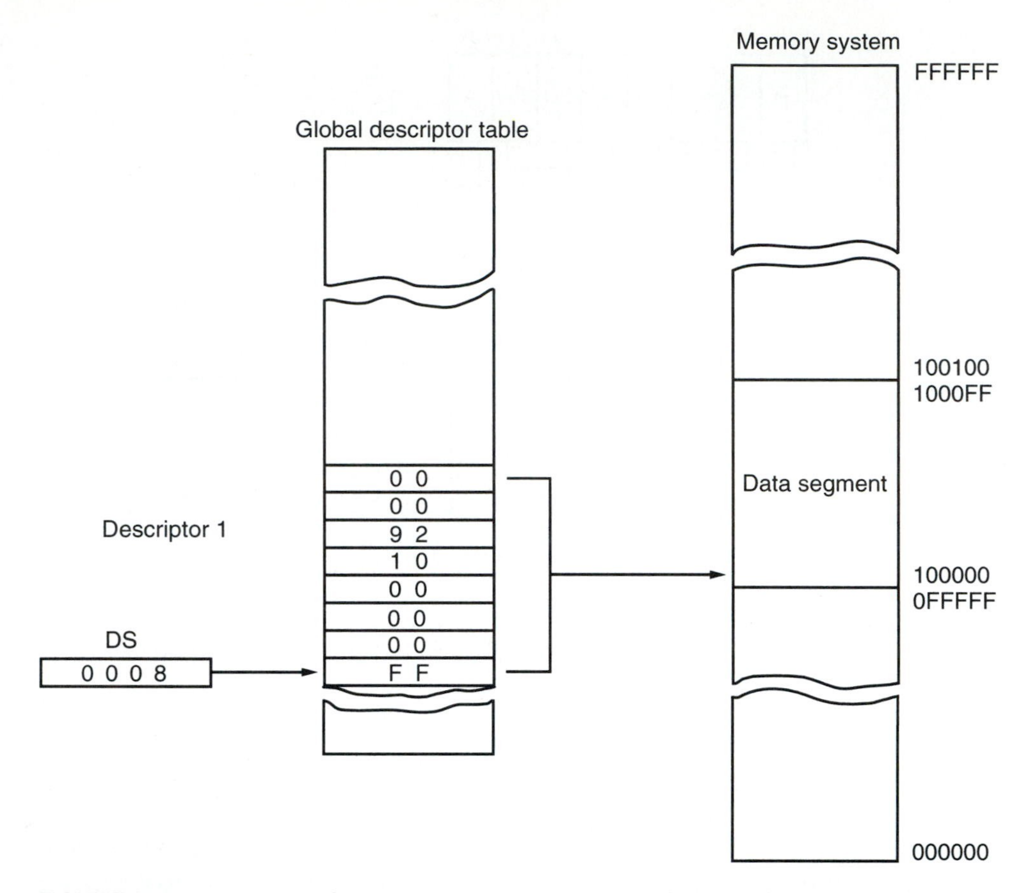

FIGURE 4–9 Using the DS register to select a descriptor from the Global Descriptor Table. In this example, the DS register accesses memory locations 100000H–1000FFH as a data segment.

gram-invisible registers. The **program-invisible registers** are not directly addressed by software, although some of them are accessed by the system software. Figure 4–10 illustrates the program-invisible registers as they appear in the 80286 through the Pentium. These registers control the microprocessor when it is operated in the protected mode.

Each of the segment registers contains a program-invisible portion used in the protected mode. The program-invisible portions of these registers are often called cache memory, because a cache is any memory that stores information. This cache is not to be confused with the Level 1 or Level 2 caches found with the microprocessor. The program-invisible portion of the segment register is loaded with the base address, limit, and access rights each time a number in the segment register is changed. When a new segment number is placed in a segment register, the microprocessor accesses a descriptor table and loads the descriptor into the program-invisible cache portion of the segment register. It is held there and used to access the memory segment until the segment number is changed again. This allows the microprocessor to repeatedly access a memory segment without referring to the descriptor table for each access, hence the term *cache*.

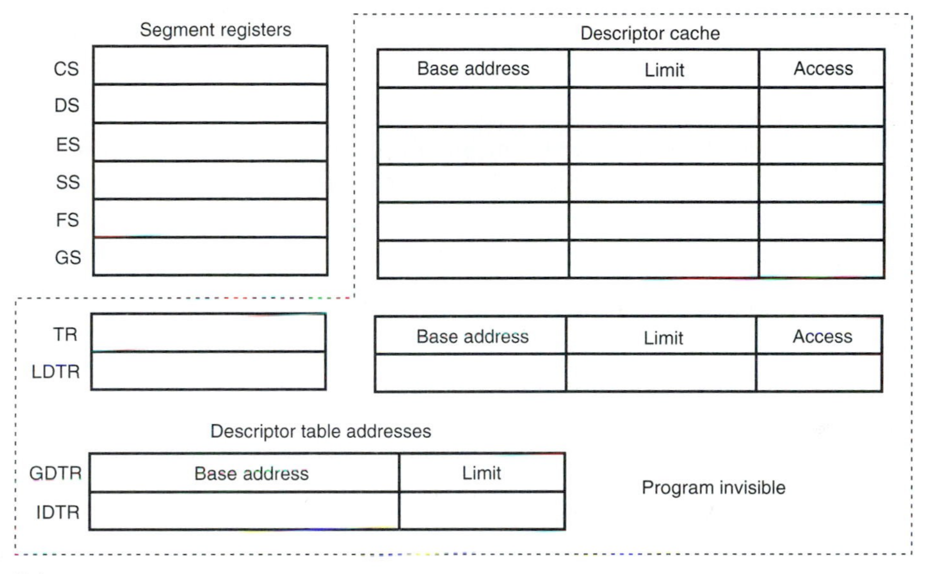

Notes:
1. The 80286 does not contain FS and GS nor the program-invisible portions of these registers.
2. The 80286 contains a base address that is 24 bits and a limit that is 16 bits.
3. The 80386/80486/Pentium contain a base address that is 32 bits and a limit that is 20 bits.
4. The access rights are 8 bits in the 80286 and 12 bits in the 80386/80486/Pentium.

FIGURE 4–10 The program-invisible register within the 80286, 80386, and Pentium microprocessor.

The GDTR (**global descriptor table register**) and IDTR (**interrupt descriptor table register**) contain the base address of the descriptor table and its limit. The limit of each descriptor table is 16 bits, because the maximum table length is 64K bytes. When protected-mode operation is desired, the address of the global descriptor table and its limit are loaded into the GDTR. Before using protected mode, the interrupt descriptor table and the IDTR must be initialized. More detail is provided on protected-mode programming in Chapter 14. At this point, the additional description of these registers is impossible.

The location of the local descriptor table is selected from the global descriptor table. One of the global descriptors is set up to address the local descriptor table. To access the local descriptor table, the LDTR (**local descriptor table register**) is loaded with a selector, just as a segment register is loaded with a selector. This selector accesses the global descriptor table and loads the base address, limit, and access rights of the local descriptor table into the cache portion of the LDTR.

The **task register** (TR) holds a selector that accesses a descriptor that defines a task. A task is most often a procedure or application program. The descriptor for the procedure or application program is stored in the global descriptor table, so access can be controlled through the privilege levels. The task register allows a context or task switch in about 17 μs. Task switching allows the microprocessor to switch between tasks in a fairly short amount of time. The task switch allows multitasking systems to switch from one task to another in a simple and orderly fashion.

4–4 MEMORY PAGING

The memory paging mechanism of the 80386 and above allows any physical memory location to be assigned to any linear address. The **linear address** is defined as the address generated by a program. With the memory paging unit, the linear address is invisibly translated into any physical address. This allows an application written to function at a specific address to be relocated through the paging mechanism. It also allows memory to be placed into areas where no memory exists. An example is the upper memory blocks provided by EMM386.EXE.

The EMM386.EXE program reassigns extended memory, in 4K blocks, to the system memory, between the video BIOS and the system BIOS ROMS. Without the paging registers, the use of this area of memory is impossible.

Paging Registers

The paging unit is controlled by the contents of the microprocessor's control registers. Refer to Figure 4–11 for the contents of control registers CR0 through CR3. Note that these registers are only available to the 80386 through Pentium microprocessors. Pentium contains an additional control register, labeled CR4, that controls extensions provided in the Pentium microprocessor. One of these features is a 4M byte page that is enabled by setting bit position 4 or CR4. As of this writing, no details on this new page size have been provided by Intel.

The registers important to the paging unit are CR0 and CR3. The leftmost bit position of CR0 (PG) selects paging when placed at a logic 1. If the PG bit is cleared (0), the linear address generated by the program becomes the physical address used to access memory. If the PG bit is set (1), the linear address is converted to a physical address through the paging mechanism. The paging mechanism functions in both the real and protected mode.

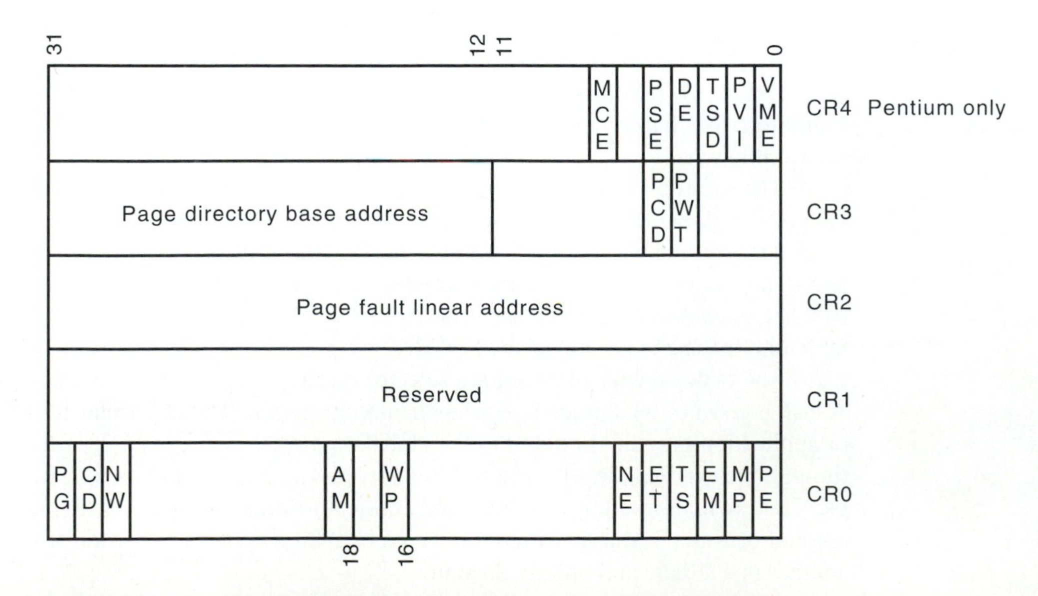

FIGURE 4–11 The control register structure of the microprocessor.

The contents of CR3 contain the page directory base address and the PCD and PWT bits. The PCD and PWT bits control the operation of the PCD and PWT pins on the microprocessor. If PCD is set (1), the PCD pin becomes a logic 1 during bus cycles that are not pages. This allows the external hardware to control the Level 2 cache memory. (The Level 2 cache memory is an external high-speed memory. It functions as a buffer between the microprocessor and the main DRAM memory system.) The PWT bit also appears on the PWT pin during bus cycles that are not pages, to control the write-through cache in the system. The page directory base address locates the page directory for the page translation unit. Note that this address locates the page directory at any 4K boundary in the memory system, because it is appended internally with a 000H. The page directory contains 1,024 directory entries of four bytes each. Each page directory entry addresses a page table that contains 1,024 entries.

The linear address, as it is generated by the software, is broken into three sections that are used to access the page directory entry, page table entry, and page offset address. Figure 4–12 shows the linear address and its makeup for paging. Notice how the leftmost 10 bits address an entry in the page directory. For linear addresses 00000000H through 003FFFFFH, the first entry of the page directory is accessed. Each page directory entry represents a 4M byte section of the memory system. The contents of the page directory select a page table that is indexed by the next 10 bits of the linear address (bit positions 12–21). This means that address 00000000H through 00000FFFH selects page directory entry 0 and page table entry 0. Notice this is a 4K byte address range. The offset part of the linear address (bit positions 0–11) next selects a byte in the 4K byte memory page. In Figure 4–12, if the page table 0 entry contains address 00100000H, then the physical

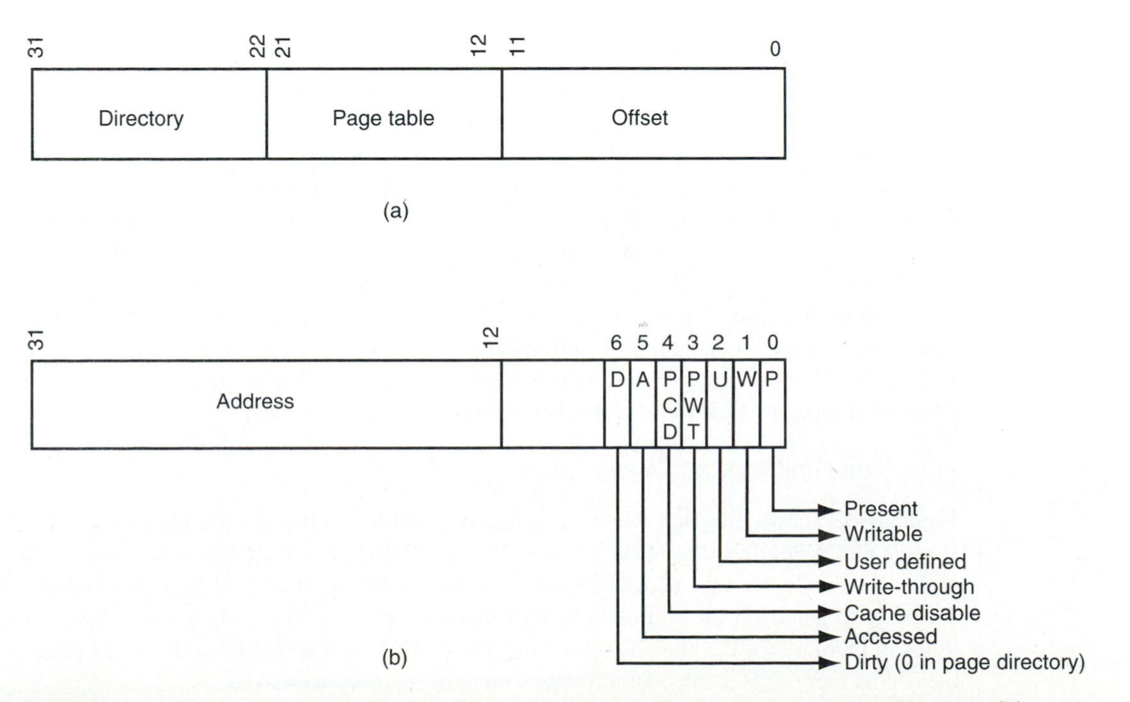

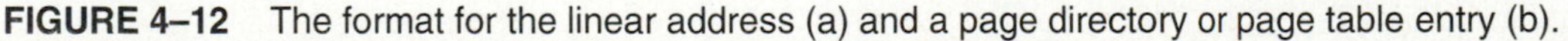
FIGURE 4–12 The format for the linear address (a) and a page directory or page table entry (b).

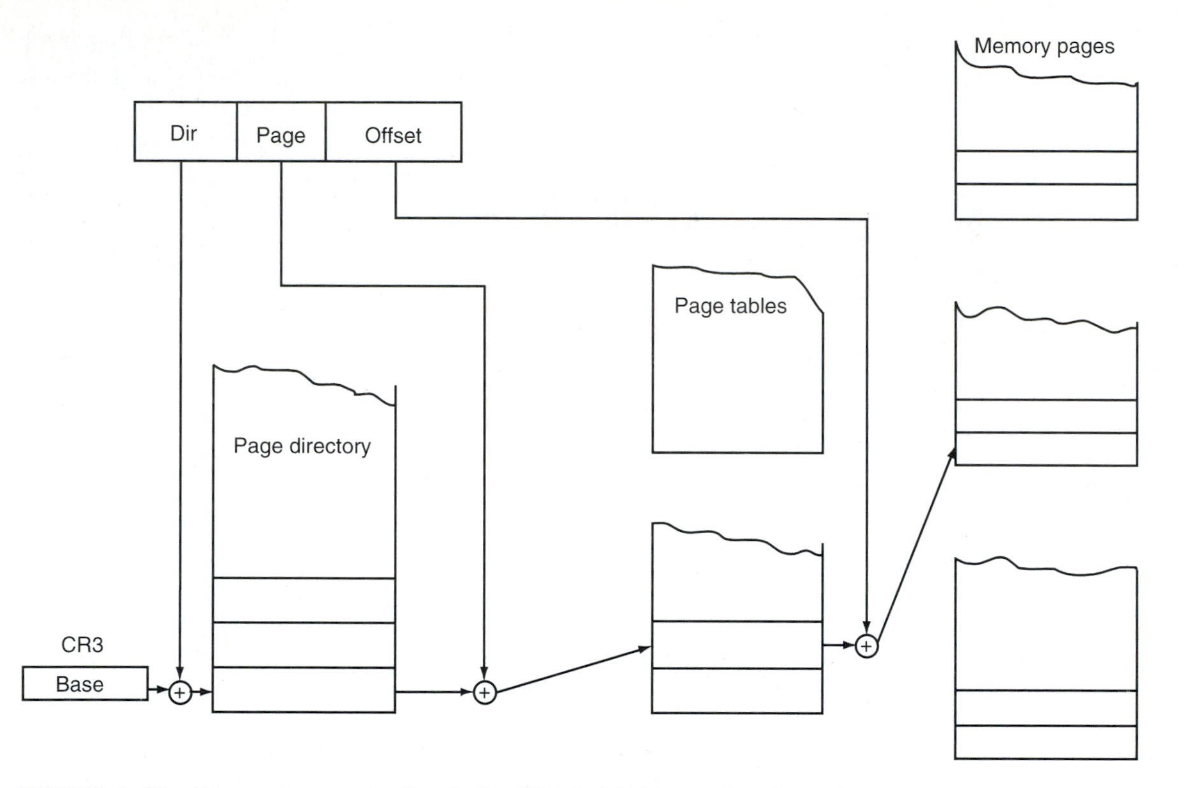

FIGURE 4–13 The paging mechanism in the 80386, 80486, and Pentium microprocessor.

address is 00100000H–00100FFFH for linear address 00000000H–00000FFFH. This means that when the program accesses a location between 00000000H and 00000FFFH, the microprocessor physically addresses location 00100000H–00100FFFH.

Because the act of repaging a 4K byte section of memory requires access to the page directory and a page table, both located in memory, Intel has incorporated a cache called the TLB **(translation lookaside buffer).** In the 80486 microprocessor, the TLB holds the 32 most recent page translation addresses. This means that the last 32 page table translations are stored in the TLB, so that if the same area of memory is accessed, the address is already present in the TLB and access to the page directory and page tables is not required. This speeds program execution. If a translation is not in the TLB, then the page directory and page table must be accessed, which requires additional execution time. The Pentium contains a separate TLB for each of its instruction and data caches.

The Page Directory and Page Table

Figure 4–13 shows the page directory, a few page tables, and some memory pages. There is only one page directory in the system. The page directory contains 1,024 doubleword addresses that locate up to 1,024 page tables. The page directory and each page table are 4K bytes in length. If the entire 4G byte of memory is paged, the system must allocate 4K bytes of memory for the page directory and 4K × 1,024, or 4M bytes, for the 1,024 page tables. This represents a considerable investment in memory resources.

The DOS system with the EMM386.EXE program uses page tables to redefine the area of memory between locations C8000H and EFFFFH as upper memory blocks. It repages extended memory to backfill this part of the conventional memory system, thus allowing DOS access to additional memory.

Suppose that the EMM386.EXE program allows access to 16M bytes of extended and conventional memory through paging, and locations C8000H through EFFFFH must be repaged to locations 110000 through 138000H with all other areas of memory paged to normal locations. Such a scheme is depicted in Figure 4–14. Here the page directory contains 4 entries. Recall that each entry in the page directory corresponds to 4M bytes of physical memory. The system also contains four page tables, with 1,024 entries each. Recall that each entry in the page table repages 4K bytes of physical memory. This scheme requires 16K of memory for the four page tables and 16 bytes of memory for the page directory.

Like DOS, Windows repages the memory system. At present, Windows version 3.11 only supports paging for 16M bytes of memory, because of the amount of memory required to store the page tables. On the Pentium microprocessor, pages can be either 4K bytes or 4M bytes in length. Although no software currently supports 4M byte memory pages, no doubt operating systems of the future will.

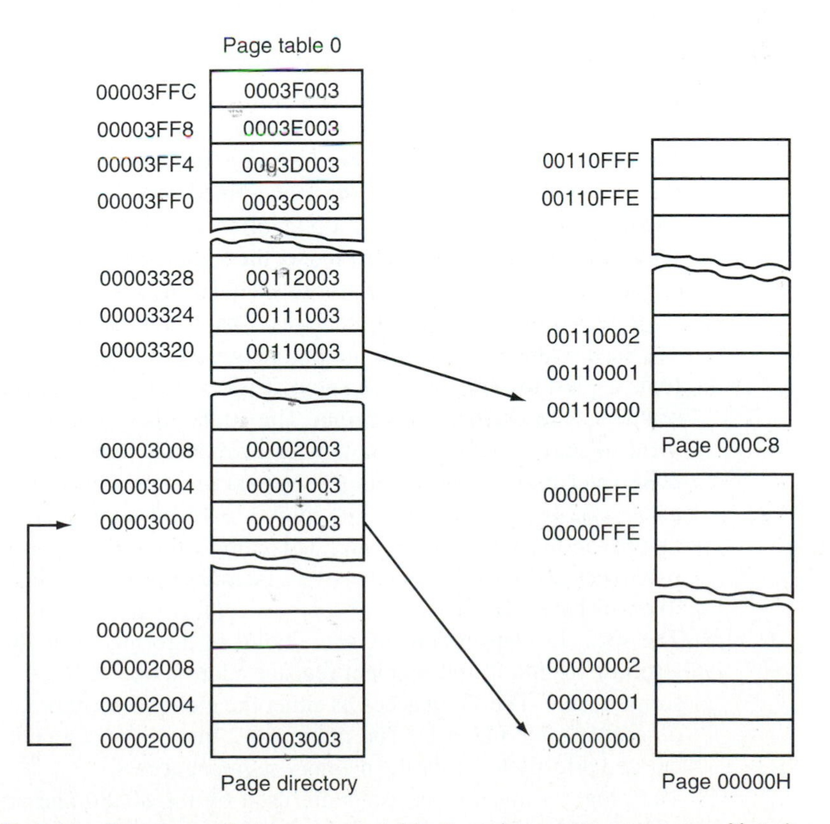

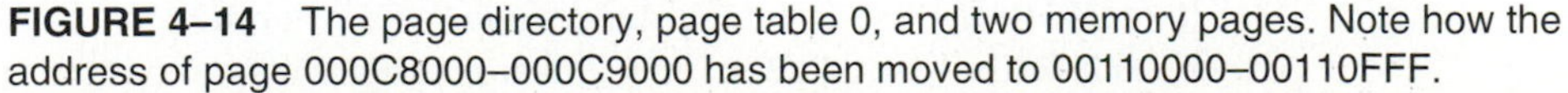
FIGURE 4–14 The page directory, page table 0, and two memory pages. Note how the address of page 000C8000–000C9000 has been moved to 00110000–00110FFF.

4–5 SUMMARY

1. The programming model of the 8086 through 80286 contains 8- and 16-bit registers. The programming model of the 80386 and above contains 8-, 16-, and the 32-bit extended registers as well as two additional 16-bit segment registers: FS and GS.
2. The 8-bit registers are AH, AL, BH, BL, CH, CL, DH, and DL. The 16-bit registers are AX, BX, CX, DX, SP, BP, DI, and SI. The segment registers are CS, DS, ES, SS, FS, and GS. The 32-bit extended registers are EAX, EBX, ECX, EDX, ESP, EBP, EDI, and ESI. In addition, the microprocessor contains an instruction pointer (IP/EIP) and flag register (FLAGS or EFLAGS).
3. All real-mode memory addresses are a combination of a segment address plus an offset address. The starting location of a segment is defined by the 16-bit number in the segment register appended with a hexadecimal zero on its rightmost end. The offset address is a 16-bit number added to the 20-bit segment address to form the real-mode memory address.
4. All instructions (code) are accessed by the combination of CS (segment address) plus IP or EIP (offset address).
5. Data are normally referenced through a data segment plus either an offset address or the contents of a register that contains the offset address. The 8086 through the Pentium use BX, DI, and SI as default offset registers for data if 16-bit registers are selected. The 80386 and above can use the 32-bit registers EAX, EBX, ECX, EDX, EDI, and ESI as default offset registers for data.
6. Protected-mode operation allows memory above the first 1M byte to be accessed by the 80286 through the Pentium microprocessors. This extended memory system (XMS) is accessed via a segment-plus-offset address, just as in the real mode. The difference is that, in the protected mode, the segment address is not held in the segment register. It is stored in a descriptor selected by the segment register.
7. A protected-mode descriptor contains a base address, limit, and access rights byte. The base address locates the starting address of the memory segment. The limit defines the last location of the segment. The access rights byte defines how the memory segment is accessed via a program. The 80286 microprocessor allows a memory segment to start at any of its 16M bytes of memory, using a 24-bit base address. The 80386 and above allow a memory segment to begin at any of its 4G bytes of memory using a 32-bit base address. The limit is a 16-bit number in the 80286 and a 20-bit number in the 80386 and above. This allows the 80286 a memory segment limit of 64K bytes and the 80386 and above a memory segment limit of either 1M bytes (G = 0) or 4G bytes (G = 1).
8. The segment register contains three fields of information in the protected mode. The leftmost 13 bits of the segment register address one of 8,192 descriptors from a descriptor table. The TI bit accesses either the global descriptor table (TI = 0) or the local descriptor table (TI = 1). The rightmost 2 bits of the segment register select the requested priority level for the memory segment access.
9. The program-invisible registers are used by the 80286 and above to access the descriptor tables. Each segment register contains a cache portion that is used in protected mode to hold the base address, limit, and access rights acquired from a descriptor. The

cache allows the microprocessor to access the memory segment without again referring to the descriptor table until the segment register's contents are changed.

10. A memory page is 4K bytes in length. The linear address, as generated by a program, can be mapped to any physical address through the paging mechanism found within the 80386 through the Pentium microprocessor.
11. Memory paging is accomplished through control registers CR0 and CR3. The PG bit of CR0 enables paging, and the contents of CR3 address the page directory. The page directory contains up to 1,024 page table addresses used to access paging tables. The page table contains 1,024 entries that locate the physical address of a 4K byte memory page.
12. The TLB (translation lookaside buffer) caches the 32 most recent page table translations. This eliminates page table translation if the translation resides in the TLB, speeding the execution of software.

4–6 QUESTIONS AND PROBLEMS

1. What are program-visible registers?
2. The 80286 addresses registers that are 8 and ______________ bits wide.
3. The extended registers are addressable by which microprocessors?
4. The extended BX register is addressed as ______________.
5. Which register holds a count for some instructions?
6. What is the purpose of the IP/EIP register?
7. The carry flag bit is set by what arithmetic operations?
8. Will an overflow occur if a signed FFH is added to a signed 01H?
9. A number that contains 3 one-bits is said to have ______________ parity.
10. Which flag bit controls the INTR pin on the microprocessor?
11. Which microprocessors contain an FS segment register?
12. What is the purpose of a segment register in real-mode operation?
13. In the real mode, show the starting and ending addresses of each segment located by the following segment register values:
    (a) 1000H
    (b) 1234H
    (c) 2300H
    (d) E000H
    (e) AB00H
14. Find the memory address of the next instruction, executed by a microprocessor in the real mode, for the following CS:IP combinations:
    (a) CS = 1000H and IP = 2000H
    (b) CS = 2000H and IP = 1000H
    (c) CS = 2300H and IP = 1A00H
    (d) CS = 1A00H and IP = B000H
    (e) CS = 3456H and IP = ABCDH
15. Real-mode memory addresses allow access to memory below which address?

16. Which register or registers are used as an offset address for string instruction destinations in the 80486 microprocessor?
17. Which 32-bit register or registers are used as an offset address for data segment data in the 80386 microprocessor?
18. The stack memory is addressed by a combination of the ______________ segment plus ______________ offset.
19. If the base pointer (BP) address memory, the ______________ segment contains the data.
20. Determine the memory location addressed by the following real-mode 80286 register combinations:
    (a) DS = 1000H and DI = 2000H
    (b) DS = 2000H and SI = 1002H
    (c) SS = 2300H and BP = 3200H
    (d) DS = A000H and BX = 1000H
    (e) SS = 2900H and SP = 3A00H
21. Determine the memory location addressed by the following real-mode 80386 register combinations:
    (a) DS = 2000H and EAX = 00003000H
    (b) DS = 1A00H and ECX = 00002000H
    (c) DS = C000H and ESI = 0000A000H
    (d) SS = 8000H and ESP = 00009000H
    (e) DS = 1239H and EDX = 0000A900H
22. Protected-mode memory addressing allows access to which area of the memory in the 80286 microprocessor?
23. Protected-mode memory addressing allows access to which area of the memory in the Pentium microprocessor?
24. What is the purpose of the segment register in protected-mode memory addressing?
25. How many descriptors are accessible in the global descriptor table in the protected mode?
26. For an 80286 descriptor that contains a base address of A00000H and a limit of 1000H, what starting and ending locations are addressed by this descriptor?
27. For an 80486 descriptor that contains a base address of 01000000H, a limit of 0FFFFH, and G = 0, what starting and ending locations are addressed by this descriptor?
28. For a Pentium descriptor that contains a base address of 00280000H, a limit of 00010H, and G = 1, what starting and ending locations are addressed by this descriptor?
29. If the DS register contains 0020H in a protected-mode system, which global descriptor table entry is accessed?
30. If DS = 0103H in a protected-mode system, the requested privilege level is ______________.
31. If DS = 0105H in a protected-mode system, which entry, table, and requested privilege level are selected?
32. What is the maximum length of the global descriptor table in the Pentium microprocessor?
33. Code a descriptor that describes a memory segment beginning at location 210000H and ending at location 21001FH. This memory segment is a code segment that can be read. The descriptor is for an 80286 microprocessor.

34. Code a descriptor that describes a memory segment that begins at location 03000000H and ends at location 05FFFFFFH. This memory segment is a data segment that grows upward in the memory system and can be written. The descriptor is for an 80386 microprocessor.
35. Which register locates the global descriptor table?
36. How is the local descriptor table addressed in the memory system?
37. Describe what happens when a new number is loaded into a segment register when the microprocessor is operated in the protected mode.
38. What are the program-invisible registers?
39. What is the purpose of the GDTR?
40. How many bytes are found in a memory page?
41. What register is used to enable the paging mechanism in the 80386, 80486, and Pentium microprocessors?
42. How many 32-bit addresses are stored in the page directory?
43. Each entry in the page directory translates how much linear memory into physical memory?
44. If the microprocessor sends linear address 00200000H to the paging mechanism, which paging directory entry is accessed, and which page table entry is accessed?
45. What value is placed in the page table to redirect linear address 20000000H to 30000000H?
46. What is the purpose of the TLB in the 80486 microprocessor?

CHAPTER 5

Addressing Modes

INTRODUCTION

Efficient software development for the microprocessor requires a complete familiarity with the addressing modes employed by each instruction. In this chapter, the **MOV** (*move data*) instruction is used to describe the data-addressing modes. The MOV instruction transfers bytes or words of data between registers or between registers and memory in the 8086 through the 80286 and transfers bytes, words, or doublewords of data in the 80386 and above. In describing the program memory-addressing modes, the call and jump instructions show how to modify the flow of the program.

The data-addressing modes include register, immediate, direct, register-indirect, base-plus-index, register-relative, and base relative-plus-index in the 8086 through the 80286 microprocessor. The 80386 and above processors also include a scaled-index mode of addressing data. The program memory-addressing modes include program relative, direct, and indirect. The operation of the stack memory is explained so the PUSH and POP instructions are understood.

OBJECTIVES

Upon completion of this chapter, you will be able to:

1. Explain the operation of each data-addressing mode.
2. Use the data-addressing modes to form assembly language statements.
3. Explain the operation of each program memory-addressing mode.
4. Use the program memory-addressing modes to form assembly and machine language statements.
5. Select the appropriate addressing mode to accomplish a given task.
6. Detail the difference between real-mode operation and protected-mode operation when addressing memory data.

7. Describe the sequence of events that place data onto the stack or remove data from the stack.
8. Explain how a data structure is placed in memory and used with software.

5–1 DATA-ADDRESSING MODES

Because the MOV instruction is common and flexible, it provides a basis for the explanation of the data-addressing modes. Figure 5–1 illustrates the MOV instruction and defines the direction of data flow. The source is to the right and the destination is to the left, next to the opcode MOV. (An **opcode** or *operation code* tells the microprocessor which operation to perform.) This direction of flow, which is applied to all instructions, is at first awkward. We naturally assume things move from left to right, but here they move from right to left. Notice that a comma *always* separates the destination from the source in an instruction. Also note that memory-to-memory transfers are not allowed by any instruction except for the MOVS instruction.

In Figure 5–1, the MOV AX,BX instruction transfers the word contents of the source register (BX) into the destination register (AX). The source *never* changes, but the destination almost always changes.[1] It is essential to remember that a MOV instruction always *copies* the source data into the destination; it never actually picks up the data and moves it. Also note that the flag register remains unaffected by most data-transfer instructions.

Figure 5–2 shows all possible variations of the data-addressing modes using the MOV instruction. This illustration helps to show how each data-addressing mode is formulated with the MOV instruction and serves as a reference. Note that these are the same data-addressing modes found in all versions of the Intel microprocessor, except for the scaled-index addressing mode, which is found only in the 80386 through Pentium. The data-addressing modes are:

Register addressing Transfers a copy of a byte or word from the source register or memory location to the detination register or memory location. (Example: the MOV CX,DX instruction copies the word-sized contents of register DX into register CX.) In the 80386 and above, a

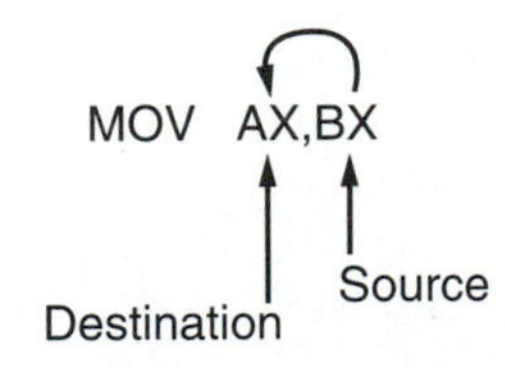

FIGURE 5–1 The MOV instruction showing the source, destination, and direction of data flow.

[1]The exceptions are the CMP and TEST instructions, which never change destinations. These instructions are described in later chapters.

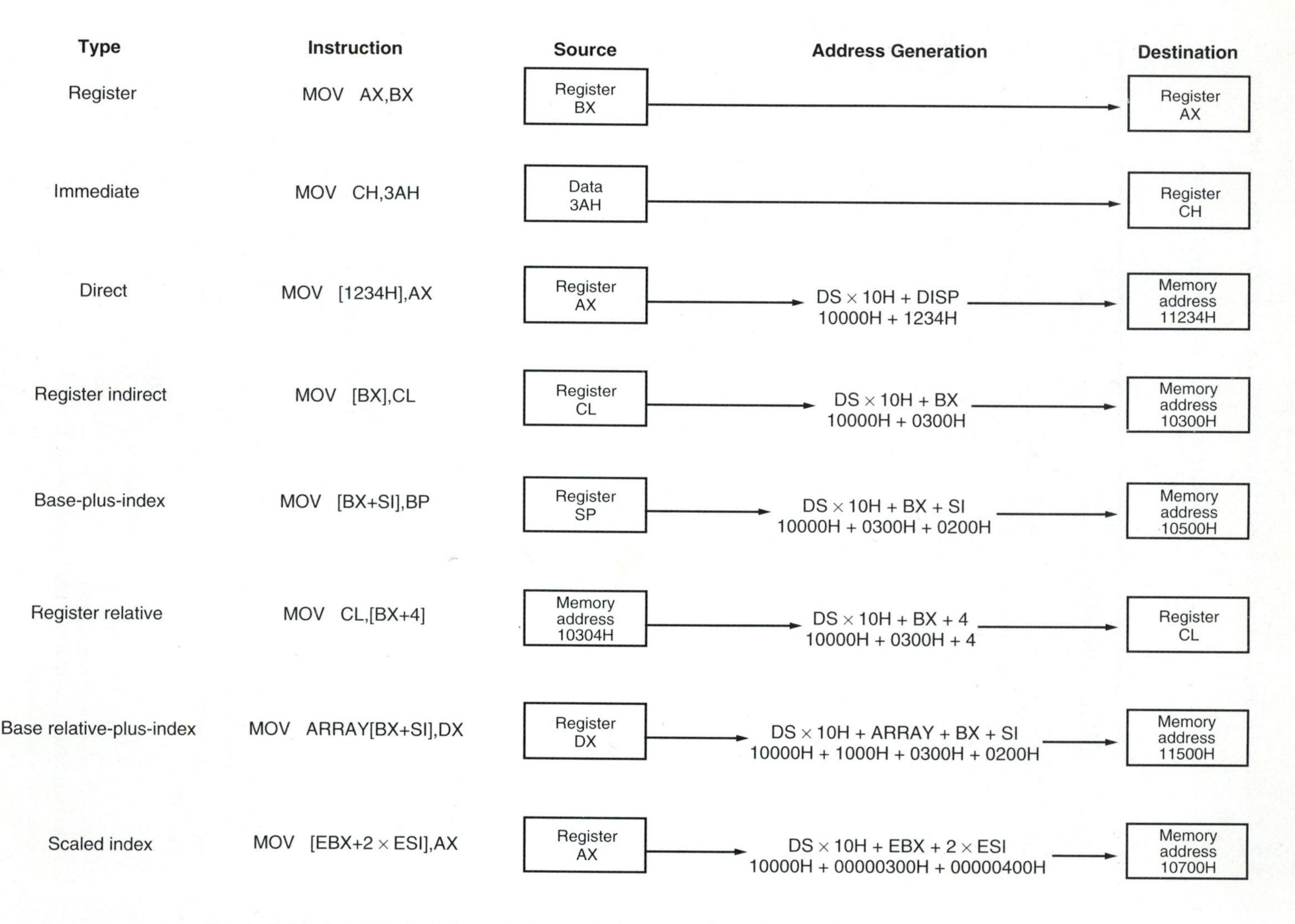
Type
Instruction
Source
Address Generation
Destination
Register
MOV AX,BX
Register BX
Register AX
Immediate
MOV CH,3AH
Data 3AH
Register CH
Direct
MOV [1234H],AX
Register AX
DS × 10H + DISP
10000H + 1234H
Memory address 11234H
Register indirect
MOV [BX],CL
Register CL
DS × 10H + BX
10000H + 0300H
Memory address 10300H
Base-plus-index
MOV [BX+SI],BP
Register SP
DS × 10H + BX + SI
10000H + 0300H + 0200H
Memory address 10500H
Register relative
MOV CL,[BX+4]
Memory address 10304H
DS × 10H + BX + 4
10000H + 0300H + 4
Register CL
Base relative-plus-index
MOV ARRAY[BX+SI],DX
Register DX
DS × 10H + ARRAY + BX + SI
10000H + 1000H + 0300H + 0200H
Memory address 11500H
Scaled index
MOV [EBX+2 × ESI],AX
Register AX
DS × 10H + EBX + 2 × ESI
10000H + 00000300H + 00000400H
Memory address 10700H
Notes: EBX = 00000300H, ESI = 00000200H, ARRAY = 1000H, and DS = 1000H

FIGURE 5–2 8086–Pentium data-addressing modes.

doubleword can also be transferred from the source register or memory location to the destination register or memory location. (Example: the MOV ECX, EDX instruction copies the doubleword-sized contents of register EDX into register ECX.)

Immediate addressing

Transfers the source-immediate byte or word of data into the destination register or memory location. (Example: the MOV AL,22H instruction copies a byte-sized 22H into register AL.) In the 80386 and above, a doubleword of source-immediate data can be transferred into a register or memory location. (Example: the MOV EBX,12345678H instruction copies a doubleword-sized 12345678H into the 32-bit wide EBX register.)

Direct addressing

Moves a byte or word between a memory location and a register. The instruction set does not support a memory-to-memory transfer, except for the MOVS instruction. (Example: the MOV CX,LIST instruction copies the word-sized contents of memory location LIST into register CX.) In the 80386 and above, a doubleword-sized memory location can also be addressed. (Example: the MOV ESI,LIST instruction copies a 32-bit number, stored in four consecutive bytes of memory, from location LIST into register ESI.)

Register-indirect addressing

Transfers a byte or word between a register and a memory location addressed by an index or base register. The index and base registers are: BP, BX, DI, and SI. (Example: the MOV AX,[BX] instruction copies the word-sized data from the data segment offset address indexed by BX into register AX.) In the 80386 and above, a byte, word, or doubleword is transferred between a register and a memory location addressed by any register: EAX, EBX, ECX, EDX, EBP, EDI, or ESI. (Example: the MOV AL,[ECX] instruction loads AL from the data segment offset address selected by the contents of ECX.)

Base-plus-index addressing

Transfers a byte or word between a register and the memory location addressed by a base register (BP or BX) plus an index register (DI or SI). (Example: the MOV [BX+DI],CL instruction copies the byte-sized contents of register CL into the data segment memory location addressed by BX plus DI.) In the 80386 and above, any register EAX, EBX, ECX, EDX, EBP, EDI, or ESI may be combined to generate the memory address. (Example: the MOV [EAX+EBX],CL instruction copies the byte-sized contents of register CL into the data segment memory location addressed by EAX plus EBX.)

Register-relative addressing

Moves a byte or word between a register and the memory location addressed by an index or base register plus a displacement. (Example: MOV AX,[BX+4] or MOV AX,ARRAY[BX]. The first instruction loads AX from the data segment address formed by BX plus 4. The second instruction loads AX from the data segment memory location in ARRAY plus the contents of BX.) The 80386 and above use any register to address memory. (Example: MOV AX,[ECX+4] or MOV AX,ARRAY[EBX]. The first instruction loads AX from the data segment address formed by ECX plus 4. The second instruction loads AX from the data segment memory location ARRAY plus the contents of EBX.)

Base relative-plus-index addressing

Transfers a byte or word between a register and the memory location addressed by a base and an index register plus a displacement. (Example: MOV AX,ARRAY[BX+DI] or MOV AX,[BX+DI+4]. These instructions both load AX from a data segment memory location. The first instruction uses an address formed by adding ARRAY, BX, and DI; the second instruction uses an address formed by adding BX, DI, and 4.) (Example: In an 80386 and above, MOV EAX,ARRAY[EBX+ECX] loads EAX from the data segment memory location accessed by the sum of ARRAY, EBX, and ECX.)

Scaled-index addressing

Available only in the 80386 through the Pentium microprocessor. The second of a pair of registers is modified by the scale factor of 2×, 4×, or 8× to generate the operand memory address. (For example: a MOV

EDX,[EAX+4*EBX] instruction loads EDX from the data segment memory location addressed by EAX plus 4 *times* EBX.) Scaling allows access to word (2×), doubleword (4×), or quadword (8×) memory array data. Note that a scaling factor of 1× also exists, but it is normally implied and does not appear in the instruction. MOV AL,[EBX+ECX] is an example where the scaling factor is 1. Alternatively, the instruction can be written as MOV AL,[EBX+1*ECX]. Another example is a MOV AL,[2*EBX] instruction, which uses only one scaled register to address memory.

Register Addressing

Register addressing is the most basic form of data addressing, once the register names are learned. The microprocessor contains the following 8-bit registers used with register addressing: AH, AL, BH, BL, CH, CL, DH, and DL. Also present are the following 16-bit registers: AX, BX, CX, DX, SP, BP, SI, and DI. In the 80386 and above, the extended 32-bit registers are: EAX, EBX, ECX, EDX, ESP, EBP, EDI, and ESI. Some MOV instructions and the PUSH and POP instructions address the 16-bit segment registers (CS, ES, DS, SS, FS, and GS) with register addressing. It is important for instructions to use registers that are the same size. *Never* mix an 8-bit register with a 16-bit register, an 8-bit register with a 32-bit register, or a 16-bit register with a 32-bit register. This is not allowed by the microprocessor and results in an error when assembled. This is true even when a MOV AX,AL or a MOV EAX,AL instruction seems to make sense. Of course, the MOV AX,AL or MOV EAX,AL instruction is *not* allowed because these registers are of different sizes. A few instructions, such as SHL DX,CL, are exceptions to this rule, as indicated in later chapters. Note that none of the MOV instructions affect the flag bits.

Table 5–1 shows many versions of register move instructions. It is impossible to show all possible combinations, because there are so many. For example, just the 8-bit subset of the MOV instruction has 64 different variations. About the only type of register MOV instruction *not* allowed is a segment-to-segment register MOV instruction. Note that the code segment register may not be changed by a MOV instruction, because the address of the next instruction is found in both IP/EIP and CS. If only CS were changed, the address of the next instruction would be unpredictable. Therefore, changing the CS register with a MOV is not allowed.

Figure 5–3 shows the operation of a MOV BX,CX instruction. Note that the source register's contents do not change, but the destination register's do. This instruction moves a 1234H from register CX into register BX. This erases the old contents (76AFH) of register BX, but the contents of CX remain unchanged. The contents of the destination register or destination memory location change for all instructions except the CMP and TEST instructions. Note that the MOV BX,CX instruction does not affect the leftmost 16 bits of register EBX.

Example 5–1 shows a short sequence of assembled instructions that copy various data between 8-, 16-, and 32-bit registers. As mentioned, the act of moving data from one register to another changes only the destination register, never the source. The last

TABLE 5–1 Examples of register-addressed instructions

Assembly Language	*Operation*
MOV AL,BL	Copies BL into AL
MOV CH,CL	Copies CL into CH
MOV AX,CX	Copies CX into AX
MOV SP,BP	Copies BP into SP
MOV DS,AX	Copies AX into DS
MOV SI,DI	Copies DI into SI
MOV BX,ES	Copies ES into BX
MOV ECX,EBX	Copies EBX into ECX
MOV ESP,EDX	Copies EDX into ESP
MOV ES,DS	Not allowed (segment to segment)
MOV BL,BX	Not allowed (mixed sizes)
MOV CS,AX	Not allowed (the code segment register may not be used as a destination register)

instruction (MOV CS,AX) in this example assembles without error, but causes problems if executed. If only the contents of CS are changed without changing IP, the next step in the program would be unknown, and the program would go awry.

EXAMPLE 5–1

```
0000 8B C3          MOV   AX,BX     ;copy contents of BX into AX
0002 8A CE          MOV   CL,DH     ;copy the contents of DH into CL
0004 8A CD          MOV   CL,CH     ;copy the contents of CH into CL
0006 66| 8B C3      MOV   EAX,EBX   ;copy the contents of EBX into EAX
0009 66| 8B D8      MOV   EBX,EAX   ;copy EAX into EBX, ECX, and EDX
000C 66| 8B C8      MOV   ECX,EAX
000F 66| 8B D0      MOV   EDX,EAX
0012 8C C8          MOV   AX,CS     ;copy CS into DS
0014 8E D8          MOV   DS,AX
0016 8E C8          MOV   CS,AX     ;assembles, but will cause problems
```

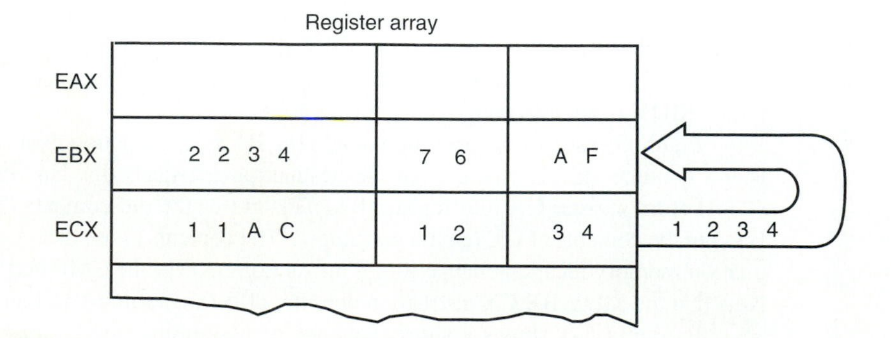

FIGURE 5–3 The effect of executing the MOV BX,CX instruction at the point just before the BX register changes. Note that only the rightmost 16 bits of register EBX change.

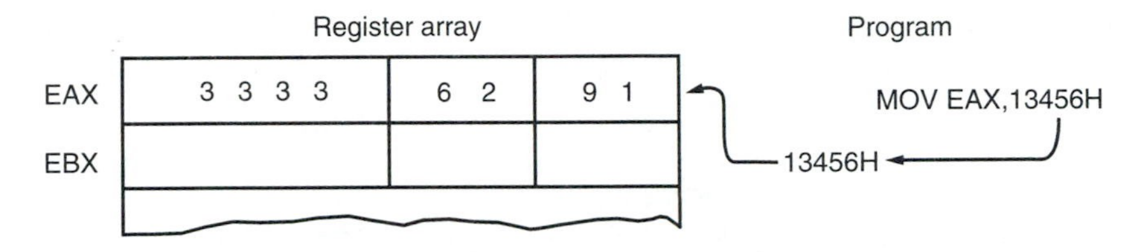

FIGURE 5–4 The operation of the MOV EAX,34546H instruction. This instruction copies the immediate data (13456H) into EAX.

Immediate Addressing

Another data-addressing mode is **immediate addressing.** The term *immediate* implies that the data immediately follow the hexadecimal opcode in the memory. Immediate data is *constant data*, while the data transferred from a register is *variable data*. Immediate addressing operates upon a byte or word of data. In the 80386 through the Pentium microprocessor, immediate addressing also operates on doubleword data. The MOV immediate instruction transfers a copy of the immediate data into a register or a memory location. Figure 5–4 shows the operation of a MOV EAX,13456H instruction. This instruction copies the 13456H from the instruction (located in the memory immediately following the hexadecimal opcode) into register EAX. As in Figure 5–3, the source data overwrites the destination data.

In symbolic assembly language, the symbol # precedes immediate data in some assemblers.[2] The MOV AX,#3456H instruction is an example. Most assemblers do not use the # symbol, but represent immediate data as in the MOV AX,3456H instruction. In this text, the # symbol is not used for immediate data. The most common assemblers—Intel ASM, Microsoft MASM,[3] and Borland TASM[4]—do not use the # symbol for immediate data, but older assemblers may.

The symbolic assembler portrays immediate data in many ways. The letter H appends hexadecimal data. If hexadecimal data begins with a letter, the assembler requires that it starts with a 0. For example, to represent a hexadecimal F2, 0F2H is used in assembly language. In some assemblers, (though not in MASM, TASM, or this text), hexadecimal data is represented with an 'h, as in MOV AX,#'h1234. Decimal data is represented as is, and requires no special codes or adjustments. An example is the 100 decimal in the MOV AL,100 instruction. An ASCII-coded character or characters may be depicted in immediate form if the ASCII data are enclosed in apostrophes. An example is the MOV BH,'A' instruction, which moves an ASCII-coded A (41H) into register BH. Be careful to use the apostrophe (') for ASCII data and not the single quotation mark ('). Binary data are represented by a letter B or, in some assemblers, the letter Y after the binary number. Table 5–2 shows many variations of MOV instructions that apply immediate data.

---

[2]This is true in the assembler provided for an HP64100 Logic Development system manufactured by Hewlett-Packard, Inc.

[3]The MASM (macro assembler) is an assembler program from Microsoft Corporation.

[4]The TASM (turbo assembler) is an assembler program from Borland Corporation.

TABLE 5–2 Examples of immediate addressing using the MOV instruction

Assembly Language	*Operation*
MOV BL,44	Moves a 44 decimal (2CH) into BL
MOV AX,44H	Moves a 44 hexadecimal into AX
MOV SI,0	Moves a 0000H into SI
MOV CH,100	Moves a 100 (64H) into CH
MOV AL,'A'	Moves an ASCII A (41H) into AL
MOV AX,'AB'	Moves an ASCII BA* (4241H) into AX
MOV CL,11001110B	Moves a binary 11001110 into CL
MOV EBX,12340000H	Moves a 12340000H into EBX
MOV ESI,12	Moves a 12 decimal into ESI
MOV EAX,100Y	Moves a 100 binary into EAX

**Note:* This is not an error. The ASCII characters are stored as a BA, so care should be exercised when using a word-sized pair of ASCII characters.

Example 5–2 shows various immediate instructions in a short program that places a 0000H into the 16-bit registers AX, BX, and CX. This is followed by instructions that use register addressing to copy the contents of AX into registers SI, DI, and BP. This is a complete program that uses programming models for assembly and execution. The .MODEL TINY statement directs the assembler to assemble the program into a single segment. The .CODE statement or directive indicates the start of the code segment, the .STARTUP statement indicates the starting instruction in the program, and the .EXIT statement causes the program to exit to DOS. The END statement indicates the end of the program file. This program can be assembled with MASM and, if executed with CodeView (CV), its execution can be viewed onscreen. To store the program into the system, use either DOS EDIT or Programmer's WorkBench (PWB). Note that a TINY program assembles as a command (.COM) program.

EXAMPLE 5–2

```
                        .MODEL TINY         ;choose single segment model
0000                    .CODE               ;indicate start of code segment

                        .STARTUP            ;indicate start of program

0100  B8 0000           MOV    AX,0         ;place 0000H into AX
0103  BB 0000           MOV    BX,0000H     ;place 0000H into BX
0106  B9 0000           MOV    CX,0         ;place 0000H into CX

0109  8B F0             MOV    SI,AX        ;copy AX into SI
010B  8B F8             MOV    DI,AX        ;copy AX into DI
010D  8B E8             MOV    BP,AX        ;copy AX into BP

                        .EXIT               ;exit to DOS
                        END                 ;end of file
```

Each statement in a program consists of four parts or fields, as illustrated in Example 5–3. The leftmost field is called the label and is used to store a symbolic name for the memory location that it represents. All labels begin with a letter or one of the following special characters: @, $, _, or ?. A label may be of any length from 1 to 35 characters. The label appears in a program to identify the name of a memory location for storing data and for other purposes that are explained as they appear. The next field is called the opcode field and is designed to hold the instruction or opcode. The MOV instruction is an example of an opcode. To the right of the opcode field is the operand field that contains information used by the opcode. For example, the MOV AL,BL instruction has the opcode MOV and operands AL and BL. Note that some instructions contain between zero and three operands. The final field, the comment field, contains a comment about an instruction or a group of instructions. A comment always begins with a semicolon (;).

EXAMPLE 5–3

```
LABEL          OPCODE  OPERAND      COMMENT

DATA1          DB      23H          ;define DATA1 as a byte of 23H
DATA2          DW      1000H        ;define DATA2 as a word of 1000H

START:         MOV     AL,BL        ;copy BL into AL
               MOV     BH,AL        ;copy AL into BH
               MOV     CX,200       ;copy 200 decimal into CX
```

When the program is assembled and the list (.LST) file is viewed, it appears as the program listed in Example 5–2. The hexadecimal number at the far left is the offset address of the instruction or data. This number is generated by the assembler. The number or numbers to the right of the offset address are the machine-coded instructions or data; they, too, are generated by the assembler. In Example 5–2, if the instruction MOV AX,0 appears in a file and the file is assembled, it appears in memory location 0100. Its hexadecimal machine language form is B8 0000. When the program was written, only MOV AX,0 was typed into the editor; the assembler generated the machine code and address and stored the program in a file ending with the extension .LST. All programs shown in this text are in the form generated by the assembler.

Direct Data Addressing

Most instructions can use the direct data addressing mode. In fact, direct data addressing is applied to many instructions in a typical program. There are two basic forms of direct data addressing: (1) direct addressing, which applies to a MOV between a memory location and AL, AX, or EAX and (2) displacement addressing, which applies to almost any instruction in the instruction set. In either case, the address is formed by adding the displacement to the default data segment address or an alternate segment address.

Direct Addressing. Direct addressing with a MOV instruction transfers data between a memory location, located within the data segment, and the AL (8-bit), AX (16-bit), or EAX (32-bit) register. This instruction is usually a 3-byte long instruction. (In the 80386 and above, a register-size prefix may appear before the instruction, causing it to exceed three bytes in length.)

The MOV AL,DATA instruction, as represented by most assemblers, loads AL from data segment memory location DATA (1234H). Memory location DATA is a **symbolic memory location,** while 1234H is the actual hexadecimal location. With many assemblers,

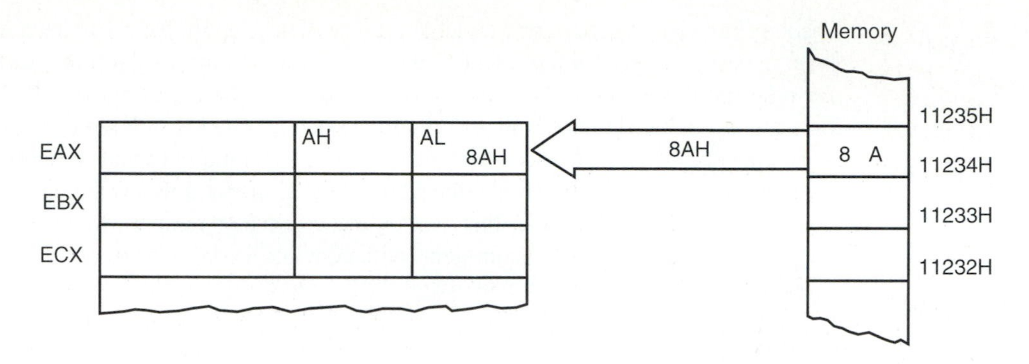

FIGURE 5–5 The operation of the MOV AL,[1234H] instruction when DS = 1000H.

this instruction is represented as MOV AL,[1234H].[5] The [1234H] is an **absolute memory location** that is not allowed by all assembler programs. Note that this may need to be formed as MOV AL,DS:[1234H] with some assemblers, to show that the address is in the data segment. Figure 5–5 shows how this instruction transfers a copy of the byte-sized contents of memory location 11234H into AL. The effective address is formed by adding 1234H (the offset address) to 10000H (the data segment address) in a system operating in the real mode.

Table 5–3 lists the three direct addressed instructions. These instructions appear often in programs, so Intel decided to make them special 3-byte long instructions to reduce the length of programs. All other instructions that move data from a memory location to a register, called displacement-addressed instructions, require four or more bytes of memory.

Displacement Addressing. Displacement addressing is almost identical to direct addressing except the instruction is four bytes wide instead of three. In the 80386 through Pentium, this instruction can be up to seven bytes wide if a 32-bit register and a 32-bit displacement is specified. This type of direct data addressing is much more flexible, because most instructions use it.

If the MOV CL,[1234H] instruction is compared to the MOV AL,[1234H] instruction, both perform basically the same operation but the destination register differs (CL versus AL). Another difference only becomes apparent upon examination of the assembled versions of these two instructions. The MOV AL,[1234H] instruction is three bytes long and the MOV CL,[1234H] instruction is four bytes long, as shown in Example 5–4. This example shows how the assembler converts these two instructions into hexadecimal machine language.

EXAMPLE 5–4

```
0000  A0 1234 R          MOV   AL,[1234H]
0003  8A 0E 1234 R       MOV   CL,[1234H]
```

[5]This form may be used with MASM, but most often appears when a program is entered or listed by DEBUG, a debugging tool provided with the disk operating system.

TABLE 5–3 Three possible direct-addressed instructions using EAX, AX, and AL

Assembly Language	*Operation*
MOV AL,NUMBER	Copies the byte contents of memory address NUMBER, located in the data segment, into AL
MOV AX,COW	Copies the word contents of memory address COW in the data segment into AX
*MOV EAX,WATER	Copies the doubleword contents of memory WATER, located in the data segment, into EAX
MOV NEWS,AL	Copies AL into memory location NEWS in the data segment
MOV THERE,AX	Copies AX into memory location THERE in the data segment
*MOV HOME,EAX	Copies EAX into data segment memory location HOME

**Note*: The 80386/80486/Pentium microprocessor will at times use more than three bytes of memory for the 32-bit move between EAX and memory.

Table 5–4 lists some MOV instructions using the displacement form of direct addressing. Not all variations are listed, because there are too many. Note that the segment registers can be stored or loaded from memory.

Example 5–5 shows a short program that addresses information in the data segment. Note that the data segment begins with a .DATA statement to inform the assembler where the data segment begins. The model size is adjusted from TINY, as in Example 5–3, to SMALL, so a data segment can be included. The SMALL model allows one data segment

TABLE 5–4 Examples of direct data addressing using a displacement

Assembly Language	*Operation*
MOV CH,DOG	Loads CH with the contents of data segment memory location DOG (the actual offset address of DOG is calculated by the assembler)
MOV CH,[1000H]	*Loads CH with the contents of data segment memory location 1000H
MOV ES,DATA6	Loads ES with the word-sized contents of data segment memory location DATA6
MOV DATAS,BP	BP is copied into data segment memory location DATAS
MOV NUMBER,SP	SP is copied into data segment memory location NUMBER
MOV DATA1,EAX	EAX is copied into data segment memory location DATA1
MOV EDI,SUM1	Loads EDI with the contents of data segment memory location SUM1

**Note*: This form of addressing is seldom used with most assemblers, because an actual numeric offset address is rarely accessed.

and one code segment. The SMALL model is often used when memory data are required for a program. A SMALL model program assembles as an execute (.EXE) program. Notice how this example allocates memory locations in the data segment using the DB and DW directives. The .STARTUP statement not only indicates the start of the code, but also loads the data segment register with the segment address of the data segment. If this program is assembled and executed with CodeView, the instructions can be viewed as they execute, changing registers and memory locations.

EXAMPLE 5–5

```
                               .MODELSMALL             ;select SMALL model
0000                           .DATA                   ;indicate start of DATA segment

0000  10              DATA1    DB      10H             ;place 10H in DATA1
0001  00              DATA2    DB      0               ;place 0 in DATA2
0002  0000            DATA3    DW      0               ;place 0 in DATA3
0004  AAAA            DATA4    DW      0AAAAH          ;place AAAAH in DATA4

0000                           .CODE                   ;indicate start of CODE segment
                               .STARTUP                ;indicate start of program

0017  A0 0000 R                MOV     AL,DATA1        ;copy DATA1 to AL
001A  8A 26 0001 R             MOV     AH,DATA2        ;copy DATA2 to AH
001E  A3 0002 R                MOV     DATA3,AX        ;save AX at DATA3
0021  8B 1E 0004 R             MOV     BX,DATA4        ;load BX with DATA4

                               .EXIT                   ;exit to DOS
                               END                     ;end file
```

Register-Indirect Addressing

Register-indirect addressing allows data to be addressed at any memory location through the offset address held in any of the following registers: BP, BX, DI and SI. For example, if register BX contains a 1000H and the MOV AX,[BX] instruction executes, the word contents of data segment memory offset address 1000H are copied into register AX. If the microprocessor is operated in the real mode and DS = 0100H, this instruction addresses a word stored at memory bytes 2000H and 2001H and transfers them into register AX (see Figure 5–6). Note that the contents of 2000H are moved into AL and the contents of 2001H are moved into AH. The [ ] symbols denote indirect addressing in assembly language. In addition to using the BP, BX, DI, and SI registers to indirectly address memory, the 80386 and above allow register-indirect addressing with any extended register except ESP. Some typical instructions using indirect addressing appear in Table 5–5.

The data segment is used by default with register-indirect addressing or any other addressing mode that uses BX, DI, or SI to address memory. If register BP addresses memory, the stack segment is used by default. These are considered the default settings for these four index and base registers. For the 80386 and above, EBP addresses memory, by default, in the stack segment while EAX, EBX, ECX, EDX, EDI, and ESI address memory, by default, in the data segment. When using a 32-bit register to address memory in the real mode, the contents of the 32-bit register must never exceed 0000FFFFH. In the protected mode, any value can be used in a 32-bit register to indirectly address memory, as long as it does not access a location outside of the segment dictated by the access rights byte. An example 80386/80486/Pentium instruction is MOV EAX,[EBX]. This instruction

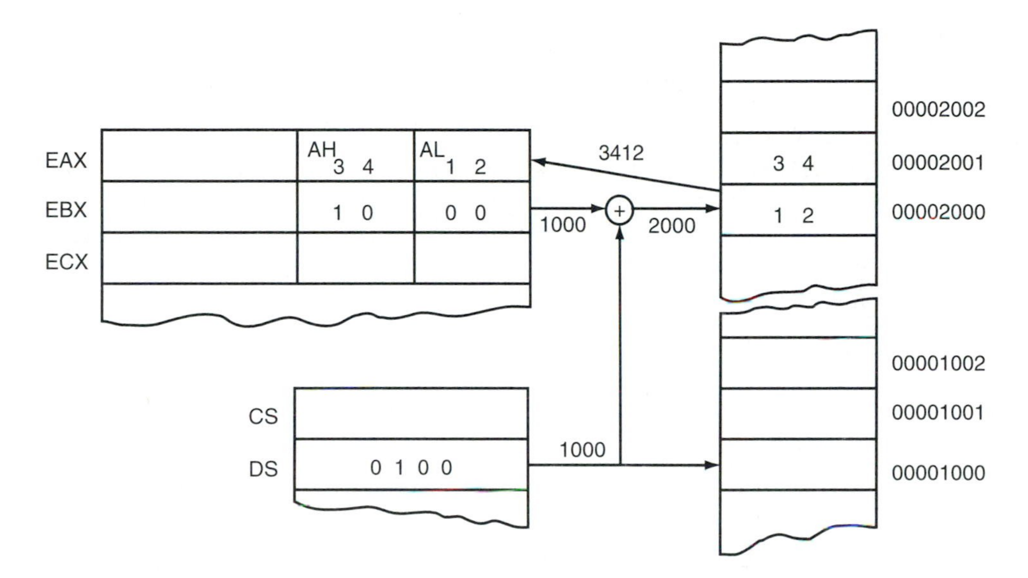

FIGURE 5–6 The operation of the MOV AX,[BX] instruction when BX = 1000H and DS = 0100H. Note that this instruction is shown after the contents of memory are transferred to AX.

loads EAX with the doubleword-sized number stored at the data segment offset address indexed by EBX.

In some cases, indirect addressing requires that the size of the data are specified with the special assembler directive BYTE PTR, WORD PTR or DWORD PTR. These directives indicate the size of the memory data addressed by the *memory pointer* (PTR). For ex-

TABLE 5–5 Example instructions using register-indirect addressing

Assembly Language	**Operation*
MOV CX,[BX]	CX is loaded with a word from the data segment memory location addressed by BX
MOV [BP],DL	A byte is copied from register DL into the stack segment memory location addressed by BP
MOV [DI],BH	A byte is copied from register BH into the data segment memory location addressed by DI
MOV [DI],[BX]	Memory-to-memory moves are not allowed except with string instructions
MOV DI,[DI]	This instruction is not allowed because the register used to indirectly address memory may not be changed by the instruction
MOV AL,[EDX]	AL is loaded from the data segment memory location addressed by EDX
MOV ECX,[EBX]	ECX is loaded with a doubleword from the data segment memory location addressed by EBX

**Note*: Data addressed by BP or EBP is, by default, located in the stack segment. All other indirect addressing modes use the data segment by default.

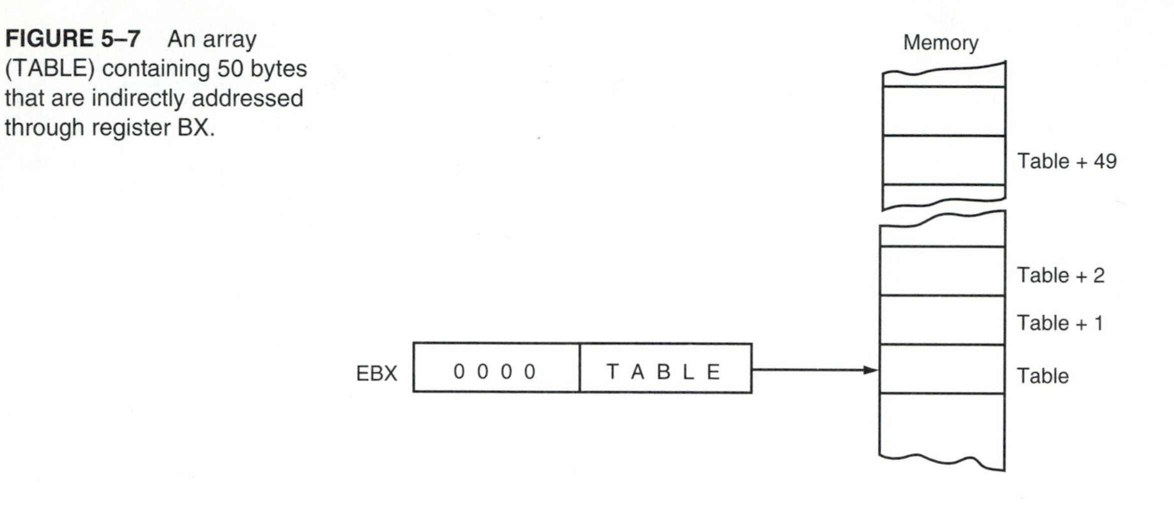

FIGURE 5–7 An array (TABLE) containing 50 bytes that are indirectly addressed through register BX.

ample, the MOV AL,[DI] instruction is clearly a byte-sized move instruction, but the MOV [DI],10H instruction is ambiguous. Does the MOV [DI],10H instruction address a byte-, word-, or doubleword-sized memory location? The assembler can't determine the size of the 10H. The instruction MOV BYTE PTR [DI],10H clearly designates the location addressed by DI as a byte-sized memory location. Likewise, the MOV DWORD PTR [DI],10H clearly identifies the memory location as doubleword-sized. The BYTE PTR, WORD PTR, and DWORD PTR directives are used only with instructions that address a memory location through a pointer or index register with immediate data and for a few other instructions that are described in subsequent chapters.

Indirect addressing often allows a program to refer to tabular data located in the memory system. For example, suppose that you must create a table of information that contains fifty samples taken from memory location 0000:046C. Location 0000:046C contains a counter that is maintained by the personal computer's real-time clock. Figure 5–7 shows the table and the BX register used to address each location in the table sequentially. To accomplish this task, load the starting location of the table into the BX register with a MOV immediate instruction. After initializing the starting address of the table, use register indirect addressing to store the fifty samples sequentially.

EXAMPLE 5–6

```
                                .MODEL SMALL                    ;select SMALL model
0000                            .DATA                           ;indicate start of DATA segment

0000  0032 [            DATAS   DW      50 DUP (?)              ;setup array of 50 bytes
                 0000
                       ]

0000                            .CODE                           ;indicate start of CODE segment
                                .STARTUP                        ;indicate start of program

0017  B8 0000                   MOV     AX,0
001A  8E C0                     MOV     ES,AX                   ;address segment 0000 with ES

001C  BB 0000 R                 MOV     BX,OFFSET DATAS         ;address DATAS array
```

```
001F  B9 0032              MOV     CX,50          ;load counter with 50
0022               AGAIN:
0022  26: A1 046C          MOV     AX,ES:[046CH]  ;get clock value
0026  89  07               MOV     [BX],AX        ;save clock value in DATAS
0028  43                   INC     BX             ;increment BX to next element
0029  E2 F7                LOOP    AGAIN          ;repeat 50 times

                           .EXIT                  ;exit to DOS
                           END                    ;end file
```

The sequence shown in Example 5–6 loads register BX with the starting address of the table and initializes the count, located in register CX, to 50. The OFFSET directive tells the assembler to load BX, with the offset address of memory location TABLE, not the contents of TABLE. For example, the MOV BX,DATAS instruction copies the contents of memory location DATAS into BX, while the MOV BX,OFFSET DATAS instruction copies the *offset address* of DATAS into BX. When the OFFSET directive is used with the MOV instruction, the assembler calculates the offset address, then uses a move immediate instruction to load the address into the specified 16-bit register.

Once the counter and pointer are initialized, a repeat-until CX = 0 loop executes. Here, data are read from extra segment memory location 46CH with the MOV AX,ES:[046CH] instruction and stored in memory indirectly addressed by the offset address located in register BX. Next, BX is incremented (*one is added to BX*) to the next table location. Finally, the LOOP instruction repeats the loop 50 times. The LOOP instruction decrements (*subtracts one from*) the counter (CX), and if CX is not zero, LOOP causes a jump to memory location AGAIN. If CX becomes zero, no jump occurs and this sequence of instructions ends. This example copies the most recent 50 values from the clock into the memory array DATAS. This program will often show the same data in each location, because the contents of the clock are changed only 18.2 times per second. To view the program and its execution, use the CodeView program. To use CodeView, type CV FILE.EXE or CV FILE.COM, or access it as DEBUG from the Programmer's Work-Bench program under the RUN menu. CodeView only functions with .EXE or .COM files. Some useful CodeView switches are /50 for a fifty-line display and /S for high-resolution video displays in an application. To debug the file TEST.COM with 50 lines, type CV /50 TEST.COM at the DOS prompt.

Base-Plus-Index Addressing

Base-plus-index addressing is similar to indirect addressing because it indirectly addresses memory data. In the 8086 through the 80286, this type of addressing uses one base register (BP or BX) and one index register (DI or SI) to indirectly address memory. Often the base register holds the beginning location of a memory array, while the index register holds the relative position of an element in the array. Remember that whenever BP addresses memory data, both the stack segment register and BP generate the effective address.

In the 80386 and above, this type of addressing allows the combination of any two 32-bit extended registers except ESP. For example, the MOV DL,[EAX+EBX] instruction is an example using EAX as the base plus EBX as the index. If the EBP register is used, the data are located in the stack segment instead of the data segment.

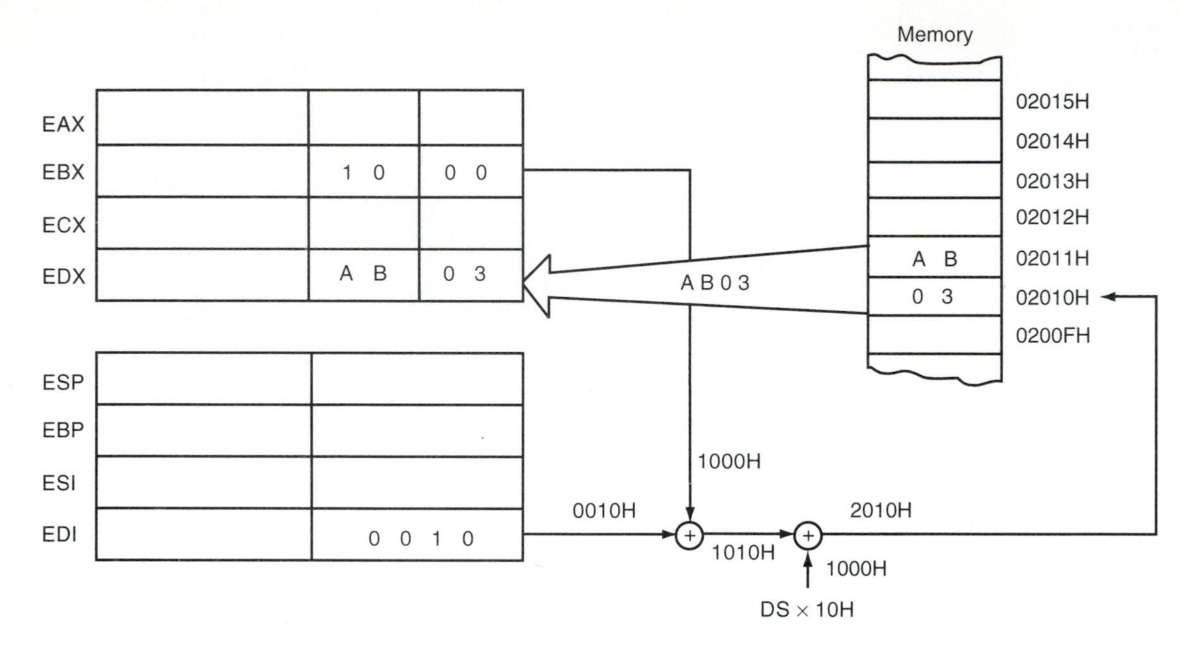

FIGURE 5–8 An example showing how the base-plus-index addressing mode functions for the MOV DX,[BX+DI] instruction. Notice that memory address 02010H is accessed because DS = 0100H, BX = 1000H, and DI = 0010H.

Locating Data with Base-Plus-Index Addressing. Figure 5–8 shows how data are addressed by the MOV DX,[BX+DI] instruction when the microprocessor operates in the real mode. In this example, BX = 1000H, DI = 0010H, and DS = 0100H, which translates into memory address 02010H. This instruction transfers a copy of the word from location 02010H into the DX register. Table 5–6 lists some instructions used for base-plus-index

TABLE 5–6 Examples of base-plus-index addressing

Assembly Language	*Operation*
MOV CX,[BX+DI]	Loads CX with the word contents of the data segment memory location addressed by BX plus DI
MOV CH,[BP+SI]	Loads CH with the byte contents of the stack segment memory location addressed by BP plus SI
MOV [BX+SI],SP	Stores the word contents of SP into the data segment memory location addressed by BX plus SI
MOV [BP+DI],CX	Stores the word contents of CX into the stack segment memory location addressed by BP plus DI
MOV CL,[EDX+EDI]	Loads CL with the byte contents of the data segment memory location addressed by EDX plus EDI
MOV [EAX+EBX],ECX	Stores the doubleword contents of ECX into the data segment memory location addressed by EAX plus EBX

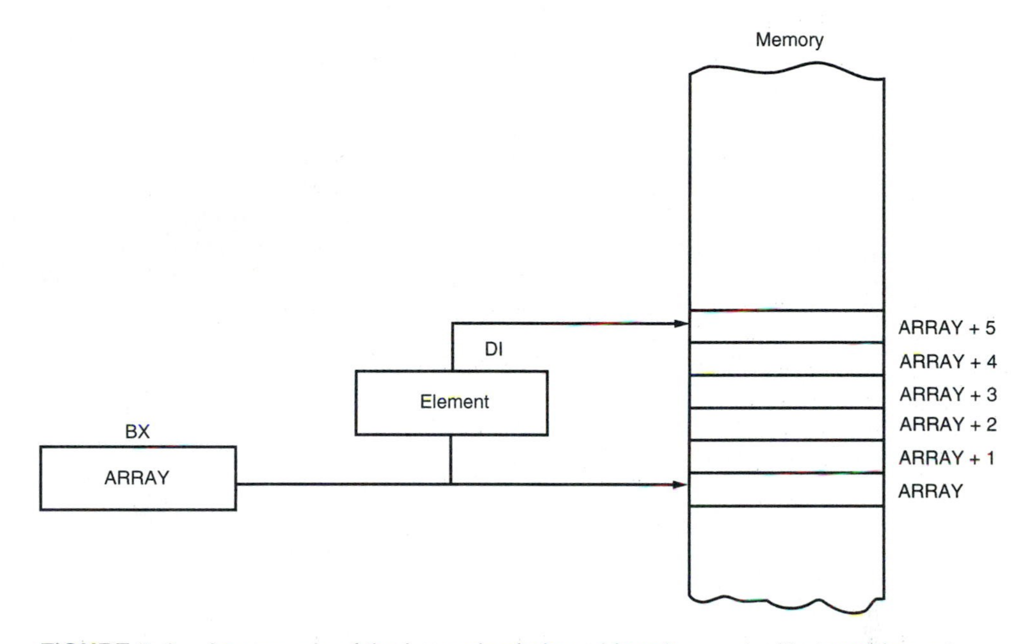

FIGURE 5–9 An example of the base-plus-index addressing mode. Here an element (DI) of an ARRAY (BX) is addressed.

addressing. Note that the Intel ASM assembler requires that this instruction appear as MOV DX, [BX][DI]. This text uses the first form [BX+DI] in all example programs, but the second form [BX][DI] can be used in many assemblers, including MASM from Microsoft.

Locating Array Data Using Base-Pus-Index Addressing. A major use of the base-plus-index addressing mode is to address elements in a memory array. Suppose that the elements in an array located in the data segment at memory location ARRAY must be accessed. To accomplish this, load the BX register (base) with the beginning address of the array and the DI register (index) with the element number to be accessed. Figure 5–9 shows the use of BX and DI to access an element in an array of data.

EXAMPLE 5–7

```
                             .MODEL SMALL              ;select SMALL model
0000                         .DATA                     ;indicate start of DATA segment

0000  0010 [       ARRAY  DB     16 DUP (?)           ;setup ARRAY
            00
               ]
0010  29                  DB     29H                   ;sample data at element 10H
0011  001E [              DB     30 DUP (?)
            00
               ]
0000                         .CODE                     ;indicate start of CODE segment
                             .STARTUP                  ;indicate start of program

0017  BB 0000 R              MOV    BX,OFFSET ARRAY   ;address ARRAY
001A  BF 0010                MOV    DI,10H            ;address element 10H
```

```
001D  8A 01           MOV    AL,[BX+DI]        ;get element 10H
001F  BF 0020         MOV    DI,20H            ;address element 20H
0022  88 01           MOV    [BX+DI],AL        ;save in element 20H

                      .EXIT                    ;exit to DOS
                      END                      ;indicate end of file
```

A short program listed in Example 5–7 moves array element 10H into array element 20H. Notice that the array element number, loaded into the DI register, addresses the array element. Also notice how the contents of the ARRAY have been initialized so element 10H contains a 29H.

Register-Relative Addressing

Register-relative addressing is similar to base-plus-index addressing and displacement addressing. In register-relative addressing, the data in a segment of memory are addressed by adding the displacement to the contents of a base or an index register (BP, BX, DI, or SI). Figure 5–10 shows the operation of the MOV AX,[BX+1000H] instruction. In this example, BX = 0100H and DS = 0200H, so the address generated is the sum of DS × 10H, BX, and the displacement of 1000H, or 03100H. Remember that BX, DI, or SI address the data segment, and BP addresses the stack segment. In the 80386 and above, the displacement can be a 32-bit number and the register can be any 32-bit register except the ESP register. Remember that a real-mode segment is 64K bytes in length. Table 5–7 lists a few instructions that use register-relative addressing.

The displacement can be a number added to the register within the [ ], as in the MOV AL,[DI+2] instruction, or it can be a displacement subtracted from the register, as in MOV AL,[SI–1]. A displacement also can be an offset address appended to the front of the [ ] as in MOV AL,DATA[DI]. Both forms of displacements can appear simultaneously as in the MOV AL,DATA[DI+3] instruction. In all cases, both forms of the displacement add to the base or base and index register within the [ ]. In the 8086–80286 microprocessor, the value of the displacement is limited to a 16-bit signed number with a value ranging between +32,767 (7FFFH) and –32,768 (8000H). In the 80386 and above, a 32-bit displacement is allowed, ranging in value from +2,147,483,647 (7FFFFFFFH) and –2,147,483,648 (80000000H).

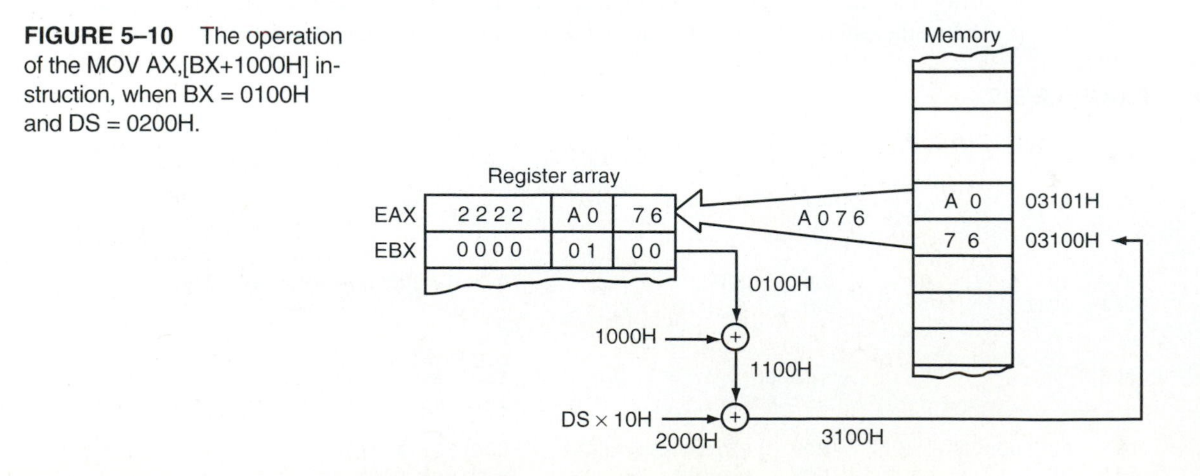

FIGURE 5–10 The operation of the MOV AX,[BX+1000H] instruction, when BX = 0100H and DS = 0200H.

TABLE 5–7 Examples of register-relative addressing

Assembly Language	*Operation*
MOV AX,[DI+100H]	Loads AX from the word contents of the data segment memory location addressed by DI plus 100H
MOV ARRAY[SI],BL	Stores the byte contents of BL into the data segment memory location addressed by ARRAY plus SI
MOV LIST[SI+2],CL	Stores the byte contents of CL into the data segment memory location addressed by the sum of LIST, SI, and 2
MOV DI,SETS[BX]	Loads DI from the data segment memory location addressed by SETS plus BX
MOV DI,[EAX+100H]	Loads DI from the data segment memory location addressed by EAX plus 100H
MOV ARRAY[EBX],AL	Stores AL into the data segment memory location addressed by ARRAY plus EBX

Addressing Array Data with Register-Relative Addressing. It is possible to address array data with register-relative addressing, much as one does with base-plus-index addressing. In Figure 5–11, register-relative addressing is illustrated with the same example used for base-plus-index addressing. This shows how the displacement ARRAY adds to index register DI to generate a reference to an array element.

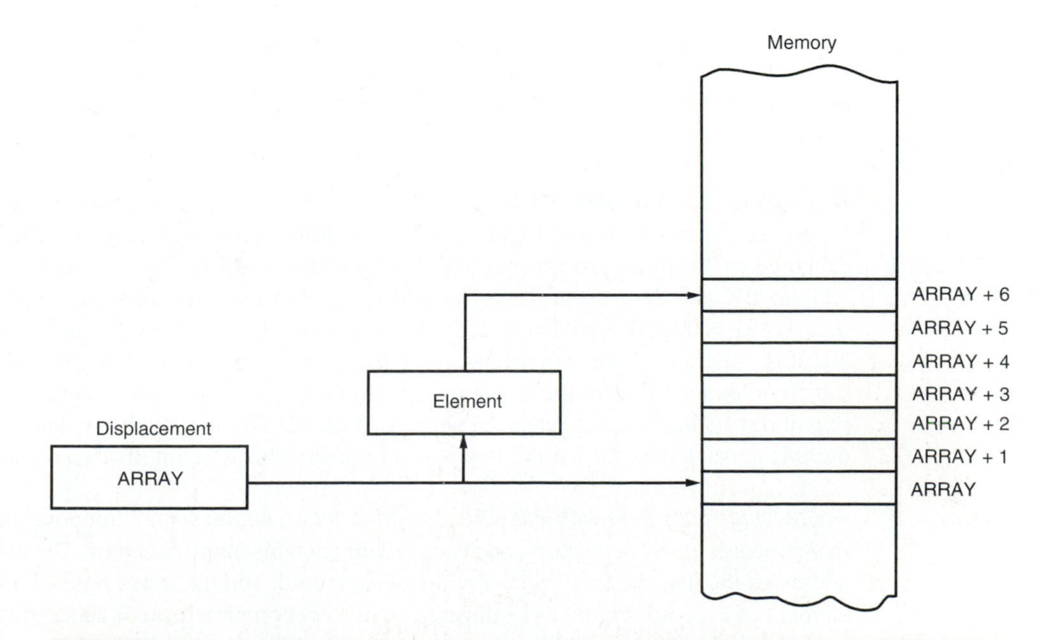

FIGURE 5–11 Register-relative addressing used to address an element of ARRAY. The displacement addresses the start of ARRAY, and DI accesses an element.

Example 5–8 shows how this new addressing mode can transfer the contents of array element 10H into array element 20H. Notice the similarity between this example and Example 5–7. The main difference is that in Example 5–8, register BX is not used to address memory area ARRAY; instead, ARRAY is used as a displacement to accomplish the same task.

EXAMPLE 5-8

```
                                .MODEL SMALL            ;select SMALL model
0000                            .DATA                   ;indicate start of DATA segment

0000  0010 [          ARRAY     DB    16 DUP (?)        ;setup ARRAY
              00
                 ]
0010  29                        DB    29H               ;sample data at element 10H
0011  001E [                    DB    30 DUP (?)
              00
                 ]
0000                            .CODE                   ;indicate start of CODE segment
                                .STARTUP                ;indicate start of program

0017  BF 0010                   MOV   DI,10H            ;address element 10H
001A  8A 85 0000 R              MOV   AL,ARRAY[DI]      ;get element 10H
001E  BF 0020                   MOV   DI,20H            ;address element 20H
0021  88 85 0000 R              MOV   ARRAY[DI],AL      ;save in element 20H

                                .EXIT                   ;exit to DOS
                                END                     ;indicate end of file
```

Base Relative-Plus-Index Addressing

The base relative-plus-index addressing mode is similar to the base-plus-index addressing mode, but it adds a displacement besides and uses a base register and index register to form the memory address. This addressing mode often addresses a two-dimensional array of memory data.

Addressing Data with Base Relative-Plus-Index. Base relative-plus-index addressing is the least-used addressing mode. Figure 5–12 shows how data are referenced if the instruction executed by the microprocessor is MOV AX,[BX+SI+100H]. The displacement of 100H adds to BX and SI to form the offset address within the data segment. Registers BX = 0020H, SI = 0010H, and DS = 1000H, so the effective address for this instruction is 10130H—the sum of these registers plus a displacement of 100H. This addressing mode is too complex for frequent use in a program. Some typical instructions using base relative-plus-index addressing appear in Table 5–8. With the 80386 and above, the effective address is generated by the sum of two 32-bit registers plus a 32-bit displacement.

Addressing Array Data with Base Relative-Plus-Index Addressing. Suppose that a file of many records exists in memory, and each record contains many elements. The displacement addresses the file, the base register addresses a record, and the index register addresses an element of a record. Figure 5–13 illustrates this very complex form of addressing.

Example 5–9 provides a program that copies Element 0 of Record A into Element 2 of Record C using the base relative-plus-index mode of addressing. This example FILE

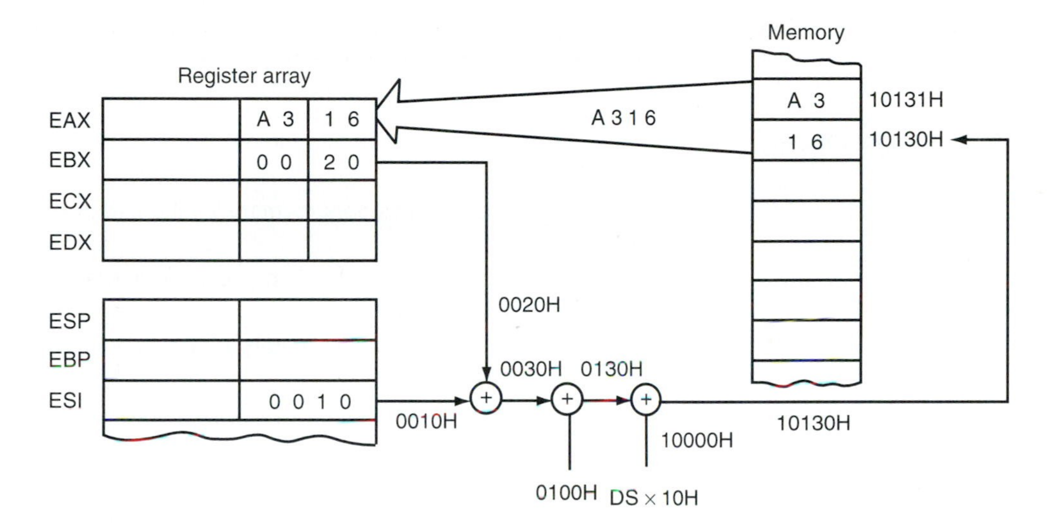

FIGURE 5–12 An example of base relative-plus-index addressing using a MOV AX,[BX+SI+100H] instruction. Note: DS = 1000H.

contains four records, and each record contains ten elements. Notice how the THIS BYTE statement is used to define the label FILE and RECA as the same memory location.

EXAMPLE 5–9

```
                    .MODEL SMALL                       ;indicate SMALL model
0000                .DATA                              ;indicate start of DATA segment

0000  = 0000        FILE  EQU   THIS BYTE              ;assign FILE to this byte

0000  000A [        RECA  DB    10 DUP (?)             ;reserve 10 bytes for RECA
           00
             ]
000A  000A [        RECB  DB    10 DUP (?)             ;reserve 10 bytes for RECB
           00
             ]
0014  000A [        RECC  DB    10 DUP (?)             ;reserve 10 bytes for RECC
           00
             ]
001E  000A [        RECD  DB    10 DUP (?)             ;reserve 10 bytes for RECD
           00
             ]

0000                      .CODE                        ;indicate start of CODE segment
                          .STARTUP                     ;indicate start of program

0017  BB 0000 R           MOV   BX,OFFSET RECA         ;address RECA
001A  BF 0000             MOV   DI,0                   ;address element 0
001D  8A 81 0000 R        MOV   AL,FILE[BX+DI]         ;get data
0021  BB 0014 R           MOV   BX,OFFSET RECC         ;address RECC
0024  BF 0002             MOV   DI,2                   ;address element 2
0027  88 81 0000 R        MOV   FILE[BX+DI],AL         ;save data

                          .EXIT                        ;exit to DOS
                          END                          ;indicate end of file
```

TABLE 5–8 Examples of base relative-plus-index addressing

Assembly Language	*Operation*
MOV DH,[BX+DI+20H]	Loads DH from the data segment memory location addressed by the sum of BX, DI, and 20H
MOV AX,FILE[BX+DI]	Loads AX from the data segment memory location addressed by the sum of FILE, BX, and DI
MOV LIST[BP+DI],CL	Stores CL at the stack segment memory location addressed by the sum of LIST, BP, and DI
MOV LIST[BP+SI+4],DH	Stores DH at the stack segment memory location addressed by the sum of LIST, BP, SI, and 4
MOV AL,FILE[EBX+ECX+2]	Loads AL from the data segment memory location addressed by the sum of FILE, EBX, ECX, and 2

Scaled-Index Addressing

Scaled-index addressing is the last data-addressing mode discussed. This data-addressing mode is unique to the 80386 through Pentium microprocessors. Scaled-index addressing uses two 32-bit registers (a base register and an index register) to access the memory. The second register (index) is multiplied by a scaling factor. The scaling factor can be 1×, 2×, 4×, or 8×. A scaling factor of 1× is implied and need not be included in the assembly language instruction (MOV AL,[EBX+ECX]). A scaling factor of 2× is used to address word-sized memory arrays, a scaling factor of 4× is used with doubleword-sized memory arrays, and a scaling factor of 8× is used with quadword-sized memory arrays.

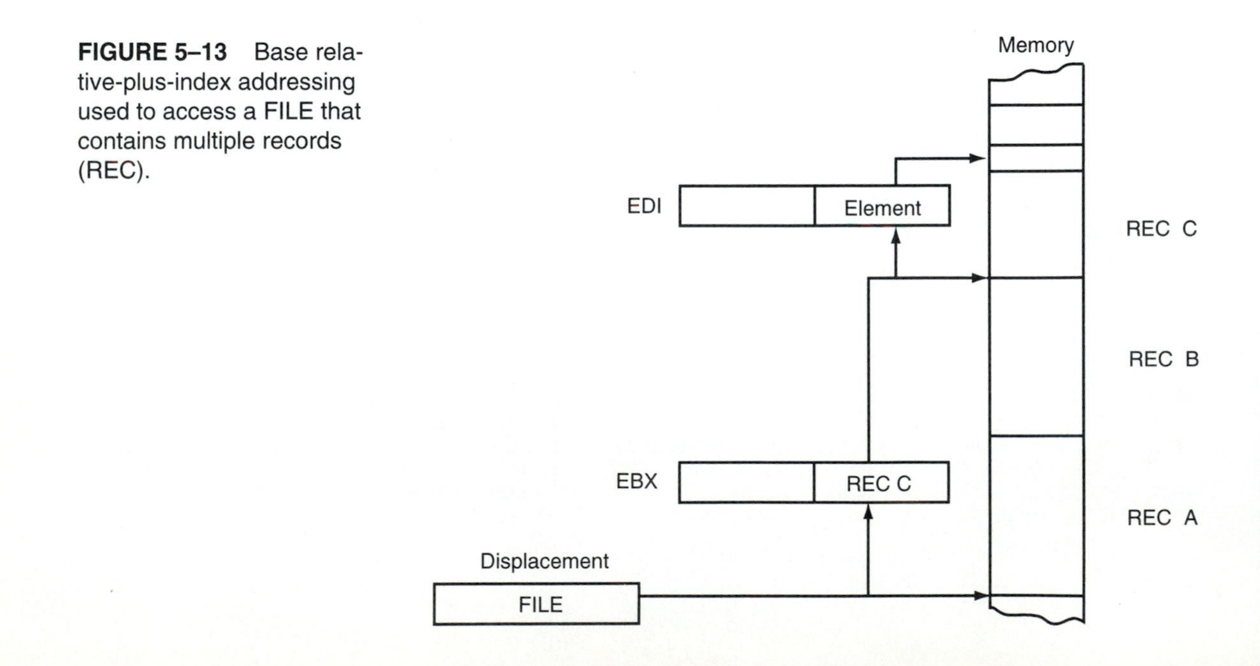

FIGURE 5–13 Base relative-plus-index addressing used to access a FILE that contains multiple records (REC).

TABLE 5–9 Examples of scaled-index addressing

Assembly Language	*Operation*
MOV EAX,[EBX+4*ECX]	Loads EAX from the data segment memory location addressed by the sum of EBX plus 4 times ECX
MOV [EAX+2*EBX],CX	Stores CX into the data segment memory location addressed by EAX plus 2 times EBX
MOV AX,[EBP+2*EDI+100H]	Loads AX from the stack segment memory location addressed by EBP plus 2 times EDI plus 100H
MOV LIST[EAX+2*EBX+10H],DX	Stores DX in the data segment memory location addressed by LIST plus EAX plus 2 times EBX plus 10H
MOV EAX,ARRAY[4*ECX]	Loads EAX with the contents of the doubleword data segment plus 4 times ECX

An example instruction is MOV AX,[EDI+2*ECX]. This instruction uses a scaling factor of 2×, which multiplies the contents of ECX by 2 before adding it to the EDI register to form the memory address. If ECX contains a 00000000H, word-sized memory Element 0 is addressed. If ECX contains a 00000001H, word-sized memory Element 1 is accessed, and so forth. This scales the index (ECX) by a factor of 2 for a word-sized memory array. Refer to Table 5–9 for some examples of scaled-index addressing. As you can imagine, there are an extremely large number of the scaled-index address register combinations.

Example 5–10 shows a sequence of instructions that use scaled-index addressing to access a word-sized array of data called LIST. Note that the offset address of LIST is loaded into register EBX with the MOV EBX,OFFSET LIST instruction. Once EBX addresses array LIST, the elements (located in ECX) of 2, 4, and 7 of this word-wide array are added using a scaling factor of 2 to access the elements. This program stores the 2 at Element 2 into Elements 4 and 7. Notice the .386 directive to select the 80386 microprocessor. This directive must follow the .MODEL statement for the assembler to process 80386 instructions for DOS. If the 80486 is in use, the .486 directive appears after the .MODEL statement. If the Pentium is in use, the .586 directive appears after .MODEL. If the microprocessor selection directive appears before the .MODEL statement, the microprocessor executes instructions in the 32-bit mode, which is not compatible with DOS.

EXAMPLE 5–10

```
                              .MODEL SMALL          ;select SMALL model
                              .386                  ;use the 80386 microprocessor
0000                          .DATA                 ;indicate the start of the DATA segment

0000  0000 0001 0002  LIST     DW    0,1,2,3,4      ;define array list
        0003 0004
000A  0005 0006 0007           DW    5,6,7,8,9
        0008 0009

0000                           .CODE                ;indicate start of the CODE segment
                               .STARTUP             ;indicate start of program
0010  66| BB 00000000 R        MOV   EBX,OFFSET LIST ;address array LIST

0016  66| B9 00000002          MOV   ECX,2          ;get element 2
001C  67& 8B 04 4B             MOV   AX,[EBX+2*ECX]
```

```
0020  66| B9 00000004        MOV  ECX,4              ;store in element 4
0026  67& 89 04 4B           MOV  [EBX+2*ECX],AX

002A  66| B9 00000007        MOV  ECX,7              ;store in element 7
0030  67& 89 04 4B           MOV  [EBX+2*ECX],AX

                             .EXIT                   ;exit to DOS
                             END                     ;end of file
```

Data Structures

A data structure is used to specify how information is stored in a memory array, and can be quite useful with applications that use arrays. It is best to think of a data structure as a template for data. The start of a structure is identified with the STRUC assembly language directive and the end of a structure is identified with the ENDS statement. A typical data structure is defined and used three times in Example 5–11. Notice that the name of the structure appears with the STRUC statement and also with the ENDS statement.

EXAMPLE 5–11

```
                         ;Define INFO data structure
                         ;
0057                     INFO    STRUC

0000  0020  [            NAMES   DB    32 DUP (?)        ;32 bytes for name
              00
                ]
0020  0020  [            STREET  DB    32 DUP (?)        ;32 bytes for street
              00
                ]
0040  0010  [            CITY    DB    16 DUP (?)        ;16 bytes for city
              00
                ]
0050  0002               STATE   DB    2 DUP (?)         ;2 bytes for state
              00
                ]
0052  0005 [             ZIP     DB    5 DUP (?)         ;5 bytes for zipcode
              00
                ]
                         INFO    ENDS

0000  42 6F 62 20 53 6D NAME1   INFO   <'Bob Smith','123 Main Street','Wanda','OH','44444'>
      69 74 68
      0017 [
              00
                ]
      31 32 33 20 4D
      61 69 6E 20 53 74
      72 65 65 74
      0011 [
              00
                ]
      57 61 6E 64 61
      000B [
              00
                ]
      4F 48 34 34 34
      34 34

0057  53 74 65 76 65 20 NAME2   INFO   <'Steve Doe','222 Mouse Lane','Miller','PA','18100'>
  44 6F 65
```

```
        0017 [
               00
                  ]
        32 32 32 20 4D
        6F 75 73 65 20 4C
        61 6E 65
        0012 [
               00
                  ]
        4D 69 6C 6C 65
        72
        000A [
               00
                  ]
        50 41 31 38 31
        30 30

00AE 42 65 6E 20 44 6F NAME3   INFO   <'Ben Dover','303 Main Street','Orender','CA','90000'>
        76 65 72
        0017 [
               00
                  ]
        33 30 33 20 4D
        61 69 6E 20 53 74
        72 65 65 74
        0011 [
               00
                  ]
        4F 72 65 6E 64
        65 72
        0009 [
               00
                  ]
        43 41 39 30 30
        30 30
```

The data structure in Example 5–11 defines five fields of information. The first is 32 bytes in length and holds a name, the second is 32 bytes in length and holds a street address, the third is 16 bytes in length and holds the city, the fourth is 2 bytes in length and holds the state, and the fifth is 5 bytes in length and holds the ZIPCode. Once the structure is defined (INFO), it can be filled as illustrated with names and addresses. Three examples are illustrated. Note that when the data structure is used to define data, literals are surrounded with apostrophes and the entire field is surrounded by < > symbols.

When data are addressed in a structure, use the structure name and the field name to select a field from the structure. For example, to address the STREET in NAME2, use the operand NAME2.STREET, where the name of the structure is followed by a period and then the name of the field. Likewise, use NAME3.CITY to refer to the city in structure NAME3.

EXAMPLE 5–12

```
                                  ;Clear names in array NAME1

0000  B9 0020                     MOV   CX,32
0003  B0 00                       MOV   AL,0
0005  BE 0000 R                   MOV   SI,OFFSET NAME1.NAMES
0008  F3/ AA                      REP   STOSB

                                  ;Clear street in array NAME2
```

```
000A  B9 0020                  MOV   CX,32
000D  B0 00                    MOV   AL,0
0010  BE 0077 R                MOV   SI,OFFSET NAME2.STREET
0013  F3/ AA                   REP   STOSB

                    ;Clear ZIPCode in array NAME3

0015  B9 0005                  MOV   CX,5
0018  B0 00                    MOV   AL,0
001A  BE 0100 R                MOV   SI,OFFSET NAME3.ZIP
001D  F3/ AA                   REP   STOSB
```

The short sequence of instructions in Example 5–12 clears the name field in structure NAME1, the address field in structure NAME2, and the ZIPCode field in structure NAME3. The function and operation of the instructions in this program are defined in later chapters. You may wish to refer back to this example once these instructions are learned.

5–2 PROGRAM MEMORY-ADDRESSING MODES

Program memory-addressing modes, used with the JMP and CALL instructions, consist of three distinct forms: direct, relative, and indirect. This section introduces these three addressing forms, using the JMP instruction to illustrate their operation.

Direct Program Memory-Addressing

Many early microprocessors used direct program memory-addressing for all jumps and calls. Direct program memory-addressing is also used in high-level languages, such as the GOTO and GOSUB instructions in BASK. The microprocessor uses this form of addressing, but not nearly as often as it uses relative and indirect program memory-addressing.

The instructions for direct program memory-addressing store the address with the opcode. For example, if a program jumps to memory location 10000H for the next instruction, the address 10000H is stored following the opcode in the memory. Figure 5–14 shows the direct intersegment JMP instruction and the four bytes required to store the address 10000H. This JMP instruction loads CS with 1000H and IP with 0000H to jump to memory location 10000H for the next instruction. An **intersegment** jump is a jump to any memory location within the entire memory system.) The direct jump is often called a *far jump* because it can jump to any memory location for the next instruction. In the real mode, a far jump accesses any location within the first 1M byte of memory by changing both CS and IP. In protected-mode operation in the 80386 and above, the far jump accesses a new code segment descriptor from the descriptor table, allowing it to jump to any memory location in the entire 4G byte address range.

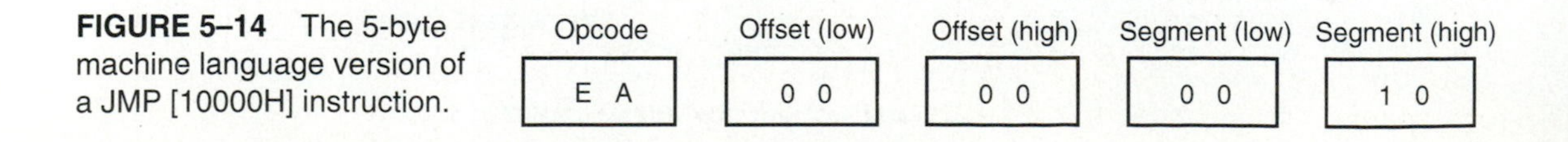

FIGURE 5–14 The 5-byte machine language version of a JMP [10000H] instruction.

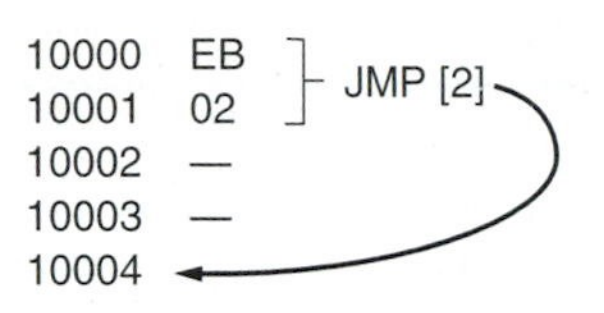

FIGURE 5–15 A JMP [2] instruction. This instruction skips over the two bytes of memory that follow the JMP instruction.

The only other instruction that uses direct program memory-addressing is the intersegment, or far CALL instruction. Usually, the name of a memory address, called a *label*, refers to the *location* that is called or jumped to, rather than the actual numeric address. When using a label with the CALL or JMP instruction, most assemblers select the best form of program addressing.

Relative Program Memory-Addressing

Relative program memory-addressing is not available in all early microprocessors, but it is available to the Intel family. The term *relative* means relative to the instruction pointer (IP). For example, if a JMP instruction skips the next two bytes of memory, the address in relation to the instruction pointer is a 2 that adds to the instruction pointer. This develops the address of the next program instruction. An example of the relative JMP instruction is shown in Figure 5–15. Notice that the JMP instruction is a one-byte instruction with a one- or two-byte displacement that adds to the instruction pointer. A one-byte displacement is used in *short jumps*; a two-byte displacement is used with *near jumps* and calls. Both types are considered intrasegment jumps. (An **intrasegment** jump is a jump anywhere within the current code segment.) In the 80386 and above, the displacement can also be a 32-bit value, allowing relative addressing to any location within a 4G byte code segment.

Relative JMP and CALL instructions contain an 8- or 16-bit signed displacement that allows forward memory reference or reverse memory reference. (The 80386 and above can have an 8-bit or 32-bit displacement.) All assemblers automatically calculate the distance for the displacement and select the proper one-, two- or, four-byte form. If the distance is too far for a two-byte displacement in an 8086 through 80286 microprocessor, some assemblers use a direct jump. An 8-bit displacement (*short*) has a jump range of between +127 and –128 bytes from the next instruction, while a 16-bit displacement (*near*) has a range of ±32K bytes. In the 80386 and above, a 32-bit displacement allows a range of ±2G bytes.

Indirect Program Memory-Addressing

The microprocessor allows several forms of indirect program memory-addressing for the JMP and CALL instructions. Table 5–10 lists some acceptable indirect program jump instructions, which can use any 16-bit register (AX, BX, CX, DX, SP, BP, DI, or SI), any relative register ([BP], [BX], [DI], or [SI]), or any relative register with a displacement. In the 80386 and above, an extended register can also be used to hold the address or indirect address of a relative JMP or CALL. For example, the JMP EAX instruction jumps to the location addressed by register EAX.

If a 16-bit register holds the address of a JMP instruction, the jump is near. For example, if the BX register contains a 1000H and a JMP BX instruction executes, the microprocessor jumps to offset address 1000H in the current code segment.

TABLE 5–10 Examples of indirect program memory-addressing

Assembly Language	*Operation*
JMP AX	Jump to the location addressed by AX in the current code segment
JMP CX	Jump to the location addressed by CX in the current code segment
JMP NEAR PTR [BX]	Jump to the current code segment location using the address stored at the data segment location plus BX
JMP NEAR PTR [DI+2]	Jump to the current code segment location using the address stored at the data segment location plus DI + 2
JMP TABLE [BX]	Jump to the current code segment location addressed by the contents of TABLE plus BX
JMP ECX	Jump to the current code segment location addressed by ECX

FIGURE 5–16 A jump table that stores addresses of various programs. The exact address chosen from the TABLE is determined by an index stored with the jump instruction.

```
TABLE  DW  LOC0
       DW  LOC1
       DW  LOC2
       DW  LOC3
```

If a relative register holds the address, the jump is considered an indirect jump. For example, JMP [BX] refers to a memory location within the data segment at the offset address contained in BX. At this offset address is a 16-bit number used as the offset address in the intrasegment jump. This type of jump is called an indirect-indirect or double-indirect jump.

Figure 5-16 shows a jump table stored beginning at memory location TABLE. This jump table is referenced by the short program shown in Example 5–13. In this example, the BX register is loaded with a 4, so when it combines in the JMP TABLE[BX] instruction with TABLE, the effective address is the contents of the second entry in the jump table.

EXAMPLE 5–13

```
                           ;Using indirect addressing for a jump
                           ;
0000  BB 0004                     MOV   BX,4               ;address LOC2
0003  FF A7 23A1 R                JMP   TABLE[BX]          ;jump to LOC2
```

5–3 STACK MEMORY-ADDRESSING

The stack plays an important role in all microprocessors. It holds data temporarily and stores return addresses for procedures. The *stack memory* is a LIFO (last-in, first-out) memory, which describes the way data are stored and removed from the stack. Data are placed onto the stack with a PUSH instruction and removed with a POP instruction. The CALL instruction uses the stack to hold the return address for procedures, and the RET (return) instruction removes the return address from the stack.

The stack memory is maintained by two registers: the *stack pointer* (SP or ESP) and the *stack segment register* (SS). Whenever a word of data is pushed onto the stack (see Figure 5–17 (a)), the high-order 8-bits are placed in the location addressed by SP–1. The low-order 8-bits are placed in the location addressed by SP–2. The SP is then decremented by 2, so the next word of data is stored in the next available stack memory location. The SP/ESP register always points to an area of memory located within the stack segment. The SP/ESP register adds to SS × 10H to form the stack memory address in the real mode. In protected-mode operation, the SS register holds a selector that accesses a descriptor for the base address of the stack segment.

Whenever data are popped from the stack (see Figure 5–17 (b)), the low-order 8-bits are removed from the location addressed by SP. The high-order 8-bits are removed from the location addressed by SP+1. The SP register is then incremented by 2. Table 5–11 lists some of the PUSH and POP instructions available to the microprocessor. Note that PUSH

TABLE 5–11 Example PUSH and POP instructions

Assembly Language	*Operation*
POPF	Removes a word from the stack and places it into the flags
POPFD	Removes a doubleword from the stack and places it into the EFLAGS
PUSHF	Stores a copy of the flag word on the stack
PUSHFD	Stores a copy of the EFLAGS on the stack
PUSH AX	Stores a copy of AX on the stack
POP BX	Removes a word from the stack and places it into BX
PUSH DS	Stores a copy of DS on the stack
PUSH 1234H	Stores a 1234H on the stack
POP CS	Illegal instruction
PUSH [BX]	Stores a copy of the word contents of the memory location addressed by BX in the data segment on the stack
PUSHA	Stores a copy of registers AX, CX, DX, BX, SP, BP, DI, and SI on the stack
PUSHAD	Stores a copy of registers EAX, ECX, EDX, EBX, ESP, EBP, EDI, and ESI on the stack
POPA	Removes data from the stack and places it into SI, DI, BP, SP, BX, DX, CX, and AX
POPAD	Removes data from the stack and places it into ESI, EDI, EBP, ESP, EBX, EDX, ECX, and EAX
POP EAX	Removes data from the stack and places into EAX
PUSH EDI	Stores a copy of EDI onto the stack

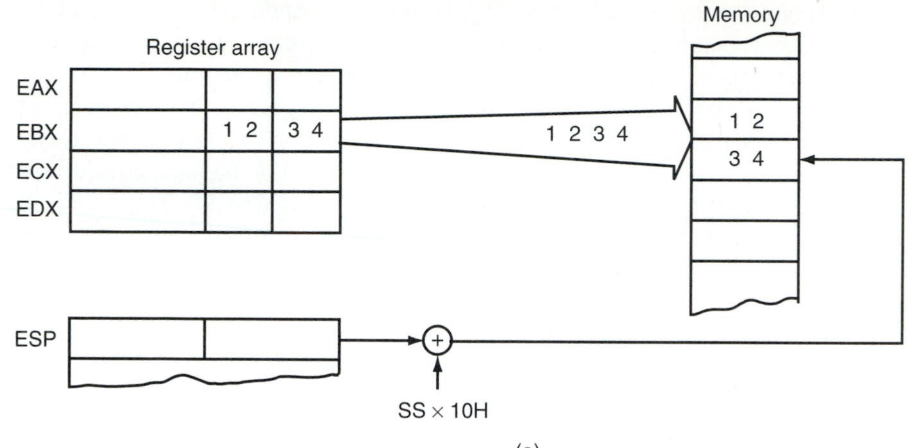

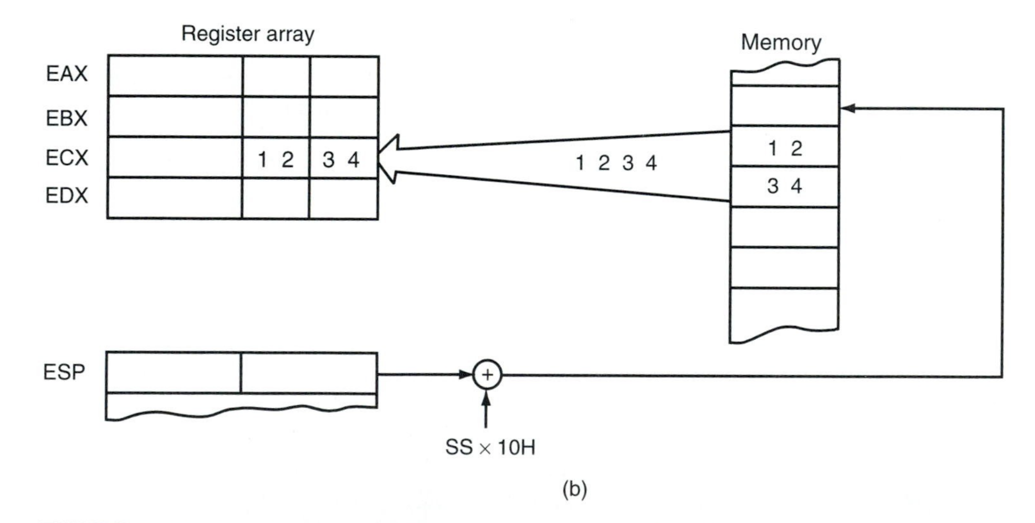

FIGURE 5–17 The PUSH and POP instructions. (a) PUSH BX places the contents of BX onto the stack. (b) POP CX removes data from the stack and places them into CX. Both instructions are shown after execution.

and POP always store or retrieve *words* of data—never bytes—in the 8086 through the 80286 microprocessor. The 80386 and above allow words or doublewords to be transferred to and from the stack. Data may be pushed onto the stack from any 16-bit register or segment register or, in the 80386 and above, any 32-bit extended register. Data may be popped off the stack into any 16-bit register or any segment register except CS. The reason data may not be popped from the stack into CS is that this only changes part of the address of the next instruction.

The PUSHA and POPA instructions either push or pop all of the registers on the stack except the segment registers. These instructions are not available on 8086/8088 microprocessors. The push immediate instruction is also new to the 80286 through Pentium microprocessors. Table 5–11 shows the order of the registers transferred by the PUSHA and POPA instructions. The 80386 and above also allow extended registers to be pushed or popped.

Example 5–14 lists a short program that pushes the contents of AX, BX, and CX onto the stack. The first POP retrieves the value that was pushed onto the stack from CX and places it into AX. The second POP places the original value of BX into CX. The last POP places the original value of AX into BX.

EXAMPLE 5–14

```
                    .MODEL TINY             ;select TINY model
0000                .CODE                   ;start CODE segment
                    .STARTUP                ;start of program
0100  B8 1000       MOV    AX,1000H         ;load test data
0103  BB 2000       MOV    BX,2000H
0106  B9 3000       MOV    CX,3000H

0109  50            PUSH   AX               ;1000H to stack
010A  53            PUSH   BX               ;2000H to stack
010B  51            PUSH   CX               ;3000H to stack

010C  58            POP    AX               ;3000H to AX
010D  59            POP    CX               ;2000H to CX
010E  5B            POP    BX               ;1000H to BX
                    .EXIT                   ;exit to DOS
                    END                     ;end of file
```

5–4 SUMMARY

1. The data-addressing modes include: register, immediate, direct, register-indirect, base-plus-index, register-relative, and base relative-plus-index addressing. In the 80386 through Pentium microprocessors, an additional addressing mode, called scaled-index addressing, exists.
2. The program memory-addressing modes include: direct, relative, and indirect addressing.
3. Table 5–12 lists all real-mode data-addressing modes available to the 8086 through the 80286 microprocessor. Note that the 80386 and above use these modes, plus many more defined throughout this chapter. In the protected mode, the function of the segment register is to address a descriptor that contains the base address of the memory segment.
4. The 80386, 80486, and Pentium microprocessors have additional addressing modes that allow the extended registers EAX, EBX, ECX, EDX, EBP, EDI, and ESI to address memory. These addressing modes are too numerous to list in tabular form but, in general, any of these registers function in the same way as those listed in Table 5–12. For example, MOV AL,TABLE[EBX+2*ECX+10H] is a valid addressing mode for the 80386/80486/Pentium microprocessor.
5. The MOV instruction copies the contents of the source operand into the destination operand. The source never changes for any instruction.

TABLE 5–12 Real-mode data-addressing modes

Assembly Language	*Address Generation*
MOV AL,BL	8-bit register addressing
MOV DI,BP	16-bit register addressing
MOV DS,BX	Segment register addressing
MOV AL,LIST	(DS × 10H) + LIST
MOV CH,DATA1	(DS × 10H) + DATA1
MOV DS,DATA2	(DS × 10H) + DATA2
MOV AL,12	Immediate data of 12 decimal
MOV AL,[BP]	(SS × 10H) + BP
MOV AL,[BX]	(DS × 10H) + BX
MOV AL,[DI]	(DS × 10H) + DI
MOV AL,[SI]	(DS × 10H) + SI
MOV AL,[BP+2]	(SS × 10H) + BP + 2
MOV AL,[BX–4]	(DS × 10H) + BX – 4
MOV AL,[DI+1000H]	(DS × 10H) + DI + 1000H
MOV AL,[SI+300H]	(DS × 10H) + SI + 0300H
MOV AL,LIST[BP]	(SS × 10H) + BP + LIST
MOV AL,LIST[BX]	(DS × 10H) + BX + LIST
MOV AL,LIST[DI]	(DS × 10H) + DI + LIST
MOV AL,LIST[SI]	(DS × 10H) + SI + LIST
MOV AL,LIST[BP+2]	(SS × 10H) + BP + LIST + 2
MOV AL,LIST[BX–6]	(DS × 10H) + BX + LIST – 6
MOV AL,LIST[DI+100H]	(DS × 10H) + DI + LIST + 100H
MOV AL,LIST[SI+20H]	(DS × 10H) + SI + LIST + 20H
MOV AL,[BP+DI]	(SS × 10H) + BP + DI
MOV AL,[BP+SI]	(SS × 10H) + BP + SI
MOV AL,[BX+DI]	(DS × 10H) + BX + DI
MOV AL,[BX+SI]	(DS × 10H) + BX + SI
MOV AL,[BP+DI+2]	(SS × 10H) + BP + DI + 2
MOV AL,[BP+SI–4]	(SS × 10H) + BP + SI – 4
MOV AL,[BX+DI+30H]	(DS × 10H) + BX + DI + 30H
MOV AL,[BX+SI+10H]	(DS × 10H) + BX + SI + 10H
MOV AL,LIST[BP+DI]	(SS × 10H) + BP + DI + LIST
MOV AL,LIST[BP+SI]	(SS × 10H) + BP + SI + LIST
MOV AL,LIST[BX+DI]	(DS × 10H) + BX + DI + LIST
MOV AL,LIST[BX+SI]	(DS × 10H) + BX + SI + LIST
MOV AL,LIST[BP+DI+2]	(SS × 10H) + BP + DI + LIST + 2
MOV AL,LIST[BP+SI–7]	(SS × 10H) + BP + SI + LIST – 7
MOV AL,LIST[BX+DI–10H]	(DS × 10H) + BX + DI + LIST – 10H
MOV AL,LIST[BX+SI+1AFH]	(DS × 10H) + BX + SI + LIST + 1AFH

6. Register addressing specifies any 8-bit register (AH, AL, BH, BL, CH, CL, DH, or DL) or any 16-bit register (AX, BX, CX, DX, SP, BP, SI, or DI). The segment registers (CS, DS, ES, or SS) are also addressable for moving data between a segment register and a 16-bit register/memory location or for PUSH and POP. In the 80386 through the Pentium microprocessors, the extended registers also are used for register

addressing and consist of: EAX, EBX, ECX, EDX, ESP, EBP, EDI, and ESI. Also available to the 80386 and above are the FS and GS segment registers.

7. The MOV immediate instruction transfers the byte or word immediately following the opcode into a register or a memory location. Immediate addressing manipulates constant data in a program. In the 80386 and above, a doubleword immediate data may also be loaded into a 32-bit register or memory location.
8. The .MODEL statement is used with assembly language to identify the start of a file and the type of memory model used with the file. If the size is TINY, the program exists in one segment—the code segment—and is assembled as a command (.COM) program. If the SMALL model is used, the program uses a code and data segment and assembles as an execute (.EXE) program. Other model sizes and their attributes are listed in Appendix A.
9. Direct addressing occurs in two forms in the microprocessor: (1) direct addressing and (2) displacement addressing. These forms of addressing are identical, except direct addressing is used to transfer data between EAX, AX, or AL and memory while displacement addressing is used with any register–memory transfer. Direct addressing requires three bytes of memory, while displacement addressing requires four bytes. In the 80386 and above, some of these instructions may require additional bytes in the form of prefixes for register and operand sizes.
10. Register-indirect addressing allows data to be addressed at the memory location pointed to by either a base (BP and BX) or index register (DI and SI). In the 80386 and above, extended registers EAX, EBX, ECX, EDX, EBP, EDI, and ESI are used to address memory data.
11. Base-plus-index addressing often addresses data in an array. The memory address for this mode is formed by adding a base register, index register, and the contents of a segment register times 10H. In the 80386 and above, the base and index registers may be any 32-bit register except EIP and ESP.
12. Register-relative addressing uses either a base or index register plus a displacement to access memory data.
13. Base relative-plus-index addressing is useful for addressing a two-dimensional memory array. The address is formed by adding a base register, an index register, displacement, and the contents of a segment register times 10H.
14. Scaled-index addressing is unique to the 80386 through Pentium. The second of two registers (index) is scaled by a factor of 2×, 4×, or 8× to access words, doublewords, or quadwords in memory arrays. MOV AX,[EBX+2*ECX] and MOV [4*ECX],EDX are examples of scaled-index instructions.
15. Data structures are templates for storing arrays of data and are addressed by array name and field. For example, Field TEN of Array NUMBER is addressed as NUMBER.TEN.
16. Direct program memory-addressing is allowed with the JMP and CALL instructions to any location in the memory system. With this addressing mode, the offset address and segment address are stored with the instruction.
17. Relative program addressing allows a JMP or CALL instruction to branch forward or backward in the current code segment by ± 32K bytes. In the 80386 and above, the 32-bit displacement allows a branch to any location in the current code segment using a displacement value of ± 2G bytes.

18. Indirect program addressing allows the JMP or CALL instructions to address another portion of the program or subroutine indirectly through a register or memory location.
19. The PUSH and POP instructions transfer a word between the stack and a register or memory location. A PUSH immediate instruction is available to place immediate data on the stack. The PUSHA and POPA instructions transfer AX, CX, DX, BX, BP, SP, SI, and DI between the stack and these registers. In the 80386 and above, the extended register and extended flags can also be transferred between registers and the stack. A PUSHFD stores the EFLAGS, while a PUSHF stores the FLAGS.
20. Example 5–15 shows many of the addressing modes presented in the chapter. This example program fills ARRAY1 from locations 0000:0000 through 0000:0009. It then fills ARRAY2 with 0 through 9. Finally, it exchanges the contents of ARRAY1, Element 2 with ARRAY2, Element 3.

EXAMPLE 5–15

```
                              .MODEL SMALL                ;select SMALL model
0000                          .DATA                       ;indicate start of DATA segment

0000  000A [         ARRAY1 DB     10 DUP (?)             ;reserve 10 bytes for ARRAY1
            00
               ]
000A  000A [         ARRAY2 DB     10 DUP (?)             ;reserve 10 bytes for  ARRAY2
            00
               ]
0000                           .CODE                      ;indicate start of CODE segment
                               .STARTUP                   ;indicate start of program

0017  B8 0000                  MOV    AX,0                ;address segment 0000 with ES
001A  8E C0                    MOV    ES,AX

001C  BF 0000                  MOV    DI,0                ;address element 0
001F  B9 000A                  MOV    CX,10               ;count of 10
0022                 LAB1:
0022  26: 8A 05                MOV    AL,ES:[DI]          ;copy 0000:0000 through 0000:0009
0025  88 85 0000 R             MOV    ARRAY1[DI],AL       ;into ARRAY1
0029  47                       INC    DI
002A  E2 F6                    LOOP   LAB1

002C  BF 0000                  MOV    DI,0                ;address element 0
002F  B9 000A                  MOV    CX,10               ;count of 10
0032  B0 00                    MOV    AL,0                ;initial value
0034                 LAB2:
0034  88 85 000A R             MOV    ARRAY2[DI],AL       ;fill ARRAY2
0038  FE C0                    INC    AL
003A  47                       INC    DI
003B  E2 F7                    LOOP   LAB2

003D  BF 0003                  MOV    DI,3                ;exchange array data
0040  8A 85 0000 R             MOV    AL,ARRAY1[DI]
0044  8A A5 000B R             MOV    AH,ARRAY2[DI+1]
0048  88 A5 0000 R             MOV    ARRAY1[DI],AH
004C  88 85 000B R             MOV    ARRAY2[DI+1],AL

                               .EXIT                      ;exit to DOS
                               END                        ;end of file
```

5–5 QUESTIONS AND PROBLEMS

1. What do the following MOV instructions accomplish?
   (a) MOV AX,BX
   (b) MOV BX,AX
   (c) MOV BL,CH
   (d) MOV ESP,EBP
   (e) MOV AX,CS
2. List the 8-bit registers that are used for register addressing.
3. List the 16-bit registers that are used for register addressing.
4. List the 32-bit registers that are used for register addressing in the 80386 through Pentium microprocessors.
5. List the 16-bit segment registers used with register addressing by MOV, PUSH, and POP.
6. What is wrong with the instruction MOV BL,CX?
7. What is wrong with the instruction MOV DS,SS?
8. Select an instruction for each of the following tasks:
   (a) copy EBX into EDX
   (b) copy BL into CL
   (c) copy SI into BX
   (d) copy DS into AX
   (e) copy AL into AH
9. Select an instruction for each of the following tasks:
   (a) move 12H into AL
   (b) move 123AH into AX
   (c) move 0CDH into CL
   (d) move 1000H into SI
   (e) move 1200A2H into EBX
10. What special symbol is sometimes used to denote immediate data?
11. What is the purpose of the .MODEL TINY statement?
12. What assembly language directive indicates the start of the CODE segment?
13. What is a label?
14. The MOV instruction is placed in what field of a statement?
15. A label may begin with what characters?
16. What is the purpose of the .EXIT directive?
17. True or false: The .MODEL TINY statement causes a program to assemble an execute program.
18. The .STARTUP directive accomplishes what tasks in the small memory model?
19. What is a displacement? How does it determine the memory address in a MOV [2000H],AL instruction?
20. What do the symbols [ ] indicate?
21. Suppose DS = 0200H, BX = 0300H, and DI = 400H. Determine the memory address accessed by each of the following instructions assuming real-mode operation:
    (a) MOV AL,[1234H]
    (b) MOV EAX,[BX]
    (c) MOV [DI],AL

22. What is wrong with the instruction MOV [BX],[DI]?
23. Choose an instruction that requires BYTE PTR.
24. Choose an instruction that requires WORD PTR.
25. Choose an instruction that requires DWORD PTR.
26. Explain the difference between the instruction MOV BX,DATA and the instruction MOV BX,OFFSET DATA.
27. Given that DS = 1000H, SS = 2000H, BP = 1000H, and DI = 0100H, determine the memory address accessed by each of the following instructions, assuming real-mode operation:
    (a) MOV AL,[BP+DI]
    (b) MOV CX,[DI]
    (c) MOV EDX,[BP]
28. What, if anything, is wrong with a MOV AL,[BX][SI] instruction?
29. Given that DS = 1200H, BX = 0100H, and SI = 0250H, determine the address accessed by each of the following, instructions assuming real-mode operation:
    (a) MOV [100H],DL
    (b) MOV [SI+100H],EAX
    (c) MOV DL,[BX+100H]
30. Given that DS = 1100H, BX = 0200H, LIST = 0250H, and SI = 0500H, determine the address accessed by each of the following instructions, assuming real-mode operation:
    (a) MOV LIST[SI],EDX
    (b) MOV CL,LIST[BX+SI]
    (c) MOV CH,[BX+SI]
31. Given that DS = 1300H, SS = 1400H, BP = 1500H, and SI = 0100H, determine the address accessed by each of the following instructions, assuming real-mode operation:
    (a) MOV EAX,[BP+200H]
    (b) MOV AL,[BP+SI-200H]
    (c) MOV AL,[SI-0100H]
32. Which base register addresses data in the stack segment?
33. Given that EAX = 00001000H, EBX = 00002000H and DS = 0010H, determine the addresses accessed by the following instructions, assuming real-mode operation:
    (a) MOV ECX,[EAX+EBX]
    (b) MOV [EAX+2*EBX],CL
    (c) MOV DH,[EBX+4*EAX+1000H]
34. Develop a data structure that has five fields of one word each, named F1, F2, F3, F4, and F5, with a structure name of FIELDS.
35. Show how field F3 of the data structure constructed in Question 34 is addressed in a program.
36. List all three program memory-addressing modes.
37. How many bytes of memory store a far direct jump instruction? What is stored in each of the bytes?
38. What is the difference between an intersegment jump and an intrasegment jump?
39. If a near jump uses a signed 16-bit displacement, how can it jump to any memory location within the current code segment?
40. The 80386 and above use a _______________-bit displacement to jump to any location within the 4G byte code segment.

41. What is a far jump?
42. If a JMP instruction is stored at memory location 100H within the current code segment, and it is jumping to memory location 200H within the current code segment, it cannot be a ______________ jump.
43. Show which JMP instruction (short, near, or far) assembles if the JMP THERE instruction is stored at memory address 10000H and the address of THERE is:
    (a) 10020H
    (b) 11000H
    (c) 0FFFEH
    (d) 30000H
44. Form a JMP instruction that jumps to the address pointed to by the BX register.
45. Select a JMP instruction that jumps to the location stored in memory at location table. Assume that it is a near JMP.
46. How many bytes are stored on the stack by PUSH instructions?
47. Explain how the PUSH [DI] instruction functions.
48. What registers are placed on the stack by the PUSHA instruction? In what order are they placed there?
49. What does the PUSHAD instruction accomplish?
50. Which instruction places the EFLAGS on the stack in the Pentium microprocessor?

CHAPTER 6

Data Movement Instructions

INTRODUCTION

This chapter concentrates on the data movement and string instructions. The data movement instructions include: MOV, MOVSX, MOVZX, PUSH, POP, BSWAP, XCHG, XLAT, IN, OUT, LEA, LDS, LES, LFS, LGS, LSS, LAHF, SAHF. The string instructions include: MOVS, LODS, STOS, INS, and OUTS. The data movement instructions are presented first, because they are more commonly used in programs and easy to understand.

The microprocessor requires an assembler program to generate machine language, because machine language instructions are too complex to efficiently generate by hand. This chapter describes the assembly language syntax and some of its directives. (This text assumes that the user is developing software on an IBM personal computer or clone. It is recommended that the assembler used be the Microsoft MACRO assembler (MASM), but the Intel Assembler (ASM), Borland Turbo assembler (TASM), or similar assembler software functions as well. This text presents information that functions with the Microsoft MASM assembler, but most programs assemble without modification with other assemblers. Appendix A explains the Microsoft assembler and provides details on the linker program and the programmer's workbench.)

OBJECTIVES

Upon completion of this chapter, you will be able to:

1. Explain the operation of each data movement instruction with applicable addressing modes.
2. Explain the purposes of the assembly language pseudo-operations and key words such as: ALIGN, ASSUME, DB, DD, DW, END, ENDS, ENDP, EQU, .MODEL, OFFSET, ORG, PROC, PTR, SEGMENT, USE16, USE32, and USES.
3. Given a specific data movement task, select the appropriate assembly language instruction to accomplish it.

4. Given a hexadecimal machine language instruction, determine the symbolic opcode, source, destination, and addressing mode.
5. Use the assembler to set up a data segment, stack segment, and a code segment.
6. Show how to set up a procedure using PROC and ENDP.
7. Explain the difference between memory models and full segment definitions for the MASM assembler.

6–1 MOV REVISITED

The MOV instruction, introduced in Chapter 5, explains the diversity of 8086–80486/Pentium addressing modes. This chapter introduces the machine language instructions available with various addressing modes and instructions. Machine code is introduced because it may occasionally be necessary to interpret machine language programs generated by an assembler. Interpretation of the machine's native language allows debugging or modification at the machine language level. Occasionally, machine language patches are made using the DOS DEBUG program, so some knowledge of machine language is useful. Conversion between machine and assembly language instructions are illustrated in Appendix B.

Machine Language

Machine language is the native binary code that the microprocessor understands and uses to control its operation. Machine language instructions for the 8086 through Pentium vary in length from 1 to as many as 13 bytes. Although machine language appears complex, there is order to it. There are well over 100,000 variations of machine language instructions, which means there is no complete list of these variations. Because of this, some binary bits in a machine language instruction are given, and the remainder are determined for each variation of the instruction.

Instructions for the 8086–80286 are 16-bit mode instructions that take the form found in Figure 6–1(a). The 16-bit mode instructions are compatible with the 80386 and above if they are programmed to operate in the 16-bit instruction mode, but may be prefixed as in Figure 6–1(b). The 80386 through Pentium assume that all instructions are 16-bit mode instructions when the machine is operated in the real mode. In the protected mode, the upper byte of the descriptor contains the D-bit that selects either the 16- or 32-bit instruction mode. At present, only Windows NT and OS/2 operate in the 32-bit instruction mode. The 32-bit mode instructions are in the form shown in Figure 6–1(b). These instructions occur in the 16-bit instruction mode by the use of prefixes, which are explained later.

The first two bytes of the 32-bit instruction mode format are called override prefixes, because they are not always present. The first modifies the size of the operand address used by the instruction and the second modifies the register size. If the 80386 through Pentium operate as 16-bit instruction mode machines (real or protected mode) and a 32-bit register is used, the register-size prefix (66H) is appended to the front of the instruction. If the microprosessor is operated in the 32-bit instruction mode (protected mode only) and a

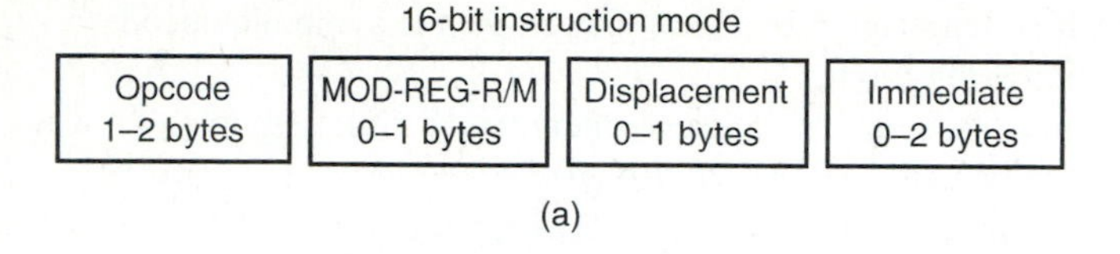

(a)

32-bit instruction mode (80386, 80486, or Pentium only)

Address size 0–1 bytes | Operand size 0–1 bytes | Opcode 1–2 bytes | MOD-REG-R/M 0–1 bytes | Scaled-index 0–1 bytes | Displacement 0–4 bytes | Immediate 0–4 bytes

(b)

FIGURE 6–1 The formats of the 8086–Pentium instructions. (a) The 16-bit form and (b) the 32-bit form.

32-bit register is used, the register-size prefix is absent. If a 16-bit register appears in an instruction in the 32-bit instruction mode, the register-size prefix is present to select a 16-bit register. The address-size prefix (67H) is used in a similar fashion, as explained later. The prefixes toggle the size of the register and operand address from 16-bit to 32-bit or 32-bit to 16-bit for the prefixed instruction. Note that, by default, the 16-bit instruction mode uses 8- and 16-bit registers and addressing modes, while the 32-bit instruction mode uses 8- and 32-bit registers and addressing modes. The prefixes override these defaults so a 32-bit register can be used in the 16-bit mode or a 16-bit register can be used in the 32-bit mode. The mode of operation (16 or 32 bits) should be selected to conform with the application at hand. If 8- and 32-bit data pervade the application, then the 32-bit mode should be selected; likewise, if 8- and 16-bit data pervade, then the 16-bit mode should be selected. Normally, mode selection is a function of the operating system.

The Opcode. The *opcode* selects the operation (addition, subtraction, move, etc.) performed by the microprocessor. The opcode is either one or two bytes in length for most machine language instructions. Figure 6–2 illustrates the general form of the first opcode byte of many, but not all, machine language instructions. Here, the first six bits of the first byte are the binary opcode. The remaining two bits indicate the *direction* (D) of data flow (not to be confused with the instruction mode bit or direction flag bit), and whether the data are a *byte* or a *word* (W). In the 80386 and above, words and doublewords are both specified when W = 1. The instruction mode and register-size prefix (66H) determine whether W represents a word or a doubleword.

If the direction bit (D) = 1, data flow to the register (REG) field from the R/M field located in the second byte of an instruction. If the D-bit = 0 in the opcode, data flow to the

FIGURE 6–2 Byte 1 of many machine language instructions, showing the position of the D and W bits.

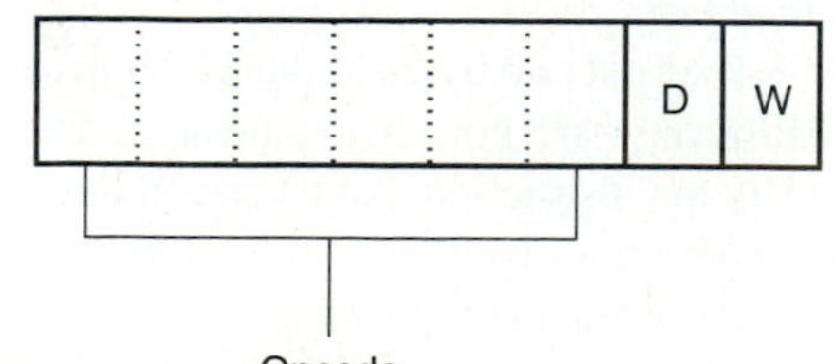

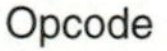

FIGURE 6–3 Byte 2 of many machine language instructions, showing the position of the MOD, REG, and R/M fields.

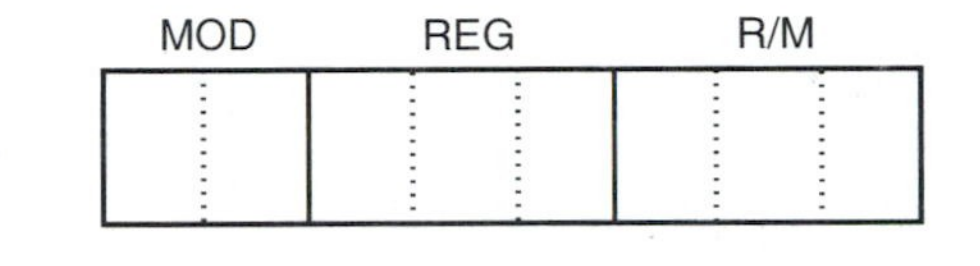

R/M field from the REG field. If the W-bit = 1, the data size is a word or doubleword. If the W-bit = 0, the data size is a byte. The W-bit appears in most instructions while the D-bit mainly appears with the MOV and some other instructions. Refer to Figure 6–3 for the binary bit pattern of the second opcode byte (reg-mod-r/m) of many instructions. This illustration shows the location of the MOD (*mode*), REG (*register*), and R/M (*register/memory*) fields.

MOD Field. The **MOD field** specifies the addressing mode (MOD) for the selected instruction. The MOD field selects the type of addressing and whether a displacement is present with the selected type. Table 6–1 lists the operand forms available to the MOD field for the 16-bit instruction mode, unless the operand address-size override prefix (67H) appears. If the MOD field contains an 11, it selects the register-addressing mode. Register addressing uses the R/M field to specify a register instead of a memory location. If the MOD field contains a 00, 01, or 10, the R/M field selects one of the data memory-addressing modes. When MOD selects a data memory-addressing mode, it indicates that the addressing mode contains no displacement (00), an 8-bit sign-extended displacement (01), or a 16-bit displacement (10). The MOV AL,[DI] instruction is an example showing no displacement, a MOV AL,[DI + 2] instruction uses an 8-bit displacement (+ 2), and a MOV AL,[DI + 1000H] instruction uses a 16-bit displacement (+ 1000H).

All 8-bit displacements are sign-extended into 16-bit displacements when the microprocessor executes the instruction. If the 8-bit displacement is 00H–7FH (positive), it is sign-extended to 0000H–007FH before adding to the offset address. If the 8-bit displacement is 80H–FFH (negative), it is sign-extended to FF80H–FFFFH. To sign-extend a number, its sign-bit is copied to the next higher-order byte, which generates either a 00H or an FFH in the higher-order byte. Note that some assembler programs do not use the 8-bit displacements.

In the 80386 through Pentium microprocessor, the MOD field may be the same as Table 6–1 or, if the instruction mode is 32 bits, it is the same as Table 6–2. The MOD field is interpreted as selected by the address-size override prefix or the operating mode of the microprocessor. This change in the interpretation of the MOD field and instruction supports many of the numerous addressing modes allowed in the 80386 through Pentium. The main difference is that when the MOD field is a 10, the 16-bit displacement becomes

TABLE 6–1 MOD field specifications for the 16-bit instruction mode

MOD	*Function*
00	No displacement
01	8-bit sign-extended displacement
10	16-bit displacement
11	R/M is a register

TABLE 6–2 MOD field specifications for the 32-bit instruction mode (80386/80486/Pentium only)

MOD	*Function*
00	No displacement
01	8-bit sign-extended displacement
10	32-bit displacement
11	R/M is a register

a 32-bit displacement, allowing any protected-mode memory location (4G bytes) to be accessed. The 80386 and above allow an 8- or 32-bit displacement only when operated in the 32-bit instruction mode, unless the address-size override prefix appears. Note that if an 8-bit displacement is selected, it is sign-extended into a 32-bit displacement by the microprocessor.

Register Assignments. Table 6–3 lists the register assignments for the REG field and the R/M field (MOD = 11). This table contains three lists of register assignments: one is used when the W-bit = 0 (bytes), and the other two are used when the W-bit = 1 (words or doublewords). Note that doubleword registers are only available to the 80386 through Pentium.

Suppose that a two-byte instruction, 8BECH, appears in a machine language program. Because neither a 67H (operand address-size override prefix) nor 66H (register-size override prefix) appears as the first byte, the first byte is the opcode. Assuming that the microprocessor is operated in the 16-bit instruction mode, this instruction is converted to binary and placed in the instruction format of Bytes 1 and 2, as illustrated in Figure 6–4. The opcode is 100010. If you refer to Appendix B, which lists the machine language instructions, you will find that this is the opcode for a MOV instruction. Also notice that both the D and W bits are a logic 1, which means that a word moves into the destination register specified in the REG field. The REG field contains a 101, indicating register BP, so the MOV instruction moves data into register BP. Because the MOD field contains a 11, the R/M field also indicates a register. Here, R/M = 100 (SP), therefore this instruction moves data from SP into BP and is written in symbolic form as a MOV BP,SP instruction.

Suppose that a 668BE8H instruction appears in an 80386 or above operated in the 16-bit instruction mode. The first byte (66H) is the register-size override prefix that selects 32-bit register operands for the 16-bit instruction mode. The remainder of the instruction indicates that the opcode is a MOV instruction with a source operand of EAX

TABLE 6–3 REG and R/M assignments (when MOD = 11)

Code	*W = 0 (Byte)*	*W = 1 (Word)*	*W = 1 (Doubleword)*
000	AL	AX	EAX
001	CL	CX	ECX
010	DL	DX	EDX
011	BL	BX	EBX
100	AH	SP	ESP
101	CH	BP	EBP
110	DH	SI	ESI
111	BH	DI	EDI

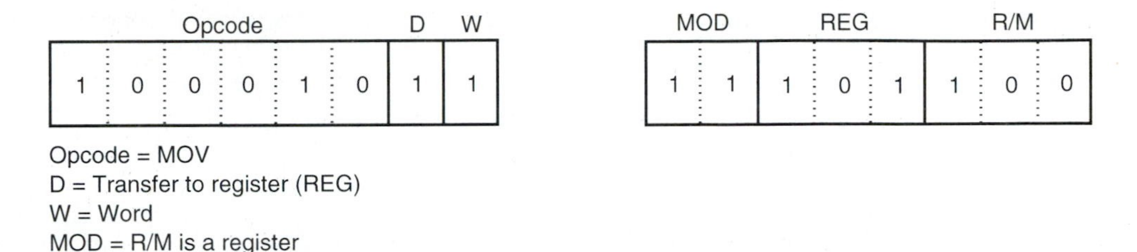

Opcode = MOV
D = Transfer to register (REG)
W = Word
MOD = R/M is a register
REG = BP
R/M = SP

FIGURE 6–4 The 8BEC instruction placed into Byte 1 and 2 formats from Figure 6–2 and 6–3. This instruction is a MOV BP,SP.

and a destination operand of EBP. This instruction is a MOV EBP,EAX. The same instruction becomes a MOV BP,AX instruction in the 80386 and above if it is operated in the 32-bit instruction mode, because the register-size override prefix selects a 16-bit register. Luckily, the assembler program keeps track of the register- and address-size prefixes and the mode of operation. Recall that if the .386 switch is placed before the .MODEL statement, the 32-bit mode is selected, and if it is placed after the .MODEL statement, the 16-bit mode is selected.

R/M Memory Addressing. If the MOD field contains a 00, 01, or 10, the R/M field takes on a new meaning. Table 6–4 lists the memory-addressing modes for the R/M field when MOD is a 00, 01, or 10 for the 16-bit instruction mode.

All of the 16-bit addressing modes presented in Chapter 5 appear in Table 6–4. The displacement, discussed in Chapter 5, is defined by the MOD field. If MOD = 00 and R/M = 101, the addressing mode is [DI]. If MOD = 01 or 10, the addressing mode is [DI + 33H] or LIST [DI + 22H] for the 16-bit instruction mode. This example uses LIST, 33H, and 22H as arbitrary values for the displacement.

Figure 6–5 illustrates the machine language version of the 16-bit instruction MOV DL,[DI] or instruction (8A15H). This instruction is two bytes long and has an opcode 100010, D = 1 (*to REG from R/M*), W = 0 (*byte*), MOD = 00 (*no displacement*), REG = 010 (*DL*) and R/M = 101 (*[DI]*). If the instruction changes to MOV DL,[DI + 1], the MOD

TABLE 6–4 16-bit R/M memory-addressing modes

Code	*Addressing Mode*
000	DS:[BX+SI]
001	DS:[BX+DI]
010	SS:[BP+SI]
011	SS:[BP+DI]
100	DS:[SI]
101	DS:[DI]
110	SS:[BP]*
111	DS:[BX]

**Note*: See text section, Special Addressing Mode.

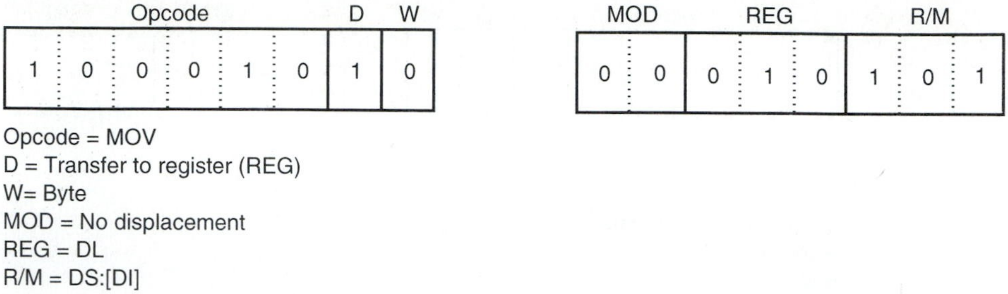

Opcode						D	W
1	0	0	0	1	0	1	0

MOD		REG			R/M		
0	0	0	1	0	1	0	1

Opcode = MOV
D = Transfer to register (REG)
W= Byte
MOD = No displacement
REG = DL
R/M = DS:[DI]

FIGURE 6–5 A MOV DL,[DI] instruction converted to its machine-language form.

field changes to 01, for an 8-bit displacement, but the first two bytes of the instruction otherwise remain the same. The instruction now becomes 8A5501H instead of 8A15H. Notice that the 8-bit displacement appends to the first two bytes of the instruction to form a 3-byte rather than a 2-byte instruction. If the instruction is again changed to a MOV DL,[DI+1000H], the machine language form becomes a 8A750010H. Here the 16-bit displacement of 1000H (coded as 0010H) appends the opcode.

Special Addressing Mode. There is a special addressing mode, which does not appear in Tables 6–2, 6–3, or 6–4, that occurs whenever memory data are referenced by only the displacement mode of addressing for 16-bit instructions. Examples are the MOV [1000H],DL and MOV NUMB,DL instructions. The first instruction moves the contents of register DL into data segment memory location 1000H. The second instruction moves register DL into symbolic data segment memory location NUMB.

Whenever an instruction has only a displacement, the MOD field is always a 00 and the R/M field is always a 110. This combination normally shows that the instruction contains no displacement and uses addressing mode [BP]. You cannot actually use addressing mode [BP] without a displacement in machine language. The assembler takes care of this by using an 8-bit displacement (MOD = 01) of 00H whenever the [BP] addressing mode appears in an instruction. This means that the [BP] addressing mode assembles as a [BP + 0], even though a [BP] is used in the instruction. The same special addressing mode is also available to the 32-bit mode.

Figure 6–6 shows the binary bit pattern required to encode the MOV [1000H],DL instruction in machine language. If the individual translating this symbolic instruction into machine language does not know about the special addressing mode, he or she would incorrectly translate to a MOV [BP],DL instruction. Figure 6–7 shows the actual form of the MOV [BP],DL instruction. Notice that this is a 3-byte instruction with a displacement of 00H.

32-Bit Addressing Modes. The 32-bit addressing modes found in the 80386 and above are obtained by either running these machines in the 32-bit instruction mode or in the 16-bit instruction mode using the address-size prefix 67H. Table 6–5 shows the R/M coding used to specify the 32-bit addressing modes. Notice that when R/M = 100, an additional byte, called a *scaled-index byte,* appears in the instruction. The scaled-index byte indicates the additional forms of scaled-index addressing that do not appear in Table 6–5. The scaled-index byte is mainly used when two registers are added to specify the memory

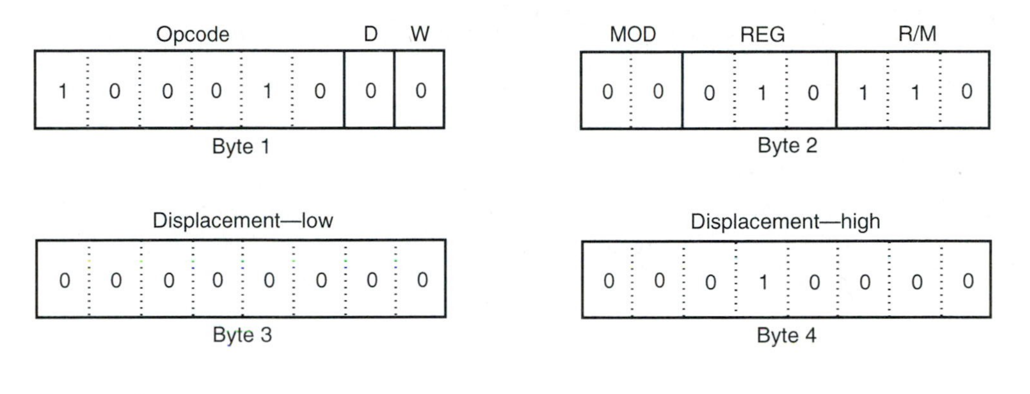

Opcode = MOV
D = Transfer from register (REG)
W = Byte
MOD = because R/M is [BP] (special addressing)
REG = DL
R/M = DS:[BP]
Displacement = 1000H

FIGURE 6–6 The MOV [1000H],DL instruction uses the special addressing mode.

address in an instruction. Because the scaled-index byte is added to the instruction, there are seven bits to define in the opcode and eight bits to define in the scaled-index byte. This means that a scaled-index instruction has 2^{15} (32K) possible combinations. There are more than 32,000 variations of the MOV instruction alone in the 80386 through Pentium microprocessor.

Figure 6–8 shows the format of the **scaled-index byte** as selected by a value of 100 in the R/M field of an instruction, when the 80386 and above use a 32-bit address. The leftmost two bits select a scaling factor (multiplier) of 1×, 2×, 4×, or 8×. Note that a scaling factor of 1× is implicit in an instruction that contains two 32-bit indirect address registers. The index and base fields both contain register numbers, as indicated in Table 6–3.

TABLE 6–5 32-bit addressing modes selected by R/M

Code	*Function*
000	DS:[EAX]
001	DS:[ECX]
010	DS:[EDX]
011	DS:[EBX]
100	Uses scaled index byte
101	SS:[EBP]*
110	DS:[ESI]
111	DS:[EDI]

**Note*: If the MOD bits are 00, this addressing mode uses a 32-bit displacement without register EBP. This is similar to the special addressing mode for the 16-bit instruction mode.

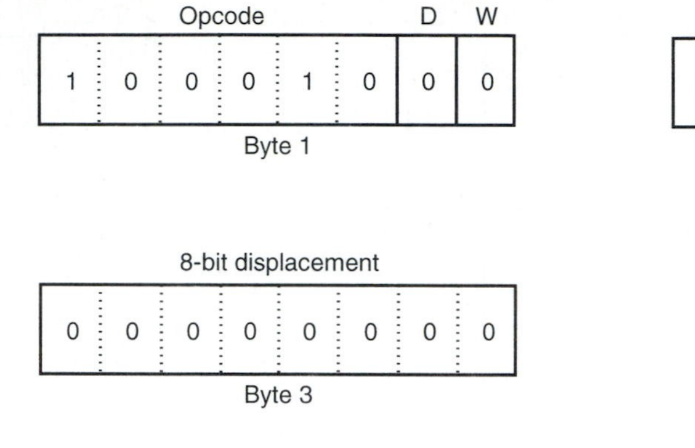

Opcode = MOV
D = Transfer from register (REG)
W = Byte
MOD = because R/M is [BP] (special addressing)
REG = DL
R/M = DS:[BP]
Displacement = 00H

FIGURE 6–7 The MOV [BP],DL instruction converted to binary machine language.

The instruction MOV EAX,[EBX+4*ECX] is encoded as 67668B048BH. Both address-size (67H) and register-size (66H) override prefixes appear for this instruction. This means that the instruction is 67668B048BH when operated with the 80386 and above in the 16-bit instruction mode. If the microprocessor is operated in the 32-bit instruction mode, both prefixes disappear, so the instruction becomes 8B048BH. The use of the prefixes depends on the mode of operation of the microprocessor. Scaled-index addressing can also use a single register multiplied by a scaling factor. An example is the MOV AL,[2*ECX] instruction. The contents of the data segment location addressed by two times ECX is copied into AL.

An Immediate Instruction. Let's examine the MOV WORD PTR [BX+1000H],1234H instruction as an example of a 16-bit instruction using immediate addressing. This instruction moves a 1234H into the word-sized memory location addressed by the sum of 1000H, BX, and DS × 10H. This 6-byte instruction uses two bytes for the opcode, W, MOD, and R/M fields. Two of the six bytes are the data, 1234H. The last two bytes

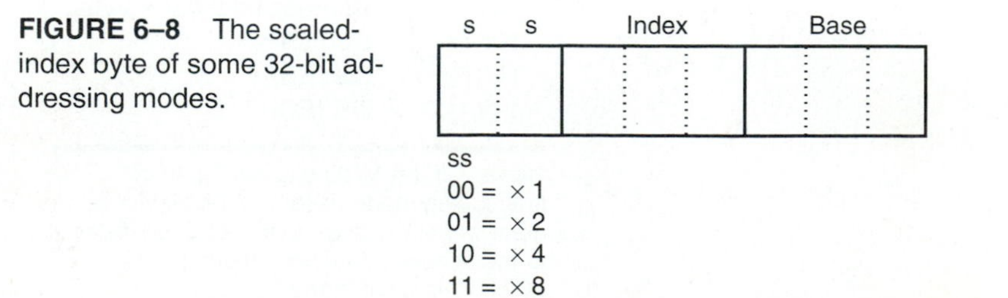

FIGURE 6–8 The scaled-index byte of some 32-bit addressing modes.

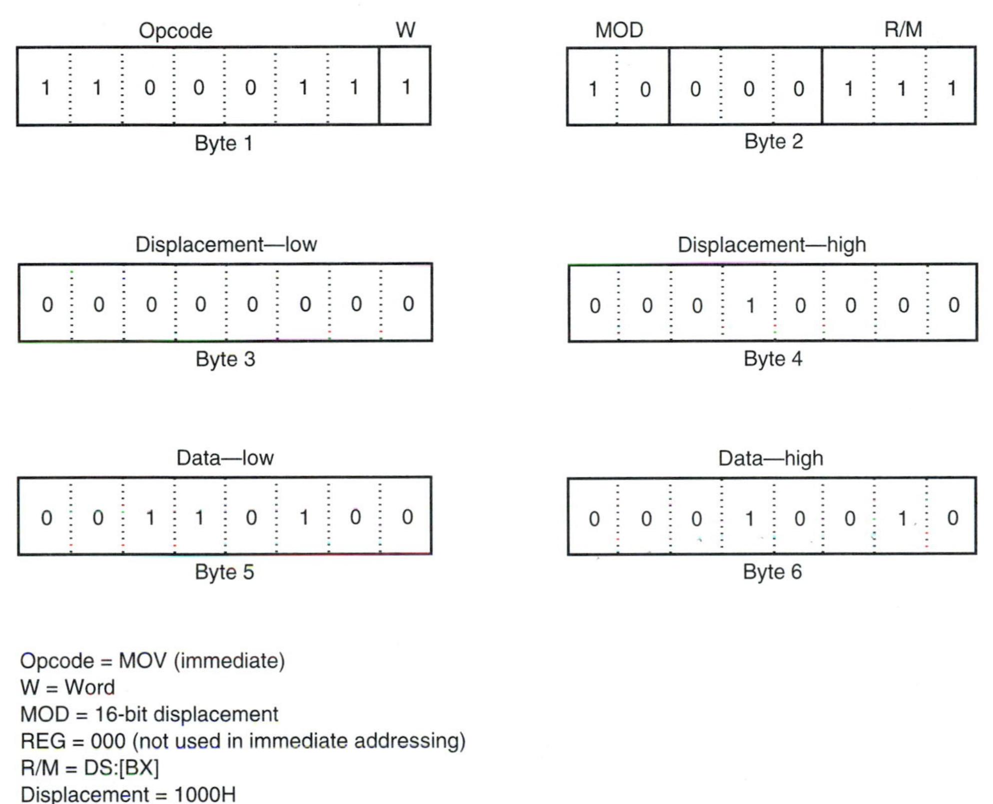

FIGURE 6–9 A MOV WORD PTR [BX+1000H],1234H instruction converted to binary machine language.

are the displacement of 1000H. Figure 6–9 shows the binary bit pattern for each byte of this instruction.

This instruction, in symbolic form, includes WORD PTR. The WORD PTR directive indicates to the assembler that the instruction uses a word-sized memory pointer. If the instruction moves a byte of immediate data, then BYTE PTR replaces WORD PTR in the instruction. Likewise, if the instruction uses a doubleword of immediate data the DWORD PTR directive replaces BYTE PTR. Most instructions that refer to memory through a pointer do not need the BYTE PTR, WORD PTR, or DWORD PTR directives. These are only necessary when it is not clear if the operation is a byte or a word. The MOV [BX],AL instruction is clearly a byte move, while the MOV [BX],1 instruction is not exact and could therefore be a byte-, word-, or doubleword-sized move. Here the instruction must be coded as MOV BYTE PTR [BX],1, MOV WORD PTR [BX],1, or MOV DWORD PTR [BX],1. If not, the assembler flags it as an error, because it cannot determine the intent of the instruction.

Segment MOV Instructions. If the contents of a segment register are moved by a MOV, PUSH, or POP instruction, a special set of register bits (REG field) selects the segment register (see Table 6–6).

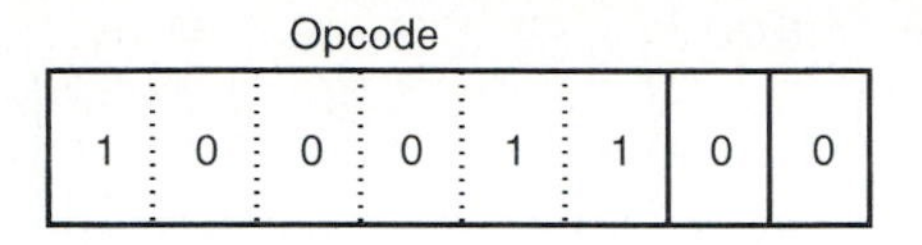

MOD REG R/M

1 1 0 0 1 0 1 1

Opcode = MOV
MOD = R/M is a register
REG = CS
R/M = BX

FIGURE 6–10 A MOV BX,CS instruction converted to binary machine language.

TABLE 6–6 Segment register selection bits

Code	*Segment Register*
000	ES
001	CS*
010	SS
011	DS
100	FS
101	GS

**Note*: MOV CS,R/M(16) and POP CS are not allowed by the microprocessor. The FS and GS segments are only available to the 80386/80486/Pentium microprocessor.

Figure 6–10 shows a MOV BX,CS instruction converted to binary. The opcode for this type of MOV instruction is different than for the prior MOV instructions. Segment registers can be moved between any 16-bit register or 16-bit memory location. For example, the MOV [DI],DS instruction stores the contents of DS into the memory location addressed by DI in the data segment. An immediate segment register MOV is not available in the instruction set. To load a segment register with immediate data, first load another register with the data and then move it to a segment register.

Although this has not been a complete coverage of machine language coding, it should give you a good start in machine language programming. Remember a program written in symbolic assembly language (assembly language), is rarely assembled by hand into binary machine language. An *assembler* converts symbolic assembly language into machine language. Because microprocessors have more than 100,000 instruction variations, an assembler is practically a necessity. Assembly by hand is very time-consuming, although not impossible.

6–2 PUSH/POP

The PUSH and POP instructions are important instructions that store and retrieve data from the LIFO (last-in, first-out) stack memory. The microprocessor has six forms of the PUSH and POP instructions: register, memory, immediate, segment register, flags, and all registers. The PUSH and POP immediate and the PUSHA and POPA (all registers) forms are not available in the earlier 8086/8088 microprocessor, but are available to the 80286 through Pentium.

Register addressing allows the contents of any 16-bit register to be transferred to or from the stack. In the 80386 and above, the 32-bit extended registers and flags (EFLAGS) can also be pushed or popped from the stack. The PUSH and POP instructions store the contents of a 16-bit memory location (or 32-bit memory location in the 80386 and above) on the stack or stack data into a memory location. Immediate addressing allows immediate data to be pushed onto the stack, but not popped off the stack. Segment register addressing allows the contents of any segment register to be pushed onto the stack or removed from the stack (CS may be pushed, but data from the stack may never be popped into CS). The flags may be pushed or popped from that stack and the contents of all the registers may be pushed or popped.

PUSH

The 8086–80286 PUSH instruction always transfers *two bytes* of data to the stack, and the 80386 and above transfer two or four bytes of data, depending on the register or size of the memory location. The source of the data may be any internal 16-bit/32-bit register, immediate data, any segment register, or any two bytes of memory data. There is also a PUSHA instruction that copies the contents of the internal register set, except the segment registers, to the stack. The PUSHA (*push all*) instruction copies the registers to the stack in the following order: AX, CX, DX, BX, SP, BP, SI, and DI. The value of SP pushed to the stack is whatever it was before the PUSHA instruction executed. The PUSHF (*push flags*) instruction copies the contents of the flag register to the stack. The PUSHAD and POPAD instructions push and pop the contents of the 32-bit register set found in the 80386 through Pentium.

Whenever data are pushed onto the stack, the first (most significant) data byte moves into the stack segment memory location addressed by SP – 1. The second (least significant) data byte moves into the stack segment memory location addressed by SP – 2. After the data are stored by a PUSH, the contents of the SP register decrement by 2. The same is true for a doubleword push, except four bytes are moved to the stack memory (most significant byte first) then the stack pointer decrements by 4. Figure 6–11 shows the operation of the PUSH AX instruction. This instruction copies the contents of AX onto the stack where address SS:[SP – 1] = AH, SS:[SP – 2] = AL, and afterwards SP = SP – 2.

The PUSHA instruction pushes all the internal 16-bit registers onto the stack as illustrated in Figure 6–12. This instruction requires 16 bytes of stack memory space to store all eight 16-bit registers. After all registers are pushed, the contents of the SP register are decremented by 16. The PUSHA instruction is very useful when the entire register set (*microprocessor environment*) of the 80286 and above must be saved during a task. Note that the PUSHAD instruction places the 32-bit register on the stack in the 80386 through Pentium. PUSHAD requires 32 bytes of stack storage space.

The PUSH immediate data instruction has two different opcodes, but in both cases, a 16-bit immediate number moves onto the stack. If PUSHD is used, a 32-bit immediate datum is pushed. If the value of the immediate data are 00H–FFH, the opcode is 6AH; if the data are 0100H–FFFFH, the opcode is 68H. The PUSH 8 instruction, which pushes an 0008H onto the stack, assembles as a 6A08H, and the PUSH 1000H instruction assembles

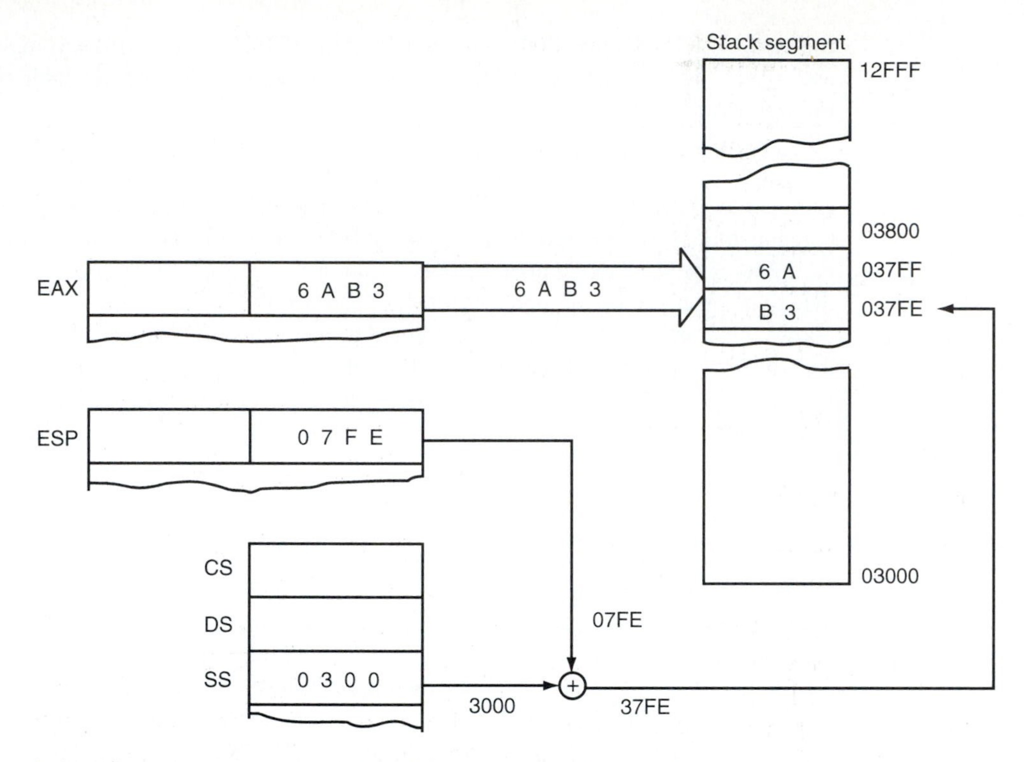

FIGURE 6–11 The effect of the PUSH AX instruction on ESP and stack memory locations 37FFH and 37FEH. This instruction is shown at the point after execution.

as 680010H. Another example of PUSH immediate is the PUSH 'A' instruction, which pushes a 0041H onto the stack. Here the 41H is the ASCII code for the letter A.

Table 6–7 lists the forms of the PUSH instruction that include PUSHA and PUSHF. Notice how the instruction set is used to specify different data sizes with the assembler.

POP

The POP instruction performs the inverse operation of a PUSH instruction. The POP instruction removes data from the stack and places it into the target 16-bit register, segment register, or a 16-bit memory location. In the 80386 and above, a POP can also remove 32-bit data from the stack and use a 32-bit address. The POP instruction is not available as an immediate POP. The POPF (*pop flags*) instruction removes a 16-bit number from the stack and places it into the flag register. The POPFD removes a 32-bit number from the stack and places it into the extended flag register. The POPA (*pop all*) instruction removes 16 bytes of data from the stack and places them into the following registers in the order shown: DI, SI, BP, SP, BX, DX, CX, and AX. This is the reverse of the order in which they are placed on the stack by the PUSHA instruction, causing the same data to return to the same registers. In the 80386 and above, a POPAD instruction reloads the 32-bit registers from the stack.

Suppose that a POP BX instruction executes. The first byte of data removed from the stack (the memory location addressed by SP in the stack segment) moves into register BL.

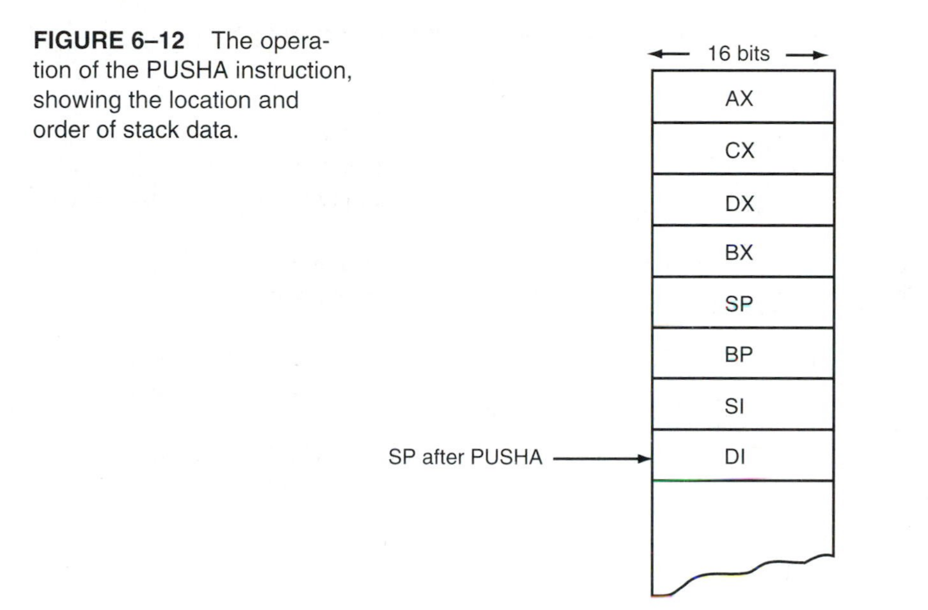

FIGURE 6–12 The operation of the PUSHA instruction, showing the location and order of stack data.

The second byte is removed from stack segment memory location SP + 1, and placed into register BH. After both bytes are removed from the stack, the SP register increments by 2. Figure 6–13 shows how the POP BX instruction removes data from the stack and places them into register BX.

The opcodes used for the POP instruction and all its variations, appear in Table 6–8. Like Table 6–7, this table also lists many example instructions. Note that POP CS is not a valid instruction. If a POP CS instruction executes, only a portion of the address (CS) of the next instruction changes. This makes the POP CS instruction unpredictable and therefore not allowed.

TABLE 6–7 The PUSH instructions

Symbolic	*Example*	*Note*
PUSH reg16	PUSH BX	16-bit register
PUSH reg32	PUSH EAX	32-bit register
PUSH mem16	PUSH [BX]	16-bit addressing mode
PUSH mem32	PUSH [EAX]	32-bit addressing mode
PUSH seg	PUSH DS	Any segment register
PUSH imm8	PUSH 12H	8-bit immediate data
PUSHW imm16	PUSHW 1000H	16-bit immediate data
PUSHD imm32	PUSHD 20	32-bit immediate data
PUSHA	PUSHA	Save 16-bit registers
PUSHAD	PUSHAD	Save 32-bit registers
PUSHF	PUSHF	Save 16-bit flag register
PUSHFD	PUSHFD	Save 32-bit flag register

Note: The 80386/80486/Pentium is required to operate with 32-bit addresses, registers, and immediate data.

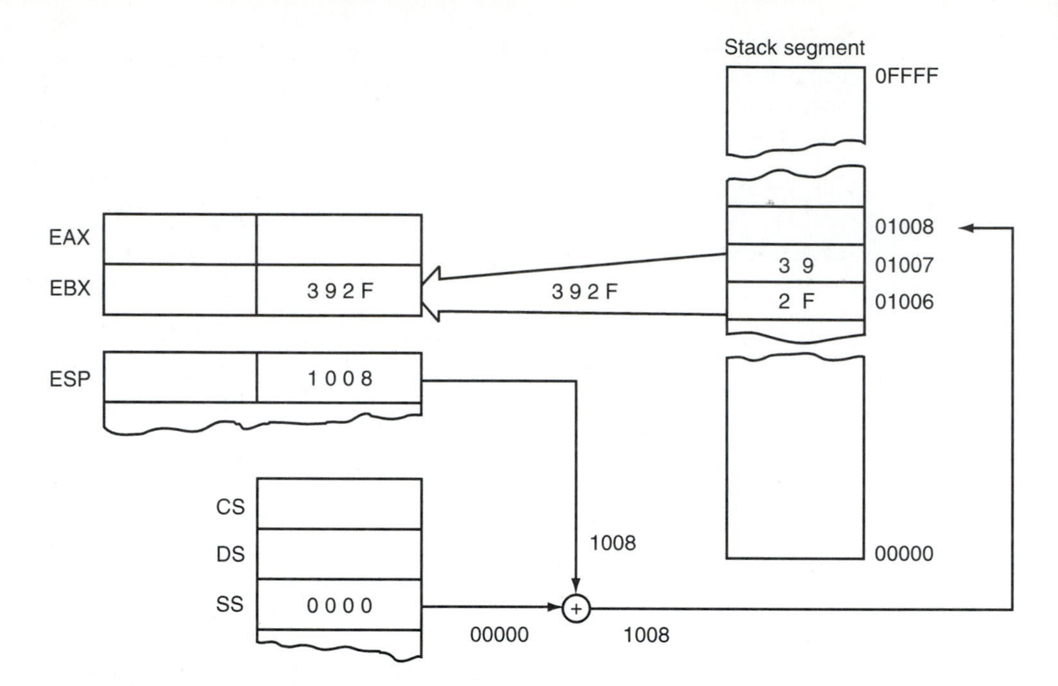

FIGURE 6–13 The POP BX instruction, showing how data are removed from the stack. This instruction is shown after execution.

Initializing the Stack

When the stack area is initialized, load both the stack segment register (SS) and the stack pointer (SP) register. It is normal to designate an area of memory as the stack segment by loading SS with the bottom location of the stack segment.

For example, if the stack segment is to reside in memory locations 10000H–1FFFFH, load SS with a 1000H. (Recall that the rightmost end of the stack segment register is appended with a 0H for real-mode addressing.) To start the stack at the top of this

TABLE 6–8 The POP instructions

Symbolic	*Example*	*Note*
POP reg16	POP DI	16-bit register
POP reg32	POP EBX	32-bit register
POP mem16	POP WORD PTR[DI+2]	16-bit memory address
POP mem32	POP DATA3	32-bit memory address
POP seg	POP GS	Any segment register
POPA	POPA	16-bit registers
POPAD	POPAD	32-bit registers
POPF	POPF	16-bit flag register
POPFD	POPFD	32-bit extended flag register

Note: The 80386/80486/Pentium is required to operate with 32-bit addresses and registers.

64K-byte stack segment, the stack pointer (SP) is loaded with a 0000H. Likewise, to address the top of the stack at location 10FFFH, use a value of 1000H in SP. Figure 6–14 shows how this value causes data to be pushed onto the top of the stack segment with a PUSH CX instruction. Remember, all segments are *cyclic* in nature, that is, the top location of a segment is *contiguous* with the bottom location of the segment.

EXAMPLE 6–1

```
0000                    STACK_SEG     SEGMENT STACK

0000 0100[                       DW   100H DUP (?)
              ????
                   ]
0200                    STACK_SEG     ENDS
```

In assembly language, a stack segment is set up as illustrated in Example 6–1. The first statement identifies the start of the stack segment and the last statement identifies the end of the stack segment. The assembler and linker program place the correct stack segment address in SS and the length of the segment (top of the stack) into SP. There is no need to load these registers in your program unless you wish to change the initial values for some reason.

An alternative method for defining the stack segment is used with one of the memory models for the MASM assembler only (refer to Appendix A). Other assemblers

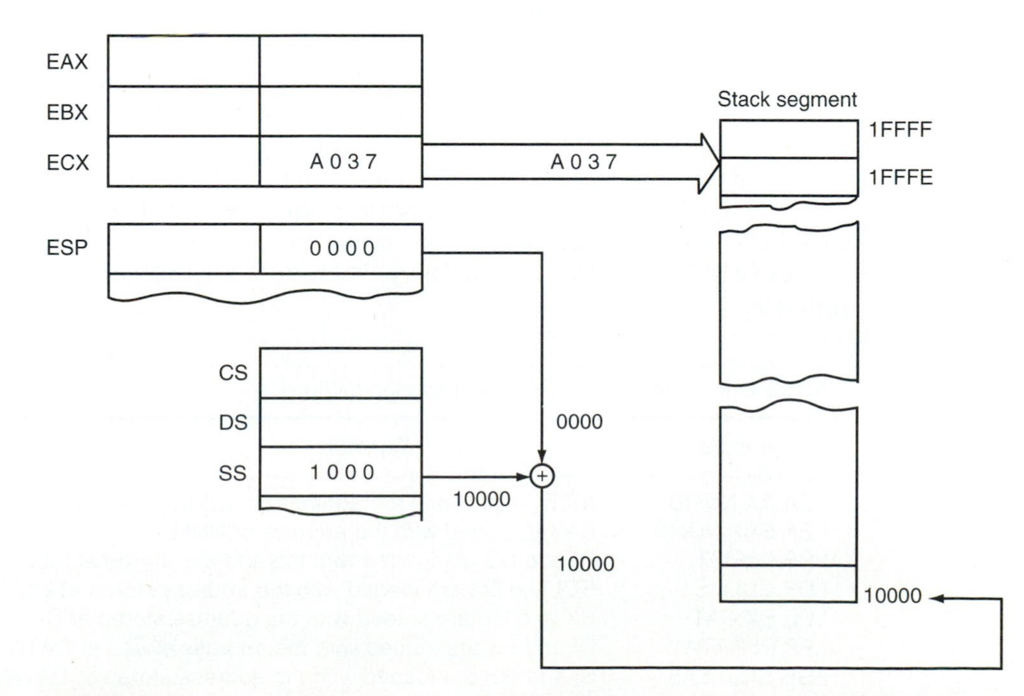

FIGURE 6–14 The PUSH CX instruction showing the cyclic nature of the stack segment. This instruction is shown just before execution, to illustrate that the stack bottom is contiguous to the top.

do not use models, or if they do, they are not exactly the same as MASM. In Example 6–2, the .STACK statement, followed by the number of *bytes* allocated to the stack, defines the stack area. The function is identical to Example 6–1. The .STACK statement initializes both SS and SP. Note: this text uses memory models designed to be used with the Microsoft Macro Assembler program MASM.

EXAMPLE 6–2

```
.MODEL SMALL
.STACK 200H          ;set stack size
```

If the stack is not specified using either method, a warning will appear when the program is linked. The warning may be ignored if the stack size is 128 bytes or less. The system automatically assigns (through DOS) at least 128 bytes of memory to the stack. This memory section is located in the program segment prefix (PSP), which is appended to the beginning of each program file. If you use more memory for the stack, you will erase information in the PSP that is critical to the operation of your program and the computer. This error often causes the computer program to crash. If the TINY memory model is used, the stack is automatically located at the very end of the segment, which allows for a larger stack area.

6–3 LOAD-EFFECTIVE ADDRESS

There are several load-effective address instructions in the microprocessor instruction set. The LEA instruction loads any 16-bit register with the address determined by the addressing mode selected for the instruction. The LDS and LES variations load any 16-bit register with the offset address retrieved from a memory location, and then load either DS or ES with a segment address retrieved from memory. In the 80386 and above, LFS, LGS, and LSS are added to the instruction set and a 32-bit register can be selected to receive a 32-bit offset from memory. Table 6–9 lists the load-effective address instructions.

TABLE 6–9 The load-effective address instructions

Symbolic	*Function*
LEA AX,NUMB	AX is loaded with the address of NUMB
LEA EAX,NUMB	EAX is loaded with the address of NUMB
LDS DI,LIST	DI and DS are loaded with the address stored at LIST
LDS EDI,LIST	EDI and DS are loaded with the address stored at LIST
LES BX,CAT	BX and ES are loaded with the address stored at CAT
LFS DI,DATA1	DI and FS are loaded with the address stored at DATA1
LGS SI,DATA5	SI and GS are loaded with the address stored at DATA5
LSS SP,MEM	SP and SS are loaded with the address stored at MEM

LEA

The LEA instruction loads a 16- or 32-bit register with the offset address of the data specified by the operand. As the first example in Table 6–9 shows, the operand address NUMB, not the contents of address NUMB, is loaded into register AX.

By comparing LEA with MOV, it is observed that: LEA BX,[DI] loads the offset address specified by DI into the BX register; MOV BX,[DI] loads the data stored at the memory location addressed by [DI] into register BX.

Earlier in the text, several examples are presented using the OFFSET directive. The **OFFSET** directive performs the same function as an LEA instruction if the operand is a displacement. For example, the MOV BX,OFFSET LIST performs the same function as LEA BX,LIST. Both instructions load the offset address of memory location LIST into the BX register. Refer to Example 6–3 for a short program that loads SI with the address of DATA1 and DI with the address of DATA2. It then exchanges the contents of these memory locations. Note that the LEA and MOV with OFFSET instructions are the same length (3 bytes).

EXAMPLE 6–3

```
                            .MODEL SMALL               ;select SMALL model
0000                        .DATA                      ;indicate start of DATA segment
0000  2000       DATA1      DW     2000H               ;define DATA1
0002  3000       DATA2      DW     3000H               ;define DATA2
0000                        .CODE                      ;indicate start of CODE segment
                            .STARTUP                   ;indicate start of program
0017  BE 0000 R             LEA    SI,DATA1            ;address DATA1 with SI
001A  BF 0002 R             MOV    DI,OFFSET DATA2     ;address DATA2 with DI

001D  8B 1C                 MOV    BX,[SI]             ;exchange DATA1 with DATA2
001F  8B 0D                 MOV    CX,[DI]
0021  89 0C                 MOV    [SI],CX
0023  89 1D                 MOV    [DI],BX
                            .EXIT                      ;exit to DOS
                            END                        ;indicate end of file
```

But why is the LEA instruction available if the OFFSET directive accomplishes the same task? First, OFFSET only functions with simple operands such as LIST. It may not be used for an operand such as [DI], LIST [SI], etc. The OFFSET directive is more efficient than the LEA instruction for simple operands. It takes the microprocessor longer to execute the LEA BX,LIST instruction than the MOV BX,OFFSET LIST. The 80486 microprocessor, for example, requires two clocks to execute the LEA BX,LIST instruction, and only one clock to execute MOV BX,OFFSET LIST. The reason the MOV BX,OFFSET LIST instruction executes faster is that the assembler calculates the offset address of LIST, while with the LEA instruction, the microprocessor does the calculation as it executes the instruction. The MOV BX,OFFSET LIST instruction is actually assembled as a move immediate instruction and is more efficient.

Suppose the microprocessor executes an LEA BX,[DI] instruction and DI contains a 1000H. Because DI contains the offset address, the microprocessor transfers a copy of DI into BX. A MOV BX,DI instruction performs this task in less time and is often preferred to the LEA BX,[DI] instruction.

Another example is LEA SI,[BX + DI]. This instruction adds BX to DI and stores the sum in the SI register. The sum generated by this instruction is a modulo-64K sum. If BX = 1000H and DI = 2000H, the offset address moved into SI is 3000H. If BX = 1000H and DI = FF00H, the offset address is 0F00H instead of 10F00H. Notice that the second result is a modulo-64K sum of 0F00H. (A modulo-64K sum drops any carry out of the 16-bit result.)

LDS, LES, LFS, LGS, and LSS

The LDS, LES, LFS, LGS, and LSS instructions load any 16-bit or 32-bit register with an offset address and the DS, ES, FS, GS, or SS segment register with a segment address. These instructions use any of the memory addressing modes to access a 32-bit or 48-bit section of memory that contains both the segment and offset address. The 32-bit section of memory contains a 16-bit offset and segment address, while the 48-bit section contains a 32-bit offset and a segment address. These instructions may not use the register addressing mode (MOD = 11). Note that the LFS, LGS, and LSS instructions are only available on 80386 and above, as are the 32-bit registers.

Figure 6–15 illustrates an example LDS BX,[DI] instruction. This instruction transfers the 32-bit number addressed by DI in the data segment into the BX and DS registers. The LDS, LES, LFS, LGS, and LSS instructions obtain a new far address from memory.

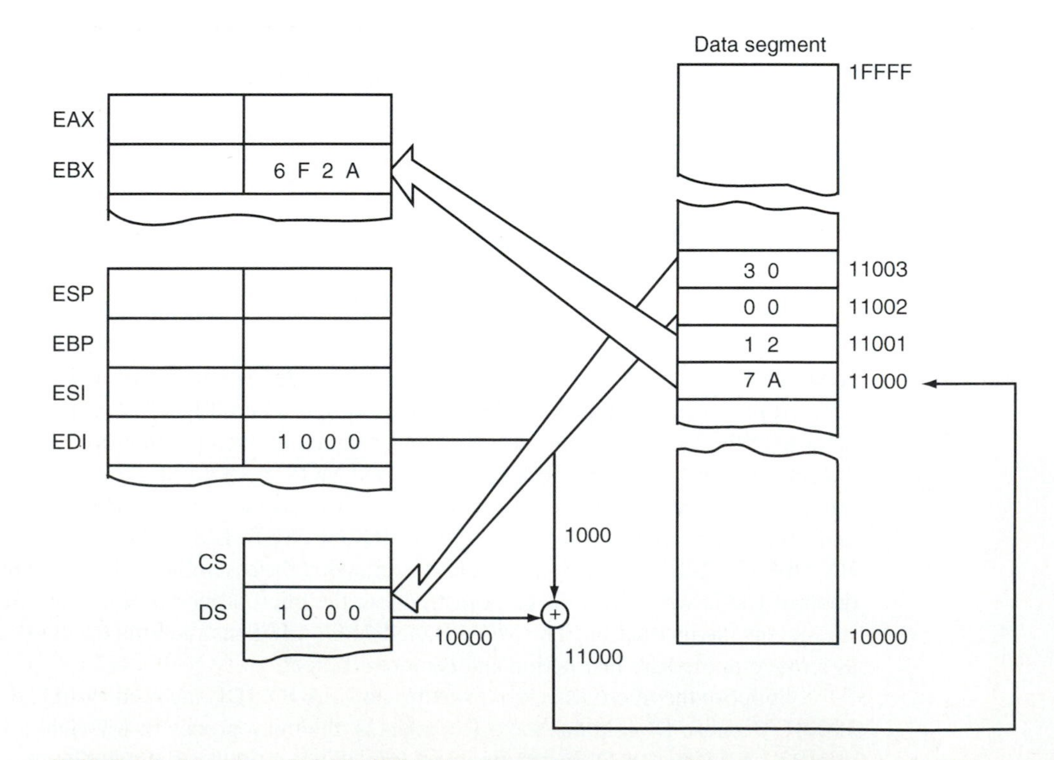

FIGURE 6–15 The LDS BX,[DI] instruction loads register BX from addresses 11000H and 11001H and register DS from locations 11002H and 11003H. This instruction is shown at the point just before DS changes to 3000H and BX changes to 127AH.

The offset address appears first, followed by the segment address. This format is used for storing all 32-bit memory addresses.

A far address can be stored in memory by the assembler. For example, the ADDR DD FAR PTR FROG instruction stores the offset and segment address (*far address*) of FROG in 32 bits of memory at location ADDR. The DD directive tells the assembler to store a *doubleword* (32-bit number) in memory address ADDR.

In the 80386 and above, an LDS EBX,[DI] instruction loads EBX from the four-byte section of memory addressed by DI in the data segment. Following this four-byte offset is a word that is loaded to the DS register. Notice that instead of addressing a 32-bit section of memory, the 80386 and above address a 48-bit section of the memory whenever a 32-bit offset address is loaded to a 32-bit register. The first four bytes contain the offset value loaded to the 32-bit register, and the last two bytes contain the segment address.

EXAMPLE 6–4

```
                                .MODEL SMALL            ;select SMALL model
                                .386                    ;select 80386 microprocessor
0000                            .DATA                   ;indicate start of DATA segment
0000  00000000        SADDR     DD     ?                ;old stack address
0004  1000 [          SAREA     DW     1000H DUP (?)    ;new stack area
            0000
                ]
2004 = 2004          STOP       EQU    THIS WORD        ;define to of new stack
0000                            .CODE                   ;indicate start of CODE segment
                                .STARTUP                ;indicate start of program

0010  FA                        CLI                     ;disable interrupt

0011  8B C4                     MOV    AX,SP            ;save old SP
0013  A3 0000 R                 MOV    WORD PTR SADDR,AX
0016  8C D0                     MOV    AX,SS            ;save old SS
0018  A3 0002 R                 MOV    WORD PTR SADDR+2,AX

001B  8C D8                     MOV    AX,DS           ;load new SS
001D  8E D0                     MOV    SS,AX
001F  B8 2004 R                 MOV    AX,OFFSET STOP  ;load new SP
0022  8B E0                     MOV    SP,AX

0024  FB                        STI                    ;enable interrupt

0025  8B C0                     MOV    AX,AX           ;do dummy instructions
0027  8B C0                     MOV    AX,AX           ;note: anything can appear here

0029  0F B2 26 0000 R           LSS    SP,SADDR        ;load old SS and SP

                                .EXIT                  ;exit to DOS
                                END                    ;end of file
```

The most useful of the load instructions is the LSS instruction. Example 6–4 shows a short program that creates a new stack area after saving the address of the old stack area. After executing some dummy instructions, the old stack area is reactivated by loading both SS and SP with the LSS instruction. Note that the CLI (disable interrupts) and STI (enable interrupts) instructions must be included to disable interrupts, a topic discussed near the end of this chapter. Because the LSS instruction functions in the 80386 or above, the .386 statement appears after the .MODEL statement to select the 80386 microprocessor. Notice how the WORD PTR directive is used to override the doubleword (DD) definition for the old stack address memory location. If an 80386 or newer microprocessor is in use, it is suggested that the .386 switch be used to develop software for the 80386 microprocessor. This is true even if the microprocessor is a Pentium. The reason is that the 80486 and Pentium offer only a few additional instructions that are seldom used in software development. If the need arises to use the CMPXCHG, CMPXCHG8 (new to the Pentium), XADD, or BSWAP instructions, then select either the .486 switch for the 80486 microprocessor or the .586 switch for the Pentium.

6–4 STRING DATA TRANSFERS

There are five string data transfer instructions—LODS, STOS, MOVS, INS, and OUTS. Each string instruction allows data transfers that are either a single byte, word, doubleword, or if repeated, a block of bytes, words, or doublewords. Before the string instructions are presented, the operation of the D flag-bit (*direction*), DI, and SI must be understood as they apply to the string instructions.

The Direction Flag

The direction flag (D) selects auto-increment (D = 0) or auto-decrement (D = 1) operation for the DI and SI registers during string operations. The direction flag is only used with the string instructions. The CLD instruction *clears* the D flag (D = 0) and the STD instruction *sets* it (D = 1). Therefore, the CLD instruction selects the auto-increment mode (D = 0) and STD selects the auto-decrement mode (D = 1).

Whenever a string instruction transfers a byte, the contents of DI and/or SI increment or decrement by 1. If a word is transferred, the contents of DI and/or SI increment or decrement by 2. Doubleword transfers cause DI and/or SI to increment or decrement by 4. Only the actual registers used by the string instruction increment or decrement. For example, the STOSB instruction uses the DI register to address a memory location. When STOSB executes, only DI increments or decrements, without effecting SI. The same is true of the LODSB instruction, which uses the SI register to address memory data. LODSB only increments/decrements SI without effecting DI.

DI AND SI

During the execution of a string instruction, memory accesses occur through either or both the DI and SI registers. The DI offset address accesses data in the extra segment for all string instructions that use it. The SI offset address accesses data, by default, in the data segment. The segment assignment of SI may be changed with a segment override prefix as described later in this chapter. The DI segment assignment is *always* in the extra segment when a string instruction executes. This assignment cannot be changed. The reason one pointer addresses data in the extra segment and the other pointer addresses data in the data segment is so the MOVS instruction can move 64K bytes of data from one segment of memory to another.

TABLE 6–10 Forms of the LODS instruction

Symbolic	*Function*
LODSB	AL = DS:[SI]; SI = SI ± 1
LODSW	AX = DS:[SI]; SI = SI ± 2
LODSD	EAX = DS:[SI]; SI = SI ± 4
LODS LIST	AL = DS:[SI]; SI = SI ± 1 (if LIST is a byte)
LODS DATA1	AX = DS:[SI]; SI = SI ± 2 (if DATA1 is a word)
LODS ES:DATA4	EAX = ES:[SI]; SI = SI ± 4 (if DATA4 is a doubleword)

Note: Only the 80386/80486/Pentium can use doublewords. Also, the segment can be overridden with a segment override prefix as in LODS ES:DATA4.

LODS

The **LODS** instruction loads AL, AX, or EAX with data stored at the data segment offset address indexed by the SI register. (Note that only the 80386, 80486, and Pentium can use EAX.) After loading AL with a byte, AX with a word, or EAX with a doubleword, the contents of SI increment if D = 0, or decrement if D = 1. A one is added to or subtracted from SI for a byte-sized LODS, a two is added or subtracted for a word-sized LODS, and a four is added or subtracted for a doubleword-sized LODS.

Table 6–10 lists the permissible forms of the LODS instruction. The LODSB (*loads a byte*) instruction causes a byte to be loaded into AL, the LODSW (*loads a word*) instruction causes a word to be loaded into AX, and the LODSD (*loads a doubleword*) instruction causes a doubleword to be loaded into EAX. The LODS instruction may also be followed by a byte-, word-, or doubleword-sized operand to select a byte, word, or doubleword transfer, although that is unusual. Operands are often defined as bytes with DB, as words with DW, and as doublewords with DD. The **DB** pseudo-operation defines byte(s), the **DW** pseudo-operation defines word(s), and the **DD** pseudo-operation defines doubleword(s).

Figure 6–16 shows the effect of executing the LODSW instruction if the D flag = 0, SI = 1000H, and DS = 1000H. Here a 16-bit number stored at memory locations 11000H and 11001H moves into AX. Because D = 0 and this is a word transfer, the contents of SI increment by 2 *after* AX loads with memory data.

STOS

The **STOS** instruction stores AL, AX, or EAX at the extra segment memory location addressed by the DI register. (Note only the 80386/80486/Pentium can use EAX and doublewords.) Table 6–11 lists all forms of the STOS instruction. Like LODS, a STOS instruction may be appended with a B, W, or D for byte, word, or doubleword transfers. The STOSB (*stores a byte*) instruction stores the byte in AL at the extra segment memory location addressed by DI. The STOSW (*stores a word*) instruction stores AX in the extra segment memory location addressed by DI. The STOSD instruction (stores a doubleword) stores EAX in the extra segment location addressed by DI. After the byte (AL), word (AX), or doubleword (EAX) is stored, the contents of DI increments or decrements.

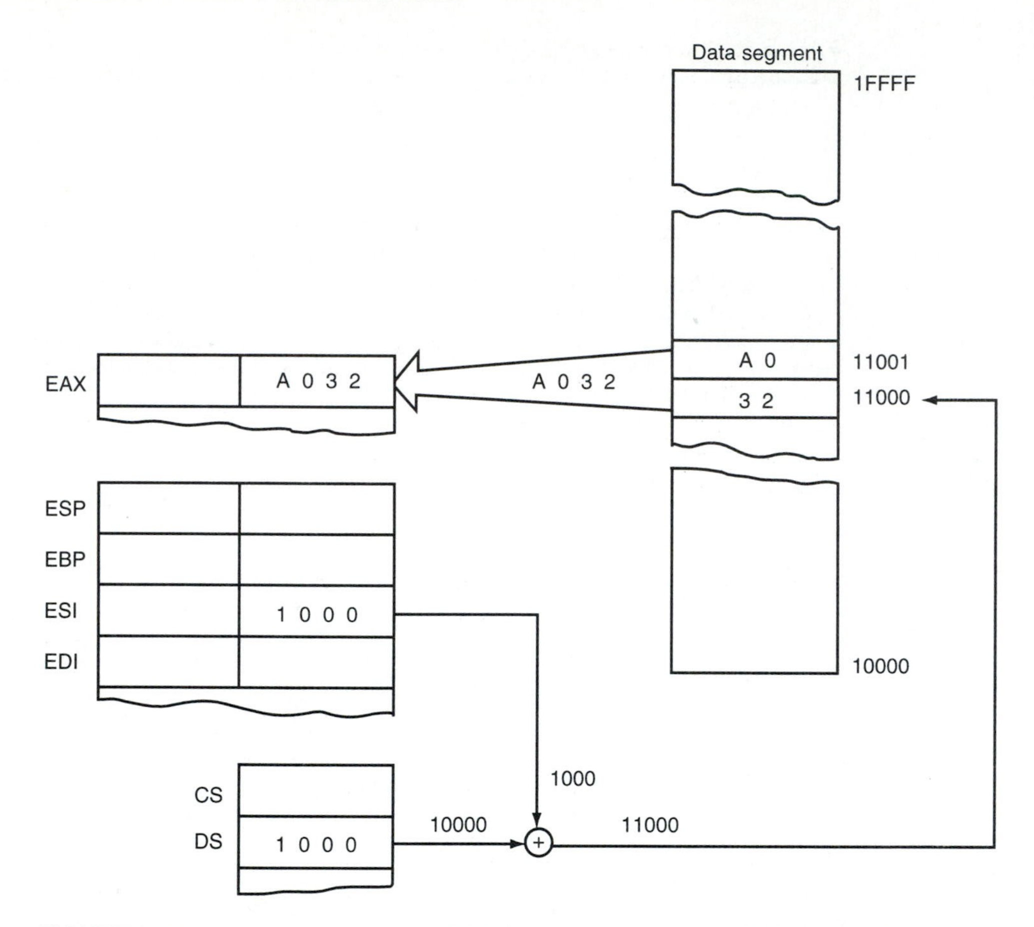

FIGURE 6–16 The operation of the LODSW instruction if DS = 1000H, SI = 1000H, D = 0, 11000H = 32 and 11001H = A0. This instruction is shown after AX is loaded from memory, but before SI increments by 2.

STOS with a REP. The repeat prefix (**REP**) is added to any string data transfer instruction except the LODS instruction. It doesn't make any sense to perform a repeated LODS operation. The REP prefix causes CX to decrement by 1 each time the string instruction executes. After CX decrements, the string instruction repeats. If CX reaches a value of 0, the instruction terminates and the program continues with the next sequential instruction. Thus, if CX is loaded with a 100, and a REP STOSB instruction executes, the microprocessor automatically repeats the STOSB instruction 100 times. Since the DI register is automatically incremented or decremented after each datum is stored, this instruction stores the contents of AL in a block of memory instead of a single byte of memory.

Suppose that the STOSW instruction is used to clear the video text display. This is accomplished by addressing video text memory that begins at memory location B800:0000. Each character position on the 25-line-by-80-character per line display comprises two bytes. The first byte contains the ASCII-coded character and the second byte contains the color and attributes of the character. In this example, AL is the ASCII-coded

TABLE 6–11 Forms of the STOS instruction

Symbolic	*Function*
STOSB	ES:[DI] = AL; DI = DI ± 1
STOSW	ES:[DI] = AX; DI = DI ± 2
STOSD	ES:[DI] = EAX; DI = DI ± 4
STOS LIST	ES:[DI] = AL; DI = DI ± 1 (if LIST is a byte)
STOS DATA1	ES:[DI] = AX; DI = DI ± 2 (if DATA1 is a word)
STOS DATA4	ES:[DI] = EAX; DI = DI ± 4 (if DATA4 is a doubleword)

Note: Doublewords are only used by the 80386, 80486, or Pentium microprocessor.

space (20H) and AH is the color—white text on a black background (07H). Notice how this program uses a count of 25*80 and the REP STOSW instruction to clear the screen with ASCII spaces.

The operands in a program can be modified by using arithmetic or logic operators such as multiplication (*). Other operators appear in Table 6–12.

EXAMPLE 6–5

```
                    .MODEL TINY          ;select TINY model
0000                .CODE                ;indicate start of CODE segment
                    .STARTUP             ;indicate start of program
0100  FC            CLD                  ;select increment mode
0101  B8 B800       MOV   AX,0B800H      ;address segment B800
0104  8E C0         MOV   ES,AX

0106  BF 0000       MOV   DI,0           ;address offset 0000
0109  B9 07D0       MOV   CX,25*80       ;load count
010C  B8 0720       MOV   AX,0720H       ;load data

010F  F3/ AB        REP   STOSW          ;clear the screen
                    .EXIT                ;exit to DOS
                    END                  ;end of file
```

TABLE 6–12 Common operand operators

Operator	*Example*	*Comment*
+	MOV AL,6+3	Copies a 9 into AL
–	MOV AL,8-2	Copies a 6 into AL
*	MOV CX,4*3	Copies a 12 into CX
/	MOV AX,12/5	Copies a 2 into AX (remainder is lost)
MOD	MOV AX,12 MOD 7	Copies a 5 into AX (quotient is lost)
AND	MOV AX,12 AND 4	Copies a 4 into AX (1100 AND 0100 = 0100)
OR	MOV AX,12 OR 1	Copies a 13 into AX (1100 OR 0001 = 1101)
NOT	MOV AL,NOT1	Copies 254 into AL (0000 0001 NOT is equal to 1111 1110 or 254)

The REP prefix precedes the STOSW instruction in both assembly language and hexadecimal machine language. In machine language, the F3H is the REP prefix and ABH is the STOSW opcode.

If the value loaded to AX is changed to 0731H, the video display fills with white 1's on a black background. If AX is changed to 0132H, the video display fills with blue 2's on a black background. By changing the value loaded to AX, the display can be filled with any character and any color combination. More information appears in a later chapter on accessing the video display.

MOVS

One of the more useful string data transfer instructions is MOVS, because it transfers data from one memory location to another. This is the only *memory-to-memory transfer* allowed in the 8086–80486 and Pentium microprocessors. The MOVS instruction transfers a byte, word, or doubleword from the data segment location addressed by SI to the extra segment location addressed by DI. The pointers then increment or decrement as dictated by the direction flag. Table 6–13 lists all permissible forms of the MOVS instruction. Note that only the source operand (SI), located in the data segment, may be overridden so another segment may be used. The destination operand (DI) must always be located in the extra segment.

Suppose that the video display needs to be scrolled up one line. Because we now know the location of the video display, a repeated MOVSW instruction can be used to scroll the video display up a line. Example 6–6 lists a short program that addresses the video text display beginning at location B800:0000 with the DS:SI register combination and at location B800:00A0 with the ES:DI register combination. Next, the REP MOVSW instruction is executed 24*80 times to scroll the display up a line. This is followed by a sequence that addresses the last line of the display so it can be cleared. The last line is cleared in this example by storing spaces on a black background. The last line could also be cleared by changing the ASCII code to a space without modifying the attribute, by reading

TABLE 6–13 Forms of the MOVS instruction

Symbolic	*Function*
MOVSB	ES:[DI] = DS:[SI]; DI = DI ± 1; SI = SI ± 1 (byte transferred)
MOVSW	ES:[DI] = DS:[SI]; DI = DI ± 2; SI = SI ± 2 (word transferred)
MOVSD	ES:[DI] = DS:[SI]; DI = DI ± 4; SI = SI ± 4 (doubleword transferred)
MOVS BYTE1,BYTE2	ES:[DI] = DS:[SI]; DI = DI ± 1; SI = SI ± 1 (if BYTE1 and BYTE2 are bytes)
MOVS WORD1,WORD2	ES:[DI] = DS:[SI]; DI = DI ± 2; SI = SI ± 2 (if WORD1 and WORD2 are words)
MOVS DWORD1,DWORD2	ES:[DI] = DS:[SI]; DI = DI ± 4; SI = SI ± 4 (if DWORD1 and DWORD2 are doublewords)

the code and attribute into a register. Once in a register, the code is modified, and both the code and attribute are stored in memory.

EXAMPLE 6–6

```
                    .MODEL TINY              ;select TINY model
0000                .CODE                    ;indicate start of CODE segment
                    .STARTUP                 ;indicate start of program
0100  FC            CLD                      ;select increment
0101  B8 B800       MOV    AX,0B800H         ;load ES and DS with B800
0104  8E C0         MOV    ES,AX
0106  8E D8         MOV    DS,AX

0108  BE 00A0       MOV    SI,160            ;address line 1
010B  BF 0000       MOV    DI,0              ;address line 0
010E  B9 0780       MOV    CX,24*80          ;load count
0111  F3/ A5        REP    MOVSW             ;scroll screen

0113  BF 0F00       MOV    DI,24*80*2        ;clear bottom line
0116  B9 0050       MOV    CX,80
0119  B8 0720       MOV    AX,0720H
011C  F3/ AB        REP    STOSW
                    .EXIT                    ;exit to DOS
                    END                      ;end of file
```

INS

The INS (*input string*) instruction (not available on the 8086/8088 microprocessor) transfers a byte, word, or doubleword of data from an I/O device into the extra segment memory location addressed by the DI register. The *I/O address* is contained in the DX register. This instruction is useful for inputting a block of data from an external I/O device directly into the memory. One application transfers data from a disk drive to memory. Disk drives are often considered to be I/O devices in a computer system and are interfaces as such.

Like other string instructions, the INS instruction has three basic forms. The INSB instruction inputs data from an 8-bit I/O device and stores it in the byte-sized memory location indexed by SI. The INSW instruction inputs data from a 16-bit I/O device and stores it in a word-sized memory location. The INSD instruction inputs a doubleword. These instructions

TABLE 6–14 Forms of the INS instruction

Symbolic	*Function*
INSB	ES:[DI] = [DX]; DI = DI ± 1 (byte transferred)
INSW	ES:[DI] = [DX]; DI = DI ± 2 (word transferred)
INSD	ES:[DI] = [DX]; DI = DI ± 4 (doubleword transferred)
INS LIST	ES:[DI] = [DX]; DI = DI ± 1 (if LIST is a byte)
INS DATA1	ES:[DI] = [DX]; DI = DI ± 2 (if DATA1 is a word)
INS DATA4	ES:[DI] = [DX]; DI = DI ± 4 (if DATA4 is a doubleword)

Note: [DX] indicates that DX contains the I/O device address. These instructions are not available on the 8086/8088 microprocessor. Only the 80386/80486/Pentium use doublewords.

can be repeated using the REP prefix. This allows an entire block of input data to be stored in the memory from an I/O device. Table 6–14 lists the various forms of the INS instruction.

Example 6–7 shows a sequence of instructions that input 50 bytes of data from an I/O device whose address is 03ACH and stores the data in the extra segment memory array LISTS. This software assumes that data are available from the I/O device at all times. Otherwise, the software must check to see if the I/O device is ready to transfer data precluding the use of a REP prefix.

EXAMPLE 6–7

```
                        ;Using the REP INSB to input data to a memory array
                        ;
0000  BF 0000 R               MOV   DI,OFFSET LISTS     ;address array
0003  BA 03AC                 MOV   DX,3ACH             ;address I/O
0006  FC                      CLD                       ;auto-increment
0007  B9 0032                 MOV   CX,50               ;load count
000A  F3/6C                   REP INSB                  ;input data
```

OUTS

The OUTS (*output string*) instruction (not available on the 8086/8088 microprocessor) transfers a byte, word, or doubleword of data from the data segment memory location addressed by SI to an I/O device. The OUTS instruction addresses the I/O device just as the INS instruction did. Table 6–15 shows the variations available for the OUTS instruction.

Example 6–8 shows a short sequence of instructions that transfer data from a data segment memory array (ARRAY) to an I/O device at I/O address 3ACH. This software assumes that the I/O device is always ready for data.

EXAMPLE 6–8

```
                        ;Using the REP OUTS to output data from a memory array
                        ;
0000  BE 0064 R               MOV   SI,OFFSET ARRAY     ;address array
0003  BA 03AC                 MOV   DX,3ACH             ;address I/O
0006  FC                      CLD                       ;auto-increment
0007  B9 0064                 MOV   CX,100              ;load count
000A  F3/6E                   REP OUTSB
```

TABLE 6–15 Forms of the OUTS instruction

Symbolic	*Function*
OUTSB	[DX] = DS:[SI]; SI = SI ± 1 (byte transferred)
OUTSW	[DX] = DS:[SI]; SI = SI ± 2 (word transferred)
OUTSD	[DX] = DS:[SI]; SI = SI ± 4 (doubleword transferred)
OUTS LIST	[DX] = DS:[SI]; SI = SI ± 1 (if LIST is a byte)
OUTS DATA1	[DX] = DS:[SI]; SI = SI ± 2 (if DATA1 is a word)
OUTS DATA4	[DX] = DS:[SI]; SI = SI ± 4 (if DATA4 is a doubleword)

Note: [DX] indicates that DX contains the I/O device address. These instructions are not available on the 8086/8088 microprocessor. Only the 80386/80486/Pentium use doublewords.

6–5 MISCELLANEOUS DATA TRANSFER INSTRUCTIONS

Don't be fooled by the term *miscellaneous*; these instructions are used in programs. The data transfer instructions detailed in this section are: XCHG, LAHF, SAHF, XLAT, IN, OUT, BSWAP, MOVSX, and MOVZX. Because the miscellaneous instructions are not used as often as MOV instructions, they have been grouped together and represented in this section.

XCHG

The exchange instruction (XCHG) exchanges the contents of a register with the contents of any other register or memory location. The XCHG instruction cannot exchange segment registers or memory-to-memory data. Exchanges are byte-, word-, or doubleword-sized and use any addressing mode discussed in Chapter 5 except immediate addressing. Table 6–16 shows the forms available for the XCHG instruction.

The XCHG instruction, using the 16-bit AX register with another 16-bit register, is the most efficient exchange. This instruction occupies one byte of memory. Other XCHG instructions require two or more bytes of memory, depending on the addressing mode selected.

When using a memory addressing mode and the assembler, it doesn't matter which operand addresses memory. To the assembler, an XCHG AL,[DI] instruction is an XCHG [DI],AL instruction.

If the 80386 through the Pentium microprocessor is available, the XCHG instruction can exchange doubleword data. For example, the XCHG EAX,EBX instruction exchanges the contents of the EAX register with the EBX register.

LAHF and SAHF

The LAHF and SAHF instructions are seldom used. They were designed as *bridge* instructions, allowing 8085 software to be translated into 8086 software by a translation program. Because any software that required translation was probably completed many years ago, these instructions have little application today. The LAHF instruction transfers the rightmost 8 bits of the flag register into the AH register. The SAHF instruction transfers the AH register into the rightmost 8 bits of the flag register.

At times, the SAHF instruction may find some application with the numeric coprocessor. The numeric coprocessor contains a status register that is copied into the AX

TABLE 6–16 Forms of the XCHG instruction

Symbolic	*Note*
XCHG reg,reg	Exchanges byte, word, or doubleword registers
XCHG reg,mem	Exchanges byte, word, or doubleword memory data with register data

Note: Only the 80386/80486/Pentium use doubleword data.

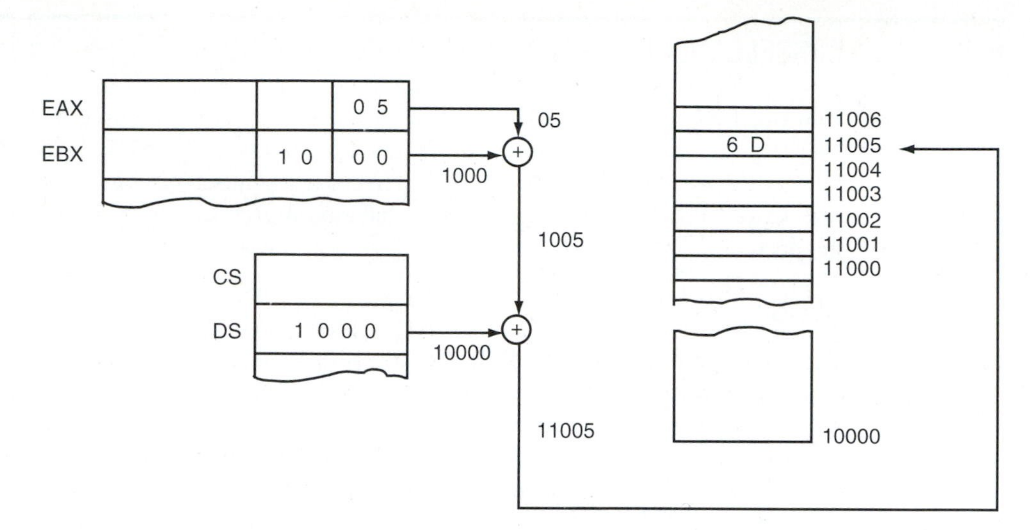

FIGURE 6–17 The operation of the XLAT instruction at the point just before 6DH is loaded into AL.

register with the FSTSW instruction. The SAHF instruction is then used to copy from AH into the flag register. The flags are then tested for some of the conditions of the numeric coprocessor. This is detailed in Chapter 13, which explains the operation and programming of the numeric coprocessor.

XLAT

The XLAT (*translate*) instruction converts the contents of the AL register into a number stored in a memory table. This instruction performs the direct table lookup technique often used to convert one code to another. An XLAT instruction first adds the contents of AL to BX to form a memory address within the data segment. It then copies the contents of this address into AL. This is the only instruction that adds an 8-bit number to a 16-bit number.

Suppose that a 7-segment LED display lookup table is stored in memory at address TABLE. The XLAT instruction then translates the BCD number in AL to a 7-segment code in AL. Example 6–9 provides a short program that converts from a BCD code to a 7-segment code. Figure 6–17 shows the operation of this example program if TABLE = 1000H, DS = 1000H, and the initial value of AL = 05H (a 5 BCD). After the translation, AL = 6DH.

EXAMPLE 6–9

```
                          ;Using an XLAT to convert from BCD to 7-segment code
                          ;
                                 .MODEL SMALL                  ;select SMALL model
0000                             .DATA                         ;indicate start of DATA segment
0000  3F 06 5B 4F         TABLE  DB     3FH,6,5BH,4FH          ;7-segment code lookup table
0004  66 6D 7D 27                DB     66H,6DH,7DH,27H
0008  7F 6F                      DB     7FH,6FH
000A  00                  CODE7  DB     ?                      ;reserve place for result
```

```
0000                  .CODE                      ;indicate start of CODE segment
                      .STARTUP                   ;indicate start of program
0017  B0 04           MOV   AL,4                 ;load test data
0019  BB 0000 R       MOV   BX,OFFSET TABLE      ;address lookup table
001C  D7              XLAT                       ;convert AL to 7-segment code
001D  A2 000A R       MOV   CODE7,AL             ;save 7-segment code
                      .EXIT                      ;exit to DOS
                      END                        ;end of file
```

IN and OUT

Table 6–17 lists the forms of the IN and OUT instructions, which perform I/O operations. Notice that only the contents of AL, AX, or EAX are transferred between the I/O device and the microprocessor. An IN instruction transfers data from an external I/O device to AL, AX, or EAX, and an OUT instruction transfers data from AL, AX, or EAX to an external I/O device. (Only the 80386 and above contain EAX.)

Two forms of I/O device (*port*) address exist for IN and OUT: fixed-port and variable-port. Fixed-port addressing allows data transfer between AL, AX, or EAX using an 8-bit I/O port address. It is called **fixed-port addressing** because the port number follows the instruction's opcode. Often instructions are stored in a ROM. A fixed-port instruction stored in a ROM has its port number permanently fixed because of the nature of read-only memory. If the fixed-port address is stored in a RAM it is possible to modify it, but such a modification does not conform to good programming practices.

The port address appears on the address bus during an I/O operation. For the 8-bit fixed-port I/O instructions, the 8-bit port address is zero-extended into a 16-bit address. For example, if the IN AL,6AH instruction executes, data from I/O address 6AH is input to AL. The address appears as a 16-bit 006AH on pins A0–A15 of the address bus. Address bus bits A16–A19 (8086/8088), A16–A23 (80286/80386SX), A16–A24 (80386SL/80386SLC/80386EX), or A16–A32 (80386/80486/Pentium) are undefined for an IN or OUT instruction. Note that Intel reserves the last 16 I/O ports for use with some of its peripheral components.

TABLE 6–17 IN and OUT instructions

Symbolic	*Function*
IN AL,p8	8-bit data are input to AL from port p8
IN AX,p8	16-bit data are input to AX from port p8
IN EAX,p8	32-bit data are input to EAX from port p8
IN AL,DX	8-bit data are input to AL from port DX
IN AX,DX	16-bit data are input to AX from port DX
IN EAX,DX	32-bit data are input to EAX from port DX
OUT p8,AL	8-bit data are sent to port p8 from AL
OUT p8,AX	16-bit data are sent to port p8 from AX
OUT p8,EAX	32-bit data are sent to port p8 from EAX
OUT DX,AL	8-bit data are sent to port DX from AL
OUT DX,AX	16-bit data are sent to port DX from AX
OUT DX,EAX	32-bit data are sent to port DX from EAX

Note: p8 = an 8-bit I/O port number. DX = the 16-bit port address held in DX.

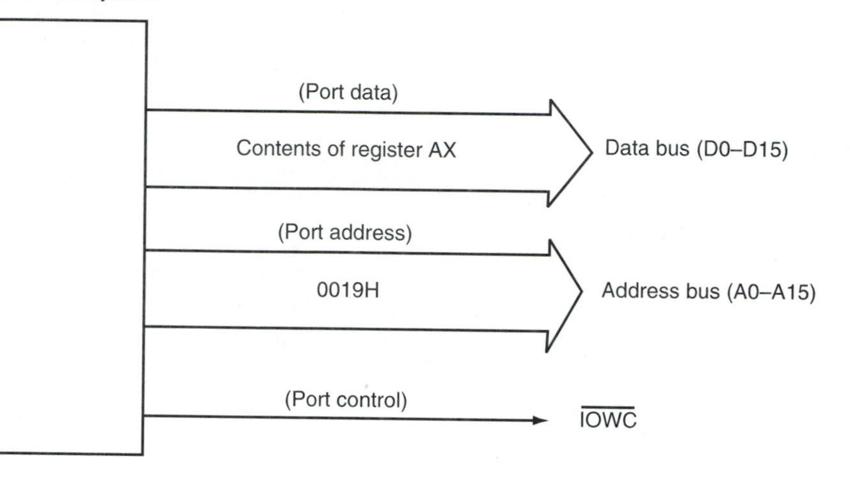

FIGURE 6–18 The signals found in the microprocessor-based system for an OUT 19H,AX instruction.

Variable-port addressing allows data transfers between AL, AX, or EAX and a 16-bit port address. It is called **variable-port addressing** because the I/O port number is stored in register DX, which can be changed (*varied*) during the execution of a program. The 16-bit I/O port address appears on the address bus pin connections A0–A15. The IBM PC uses a 16-bit port address to access its I/O space. The I/O space for a PC is located at I/O port 0000H–03FFH. Note that some plug-in adapter cards may use I/O addresses above 03FFH.

Figure 6–18 illustrates the execution of the OUT 19H,AX instruction, which transfers the contents of AX to I/O port 19H. Notice that the I/O port number appears as a 0019H on the 16-bit address bus and that the data from AX appears on the data bus of the microprocessor. The system control signal, $\overline{\text{IOWC}}$ (I/O write control) is a logic zero to enable the I/O device.

A short program that clicks the speaker in the personal computer appears in Example 6–10. The speaker is controlled by accessing I/O port 61H. If the rightmost two bits of this port are set (11) and then cleared (00), a click is heard on the speaker. Note that this program uses a logical OR instruction to set these two bits and a logical AND instruction to clear them. These logic operation instructions are described in the next chapter. The MOV CX,1000H instruction followed by the LOOP L1 instruction is used as a time delay. If the count is increased, the click will become longer; if the count is decreased, the click will become shorter.

EXAMPLE 6–10

```
                    .MODEL TINY          ;select TINY model
0000                .CODE                ;indicate start of code segment
                    .STARTUP             ;indicate start of program
0100   E4 61        IN      AL,61H       ;read port 61H
0102   0C 03        OR      AL,3         ;set rightmost two bits
0104   E6 61        OUT     61H,AL       ;speaker is on
```

```
0106  B9 1000            MOV    CX,1000H      ;delay count
0109             L1:
0109  E2 FE              LOOP   L1            ;time delay

010B  E4 61              IN     AL,61H        ;read port 61H
010D  24 FC              AND    AL,0FCH       ;clear rightmost two bits
010F  E6 61              OUT    61H,AL        ;speaker is off
                         .EXIT                ;exit to DOS
                         END                  ;end of file
```

MOVSX and MOVZX

The MOVSX (move and sign-extend) and MOVZX (move and zero-extend) instructions are found in the 80386, 80486, and Pentium instruction sets. These instructions move data and at the same time either sign- or zero-extend it. Table 6–18 illustrates these instructions with several examples of each.

When a number is zero-extended, the most significant part fills with zeros. For example, if an 8-bit 34H is zero-extended into a 16-bit number, it becomes 0034H. Zero-extension is often used to convert unsigned 8- or 16-bit numbers into unsigned 16- or 32-bit numbers using the MOVZX instruction.

A number is sign-extended when its sign-bit is copied into the most significant part. For example, if an 8-bit 84H is sign-extended into a 16-bit number, it becomes FF84H. The sign-bit of an 84H is a one, which is copied into the most significant part of the sign-extended result. Sign-extension is most often used to convert 8- or 16-bit signed numbers into 16- or 32-bit signed numbers using the MOVSX instruction.

TABLE 6–18 The MOVSX and MOVZX instructions

Symbolic	*Example*	*Note*
MOVSX reg,reg	MOVSX CX,BL	Converts the 8-bit contents of BL into a 16-bit number in CX by sign-extension
	MOVSX ECX,AX	Converts the 16-bit contents of AX into a 32-bit number in ECX by sign-extension
MOVSX reg,mem	MOVSX BX,DATA	Converts the 8-bit contents of DATA into a 16-bit number in BX by sign-extension
	MOVSX EAX,[EDI]	Converts the 16-bit contents of the data segment memory location addressed by EDI into a 32-bit number in EAX by sign-extension
MOVZX reg,reg	MOVZX DX,AL	Converts the 8-bit contents of AL into a 16-bit number in DX by zero-extension
	MOVZX EBP,DI	Converts the 16-bit contents of DI into a 32-bit number in EBP by zero-extension
MOVZX reg,mem	MOVZX DX,DATA1	Converts the 8-bit contents of DATA1 into a 16-bit number in DX by zero-extension
MOVZX reg,mem	MOVZX EAX,DATA2	Converts the 16-bit contents of DATA2 into a 32-bit number in EAX by zero-extension

BSWAP

The byte swap instruction (BSWAP) is available only in the 80486 and Pentium microprocessor. This instruction takes the contents of any 32-bit register and swaps the first byte with the fourth and the second byte with the third. For example, the BSWAP EAX instruction with EAX = 00112233H swaps bytes in EAX so it results in EAX = 33221100H. Notice that the order of all four bytes is reversed by this instruction. This instruction is used to convert data from big endian form to little endian form or vice versa.

6–6 SEGMENT OVERRIDE PREFIX

The segment override prefix, which may be added to almost any instruction in any memory addressing mode, allows the programmer to deviate from the default segment. The segment override prefix is an additional byte that appends the front of an instruction to select an alternate segment register. About the only instructions that cannot be prefixed are the jump and call instructions, which must use the code segment register for address generation. The segment override is also used to select the FS and GS segments in the 80386 through the Pentium microprocessor.

For example, the MOV AX,[DI] instruction accesses data within the data segment by default. If required by a program, this can be changed by prefixing the instruction. Suppose that the data are in the extra segment instead of the data segment. This instruction addresses the extra segment if changed to MOV AX,ES:[DI].

Table 6–19 shows some altered instructions that address different memory segments than normal. Each time an instruction is prefixed with a segment override prefix, the instruction becomes one byte longer. Although this is no serious change to the length of the instruction, it does add to its execution time. It is customary to limit the use of the segment override prefixes in order to write shorter and more efficient software.

TABLE 6–19 Instructions that include segment override prefixes

Symbolic	*Segment Accessed*	*Normal Segment*
MOV AX,DS:[BP]	Data segment	Stack segment
MOV AX,ES:[BP]	Extra segment	Stack segment
MOV AX,SS:[DI]	Stack segment	Data segment
MOV AX,CS:[SI]	Code segment	Data segment
MOV AX,ES:LIST	Extra segment	Data segment
LODS ES:DATA	Data segment	Extra segment
MOV EAX,FS:DATA2	FS segment	Data segment
MOV BL,GS:[ECX]	GS segment	Data segment

Note: Only the 80386, 80486, and Pentium allow the use of the FS and GS segments.

6–7 ASSEMBLER DETAIL

The assembler[1] for the microprocessor can be used in two ways: (1) with models that are unique to a particular assembler (used in most examples in this text) and (2) with full-segment definitions that allow complete control over the assembly process and are universal to all assemblers. This section of the text presents both methods, and explains how to organize a program's memory space using the assembler. It also explains the purpose and use of some of the more important directives used with this assembler. Appendix A provides additional details about the assembler.

Directives

Before the format of an assembly language program is discussed, some details about the directives (*pseudo-operations*) that control the assembly process must be learned. Some common assembly language directives appear in Table 6–20. Directives indicate how an operand or section of a program is to be processed by the assembler. Some directives generate and store information in the memory; others do not. The DB (define byte) directive stores bytes of data in the memory, while the BYTE PTR directive never stores data. The BYTE PTR directive indicates the size of the data referenced by a pointer or index register.

Note that, by default, the assembler accepts only 8086/8088 instructions, unless a program is preceded by the .386 or .386P directive or one of the other microprocessor selection switches. The .386 directive tells the assembler to use the 80386 instruction set in the real mode, while the .386P directive tells the assembler to use the 80386 protected-mode instruction set.

Storing Data in a Memory Segment. The DB (define byte), DW (define word), and DD (define doubleword) directives, first presented in Chapter 1, are most often used with the microprocessor to define and store memory data. If a numeric coprocessor is present in the system, the DQ (define quadword) and DT (define ten bytes) directives are also common. These directives identify a memory location and indicate the size of the location.

Example 6–11 shows a memory segment that contains various forms of data definition directives. It also shows the full segment definition, with the first SEGMENT statement to indicate the start of the segment and its symbolic name. Alternatively, the SMALL model can be used with the .DATA statement. The last statement in this example contains the ENDS directive, which indicates the end of the segment. The name of the segment (LIST_SEG), can be anything the programmer desires. This allows a program to contain as many segments as required.

EXAMPLE 6–11

```
                            ;Using the DB, DW, and DD directives
                            ;
0000                        LIST_SEG   SEGMENT

0000  01 02 03              DATA1    DB      1,2,3              ;define bytes
```

[1]The assembler used in this text is the Microsoft macro assembler, called MASM.

TABLE 6–20 Common assembler directives

Directive	*Function*
.286	Selects the 80286 instruction set
.286P	Selects the protected-mode 80286 instruction set
.386	Selects the 80386 instruction set
.386P	Selects the protected-mode 80386 instruction set
.486	Selects the 80486 instruction set
.486P	Selects the 80486 protected-mode instruction set
.586	Selects the Pentium instruction set*
.586P	Selects the Pentium protected-mode instruction set*
.287	Selects the 80287 numeric coprocessor
.387	Selects the 80387 numeric coprocessor
.EXIT	Used with programming models to exit to DOS
.MODEL	Selects the programming model
.STARTUP	Used with programming models to indicate the start of the program
ALIGN 2	Starts the data in a segment at word or doubleword boundaries
ASSUME	Indicates the names of each segment to the assembler. Does not load the segment registers
BYTE	Indicates a byte-sized operand, as in BYTE PTR or THIS BYTE
DB	Defines byte(s) (8 bits)
DD	Defines doubleword(s) (32 bits)
DQ	Defines quadword(s) (64 bits)
DT	Defines ten bytes (80 bits)
DUP	Generates duplicates of characters or numbers
DW	Defines word(s) (16 bits)
DWORD	Indicates a doubleword-sized operand, as in THIS DWORD

```
0003  45                         DB     45H             ;hexadecimal
0004  41                         DB     'A'             ;ASCII
0005  F0                         DB     11110000B       ;binary
0006  000C 000D         DATA2    DW     12,13           ;define words
000A  0200                       DW     LIST1           ;symbolic
000C  2345                       DW     2345H           ;hexadecimal
000E  00000300          DATA3    DD     300H            ;hexadecimal
0012  4007DF3B                   DD     2.123           ;real
0016  544269E1                   DD     3.34E+12        ;real
001A  00                LISTA    DB     ?               ;reserve 1 byte
001B  000A[             LISTB    DB     10 DUP (?)      ;reserve 10 bytes
              ??
                    ]
0025   00                        ALIGN  2               ;set word boundary

0026   0100[            LISTC    DW     100H DUP (0)    ;word array
```

TABLE 6–20 *(continued)*

Directive	*Function*
END	Indicates the end of the program
ENDM	Indicates the end of a macro sequence
ENDP	Indicates the end of a procedure
ENDS	Indicates the end of a segment or structure
EQU	Equates data to a label
FAR	Specifies a far address as in JMP FAR PTR LISTS
MACRO	Defines the name, parameters, and start of a macro
NEAR	Specifies a near address as in JMP NEAR PTR HELP
OFFSET	Specifies an offset address
ORG	Sets the origin within a segment
PROC	Defines the beginning of a procedure
PTR	Indicates a memory pointer
SEGMENT	Defines the start of a memory segment
STACK	Indicates that a segment is a stack segment
STRUC	Defines the start of a data structure
THIS	Used with EQU to set a label to a byte, word, or doubleword
USES	An MASM version 6.X directive that automatically saves registers used by a procedure
USE16	Directs the assembler to use the 16-bit instruction mode and data sizes for the 80386, 80486, and Pentium microprocessors
USE32	Directs the assembler to use the 32-bit instruction mode and data sizes for the 80386, 80486, and Pentium microprocessors
WORD	Acts as a word operand as in WORD PTR or THIS WORD

Note: *For version 6.11 of MASM. Most of these directives function with most versions of the assembler.

```
                      0000
                                ]
0226    0016[              LIST_9   DD     22 DUP (?)        ;doubleword array
                      ????????
                                      ]
027E    0064[              SIXES    DB     100 DUP (6)       ;byte array
                      06
                           ]
02E2                       LIST_SEG ENDS
```

This example shows various forms of data storage for bytes at DATA1. More than one byte can be defined on a line in binary, hexadecimal, decimal, or ASCII code. The DATA2 label shows how to store various forms of word data. Doublewords are stored at DATA3. They include floating-point single-precision real numbers.

Memory can be reserved for use in the future by using a ? as an operand for a DB, DW, or DD directive. When a ? is used in place of a numeric or ASCII value, the assembler sets aside a location and does not initialize it to any specific value. (Actually, the assembler stores a zero into locations specified with a ?). The **DUP** (*duplicate*) directive creates an array, as shown in several ways in Example 6–11. A 10 DUP (?) reserves 10 locations of memory, but stores no specific value in any of the 10 locations. If a number appears within the ( ) part of the DUP statement, the assembler initializes the reserved section of memory with the data indicated. For example, the DATA1 DB 10 DUP (2) instruction reserves 10 bytes of memory for array DATA1 and initializes each location with a 02H.

The ALIGN directive, used in this example, makes sure that the memory arrays are stored on word boundaries. An ALIGN 2 places data on word boundaries and an ALIGN 4 places them on doubleword boundaries. In the Pentium, quadword data for a double-precision floating-point number should use ALIGN 8. It is important that word-sized data is placed at word boundaries and doubleword-sized data is placed at doubleword boundaries. If not, the microprocessor spends additional time accessing these data types. A word stored at an odd-numbered memory location takes twice as long to access as a word stored at an even-numbered memory location. Note that the ALIGN directive cannot be used with memory models, because the size of the model determines the data alignment.

EQU and THIS. The **EQU** (*equate*) directive equates a numeric value, ASCII character, or a label to another label. Equates make a program clearer and simplify debugging. Example 6–12 shows several equate statements and a few instructions that show how they function in a program.

EXAMPLE 6–12

```
                        ;Using equate directive
                        ;
= 000A                  TEN     EQU   10
= 0009                  NINE    EQU   9

0000   B0 0A                    MOV   AL,TEN
0002   04 09                    ADD   AL,NINE
```

The THIS directive always appears as THIS BYTE, THIS WORD, or THIS DWORD. In certain cases, datum must be referred to as both a byte and a word. The assembler can only assign either a byte address or a word address to a label. To assign a byte label to a word, use the software listed in Example 6–13.

EXAMPLE 6–13

```
                        ;Using the THIS and ORG directives
                        ;
0000                    DATA_SEG      SEGMENT

0100                           ORG    100H

= 0100                  DATA1   EQU   THIS BYTE
0100  0000              DATA2   DW    ?

0102                    DATA_SEG      ENDS
```

```
0000                    CODE_SEG     SEGMENT 'CODE'

                                 ASSUME CS:CODE_SEG,DS:DATA_SEG

0000  8A 1E 0100 R               MOV    BL,DATA1
0004  A1 0100 R                  MOV    AX,DATA2
0007  8A 3E 0101 R               MOV    BH,DATA1+1

000B                    CODE_SEG   ENDS
```

This example illustrates how the ORG (origin) statement changes the starting offset address of the data in the data segment to location 100H. At times, the origin of data or the code must be assigned to an absolute offset address with the ORG statement. The ASSUME statement tells the assembler what names have been chosen for the code, data, extra, and stack segments. Without the assume statement, the assembler assumes *nothing* and automatically uses a segment override prefix on all instructions that address memory data. The ASSUME statement is only used with full-segment definitions, as described later in this section of the text.

PROC and ENDP. The PROC and ENDP directives indicate the start and end of a **procedure** (subroutine). These directives force structure, because the procedure is clearly defined. If structure is to be violated, use the CALLF, CALLN, RETF, and RETN instructions. Both the PROC and ENDP directives require a label to indicate the name of the procedure. The PROC directive, which indicates the start of a procedure, must be followed with either NEAR or FAR. A NEAR procedure is one that resides in the same code segment as the program. A FAR procedure may reside at any location in the memory system. A call NEAR procedure is considered local, and a call FAR procedure is considered global. The term *global* denotes a procedure that can be used by any program, while *local* defines a procedure that is only used by the current program. Any labels that are defined within the procedure block are defined as either local (NEAR) or global (FAR).

EXAMPLE 6–14

```
             ;A procedure that adds BX, CX, and DX with the sum
             ;stored in AX
             ;
0000         ADDEM    PROC    FAR               ;start procedure

0000  03 D9           ADD     BX,CX
0002  03 DA           ADD     BX,DX
0004  8B C3           MOV     AX,BX
0006  CB              RET

0007         ADDEM    ENDP                      ;end procedure
```

Example 6–14 shows a procedure that adds BX, CX, and DX, and stores the sum in register AX. Although this procedure is short, and may not be particularly useful, it does illustrate how to use the PROC and ENDP directives to delineate the procedure. Note that information about the operation of the procedure should appear as a grouping of comments that show the registers changed by the procedure and the result of the procedure.

If the Microsoft MASM assembler program version 6.X is available, the PROC directive specifies and automatically saves any registers used within the procedure. The *USES* statement indicates which registers are used by the procedure, so the assembler can automatically save them before the procedure begins and restore them before the procedure ends with the RET instruction. For example, the ADDS PROC USES AX BX CX statement automatically pushes AX, BX, and CX on the stack before the procedure begins and pops them from the stack before the RET instruction executes at the end of the procedure. Example 6–15 illustrates a procedure written using MASM 6.X that shows the USES statement. The registers in the list are separated not by commas, but spaces. The PUSH and POP instructions are displayed in the procedure listing because it was assembled with the .LISTALL directive. The instructions prefaced with an asterisk are inserted by the assembler. The USES statement does not appear elsewhere in this text, so compatibility with MASM version 5.10 is maintained. If version 6.X is available, it is suggested that the USES statement be included with most procedures.

EXAMPLE 6–15

```
                    ;A procedure that includes the USES directive to save
                    ;BX, CX, and DX on the stack and restore them before
                    ;the RET instruction.
                    ;
0000                ADDS    PROC    NEAR USES BX CX DX

0000    53          *       push    bx
0001    51          *       push    cx
0002    52          *       push    dx
0003    03 D8               ADD     BX,AX
0005    03 CB               ADD     CX,BX
0007    03 D1               ADD     DX,CX
0009    8B C2               MOV     AX,DX
                            RET
000B    5A          *       pop     dx
000C    59          *       pop     cx
000D    5B          *       pop     bx
000E    C3          *       ret     00000h

000F                ADDS    ENDP
```

Memory Organization

The assembler uses two basic formats for developing software. One method uses models and the other uses full-segment definitions. Memory models, as presented in this section and also briefly in Chapter 5, are unique to the MASM assembler program. The TASM assembler uses memory models, but they differ somewhat from the MASM models. Full-segment definitions are common to most assemblers, including the Intel assembler, and are often used for software development. The models are easier to use for simple tasks. The full-segment definitions offer better control over the assembly language task and are recommended for complex programs. The model was used in early chapters because it is easier to understand. Models are also used with assembly language procedures that are used by high-level languages, such as C/C++. This text fully develops and uses the memory model definitions for its programming examples, but realize that full-segment definitions offer some advantages over memory models, as discussed later in this section.

Models. There are many models available to the MASM assembler, from tiny to huge. Appendix A contains a table that lists all the models available for use with the assembler. To designate a model, use the .MODEL statement followed by the size of the memory system. The tiny model, useful for many small programs, requires that all software and data fit into one 64K byte memory segment. The small model requires that only one data segment with one code segment is used, for a total of 128K bytes of memory. Other models are available up to the huge model.

Example 6–16 illustrates how the .MODEL statement defines the parameters of a short program that copies the contents of a 100-byte block of memory (LISTA) into a second 100-byte block of memory (LISTB). It also shows how to define the stack, data, and code segments. The .EXIT 0 directive returns to DOS with an error code of 0 (no error). If no parameter is added to .EXIT, it still returns to DOS, but the error code is not defined. Also note that special directives such as @DATA (see Appendix A) are used to identify various segments. If the .STARTUP directive is used (MASM version 6.X), the MOV AX,@DATA followed by MOV DS,AX statements can be eliminated. Models are important with both Microsoft C/C++ and Borland C/C++ development systems if assembly language is included with C/C++ programs. Both development systems use in-line assembly programming for adding assembly language instructions and require an understanding of programming models. Refer to the respective C/C++ language reference for each system to determine the model protocols.

EXAMPLE 6–16

```
                                .MODEL SMALL
                                .STACK 100H            ;define stack
                                .DATA                  ;define data segment

0000  0064[          LISTA       DB     100 DUP (?)
              ??
                         ]
0064  0064[          LISTB       DB     100 DUP (?)
              ??
                         ]

                                .CODE                  ;define code segment

0000  B8 ---- R      HERE:      MOV    AX,@DATA        ;load ES, DS
0003  8E C0                     MOV    ES,AX
0005  8E D8                     MOV    DS,AX

0007  FC                        CLD                    ;move data
0008  BE 0000 R                 MOV    SI,OFFSET LISTA
000B  BF 0064 R                 MOV    DI,OFFSET LISTB
000E  B9 0064                   MOV    CX,100
0011  F3/A4                     REP    MOVSB

0013                            .EXIT  0               ;exit to DOS

                                END    HERE
```

Full-Segment Definitions. Example 6–17 illustrates the same program using full-segment definitions. Full-segment definitions are also used with the Borland and Microsoft C/C++ environments for procedures developed in assembly language. The program in Example 6–17 appears longer than the one pictured in 6–16, but more structured than the model method of setting up a program.

EXAMPLE 6–17

```
0000                      STACK_SEG   SEGMENT STACK

0000  0100[                       DW    100H DUP (?)
                  ????
                      ]

0200                      STACK_SEG   ENDS

0000                      DATA_SEG   SEGMENT 'DATA'

0000  0064[               LISTA   DB    100 DUP (?)
                  ??
                          ]
0064  0064[               LISTB   DB    100 DUP (?)
                  ??
                          ]

00C8                      DATA_SEG   ENDS

0000                      CODE_SEG   SEGMENT 'CODE'

                                  ASSUME CS:CODE_SEG,DS:DATA_SEG
                                  ASSUME SS:STACK_SEG

0000                      MAIN    PROC   FAR

0000  B8 ---- R                   MOV    AX,DATA_SEG        ;load DS and ES
0003  8E C0                       MOV    ES,AX
0005  8E D8                       MOV    DS,AX

0007  FC                          CLD                       ;move data
0008  BE 0000 R                   MOV    SI,OFFSET LISTA
000B  BF 0064 R                   MOV    DI,OFFSET LISTB
000E  B9 0064                     MOV    CX,100
0011  F3/A4                       REP MOVSB

013   B4 4C                       MOV    AH,4CH             ;exit to DOS
0015  CD 21                       INT    21H

0017                      MAIN    ENDP

0017                      CODE_SEG   ENDS

                                  END   MAIN
```

The first segment defined is the STACK_SEG, which is clearly delineated with the SEGMENT and ENDS directives. Within these directives, a DW 100 DUP (?) sets aside 100H words for the stack segment. Because the word STACK appears next to SEGMENT, the assembler and linker automatically load both the stack segment register (SS) and stack pointer (SP).

Next the data are defined in the DATA_SEG. Here two arrays of data appear as LISTA and LISTB. Each array contains 100 bytes of space for the program. The names of the segments in this program can be changed to any name. Always include the group name 'DATA' so the Microsoft program CodeView can be effectively used to symbolically debug this software. CodeView is part of the MASM package. To access CodeView, type CV followed by the file name at the DOS command line, or if operating from Programmer's WorkBench, select Debug under the Run menu. If the group name is not placed in a program, CodeView can still be used to debug a program, but the program will not be debugged in symbolic form. Other group names such as 'STACK', 'CODE', and so forth are listed in Appendix A. You must at least place the word 'CODE' next to the code segment SEGMENT statement if you want to view the program symbolically with CodeView.

The CODE_SEG is organized as a far procedure, because most software is procedure oriented. Before the program begins, the code segment contains the ASSUME statement. The ASSUME statement tells the assembler and linker the name used for the code segment (CS) is CODE_SEG. It also tells the assembler and linker that the data segment is DATA_SEG and the stack segment is STACK_SEG. The group name 'CODE' is used for the code segment to run CodeView. Other group names appear in Appendix A with the models.

After the program loads both the extra segment register and data segment register with the location of the data segment, it transfers 100 bytes from LISTA to LISTB. Following this is a sequence of two instructions that return control to DOS. Note that the program loader does not automatically initialize DS and ES. These registers must be loaded with the desired segment addresses in the program.

The last statement in the program is END MAIN. The END statement indicates the end of the program and the location of the first instruction executed. Here we want the machine to execute the main procedure, so a label follows the END directive.

In the 80386 through Pentium, an additional directive is attached to the code segment. The USE16 or USE32 directive tells the assembler to use either the 16- or 32-bit instruction modes for the microprocessor. Software developed for the DOS environment must use the USE16 directive for the 80386 through Pentium program to function correctly, because MASM assumes by default that all segments are 32 bits and all instruction modes are 32 bits. In fact, any program designed to execute in the real mode must include the USE16 directive to deviate from the default 8086/8088. Example 6–18 shows how the same software listed in Example 6–17 is formed for the 80386 microprocessor.

EXAMPLE 6–18

```
                          .386                                  ;select the 80386
0000                      STACK_SEG SEGMENT STACK

0000  0100[                         DW      100H DUP (?)
                  ????
                      ]

0200                      STACK_SEG ENDS

0000                      DATA_SEG SEGMENT 'DATA'

0000  0064[               LISTA     DB      100 DUP (?)
```

```
                ??
                         ]
0064  0064[          LISTB    DB     100 DUP (?)
                ??
                         ]

00C8                 DATA_SEG   ENDS

0000                 CODE_SEG   SEGMENT USE16 'CODE'

                            ASSUME CS:CODE_SEG,DS:DATA_SEG
                            ASSUME SS:STACK_SEG

0000                 MAIN   PROC   FAR

0000  B8 ---- R             MOV    AX,DATA_SEG        ;load DS and ES
0003  8E C0                 MOV    ES,AX
0005  8E D8                 MOV    DS,AX

0007  FC                    CLD                       ;move data
0008  BE 0000 R             MOV    SI,OFFSET LISTA
000B  BF 0064 R             MOV    DI,OFFSET LISTB
000E  B9 0064               MOV    CX,100
0011  F3/A4                 REP MOVSB

0013  B4 4C                 MOV    AH,4CH             ;exit to DOS
0015  CD 21                 INT    21H

0017                        MAIN   ENDP

0017                        CODE_SEG   ENDS
                                   END   MAIN
```

A Sample Program

EXAMPLE 6–19

```
                ;An example program that reads a key and displays it.
                ;Note that an @ key ends the program
                ;
0000            CODE_SEG   SEGMENT 'CODE'

                           ASSUME CS:CODE_SEG

0000            MAIN       PROC   FAR

0000  B4 06                MOV    AH,6            ;read key
0002  B2 FF                MOV    DL,0FFH
0004  CD 21                INT    21H
0006  74 F8                JE     MAIN            ;if no key

0008  3C 40                CMP    AL,'@'          ;test for @
000A  74 08                JE     MAIN1           ;if @

000C  B4 06                MOV    AH,6            ;display key
000E  8A D0                MOV    DL,AL
0010  CD 21                INT    21H
0012  EB EC                JMP    MAIN            ;repeat
0014            MAIN1:
0014  B4 4C                MOV    AH,4CH          ;exit to DOS
0016  CD 21                INT    21H
```

```
0018                 MAIN      ENDP

0018                 CODE_SEG     ENDS

                               END     MAIN
```

Example 6–19 provides a sample program, using full-segment definitions, that reads a character from the keyboard and displays it on the CRT screen. Although this program is trivial, it illustrates a complete, workable program that functions on any personal computer using DOS, from the earliest 8088-based system to the latest Pentium-based system. This program also illustrates the use of a few DOS function calls. Appendix A lists the DOS function calls with their parameters. The BIOS function calls allow the use of the keyboard, printer, disk drives, and everything else available in your computer system.

This example program uses only a code segment, because there is no data. A stack segment should appear, but has been left out because DOS automatically allocates a 128-byte stack for all programs. The only time the stack is used in this example is for the INT 21H instructions, which call a procedure in DOS. When this program is linked, the linker signals that no stack segment is present. This warning may be ignored in this example, because the stack is less than 128 bytes.

The entire program is placed into a far procedure called MAIN. It is good programming practice to write all software in procedural form. This allows the program to be used as a procedure if necessary at some future time. It is also important to document register use and any parameters required for the program in the program header. The program header is a section of comments that appear at the start of the program.

The program uses DOS functions 06H and 4CH. The function number is placed in AH before the INT 21H instructions execute. If DL = 0FFH, 06H function reads the keyboard. The 06H function displays the ASCII contents of DL if DL does not equal 0FFH. The first section of the program moves a 06H into AH and a 0FFH into DL, so a key is read from the keyboard. The INT 21H tests the keyboard and if no key is typed, it returns equal. The JE instruction tests the equal condition and jumps to MAIN if no key is typed.

When a key is typed, the program continues to the next step. This step compares the contents of AL with an @ symbol. Upon return from INT 21H, the ASCII character of the typed key is found in AL. In this program, if an @ symbol is typed, the program ends. If the @ symbol is not typed, the program continues by displaying the character typed on the keyboard with the next INT 21H instruction.

The second INT 21H instruction moves the ASCII character into DL, so it can be displayed on the CRT screen. After displaying the character, a JMP executes. This causes the program to continue at MAIN, where it repeats reading a key.

If the @ symbol is typed, the program continues at MAIN1, where it executes the DOS function code number 4CH. This causes the program to return to the DOS prompt (A>) so the computer can be used for other tasks.

More information about the assembler and its applications appears in Appendix A and in the next several chapters. The Appendix provides a complete overview of the assembler, linker, and DOS functions. It also provides a list of the **BIOS** (basic I/O system) functions. The information provided in the following chapters clarifies how to use the assembler for certain tasks.

6–8 SUMMARY

1. Data movement instructions transfer data between registers, a register and memory, a register and the stack, memory and the stack, the accumulator and I/O, and the flags and the stack. Memory-to-memory transfers are only allowed with the MOVS instruction.
2. Data movement instructions include MOV, PUSH, POP, XCHG, XLAT, IN, OUT, LEA, LDS, LES, LSS, LGS, LFS, LAHF, SAHF, and the string instructions: LODS, STOS, MOVS, INS, and OUTS.
3. The first byte of an instruction contains the opcode. The opcode specifies the operation performed by the microprocessor. The opcode may be preceded by one or more override prefixes in some forms of instructions.
4. The D bit, located in many instructions, selects the direction of data flow. If D = 0, the data flow from the REG field to the R/M field of the instruction. If D = 1, the data flow from the R/M field to the REG field.
5. The W bit, found in most instructions, selects the size of the data transfer. If W = 0, the data are byte-sized; if W = 1, the data are word-sized. In the 80386 and above, W = 1 specifies either a word or doubleword register.
6. MOD selects the addressing mode of operation for a machine language instruction's R/M field. If MOD = 00, there is no displacement. If MOD = 01, an 8-bit sign-extended displacement appears. If MOD = 10, a 16-bit displacement occurs. If MOD = 11, a register is used instead of a memory location. In the 80386 and above, the MOD bits also specify a 32-bit displacement.
7. A 3-bit binary register code specifies the REG and R/M fields when MOD = 11. The 8-bit registers are AH, AL, BH, BL, CH, CL, DH, and DL. The 16-bit registers are AX, BX, CX, DX, SP, BP, DI, and SI. The 32-bit registers are EAX, EBX, ECX, EDX, ESP, EBP, EDI, and ESI.
8. When the R/M field depicts a memory mode, a 3-bit code selects one of the following modes: [BX+DI], [BX+SI], [BP+DI], [BP+SI], [BX], [BP], [DI], or [SI] for 16-bit instructions. In the 80386 and above, the R/M field specifies EAX, EBX, ECX, EDX, EBP, EDI, ESI or one of the scaled-index modes of addressing memory data. If the scaled-index mode is selected (R/M = 100), an additional byte (scaled-index byte) is added to the instruction to specify the base register, index register, and the scaling factor.
9. All memory addressing modes, by default, address data in the data segment, unless BP or EBP addresses memory. The BP or EBP register addresses data in the stack segment.
10. The segment registers are only addressed by the MOV, PUSH, or POP instruction. The MOV instruction may transfer a segment register to a 16-bit register or vice versa. A MOV CS,reg or POP CS instruction is not allowed, because these instructions only change part of the address. In the 80386 through the Pentium, two additional segment registers, FS and GS, exist.
11. Data are transferred between a register or a memory location and the stack by the PUSH and POP instructions. Variations of these instructions allow immediate data to be pushed onto the stack, the flags to be transferred between the stack, and all 16-bit

registers to be transferred between the stack and the registers. When data are transferred to the stack, two bytes (8086–80286) always move, with the least significant byte placed at the SP location –1 byte and the most significant byte placed at the SP location –2 byte. After placing the data on the stack, SP decrements by 2. In the 80386/80486/Pentium, four bytes of data from a memory location or register may also be transferred to the stack.

12. Opcodes that transfer data between the stack and the flags are PUSHF and POPF. Opcodes that transfer all the 16-bit registers between the stack and the registers are PUSHA and POPA. In the 80386 and above, a PUSHFD and POPFD transfers the contents of the EFLAGS between the microprocessor and the stack.
13. LEA, LDS, and LES instructions load a register or registers with an effective address. The LEA instruction loads any 16-bit register with an effective address, while LDS and LES load any 16-bit register and either DS or ES with the effective address. In the 8038 and above, additional instructions include LFS, LGS, and LSS, which load a 16-bit register and FS, GS, or SS.
14. String data transfer instructions use DI and/or SI to address memory. The DI offset address is located in the extra segment, and the SI offset address is located in the data segment.
15. The direction flag (D) chooses the auto-increment or auto-decrement mode of operation for DI and SI for string instructions. To clear D to 0, use the CLD instruction to select the auto-increment mode. To set D to 1, use the STD instruction to select the auto-decrement mode. DI and/or SI increment/decrement by 1 for a byte operation, by 2 for a word operation, and by 4 for a doubleword operation.
16. LODS loads AL, AX, or EAX with data from the memory location addressed by SI. STOS stores AL, AX, or EAX in the memory location addressed by DI. MOVS transfers a byte or a word from the memory location addressed by SI into the location addressed by DI.
17. INS inputs data from an I/O device addressed by DX and stores it in the memory location addressed by DI. OUTS outputs the contents of the memory location addressed by SI and sends it to the I/O device addressed by DX.
18. The REP prefix may be attached to any string instruction to repeat it. The REP prefix repeats the string instruction the number of times found in register CX.
19. Arithmetic and logic operators can be used in assembly language. An example is MOV AX,34*3, which loads AX with 102.
20. Translate (XLAT) converts the data in AL into a number stored at the memory location addressed by BX plus AL.
21. IN and OUT transfer data between AL, AX, or EAX and an external I/O device. The address of the I/O device is either stored with the instruction (fixed port) or in register DX (variable port).
22. The segment override prefix selects a different segment register for a memory location than the default segment. For example, the MOV AX,[BX] instruction uses the data segment, but the MOV AX,ES:[BX] instruction uses the extra segment because of the ES: prefix. The segment override prefix is the only way the FS and GS segments are addressed in the 80386 through the Pentium.
23. The MOVZX (move and zero-extend) and MOVSX (move and sign-extend) instructions found in the 80386 and above, increase the size of a byte to a word or the size of

a word to a doubleword. The zero-extend instruction increases the size of the number by inserting leading zeros. The sign-extend instruction increases the size of the number by copying the sign bit into the more significant bits of the number.

24. Assembler directives DB, (define byte), DW (define word), DD (define doubleword) and DUP (duplicate) store data in the memory system.
25. The EQU (equate) directive allows data or labels to be equated to labels.
26. When full-segment definitions are in use, the SEGMENT directive identifies the start of a memory segment and ENDS identifies the end of a segment.
27. When full-segment definitions are in effect, the ASSUME directive tells the assembler what segment names you have assigned to CS, DS, ES, and SS. In the 80386 and above, ASSUME also indicates the segment name for FS and GS.
28. The PROC and ENDP directives indicate the start and end of a procedure. The USES directive (MASM version 6.X) automatically saves and restores any number of registers on the stack, if it appears with the PROC directive.
29. The assembler assumes that software is being developed for the 8086/8088 microprocessor unless the .286, .386, .486, or .586 directive is used to select one of these other microprocessors. This directive follows the .MODEL statement to use the 16-bit instruction mode, and precedes it for the 32-bit instruction mode.
30. Memory models can be used to shorten the program slightly, but they can cause problems for very large programs. Memory models are not compatible with all assembler programs.

6–9 QUESTIONS AND PROBLEMS

1. The first byte of an instruction is the _____________ unless it contains one of the override prefixes.
2. Describe the purpose of the D and W bits found in some machine language instructions.
3. The MOD field, in a machine language instruction, specifies what information?
4. If the register field (REG) of an instruction contains a 010 and W = 0, what register is selected, assuming the instruction is a 16-bit mode instruction?
5. How are the 32-bit registers selected for the 80486 microprocessor?
6. What memory-addressing mode is specified by R/M = 001 with MOD = 00 for a 16-bit instruction?
7. Identify the default segment register assigned to:
   (a) SP
   (b) EBX
   (c) DI
   (d) EBP
   (e) SI
8. Convert an 8B07H from machine language to assembly language.
9. Convert an 8B1E004CH from machine language to assembly language.
10. If a MOV SI,[BX+2] instruction appears in a program, what is its machine language equivalent?

11. If a MOV ESI,[EAX] instruction appears in a program for the Pentium microprocessor operated in the 16-bit instruction mode, what is its machine language equivalent?
12. What is wrong with a MOV CS,AX instruction?
13. Form a short sequence of instructions to load the data segment register with a 1000H.
14. The PUSH and POP instructions always transfer a ______________-bit number between the stack and a register or memory location in the 8086–80286 microprocessor.
15. What segment register may not be popped from the stack?
16. Which registers move onto the stack with the PUSHA instruction?
17. Which registers move onto the stack for a PUSHAD instruction?
18. Describe the operation of each of the following instructions:
    (a) PUSH AX
    (b) POP ESI
    (c) PUSH [BX]
    (d) PUSHFD
    (e) POP DS
    (f) PUSHD 4
19. Explain what happens when the PUSH BX instruction executes. Be sure to show where BH and BL are stored. Assume that SP = 0100H and SS = 0200H.
20. Repeat Question 19 for the PUSH EAX instruction.
21. The 16-bit POP instruction (except for POPA) increments SP by ______________.
22. What values appear in SP and SS, if the stack is addressed at memory location 02200H?
23. Compare the operation of a MOV DI,NUMB instruction with an LEA DI,NUMB instruction.
24. What is the difference between an LEA SI,NUMB instruction and a MOV SI,OFFSET NUMB instruction?
25. Which is more efficient, a MOV with an OFFSET or an LEA instruction?
26. Describe how the LDS BX,NUMB instruction operates.
27. What is the difference between the LDS and LSS instructions?
28. Develop a sequence of instructions that move the contents of data segment memory locations NUMB and NUMB+1 into BX, DX, and SI.
29. What is the purpose of the direction flag?
30. Which instructions set and clear the direction flag?
31. The string instructions use DI and SI to address memory data in which memory segments?
32. Explain the operation of the LODSB instruction.
33. Explain the operation of the STOSW instruction.
34. Explain the operation of the OUTSB instruction.
35. What does the REP prefix accomplish and what type of instruction is it used with?
36. Develop a sequence of instructions to copy 12 bytes of data from an area of memory addressed by SOURCE into an area of memory addressed by DEST.
37. Where is the I/O address (port number) stored for an INSB instruction?
38. Select an assembly language instruction that exchanges the contents of the EBX register with the ESI register.
39. Would the LAHF and SAHF instructions normally appear in software?
40. Explain how the XLAT instruction transforms the contents of the AL register.

41. Write a short program that uses the XLAT instruction to convert BCD numbers 0–9 into ASCII-coded numbers 30H–39H. Store the ASCII-coded data in a TABLE located within the data segment.
42. Explain what the IN AL,12H instruction accomplishes.
43. Explain how the OUT DX,AX instruction operates.
44. What is a segment override prefix?
45. Select an instruction that moves a byte of data from the memory location addressed by the BX register in the extra segment into the AH register.
46. Develop a sequence of instructions to exchange the contents of AX with BX, ECX with EDX, and SI with DI.
47. What is an assembly language directive?
48. Describe the purpose of the following assembly language directives: DB, DW, and DD.
49. Select an assembly language directive that reserves 30 bytes of memory for array LIST1.
50. Describe the purpose of the EQU directive.
51. What is the purpose of the .386 directive?
52. What is the purpose of the .MODEL directive?
53. If the start of a segment is identified with .DATA, what type of memory organization is in effect?
54. If the SEGMENT directive identifies the start of a segment, what type of memory organization is in effect?
55. What does the INT 21H accomplish, if AH contains a 4CH?
56. What directives indicate the start and end of a procedure?
57. Explain the purpose of the USES statement as it applies to a procedure with MASM version 6.X.
58. How is the 80486 microprocessor instructed to use the 16-bit instruction mode?
59. Develop a near procedure that stores AL in four consecutive memory locations within the data segment, as addressed by the DI register.
60. Develop a far procedure that copies the contents of word-sized memory location CS:DATA1 into AX, BX, CX, DX, and SI.

CHAPTER 7

Arithmetic and Logic Instructions

INTRODUCTION

In this chapter, arithmetic and logic instructions are examined. Arithmetic instructions include addition, subtraction, multiplication, division, comparison, negation, incrementation, and decrementation. Logic instructions include AND, OR, Exclusive-OR, NOT, shifts, rotates, and the logical compare (TEST). Also presented are the 80386 through Pentium instructions XADD, SHRD, SHLD, bit tests, and bit scans.

Also introduced are string comparison instructions, which are used for scanning tabular data and for comparing sections of memory data. Both tasks perform efficiently with the string scan (SCAS) and string compare (CMPS) instructions.

If you are familiar with an 8-bit microprocessor, you will recognize that the 8086 through Pentium instruction set is superior to most 8-bit microprocessors, because most of the instructions have two operands instead of one. Even if this is your first microprocessor, you will quickly learn that the 8086 through Pentium microprocessors possess a powerful and easy-to-use set of arithmetic and logic instructions.

OBJECTIVES

Upon completion of this chapter, you will be able to:

1. Use the arithmetic and logic instructions to accomplish simple binary, BCD, and ASCII arithmetic.
2. Use AND, OR, and Exclusive-OR to accomplish binary bit manipulation.
3. Use the shift and rotate instructions.
4. Explain the operation of the 80386 through the Pentium exchange and add, compare and exchange, double-precision shift, bit test, and bit scan instructions.
5. Check the contents of a table for a match with the string instructions.

7–1 ADDITION, SUBTRACTION, AND COMPARISON

The bulk of the arithmetic instructions found in any microprocessor includes addition, subtraction, and comparison. In this section, addition, subtraction, and comparison instructions are illustrated. Also shown is their use in manipulating register and memory data.

Addition

Addition appears in many forms in the microprocessor. This section details the use of the ADD instruction for 8-, 16-, and 32-bit binary addition. Another form of addition, called *add-with-carry,* is introduced with the **ADC** instruction. Finally, the increment instruction (INC), a special type of addition that adds a one to a number, is presented. In Section 7–3, other forms of addition are examined, such as BCD and ASCII. Also described is the XADD instruction found in the 80486 and Pentium microprocessors.

Table 7–1 illustrates the addressing modes available to the ADD instruction. (These addressing modes include almost all those mentioned in Chapter 5.) However, since there are over 32,000 variations of the ADD instruction, it is impossible to list them all in this table. The only types of addition not allowed are memory-to-memory and segment-register. The segment registers can only be moved, pushed, or popped. The 32-bit registers are only available with the 80386 through the Pentium microprocessor.

Register Addition. Example 7–1 shows a simple procedure that uses register addition to add the contents of several registers. In this example, the contents of AX, BX, CX, and DX are added to form a 16-bit result stored in the AX register. A procedure is used because assembly language, like most languages, is procedure-oriented.

EXAMPLE 7–1

```
                    ;procedure that sums AX, BX, CD, and DX
                    ;result is returned in AX
                    ;
0000                ADDS    PROC    NEAR

0000  03 C3                 ADD     AX,BX
0002  03 C1                 ADD     AX,CX
0004  03 C2                 ADD     AX,DX
0006  C3                    RET

0007                ADDS    ENDP
```

Whenever arithmetic and logic instructions execute, the contents of the flag register change. Note that the contents of the interrupt, trap, and other flags do not change due to arithmetic and logic instructions. Only the flags located in the rightmost eight bits of the flag register and the overflow flag change. These rightmost flags denote the result of the arithmetic or a logic operation. Any ADD instruction modifies the contents of the sign, zero, carry, auxiliary carry, parity, and overflow flags. The flag bits never change for most of the data transfer instructions presented in Chapter 6.

TABLE 7–1 Addition instructions

Instruction	*Comment*
ADD AL,BL	AL = AL + BL
ADD CX,DI	CX = CX + DI
ADD EBP,EAX	EBP = EBP + EAX
ADD CL,44H	CL = CL + 44H
ADD BX,35AFH	BX = BX + 35AFH
ADD EDX,12345H	EDX = EBX + 00012345H
ADD [BX],AL	AL adds to the contents of the data-segment offset location addressed by BX, and the result is stored in the same memory location
ADD CL,[BP]	The contents of the stack-segment offset location addressed by BP adds to CL, and the result is stored in CL
ADD AL,[EBX]	The contents of the data-segment offset location addressed by EBX adds to AL, and the result is stored in AL
ADD BX,[SI + 2]	The word-sized contents of the data-segment location addressed by SI plus 2 adds to BX, and the result is stored in the same memory location
ADD CL,TEMP	The contents of data-segment location TEMP add to CL, with the result stored in CL
ADD BX,TEMP[DI]	The word-sized contents of the data-segment location addressed by TEMP plus DI adds to BX, and the result is stored in the same memory location
ADD [BX + DI],DL	The data-segment memory byte addressed by BX + DI is the sum of that byte plus DL
ADD BYTE PTR [DI],3	Adds a 3 to the contents of the byte-sized memory location addressed by DI within the data segment
ADD BX,[EAX+2*ECX]	The data-segment memory word addressed by the sum of 2 times ECX plus EAX, adds to BX

Immediate Addition. Immediate addition is employed whenever constant or known data are added. An 8-bit immediate addition appears in Example 7–2. In this example, DL is loaded with a 12H by using an immediate move instruction. Next, a 33H is added to the 12H in DL, using an immediate addition instruction.

EXAMPLE 7–2

```
0006    B2 12       MOV   DL,12H
0008    80 C2 33    ADD   DL,33H
```

After the addition, the sum (45H) moves into register DL and the flags change as follows:

Z = 0 (result not zero)

C = 0 (no carry)

A = 0 (no half-carry)

S = 0 (result positive)

P = 0 (odd parity)

O = 0 (no overflow)

Memory-to-Register Addition. Suppose an application requires that memory data add to the AL register. Example 7–3 shows an example that adds 2 consecutive bytes of data, stored at data-segment offset locations NUMB and NUMB+1, to the AL register.

EXAMPLE 7–3

```
                    ;procedure that sums data in locations NUMB and NUMB+1
                    ;the result is returned in AX
                    ;
0000                SUMS    PROC    NEAR

0000  BF 0000 R             MOV     DI,OFFSET NUMB        ;address NUMB
0003  B0 00                 MOV     AL,0                  ;clear sum
0005  02 05                 ADD     AL,[DI]               ;add NUMB
0007  02 45 01              ADD     AL,[DI+1]             ;add NUMB+1
000A  C3                    RET

000B                SUMS    ENDP
```

The procedure first loads the contents of the destination index register (DI) with offset address NUMB. The DI register, used in this example, addresses data in the data segment beginning at memory location NUMB. In most cases, loading the address inside a procedure is poor programming practice. It is usually better to load the address outside the procedure and then CALL the procedure with the address in place. Next, the ADD AL,[DI] instruction adds the contents of memory location NUMB to AL. Note that AL is initialized to zero. This occurs because DI addresses memory location NUMB, and the instruction adds its contents to AL. Finally, the ADD AL,[DI+1] instruction adds the contents of memory location NUMB plus one byte to the AL register. After both ADD instructions execute, the result appears in the AL register as the sum of the contents of NUMB plus the contents of NUMB+1.

Array Addition. Memory arrays are sequential lists of data. Suppose that an array of data (ARRAY) contains 10 bytes numbered from element 0 through element 9. Example 7–4 shows a procedure that adds the contents of array elements 3, 5, and 7. (The procedure and the array elements it adds are chosen to demonstrate the use of some of the addressing modes for the microprocessor.)

EXAMPLE 7–4

```
                    ;procedure that sums ARRAY elements 3, 5, and 7
                    ;result is returned in AL
                    ;
                    ;Note this procedure destroys the contents of SI
                    ;
0000                SUM     PROC    NEAR

0000   B0 00                MOV     AL,0                  ;clear sum
```

```
0002    BE 0003                MOV    SI,3               ;address element 3
0005    02 84 0002 R           ADD    AL,ARRAY[SI]       ;add element 3
0009    02 84 0004 R           ADD    AL,ARRAY[SI+2]     ;add element 5
000D    02 84 0006 R           ADD    AL,ARRAY[SI+4]     ;add element 7
0011    C3                     RET

0012                    SUM    ENDP
```

This example first clears AL to zero, so it can be used to accumulate the sum. Next, register SI is loaded with a 3 to initially address array element 3. The ADD AL,ARRAY[SI] instruction adds the contents of array element 3 to the sum in AL. The instructions that follow add array elements 5 and 7 to the sum in AL, using a 3 in SI plus a displacement of 2 to address element 5, and a displacement of 4 to address element 7.

Suppose that an array of data contains 16-bit numbers used to form a 16-bit sum in register AX. Example 7–5 shows a procedure, written for the 80386 and above, using scaled-index addressing to add elements 3, 5, and 7 of an area of memory called ARRAY. In this example, EBX is loaded with the address ARRAY, and ECX holds the array element number. Note how the scaling factor is used to multiply the contents of the ECX register by 2 to address words of data. Recall that words are two bytes in length.

EXAMPLE 7–5

```
                    ;procedure that sums ARRAY elements 3, 5 and 7
                    ;result is returned in AX
                    ;
                    ;note that the contents of registers EBX and ECX are destroyed
                    ;
0000                SUM     PROC    NEAR

0000  66| BB 00000000 R     MOV     EBX,OFFSET ARRAY    ;address ARRAY
0006  66| B9 00000003       MOV     ECX,3               ;address element 3
000C  67& 8B 04 4B          MOV     AX,[EBX+2*ECX]      ;get element 3
0010  66| B9 00000005       MOV     ECX,5               ;address element 5
0016  67& 03 04 4B          ADD     AX,[EBX+2*ECX]      ;add element 5
001A  66| B9 00000007       MOV     ECX,7               ;address element 7
0020  67& 03 04 4B          ADD     AX,[EBX+2*ECX]      ;add element 7
0024  C3                    RET

0025                SUM     ENDP
```

Increment Addition. **Increment addition** (INC) adds 1 to a register or a memory location. The INC instruction can add 1 to any register or memory location except a segment register. Table 7–2 illustrates some of the possible forms of the increment instruction available to the 8086–80486 and Pentium microprocessor. It is impossible to show all variations of the INC instruction because of the large number available.

With indirect memory increments, the size of the data must be described using the BYTE PTR, WORD PTR, or DWORD PTR directives. This is because the assembler program cannot determine whether, for example, the INC [DI] instruction is a byte-, word-, or doubleword-sized increment. The INC BYTE PTR [DI] instruction clearly indicates byte-sized memory data, the INC WORD PTR [DI] instruction unquestionably indicates a word-sized memory data, and the INC DWORD PTR [DI] instruction indicates doubleword-sized data.

TABLE 7–2 Increment instructions

Instruction	*Comment*
INC BL	BL = BL + 1
INC SP	SP = SP + 1
INC EAX	EAX = EAX + 1
INC BYTE PTR [BX]	The byte contents of the memory location addressed by BX in the data-segment increment
INC WORD PTR [SI]	The word contents of the memory location addressed by SI in the data-segment increment
INC DWORD PTR [ECX]	The doubleword contents of the data-segment memory location addressed by ECX increment
INC DATA1	The contents of data-segment location DATA1 increment

EXAMPLE 7–6

```
                        ;procedure that sums NUMB and NUMB+1
                        ;result is returned in AL
                        ;
                        ;Note that the contents of DI are destroyed
                        ;
0000                    SUMS    PROC    NEAR

0000    BF 0000 R               MOV     DI,OFFSET NUMB      ;address NUMB
0003    B0 00                   MOV     AL,0                ;clear sum
0005    02 05                   ADD     AL,[DI]             ;add NUMB
0007    47                      INC     DI                  ;address NUMB+1
0008    02 05                   ADD     AL,[DI]             ;add NUMB+1
000A    C3                      RET

000B                    SUMS    ENDP
```

Example 7–6 shows how Example 7–3 is modified to use the increment instruction for addressing NUMB and NUMB+1. Here, an INC DI instruction changes the contents of register DI from offset address NUMB to offset address NUMB+1. Examples 7–3 and 7–6 both add the contents of NUMB and NUMB+1. The difference between these programs is the way that this data's address is formed, through the contents of the DI register, using the increment instruction.

Increment instructions affect the flag bits, as do most other arithmetic and logic operations. The difference is that increment instructions do not affect the carry flag bit. Carry doesn't change because we often use increments in programs that depend upon the contents of the carry flag. Note that increment is used to point to the next memory element only in a byte-sized array of data. If word-sized data are addressed, it is better to use an ADD DI,2 instruction to modify the DI pointer in place of two INC DI instructions. For doubleword arrays, use the ADD DI,4 instruction to modify the DI pointer. In some cases, the carry flag must be preserved, which may mean that a pair of, or four, INC instructions might appear in a program to modify a pointer.

TABLE 7–3 Add-with-carry instructions

Instruction	*Comment*
ADC AL,AH	AL = AL + AH + carry
ADC CX,BX	CX = CX + BX + carry
ADC EBX,EDX	EBX = EBX + EDX + carry
ADC DH,[BX]	The byte contents of the data-segment memory location addressed by BX are added to DH and carry; the result is stored in DH
ADC BX,[BP + 2]	BX and the word contents of the stack-segment memory location addressed by BP are added to carry; the result is stored in BX
ADC ECX,[EBX]	ECX and the doubleword contents of the data-segment memory location addressed by EBX are added to carry; the result is stored in ECX

Note: Only the 80386/80486/Pentium use 32-bit registers and addressing modes.

Addition with Carry. An addition-with-carry instruction (ADC) adds the bit in the carry flag (C) to the operand data. This instruction mainly appears in software that adds numbers wider than 16 bits in the 8086–80286, or wider than 32-bits in the 80386 through Pentium microprocessor.

Table 7–3 lists several add-with-carry instructions, with comments that explain their operation. Like the ADD instruction, ADC affects the flags after the addition.

Suppose a program is written for the 8086–80286 to add the 32-bit number in BX and AX to the 32-bit number in DX and CX. Figure 7–1 illustrates this addition so the placement and function of carry flag can be understood. This addition cannot be easily performed without adding the carry flag bit, because the 8086–80286 only add 8- or 16-bit numbers. Example 7–7 shows how the addition occurs with a procedure. Here the contents of registers AX and CX add to form the least significant 16 bits of the sum. This addition may or may not generate a carry. A carry appears in the carry flag if the sum is greater than FFFFH. Because it is impossible to predict a carry, the most significant 16 bits of this

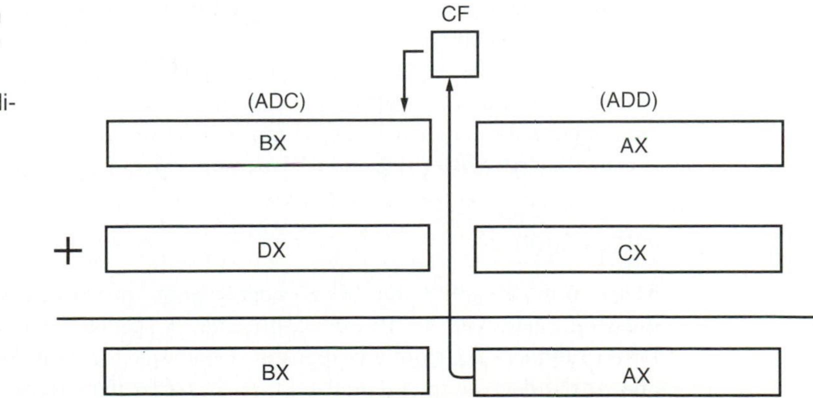

FIGURE 7–1 Addition with carry showing how the carry flag (C) links the two 16-bit additions into one 32-bit addition.

addition are added with the carry flag using the ADC instruction. The ADC instruction adds the one or the zero in the carry flag to the most significant 16 bits of the result. This program adds BX–AX to DX–CX, with the sum appearing in BX–AX.

EXAMPLE 7–7

```
                        ;procedure that sums BX-AX and DX-CX
                        ;result is returned in BX-AX
                        ;
0000                    SUM32   PROC    NEAR

0000  03 C1                     ADD     AX,CX
0002  13 DA                     ADC     BX,DX
0004  C3                        RET

0005                    SUM32   ENDP
```

Suppose the same procedure is rewritten for the 80386 through the Pentium microprocessor, but modified to add two 64-bit numbers. This requires modification of the instructions and the use of the extended registers to hold the data. These changes are shown in Example 7–8, which adds two 64-bit numbers.

EXAMPLE 7–8

```
                          ;procedure that sums EBX-EAX and EDX-ECX
                          ;result is returned in EBX-EAX
                          ;
0000                      SUM64   PROC    NEAR

0000  66| 03 C1                   ADD     EAX,ECX
0003  66| 13 DA                   ADC     EBX,EDX
0006  C3                          RET

0007                      SUM64   ENDP
```

Exchange and Add for the 80486/Pentium Microprocessor. A new type of addition, called *exchange and add* (XADD), appears in the 80486 instruction set and continues with the Pentium. The XADD instruction adds the source to the destination and stores the sum in the destination, like any addition. The difference is that after the addition takes place, the original value of the destination is copied into the source operand. This is one of the few instructions that change the source.

For example, if BL = 12H and DL = 02H, and the XADD BL,DL instruction executes, the BL register contains the sum of 14H and DL becomes 12H. The sum of 14H is generated and the original destination of 12H replaces the source. This instruction functions with any register size and any memory operand, just like the ADD instruction.

Subtraction

Many forms of subtraction (SUB) appear in the instruction set. These forms use any addressing mode with 8-, 16-, or 32-bit data. A special form of subtraction (decrement or DEC) subtracts a 1 from any register or memory location. Section 7–3 shows how BCD and ASCII data subtract. Numbers that are wider than 16 or 32 bits must occasionally be subtracted. The subtract-with-borrow instruction (SBB) performs this type of subtraction. In the 80486 and Pentium, the instruction set also includes a compare and exchange.

Table 7–4 lists some of the many addressing modes allowed with the subtract instruction (SUB). There are well over a thousand possible subtraction instructions, far too many to list. About the only types of subtraction not allowed are memory-to-memory and segment-register subtractions. Like other arithmetic instructions, the subtract instruction affects the flag bits.

Register Subtraction. Example 7–9 shows a sequence of instructions that perform register subtraction. This example subtracts the 16-bit contents of registers CX and DX from the contents of register BX. After each subtraction, the microprocessor modifies the contents of the flag register. The flags change for most arithmetic and logic operations.

EXAMPLE 7–9

```
0000  2B D9          SUB   BX,CX
0002  2B DA          SUB   BX,DX
```

Immediate Subtraction. The microprocessor allows immediate operands for the subtraction of constant data. Example 7–10 presents a short sequence of instructions that subtract a 44H from a 22H. Here, we first load the 22H into CH using an immediate move instruction. Next, the SUB instruction, using immediate data 44H, subtracts a 44H from the 22H.

TABLE 7–4 Subtraction instructions

Instruction	*Comment*
SUB CL,BL	CL = CL – BL
SUB AX,SP	AX = AX – SP
SUB ECX,EBP	ECX = ECX – EBP
SUB DH,6FH	DH = DH – 6FH
SUB AX,0CCCCH	AX = AX – 0CCCCH
SUB EAX,23456H	EAX = EAX – 00023456H
SUB [DI],CH	CH subtracts from the byte contents of the data-segment memory location addressed by DI
SUB CH,[BP]	The byte contents of the stack-segment memory location addressed by BP subtracts from CH
SUB AH,TEMP	The byte contents of the data-segment memory location TEMP subtracts from AH
SUB DI,TEMP[BX]	The word contents of the data-segment memory location addressed by TEMP + BX subtracts from DI
SUB ECX,DATA1	The doubleword contents of the data-segment memory location addressed as DATA1 subtracts from ECX

Note: Only the 80386/80486/Pentium use 32-bit registers and addressing modes.

TABLE 7–5 Decrement instructions

Instruction	*Comment*
DEC BH	BH = BH – 1
DEC SP	SP = SP – 1
DEC ECX	ECX = ECX – 1
DEC BYTE PTR [DI]	The byte contents of the data-segment memory location addressed by DI decrements
DEC WORD PTR [BP]	The word contents of the stack-segment memory location addressed by BP decrements
DEC DWORD PTR [EBX]	The doubleword contents of the data-segment memory location addressed by EBX decrements
DEC NUMB	The contents of data-segment memory location NUMB decrement; the way NUMB is defined determines whether this is a byte or word decrement

Note: Only the 80386/80486/Pentium use 32-bit registers and addressing modes.

EXAMPLE 7–10

```
0000  B5 22                MOV   CH,22H
0002  80 ED 44             SUB   CH,44H
```

After the subtraction, the difference (DEH) moves into the CH register. The flags change as follows for this subtraction:

Z = 0 (result not zero)
C = 1 (borrow)
A = 1 (half-borrow)
S = 1 (result negative)
P = 1 (even parity)
O = 0 (no overflow)

Both carry flags (C and A) hold borrows instead of carries after a subtraction. Notice, in Example 7–10 there was no overflow. Example 7–10 subtracted a 44H (+68) from a 22H (+34), resulting in a DEH (–34). Because the correct 8-bit signed result is –34, there is no overflow. An 8-bit overflow only occurs if the signed result is greater than +127 or less than –128.

Decrement Subtraction. **Decrement subtraction** (DEC) subtracts a 1 from a register or the contents of a memory location. Table 7–5 lists some decrement instructions that illustrate register and memory decrements.

The decrement indirect memory data instructions require BYTE PTR, WORD PTR, or DWORD PTR because the assembler cannot distinguish a byte from a word when an index register addresses memory. For example, DEC [SI] is vague because the assembler cannot determine whether the location addressed by SI is a byte or a word. Using DEC BYTE PTR [SI], DEC WORD PTR [DI], or DEC DWORD PTR [SI] reveals the size of the data.

TABLE 7–6 Subtract-with-borrow instructions

Instruction	*Comment*
SBB AH,AL	AH = AH – AL – carry
SBB AX,BX	AX = AX – BX – carry
SBB EAX,EBX	EAX = EAX – EBX – carry
SBB CL,3	CL = CL – 3 – carry
SBB BYTE PTR [DI],3	3 and carry subtract from the byte contents of the data-segment memory location addressed by DI
SBB [DI],AL	AL and carry subtract from the byte contents of the data-segment memory location addressed by DI
SBB DI,[BP + 2]	The word contents of the stack-segment memory location addressed by BP plus 2 and carry subtract from DI
SBB AL,[EBX+ECX]	The byte contents of the data-segment memory location addressed by the sum of EBX and ECX and carry subtract from AL

Note: Only the 80386/80486/Pentium use 32-bit registers and addressing modes.

Subtract with Borrow. The **SBB** instruction (*subtraction-with-borrow*) functions as a regular subtraction, except the carry flag (C), which holds the borrow, also subtracts from the difference. The most common use for this instruction is for subtractions wider than 16 bits in the 8086–80286 microprocessor, 32 bits in the 80386 through the Pentium. Wide subtractions require that borrows propagate through the subtraction, just as wide additions propagate the carry.

Table 7–6 lists many SBB instructions with comments that define their operation. Like the SUB instruction, SBB affects the flags. Notice that the subtract from memory immediate instruction in this table requires a BYTE PTR, WORD PTR, or DWORD PTR directive.

When the 32-bit number held in BX and AX is subtracted from the 32-bit number held in SI and DI, the carry flag propagates the borrow between the two 16-bit subtractions required to perform this operation in the microprocessor. Figure 7–2 shows how the borrow propagates through the carry flag (C) for this task. Example 7–11 shows how this subtraction is performed by a program. With wide subtraction, the least significant 16- or 32-bit data are subtracted with the SUB instruction. All subsequent, and more significant data are subtracted using the SBB instruction. The example uses the SUB instruction to subtract DI from AX then SBB to subtract-with-borrow SI from BX.

EXAMPLE 7–11

```
0004  2B C7            SUB   AX,DI
0006  1B DE            SBB   BX,SI
```

Comparison

The **CMP** instruction (*comparison*) is a subtraction that only changes the flag bits. A comparison is useful for checking the entire contents of a register or a memory location against

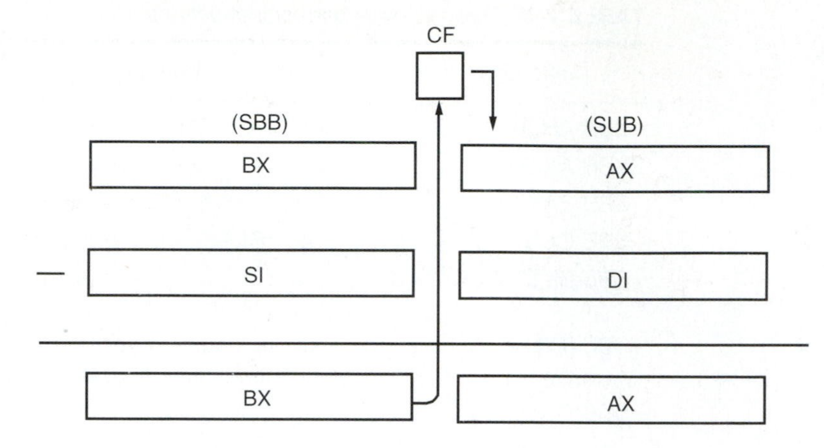

FIGURE 7–2 Subtraction with borrow showing how the carry flag (C) propagates the borrow.

another value. A CMP is normally followed by a conditional jump instruction, which tests the condition of the flag bits.

Table 7–7 lists a variety of comparison instructions that use the same addressing modes as the addition and subtraction instructions already presented. The only disallowed forms of comparison are memory-to-memory and segment-register comparisons.

EXAMPLE 7–12

```
0000  3C 10         CMP  AL,10H        ;compare with 10H
0002  73 1C         JAE  SUBER         ;if 10H or above
```

Example 7–12 shows a comparison followed by a conditional jump instruction. In this example, the contents of AL are compared with a 10H. Conditional jump instructions that often follow the comparison are JA (jump above) or JB (jump below). If the JA follows the comparison, the jump occurs if the value in AL is above 10H. If the JB follows the comparison, the jump occurs if the value in AL is below 10H. In this example, the JAE instruction follows the comparison. This instruction causes the program to continue at memory location SUBER if the value in AL is 10H or above. There is also a JBE (jump below or equal) instruction that could follow the comparison to jump if the outcome is below or equal to 10H. Chapter 8 provides more detail on the comparison and conditional jump instructions.

Compare and Exchange (80486/Pentium Only). The *compare and exchange* instruction (CMPXCHG), found only in the 80486 and Pentium instruction sets, compares the destination operand with the accumulator. If they are equal, the source operand is copied into the destination. If they are not equal, the destination operand is copied into the accumulator. This instruction functions with 8-, 16-, or 32-bit data.

The CMPXCHG CX,DX instruction is an example of the compare and exchange instruction. This instruction first compares the contents of CX with AX. If CX equals AX, DX is copied into AX. If CX is not equal to AX, CX is copied into AX. This instruction also compares AL with 8-bit data and EAX with 32-bit data, if the operands are either 8- or 32-bit.

In the Pentium only, a CMPXCHG8B instruction that compares two quad words is available. This is the only new data manipulation instruction provided to the Pentium. The compare-and-exchange-8-bytes instruction compares the 64-bit value located in EDX:EAX

TABLE 7–7 Comparison instructions

Instruction	*Comment*
CMP CL,BL	Subtracts BL from CL; neither BL nor CL changes, but the flags change
CMP AX,SP	Subtracts SP from AX; neither AX nor SP changes, but the flags change
CMP EBP,ESI	Subtracts ESI from EBP; neither EBP nor ESI changes, but the flags change
CMP AX,0CCCCH	Subtracts 0CCCCH from AX; AX does not change, but the flags change
CMP [DI], CH	Subtracts CH from the byte contents of the data-segment memory location addressed by DI; neither CH nor memory changes, but the flags change
CMP CL,[BP]	Subtracts the byte contents of the stack-segment memory location addressed by BP from CL; neither CL nor memory changes, but the flags change
CMP AH,TEMP	Subtracts the byte contents of the data-segment memory location addressed by TEMP from AH; neither AH nor memory changes, but the flags change
CMP DI,TEMP [BX]	Subtracts the word contents of the data-segment memory location addressed by TEMP plus BX from DI; neither DI nor memory changes, but the flags change
CMP AL,[EDI+ESI]	Subtracts the byte contents of the data-segment memory location addressed by the sum of EDI and ESI; neither AL nor memory changes, but the flags change

Note: Only the 80386/80486/Pentium use 32-bit registers and addressing modes.

with a 64-bit number located in memory. An example is CMPXCHG8B TEMP. If TEMP equals EDX:EAX, TEMP is replaced with the value found in ECX:EBX. If TEMP does not equal EDX:EAX, then the number found in TEMP is loaded into EDX:EAX. The zero flag bit indicates that the values are equal after the comparison.

7–2 MULTIPLICATION AND DIVISION

Only modern microprocessors contain multiplication and division instructions. Earlier 8-bit microprocessors could not multiply or divide without the use of a program that used a series of shifts and additions or subtractions. Because microprocessor manufacturers were aware of this inadequacy, they incorporated multiplication and division instructions into the instruction sets of the newer microprocessors. In fact, the Pentium contains special circuitry that performs a multiplication in as little as one clocking period. It took more than 40 clocking periods to perform the same multiplication in earlier Intel microprocessors.

TABLE 7–8 8-bit multiplication instructions

Instruction	*Comment*
MUL CL	AL is multiplied by CL; this unsigned multiplication leaves the product in AX
IMUL DH	AL is multiplied by DH; this signed multiplication leaves the product in AX
IMUL BYTE PTR [BX]	AL is multiplied by the byte contents of the data-segment memory location addressed by BX; this unsigned multiplication leaves the product in AX
MUL TEMP	AL is multiplied by the contents of the data-segment memory location TEMP; if TEMP is defined as an 8-bit number, the unsigned product is found in AX

Multiplication

Multiplication is performed on bytes, words, or doublewords and can be signed integer (IMUL) or unsigned (MUL). Only the 80386 through the Pentium multiply 32-bit doublewords. The product after a multiplication is always a double-width product. If two 8-bit numbers are multiplied, they generate a 16-bit product; if two 16-bit numbers are multiplied, they generate a 32-bit product; and if two 32-bit numbers are multiplied, a 64-bit product is generated.

Some flag bits (O and C) change when the multiply instruction executes, producing predictable outcomes. Other flags change too, but their results are unpredictable and therefore they are unused. In an 8-bit multiplication, if the most significant 8 bits of the result are 0, both C and O flag bits equal 0. These flag bits show that the result is 8 bits wide (C = 0) or 16 bits wide (C = 1). In a 16-bit multiplication, if the most significant 16-bits of the product are 0, both C and O clear to 0. In a 32-bit multiplication, both C and O indicate that the most significant 32 bits of the product are zero.

8-bit Multiplication. With 8-bit multiplication, whether signed or unsigned, the multiplicand is always in the AL register. The multiplier can be any 8-bit register or any memory location. Immediate multiplication is not allowed unless the special signed immediate multiplication instruction, discussed later in this section, appears in a program. The multiplication instruction contains one operand because it always multiplies the operand by the contents of register AL. An example is the MUL BL instruction, which multiplies the unsigned contents of AL by the unsigned contents of BL. After the multiplication, the unsigned product—a double-width product—is placed in AX. Table 7–8 illustrates some 8-bit multiplication instructions.

Suppose that BL and CL each contain two 8-bit unsigned numbers, and these numbers must be multiplied to form a 16-bit product stored in DX. This procedure cannot be accomplished by a single instruction, because we can only multiply a number by the AL register for an 8-bit multiplication. Example 7–13 shows a short program that generates DX = BL × CL. This example loads register BL and CL with example data 5 and 10. The

TABLE 7–9 16-bit multiplication instructions

Instruction	*Comment*
MUL CX	AX is multiplied by CX; the unsigned product is found in DX–AX
IMUL DI	AX is multiplied by DI; the signed product is found in DX–AX
MUL WORD PTR [SI]	AX is multiplied by the word contents of the data-segment memory location addressed by SI; the unsigned product is found in DX–AX

product, a 50, moves into DX from AX after the multiplication, by using the MOV DX,AX instruction.

EXAMPLE 7–13

```
0000  B3 05          MOV   BL,5        ;load data
0002  B1 0A          MOV   CL,10
0004  8A C1          MOV   AL,CL       ;position data
0006  F6 E3          MUL   BL          ;multiply
0008  8B D0          MOV   DX,AX       ;position product
```

For signed multiplication, the product is in true binary form if positive, and in two's complement form if negative. These are the same forms used to store all positive and negative signed numbers used by the microprocessor. If the program of Example 7–13 multiplies two signed numbers, only the MUL instruction is changed to IMUL.

16-bit Multiplication. Word multiplication is very similar to byte multiplication. The difference is that AX contains the multiplicand instead of AL and the product appears in DX–AX instead of AX. The DX register always contains the most significant 16 bits of the product and the AX register always contains the least significant 16 bits. As with 8-bit multiplication, the choice of the multiplier is up to the programmer. Table 7–9 shows several different 16-bit multiplication instructions.

Immediate 16-bit Multiplication. The 8086/8088 microprocessor could not perform immediate multiplication but the 80286 through Pentium can, by using a special version of the multiply instruction. Immediate multiplication must be signed. The instruction format is different because it contains three operands. The first operand is the 16-bit destination register, the second operand is a register or memory location that contains the 16-bit multiplicand, and the third operand is either an 8-bit or 16-bit immediate data used as the multiplier.

The IMUL CX,DX,12H instruction multiplies 12H by DX and leaves a 16-bit signed product in CX. If the immediate data are 8 bits, they sign-extend into a 16-bit number before the multiplication occurs. Another example is IMUL BX,NUMBER,1000H, which multiplies NUMBER times 1000H and leaves the product in BX. Both the destination and multiplicand must be 16-bit numbers. The restrictions placed upon immediate multiplication—especially the fact that it is a signed multiplication and the product is 16 bits wide—limit its utility.

TABLE 7–10 32-bit multiplication instructions

Instruction	*Comment*
MUL ECX	EAX is multiplied by ECX; the unsigned product is found in EDX–EAX
IMUL EDI	EAX is multiplied by EDI; the signed product is found in EDX–EAX
MUL DWORD PTR[ECX]	EAX is multiplied by the doubleword contents of the data-segment memory location addressed by ECX; the unsigned product is found in EDX–EAX

32-Bit Multiplication. In the 80386 and above, 32-bit multiplication is allowed, because these microprocessors contain 32-bit registers. 32-bit multiplication can be signed or unsigned by using the IMUL and MUL instructions. With 32-bit multiplication, the contents of EAX are multiplied by the operand specified with the instruction. The product (64 bits wide) is found in EDX–EAX when EAX contains the least-significant 32 bits of the product. Table 7–10 lists some of the 32-bit multiplication instructions found in the 80386 and above instruction set.

Division

Like multiplication, division occurs on 8- or 16-bit numbers (and also on 32-bit numbers in the 80386 through Pentium). These numbers are signed (IDIV) or unsigned (DIV) integers. The dividend is always a double-width dividend that is divided by the operand. This means that an 8-bit division divides a 16-bit number by an 8-bit number, a 16-bit division divides a 32-bit number by a 16-bit number, and a 32-bit division divides a 64-bit number by a 32-bit number. There is no immediate division instruction available to any microprocessor.

None of the flag bits change predictably for a division. A division can result in two different types of errors. One of these is an attempt to divide by zero and the other is a divide overflow. A divide overflow occurs when a small number divides into a large number. For example, suppose that AX = 3,000 and that it is divided by 2. Because the quotient for an 8-bit division appears in AL, the result, 1,500, causes a divide overflow. With either type of error, the microprocessor generates an interrupt. In most cases, a divide error interrupt displays an error message on the video screen. The divide error interrupt and all other interrupts for the microprocessor are explained in Chapter 8.

8-Bit Division. An 8-bit division uses the AX register to store the dividend that is divided by the contents of any 8-bit register or memory location. The quotient moves into AL after the division, with AH containing a whole number remainder. For a signed division, the quotient is positive or negative and the remainder *always* assumes the sign of the dividend and is always an integer. For example, if AX = 0010H (+16) and BL = FDH (–3) and the IDIV BL instruction executes, AX = 01FBH, this represents a quotient of –5 (AL) with a remainder of 1 (AH). If, on the other hand, –16 is divided by +3, the result will be a quotient of –5 (AL) with a remainder of –1 (AH). Table 7–11 lists some of the 8-bit division instructions.

TABLE 7–11 8-bit division instructions

Instruction	*Comment*
DIV CL	AX is divided by CL; the unsigned quotient is in AL and the remainder is in AH
IDIV BL	AX is divided by BL; the signed quotient is in AL and the remainder is in AH
DIV BYTE PTR [BP]	AX is divided by the byte contents of the stack-segment memory location addressed by BP; the unsigned quotient is in AL and the remainder is in AH

With 8-bit division, the numbers are usually 8 bits wide. This means that one of them, the dividend, must be converted to a 16-bit wide number in AX. This is accomplished differently for signed and unsigned numbers. For an unsigned number, the most significant 8 bits must be cleared to zero (*zero-extended*). The MOVZX instruction described in Chapter 6 can be used to zero-extend a number in the 80386 through the Pentium microprocessor. For the signed number, the least significant 8 bits are sign-extended into the most significant 8 bits. In the microprocessor a special instruction exists that sign-extends AL into AH, or converts an 8-bit signed number in AL into a 16-bit signed number in AX. The CBW (*convert byte to word*) instruction performs this conversion. In the 80386 through the Pentium microprocessor, a MOVSX instruction can sign-extend a number (see Chapter 6).

EXAMPLE 7–14

```
0000  A0 0000 R               MOV   AL,NUMB        ;get NUMB
0003  B4 00                   MOV   AH,0           ;zero-extend
0005  F6 36 0002 R            DIV   NUMB1          ;divide by NUMB1
0009  A2 0003 R               MOV   ANSQ,AL        ;save quotient
000C  88 26 0004 R            MOV   ANSR,AH        ;save remainder
```

Example 7–14 illustrates a short program that divides the unsigned byte contents of memory location NUMB by the unsigned contents of memory location NUMB1. Here the quotient is stored in location ANSQ and the remainder in location ANSR. Notice how the contents of location NUMB are retrieved from memory and then zero-extended to form a 16-bit unsigned number for the dividend.

EXAMPLE 7–15

```
0000  A0 0000 R               MOV    AL,NUMB       ;get NUMB
0003  98                      CBW                  ;sign-extend
0004  F6 3E 0002 R            IDIV   NUMB1         ;divide by NUMB1
0008  A2 0003 R               MOV    ANSQ,AL       ;save quotient
000B  88 26 0004 R            MOV    ANSR,AH       ;save remainder
```

Example 7–15 shows the same basic program, except the numbers are signed numbers. This means that instead of zero-extending AL into AH, it is sign-extended with the CBW instruction.

TABLE 7–12 16-bit division instructions

Instruction	*Comment*
DIV CX	DX–AX is divided by CX; the unsigned quotient is in AX and the remainder is in DX
IDIV SI	DX–AX is divided by SI; the signed quotient is in AX and the remainder is in DX
DIV NUMB	DX–AX is divided by the word contents of the data-segment memory location NUMB; the unsigned quotient is in AX and the remainder is in DX

16-bit Division. Sixteen-bit division is similar to 8-bit division, except the 16-bit number is divided into DX–AX, a 32-bit dividend. After a 16-bit division, the quotient appears in AX and the remainder in DX. Table 7–12 lists some of the 16-bit division instructions.

In 16-bit division, numbers must often be converted to the proper form for the dividend. If a 16-bit unsigned number is placed in AX, then DX must be cleared to 0. In the 80386 and above, the number is zero-extended using the MOVZX instruction. If AX is a 16-bit signed number, the CWD (*convert word to doubleword*) instruction sign-extends it into a signed 32-bit number. If the 80386 and above is available, the MOVSX instruction can also be used to sign-extend a number.

EXAMPLE 7–16

```
0000  B8 FF9C            MOV    AX,-100        ;load -100
0003  B9 0009            MOV    CX,9           ;load +9
0006  99                 CWD                   ;sign-extend
0007  F7 F9              IDIV   CX
```

Example 7–16 shows the division of two 16-bit signed numbers. Here a –100 in AX is divided by a +9 in CX. The CWD instruction converts the –100 in AX to a –100 in DX–AX before the division. After the division, the results appear in DX–AX as a quotient of –11 in AX and a remainder of –1 in DX.

32-bit Division. The 80386 through the Pentium microprocessor performs 32-bit division on signed or unsigned numbers. The 64-bit contents of EDX–EAX are divided by the operand specified by the instruction, leaving a 32-bit quotient in EAX and a 32-bit remainder in EDX. Other than the size of the registers, this instruction functions in the same manner as the 8- and 16-bit divisions. Table 7–13 shows some example 32-bit division instructions. The convert-doubleword-to-quadword instruction (CDQ) is used before a signed division to convert the 32-bit contents of EAX into a 64-bit signed number in EDX–EAX.

The Remainder. What is done with the remainder after a division? There are a few possible choices. The remainder could be used to round the result or dropped to truncate the result. If the division is unsigned, rounding requires the remainder to be compared with half the divisor, to decide whether to round the quotient up or down. The remainder could also be stated as a fraction.

TABLE 7–13 32-bit division instructions

Instruction	*Comment*
DIV ECX	EDX–EAX is divided by ECX; the unsigned quotient is in EAX and the remainder is in EDX
DIV DATA2	EDX–EAX is divided by the doubleword contents of the data-segment memory location addressed by DATA2; the unsigned quotient is in EAX and the remainder is in EDX
IDIV DWORD PTR [EDI]	EDX–EAX is divided by the doubleword contents of the data-segment memory location addressed by EDI; the signed quotient is in EAX and the remainder is in EDX

EXAMPLE 7–17

```
0000  F6 F3              DIV   BL         ;divide
0002  02 E4              ADD   AH,AH      ;double remainder
0004  3A E3              CMP   AH,BL      ;test for rounding
0006  72 02              JB    NEXT
0008  FE C0              INC   AL         ;round
000A            NEXT:
```

Example 7–17 shows a sequence of instructions that divide AX by BL and round the result. This program doubles the remainder before comparing it with BL to decide whether or not to round the quotient. Here, an INC instruction rounds the contents of AL after the comparison.

Suppose that a fractional remainder is required, rather than an integer remainder. A fractional remainder is obtained by saving the quotient. Next the AL register is cleared to zero. The number remaining in AX is now divided by the original operand to generate a fractional remainder.

EXAMPLE 7–18

```
0000  B8 000D            MOV   AX,13      ;load 13
0003  B3 02              MOV   BL,2       ;load 2
0005  F6 F3              DIV   BL         ;13/2
0007  A2 0003 R          MOV   ANSQ,AL    ;save quotient
000A  B0 00              MOV   AL,0       ;clear AL
000C  F6 F3              DIV   BL         ;generate remainder
000E  A2 0004 R          MOV   ANSR,AL    ;save remainder
```

Example 7–18 shows how 13 is divided by 2. The 8-bit quotient is saved in memory location ANSQ and AL is cleared. Next, the contents of AX are again divided by 2 to generate a fractional remainder. After the division, the AL register equals 80H. This is a 10000000_2. If the binary point (radix) is placed before the leftmost bit of AL, the fractional remainder in AL is 0.10000000_2 or 0.5 decimal. The remainder is saved in memory location ANSR.

7–3 BCD AND ASCII ARITHMETIC

The microprocessor allows arithmetic manipulation of both binary-coded decimal (BCD) and American Standard Code for Information Interchange (ASCII) data. This is accomplished by instructions that adjust the numbers for BCD and ASCII arithmetic.

The BCD operations occur in systems, like point-of-sales terminals (cash registers), that seldom require arithmetic. The ASCII operations are performed on ASCII data used by many programs. BCD or ASCII arithmetic is rarely used today.

BCD Arithmetic

Two arithmetic techniques operate with BCD data: addition and subtraction. The instruction set provides two instructions that correct the result of a BCD addition and a BCD subtraction. The DAA instruction (*decimal adjust after addition*) follows BCD addition and the DAS instruction (*decimal adjust after subtraction*) follows BCD subtraction. Both instructions correct the result of the addition or subtraction so it is a BCD number.

For BCD data, the numbers always appear in packed BCD form and are stored as two BCD digits per byte. The adjust instructions only function with the AL register after BCD addition and subtraction.

DAA Instruction. The DAA instruction follows the ADD or ADC instruction, to adjust the result into a BCD result. Suppose that DX and BX each contain 4-digit packed BCD numbers. Example 7–19 provides a short sample program that adds the BCD numbers in DX and BX and stores the result in CX.

EXAMPLE 7–19

```
0000  BA 1234        MOV   DX,1234H      ;load 1,234
0003  BB 3099        MOV   BX,3099H      ;load 3,099
0006  8A C3          MOV   AL,BL         ;sum BL with DL
0008  02 C2          ADD   AL,DL
000A  27             DAA                 ;adjust
000B  8A C8          MOV   CL,AL         ;answer to CL
000D  8A C7          MOV   AL,BH         ;sum BH, DH, and carry

000F  12 C6          ADC   AL,DH
0011  27             DAA                 ;adjust
0012  8A E8          MOV   CH,AL         ;answer to CH
```

Because the DAA instruction only functions with the AL register, this addition must occur 8 bits at a time. After adding the BL and DL registers, the result is adjusted with a DAA instruction before being stored in CL. Next, the BH and DH registers add with carry, and the result is adjusted with DAA before being stored in CH. In this example, 1,234 adds to 3,099 to generate a sum, 4,333, that moves into CX after the addition. Note that 1234 BCD is the same as 1234H.

DAS Instruction. The DAS instruction functions just like the DAA instruction, except it follows a subtraction instead of an addition. Example 7–20 is basically the same as

Example 7–19, except it subtracts instead of adds DX and BX. The main difference in these programs is that the DAA instructions change to DAS and the ADD and ADC instructions change to SUB and SBB instructions.

EXAMPLE 7–20

```
0000  BA 1234          MOV   DX,1234H        ;load 1,234
0003  BB 3099          MOV   BX,3099H        ;load 3,099
0006  8A C3            MOV   AL,BL           ;subtract DL from BL
0008  2A C2            SUB   AL,DL
000A  2F               DAS                   ;adjust
000B  8A C8            MOV   CL,AL           ;answer to CL
000D  8A C7            MOV   AL,BH           ;subtract DH
000F  1A C6            SBB   AL,DH
0011  2F               DAS                   ;adjust
0012  8A E8            MOV   CH,AL           ;answer to CH
```

ASCII Arithmetic

The ASCII arithmetic instructions function with ASCII-coded numbers. These numbers range in value from 30H to 39H for the numbers 0–9. There are four instructions used with ASCII arithmetic operations: AAA (*ASCII adjust after addition*), AAD (*ASCII adjust before division*), AAM (*ASCII adjust after multiplication*), and AAS (*ASCII adjust after subtraction*). These instructions use register AX as the source and the destination.

AAA Instruction. The addition of two one-digit ASCII-coded numbers will not result in any useful data. For example, if 31H and 39H are added, the result is 6AH. This ASCII addition (1 + 9) should produce a two-digit ASCII result equivalent to 10 decimal, (31H and 30H in ASCII code). If the AAA instruction is executed after this addition, the AX register will contain a 0100H. Although this is not ASCII code, it can be converted to ASCII code by adding 3030H, which generates 3130H. The AAA instruction clears AH if the result is less than 10 and *adds* 1 to AH if the result is greater than 10.

EXAMPLE 7–21

```
0000  B8 0031          MOV   AX,31H          ;load ASCII 1
0003  04 39            ADD   AL,39H          ;add ASCII 9
0005  37               AAA                   ;adjust
0006  05 3030          ADD   AX,3030H        ;answer to ASCII
```

Example 7–21 shows how ASCII addition functions in the microprocessor. AH is cleared before the addition by using the MOV AX,31H instruction. The operand, 0031H, places 00H in AH and 31H into AL.

AAD Instruction. Unlike all the other adjust instructions, the AAD instruction appears *before* a division. The AAD instruction requires that the AX register contains a two-digit unpacked BCD (not ASCII) number before executing. After adjusting the AX register with AAD, it is divided by an unpacked BCD number to generate a single-digit result in AL, with any remainder in AH.

Example 7–22 illustrates how 72 in unpacked BCD is divided by 9 to produce a quotient of 8. The 0702H loaded into the AX register is adjusted by the AAD instruction to 0048H. This instruction converts any two-digit unpacked BCD number between 00 and 99 into a binary number, so it can be divided with the binary division instruction (DIV).

EXAMPLE 7–22

```
0000  B3 09         MOV   BL,9          ;load divisor
0002  B8 0702       MOV   AX,0702H      ;load dividend
0005  D5 0A         AAD                 ;adjust
0007  F6 F3         DIV   BL
```

AAM Instruction. The AAM instruction follows the multiplication instruction after multiplying two one-digit unpacked BCD numbers. Example 7–23 shows a short program that multiplies 5 times 5. The result after the multiplication is 0019H in the AX register. After adjusting the result with the AAM instruction, AX contains a 0205H. This is an unpacked BCD result of 25. If 3030H adds to 0205H, this becomes an ASCII result of 3235H.

EXAMPLE 7–23

```
0000  B0 05         MOV   AL,5     ;load multiplicand
0002  B1 05         MOV   CL,5     ;load multiplier
0004  F6 E1         MUL   CL
0006  D4 0A         AAM            ;adjust
```

The AAM instruction accomplishes this conversion by dividing AX by 10. The remainder is found in AL and the quotient is in AH. It has been noted that the second byte of the instruction contains a 0AH. If the 0AH is changed to another value, AAM divides by the new value. For example, if the second byte is changed to a 0BH, the AAM instruction divides by 11.

One side benefit of the AAM instruction is that it converts from binary to unpacked BCD. If a binary number between 0000H and 0063H appears in the AX register, the AAM instruction converts it to BCD. For example, if AX contains a 0060H before AAM, it will contain a 0906H after AAM executes. This is the unpacked BCD equivalent of 96 decimal. If 3030H is added to 0906H, the result changes to ASCII code.

Example 7–24 shows how the 16-bit binary contents of AX are converted to a four-digit ASCII character string by using division and the AAM instruction. This works for numbers between 0 and 9,999. First DX is cleared and then DX–AX is divided by 100. For example, AX = 245 after the division, AX = 2, and DX = 45. These separate halves are converted to BCD using AAM, then a 3030H is added to convert to ASCII code.

EXAMPLE 7–24

```
0000  33 D2         XOR   DX,DX         ;clear DX register
0002  B9 0064       MOV   CX,100        ;divide DX--AX by 100
0005  F7 F1         DIV   CX            ;AX=quotient and DX=remainder
0007  D4 0A         AAM                 ;convert quotient to BCD
0009  05 3030       ADD   AX,3030H      ;convert to ASCII
000C  92            XCHG  AX,DX         ;repeat for remainder
000D  D4 0A         AAM
000F  05 3030       ADD   AX,3030H
```

Example 7–25 uses the DOS 21H function AH = 02H to display a sample number in decimal on the video display using the AAM instruction. Notice how AAM is used to convert AL into BCD. Next, the ADD AX,3030H instruction converts the BCD code in AX into ASCII, for display with DOS INT 21H. Once the data are converted to ASCII code, they are displayed by loading DL with the most significant digit from AH. Next, the least significant digit is displayed from AL. Note that the DOS INT 21H function calls change AL.

EXAMPLE 7–25

```
                    ;a program that displays the number loaded into AL,
                    ;with the first instruction (48H), as a decimal number.
                    ;
                            .MODEL TINY             ;select TINY model
0000                        .CODE                   ;indicate start of CODE segment
                            .STARTUP                ;indicate start of program
0100  B0 48                 MOV     AL,48H          ;load AL with test data
0102  B4 00                 MOV     AH,0            ;clear AH
0104  D4 0A                 AAM                     ;convert to BCD
0106  05 3030               ADD     AX,3030H        ;convert to ASCII
0109  8A D4                 MOV     DL,AH           ;display most significant digit
010B  B4 02                 MOV     AH,2
010D  50                    PUSH    AX              ;save least significant digit
010E  CD 21                 INT     21H
0110  58                    POP     AX              ;restore AL
0111  8A D0                 MOV     DL,AL           ;display least significant digit
0113  CD 21                 INT     21H
                            .EXIT                   ;exit to DOS
                            END                     ;end of file
```

AAS Instruction. Like other ASCII adjust instructions, AAS adjusts the AX register after an ASCII subtraction. For example, suppose 35H subtracts from a 39H. The result will be 04H, which requires no correction. Here AAS will modify neither AH nor AL. On the other hand, if 38H subtracts from 37H, then AL will equal 09H and the number in AH will decrement by 1. This decrement allows multiple-digit ASCII numbers to be subtracted from each other.

7–4 BASIC LOGIC INSTRUCTIONS

The basic logic instructions include AND, OR, Exclusive-OR, and NOT. Another logic instruction, TEST, is explained in this section of the text because it is a special form of the AND instruction. Also explained is the NEG instruction, which is similar to the NOT instruction.

Logic operations provide binary bit control in *low-level software*. The logic instructions allow bits to be set, cleared, or complemented. Low-level software appears in machine language or assembly language form, and often controls the I/O devices in a system. All logic instructions affect the flag bits. Logic operations always clear the carry and overflow flags, while the other flags change to reflect the condition of the result.

When binary data are manipulated in a register or a memory location, the rightmost bit position is always numbered Bit 0. Bit position numbers increase from Bit 0 toward the left to Bit 7 for a byte and to Bit 15 for a word. A doubleword (32 bits) uses Bit position 31 as its leftmost bit.

FIGURE 7–3 (a) The truth table for the AND operation and (b) the logic symbol of an AND gate.

A	B	T
0	0	0
0	1	0
1	0	0
1	1	1

(a)

(b)

AND

The AND operation performs logical multiplication as illustrated by the truth table in Figure 7–3. Two bits, A and B, are ANDed to produce the result X. As indicated by the truth table, X is a logic 1 only when both A and B are logic 1's. For all other input combinations of A and B, X is a logic 0. It is important to remember that 0 AND anything is always 0 and 1 AND 1 is always 1.

The AND instruction can replace discrete AND gates if the speed required is not too great, although this is normally reserved for embedded control applications. (Intel has released the 80386EX embedded controller, which embodies the basic structure of the personal computer system.) With the 8086 microprocessor, the AND instruction executes in about a microsecond. With newer versions, the execution speed is greatly increased. If the circuit the AND instruction is to replace operates at a much slower speed than the microprocessor, the AND instruction is a logical replacement. This replacement can save a considerable amount of money. A single AND gate integrated circuit (7408) costs approximately 40¢, while it costs less than 1/100¢ to store the AND instruction in a read-only memory. Note that logic circuit replacements such as this appear only in control systems based on microprocessors, and do not generally find application in the personal computer.

The AND operation also clears bits of a binary number. The task of clearing a bit of a binary number is called **masking.** Figure 7–4 illustrates the process of masking. Notice that the leftmost four bits clear to 0, because 0 AND anything is 0. The bit positions of AND with 1's do not change. This occurs because if a 1 ANDs with a 1, a 1 results, and if a 1 ANDs with a 0, a 0 results.

The AND instruction uses any addressing mode except memory-to-memory and segment-register addressing. Refer to Table 7–14 for a list of some AND instructions and a comment about their operation.

An ASCII-coded number can be converted to BCD by using the AND instruction to mask off the four leftmost binary bit positions. This converts the ASCII 30H to 39H to 0–9. Example 7–26 shows a short program that converts the ASCII contents of BX into

FIGURE 7–4 The operation of the AND function showing how bits of a number are cleared to zero.

```
  x x x x  x x x x   Unknown number
• 0 0 0 0  1 1 1 1   Mask
-------------------
  0 0 0 0  x x x x   Result
```

TABLE 7–14 AND instructions

Instruction	*Comment*
AND AL,BL	AL = AL AND BL
AND CX,DX	CX = CX AND DX
AND ECX,EDI	ECX = ECX AND EDI
AND CL,33H	CL = CL AND 33H
AND DI,4FFFH	DI = DI AND 4FFFH
AND ESI,34H	ESI = ESI AND 00000034H
AND AX, [DI]	AX is ANDed with the word contents of the data-segment memory location addressed by DI; the result moves into AX
AND ARRAY [SI],AL	The byte contents of the data-segment memory location addressed by ARRAY plus SI are ANDed to AL; the result moves to memory
AND [EAX],CL	The byte contents of CL are ANDed with the byte contents of the data-segment memory location addressed by register EAX; the result moves to memory

BCD. The AND instruction in this example converts two digits from ASCII to BCD simultaneously.

EXAMPLE 7–26

```
0000  BB 3135                MOV   BX,3135H      ;load ASCII
0003  81 E3 0F0F             AND   BX,0F0FH      ;mask BX
```

OR

The OR operation performs logical addition and is often called the *inclusive-OR function*. The OR function generates a logic 1 output if any inputs are 1. A 0 appears at the output only when all inputs are 0. The truth table for the OR function appears in Figure 7–5. Here, the inputs, A and B, OR together to produce the X output. It is important to remember that 1 ORed with anything yields a 1.

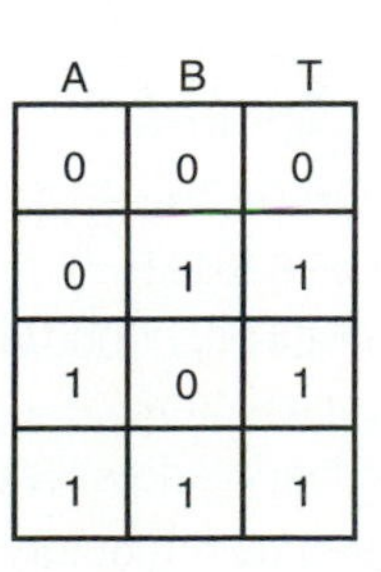

A	B	T
0	0	0
0	1	1
1	0	1
1	1	1

(a)

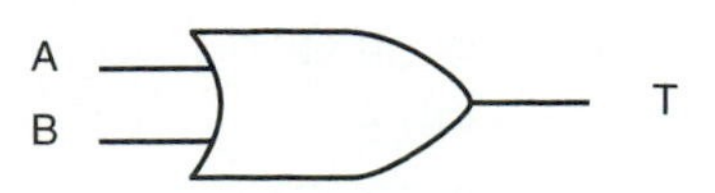

(b)

FIGURE 7–5 (a) The truth table for the OR operation and (b) the logic symbol of an OR gate.

FIGURE 7–6 The operation of the OR function showing how bits of a number are set to one.

```
  xxxx xxxx   Unknown number
+ 0000 1111   Mask
-----------
  xxxx 1111   Result
```

In embedded controller applications, the OR instruction can also replace discrete OR gates. This results in a considerable savings, because a quad, 2-input OR gate (7432) costs about 40¢, while the OR instruction costs less than 1/100¢ to store in a read-only memory.

Figure 7–6 shows how the OR gate sets (1) any bit of a binary number. Here, an unknown number (XXXX XXXX) ORs with a 0000 1111 to produce a result of XXXX 1111. The four rightmost bits set, while the four leftmost bits remain unchanged. The OR operation sets any bit and the AND operation clears any bit.

The OR instruction uses any of the addressing modes allowed to any other instruction except segment register addressing. Table 7–15 illustrates several OR instructions with comments about their operation.

Suppose two BCD numbers are multiplied and adjusted with the AAM instruction. The result appears in AX as a two-digit unpacked BCD number. Example 7–27 illustrates this multiplication and shows how to change the result into a two-digit ASCII-coded number using the OR instruction. Here, the instruction OR AX,3030H converts the 0305H found in AX to 3335H. The OR operation can be replaced with an ADD AX,3030H instruction to obtain the same results.

EXAMPLE 7–27

```
0000  B0 05        MOV   AL,5        ;load data
0002  B3 07        MOV   BL,7
0004  F6 E3        MUL   BL
0006  D4 0A        AAM               ;adjust
0008  0D 3030      OR    AX,3030H    ;to ASCII
```

TABLE 7–15 OR instructions

Instruction	*Comment*
OR AH,BL	AH = AH OR BL
OR SI,DX	SI = SI OR DX
OR EAX,EBX	EAX = EAX OR EBX
OR DH,0A3H	DH = DH OR 0A3H
OR SP,990DH	SP = SP OR 990DH
OR EBP,10	EBP = EBP OR 0000000AH
OR DX,[BX]	DX is ORed with the word contents of the data-segment memory location addressed by BX
OR DATES [DI + 2],AL	The data-segment memory location addressed by DATES plus DI plus 2 is ORed with AL

FIGURE 7–7 (a) The truth table for the exclusive-OR operation and (b) the logic symbol of an exclusive-OR gate.

A	B	T
0	0	0
0	1	1
1	0	1
1	1	0

(a)

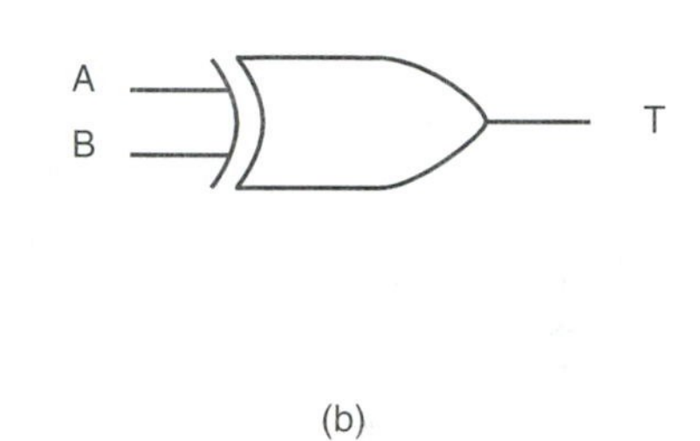

(b)

Exclusive-OR

The exclusive-OR instruction (XOR) differs from inclusive-OR (OR). The 1,1 condition of the OR function produces a 1, while the 1,1 condition of the exclusive-OR operation produces a 0. The exclusive-OR operation excludes this condition, while the inclusive-OR includes it.

Figure 7–7 shows the truth table of the exclusive-OR function. (Compare this with Figure 7–5 to appreciate the difference between these two OR functions.) If the inputs of the exclusive-OR function are both 0 or both 1, the output is 0. If the inputs are different, the output is a 1. Because of this, the exclusive-OR is sometimes called a *comparator*.

The XOR instruction uses any addressing mode except segment-register addressing. Table 7–16 lists various forms of the exclusive-OR instruction with comments about their operation.

Like the AND and OR functions, exclusive-OR can replace discrete logic circuitry in embedded applications. The 7486 quad, two-input exclusive-OR gate is replaced by one XOR instruction. The 7486 costs about 40¢, while the instruction costs less than 1/100¢ to

TABLE 7–16 Exclusive-OR instructions

Instruction	*Comment*
XOR CH,DL	CH = CH XOR DL
XOR SI,BX	SI = SI XOR BX
XOR EBX,EDI	EBX = EBX XOR EDI
XOR CH,0EEH	CH = CH XOR 0EEH
XOR DI,00DDH	DI = DI XOR 00DDH
XOR ESI,100	ESI = ESI XOR 00000064H
XOR DX,[SI]	DX is XORed with the word contents of the data-segment memory location addressed by SI; the result is left in DX
XOR DATES[DI + 2],AL	The byte contents of the data-segment memory location addressed by DATES plus DI plus 2 are XORed with AL; the result is left in memory

FIGURE 7–8 The operation of the exclusive-OR function showing how bits of a number are inverted.

```
 xxxx xxxx   Unknown number
⊕0000 1111   Mask
 ---------
 xxxx x̄x̄x̄x̄   Result
```

store in the memory. Replacing just one 7486 saves a considerable amount of money, especially if many systems are built.

The exclusive-OR instruction is useful if some bits of a register or memory location must be inverted. This instruction allows part of a number to be inverted or complemented. Figure 7–8 shows how just part of an unknown quantity can be inverted by XOR. Notice that when a 1 exclusive-ORs with X, the result is $\overline{X}$. If a 0 exclusive-ORs with X, the result is X.

Suppose that the 10 leftmost bits of the BX register must be inverted without changing the 6 rightmost bits. The XOR BX,0FFC0H instruction accomplishes this task. The AND instruction clears bits, the OR instruction sets bits, and the exclusive-OR instruction inverts bits. These three instructions allow a program to gain complete control over any bit stored in any register or memory location. This is ideal for control system applications where equipment must be turned on (1), turned off (0), or toggled from on to off or off to on.

A fairly common use for the exclusive-OR instruction is to clear a register to zero. For example, the XOR CH,CH instruction clears register CH to 00H and requires two bytes of memory. The MOV CH,00H instruction also clears CH to 00H, but requires three bytes of memory. Because of this savings, the XOR instruction is used in place of a move immediate to clear a register.

Example 7–28 shows a short sequence of instructions that clear Bits 0 and 1 of CX, set Bits 9 and 10 of CX, and invert Bit 12 of CX.

EXAMPLE 7–28

```
0000  81 C9 0600        OR    CX,0600H       ;set bits 9 and 10
0004  83 E1 FC          AND   CX,0FFFCH      ;clear bits 0 and 1
0007  81 F1 1000        XOR   CX,1000H       ;invert bit 12
```

Test and Bit Test Instructions

The TEST instruction performs the AND operation. The difference is that the AND instruction changes the destination operand, while the TEST instruction does not. A TEST only affects the condition of the flag register, which indicates the result of the test. The TEST instruction uses the same addressing modes as the AND instruction. Table 7–17 illustrates some forms of the TEST instruction, with comments about the operation of each form.

The TEST instruction functions the same way as a CMP instruction. The difference is that the TEST instruction normally tests a single bit (or occasionally multiple bits), while the CMP instruction tests the entire byte or word. The zero flag (Z) is a logic 1 (indicating a zero result) if the bit under test is a zero, and a logic 0 (indicating a nonzero result) if the bit under test is not zero.

Usually the TEST instruction is followed by either a JZ (*jump zero*) or JNZ (*jump not zero*) instruction. The destination operand is normally tested against immediate data. The value of immediate data are 1 to test the rightmost bit position, 2 to test the next bit, 4 for the next, etc.

TABLE 7–17 TEST instructions

Instruction	*Comment*
TEST DL,DH	DL is ANDed with DH; neither DL nor DH changes; only the flags change
TEST CX,BX	CX is ANDed with BX; neither CX nor BX changes; only the flags change
TEST EDX,ECX	EDX is ANDed with ECX; neither EDX nor ECX changes; only the flags change
TEST AH,4	AH is ANDed with 4; AH does not change; only the flags change
TEST EAX,256	EAX is ANDed with 256; EAX does not change; only the flags change

Example 7–29 lists a short program that tests the rightmost and leftmost bit positions of the AL register. Here, 1 selects the rightmost bit and 128 selects the leftmost bit. The JNZ instruction follows each test to jump to different memory locations depending on the outcome of the tests. The JNZ instruction jumps to the operand address (RIGHT or LEFT in the example) if the bit under test is not zero.

EXAMPLE 7–29

```
0000  A8 01           TEST   AL,1        ;test right bit
0002  75 1C           JNZ    RIGHT       ;if set
0004  A8 80           TEST   AL,128      ;test left bit
0006  75 38           JNZ    LEFT        ;if set
```

The 80386/80486/Pentium microprocessors also contain additional test instructions that test bit positions. Table 7–18 lists the four bit test instructions available to these microprocessors.

TABLE 7–18 Bit test instructions

Instruction	*Comment*
BT	Test a bit in the destination operand specified by the source operand
BTC	Test and complement a bit in the destination operand specified by the source operand
BTR	Test and reset a bit in the destination operand specified by the source operand
BTS	Test and set a bit in the destination operand specified by the source operand

TABLE 7–19 NOT and NEG instructions

Instruction	*Comment*
NOT CH	CH is one's complemented
NEG CH	CH is two's complemented
NEG AX	AX is two's complemented
NOT EBX	EBX is one's complemented
NEG ECX	ECX is two's complemented
NOT TEMP	The contents of the data-segment memory location addressed by TEMP are one's complemented (the size of TEMP is determined by how TEMP is defined)
NOT BYTE PTR [BX]	The byte contents of the data-segment memory location addressed by BX are one's complemented

All four forms of the bit test instruction test the bit position in the destination operand selected by the source operand. For example, the BT AX,4 instruction tests bit position 4 in AX. The result of the test is located in the carry flag bit. If bit position 4 is a 1, carry is set; if bit position 4 is a zero, carry is cleared.

The remaining three bit test instructions also place the bit under test into the carry flag, and, afterwards, change the bit under test. The BTC AX,4 instruction complements bit position 4 after testing it, the BTR AX,4 instruction clears it (0) after the test, and the BTS AX,4 instruction sets it (1) after the test.

Example 7–30 repeats the sequence of instructions listed in Example 7–28. Here, the BTR instruction clears bits in CX, BTS sets bits in CX, and BTC inverts bits in CX.

EXAMPLE 7–30

```
0000  0F BA E9 09          BTS   CX,9       ;set bit 9
0004  0F BA E9 0A          BTS   CX,10      ;set bit 10
0008  0F BA F1 00          BTR   CX,0       ;clear bit 0
000C  0F BA F1 01          BTR   CX,1       ;clear bit 1
0010  0F BA F9 0C          BTC   CX,12      ;invert bit 12
```

NOT and NEG

Logical inversion (the *one's complement* or NOT) and arithmetic sign inversion (the *two's complement* or NEG) are the next two logic functions presented. These two instructions contain only one operand. Table 7–19 lists some variations of the NOT and NEG instructions. Like most other instructions, NOT and NEG can use any addressing mode except segment-register addressing.

The NOT instruction inverts all bits of a byte, word, or doubleword. The NEG instruction two's complements a number, which means that the arithmetic sign of a signed number changes from positive to negative or negative to positive. The NOT function is considered logical and the NEG function is considered arithmetic.

7–5 SHIFT AND ROTATE INSTRUCTIONS

Shift and rotate instructions manipulate binary numbers at the binary bit level, just as did the AND, OR, exclusive-OR, and NOT instructions. Shifts and rotates find their most common application in low-level software used to control I/O devices. The microprocessor contains a complete set of shift and rotate instructions used to shift or rotate any memory, data, or register.

Shifts

Shift instructions position or move numbers to the left or right within a register or memory location. They also perform simple arithmetic such as multiplication by powers of 2^{+n} (*left shift*) and division by powers of 2^{-n} (*right shift*). The microprocessor's instruction set contains four different shift instructions: two are *logical shifts* and two are *arithmetic shifts*. All four shift operations appear in Figure 7–9.

Notice in Figure 7–9 that there are two right shifts and two left shifts. The logical shifts move a 0 into the rightmost bit position for a logical left shift and a 0 into the leftmost bit position for a logical right shift. There are also two arithmetic shifts. The arithmetic and logical left shifts are identical. The arithmetic and logical right shifts are different, because the arithmetic right shift copies the sign bit through the number, while the logical right shift copies a 0 through the number.

Logical shift operations function with unsigned numbers and arithmetic shifts function with signed numbers. Logical shifts multiply or divide unsigned data, and arithmetic

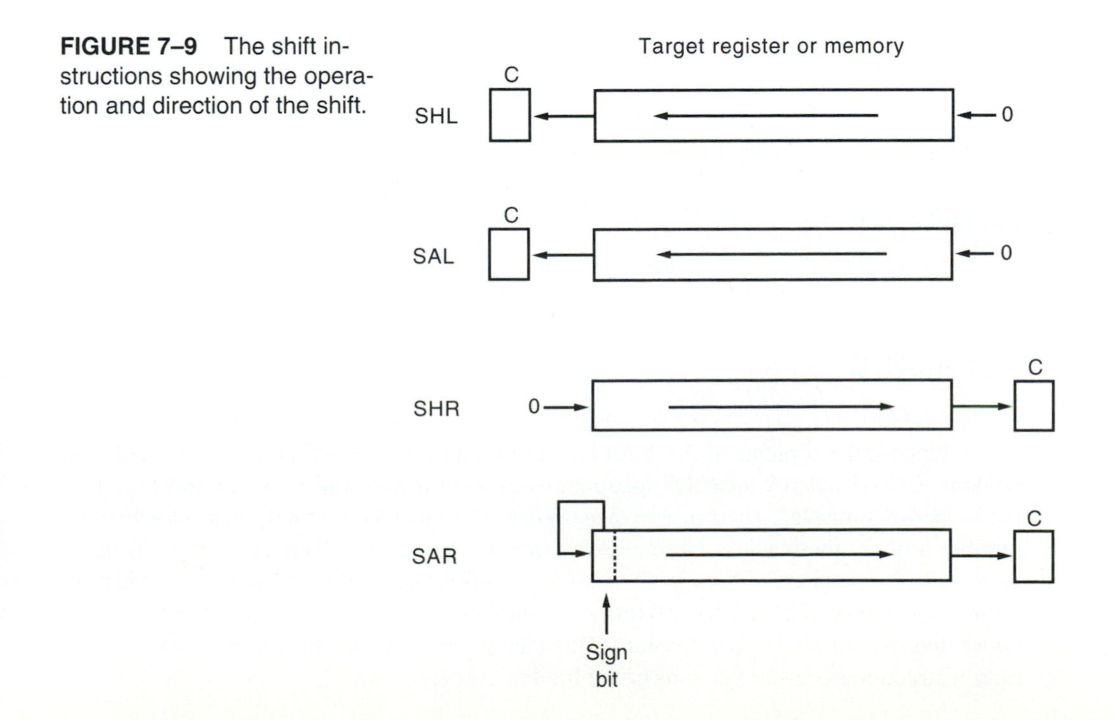

FIGURE 7–9 The shift instructions showing the operation and direction of the shift.

TABLE 7–20 Shift instructions

Instruction	*Comment*
SHL AX,1	Logically shifts AX left one place
SHR BX,12	Logically shifts BX right 12 places
SHR ECX,10	Logically shifts ECX right 10 places
SAL DATA1,CL	Arithmetically shifts DATA1 in the data segment left the number of places contained in CL
SAR SI,2	Arithmetically shifts SI right two places
SAR EDX,14	Arithmetically shifts EDX right 14 places

Note: The 8086/8088 microprocessor allows an immediate shift count of 1 only.

shifts multiply or divide signed data. A shift left always multiplies by 2 for each bit position shifted and a shift right always divides by 2 for each bit position shifted. Shifting a number two places multiplies or divides by 4.

Table 7–20 illustrates some addressing modes allowed for the various shift instructions. Two different types of shifts allow any register (except the segment register) or memory location to be shifted. One mode uses an immediate shift count, and the other uses register CL to hold the shift count. Note that CL must hold the shift count. When CL is the shift count, it does not change when the shift instruction executes. Note that the shift count is a modulo-32 count. This means that a shift count of 33 will shift the data one place (33 / 32 = remainder of 1).

Example 7–31 shows how to shift the DX register left 14 places in two different ways. The first method uses an immediate shift count of 14. The second method loads a 14 into CL and then uses CL as the shift count. Both instructions shift the contents of the DX register logically to the left 14 binary bit positions or places.

EXAMPLE 7–31

```
0000  C1 E2 0E          SHL   DX,14

                        or

0003  B1 0E             MOV   CL,14
0005  D3 E2             SHL   DX,CL
```

Suppose the contents of AX must be multiplied by 10, as in Example 7–32. This can be done in two ways: by the MUL instruction or by shifts and additions. A number is doubled when it shifts left one binary place. When a number is doubled, then added to the number times 8, the result is 10 times the number. The number 10 decimal is 1010 in binary. A logic 1 appears in both the 2's and 8's positions. If 2 times the number is added to 8 times the number, the result is 10 times the number. Using this technique, a program can be written to multiply by any constant. This technique often executes faster than the multiply instruction in earlier versions of the Intel microprocessor.

EXAMPLE 7–32

```
                        ;Multiply AX by 10 (1010)
                        ;
0000  D1 E0                       SHL   AX,1         ;AX times 2
0002  8B D8                       MOV   BX,AX
0004  C1 E0 02                    SHL   AX,2         ;AX times 8
0007  03 C3                       ADD   AX,BX        ;10 times AX
                        ;
                        ;Multiply AX by 18 (10010)
                        ;
0009  D1 E0                       SHL   AX,1         ;AX times 2
000B  8B D8                       MOV   BX,AX
000D  C1 E0 03                    SHL   AX,3         ;AX times 16
0010  03 D8                       ADD   BX,AX        ;18 times AX
```

Double-Precision Shifts (80386/80486/Pentium Only). The 80386 and above contain two double-precision shifts: SHLD (*shift left*) and SHRD (*shift right*). Each instruction contains three operands instead of the two found with the other shift instructions. Both instructions function with two 16- or 32-bit registers or with one 16- or 32-bit memory location and a register.

The SHRD AX,BX,12 instruction is an example of the double-precision shift right instruction. This instruction logically shifts AX right by 12 bit positions. The rightmost 12 bits of BX shift into the leftmost 12 bits of AX. The contents of BX remain unchanged by this instruction. The shift count can be an immediate count, as in this example, or can be found in register CL, like other shift instructions.

The SHLD EBX,ECX,16 instruction shifts EBX left. The leftmost 16 bits of ECX fill the rightmost 16 bits of EBX after the shift. As before, the contents of ECX, the second operand, remain unchanged. This instruction as well as SHRD affect the flag bits.

Rotates

Rotate instructions position binary data by rotating the information in a register or memory location either from one end to another or through the carry flag. They are often used to shift or position numbers that are wider than 16 bits in the 8086–80286 microprocessor or wider than 32 bits in the 80386 through Pentium. The four rotate instructions available appear in Figure 7–10.

Numbers rotate through a register or a memory location and the C flag (carry) or through a register or memory location only. With either type of rotate instruction, the programmer can select either a left or a right rotate. Addressing modes used with rotates are the same as those used with shifts. A rotate count can be immediate or located in register CL. Table 7–21 lists some possible rotate instructions. If CL is used for a rotate count, it does not change. The count in CL is a modulo-32 count.

EXAMPLE 7–33

```
0000  D1 E0                 SHL   AX,1
0002  D1 D3                 RCL   BX,1
0004  D1 D2                 RCL   DX,1
```

Rotates are often used to shift wide numbers to the left or right. The program listed in Example 7–33 shifts the 48-bit number in registers DX, BX, and AX left one binary

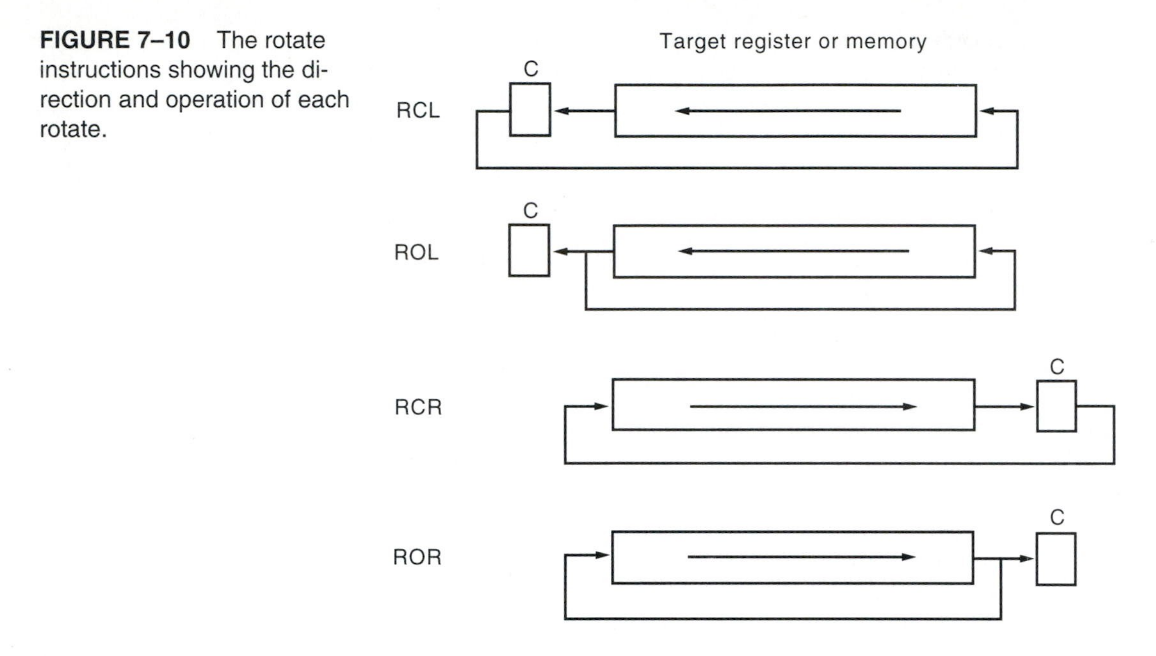

FIGURE 7–10 The rotate instructions showing the direction and operation of each rotate.

place. Notice that the least significant 16 bits (AX) are shifted left first. This moves the leftmost bit of AX into the carry flag bit. Next the rotate BX instruction rotates carry into BX and its leftmost bit moves into carry. The last instruction rotates carry into DX and the shift is complete.

Bit Scan Instructions

Although bit scan instructions don't shift or rotate numbers, they do scan through a number searching for a one bit. Because this is accomplished within the microprocessor by shifting the number, bit scan instructions are included in this section of the text.

The bit scan instructions BSF (bit scan forward) and BSR (bit scan reverse) are only available in the 80386/80486/Pentium microprocessors. Both forms scan through a number

TABLE 7–21 Rotate instructions

Instruction	*Comment*
ROL SI,4	SI rotates left four places
RCL BL,6	BL rotates left through carry six places
ROL ECX,18	ECX rotates left 18 places
RCR AH,CL	AH rotates right through carry the number of places contained in CL
ROR WORD PTR [BP],2	The word contents of the stack-segment memory location addressed by BP rotates right two places

Note: The 8086/8088 microprocessor can only use an immediate rotate count of 1.

searching for the first one bit encountered. The BSF instruction scans the number from the rightmost bit towards the left and BSR scans the number from the leftmost bit towards the right. If a one bit is encountered by either bit scan instruction, the zero flag is set and the bit position of the one bit is placed into the destination operand. If no one bit is encountered (i.e., the number contains all zeros), the zero flag is cleared. This means that the result is not-zero if no one bit is encountered.

For example, if EAX = 60000000H and the BSF EBX,EAX instruction executes, the number is scanned from the rightmost bit toward the left. The first one bit encountered is at bit position 29, which is placed into EBX, and the zero flag bit is set. If the same value for EAX is used for the BSR instruction, the EBX register is loaded with a 30 and the zero flag bit is set.

7–6 STRING COMPARISONS

String instructions are very powerful because they allow the programmer to manipulate large blocks of data with relative ease. Block data manipulation occurs with the string instructions MOVS, LODS, STOS, INS, and OUTS, which were discussed in Chapter 6.

In this section, additional string instructions that allow a section of memory to be tested against a constant or against another section of memory are discussed. To accomplish these tasks use the SCAS (*string scan*) or CMPS (*string compare*) instructions.

SCAS

The string scan instruction (SCAS) compares the AL register with a byte block of memory, the AX register with a word block of memory, or the EAX register (80386/80486/Pentium only) with a doubleword block of memory. The SCAS instruction subtracts memory from AL, AX, or EAX without affecting either the register or the memory location. The opcode used for byte comparison is *SCASB*, the opcode used for the word comparison is *SCASW*, and the opcode used for a doubleword comparison is *SCASD*. In all cases, the contents of the extra segment memory location addressed by DI is compared with AL, AX, or EAX. This default segment (ES) cannot be changed with a segment override prefix.

Like the other string instructions, SCAS instructions use the direction flag (D) to select either auto-increment or auto-decrement operation for DI. SCAS instructions repeat if prefixed by a *conditional* repeat prefix.

EXAMPLE 7–34

```
0000  BF 0011 R      MOV    DI,OFFSET BLOCK    ;address data
0003  FC             CLD                       ;auto-increment
0004  B9 0064        MOV    CX,100             ;load counter
0007  32 C0          XOR    AL,AL              ;clear AL
0009  F2/AE          REPNE SCASB               ;search
```

Suppose a section of memory 100 bytes in length begins at location BLOCK. This section of memory must be tested to see if any location contains a 00H. The program in

Example 7–34 shows how to search this part of memory for a 00H using the SCASB instruction. In this example, the SCASB instruction is prefixed with an **REPNE** (*repeat while not equal*). The REPNE prefix causes the SCASB instruction to repeat until *either* the CX register reaches 0, or until an equal condition exists as the outcome of the SCASB instruction's comparison. Another conditional repeat prefix is **REPE** (*repeat while equal*). With either repeat prefix, the contents of CX decrement without affecting the flag bits. The SCASB instruction and the comparison it makes change the flags.

Suppose you must develop a program that skips ASCII-coded spaces in a memory array. This task appears in the procedure listed in Example 7–35. This procedure assumes that the DI register already addresses the ASCII-coded character string, and that the length of the string is 256 bytes or less. Because this program is to skip spaces (20H), the REPE (repeat while equal) prefix is used with a SCASB instruction. The SCASB instruction repeats the comparison, searching for a 20H, as long as an equal condition exists.

EXAMPLE 7–35

```
0000                 SKIP    PROC  FAR

0000  FC                     CLD                  ;auto-increment
0001  B9 0100                MOV   CX,256         ;counter
0004  B0 20                  MOV   AL,20H         ;get space
0006  F3/AE                  REPE SCASB           ;search
0008  CB                     RET

0009                 SKIP    ENDP
```

CMPS

The *compare strings* instruction (CMPS) always compares two sections of memory data as bytes (CMPSB), words (CMPSW), or doublewords (CMPSD). Only the 80386/80486/Pentium can use doublewords. The contents of the data-segment memory location addressed by SI are compared with the contents of the extra segment memory location addressed by DI. The CMPS instruction increments or decrements both SI and DI. The CMPS instruction is normally used with either the REPE or REPNE prefix. Alternates to these prefixes are REPZ (*repeat while zero*) and REPNZ (*repeat while not zero*), but usually REPE or REPNE are used in programming.

EXAMPLE 7–36

```
0000                 MATCH    PROC   FAR

0000  BE 0075 R               MOV    SI,OFFSET LINE      ;address LINE
0003  BF 007F R               MOV    DI,OFFSET TABLE     ;address TABLE
0006  FC                      CLD                        ;auto-increment
0007  B9 000A                 MOV    CX,10               ;counter
000A  F3/A6                   REPE   CMPSB               ;search
000C  CB                      RET

000D                 MATCH    ENDP
```

Example 7–36 illustrates a short procedure that compares two sections of memory searching for a match. The CMPSB instruction is prefixed with a REPE. This causes the search to continue as long as an equal condition exists. When the CX register becomes 0, or an unequal condition exists, the CMPSB instruction stops execution. After the CMPSB instruction ends, the CX register is zero or the flags indicate an equal condition when the two strings match. If CX is not zero or the flags indicate a not-equal condition, the strings do not match.

7–7 SUMMARY

1. Addition (ADD) can be 8-, 16-, or 32-bit. The ADD instruction allows any addressing mode except segment-register addressing. Most flags (C, A, S, Z, P, and O) change when the ADD instruction executes. A different type of addition, add-with-carry (ADC), adds two operands and the contents of the carry flag bit (C). The 80486 and Pentium microprocessors have an additional instruction (XADD) that combines an addition with an exchange.
2. The increment instruction (INC) adds 1 to the byte, word, or doubleword contents of a register or memory location. The INC instruction affects the same flag bits as ADD except the carry flag. The BYTE PTR, WORD PTR, and DWORD PTR directives appear with the INC instruction when the contents of a memory location are addressed by a pointer.
3. Subtraction (SUB) is a byte, word, or doubleword and is performed on a register or a memory location. The only form of addressing not allowed by the SUB instruction is segment-register addressing. The subtract instruction affects the same flags as ADD and carries if the SBB form is used.
4. The decrement (DEC) instruction subtracts one from the contents of a register or a memory location. The only addressing modes not allowed with DEC are immediate or segment-register addressing. The DEC instruction does not affect the carry flag and is often used with BYTE PTR, WORD PTR, or DWORD PTR.
5. The compare (CMP) instruction is a special form of subtraction that does not store the difference; instead, the flags change to reflect the difference. Compare is used to compare an entire byte or word located in any register (except segment) or memory location. An additional compare instruction (CMPXCHG), which is a combination of compare and exchange instructions, is available in the 80486/Pentium microprocessor. In the Pentium microprocessor, the CMPXCHG8B instruction compares and exchanges quadword data.
6. Multiplication is byte, word, or doubleword and can be signed (IMUL) or unsigned (MUL). An 8-bit multiplication always multiplies register AL by an operand, with the product found in AX. A 16-bit multiplication always multiplies register AX by an operand, with the product found in DX–AX. A 32-bit multiplication always multiplies register EAX by an operand, with the product found in EDX–EAX. A special IMUL immediate instruction that contains three operands exists on the 80286–80486 and Pentium. For example, IMUL BX,CX,3 multiplies CX by 3 and leaves the product in BX.
7. Division is byte, word, or doubleword and can be signed (IDIV) or unsigned (DIV). For 8-bit division, the AX register divides by the operand, after which the quotient appears in AL and remainder appears in AH. In 16-bit division, the DX–AX register

divides by the operand, after which the AX register contains the quotient and DX register contains the remainder. In 32-bit division, the EDX–EAX register is divided by the operand, after which the EAX register contains the quotient and the EDX register contains the remainder. Note that the remainder after a signed division always assumes the sign of the dividend.

8. BCD data can be added or subtracted in packed form by adjusting the result of the addition with DAA or the result of the subtraction with DAS. ASCII data can be added, subtracted, multiplied, or divided by adjusting the operations with AAA, AAS, AAM, and AAD.
9. The AAM instruction has an interesting added feature that allows it to convert a binary number into unpacked BCD. This instruction converts a binary number between 00H–63H into unpacked BCD in AX. The AAM instruction divides AX by 10, and leaves the remainder in AL and the quotient in AH.
10. The AND, OR, and exclusive-OR instructions perform logic functions on a byte, word, or doubleword stored in a register or memory location. All flags change with these instructions with carry (C) and overflow (O) cleared.
11. The TEST instruction performs the AND operation, but the logical product is lost. This instruction changes the flag bits to indicate the outcome of the test.
12. The NOT and NEG instructions perform logical inversion and arithmetic inversion. The NOT instruction one's complements an operand, and the NEG instruction two's complements an operand.
13. There are eight different shift and rotate instructions. Each of these instructions shifts or rotates a byte, word, or doubleword register or memory data. These instructions have two operands: the first is the location of the data shifted or rotated, and the second is an immediate shift or rotate count or CL. If the second operand is CL, the CL register holds the shift or rotate count. In the 80386/80486/Pentium microprocessor, two additional double-precision shift instructions (SHRD and SHLD) exist.
14. The scan string (SCAS) instruction compares AL, AX, or EAX with the contents of the extra segment memory location addressed by DI.
15. The string compare (CMPS) instruction compares the byte, word, or doubleword contents of two sections of memory. One section is addressed by DI in the extra segment, and the other section is addressed by SI in the data segment.
16. The SCAS and CMPS instructions repeat with the REPE or REPNE prefixes. The REPE prefix repeats the string instruction while an equal condition exists, and the REPNE repeats the string instruction while a not-equal condition exists.
17. Example 7–36 illustrates a program that uses some of the instructions in this chapter to search the video display (beginning at address B800:000) to find if it contains the word BUG. If the word BUG is found, the program displays a Y. If BUG is not found it displays an N. Notice how the CMPSB instruction is used to search for BUG.

EXAMPLE 7–37

```
                  ;a program that tests the video display for the word BUG
                  ;if BUG appears anywhere on the display the program display Y
                  ;if BUG does not appear, the program displays N
                  ;
                          .MODEL SMALL               ;select SMALL model
0000                      .DATA                      ;indicate start of DATA segment
```

```
0000 42 55 47     DATA1  DB      'BUG'              ;define BUG
0000                     .CODE                      ;indicate start of CODE segment
                         .STARTUP                   ;indicate start of program
0017 B8 B800             MOV     AX,0B800H          ;address segment B800 with ES
001A 8E C0               MOV     ES,AX
001C B9 07D0             MOV     CX,25*80           ;set count
001F FC                  CLD                        ;select increment
0020 BF 0000             MOV     DI,0               ;address first display position
0023              L1:
0023 BE 0000 R           MOV     SI,OFFSET DATA1    ;address BUG
0026 57                  PUSH    DI                 ;save display address
0027 A6                  CMPSB                      ;test for B
0028 75 0A               JNE     L2                 ;if display is not B
002A 47                  INC     DI                 ;address next display position
002B A6                  CMPSB                      ;test for U
002C 75 06               JNE     L2                 ;if display is not U
002E 47                  INC     DI                 ;address next display position
002F A6                  CMPSB                      ;test for G
0030 B2 59               MOV     DL,'Y'             ;load Y for possible BUG
0032 74 09               JE      L3                 ;if BUG is found
0034              L2:
0034 5F                  POP     DI                 ;restore display address
0035 83 C7 02            ADD     DI,2               ;point to next display position
0038 E2 E9               LOOP    L1                 ;repeat until entire screen is tested
003A 57                  PUSH    DI                 ;save display address
003B B2 4E               MOV     DL,'N'             ;indicate N if BUG not found
003D              L3:
003D 5F                  POP     DI                 ;clear stack
003E B4 02               MOV     AH,2               ;display DL function
0040 CD 21               INT     21H                ;display ASCII from DL
                         .EXIT                      ;exit to DOS
                         END                        ;end of file
```

7–8 QUESTIONS AND PROBLEMS

1. Select an ADD instruction that will:
   (a) add BX to AX
   (b) add 12H to AL
   (c) add EDI and EBP
   (d) add 22H to CX
   (e) add the data addressed by SI to AL
   (f) add CX to the data stored at memory location FROG
2. What is wrong with the ADD ECX,AX instruction?
3. Is it possible to add CX to DS with the ADD instruction?
4. If AX = 1001H and DX = 20FFH, list the sum and the contents of each flag register bit (C, A, S, Z, and O) after the ADD AX,DX instruction executes.
5. Develop a short sequence of instructions that add AL, BL, CL, DL, and AH. Save the sum in the DH register.
6. Develop a short sequence of instructions that add AX, BX, CX, DX, and SP. Save the sum in the DI register.
7. Develop a short sequence of instructions that add ECX, EDX, and ESI. Save the sum in the EDI register.

8. Select an instruction that adds BX to DX and that also adds the contents of the carry flag (C) to the result.
9. Choose an instruction that adds a 1 to the contents of the SP register.
10. What is wrong with the INC [BX] instruction?
11. Select a SUB instruction that will:
    (a) subtract BX from CX
    (b) subtract 0EEH from DH
    (c) subtract DI from SI
    (d) subtract 3322H from EBP
    (e) subtract the data address by SI from CH
    (f) subtract the data stored ten words after the location addressed by SI from DX
    (g) subtract AL from memory location FROG
12. If DL = 0F3H and BH = 72H, list the difference after BH subtracts from DL, and show the contents of the flag register bits.
13. Write a short sequence of instructions that subtract the numbers in DI, SI, and BP from the AX register. Store the difference in register BX.
14. Choose an instruction that subtracts 1 from register EBX.
15. Explain what the SBB [DI –4],DX instruction accomplishes.
16. Explain the difference between the SUB and CMP instruction.
17. When two 8-bit numbers are multiplied, where is the product found?
18. When two 16-bit numbers are multiplied, what two registers hold the product? Show which register contains the most and least significant portions of the product.
19. When two numbers multiply, what happens to the O and C flag bits?
20. Where is the product stored for the MUL EDI instruction?
21. What is the difference between the IMUL and MUL instructions?
22. Write a sequence of instructions that cube the 8-bit number found in DL. Load DL with a 5 initially, and make sure your result is a 16-bit number.
23. Describe the operation of the IMUL BX,DX,100H instruction.
24. When 8-bit numbers are divided, in which register is the dividend found?
25. When 16-bit numbers are divided, in which register is the quotient found?
26. What type of errors are detected during a division?
27. Explain the difference between the IDIV and DIV instructions.
28. Where is the remainder found after an 8-bit division?
29. Write a short sequence of instructions that divide the number in BL by the number in CL and then multiply the result by 2. The final answer must be a 16-bit number stored in the DX register.
30. Which instructions are used with BCD arithmetic operations?
31. Which instructions are used with ASCII arithmetic operations?
32. Explain how the AAM instruction converts from binary to BCD.
33. Develop a sequence of instructions that convert the unsigned number in AX (values of 0–65535) into a 5-digit BCD number stored in memory beginning at the location addressed by the BX register in the data segment. Note that the most significant character is stored first, and no attempt is made to blank leading zeros.
34. Develop a sequence of instructions that add the 8-digit BCD number in AX and BX to the 8-digit BCD number in CX and DX. (AX and CX are the most significant registers.) The result must be found in CX and DX after the addition.

35. Select an AND instruction that will:
    (a) AND BX with DX and save the result in BX
    (b) AND 0EAH with DH
    (c) AND DI with BP and save the result in DI
    (d) AND 1122H with EAX
    (e) AND the data addressed by BP with CX and save the result in memory
    (f) AND the data stored in four words before the location addressed by SI with DX and save the result in DX
    (g) AND AL with memory location WHAT and save the result at location WHAT
36. Develop a short sequence of instructions that clear (0) the three leftmost bits of DH without changing the remainder DH and store the result in BH.
37. Select an OR instruction that will:
    (a) OR BL with AH and save the result in AH
    (b) OR 88H with ECX
    (c) OR DX with SI and save the result in SI
    (d) OR 1122H with BP
    (e) OR the data addressed by BX with CX and save the result in memory
    (f) OR the data stored 40 bytes after the location addressed by BP with AL and save the result in AL
    (g) OR AH with memory location WHEN and save the result in WHEN
38. Develop a short sequence of instructions that set (1) the rightmost five bits of DI without changing the remaining bits of DI. Save the result in SI.
39. Select the XOR instruction that will:
    (a) XOR BH with AH and save the result in AH
    (b) XOR 99H with CL
    (c) XOR DX with DI and save the result in DX
    (d) XOR 1A23H with ESP
    (e) XOR the data addressed by EBX with DX and save the result in memory
    (f) XOR the data stored 30 words after the location addressed by BP with DI and save the result in DI
    (g) XOR DI with memory location WELL and save the result in DI
40. Develop a sequence of instructions that set (1) the rightmost four bits of AX, clear (0) the leftmost three bits of AX, and invert bits 7, 8, and 9 of AX.
41. Describe the difference between the AND and TEST instructions.
42. Select an instruction that tests bit position 2 of register CH.
43. What is the difference between the NOT and NEG instructions?
44. Select the correct instruction to perform each of the following tasks:
    (a) shift DI right 3 places with zeros moved into the leftmost bit
    (b) move all bits in AL left one place, making sure that a 0 moves into the rightmost bit position
    (c) rotate all the bits of AL left 3 places
    (d) rotate carry right one place through EDX
    (e) move the DH register right one place, making sure that the sign of the result is the same as the sign of the original number
45. What does the SCASB instruction accomplish?
46. For string instructions, DI always addresses data in the ______________ segment.

47. What is the purpose of the D flag bit?
48. Explain what the REPE prefix does when coupled with the SCASB instruction.
49. What condition or conditions will terminate the repeated string instruction REPNE SCASB?
50. Describe what the CMPSB instruction accomplishes.
51. Develop a sequence of instructions that scan through a 300H byte section of memory called LIST located in the data segment, searching for a 66H.
52. What happens if AH = 02H and DL = 43H when the INT 21H instruction is executed?

CHAPTER 8

Program Control Instructions

INTRODUCTION

The program control instructions direct the flow of a program and allow the flow to change. A change in flow often occurs after a decision, made with the CMP or TEST instruction, is followed by a conditional jump instruction. This chapter explains the program control instructions including jumps, calls, returns, interrupts, and machine control instructions.

OBJECTIVES

Upon completion of this chapter, you will be able to:

1. Use both conditional and unconditional jump instructions to control the flow of a program.
2. Use the call and return instructions to include procedures in the program structure.
3. Explain the operation of the interrupts and interrupt control instructions.
4. Use machine control instructions to modify the flag bits.
5. Use ENTER and LEAVE to enter and leave programming structures.

8–1 THE JUMP GROUP

The main program control instruction, jump (JMP), allows the programmer to skip sections of a program and branch to any part of the memory for the next instruction. A conditional jump instruction allows the programmer to make decisions based upon numerical tests. The results of these numerical tests are held in the flag bits, which are then tested by conditional jump instructions. Another instruction similar to the conditional jump, the conditional set, is also explained in this section.

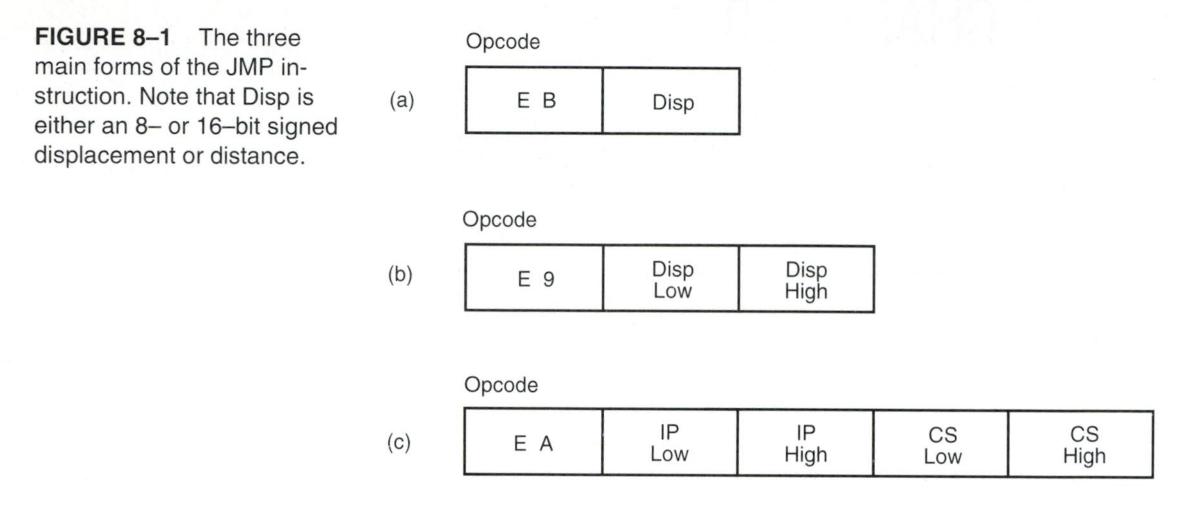

FIGURE 8–1 The three main forms of the JMP instruction. Note that Disp is either an 8– or 16–bit signed displacement or distance.

In this section of the text, all jump instructions are illustrated in sample programs. Also revisited are the LOOP and conditional LOOP instructions, first presented in Chapter 5, because they are also forms of the jump instruction.

Unconditional Jump (JMP)

Three types of unconditional jump instructions (refer to Figure 8–1) are available to the microprocessor: short jump, near jump, and far jump. The *short jump* is a 2-byte instruction that allows jumps or branches to memory locations within +127 and –128 bytes from the address following the jump. The 3-byte *near jump* allows a branch or jump within ±32K bytes (or anywhere in the current code segment) from the instruction in the current code segment. Remember, segments are cyclic in nature, which means that one location above offset address FFFFH is offset address 0000H. For this reason, if you jump 2 bytes ahead in memory, and the instruction pointer addresses offset address FFFFH, the flow continues at offset address 0001H. Thus a displacement of ±32K bytes allows a jump to any location within the current code segment. The 5-byte *far jump* allows a jump to any memory location within the entire real memory system. Short and near jumps are often called *intrasegment jumps,* and far jumps are often called *intersegment jumps.*

In the 80386 through Pentium microprocessor, the near jump is within ±2G if the machine is operated in the protected mode with a code segment 4G bytes in length, and ±32K bytes if operated in the real mode. In the protected mode, the 80386 and above use a 32-bit displacement that is not shown in Figure 8–1. The 80386 through Pentium far jump allows a jump to any location within the 4G byte address range of these microprocessors.

Short Jump. Short jumps are called *relative jumps* because they can be moved, along with their related software, to any location in current code segment without a change. This is because the jump address is *not* stored with the opcode. Instead of a jump address, a distance or *displacement* follows the opcode. The short jump displacement is a *distance* represented by a 1-byte signed number whose value ranges between +127 and –128. The short jump instruction appears in Figure 8–2. When the microprocessor executes a short jump, the displacement is sign-extended and added to the instruction pointer (IP/EIP), to gen-

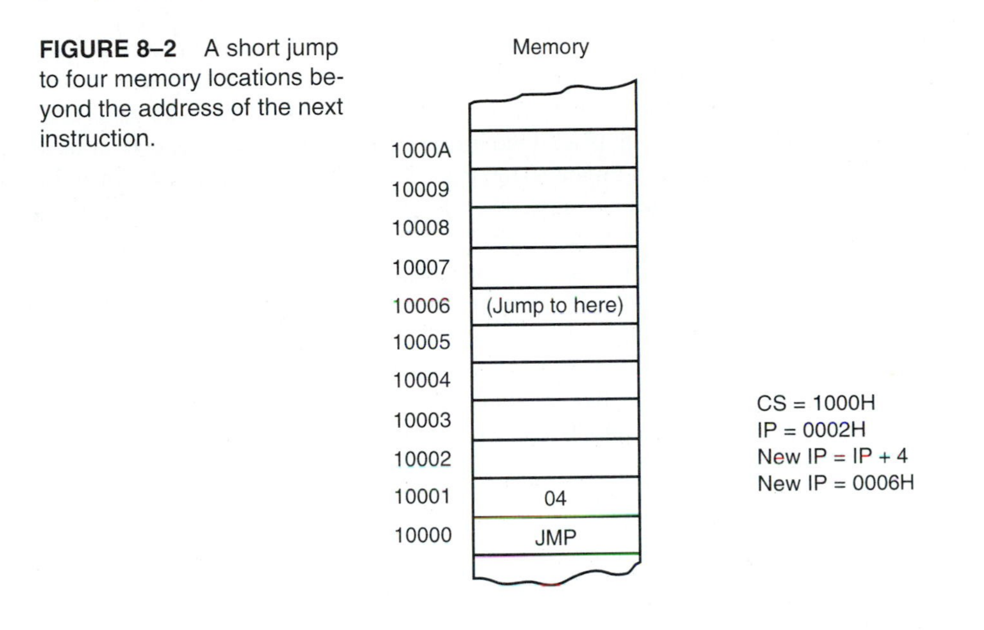

FIGURE 8–2 A short jump to four memory locations beyond the address of the next instruction.

erate the jump address within the current code segment. The short jump instruction branches to this new address for the next instruction in the program.

EXAMPLE 8–1

```
0000  33 DB                    XOR   BX,BX

0002  B8 0001        START:    MOV   AX,1
0005  03 C3                    ADD   AX,BX
0007  EB 17                    JMP   SHORT NEXT

0020  8B D8          NEXT:     MOV   BX,AX
0022  EB DE                    JMP   START
```

Example 8–1 shows how short jump instructions pass control from one part of the program to another. It also illustrates the use of a label with the jump instruction. Notice how one jump (JMP SHORT NEXT) uses the SHORT directive to force a short jump, while the other does not. *Most* assembler programs choose the best form of the jump instruction, so the second jump instruction (JMP START) also assembles as a short jump. If the address of the next instruction (0009H) is added to the sign-extended displacement (0017H) of the first jump, the address of NEXT is at location 0017H + 0009H, or 0020H.

Whenever a jump instruction references an address, a *label* identifies the address. The JMP NEXT is an example, which jumps to label NEXT for the next instruction. It is very rare to use an actual hexadecimal address with a jump instruction, but the assembler supports addressing in relation to the instruction pointer by using the $ ± a displacement. For example, JMP $+2 jumps over the next two memory locations following the JMP instruction. The label NEXT must be followed by a colon (NEXT:) to allow an instruction to reference it for a jump. If a colon does not follow a label, you *cannot* jump to it. The only time a colon is used after a label is when the label is used with a jump or call instruction.

Near Jump. A near jump is similar to a short jump except the distance is greater. A near jump passes control to an instruction in the current code segment located within ±32K bytes from the near jump instruction, or ±2G in the 80386 and above operated in protected mode. The near jump is a 3-byte instruction that contains an opcode followed by a signed 16-bit displacement. In the 80386 through the Pentium, the displacement is 32 bits and the near jump is 5 bytes in length. The signed displacement adds to the instruction pointer (IP) to generate the jump address. Because the signed displacement is in the range of ±32K, a near jump can jump to *any* memory location within the current real-mode code segment. The protected-mode code segment in the 80386 and above can be 4G bytes in length, so the 32-bit displacement allows a near jump to any location within ±2G bytes. Figure 8–3 illustrates the operation of the real-mode near jump instruction.

The near jump, like the short jump, is relocatable, because it is also a *relative* jump. If the code segment moves to a new location in the memory, the distance between the jump instruction and the operand address remains the same. This allows a code segment to be relocated by simply moving it. This feature, along with the relocatable data segments, makes the Intel family of microprocessors ideal for use in a general-purpose computer system. Software can be written and loaded anywhere in the memory, and function without modification, because of the relative jumps and relocatable data segments.

EXAMPLE 8–2

```
0000  33 DB                          XOR   BX,BX

0002  B8 0001         START:         MOV   AX,1
0005  03 C3                          ADD   AX,BX
0007  E9 0200 R                      JMP   NEXT

0200  8B D8           NEXT:          MOV   BX,AX
0202  E9 0002 R                      JMP   START
```

FIGURE 8–3 A near jump that adds the displacement (0002H) to the contents of IP.

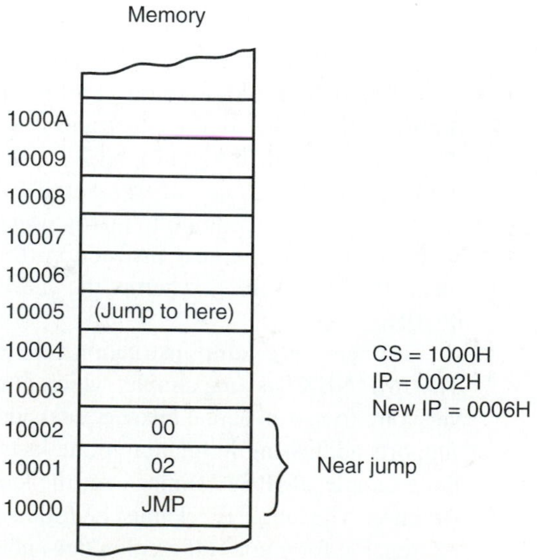

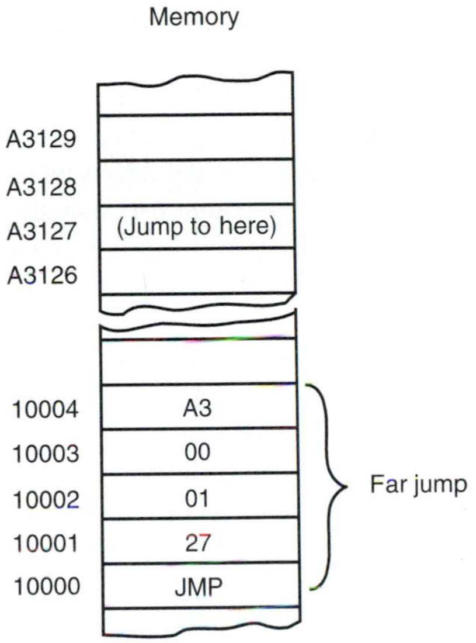

FIGURE 8–4 A far jump instruction replaces the contents of both CS and IP with the four bytes following the opcode.

Example 8–2 shows the same basic program that appeared in Example 8–1, except the jump distance is greater. The first jump (JMP NEXT) passes control to the instruction at offset memory location 0200H within the code segment. Notice that the instruction assembles as an E9 0200 R. The letter R denotes a relocatable jump address of 0200H. The relocatable address of 0200H is for the assembler program's internal use only. The actual machine language instruction assembles as an E9 F6 01, which *does not* appear in the assembler listing. The actual displacement for this jump is a 01F6H instruction. The assembler lists the jump address as 0200 R, so the address is easier to interpret as software is developed. If the linked execution file (.EXE) is displayed in hexadecimal, the jump instruction appears as an E9 F6 01.

Far Jump. A far jump instruction (see Figure 8–4) obtains a new segment and offset address from memory bytes 2–5 to accomplish the jump. Bytes 2 and 3 of this 5-byte instruction contain the new offset address, and Bytes 4 and 5 contain the new segment address. If the microprocessor is operated in the protected mode, the segment address accesses a descriptor which contains the base address of the far jump segment. The offset address, which is either 16 or 32 bits, contains the offset location within the new code segment.

EXAMPLE 8–3

```
                                 EXTRN   UP:FAR

0000  33 DB                      XOR     BX,BX

0002  B8 0001          START:    MOV     AX,1
0005  03 C3                      ADD     AX,BX
0007  E9 0200 R                  JMP     NEXT

0200  8B D8            NEXT:     MOV     BX,AX
0202  EA 0002 ---- R             JMP     FAR PTR START

0207  EA 0000 ---- E             JMP     UP
```

Example 8–3 lists a short program that uses a far jump instruction. The far jump instruction sometimes appears with the FAR PTR directive, as illustrated. Another way to obtain a far jump is to define a label as a far label. A label is far only if it is external to the current code segment or procedure. The JMP UP instruction in the example references a far label. The label UP is defined as a far label by the EXTRN UP:FAR directive. External labels appear in programs that contain more than one program file. Another way of defining a label as global is to use a double colon (LABEL::) after the label instead of a single colon. This is required inside procedure blocks that are defined as near, if the label is accessed from outside the procedure block.

When the program files are joined, the linker inserts the address for the UP label into the JMP UP instruction. It also inserts the segment address in the JMP START instruction. The segment address in JMP FAR PTR START is listed as ---- *R* for relocatable, and the segment address in JMP UP is listed as ---- *E* for external. In both cases the ---- is filled in by the linker when it links or joins the program files.

Jumps with Register Operands. The jump instruction can also use a 16- or 32-bit register as an operand. This automatically sets up the instruction as an indirect jump. The address of the jump is in the register specified by the jump instruction. Unlike the displacement associated with a near jump, the contents of the register are transferred directly into the instruction pointer. It does not add to the instruction pointer, like short and near jumps do. The JMP AX instruction, for example, copies the contents of the AX register into IP when the jump occurs. This allows a jump to any location within the current code segment. In the 80386 and above, a JMP EAX instruction jumps to any location within the current code segment. The difference is that, in the protected mode, the code segment can be 4G bytes in length, so a 32-bit offset address is needed.

EXAMPLE 8–4

```
                        ;a program that reads only the number 1, 2, or 3 from the keyboard.
                        ;if a 1, 2, or 3 are typed, the program displays a 1, 2, or 3.
                        ;
                                .MODEL SMALL              ;select SMALL model
0000                            .DATA                     ;indicate start of DATA segment
0000  0030 R            TABLE   DW      ONE               ;define lookup table
0002  0034 R                    DW      TWO
0004  0038 R                    DW      THREE
0000                            .CODE                     ;indicate start of CODE segment
                                .STARTUP                  ;indicate start of program
0017                    TOP:
0017  B4 01                     MOV     AH,1              ;read key into AL
0019  CD 21                     INT     21H

001B  2C 31                     SUB     AL,31H            ;convert '1' to 0, '2' to 1, and '3' to 2
001D  72 F8                     JB      TOP               ;if below '1'
001F  3C 02                     CMP     AL,2
0021  77 F4                     JA      TOP               ;if above '3'

0023  B4 00                     MOV     AH,0              ;double 0 to 0, 1 to 2, and 2 to 4
0025  03 C0                     ADD     AX,AX
0027  BE 0000 R                 MOV     SI,OFFSET TABLE   ;address lookup table
002A  03 F0                     ADD     SI,AX             ;add 0, 2, or 4 to lookup table address
002C  8B 04                     MOV     AX,[SI]           ;get ONE, TWO, or THREE from table
002E  FF E0                     JMP     AX                ;jump to ONE, TWO, or THREE
0030                    ONE:
0030  B2 31                     MOV     DL,'1'            ;load '1' for display
```

```
0032    EB 06              JMP    BOT               ;go display '1'
0034              TWO:
0034    B2 32              MOV    DL,'2'            ;load '2' for display
0036    EB 02              JMP    BOT               ;go display '2'
0038              THREE:
0038    B2 33              MOV    DL,'3'            ;load '3' for display
003A              BOT:
003A    B4 02              MOV    AH,2              ;display '1', '2', or '3'
003C    CD 21              INT    21H
                           .EXIT                    ;exit to DOS
                           END                      ;end of file
```

Example 8–4 shows how the JMP AX instruction accesses a jump table in the code segment. This program reads a key from the keyboard and then modifies the ASCII code to 00H in AL for a '1', 01H for a '2', and 02H for a '3'. If a '1', '2', or '3' is typed, AH is cleared to 00H. Because the jump table contains 16-bit offset addresses, the contents of AX are doubled to 0, 2, or 4, so a 16-bit entry in the table can be accessed. Next, the offset address of the start of the jump table is loaded to SI, and AX is added to form the reference to the jump address. The MOV AX,[SI] instruction then fetches an address from the jump table so the JMP AX instruction jumps to the addresses stored in the jump table (ONE, TWO, or THREE).

Indirect Jumps Using an Index. The jump instruction may also use the [ ] form of addressing to directly access the jump table. The jump table can contain offset addresses for near indirect jumps, or segment and offset addresses for far indirect jumps. (This type of jump is also known as a double-indirect jump if the register jump is called an indirect jump.) The assembler assumes that the jump is near unless the FAR PTR directive indicates a far jump instruction. Example 8–4 repeats Example 8–5 by using the JMP TABLE [SI] instead of JMP AX. This reduces the length of the program.

EXAMPLE 8–5

```
                              .MODEL SMALL              ;select SMALL model
0000                          .DATA                     ;indicate start of DATA segment
0000    002D R         TABLE  DW     ONE                ;lookup table
0002    0031 R                DW     TWO
0004    0035 R                DW     THREE
0000                          .CODE                     ;indicate start of CODE segment
                              .STARTUP                  ;indicate start of program
0017                   TOP:
0017    B4 01                 MOV       AH,1            ;read key to AL
0019    CD 21                 INT       21H

001B    2C 31                 SUB       AL,31H          ;test for below '1'
001D    72 F8                 JB        TOP             ;if below '1'
001F    3C 02                 CMP       AL,2
0021    77 F4                 JA        TOP             ;if above '3'
0023    B4 00                 MOV    AH,0               ;calculate table address
0025    03 C0                 ADD    AX,AX
0027    03 F0                 ADD    SI,AX
0029    FF A4 0000 R          JMP    TABLE [SI]         ;jump to ONE, TWO, or THREE
002D                   ONE:
002D    B2 31                 MOV    DL,'1'             ;load DL with '1'
002F    EB 06                 JMP    BOT
0031                   TWO:
```

```
0031  B2 32              MOV   DL,'2'          ;load DL with '2'
0033  EB 02              JMP   BOT
0035             THREE:
0035  B2 33              MOV   DL,'3'          ;load DL with '3'
0037             BOT:
0037  B4 02              MOV   AH,2            ;display ONE, TWO, or THREE
0039  CD 21              INT   21H
                         .EXIT                 ;exit to DOS
                         END                   ;end of file
```

The mechanism used to access the jump table is identical with a normal memory reference. The JMP TABLE [SI] instruction points to a jump address stored at the code segment offset location addressed by SI. It jumps to the address stored in the memory at this location. Both the register and indirect indexed jump instructions usually address a 16-bit offset. This means that both types of jumps are near jumps. If a JMP FAR PTR [SI] or JMP TABLE [SI], with TABLE data defined with the DD directive, appear in a program, the microprocessor assumes that the jump table contains doubleword 32-bit addresses (IP and CS).

Conditional Jumps and Conditional Sets

Conditional jump instructions are always *short jumps* in the 8086 through the 80286 microprocessor. This limits the range of the jump to within +127 to –128 bytes from the location following the conditional jump. In the 80386 and above, conditional jumps are either short or near jumps. This allows these microprocessors to use a conditional jump to any location within the current code segment. Table 8–1 lists all the conditional jump instructions with their test conditions. Note that the Microsoft MASM version 6.X assembler automatically adjusts conditional jumps if the distance is too great.

The conditional jump instructions test the following flag bits: sign (S), zero (Z), carry (C), parity (P), and overflow (O). If the condition under test is true, a branch to the label associated with the jump instruction occurs. If the condition is false, the next sequential step in the program executes. For example, a JC will jump if the carry bit is set.

The operation of most conditional jump instructions is straightforward. They often test just one flag bit, but some test more than one. Relative magnitude comparisons require more complicated conditional jump instructions that test more than one flag bit.

Because both signed and unsigned numbers are used in programming, and because the order of these numbers is different, there are two sets of conditional jump instructions for magnitude comparisons. Figure 8–5 shows the order of both signed and unsigned 8-bit numbers. The 16- and 32-bit numbers follow the same order as the 8-bit numbers, except they are larger. Notice that an FFH (255) is above the 00H in the set of unsigned numbers, but an FFH (–1) is less than 00H for signed numbers. Therefore, an unsigned FFH is *above* 00H, but a signed FFH is *less than* 00H.

When signed numbers are compared, use the JG, JL, JGE, JLE, JE, and JNE instructions. The terms *greater than* and *less than* refer to signed numbers. When unsigned numbers are compared, use the JA, JB, JAE, JBE, JE, and JNE instructions. The terms *above* and *below* refer to unsigned numbers.

The remaining conditional jumps test individual flag bits such as overflow and parity. Notice that JE has an alternative opcode, JZ. All instructions have alternates, but

TABLE 8–1 Conditional jump instructions

Instruction	*Condition Tested*	*Comment*
JA	C = 0 and Z = 0	Jump if above
JAE	C = 0	Jump if above or equal to
JB	C = 1	Jump if below
JBE	C = 1 or Z = 1	Jump if below or equal to
JC	C = 1	Jump if carry set
JE or JZ	Z = 1	Jump if equal to or jump zero
JG	Z = 0 and S = O	Jump if greater than
JGE	S = O	Jump if greater than or equal to
JL	S <> O	Jump if less than
JLE	Z = 1 or S <> O	Jump if less than or equal to
JNC	C = 0	Jump if carry cleared
JNE or JNZ	Z = 0	Jump if not equal or jump not zero
JNO	O = 0	Jump if no overflow
JNS	S = 0	Jump if no sign
JNP/JPO	P = 0	Jump if no parity/jump parity odd
JO	O = 1	Jump if overflow is set
JP/JPE	P = 1	Jump if parity/jump parity even
JS	S = 1	Jump if sign is set
JCXZ	CX = 0	Jump if CX = 0
*JECXZ	ECX = 0	Jump if ECX = 0

Note: * = 80386/80486/Pentium only.

many aren't used in programming, because they don't usually fit the condition under test. (The alternates appear in Appendix B with the instruction set listing.) For example, the JA instruction (jump if above) has the alternative JNBE (jump if not below or equal). A JA functions exactly like a JNBE, but JNBE is awkward when compared to JA.

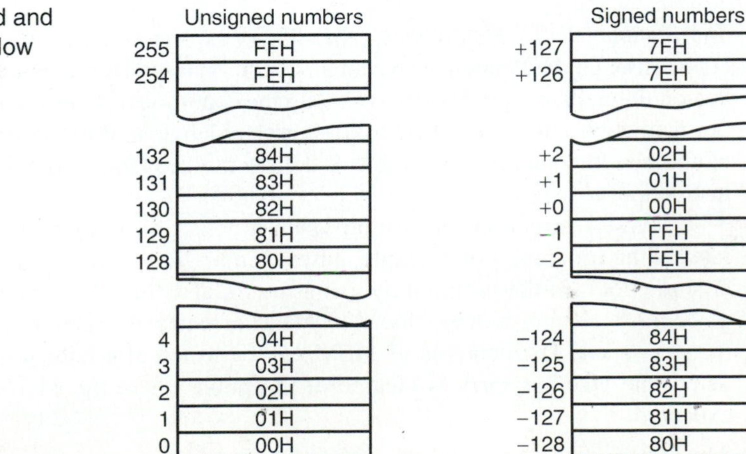

FIGURE 8–5 Signed and unsigned numbers follow different orders.

All conditional jump instructions test flag bits, except for JCXZ (jump if CX = 0) and JECXZ (jump if ECX = 0). Instead of testing flag bits, JCXZ directly tests the contents of the CX register without affecting the flag bits. JECXZ tests the contents of the ECX register. For the JCXZ instruction, if CX = 0, a jump occurs, and if CX <> 0, no jump occurs. Likewise for the JECXZ instruction: if ECX = 0, a jump occurs and if CX <> 0, no jump occurs.

EXAMPLE 8–6

```
                        ;A procedure that searches a table of 100 bytes for 0AH
                        ;The address of the TABLE is transferred to the procedure
                        ;through the SI register.
                        ;
0017                    SCAN  PROC    NEAR

0017  B9 0064                 MOV     CX,100          ;load count of 100
001A  B0 0A                   MOV     AL,0AH          ;load AL with 0AH
001C  FC                      CLD                     ;select increment
001D  F2/ AE                  REPNE   SCASB           ;test 100 bytes for 0AH
001F  F9                      STC                     ;set carry for not found
0020  E3 01                   JCXZ    NOT_FOUND       ;if not found
0022  F8                      CLC                     ;clear carry if found

0023                    NOT_FOUND:

0023  C3                      RET                     ;return from procedure

0024                    SCAN  ENDP
```

A program that uses JCXZ appears in Example 8–6. Here the SCASB instruction searches a table for a 0AH. Following the search, a JCXZ instruction tests CX to see if the count has reached zero. If the count is zero, the 0AH is not found in the table. The carry flag is used in this example to pass the not-found condition back to the calling program. Another method used to test to see if the data are found is the JNE instruction. If JNE replaces JCXZ, it performs the same function. After the SCASB instruction executes, the flags indicate a not-equal condition if the data was not found in the table.

The Conditional Set Instructions. In addition to the conditional jump instructions, the 80386 through the Pentium also contain conditional set instructions. The conditions tested by conditional jumps are put to work with the conditional set instructions. The conditional set instructions either set a byte to 01H or clear a byte to 00H, depending on the outcome of the condition under test. Table 8–2 lists the available forms of the conditional set instructions.

These instructions are useful when a condition must be tested at a point much later in the program. For example, a byte can be set to indicate that the carry is cleared at some point in the program by using the SETNC MEM instruction. This instruction places a 01H into memory location MEM if carry is cleared and a 00H into MEM if carry is set. The contents of MEM can be tested at a later point in the program to determine whether carry is cleared at the point where the SETNC MEM instruction executed.

TABLE 8–2 The conditional set instructions

Instruction	*Condition Tested*	*Comment*
SETB	C = 1	Set byte if below
SETAE	C = 0	Set byte if above or equal to
SETBE	C = 1 or Z = 1	Set byte if below or equal to
SETA	C = 0 and Z = 0	Set byte if above
SETE/SETZ	Z = 1	Set byte if equal/set byte if zero
SETNE/SETNZ	Z = 0	Set byte if not equal/set byte if not zero
SETL	S <> O	Set byte if less than
SETLE	Z = 1 or S <> O	Set byte if less than or equal to
SETG	Z = 0 and S = O	Set byte if greater than
SETGE	S = O	Set byte if greater than or equal to
SETS	S = 1	Set byte if sign (negative)
SETNS	S = 0	Set byte if no sign (positive)
SETC	C = 1	Set byte if carry
SETNC	C = 0	Set byte if no carry
SETO	O = 1	Set byte if overflow
SETNO	O = 0	Set byte if no overflow
SETP	P = 1	Set byte if parity (even)
SETNP	P = 0	Set byte if no parity (odd)

LOOP

The LOOP instruction is a combination of a decrement CX and the JNZ conditional jump. In the 8086 through the 80286, LOOP decrements CX. If CX <> 0, it jumps to the address indicated by the label. If CX becomes a 0, the next sequential instruction executes. In the 80386 and above, LOOP decrements either CX or ECX, depending upon the instruction mode. If the 80386 through Pentium operate in the 16-bit instruction mode, LOOP uses CX, and if they operate in the 32-bit instruction mode, LOOP uses ECX. This default is changed by the LOOPW (using CX) and LOOPD (using ECX) instructions in the 80386 through Pentium.

EXAMPLE 8–7

```
                        ;a program that sums the contents of BLOCK1 and BLOCK2
                        ;and stores the results over top of the data in BLOCK2
                        ;
                                .MODEL SMALL              ;select SMALL model
0000                            .DATA                     ;indicate start of DATA segment
0000  0064 [            BLOCK1 DW     100 DUP (?)         ;reserve 100 bytes for BLOCK1
                    0000
                        ]
00C8  0064 [            BLOCK2 DW     100 DUP (?)         ;reserve 100 bytes for BLOCK2
                    0000
                        ]
0000                            .CODE                     ;indicate start of CODE segment
                                .STARTUP                  ;indicate start of program
```

```
0017  8C D8                  MOV   AX,DS              ;overlap DS and ES
0019  8E C0                  MOV   ES,AX

001B  FC                     CLD                      ;select increment
001C  B9 0064                MOV   CX,100             ;load count of 100
001F  BE 0000 R              MOV   SI,OFFSET BLOCK1   ;address BLOCK1
0022  BF 00C8 R              MOV   DI,OFFSET BLOCK2   ;address BLOCK2

0025               L1:
0025  AD                     LODSW                   ;load AX with BLOCK1 data
0026  26: 03 05              ADD   AX,ES:[DI]        ;add BLOCK2 data to AX
0029  AB                     STOSW                   ;store sum in BLOCK2
002A  E2 F9                  LOOP  L1                ;repeat 100 times
                             .EXIT                   ;exit to DOS
                             END                     ;end of file
```

Example 8–7 shows how data in one block of memory (BLOCK1) adds to data in a second block of memory (BLOCK2) using LOOP to control how many numbers add. The LODSW and STOSW instructions access the data in BLOCK1 and BLOCK2. The ADD AX,ES:[DI] instruction accesses the data in BLOCK2 located in the extra segment. The only reason BLOCK2 is in the extra segment is that DI addresses extra segment data for the STOSW instruction. The .STARTUP directive only loads DS with the address of the data segment. In this example, the extra segment also addresses data in the data segment, so the contents of DS are copied to ES through the accumulator. Unfortunately, there is no direct instruction for segment register to segment register moves.

Conditional LOOPs. Like REP, the LOOP instruction has conditional forms: LOOPE and LOOPNE. The LOOPE (loop while equal) instruction jumps if CX <> 0 while an equal condition exists. It will exit the loop if the condition is not equal, or if the CX register decrements to 0. The LOOPNE (loop while not equal) instruction jumps if CX <> 0 while a not-equal condition exists. It will exit the loop if the condition is equal, or if the CX register decrements to 0. In the 80386/80486/Pentium the conditional LOOP instruction can use either CX or ECX as the counter. The LOOPEW, LOOPED, LOOPNEW, or LOOPNED instructions override the LOOP instruction if needed.

Alternates exist for LOOPE and LOOPNE: The LOOPE instruction is the same as LOOPZ and LOOPNE is the same as LOOPNZ. In most programs, only LOOPE and LOOPNE apply.

8–2 PROCEDURES

The procedure or **subroutine** is an important part of any computer system's architecture. A **procedure** is a group of instructions that usually perform one task. A *procedure* is a reusable section of the software that is stored in memory once, but used as often as necessary. This saves memory space and makes it easier to develop software. The only disadvantage of a procedure is that it takes the computer a small amount of time to link to the procedure and return from it. The CALL instruction links to the procedure and the **RET** (return) instruction returns from the procedure.

The stack stores the return address whenever a procedure is called during the execution of a program. The **CALL** instruction pushes the address of the instruction following the CALL (return address) onto the stack. The RET instruction removes an address from the stack, so the program returns to the instruction following the CALL.

EXAMPLE 8–8

```
0000                SUMS    PROC    NEAR

0000  03 C3                 ADD     AX,BX
0002  03 C1                 ADD     AX,CX
0004  03 C2                 ADD     AX,DX
0006  C3                    RET

0007                SUMS    ENDP

0007                SUMS1   PROC    FAR

0007  03 C3                 ADD     AX,BX
0009  03 C1                 ADD     AX,CX
000B  03 C2                 ADD     AX,DX
000D  CB                    RET

000E                SUMS1   ENDP

000E                SUMS2   PROC    NEAR USES BX CX DX

0011  03 C3                 ADD     AX,BX
0013  03 C1                 ADD     AX,CX
0015  03 C2                 MOV     AX,DX
                            RET

001B                SUMS2   ENDP
```

With the assembler, there are specific rules for the storage of procedures. A procedure begins with the PROC directive and ends with the ENDP directive. Each directive appears with the name of the procedure. This programming structure makes it easy to locate a procedure in a program listing. The PROC directive is followed by the type of procedure: NEAR or FAR. Example 8–8 shows how the assembler uses the definition of both a near (intrasegment) and far (intersegment) procedure. In MASM version 6.X, a NEAR or FAR procedure can be followed by the USES statement. The USES statement allows any number of registers to be automatically pushed to the stack and popped from the stack within the procedure. The USES statement is also illustrated in Example 8–8.

When these two procedures are compared, the only difference is the opcode of the return instruction. The near return instruction uses opcode C3H and the far return uses opcode CBH. A near return removes a 16-bit number from the stack and places it into the instruction pointer to return from the procedure in the current code segment. A far return removes a 32-bit number from the stack and places it into both IP and CS to return from the procedure to any memory location.

Procedures that are to be used by all software (*global*) should be written as far procedures. Procedures that are used by a given task (*local*) are normally defined as near procedures.

CALL

The CALL instruction transfers the flow of the program to the procedure. The CALL instruction differs from the jump instruction in that a CALL saves a return address on the stack. The RET instruction returns control to the instruction that immediately follows the CALL.

Near CALL. A near CALL instruction is three bytes long, with the first byte containing the opcode and the second and third bytes containing the displacement or distance of ±32K in the 8086 through the 80286. This is identical to the form of a near jump instruction. The 80386 and above use a 32-bit displacement when operated in the protected mode, to allow a distance of ±2G bytes. When a near CALL executes, it first pushes the offset address of the next instruction on the stack. The offset address of the next instruction appears in the instruction pointer (IP or EIP). After saving this return address, it then adds the displacement from Bytes 2 and 3 to the IP to transfer control to the procedure. There is no short CALL instruction. A variation on the opcode, CALLN, exists, but should be avoided. Instead use the PROC statement to define the CALL as near.

Why save the IP or EIP on the stack? The instruction pointer always points to the next instruction in the program. For the CALL instruction, the contents of IP/EIP are pushed onto the stack so program control passes to the instruction following the CALL after a procedure ends. Figure 8–6 shows the return address (IP) stored on the stack, and the call to the procedure.

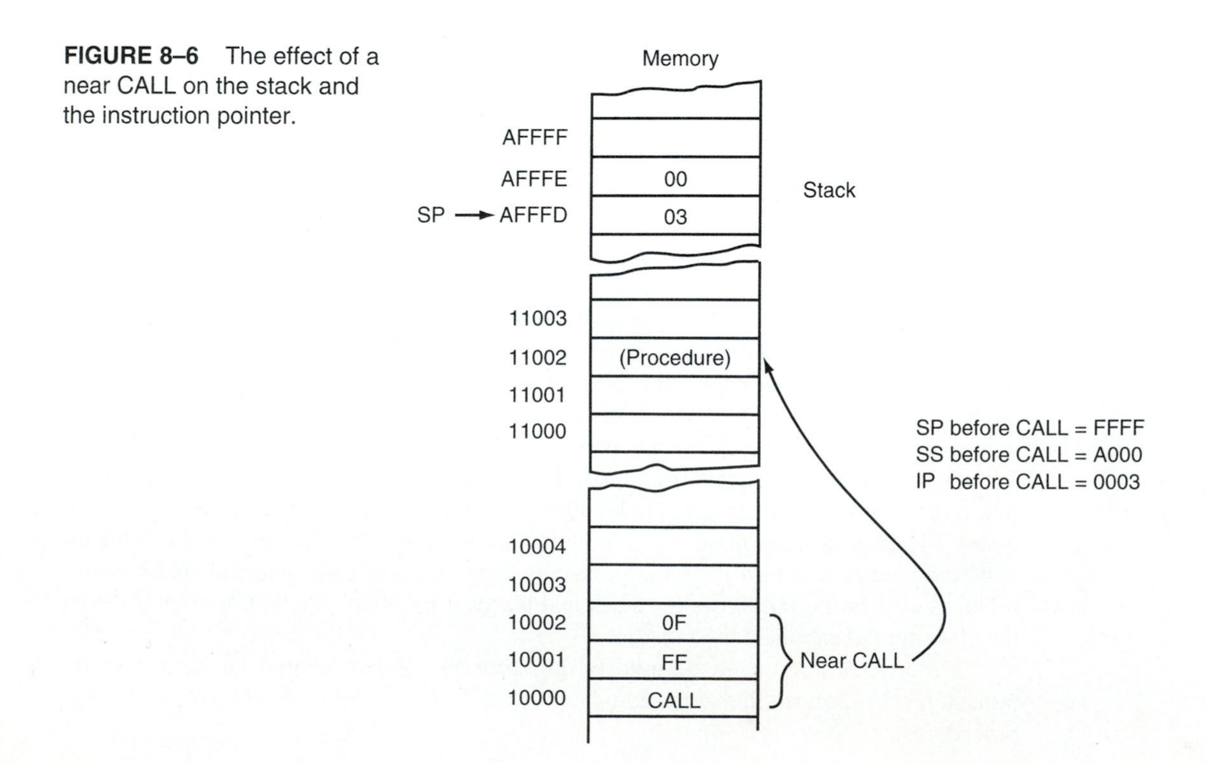

FIGURE 8–6 The effect of a near CALL on the stack and the instruction pointer.

FIGURE 8–7 The effect of a far CALL instruction.

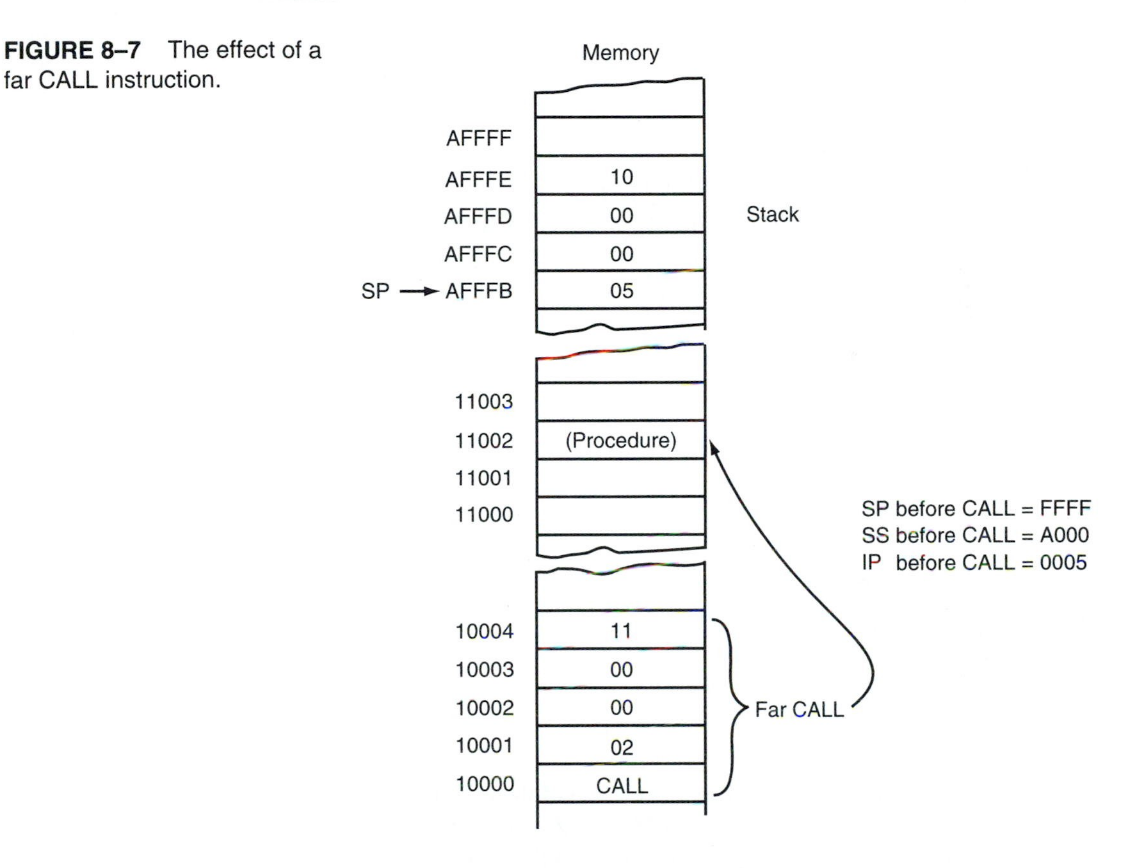

Far CALL. The far CALL instruction is like a far jump, because it can CALL a procedure stored in any memory location in the system. The far CALL is a 5-byte instruction that contains an opcode followed by the next value for the IP and CS registers. Bytes 2 and 3 contain the new contents of the IP and Bytes 4 and 5 contain the new contents for CS.

The far CALL instruction places the contents of both IP and CS on the stack before jumping to the address indicated by Bytes 2–5 of the instruction. This allows the far CALL to call a procedure located anywhere in the memory and return from that procedure.

Figure 8–7 shows how the far CALL instruction calls a far procedure. Here the contents of IP and CS are pushed onto the stack. Next the program branches to the procedure. A variant of the far call, CALLF, exists, but should be avoided. Define the type of call instruction with the PROC statement instead.

CALLs with Register Operands. Like jumps, CALLs also may contain a register operand. An example is the CALL BX instruction. This instruction pushes the contents of IP onto the stack. It then jumps to the offset address, located in register BX, in the current code segment. This type of CALL always uses a 16-bit offset address stored in any 16-bit register except the segment registers.

EXAMPLE 8–9

```
                  ;a program that displays OK on the monitor screen
                  ;using procedure DISP
                  ;
                         .MODEL TINY                 ;select TINY model
0000                     .CODE                       ;indicate start of CODE segment
                         .STARTUP                    ;indicate start of program

0100  BB 0110 R          MOV    BX,OFFSET DISP       ;address DISP with BX
0103  B2 4F              MOV    DL,'O'               ;display 'O'
0105  FF D3              CALL   BX
0107  B2 4B              MOV    DL,'K'               ;display 'K'
0109  FF D3              CALL   BX

                         .EXIT                       ;exit to DOS
                  ;
                  ;a procedure that displays the ASCII contents of DL on the
                  ;monitor screen.
                  ;
0110              DISP   PROC   NEAR

0110  B4 02              MOV    AH,2                  ;select function 02H
0112  CD 21              INT    21H                   ;execute DOS function
0114  C3                 RET                         ;return from procedure

0115              DISP   ENDP

                         END                         ;end of file
```

Example 8–9 illustrates the use of the CALL register instruction to call a procedure that begins at offset address DISP. (This call could also directly call the procedure by using the CALL DISP instruction.) The OFFSET address DISP is placed into the BX register, then the CALL BX instruction calls the procedure beginning at address DISP. This program displays an "OK" on the monitor screen.

CALLs with Indirect Memory Addresses. A CALL with an indirect memory address is particularly useful when different subroutines need to be chosen in a program. This selection process is often keyed with a number that addresses a CALL address in a lookup table.

EXAMPLE 8–10

```
                  ;program that uses a CALL lookup table to access one of three
                  ;different procedures: ONE, TWO, or THREE.
                  ;
                         .MODEL SMALL            ;select SMALL model
0000                     .DATA                   ;indicate start of DATA segment
0000  0000 R      TABLE  DW     ONE              ;define lookup table
0002  0007 R             DW     TWO
0004  000E R             DW     THREE
0000                     .CODE                   ;indicate start of CODE segment
0000              ONE    PROC   NEAR

0000  B4 02              MOV    AH,2             ;display a letter A
0002  B2 41              MOV    DL,'A'
0004  CD 21              INT    21H
0006  C3                 RET
```

```
0007                 ONE    ENDP

0007                 TWO    PROC   NEAR

0007  B4 02                 MOV    AH,2          ;display letter B
0009  B2 42                 MOV    DL,'B'
000B  CD 21                 INT    21H
000D  C3                    RET

000E                 TWO    ENDP

000E                 THREE  PROC   NEAR

000E  B4 02                 MOV    AH,2          ;display letter C
0010  B2 43                 MOV    DL,'C'
0012  CD 21                 INT    21H
0014  C3                    RET

0015                 THREE  ENDP

                            .STARTUP             ;indicates start of program

002C                 TOP:
002C  B4 01                 MOV    AH,1          ;read key into AL
002E  CD 21                 INT    21H

0030  2C 31                 SUB    AL,31H        ;convert from ASCII to 0, 1, or 2
0032  72 F8                 JB     TOP           ;if below 0
0034  3C 02                 CMP    AL,2
0036  77 F4                 JA     TOP           ;if above 2

0038  B4 00                 MOV    AH,0          ;form lookup address
003A  8B D8                 MOV    BX,AX
003C  03 DB                 ADD    BX,BX
003E  FF 97 0000 R          CALL   TABLE [BX]    ;call procedure ONE, TWO, or  THREE

                            .EXIT                ;exit to DOS
                            END                  ;end of file
```

Example 8–10 shows three separate subroutines referenced by Number 1, 2, and 3 as read from the keyboard on the personal computer. The calling sequence adjusts the value of AL and extends it to a 16-bit number before adding it to the location of the lookup table. This references one of the three subroutines using the CALL TABLE [BX] instruction. When this program executes, the letter A is displayed when a 1 is typed, the letter B is displayed when a 2 is typed, and the letter C is displayed when a 3 is typed.

The CALL instruction can also reference far pointers if the data in the table are defined as doubleword data with the DD directive, using the CALL FAR PTR [SI] or CALL TABLE [SI] instructions. These instructions retrieve a 32-bit address from the data segment memory location addressed by SI and use it as the address of a far procedure.

RET

The return instruction (RET) removes either a 16-bit number (near return) from the stack and places it into IP or a 32-bit number (far return) and places it into IP and CS. The near and far return instructions are both defined in the procedure's PROC directive. This

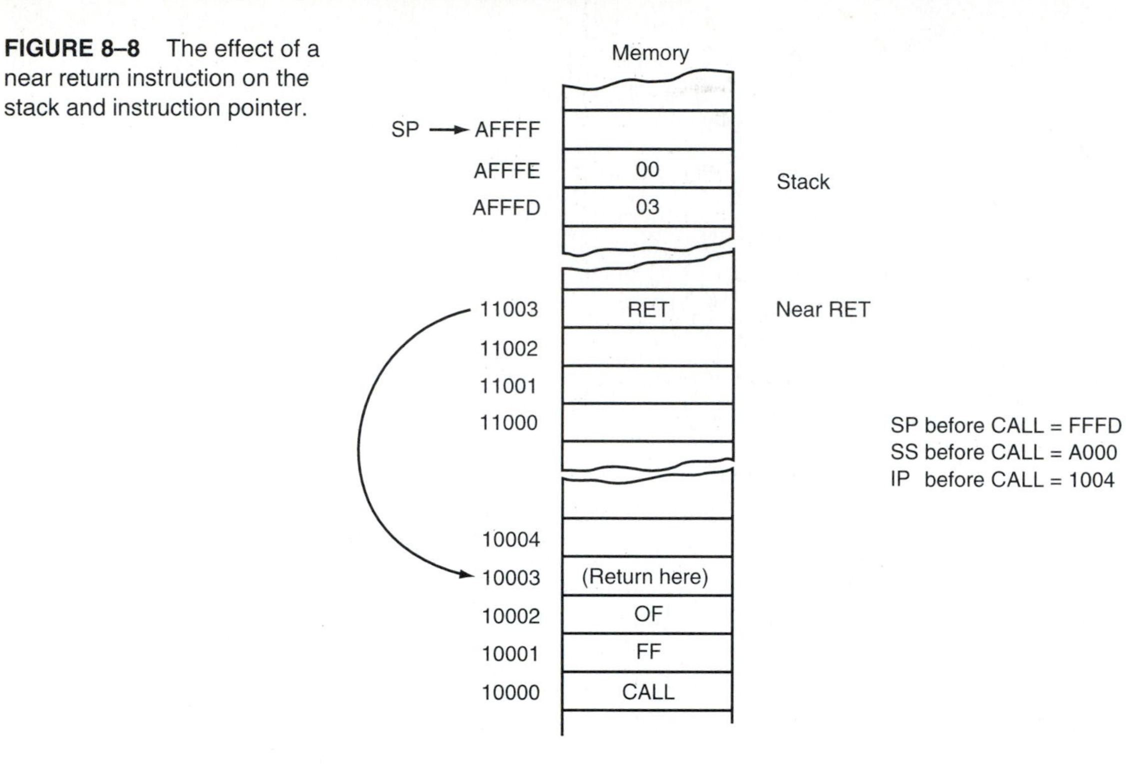

FIGURE 8–8 The effect of a near return instruction on the stack and instruction pointer.

automatically selects the proper return instruction. In the 80386 through the Pentium operated in the protected mode, the far return removes six bytes from the stack; the first four contain the new value for EIP, and the last two contain the new value for CS. In the 80386 and above, a protected-mode near return removes four bytes from the stack and places them into EIP.

When IP/EIP or IP/EIP and CS are changed, the address of the next instruction is at a new memory location. This new location is the address of the instruction that immediately follows the most recent CALL to a procedure. Figure 8–8 shows how the CALL instruction links to a procedure and how the RET instruction returns in the 8086–80286.

There is one other form of the return instruction. This form adds a number to the contents of the stack pointer (SP) after the return address is removed from the stack. A return that uses an immediate operand is ideal for use in a system that uses a C or Pascal calling convention. (This is true even though the C and PASCAL calling conventions require the caller to remove stack data for many functions.) These conventions push parameters on the stack before calling a procedure. If the parameters are to be discarded upon return, the return instruction contains a number that represents the number of bytes pushed to the stack as parameters.

EXAMPLE 8–11

```
0000  B8 001E          MOV    AX,30
0003  BB 0028          MOV    BX,40
0006  50               PUSH   AX          ;stack parameter 1
0007  53               PUSH   BX          ;stack parameter 2
```

```
0008  E8 0066                  CALL   ADDM          ;add parameters from stack
                               .      .
                               .      .             ;program continues here
                               .      .
0071                    ADDM   PROC   NEAR

0071  55                       PUSH   BP            ;save BP
0072  8B EC                    MOV    BP,SP         ;address stack with BP
0074  8B 46 04                 MOV    AX,[BP+4]     ;get parameter 1
0077  03 46 06                 ADD    AX,[BP+6]     ;add parameter 2
007A  5D                       POP    BP            ;restore BP

007B  C2 0004                  RET    4             ;return and dump parameters

007E                    ADDM   ENDP
```

Example 8–11 shows how this type of return erases the data placed on the stack by a few pushes. The RET 4 adds a 4 to SP, after removing the return address from the stack. Since the PUSH AX and PUSH BX together place four bytes of data on the stack, this return effectively deletes AX and BX from the stack. This type of return rarely appears in assembly language programs, but is used in higher level programs to clear stack data after a procedure. Notice how parameters are addressed on the stack using the BP register, which addresses the stack segment by default. Parameter stacking is common in procedures written for C or PASCAL using the C or PASCAL calling conventions. More detail on C and PASCAL interfacing is provided in Chapter 14.

There are also variants of the return instruction: RETN and RETF. Like the CALLN and CALLF instructions, these should be avoided in favor of using the PROC statement to define the type of call and return.

8–3 INTRODUCTION TO INTERRUPTS

An **interrupt** is either a hardware-generated CALL (externally derived from a hardware signal) or a software-generated CALL (internally derived from the execution of an instruction or by some other internal event). Sometimes, an internal interrupt is called an *exception*. Either type interrupts the program by calling an *interrupt service procedure* or **interrupt handler.**

This section explains software interrupts, which are special types of CALL instructions. In this section are the three types of software interrupt instructions (INT, INTO, and INT 3) are detailed. A map of the interrupt vectors is provided, and the purpose of the special interrupt return instruction (IRET) is explained.

Interrupt Vectors

An **interrupt vector** is a 4-byte number stored in the first 1,024 bytes of memory (000000H–0003FFH) when the microprocessor operates in the real mode. In the protected mode, the vector table is replaced by an interrupt descriptor table that uses 8-byte descriptors to describe each of the interrupts. There are 256 different interrupt vectors. Each

TABLE 8–3 Interrupt vectors

Number	*Address*	*Microprocessor*	*Function*
0	0H–3H	8086–80486/Pentium	Divide error
1	4H–7H	8086–80486/Pentium	Single step
2	8H–BH	8086–80486/Pentium	NMI (hardware interrupt)
3	CH–FH	8086–80486/Pentium	Breakpoint
4	10H–13H	8086–80486/Pentium	Interrupt on overflow
5	14H–17H	80286–80486/Pentium	BOUND interrupt
6	18H–1BH	80286–80486/Pentium	Invalid opcode
7	1CH–1FH	80286–80486/Pentium	Coprocessor emulation interrupt
8	20H–23H	80386–80486/Pentium	Double fault
9	24H–27H	80386	Coprocessor segment overrun
10	28H–2BH	80386–80486/Pentium	Invalid task state segment
11	2CH–2FH	80386–80486/Pentium	Segment not present
12	30H–33H	80386–80486/Pentium	Stack fault
13	34H–37H	80386–80486/Pentium	General protection fault
14	38H–3BH	80386–80486/Pentium	Page fault
15	3CH–3FH		Reserved*
16	40H–43H	80286–80486/Pentium	Floating-point error
17	44H–47H	80486SX	Alignment check interrupt
18	48H–4FH	Pentium	Machine check exception
19–31	50H–7FH	8086–80486/Pentium	Reserved*
32–255	80H–3FFH	8086–80486/Pentium	User interrupts

**Note*: Some of these interrupts will appear on newer versions of the 8086–80486/Pentium when they become available.

vector contains the address of an interrupt service procedure. Table 8–3 lists the interrupt vectors with a brief description and the memory location of each vector for the real mode. Each vector contains a value for IP and CS that forms the address of the interrupt service procedure. The first two bytes contain the IP, and the last two bytes contain the CS.

Intel reserves the first 32 interrupt vectors for present and future microprocessor products. The remaining interrupt vectors (32–255) are available for the user. Some of the reserved vectors are for errors that occur during the execution of software, like the divide error interrupt. Some vectors are reserved for the coprocessor. Still others occur for normal events in the system. In a personal computer, the reserved vectors are used for system functions as detailed later in this section. Vectors 1–6, 7, 9, 16, and 17 function in the real mode and protected mode; the remaining vectors only function in the protected mode.

INTERRUPT INSTRUCTIONS

Three different interrupt instructions are available to the programmer: INT, INTO, and INT 3. In the real mode, each of these instructions fetches a vector from the vector table and then calls the procedure stored at the location addressed by the vector. In the protected mode, each of these instructions fetches an interrupt descriptor from the interrupt descriptor table. The descriptor specifies the address of the interrupt service procedure. The interrupt call is similar to a far CALL instruction because it places the return address (IP/EIP and CS) on the stack.

INTs. There are 256 different software interrupt instructions (INT) available to the programmer. Each INT instruction has a numeric operand whose range is 0 to 255 (00H–FFH). For example, INT 100 uses interrupt vector 100, which appears at memory address 190H–193H. The address of the interrupt vector is determined by multiplying the interrupt type number by four. For example, the INT 10H instruction calls the interrupt service procedure whose address is stored beginning at memory location 40H (10H × 4) in the real mode. In protected mode, the interrupt descriptor is located by multiplying the type number by 8, because each descriptor is 8 bytes long.

Each INT instruction is two bytes long. The first byte contains the opcode and the second byte contains the vector type number. The only exception to this is INT 3, a 1-byte special software interrupt used for breakpoints.

Whenever a software interrupt instruction executes, it (1) pushes the flags onto the stack, (2) clears the T and I flag bits, (3) pushes CS onto the stack, (4) fetches the new value for CS from the vector, (5) pushes IP/EIP onto the stack, (6) fetches the new value for IP/EIP from the vector, and (7) jumps to the new location addressed by CS and IP/EIP. The INT instruction performs like a far CALL, except it pushes the flags onto the stack as well as CS and IP. The INT instruction performs the operation of a PUSHF followed by a far CALL instruction.

When the INT instruction executes, it clears the interrupt flag (I), which controls the external hardware interrupt input pin INTR (interrupt request). When I = 0, the microprocessor disables the INTR pin; when I = 1, the microprocessor enables the INTR pin.

Software interrupts are commonly used to call system procedures, because the address of the system function need not be known. The system procedures are common to all system and application software. The interrupts often control printers, video displays, and disk drives. Besides relieving the program from the need to remember the address of the system call, the INT instruction replaces a far CALL that would otherwise be used to call a system function. The INT instruction is two bytes long, whereas the far CALL is five bytes long. Each time the INT instruction replaces a far CALL, three bytes of memory are saved. This can amount to a sizable savings if the INT instruction appears often in a program, as it does for system calls.

IRET/IRETD. The interrupt return instruction (IRET) is used only with software or hardware interrupt service procedures. Unlike a simple return instruction (RET), the IRET instruction will (1) pop stack data back into the IP, (2) pop stack data back into CS, and (3) pop stack data back into the flag register. The IRET instruction accomplishes the same tasks as the POPF, followed by a far RET instruction.

Whenever an IRET instruction executes, it restores the contents of I and T from the stack. This is important because it preserves the state of these flag bits. If interrupts are enabled before an interrupt service procedure, they are *automatically* re-enabled by the IRET instruction, because it restores the flag register.

In the 80386 through the Pentium, the IRETD instruction is used to return from an interrupt service procedure called in the protected mode. It differs from the IRET in that it pops a 32-bit instruction pointer (EIP) from the stack. IRET is used in the real mode, and IRETD is used in the protected mode.

INT 3. An INT 3 instruction is a special software interrupt designed to be used as a breakpoint. The difference between it and the other software interrupts, is that INT 3 is a one-byte instruction, while the others are two-byte instructions.

It is common to insert an INT 3 instruction in software to interrupt or break the flow of the software. This function is called a *breakpoint*. A breakpoint occurs for any software interrupt, but because INT 3 is one byte long, it is easiest to use. Breakpoints help debug faulty software.

INTO. Interrupt on overflow (INTO) is a conditional software interrupt that tests the overflow flag (O). If O = 0, the INTO instruction performs no operation, but if O = 1 and an INTO instruction executes, an interrupt occurs via vector type number 4.

The INTO instruction appears in software that adds or subtracts signed binary numbers. With these operations it is possible to have an overflow. Either the JO instruction or INTO instruction detect the overflow condition.

An Interrupt Service Procedure. Suppose a procedure is required to add the contents of DI, SI, BP, and BX and save the sum in AX. Because this is a common task in this particular system, it may occasionally be worthwhile to develop the task as a software interrupt. Interrupts are usually reserved for system events; this example merely shows how an interrupt service procedure appears. Example 8–12 shows this software interrupt. The main difference between this procedure and a normal far procedure is that it ends with the IRET instruction instead of the RET instruction, and the contents of the flag register are saved on the stack during its execution.

EXAMPLE 8–12

```
0000                INTS     PROC    FAR

0000    03 C3                ADD     AX,BX
0002    03 C5                ADD     AX,BP
0004    03 C7                ADD     AX,DI
0006    03 C6                ADD     AX,SI
0008    CF                   IRET

0009                INTS     ENDP
```

Interrupt Control

This section introduces two instructions that control the INTR pin. The set interrupt flag instruction (STI) places a 1 into the I flag bit, which enables the INTR pin. The clear interrupt flag instruction (CLI) places a 0 into the I flag bit, which disables the INTR pin. The STI instruction enables INTR and the CLI instruction disables INTR. In a software interrupt service procedure, hardware interrupts are enabled as one of the first steps. This is accomplished by the STI instruction. The reason interrupts are enabled early in an interrupt service procedure is that just about all of the I/O devices in the personal computer are interrupt-processed. If the interrupts are disabled too long, severe system problems result.

Interrupts in the Personal Computer

The interrupts found in the personal computer differ somewhat from the ones presented in Table 8–3. The original personal computers were 8086/8088-based systems. They contained only Intel specified interrupts 0–4. This design was carried forward, so newer systems would be compatible with the early personal computers.

Because the personal computer is operated in the real mode, the interrupt vector table is located at addresses 00000H–003FFH. The assignments used by the computer system are listed in Table 8–4. Notice that these differ somewhat from the assignments in Table 8–3. Some of the interrupts shown in this table are used in example programs in later chapters. An example is the clock tick, which is extremely useful for timing events because it occurs a constant 18.2 times per second in all personal computers.

Interrupts 00H–1FH and 70H–77H are present in the computer no matter what operating system is installed. If DOS is installed, interrupts 20H–2FH are also present. The BIOS uses interrupts 11H through 1FH, the video BIOS uses INT 10H, and the system hardware uses interrupts 00H through 0FH and 70H through 77H.

8–4 MACHINE CONTROL AND MISCELLANEOUS INSTRUCTIONS

The last category of real-mode instructions found in the microprocessor are the machine control and miscellaneous group. These instructions provide control of the carry bit, sample the $\overline{\text{BUSY}/\text{TEST}}$ pin, and perform various other functions. Because many of these instructions are used in hardware control, they need only be explained briefly at this point.

Controlling the Carry Flag Bit

The carry flag (C) propagates the carry or borrow in multiple-word/doubleword addition and subtraction. It also indicates errors in procedures. There are three instructions that control the contents of the carry flag: STC (*set carry*), CLC (*clear carry*), and CMC (*complement carry*).

Because the carry flag is seldom used except with multiple word addition and subtraction, it is available for other uses. The most common task for the carry flag is to indicate error upon return from a procedure. Suppose that a procedure reads data from a disk memory file. This operation can be successful or an error can occur, such as file-not-found. Upon return from this procedure, if C = 1, an error has occurred and if C = 0, no error has occurred. Most DOS and BIOS procedures use the carry flag to indicate error conditions.

WAIT

The WAIT instruction monitors the hardware $\overline{\text{BUSY}}$ pin on the 80286 and 80386 and the $\overline{\text{TEST}}$ pin on the 8086/8088. The name of this pin was changed in the 80286 microprocessor from $\overline{\text{TEST}}$ to $\overline{\text{BUSY}}$. If the WAIT instruction executes while the $\overline{\text{BUSY}}$ pin = 1, nothing happens and the next instruction executes. If the $\overline{\text{BUSY}}$ pin = 0 when the WAIT instruction executes, the microprocessor waits for the $\overline{\text{BUSY}}$ pin to return to a logic 1. This pin indicates a busy condition when at logic zero.

TABLE 8–4 The hexadecimal interrupt assignments for the personal computer

Number	*Function*
0	Divide error
1	Single step (debug)
2	Nonmaskable interrupt pin
3	Breakpoint
4	Arithmetic overflow
5	Print screen key and BOUND instruction
6	Illegal instruction error
7	Coprocessor not present interrupt
8	Clock tick (hardware) (Approximately 18.2 Hz)
9	Keyboard (hardware)
A	Hardware interrupt 2 (system bus) (cascade in AT)
B–F	Hardware interrupts 3–7 (system bus)
10	Video BIOS
11	Equipment environment
12	Conventional memory size
13	Direct disk service
14	Serial COM port service
15	Miscellaneous service
16	Keyboard service
17	Parallel port LPT service
18	ROM BASIC
19	Reboot
1A	Clock service
1B	Control-break handler
1C	User timer service
1D	Pointer for video parameter table
1E	Pointer for disk drive parameter table
1F	Pointer for graphics character pattern table
20	Terminate program
21	DOS services
22	Program termination handler
23	Control C handler
24	Critical error handler
25	Read disk
26	Write disk
27	Terminate and stay resident
28	DOS idle
2F	Multiplex handler
31	DPMI (DOS protected-mode interface) assessed through Windows or a DOS extender program
33	Mouse driver
67	Expanded memory manager interrupt and VCPI (virtual control program interface)
70–77	Hardware interrupts 8–15 (AT-style computer)

The $\overline{\text{BUSY/TEST}}$ pin of the microprocessor is usually connected to the $\overline{\text{BUSY}}$ pin of an 8087 through an 80387 numeric coprocessor. This connection allows the microprocessor to wait until the coprocessor finishes a task. Because the coprocessor is inside an 80486 or Pentium, the $\overline{\text{BUSY}}$ pin is not present in these microprocessors.

HLT

The halt instruction (HLT) stops the execution of software. There are three ways to exit a halt: by an interrupt, by a hardware reset, or during a DMA operation. This instruction normally appears in a program to wait for an interrupt. It often synchronizes external hardware interrupts with the software system.

NOP

When the microprocessor encounters a no operation instruction (NOP), it takes a short time to execute. In early years a NOP, which performs absolutely no operation, was often used to pad software with space for future machine language instructions. If you are developing machine language programs, *which is extremely rare*, it is recommended that you place 10 or so NOPs in your program at 50-byte intervals. This is done in case you need to add instructions at some future point. A NOP may also find application in time delays that waste short periods of time. A NOP is not very accurate when used for timing because of the cache and pipelines in modern microprocessors.

LOCK Prefix

The LOCK prefix appends an instruction and causes the $\overline{\text{LOCK}}$ pin to become a logic 0. The $\overline{\text{LOCK}}$ pin often disables external bus masters or other system components. The LOCK prefix causes the lock pin to activate for the duration of a locked instruction. If more than one sequential instruction is locked, the $\overline{\text{LOCK}}$ pin remains a logic 0 for the duration of the sequence of locked instructions. The LOCK:MOV AL,[SI] instruction is an example of a locked instruction.

ESC

The escape (ESC) instruction passes information to the 8087–80387 numeric coprocessor. Whenever an ESC instruction executes, the microprocessor provides the memory address, if required, but otherwise performs a NOP. The 8087–80387 uses six bits of the ESC instruction to obtain its opcode and begin executing a coprocessor instruction.

The ESC opcode never appears in a program as ESC. In its place are a set of coprocessor instructions (FLD, FST, FMUL, etc.) that assemble as ESC instructions for the coprocessor. More detail is provided in Chapter 13. This instruction is considered obsolete except as a command to the coprocessor.

BOUND

The BOUND instruction, first made available in the 80286 microprocessor, is a compare instruction that may cause an interrupt (vector type number 5). This instruction compares the contents of any 16-bit or 32-bit register against the contents of two words or doublewords of memory: an upper and a lower boundary. If the value in the register compared with memory is *not* within the upper and lower boundary, a type 5 interrupt ensues. If it is within the boundary, the next instruction in the program executes.

For example, if the BOUND SI,DATA instruction executes, word-sized location DATA contains the lower boundary and word-sized location DATA +2 bytes contains the upper boundary. If the number contained in SI is less than memory location DATA or greater than memory location DATA +2 bytes, a type 5 interrupt occurs. When this interrupt occurs, the return address points to the BOUND instruction, not the instruction following BOUND. This differs from a normal interrupt, where the return address points to the next instruction in the program.

ENTER and LEAVE

The ENTER and LEAVE instructions, first made available to the 80286 microprocessor, are used with stack frames. A stack frame is a mechanism used to pass parameters to a procedure through the stack memory. The stack frame also holds local memory variables for the procedure. Stack frames provide dynamic areas of memory for procedures in multiuser environments.

The ENTER instruction creates a stack frame by pushing BP onto the stack, then loading BP with the uppermost address of the stack frame. This allows stack frame variables to be accessed through the BP register. The ENTER instruction contains two operands: the first operand specifies the number of bytes to reserve for variables on the stack frame, and the second operand specifies the level of the procedure.

Suppose that an ENTER 8,0 instruction executes. This instruction reserves 8 bytes of memory for the stack frame. The zero specifies level 0. Figure 8–9 shows the stack frame set up by this instruction. This instruction stores BP onto the top of the stack. It

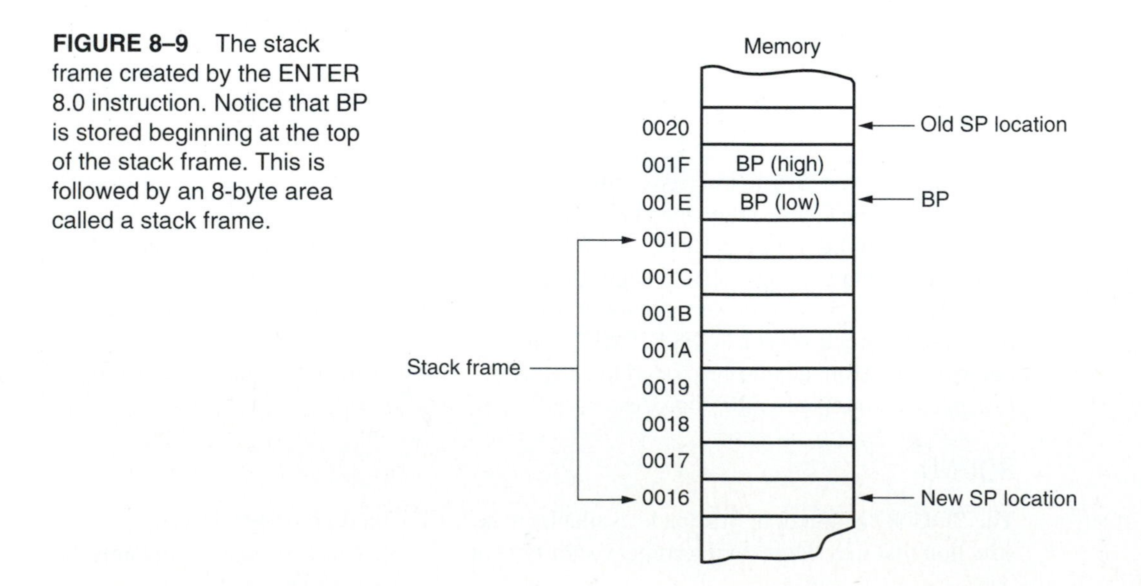

FIGURE 8–9 The stack frame created by the ENTER 8.0 instruction. Notice that BP is stored beginning at the top of the stack frame. This is followed by an 8-byte area called a stack frame.

then subtracts 8 from the stack pointer leaving 8 bytes of memory space for temporary data storage. The uppermost location of this 8-byte temporary storage area is addressed by BP. The LEAVE instruction reverses this process by reloading both SP and BP with their prior values.

EXAMPLE 8–13

```
                              ;A sequence used to call system software that
                              ;uses parameters stored in a stack frame
                              ;
0000  C8 0004 00                      ENTER  4,0                  ;create 4 byte frame

0004  A1 00C8 R                       MOV    AX,DATA1
0007  89 46 FC                        MOV    [BP-4],AX            ;save para 1
000A  A1 00CA R                       MOV    AX,DATA2
000D  89 46 FE                        MOV    [BP-2],AX            ;save para 2

0010  E8 0100 R                       CALL   SYS                  ;call subroutine

0013  8B 46 FC                        MOV    AX,[BP-4]            ;get result 1
0016  A3 00C8 R                       MOV    DATA1,AX             ;save result 1
0019  8B 46 FE                        MOV    AX,[BP-2]            ;get result 2
001C  A3 00CA R                       MOV    DATA2,AX             ;save result 2

001F  C9                              LEAVE
                                      .      .
                                      .      .
                                      (other software continues here)
                                      .      .
                                      .      .
                              ;system subroutine that uses the stack frame parameters
                              ;
0100                          SYS     PROC   NEAR

0100  60                              PUSHA

0101  8B 46 FC                        MOV    AX,[BP-4]            ;get para 1
0104  8B 5E FE                        MOV    BX,[BP-2]            ;get para 2
                                      .   .
                                      .   .
                                      (software that uses the parameters)
                                      .   .
                                      .   .
0130  89 46 FC                        MOV    [BP-4],AX            ;save result 1
0133  89 5E FE                        MOV    [BP-2],BX            ;save result 2

0136  61                              POPA
0137  C3                              RET

0138                          SYS     ENDP
```

Example 8–13 shows how the ENTER instruction creates a stack frame so two 16-bit parameters are passed to a system level procedure. Notice how the ENTER and LEAVE instructions appear in this program, and how the parameters pass through the stack frame to and from the procedure. This procedure uses two parameters that pass to it and returns two results through the stack frame.

8–5 SUMMARY

1. There are three types of unconditional jump instructions: short, near, and far. The short jump allows a branch to within +127 and −128 bytes. The near jump (using a displacement of ±32K) allows a jump to anywhere in the current code segment (intrasegment jump). The far jump allows a jump to any location in the memory system (intersegment jump). The near jump in an 80386 through a Pentium is within ±2G bytes, because these microprocessors can use a 32-bit signed displacement.
2. Whenever a label appears with a JMP instruction or conditional jump, it must be followed by a colon (LABEL:). The JMP DOGGY instruction jumps to memory location DOGGY:.
3. The displacement that follows a short or near jump is the distance from the next instruction to the jump location.
4. Indirect jumps are available in two forms: (1) jump to the location stored in a register and (2) jump to the location stored in a memory word (near indirect) or doubleword (far indirect).
5. Conditional jumps are all short jumps that test one or more of the flag bits: C, Z, O, P, or S. If the condition is true, a jump occurs, and if the condition is false, the next sequential instruction executes. Note that the 80386 and above also allow a 16-bit signed displacement for the conditional jump instructions.
6. A special conditional jump instruction (LOOP) decrements CX and jumps to the label when CX is not 0. Other forms of loop include: LOOPE, LOOPNE, LOOPZ, and LOOPNZ. The LOOPE instruction jumps if CX is not 0, and if an equal condition exists. In the 80386 through the Pentium, the LOOPD, LOOPED, and LOOPNED instructions also use the ECX register as a counter.
7. In the 80386 through the Pentium, a group of set according to conditions instructions exists that either set a byte to 01H or clear it to 00H. If the condition under test is true, the operand byte is set to 01H. If the condition under test is false, the operand byte is cleared to 00H.
8. Procedures are groups of instructions that perform one task and are used from any point in a program. The CALL instruction links to a procedure and the RET instruction returns from a procedure. In assembly language, the PROC directive defines the name and type of procedure. The ENDP directive declares the end of the procedure.
9. The CALL instruction is a combination of a PUSH and a JMP instruction. When CALL executes, it pushes the return address on the stack and then jumps to the procedure. A near CALL places the contents of IP on the stack and a far CALL places both IP and CS on the stack.
10. The RET instruction returns from a procedure by removing the return address from the stack and placing it into IP (near return) or IP and CS (far return).
11. Interrupts are either software instructions similar to CALL or hardware signals used to call procedures. This process interrupts the current program and calls a procedure. After the procedure, a special IRET instruction returns control to the interrupted software.
12. Real-mode interrupt vectors are four bytes long and contain the address (IP and CS) of the interrupt service procedure. The microprocessor contains 256 interrupt vectors in the first 1K byte of memory. The first 32 are defined by Intel, the remaining 224 are

user interrupts. In protected-mode operation, the interrupt vector is 8 bytes long and the interrupt vector table may be relocated to any section of the memory system.

13. Whenever an interrupt is accepted by the microprocessor, the flags, IP, and CS are pushed on the stack. Besides pushing the flags, the T and I flag bits are cleared to disable both the trace function and the INTR pin. The final event that occurs for the interrupt is that the interrupt vector is fetched from the vector table and a jump to the interrupt service procedure occurs.
14. Software interrupt instructions (INT) often replace system calls. Software interrupts save three bytes of memory each time they replace CALL instructions.
15. A special return instruction (IRET) must be used to return from an interrupt service procedure. The IRET instruction not only removes IP and CS from the stack, it also removes the flags from the stack.
16. Interrupt on an overflow (INTO) is a conditional interrupt that calls an interrupt service procedure if the overflow flag (O) = 1.
17. The interrupt enable flag (I) controls the INTR pin connection on the microprocessor. If the STI instruction executes, it sets I to enable the INTR pin. If the CLI instruction executes, it clears I to disable the INTR pin.
18. The carry flag bit (C) is cleared, set, and complemented by the CLC, STC, and CMC instructions.
19. The WAIT instruction tests the condition of the $\overline{\text{BUSY}}$ or $\overline{\text{TEST}}$ pin on the microprocessor. If $\overline{\text{BUSY}}$ or $\overline{\text{TEST}}$ = 1, WAIT does not wait, but if $\overline{\text{BUSY}}$ or $\overline{\text{TEST}}$ = 0, WAIT continues testing the $\overline{\text{BUSY}}$ or $\overline{\text{TEST}}$ pin until it becomes a logic 1. Note that the 8086/8088 contain the $\overline{\text{TEST}}$ pin, while the 80286–80386 contain the $\overline{\text{BUSY}}$ pin. The 80486 and Pentium do not contain a $\overline{\text{BUSY}}$ or $\overline{\text{TEST}}$ pin.
20. The LOCK prefix causes the $\overline{\text{LOCK}}$ pin to become a logic 0 for the duration of the locked instruction. The ESC instruction passes instructions to the numeric coprocessor.
21. The BOUND instruction compares the contents of any 16-bit register against the contents of two words of memory: an upper and a lower boundary. If the value in the register compared with memory is *not* within the upper and lower boundary, a type 5 interrupt ensues.
22. The ENTER and LEAVE instructions are used with stack frames. A stack frame is a mechanism used to pass parameters to a procedure through the stack memory. The stack frame also holds local memory variables for the procedure. The ENTER instruction creates the stack frame, and the LEAVE instruction removes the stack frame from the stack. The BP register addresses stack frame data.
23. Example 8–14 lists a program that uses some of the instructions presented in this chapter as well as some from prior chapters. This example contains a procedure that displays a character string on the monitor. As a test of the program, a few sample lines are displayed. Note that the character string is called a *null string* because it ends with a null (00H).

EXAMPLE 8–14

```
;a program that displays a string of characters using the
;procedure STRING.
;
        .MODEL SMALL            ;select SMALL model
```

```
0000                          .DATA                        ;indicate start of DATA segment
0000  0D 0A 0A 00            MES1DB   13,10,10,0
0004  54 68 69 73 20 MES2     DB      'This is a sample line.',0
         69 73 20 61 20
         73 61 6D 70 6C
         65 20 6C 69 6E
         65 2E 00
0000                          .CODE                        ;indicate start of CODE segment
                      ;
                      ;procedure that displays the character string address by
                      ;SI in the data segment. The character string must end with a
                      ;null.
                      ;
                      ;This procedure changes AX, DX, and SI.
                      ;
0000                  STRING PROC   NEAR

0000  AC                      LODSB                        ;get character from string
0001  3C 00                   CMP     AL,0                 ;test for null
0003  74 08                   JE      STRING1              ;if null
0005  8A D0                   MOV     DL,AL                ;move ASCII code to DL
0007  B4 02                   MOV     AH,2                 ;select function 02H
0009  CD 21                   INT     21H                  ;access DOS
000B  EB F3                   JMP     STRING               ;repeat until null
000D                  STRING1:
000D  C3                      RET                          ;return from procedure

000E                  STRING ENDP
                              .STARTUP                     ;indicate start of program
0025  FC                      CLD                          ;select increment
0026  BE 0000 R               MOV     SI,OFFSET MES1       ;address MES1
0029  E8 FFD4                 CALL    STRING               ;display MES1
002C  BE 0004 R               MOV     SI,OFFSET MES2       ;address MES2
002F  E8 FFCE                 CALL    STRING               ;display MES2
                              .EXIT                        ;exit to DOS
                              END                          ;end of file
```

8–6 QUESTIONS AND PROBLEMS

1. What is a short JMP?
2. Which type of JMP is used when jumping to any location within the current code segment?
3. Which JMP instruction allows the program to continue execution at any memory location in the system?
4. Which JMP instruction is five bytes long?
5. What is the range of a near jump in the 80386/80486/Pentium microprocessor?
6. Which type of JMP instruction (short, near, or far) assembles for the following:
   (a) if the distance is 0210H bytes
   (b) if the distance is 0020H bytes
   (c) if the distance is 10000H bytes
7. What can be said about a label followed by a colon?
8. The near jump modifies the program address by changing which register or registers?

9. The far jump modifies the program address by changing which register or registers?
10. Explain what the JMP AX instruction accomplishes. Identify it as a near or a far jump instruction.
11. Contrast the operation of a JMP DI with a JMP [DI].
12. Contrast the operation of a JMP [DI] with a JMP FAR PTR [DI].
13. List the five flag bits tested by the conditional jump instructions.
14. Describe how the JA instruction operates.
15. When will the JO instruction jump?
16. Which conditional jump instructions follow the comparison of signed numbers?
17. Which conditional jump instructions follow the comparison of unsigned numbers?
18. Which conditional jump instructions test both the Z and C flag bits?
19. When does the JCXZ instruction jump?
20. Which SET instruction is used to set AL if the flag bits indicate a zero condition?
21. The 8086 LOOP instruction decrements register ______________ and tests it for a 0 to decide whether a jump occurs.
22. The 80486 LOOPD instruction decrements register ______________ and tests it for a 0 to decide whether a jump occurs.
23. Explain how the LOOPE instruction operates.
24. Develop a short sequence of instructions that stores a 00H in 150H bytes of memory beginning at extra segment memory location DATA. You must use the LOOP instruction to help perform this task.
25. Develop a sequence of instructions that searches through a block of 100H bytes of memory. This program must count all the unsigned numbers above 42H and below 42H. Byte-sized data segment memory location UP must contain the count of numbers above 42H and data segment location DOWN must contain the count of numbers below 42H.
26. What is a procedure?
27. Explain how the near and far CALL instructions function.
28. How does the near RET instruction function?
29. The last executable instruction in a procedure must be a ______________.
30. Which directive identifies the start of a procedure?
31. How is a procedure identified as near or far?
32. Explain what the RET 6 instruction accomplishes.
33. Write a near procedure that cubes the contents of the CX register. This procedure may not affect any register except CX.
34. Write a procedure that multiplies DI by SI and then divides the result by 100H. Make sure that the result is left in AX upon returning from the procedure. This procedure may not change any register except AX.
35. Write a procedure that sums EAX, EBX, ECX, and EDX. If a carry occurs, place a logic 1 in EDI. If no carry occurs, place a 0 in EDI. The sum should be found in EAX after the execution of your procedure.
36. What is an interrupt?
37. Which software instructions call an interrupt service procedure?
38. How many different interrupt types are available in the microprocessor?
39. What is the purpose of interrupt vector type number 0?
40. Illustrate the contents of an interrupt vector and explain the purpose of each part.

41. How does the IRET instruction differ from the RET instruction?
42. What is the IRETD instruction?
43. The INTO instruction only interrupts the program for what condition?
44. The interrupt vector for an INT 40H instruction is stored at what memory locations?
45. What instructions control the function of the INTR pin?
46. Which personal computer interrupt services the parallel LPT port?
47. Which personal computer interrupt services the keyboard?
48. What instruction tests the $\overline{\text{BUSY}}$ pin?
49. When will the BOUND instruction interrupt a program?
50. An ENTER 16,0 instruction creates a stack frame that contains ______________ bytes.
51. Which register moves to the stack when an ENTER instruction executes?
52. Which instruction passes opcodes to the numeric coprocessor?
53. What is a null string?
54. Explain how the STRING procedure operates in Example 8–14.
55. Rewrite Example 8–14 so it displays your name.

CHAPTER 9

Keyboard/Display DOS and BIOS Functions

INTRODUCTION

This chapter develops programs and programming techniques using the MASM macro assembler, the DOS function calls, and the BIOS function calls. Some of the DOS and BIOS function calls are used in this chapter, but all are explained in complete detail in Appendix A. Please review the function calls as you read this chapter. The MASM assembler has been explained and demonstrated in prior chapters, yet there are many more features to learn at this point.

Some programming techniques explained in this chapter include: macro sequences, keyboard and display manipulation, program modules, library files, and other programming techniques. This chapter introduces important programming techniques so programs for the personal computer can be efficiently developed using either PCDOS[1] or MSDOS[2].

OBJECTIVES

Upon completion of this chapter, you will be able to:

1. Use the MASM assembler and linker program to create programs that contain more than one module.
2. Explain the use of EXTRN and PUBLIC as they apply to modular programming.
3. Set up a library file that contains commonly used subroutines.
4. Write and use MACRO and ENDM to develop macro sequences used with linear programming.
5. Develop programs using DOS function calls.
6. Differentiate a DOS function call from a BIOS function call.

[1]PCDOS is a registered trademark of IBM Corporation.

[2]MSDOS is a registered trademark of Microsoft Corporation.

9–1 MODULAR PROGRAMMING

Most programs are too large to be developed by one person; instead, they are often developed by teams of programmers. The linker program is provided with MSDOS or PCDOS so programming modules can be linked together into a complete program. This section describes the linker, the linking task, library files, EXTRN, and PUBLIC as they apply to program modules and modular programming.

The Assembler and Linker

The **assembler program** converts a symbolic source module or file into a hexadecimal object file. We have seen many examples of symbolic source files in previous chapters. Example 9–1 shows how assembler dialog that appears as a source module named FILE.ASM is assembled. When you create a source file, it usually contains the extension ASM to designate the file as an assembly language source file. Note that the extension is not typed into the assembler prompt when assembling a file. Source files are created using the editor that comes with the assembler or by almost any other editor or word processor capable of generating an ASCII file. If MASM version 6.X is in use, the Programmer's WorkBench editor is used to create the source file.

EXAMPLE 9–1

```
A:\>MASM

Microsoft (R) Macro Assembler Version 5.10
Copyright (C) Microsoft Corp 1981, 1989. All rights reserved.

Source filename [.ASM]: FILE
Object filename [FILE.OBJ]: FILE
Source listing [NUL.LST]: FILE
Cross reference [NUL.CRF]: FILE
```

If version 5.10 of MASM is in use, the assembler program (MASM) asks for the source file name, the object file name, the list file name, and a cross-reference file name. In most cases the name for each of these will be the same as the source file. The object file (.OBJ) is not executable, but is designed as an input file to the linker. The source listing file (.LST) contains the assembled version of the source file and its hexadecimal machine language equivalent. The cross-reference file (.CRF) lists all labels and pertinent information required for cross-referencing.

If version 6.X of MASM is in use, the project is built using the build option after the file is saved as FILE.ASM and the option menu has selected the type of project to build. In many cases, for simple assembly language programs, the standard DOS .EXE file option is selected.

The **linker program** reads the object files created by the assembler program and links them together into a single execution file. With MASM version 6.X, the linker is included and executed when a program is built using the build option. Either the linker program (MASM version 5.10) to build generates the **execution file** with the filename extension EXE. Execution files are executed by typing the file name at the DOS prompt

(A:\>). For example, execution file FROG.EXE is executed by typing FROG at the DOS command prompt.

If a file is less than 64K bytes in length, it can be converted from an execution file to a **command file** (.COM). A command file is slightly different from an execution file, in that the program must be originated at location 100H before it can execute. If MASM version 5.10 is in use, the program EXE2BIN is used for converting an execution file into a command file. The main advantage of a command file is that it loads off the disk into the computer much more quickly than an execution file. It also requires slightly less disk storage space than the execution file. If MASM version 6.X is in use, the EXE2BIN program is not needed. The Programmer's WorkBench editor program for version 6.X creates the command file directly if the command file (.COM) option is selected. WorkBench also builds the file by assembling it and linking it.

EXAMPLE 9–2

```
A:\>LINK

Microsoft (R) Overlay Linker Version 3.64
Copyright (C) Microsoft Corp 1983-1988. All rights reserved.

Object Modules [.OBJ]: FROG+WHAT+DONUT
Run File [FROG.EXE]: FROG
List File [NUL.MAP]: FROG
Libraries [.LIB]: LIBS
```

Example 9–2 shows the protocol involved with the linker program to link the files FROG, WHAT, and DONUT. The linker also links the library files (LIBS) so the procedures located within LIBS can be used with the linked execution file. To invoke the linker, type LINK at the DOS command prompt, as illustrated in Example 9–2. Note that before files can be linked, they must be assembled and they must be *error free*. Version 6.X of MASM allows a project with multiple files to be built using the build option.

In this example, after typing LINK, the linker program asks for the Object Modules, which are created by the assembler. This example has three object modules: FROG, WHAT, and DONUT. If more than one object file exists, the main program file (FROG in this example) is typed first followed by any other supporting modules. (A plus sign is used to separate module names.)

After the program module names are typed, the linker suggests that the execution (*run-time*) file name is FROG.EXE. This may be selected by typing the same name or hitting the enter key. It may also be changed to any other name at this point.

The list file is where a map of the program segments appear as created by the linking. If enter is typed, no list file is created, but if a name is typed, the list file appears on the disk.

Library files are entered in the last line. This example uses the library file name LIBS. This library contains procedures used by the other program modules.

If the Programmer's WorkBench is *not* used to assemble and link software with version 6.X of the MASM macro assembler, the sequence listed in Example 9–3 is used to assemble and link the program file. The example shows how the file TEST.ASM is assembled and linked using the ML program provided with version 6.X of MASM. The /Fl switch (must be upper case F, lower case l) generates the list file TEST.LST in this example.

EXAMPLE 9–3

```
C:\>ML /FiTEST.LST TEST.ASM

Microsoft (R) Macro Assembler Version 6.11
Copyright (C) Microsoft Corp 1981-1993. All rights reserved.

  Assembling: TEST.ASM

Microsoft (R) Segmented-Executable Linker Version 5.13
Copyright (C) Microsoft Corp 1984-1993. All rights reserved.

Object Modules [.OBJ]: TEST.obj/t
Run File [TEST.com]: "TEST.com"
List File [NUL.MAP]: NUL
Libraries [.LIB]:
Definitions File [NUL.DEF]: ;
```

PUBLIC and EXTRN

The PUBLIC and EXTRN directives are very important to modular programming. The PUBLIC directive declares that labels of code, data, or entire segments are available to other program modules. The EXTRN (external) directive (EXTERN is used as an alternate in version 6.X of the assembler) declares that labels are external to a module. Without these statements, modules could not be linked together to create a program using modular programming techniques.

EXAMPLE 9–4

```
                    DAT1    SEGMENT PUBLIC              ;declare entire segment public

                            PUBLIC  DATA1               ;declare DATA1, DATA2 public
                            PUBLIC  DATA2

0000  0064[         DATA1   DB     100 DUP (?)          ;global
          ??
            ]
0064  0064[         DATA2   DB     100 DUP (?)          ;global
          ??
            ]

00C8                DAT1    ENDS

0000                CODES   SEGMENT 'CODE'

                            ASSUME CS:CODES,DS:DAT1

                            PUBLIC  READ                ;declare READ public

0000                READ    PROC   FAR

0000  B4 06                 MOV    AH,6                 ;read keyboard
0002  B2 FF                 MOV    DL,0FFH              ;no echo
0004  CD 21                 INT    21H
0006  74 F8                 JE     READ
0008  CB                    RET
```

```
0009                READ    ENDP

0009                CODES   ENDS

                            END
```

The PUBLIC directive is normally placed in the opcode field of an assembly language statement to define a label as public so it can be used by other modules. This label can be a jump address, a data address, or an entire segment. Example 9–4 shows how the PUBLIC statement, used to define some labels as public to other modules, is a program that uses full-segment definitions. When segments are made public, they are combined with other public segments that contain data with the same segment name. With models, the segments are always public and always combined.

Example 9–4 shows how the procedure READ is made public (that is, made available to other program modules). The READ procedure reads a key without echo to the video display. Example 9–4 also shows how DATA1 and DATA2 are made public.

EXAMPLE 9–5

```
0000                      DAT1    SEGMENT PUBLIC          ;declare entire segment public

                                  EXTRN   DATA1:BYTE
                                  EXTRN   DATA2:BYTE
                                  EXTRN   DATA3:WORD
                                  EXTRN   DATA4:DWORD

0000                      DAT1    ENDS

0000                      CODES   SEGMENT 'CODE'

                                  ASSUME CS:CODES,ES:DAT1

                                  EXTRN READ:FAR

0000                      MAIN    PROC    FAR

0000  B8 ---- R                   MOV     AX,DAT1
0003  8E C0                       MOV     ES,AX

0005  BF 0000 E                   MOV     DI,OFFSET DATA1
0008  B9 000A                     MOV     CX,10
000B                      MAIN1:
000B  9A 0000 ---- E              CALL    READ
0010  AA                          STOSB
0011  E2 F8                       LOOP    MAIN1
0013  CB                          RET

0014                      MAIN    ENDP

0014                      CODES   ENDS

                                  END     MAIN
```

The EXTRN statement appears in both data and code segments to define labels as external to the segment. If data are defined as external, their size must be represented as

BYTE, WORD, or DWORD. If a jump or call address is external, it must be represented as NEAR or FAR. Example 9–5 shows how the external statement is used to indicate that several labels are external to the program listed using full-segment definitions. In this example, any external address or data is defined with the letter E in the hexadecimal assembled listing. In this example, the READ procedure from Example 9–4 is called to read ten keys from the keyboard and store them into an array called DATA1.

PUBLIC and EXTRN are also used with models. Example 9–6 shows how to declare variables public and how to use external variables with models. Note that this is a repeat of Example 9–5.

EXAMPLE 9–6

```
                                  .MODEL SMALL        ;select SMALL model
0000                              .DATA               ;indicate start of DATA segment
                                  EXTRN  DATA1:BYTE   ;declare DATA1 as an external byte
                                  EXTRN  DATA2:BYTE   ;declare DATA2 as an external byte
                                  EXTRN  DATA3:WORD   ;declare DATA3 as an external word
                                  EXTRN  DATA4:DWORD  ;declare DATA4 as an external doubleword
0000                              .CODE               ;indicate start of CODE segment
                                  EXTRN  READ:FAR     ;declare READ a far procedure
                                  .STARTUP            ;indicate start of program
0017  8C D8                       MOV    AX,DS        ;load ES with DS
0019  8E C0                       MOV    ES,AX

001B  BF 0000 E                   MOV    DI,OFFSET DATA1 ;address DATA1
001E  B9 000A                     MOV    CX,10        ;load count with 10
0021                       MAIN1:
0021  9A ---- 0000 E              CALL   READ         ;read a key without echo
0026  AA                          STOSB               ;store key code in DATA1
0027  E2 F8                       LOOP   MAIN1        ;repeat 10 times
                                  .EXIT               ;exit to DOS
                                  END                 ;end of file
```

Libraries

Library files are collections of procedures that can be used by many different programs. These procedures are assembled and compiled into a library file by the LIB program that accompanies the MASM assembler program. Libraries allow common procedures to be collected into one place so they can be used by many different applications. The library file (FILENAME.LIB) is invoked when a program is linked with the linker program.

Why bother with library files? A library file is a good place to store a collection of related procedures. When the library file is linked with a program, only the procedures required by the program are removed from the library file and added to the program. If any amount of assembly language programming is to be accomplished efficiently, a good set of library files are essential.

Creating a Library File. A library file is created with the LIB command typed at the DOS prompt. A library file is a collection of assembled .OBJ files that each perform one procedure. Example 9–7 shows two separate files that contain the READ_KEY and ECHO procedures. When these files are edited, the one containing READ_KEY is named READ_KEY.ASM and the one containing ECHO is named ECHO.ASM. Both files are used to structure a library file. Note that the name of the procedure must be declared

PUBLIC in a library file and does not necessarily need to match the file name, although it does in this example. The READ_KEY procedure uses DOS INT 21H function number 06H to read a key from the keyboard without echo. The ECHO procedure also uses DOS INT 21H function number 06H to display the contents of the AL register.

EXAMPLE 9–7

```
                    ;The first library module is called READ_KEY.
                    ;This procedure reads a key from the keyboard
                    ;and returns with the ASCII character in AL.
                    ;
0000                LIB     SEGMENT 'CODE'          ;indicate start of LIB segment
                            ASSUME  CS:LIB          ;indicate to assembler that CS is LIB

                            PUBLIC  READ_KEY        ;define READ_KEY as public

0000                READ_KEY PROC FAR               ;indicate start of procedure

0000  52                    PUSH    DX              ;save DX

0001                READ_KEY1:

0001  B4 06                 MOV     AH,6            ;select DOS function 06H
0003  B2 FF                 MOV     DL,0FFH         ;read key without echo
0005  CD 21                 INT     21H             ;access DOS
0007  74 F8                 JE      READ_KEY1       ;if no key is typed
0009  5A                    POP     DX              ;restore DX
000A  CB                    RET                     ;return from procedure

000B                READ_KEY ENDP                   ;indicate end of procedure

000B                LIB     ENDS                    ;indicate end of LIB segment
                            END

                    ;
                    ;This second library module is called ECHO
                    ;This procedure displays the ASCII character
                    ;in AL on the CRT screen.
                    ;
0000                LIB     SEGMENT 'CODE'          ;indicate start of LIB segment
                            ASSUME  CS:LIB          ;indicate to the assembler that CS is LIB

                            PUBLIC  ECHO            ;indicate ECHO is public

0000                ECHO    PROC    FAR             ;indicate start of procedure

0000  52                    PUSH    DX              ;save DX
0001  B4 06                 MOV     AH,6            ;select function 06H
0003  8A D0                 MOV     DL,AL           ;display AL
0005  CD 21                 INT     21H             ;access DOS
0007  5A                    POP     DX              ;restore DX
0008  CB                    RET                     ;return from procedure

0009                ECHO    ENDP                    ;indicate end of procedure

0009                LIB     ENDS                    ;indicate end of LIB segment
                            END                     ;end of file
```

After each file is assembled, the LIB program is used to combine the files into a library file. The LIB program prompts for information as illustrated in Example 9–8, where these files are combined to form the library IO.

EXAMPLE 9–8

```
A:\>LIB

Microsoft (R) Library Manager Version 3.10
Copyright (C) Microsoft Corp 1983-1988. All rights reserved.

Library name:IO
Library file does not exist. Create? Y
Operations:READ_KEY+ECHO
List file:IO
```

The LIB program begins with the copyright message from Microsoft, followed by the prompt *Library name:*. The library name chosen is IO for the IO.LIB file. Because this is a new file, the library program asks if we wish to create the library file. The *Operations:* prompt is where the library module names are typed. In this case, a library is created using two procedure files (READ_KEY and ECHO). The list file shows the contents of the library, it is illustrated in Example 9–9. The list file shows the size and names of the files used to create the library and also the public label (procedure name) used in the library file.

EXAMPLE 9–9

```
ECHO..............ECHO    READ_KEY...........READ_KEY

READ_KEY     Offset: 00000010H Code and data size: BH
 READ_KEY

ECHO         Offset: 00000070H Code and data size: 9H
 ECHO
```

If you must add additional library modules at a later time, type the name of the library file after invoking LIB. At the *Operations:* type the new module name, preceded by a *plus sign,* to add a new procedure. If you must delete a library module, use a *minus sign* before the operation file name.

Once the library file is linked to your program file only the library procedures actually used by your program are placed in the execution file. Don't forget to use the label EXTRN when specifying library calls from your program module.

Macros

A macro is a group of instructions that perform one task, just like a procedure performs one task. The difference is that a procedure is accessed via a CALL instruction, while a macro is inserted in the program by using the name of the macro, which represents a sequence of instructions. A *macro* creates a new instruction or directive for the assembler. Macro sequences execute faster than procedures because there are no CALL and RET instructions to execute. The instructions of the macro sequence are placed in the program at the point where they are invoked. When software is developed using macro sequences, it flows from the top to the bottom.

Macro sequences are ideal for systems that contain cache memory or are required to execute software with maximum speed and efficiency. Programs that use macro sequences in place of procedures are often called **linear programs** because they flow from the top to the bottom without any calls. If a cache memory is used in the computer system, it is desirable to develop programs using macro sequences, because the entire program can often be executed from the cache. This significantly increases execution speed. If a procedure is called, the cache must often be reloaded, requiring additional time. If a macro is invoked, no additional time is needed, because no jump to the procedure occurs. Cache memory looks ahead, in a sense, and loads instructions from the memory system before they are executed by the microprocessor. It does this by loading the cache with the next sequential section of four 32-bit doublewords. In a macro sequence, the next four 32-bit doublewords will often contain the entire macro if it is short. If a procedure is called, it is usually not in the next four 32-bit doublewords of memory, which would require access to another section of the memory as well as additional time.

The MACRO and ENDM directives are used to delineate a macro sequence. The first statement of a macro is the MACRO statement, which contains the name of the macro and any parameters associated with it. An example is MOVE MACRO A,B, which defines the macro as MOVE. This new macro name is used like an opcode or directive that contains two parameters: A and B. These parameters are often called replaceable parameters, because as the macro is expanded for storage in the program, they are replaced. Macros can contain up to 32 replaceable parameters. The last statement of a macro is the ENDM instruction, on a line by itself without a label.

EXAMPLE 9–10

```
                              MOVE   MACRO    A,B        ;;moves word from B to A

                              PUSH   AX
                              MOV    AX,B
                              MOV    A,AX
                              POP    AX

                              ENDM

                              MOVE   VAR1,VAR2           ;use macro MOVE

0000  50             1        PUSH   AX
0001  A1  0002  R    1        MOV    AX,VAR2
0004  A3  0000  R    1        MOV    VAR1,AX
0007  58             1        POP    AX

                              MOVE   VAR3,VAR4           ;use macro MOVE

0008  50             1        PUSH   AX
0009  A1  0006  R    1        MOV    AX,VAR4
000C  A3  0004  R    1        MOV    VAR3,AX
000F  58             1        POP    AX
```

Example 9–10 shows how a macro is created and used in a program. Note, this is not a complete program and will not execute as it is printed. It is included to show how macro sequences are created and generate codes when invoked. This macro moves the word-sized contents of memory location B into word-sized memory location A. After the macro is defined with the first six statements, it is used twice. The macro is *expanded* in this example to show how it assembles and generates the move. Notice how parameters A and B are replaced by VAR1 and VAR2 the first time the macro is used, and by VAR3 and

VAR4 the second time the macro is used. Any hexadecimal machine language statement followed by a 1 is a macro expansion statement. If a macro uses a second macro (nesting), the statement is followed by a 2, and so forth. The expansion statements were not typed in the source program. The expansion statements are generated each time a macro sequence is invoked. The comment in the macro is preceded by ;; instead of ; as is customary. To cause the macro expansion statements to list, place the .LISTALL statement at the start of the program.

Local Variable in a Macro. Sometimes macros contain local variables. A *local variable* appears in the macro, but is not available outside the macro. To define a local variable, use the LOCAL directive. Example 9–11 shows how a local variable, used as a jump address, appears in a macro definition. If this jump address is not defined as local, the assembler will flag it as a duplicate label error.

Example 9–11 reads two characters from the keyboard and stores them into the byte-sized memory locations indicated as parameters with the macro. Notice how the local label READ1 is treated in the expanded macros. Local variables always appear as ??nnnn, where nnnn is a number identifying it as a unique label.

The LOCAL directive must always immediately follow the MACRO directive without any intervening blank lines or comments. If a comment or blank line appears between the MACRO and LOCAL statements, the assembler indicates an error and will not accept the variable as local.

EXAMPLE 9–11

```
                          ;a program that uses a macro to read 2 keys from the keyboard
                          ;and echo them to the display. The first key code is stored at
                          ;VAR5 and the second at VAR6.
                          ;
                                  .MODEL SMALL            ;select SMALL model
0000                              .DATA                   ;indicate start of DATA segment
0000  00                  VAR5    DB      0               ;define VAR5
0001  00                  VAR6    DB      0               ;define VAR6
0000                              .CODE                   ;indicate start of CODE segment

                          READ    MACRO   A               ;;reads keyboard with echo
                                  LOCAL   READ1           ;;define READ1 as local

                                  PUSH    DX
                          READ1:
                                  MOV     AH,6            ;;read key
                                  MOV     DL,0FFH
                                  INT     21H
                                  JE      READ1
                                  MOV     A,AL            ;;save key code
                                  MOV     DL,AL           ;;echo key
                                  INT     21H
                                  POP     DX

                                  ENDM
                                  .STARTUP                ;start of program

                                  READ    VAR5            ;read key and store it at VAR5

0000  52             1            PUSH    DX
0001                 1    ??0000:
0001  B4 06          1            MOV     AH,6
0003  B2 FF          1            MOV     DL,0FFH
```

```
0005  CD 21      1         INT    21H
0007  74 F8      1         JE     ??0000
0009  A2 0000 R  1         MOV    VAR5,AL
000C  8A D0      1         MOV    DL,AL
000E  CD 21      1         INT    21H
0010  5A         1         POP    DX

                           READ  VAR6              ;read key and store it at VAR6

0011  52         1         PUSH   DX
0012             1  ??0001:
0012  B4 06      1         MOV    AH,6
0014  B2 FF      1         MOV    DL,0FFH
0016  CD 21      1         INT    21H
0018  74 F8      1         JE     ??0001
001A  A2 0001 R  1         MOV    VAR6,AL
001D  8A D0      1         MOV    DL,AL
001F  CD 21      1         INT    21H
0021  5A         1         POP    DX
                           .EXIT                   ;exit to DOS
                           END                     ;end of file
```

Placing MACRO Definitions in their Own Module. Macro definitions are placed in the program file as shown, or are placed in their own macro module. A file that contains only macros that are to be included with other program files is called an INCLUDE file. The INCLUDE directive indicates that a program file will include a module that contains external macro definitions. Although this is not a library file, it functions as a library of macro sequences. A macro INCLUDE file is an unassembled ASCII file generated by an editor or word processor. A good collection of macro statements is invaluable when programming in assembly language because it eliminates the task of writing the same sequence of instructions many times in a program.

When macro sequences are placed in a file (often with the extension INC or MAC) they do not contain PUBLIC statements. If a file called MACRO.MAC contains macro sequences, the INCLUDE statement is placed in the program file as INCLUDE C:\ASSM\MACRO.MAC. Notice that the macro file is on drive C, subdirectory ASSM in this example. The INCLUDE statement includes these macros just as if they had been typed into the file. No EXTRN statement is needed to access the macro statements that have been included.

The Modular Programming Approach

The modular programming approach often involves a team of people, each assigned to a different programming task. The team manager assigns various portions of the program to different team members. Often the team manager develops the system flowchart or shell, then divides it into modules for team members.

One team member might be assigned the task of developing a macro definition file. This file might contain macro definitions that handle the I/O operations for the system. Another team member might be assigned the task of developing the procedures used for the system. In most cases, the procedures are organized as a library file linked to the program modules. Several program files or modules might be used for the final system, each developed by different team members.

This approach requires considerable communications between team members and good documentation, so that modules interface correctly.

9–2 USING THE KEYBOARD AND VIDEO DISPLAY

This section of the text explains how to use the keyboard and video display connected to the personal computer, running under either MSDOS or PCDOS, for programming. Some examples of using the keyboard and video display have appeared earlier.

Reading the Keyboard with DOS Functions

The keyboard in the personal computer is often read via a DOS function call. A complete listing of the DOS function calls appears in Appendix A. This section uses INT 21H with various DOS function calls to read the keyboard. Data read from the keyboard is either in ASCII-coded or extended ASCII-coded form. The exact form depends on which keys are typed on the keyboard.

The ASCII-coded data appears as outlined in Tables 1–6 and 1–7. These codes correspond to most of the keys on the keyboard. The extended ASCII codes listed in Table 1–4 apply to the printer or video screen. Also available through the keyboard are extended ASCII-coded *keyboard* data. Table 9–1 lists most of the extended ASCII codes. It also lists scan codes read directly from the keys before conversion to ASCII. These codes are different from the extended ASCII printer/video display codes in Chapter 1. Notice that most keys on the keyboard have alternative key codes, selected by the function keys, the shift function keys, alternate function keys, and the control function keys.

There are three ways to read the keyboard with DOS. The first method reads a key and echoes (or displays) the key on the video screen. A second way tests to see whether a key is pressed and if it *is* pressed, it reads the key. The third way allows an entire character line to be read from the keyboard.

Reading a Key with an Echo. Example 9–12 shows a procedure that reads a key from the keyboard and **echoes** it (sends it back out) to the video display. Although this is the easiest way to read a key, it is the most limited, because it always echoes (displays) the character to the screen even if it is an unwanted character. DOS function number 01H responds to the control C key and exits the program to DOS if it is typed.

EXAMPLE 9–12

```
0000                KEY      PROC    FAR

0000 B4 01                   MOV     AH,1        ;function 01H
0002 CD 21                   INT     21H         ;read key
0004 0A C0                   OR      AL,AL       ;test for 00H, clear carry
0006 75 03                   JNZ     KEY1
0008 CD 21                   INT     21H         ;get extended
000A F9                      STC                 ;indicate extended
000B                KEY1:
000B CB                      RET

000C                KEY      ENDP
```

To read and echo a character, the AH register is loaded with DOS function number 01H. This is followed by the INT 21H instruction. (All DOS functions use AH to hold the

TABLE 9–1 The keyboard scanning and extended ASCII codes as returned from the keyboard (pp. 305–307)

		Extended ASCII Code with . . .			
Key	*Scan Code*	*Nothing*	*Shift*	*Control*	*Alternate*
ESC	01				01
1	02				78
2	03			03	79
3	04				7A
4	05				7B
5	06				7C
6	07				7D
7	08				7E
8	09				7F
9	0A				80
0	0B				81
-	0C				82
+	0D				83
BKSP	0E				0E
TAB	0F		0F	94	A5
Q	10				10
W	11				11
E	12				12
R	13				13
T	14				14
Y	15				15
U	16				16
I	17				17
O	18				18
P	19				19
[	1A				1A
]	1B				1B
Enter	1C				1C
Enter	1C				A6
L CTRL	1D				
R CTRL	1D				
A	1E				1E
S	1F				1F
D	20				20
F	21				21

TABLE 9–1 (continued)

Key	Scan Code	Extended ASCII Code with . . . Nothing	Shift	Control	Alternate
G	22				22
H	23				23
J	24				24
K	25				25
L	26				26
:	27				27
'	28				28
'	29				29
L SHFT	2A				
\	2B				
Z	2C				2C
X	2D				2D
C	2E				2E
V	2F				2F
B	30				30
N	31				31
M	32				32
,	33				33
.	34				34
/	35				35
Gray /	35			95	A4
R SHFT	36				
PRTSC	E0 2A E0 37				
L ALT	38				
R ALT	38				
Space	39				
Caps	3A				
F1	3B	3B	54	5E	68
F2	3C	3C	55	5F	69
F3	3D	3D	56	60	6A
F4	3E	3E	57	61	6B
F5	3F	3F	58	62	6C
F6	40	40	59	63	6D
F7	41	41	5A	64	6E

TABLE 9–1 (continued)

		Extended ASCII Code with . . .			
Key	*Scan Code*	*Nothing*	*Shift*	*Control*	*Alternate*
F8	42	42	5B	65	6F
F9	43	43	5C	66	70
F10	44	44	5D	67	71
F11	57	85	87	89	8B
F12	58	86	88	8A	8C
Num	45				
Scroll	46				
Home	E0 47	47	47	77	97
Up	48	48	48	8D	98
PGUP	E0 49	49	49	84	99
Gray -	4A				
Left	4B	4B	4B	73	9B
Center	4C				
Right	4D	4D	4D	74	9D
Gray +	4E				
End	E0 4F	4F	4F	75	9F
Down	E0 50	50	50	91	A0
PGDN	E0 51	51	51	76	A1
Ins	E0 52	52	52	92	A2
Del	E0 53	53	53	93	A3
Pause	E0 10 45				

function number before the INT 21H DOS function call.) Upon return from INT 21H, the AL register contains the typed ASCII character and the video display shows the typed character. The return from this DOS function call does not occur until a key is typed. If AL = 0 after the return, the INT 21H instruction must again be executed to obtain the extended ASCII-coded character. This procedure in Example 9–12 returns with carry set (1) to indicate an extended ASCII character and carry cleared (0) to indicate a normal ASCII character.

Reading a Key with No Echo. The best single character key reading function is function number 06H. This function reads a key without an echo to the screen. It also returns with extended ASCII characters and *does not* respond to the control C key. This function uses AH for the function number (06H) and DL = 0FFH to indicate that the function call (INT 21H) will read the keyboard without an echo.

EXAMPLE 9–13

```
0000                KEYS      PROC    FAR

0000  B4 06                   MOV     AH,6            ;function 06H
0002  B2 FF                   MOV     DL,0FFH
0004  CD 21                   INT     21H             ;read key
0006  74 F8                   JE      KEYS            ;if no key
0008  0A C0                   OR      AL,AL           ;test for 00H, clear carry
000A  75 03                   JNE     KEYS1
000C  CD 21                   INT     21H             ;get extended
000E  F9                      STC                     ;indicate extended
000F                KEYS1:
000F  CB                      RET

0010                KEYS      ENDP
```

Example 9–13 shows a procedure that uses function number 06H to read the keyboard. This performs the same way as Example 9–12, except no character is echoed to the video display.

If you examine Example 9–13, you will note one other difference between it and Example 9–12. Function call number 06H returns from the INT 21H even if no key is typed, while function call 01H waits for a key to be typed. This is an important difference that should be noted. This feature allows software to perform other tasks between checking the keyboard for a character.

Read an Entire Line with Echo. Sometimes it is advantageous to read an entire line of data with one function call. Function call number 0AH reads an entire line of information—up to 255 characters—from the keyboard. It continues to acquire keyboard data until the enter key (0DH) is typed. This function requires that AH = 0AH and DS:DX address the keyboard buffer (a memory area where the ASCII data are stored). The first byte of the buffer area contains the maximum number of keyboard characters read by this function. If the number typed exceeds this maximum number, the DOS function returns just as if the enter key were typed. The second byte of the buffer contains the count of the actual number of characters typed and the remaining locations in the buffer contain the ASCII keyboard data.

EXAMPLE 9–14

```
                        ;a program that reads two lines of data from the keyboard
                        ;using DOS INT 21H function number 0AH
                        ;***uses***
                        ;LINE procedure to read a line.
                        ;
                               .MODEL SMALL                     ;select SMALL model
0000                           .DATA                            ;start DATA segment
0000  0101 [           BUF1    DB      257 DUP (?)              ;define BUF1
              00
                 ]
0101  0101 [           BUF2    DB      257 DUP (?)              ;define BUF2
              00
                 ]
0000                           .CODE                            ;start CODE segment
                               .STARTUP                         ;start program
0017  C6 06 0000 R FF          MOV     BUF1,255                 ;character count of 255
001C  BA 0000 R                MOV     DX,OFFSET BUF1           ;address BUF1
001F  E8 000F                  CALL    LINE                     ;read a line
```

```
0022  C6 06 0101 R FF       MOV    BUF2,255                    ;character count of 255
0027  BA 0101 R             MOV    DX,OFFSET BUF2              ;address BUF2
002A  E8 0004               CALL   LINE                        ;read a line
                            .EXIT                              ;exit to DOS
                     ;
                     ;the LINE procedure uses DOS INT 21H function 0AH to read
                     ;and echo an entire line from the keyboard.
                     ;***parameters***
                     ;DX must contain the data segment offset address of the buffer
                     ;the first location in the buffer contains the number of characters
                     ;to be read for the line.
                     ;upon return the second location in the buffer contains the line length.
                     ;
0031                 LINE   PROC   NEAR

0031  B4 0A                 MOV    AH,0AH                      ;select function 0AH
0033  CD 21                 INT    21H                         ;access DOS
0035  C3                    RET                                ;return from procedure

0036                 LINE   ENDP
                            END                                ;end of file
```

Example 9–14 shows a program that uses function 0AH and reads two lines of information into two memory buffers (BUF1 and BUF2). Before the call to the DOS function through procedure LINE, the first byte of the buffer is loaded with a 255, so up to 255 characters can be typed. If you assemble and execute this program, the first line is accepted and so is the second. The only problem is that the second line appears on top of the first line. The next section of the text explains how to solve this problem and also display memory data on the video screen.

Reading a Key with BIOS INT 16H. In addition to DOS, the BIOS can be used to read a key from the keyboard through BIOS INT 16H. If AH = 00H (83-key keyboard) or 10H (101-key keyboard), the INT 16H instruction waits for a key to be typed and returns the ASCII character in AL and scan code in AH. (The scan codes are illustrated in Table 9–1.) If the status of the keyboard is checked, function 01H (83 key) or 11H (101 key) is used to test the keyboard. If the zero flag is set (1), no key was typed; if the zero flag is cleared (0) a key was typed that can be read with function 00H (10H). For additional details on INT 16H, refer to Appendix A. Refer to Example 9–15 for two procedures that illustrate the use of INT 16H. Note that no echo is provided by INT 16H.

EXAMPLE 9–15

```
0000              KEYS   PROC   NEAR

0000  B4 10              MOV    AH,10H          ;read 101 key keyboard
0002  CD 16              INT    16H             ;wait for key, no echo
0004  C3                 RET

0005              KEYS   ENDP

0005              STAT   PROC   NEAR

0005  B4 11              MOV    AH,11H          ;check 101 key keyboard
0007  CD 16              INT    16H             ;get status
0009  C3                 RET

000A              STAT   ENDP
```

Writing to the Video Display with DOS Functions

With almost any program, data must be displayed on the video display. Video data are displayed in a number of different ways with DOS function calls. Function 02H or 06H display one character at a time; function 09H displays an entire string of characters. Because function 02H and 06H are identical, function 06H can also be used to read a key.

Displaying One ASCII Character. DOS functions 02H and 06H are explained together, because they are identical for displaying ASCII data. Example 9–16 shows how this function displays a carriage return or enter (0DH) and a line feed (0AH). Here a macro called DISP (display) displays the carriage return and line feed. The combination of a carriage return and a line feed moves the cursor to the next line at the left margin of the video screen. This two-step process corrects the problem that occurred between the lines typed through the keyboard in Example 9–14.

EXAMPLE 9–16

```
                         ;a program that displays a carriage return and a line feed
                         ;using the DISP macro
                         ;
                         .MODEL TINY                          ;select TINY model
                         .CODE                                ;start CODE segment
                  DISP   MACRO    A                           ;;display A macro

                         MOV  AH,06H                          ;;DOS function 06H
                         MOV  DL,A                            ;;place parameter A in DL
                         INT  21H                             ;;display parameter A

                         ENDM

                         .STARTUP                             ;start program

                  DISP   0DH                                  ;display carriage return

0100  B4 06    1         MOV    AH,06H
0102  B2 0D    1         MOV    DL,0DH
0104  CD 21    1         INT    21H

                  DISP   0AH                                  ;display line feed

0106  B4 06    1         MOV    AH,06H
0108  B2 0A    1         MOV    DL,0AH
010A  CD 21    1         INT    21H
                         .EXIT                                ;exit to DOS
                         END                                  ;end of file
```

Display a Character String. A character string is a series of ASCII-coded characters that end with a $ (24H) when used with DOS function call number 09H. Other endings are possible, such as a null (00H) for a null character string. Example 9–17 shows a program that displays a message at the current cursor position on the video display. Function call number 09H requires that DS:DX addresses the character string before executing the INT 21H.

EXAMPLE 9–17

```
                              .MODEL SMALL              ;select SMALL model
0000                          .DATA                     ;start DATA segment
0000   0D 0A 0A 54     MES    DB     13,10,10,'This is a test line.$'
          68 69 73 20
          69 73 20 61
          20 74 65 73
          74 20 6C 69
          6E 65 2E 24
0000                          .CODE                     ;start CODE segment
                              .STARTUP                  ;start program

0017   B4 09                  MOV    AH,9               ;select function 09H
0019   BA 0000 R              MOV    DX,OFFSET MES      ;address character string
001C   CD 21                  INT    21H                ;access DOS

                              .EXIT                     ;exit to DOS
                              END                       ;end of file
```

This program can be entered into the assembler, linked and executed to produce *"This is a test line."* on the video display. As always, enter the source code, not the hexadecimal addresses or data.

Consolidated Read Key and Echo Macros. Example 9–18 illustrates two macro sequences that can be added to an INCLUDE file or to any program. The READ macro reads a keyboard character and returns with either the ASCII or extended ASCII character in AL. If carry is set, AL contains the extended ASCII code; if carry is cleared, AL contains the standard ASCII code. The ECHO macro displays the ASCII-coded character located next to the word echo as a parameter at the current cursor position.

EXAMPLE 9–18

```
;
;a macro that reads a key from the keyboard without echo
;if carry = 0, AL contains the standard ASCII key code
;if carry = 1, AL contains the extended ASCII key code
;
READ    MACRO                       ;;read key macro
        LOCAL  READ1,READ2
        PUSH   DX                   ;;save DX
        MOV    AH,6                 ;;select function 06H
        MOV    DL,0FFH              ;;select read key
READ1:
        INT    21H                  ;;access DOS
        JE     READ1                ;;if no key typed
        OR     AL,AL                ;;test for extended key
        JNZ    READ2                ;;if standard ASCII key
        INT    21H                  ;;access DOS
        STC                         ;;indicate extended key code
READ2:
        POP    DX                   ;;restore DX
        ENDM
;
;a macro that displays an ASCII character from parameter CHAR
;
ECHO    MACRO CHAR                  ;;display CHAR macro
```

TABLE 9–2 BIOS function INT 10H, subfunctions 02H and 03H

AH	*Description*	*Parameters*
02H	Sets cursor position	DH = Row DL = Column BH = Page number
03H	Reads cursor position	BH = Page number DH = Row DL = Column

```
PUSH    DX              ;;save DX
MOV     AH,6            ;;select function 06H
MOV     DL,CHAR         ;;character to DL
INT     21H             ;;access DOS
ENDM
```

Using the Video BIOS Function Calls

In addition to DOS function call INT 21H, the video **BIOS** (basic I/O system) uses function calls to INT 10H. The DOS function calls allow a key to be read and a character to be displayed with ease, but the cursor is difficult to position at the desired screen location. The video BIOS function calls allow more complete control over the video display than the DOS function calls. The video BIOS function calls also require less time to execute than the DOS function calls.

Cursor Position. Before any information is placed on the video screen, the position of the cursor should be known. This allows the screen to be cleared and the displayed information to be placed at any location on the video screen. The BIOS INT 10H function number 03H allows the cursor position to be read from the video interface. The BIOS INT 10H function number 02H allows the cursor to be placed at any screen position. Table 9–2 shows the contents of various registers for both video BIOS INT 10H function calls 02H and 03H.

The page number in register BH should be 0 before setting the cursor position. Most software does not access the other pages (1–7) for the video display. If the display adaptor is a VGA display, always use page 0. The page number is often ignored after a cursor read. The 0 page is available in the **CGA** (*color graphics adaptor*), **EGA** (*enhanced graphics adapter*), and **VGA** (*variable graphics array*) text modes of operation. Other pages are available in some VGA and EGA modes and all CGA modes of operation.

The cursor position assumes that the left-hand column is column 0, progressing across a line to column 79. The row number corresponds to the character line number on the screen. Row 0 is the uppermost line; row 24 is the last line on the screen. This assumes that the text mode selected for the video adapter is 80 characters per line by 25 lines (80×25). Other text modes are available, such as 40×25 and 96×43.

EXAMPLE 9–19

```
;a program that clears the screen and homes the
;cursor to the upper left-hand corner of the screen.
;
```

```
                            .MODEL TINY             ;select TINY model
0000                        .CODE                   ;start CODE segment
                     HOME   MACRO                   ;;home cursor macro
                            MOV    AH,2             ;;function 02H
                            MOV    BH,0             ;;page 0
                            MOV    DX,0             ;;row 0, line 0
                            INT    10H              ;;home cursor
                            ENDM

                            .STARTUP                ;start program
                            HOME                    ;home cursor
0100  B4 02       1         MOV    AH,2
0102  B7 00       1         MOV    BH,0
0104  BA 0000     1         MOV    DX,0
0107  CD 10       1         INT    10H
0109  B9 07D0               MOV    CX,25*80         ;load character count
010C  B4 06                 MOV    AH,6             ;select function 06H
010E  B2 20                 MOV    DL,' '           ;select a space
0110                 MAIN1:
0110  CD 21                 INT    21H              ;display a space
0112  E2 FC                 LOOP   MAIN1            ;repeat 2000 times
                            HOME                    ;home cursor
0114  B4 02       1         MOV    AH,2
0116  B7 00       1         MOV    BH,0
0118  BA 0000     1         MOV    DX,0
011B  CD 10       1         INT    10H
                            .EXIT                   ;exit to DOS
                            END                     ;end of file
```

Example 9–19 shows how the INT 10H BIOS function call is used to clear the video screen. This is just one method of clearing the screen. Notice that the first function call positions the cursor to row 0, column 0—the **home position**. Next, use the DOS function call to write 2,000 (80 characters per line × 25 character lines) blank spaces (20H) on the video display, and move the cursor back to the home position.

If this program is assembled, linked, and executed, a problem surfaces: it is far too slow to be useful in most cases. To correct this situation, another video BIOS function call is used. The scroll function (06H) clears the screen at a much higher speed.

Function 06H is used with a 00H in AL to blank the entire screen. This allows the screen to clear at a much higher speed. See Example 9–20 for a better clear-and-home-cursor program. Here function call number 08H reads the character attributes for blanking the screen. Next, they are positioned in the correct registers and DX is loaded with the screen size, 4FH (79) and 19H (25). If this program is assembled, linked, and executed, the screen is cleared much more quickly. Refer to Appendix A for other video BIOS INT 10H function calls that may prove useful in your applications. Also listed in Appendix A are all the INT functions available in most computers.

EXAMPLE 9–20

```
                     ;a program that clears the screen and homes the cursor
                     ;
                          .MODEL TINY               ;select TINY model
0000                      .CODE                     ;start code segment
                     HOME MACRO                     ;;home cursor
                          MOV    AH,2
                          MOV    BH,0
```

```
                              MOV     DX,0
                              INT     10H
                              ENDM

                              .STARTUP                    ;start program

0100  B7 00                   MOV     BH,0
0102  B4 08                   MOV     AH,8
0104  CD 10                   INT     10H                 ;read video attribute

0106  8A DF                   MOV     BL,BH               ;load page number
0108  8A FC                   MOV     BH,AH
010A  B9 0000                 MOV     CX,0                ;load attributes
010D  BA 194F                 MOV     DX,194FH            ;line 25, column 79
0110  B8 0600                 MOV     AX,600H             ;select scroll function
0113  CD 10                   INT     10H                 ;scroll screen

                              HOME                        ;home cursor
0115  B4 02       1           MOV     AH,2
0117  B7 00       1           MOV     BH,0
0119  BA 0000     1           MOV     DX,0
011C  CD 10       1           INT     10H

                              .EXIT                       ;exit to DOS
                              END                         ;end program
```

Display Macro

One of the more useful macro sequences is the one illustrated in Example 9–21. Although it is simple and has been presented before, it saves a lot of typing. With this macro, a register, an ASCII character in quotes, or the numeric value for an ASCII character, can be specified as the argument.

EXAMPLE 9–21

```
                    ;a program that displays AB followed by a carriage return
                    ;and line feed combination using the DISP macro
                    ;
                           .MODEL TINY                ;select TINY model
                           .CODE                      ;start CODE segment
                    DISP   MACRO  VAR                 ;;display VAR macro
                           MOV    DL,VAR
                           MOV    AH,6
                           INT    21H
                           ENDM
                           .STARTUP                   ;start program

                           DISP   'A'                 ;display 'A'
0100  B2 41  1             MOV    DL,'A'
0102  B4 06  1             MOV    AH,6
0104  CD 21  1             INT    21H

0106  B0 42                MOV    AL,'B'              ;load AL with 'B'
                           DISP   AL                  ;display 'B'
0008  8A D0  1             MOV    DL,AL
000A  B4 06  1             MOV    AH,6
000C  CD 21  1             INT    21H

                           DISP   13                  ;display carriage return
```

```
000E  B2 0D  1          MOV    DL,13
0010  B4 06  1          MOV    AH,6
0012  CD 21  1          INT    21H

                        DISP   10                  ;display line feed
0014  B2 0A  1          MOV    DL,10
0016  B4 06  1          MOV    AH,6
0018  CD 21  1          INT    21H

                        .EXIT                      ;exit to DOS
                        END                        ;end of file
```

9–3 DATA CONVERSIONS

In computer systems, data is seldom in the correct form. In fact, one of the primary tasks of a computer system is to convert data from one form to another. This section of the chapter describes conversions between binary and ASCII. Binary data are removed from a register or memory and converted to ASCII for the video display. In many cases, ASCII data are converted to binary as they are typed on the keyboard. Also explained are conversions between ASCII and hexadecimal data.

Converting from Binary to ASCII

Conversion from binary to ASCII is accomplished in two ways: (1) by the AAM instruction if the number is less than 100, or (2) by a series of decimal divisions. Both techniques are presented in this section.

The AAM instruction converts the value in AX into a two-digit unpacked BCD number in AX. If the number in AX is 0062H (98 decimal) before AAM executes, AX contains a 0908H after AAM executes. This is not ASCII code, but it is converted to ASCII code by adding a 3030H to AX. Example 9–22 illustrates a procedure that processes the binary value in AL (0–99) and displays it on the video screen as decimal. This procedure blanks a leading zero for the numbers 0–9 with an ASCII space code. The program displays the hexadecimal contents of AL, in this case a 4AH, as a decimal (74) on the video display.

EXAMPLE 9–22

```
                  ;a program that uses the DISP procedure to display 74 decimal
                  ;on the video display.
                  ;
                          .MODEL TINY                    ;select TINY mode
0000                      .CODE                          ;start code segment
                          .STARTUP                       ;start program
0100  B0 4A               MOV    AL,4AH                  ;load test data to AL
0102  E8 0004             CALL   DISP                    ;display test data in decimal
                          .EXIT                          ;exit to DOS
                  ;
                  ;DISP procedure displays AL (0 to 99) as a decimal number
                  ;AX is destroyed by this procedure
                  ;
0109              DISP    PROC   NEAR
```

```
0109  52                  PUSH  DX              ;save DX
010A  B4 00               MOV   AH,0            ;clear AH
010C  D4 0A               AAM                   ;convert to BCD
010E  80 C4 20            ADD   AH,20H
0111  80 FC 20            CMP   AH,20H          ;test for leading zero
0114  74 03               JE    DISP1           ;if leading zero
0116  80 C4 10            ADD   AH,10H          ;convert to ASCII
0119            DISP1:
0119  8A D4               MOV   DL,AH           ;display first digit
011B  B4 06               MOV   AH,6
011D  50                  PUSH  AX
011E  CD 21               INT   21H
0120  58                  POP   AX
0121  8A D0               MOV   DL,AL
0123  80 C2 30            ADD   DL,30H          ;convert second digit to ASCII
0126  CD 21               INT   21H             ;display second digit
0128  5A                  POP   DX              ;restore DX
0129  C3                  RET

012A            DISP      ENDP
                          END                   ;end of file
```

The reason AAM converts any number between 0 and 99 to a two-digit unpacked BCD number is because it divides AX by 10. The result is left in AX, so AH contains the quotient and AL the remainder. This same scheme of dividing by ten can be expanded to convert any whole number from binary to an ASCII-coded character string that can be displayed on the video screen. The algorithm for converting from binary to ASCII is:

1. Divide by 10 and save the remainder on the stack as a significant BCD digit.
2. Repeat Step 1 until the quotient is 0.
3. Retrieve each remainder and add 30H to convert to ASCII before displaying or printing.

Example 9–23 shows how the unsigned 16-bit contents of AX are converted to ASCII and displayed on the video screen as an unsigned integer. Here, AX is divided by 10 and the remainder is saved on the stack after each division for later conversion to ASCII. Data are stored on the stack because the least significant digit is the one returned first by the division. The stack is used to reverse the order of the data, so it can be displayed correctly. After all the digits have been converted by division, the result is displayed on the video screen by removing the remainders from the stack and converting them to ASCII. This procedure also blanks any leading zeros that occur.

EXAMPLE 9–23

```
                ;a program that uses DISPX to display AX in decimal.
                ;
                        .MODEL TINY             ;select TINY model
0000                    .CODE                   ;start CODE segment
                        .STARTUP                ;start program
0100  B8 04A3           MOV    AX,4A3H          ;load AX with test data
0103  E8 0004           CALL   DISPX            ;display AX in decimal
                        .EXIT                   ;exit to DOS
                ;
                ;the DISPX procedure displays AX in decimal
                ;AX is destroyed
```

```
                    ;
010A                DISPX   PROC    NEAR

010A  52                    PUSH    DX                  ;save DX, CX, and BX
010B  51                    PUSH    CX
010C  53                    PUSH    BX
010D  B9 0000               MOV     CX,0                ;clear digit counter
0110  BB 000A               MOV     BX,10               ;set for decimal
0113                DISPX1:
0113  BA 0000               MOV     DX,0                ;clear DX
0116  F7 F3                 DIV     BX                  ;divide DX:AX by 10
0118  52                    PUSH    DX                  ;save remainder
0119  41                    INC     CX                  ;count remainder
011A  0B C0                 OR      AX,AX               ;test for quotient of zero
011C  75 F5                 JNZ     DISPX1              ;if quotient is not zero
011E                DISPX2:
011E  5A                    POP     DX                  ;get remainder
011F  B4 06                 MOV     AH,6                ;select function 06H
0121  80 C2 30              ADD     DL,30H              ;convert to ASCII
0124  CD 21                 INT     21H                 ;display digit
0126  E2 F6                 LOOP    DISPX2              ;repeat for all digits

0128  5B                    POP     BX                  ;restore BX, CX, and DX
0129  59                    POP     CX
012A  5A                    POP     DX
012B  C3                    RET

012C                DISPX   ENDP
                            END                         ;end of file
```

It is interesting to note that the same steps outlined to convert from binary to ASCII-coded decimal can also be used to convert from binary to any other number base. For example, if Step 1 divides by 7, the converted ASCII data will be in base 7. Likewise, if Step 1 divides by 3, the output will be in base 3. To convert to a number base larger than 10, the technique must include changing the remainders larger than 9 into the letters A, B , and so forth. E.g., add 37H in place of 30H. The technique required to convert from hexadecimal data to binary is covered next.

Converting from ASCII to Binary

Conversions from ASCII to binary usually start with keyboard data entry. If a single key is typed, the conversion is accomplished by subtracting a 30H from the number. If more than one key is typed, an additional step is required. After subtracting 30H, the number is added to the prior result multiplied by 10. The algorithm used to convert ASCII to binary is:

1. Begin with a binary result of 0.
2. Subtract 30H from the character typed on the keyboard to convert it to BCD.
3. Multiply the binary result by 10 and add the new BCD digit.
4. Repeat Steps 2 and 3 until the character typed is not an ASCII-coded number of 30H–39H.

EXAMPLE 9–24

```
;a program that reads one decimal number from the keyboard
;and stores the binary value at memory word TEMP
```

```
                             ;
                                    .MODEL SMALL                ;select TINY model
0000                                .DATA                       ;start DATA segment
0000   0000                  TEMP    DW     ?                   ;define TEMP
0000                                .CODE                       ;start CODE segment
                                    .STARTUP                    ;start program
0017   E8 0007                      CALL   READN                ;read a number
001A   A3 0000 R                    MOV    TEMP,AX              ;save it in TEMP
                                    .EXIT                       ;exit to DOS
                             ;
                             ;READN procedure reads a decimal number from the keyboard
                             ;and returns its binary value in AX.
                             ;
0021                          READN PROC   NEAR

0021   53                           PUSH   BX                   ;save BX and CX
0022   51                           PUSH   CX
0023   B9 000A                      MOV    CX,10                ;load 10 for decimal conversions
0026   BB 0000                      MOV    BX,0                 ;clear result
0029                          READN1:
0029   B4 01                        MOV    AH,1                 ;read key with echo
002B   CD 21                        INT    21H

002D   3C 30                        CMP    AL,'0'
002F   72 14                        JB     READN2               ;if below '0'
0031   3C 39                        CMP    AL,'9'
0033   77 10                        JA     READN2               ;if above '9'

0035   2C 30                        SUB    AL,'0'               ;convert to ASCII

0037   50                           PUSH   AX                   ;save digit
0038   8B C3                        MOV    AX,BX                ;multiply result by 10
003A   F7 E1                        MUL    CX
003C   8B D8                        MOV    BX,AX
003E   58                           POP    AX
003F   B4 00                        MOV    AH,0
0041   03 D8                        ADD    BX,AX                ;add digit value to result
0043   EB E4                        JMP    READN1               ;repeat
0045                          READN2:
0045   8B C3                        MOV    AX,BX                ;get binary result into AX
0047   59                           POP    CX                   ;restore CX and BX
0048   5B                           POP    BX
0049   C3                           RET

004A                          READN ENDP
                                    END                         ;end of file
```

Example 9–24 illustrates a procedure that implements the ASCII-to-binary conversion algorithm. Here, the binary number returns in the AX register as a 16-bit result. If a larger result is required, the procedure must be reworked for 32-bit arithmetic. Each time this procedure is called, it reads a number from the keyboard until any key other than 0 through 9 is typed. It then returns with the binary equivalent in the AX register.

This procedure also converts from number bases other than base 10. If Step 3 is changed to multiply by 8, the result will be octal to binary conversion. The checks for numbers between 0 and 9 must also be changed to convert a number between 0 and 7 from octal.

Displaying and Reading Hexadecimal Data

Hexadecimal data is easier to read from the keyboard and display than decimal data. Hexadecimal data is used, not at the applications level, but at the system level. System-level data is often hexadecimal and must be either displayed in hexadecimal form or read from the keyboard as hexadecimal data.

Reading Hexadecimal Data. Hexadecimal data appears as 0 to 9 and A to F. The ASCII codes obtained from the keyboard for hexadecimal data are 30H to 39H for the numbers 0 through 9 and 41H to 46H (A–F) or 61H to 66H (a–f) for the letters. To be useful, a procedure that reads hexadecimal data must be able to accept both lowercase (a–f) and uppercase (A–F) letters.

EXAMPLE 9–25

```
                        ;a program that reads a 4-digit hexadecimal number from the
                        ;keyboard and stores the result in word-sized location TEMP.
                        ;
                                .MODEL SMALL                ;select SMALL model
0000                            .DATA                       ;start DATA segment
0000  0000               TEMP DW      ?                     ;define TEMP
0000                            .CODE                       ;start CODE segment
                                .STARTUP                    ;start program
0017  E8 0007                   CALL   READH                ;read hexadecimal number
001A  A3 0000 R                 MOV    TEMP,AX              ;save it at TEMP
                                .EXIT                       ;exit to DOS
                        ;
                        ;The READH procedure that reads a 4-digit hexadecimal number
                        ;from the keyboard and returns it in AX.
                        ;This procedure does next check for errors and uses CONV
                        ;
0021                    READH PROC   NEAR

0021  51                        PUSH   CX                   ;save BX and CX
0022  53                        PUSH   BX
0023  B9 0004                   MOV    CX,4                 ;load CX and SI with 4
0026  8B F1                     MOV    SI,CX
0028  BB 0000                   MOV    BX,0                 ;clear result
002B                    READH1:
002B  B4 01                     MOV    AH,1                 ;read a key with echo
002D  CD 21                     INT    21H
002F  E8 000A                   CALL   CONV                 ;convert to binary
0032  D3 E3                     SHL    BX,CL
0034  02 D8                     ADD    BL,AL                ;form result in BX
0036  4E                        DEC    SI
0037  75 F2                     JNZ    READH1               ;repeat 4 times
0039  8B C3                     MOV    AX,BX                ;move result to AX
003B  5B                        POP    BX                   ;restore BX and CX
003C  59                        POP    CX
003D  C3                        RET
003E                    READH ENDP
                        ;
                        The CONV procedure converts AL into hexadecimal
                        ;
003E                    CONV  PROC   NEAR

003E  3C 39                     CMP    AL,'9'
0040  76 08                     JBE    CONV2                ;if 0 through 9
0042  3C 61                     CMP    AL,'a'
```

```
0044  72 02          JB    CONV1          ;if uppercase A through F
0046  2C 20          SUB   AL,20H         ;convert to uppercase
0048          CONV1:
0048  2C 07          SUB   AL,7
004A          CONV2:
004A  2C 30          SUB   AL,30H
004C  C3             RET
004D          CONV   ENDP
                     END                  ;end of file
```

Example 9–25 shows two procedures: one converts the contents of the data in AL from an ASCII-coded character to a single hexadecimal digit, while the other reads a 4-digit hexadecimal number from the keyboard and returns with it in register AX. The second procedure can be modified to read any size hexadecimal number from the keyboard. Notice how the letters a through f are converted to uppercase (subtract 20H) in the first procedure. The program that uses these procedures stores the result into memory location TEMP.

The CONV procedure first tests AL for a 0 through 9. If the value of AL is 0 through 9, the procedure subtracts 30H and returns. If the value in AL is A through F, the CMP AL,'a' instruction returns a below result, after which a 07H followed by a 30H are subtracted from AL to produce 0AH through 0FH. If the letters a through f appear in AL, an additional 20H is subtracted.

Displaying Hexadecimal Data. To display hexadecimal data, a number is separated into 4-bit segments that are converted into hexadecimal digits. Conversion is accomplished by adding a 30H to the numbers 0 to 9 and a 37H to the letters A to F.

EXAMPLE 9–26

```
               ;a program that displays the hexadecimal value loaded into AX
               ;this program uses DISPH to display a 4-digit value
               ;
                      .MODEL TINY                 ;select TINY model
0000                  .CODE                       ;start CODE segment
                      .STARTUP                    ;start program
0100  B8 0ABC         MOV    AX,0ABCH             ;load AX with test data
0103  E8 0004         CALL   DISPH                ;display AX in hexadecimal format
                      .EXIT                       ;exit to DOS
               ;
               ;The DISPH procedure displays AX as a 4-digit hexadecimal number
               ;
010A           DISPH  PROC   NEAR

010A  53              PUSH   BX                   ;save BX and CX
010B  51              PUSH   CX
010C  B1 04           MOV    CL,4                 ;load rotate count
010E  B5 04           MOV    CH,4                 ;load digit count
0110           DISPH1:
0110  D3 C0           ROL    AX,CL                ;position digit
0112  50              PUSH   AX
0113  24 0F           AND    AL,0FH               ;convert it to ASCII
0115  04 30           ADD    AL,30H
0117  3C 39           CMP    AL,'9'
0119  76 02           JBE    DISPH2
011B  04 07           ADD    AL,7
011D           DISPH2:
011D  B4 02           MOV    AH,2                 ;display hexadecimal digit
```

```
011F  8A D0               MOV   DL,AL
0121  CD 21               INT   21H
0123  58                  POP   AX
0124  FE CD               DEC   CH
0126  75 E8               JNZ   DISPH1                ;repeat for 4 digits
0128  59                  POP   CX                    ;restore registers
0129  5B                  POP   BX
012A  C3                  RET

012B            DISPH     ENDP
                          END                         ;end of file
```

A procedure that displays the contents of the AX register on the video display appears in Example 9–26. Here, the number is rotated left so the leftmost digit is displayed first. Because AX contains a 4-digit hexadecimal number, the procedure displays 4 hexadecimal digits. This procedure can be modified to display wider hexadecimal numbers. The program that uses this procedure displays a test value loaded into AX before calling DISPH.

Using Lookup Tables for Data Conversions

Lookup tables are often used to convert from one data form to another. A lookup table is formed in the memory as a list of data referenced by a procedure to perform conversions. The XLAT instruction can often be used to look up data in a table, provided the table contains 8-bit wide data and its length is less than or equal to 256 bytes.

Converting from BCD to 7-Segment Code. One simple application that uses a lookup table is BCD to 7-segment code conversion. Example 9–27 illustrates a lookup table that contains the 7-segment codes for the numbers 0 to 9. These codes are used with the 7-segment display pictured in Figure 9–1. This 7-segment display uses active high (logic 1) inputs to light a segment. The code is arranged so that the *a* segment is in bit position 0 and the *g* segment is in bit position 6. Bit position seven is zero in this example, but can be used for displaying a decimal point.

EXAMPLE 9–27

```
0000               SEG7    PROC   FAR

0000  53                   PUSH   BX
0001  BB 0008 R            MOV    BX,OFFSET TABLE
0004  2E: D7               XLAT   CS:TABLE                  ;see text
0006  5B                   POP    BX
0007  CB                   RET

0008  3F           TABLE   DB     3FH                       ;0
0009  06                   DB     6                         ;1
000A  5B                   DB     5BH                       ;2
000B  4F                   DB     4FH                       ;3
000C  66                   DB     66H                       ;4
000D  6D                   DB     6DH                       ;5
000E  7D                   DB     7DH                       ;6
000F  07                   DB     7                         ;7
0010  7F                   DB     7FH                       ;8
0011  6F                   DB     6FH                       ;9

0012               SEG7    ENDP
```

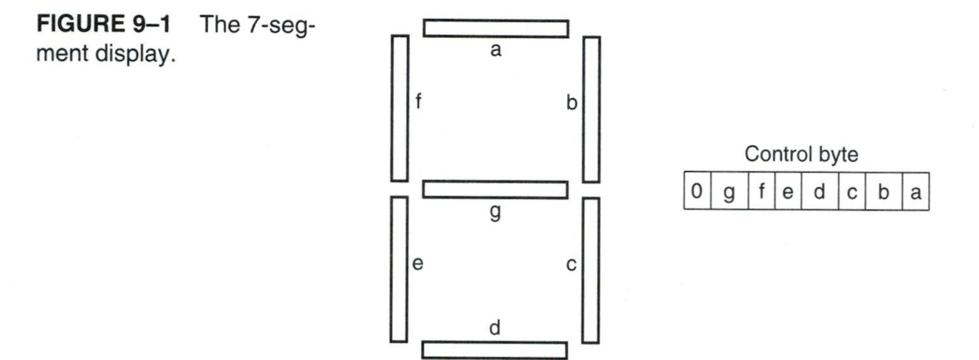

FIGURE 9–1 The 7-segment display.

The procedure that performs the conversion contains only two instructions, and assumes that AL contains the BCD digit to be converted to 7-segment code. One of the instructions addresses the lookup table by loading its address into BX, and the other performs the conversion and returns the 7-segment code in AL.

Because the lookup table is located in the code segment, and the XLAT instruction accesses the data segment by default, the XLAT instruction includes a segment override. Notice that a dummy operand (TABLE) is added to the XLAT instruction so the code segment override prefix (CS:) can be added to the instruction. Normally, XLAT does not contain an operand unless its default segment must be overridden. The LODS and MOVS instructions are overridden in the same manner as XLAT, by using a dummy operand.

Using a Lookup Table to Access ASCII Data. Some programming techniques require that numeric codes be converted to ASCII character strings. For example, suppose that you need to display the days of the week for a calendar program. Because the number of ASCII characters in each day is different, a lookup table must be used to reference the ASCII-coded days of the week.

Example 9–28 shows a table that references ASCII-coded character strings located in the code segment. Each character string contains an ASCII-coded day of the week. The table references each day of the week. The procedure that accesses the day of the week uses the AL register and the numbers 0 to 6 to refer to Sunday through Saturday. If AL contains a 2 when this procedure is called, the word *Tuesday* is displayed on the video screen.

This procedure first accesses the table by loading the table address into the SI register. Next, the number in AL is converted into a 16-bit number and doubled, because each table entry is two bytes in length. This index is then added to SI to address the correct entry in the lookup table. The address of the ASCII character string is now loaded into DX by the MOV DX,[SI] instruction.

EXAMPLE 9–28

```
                              ;a program that displays the current day of the week
                              ;by using the system clock/calendar.
                              ;
                                    .MODEL SMALL                   ;select SMALL model
0000                                .DATA                          ;start DATA segment
0000  000E R 0015 R           DTAB  DW    SUN,MON,TUE,WED,THU,FRI,SAT
```

```
001C  R 0024 R
002E  R 0037 R
003E  R
000E  53 75 6E 64 61 79    SUN   DB    'Sunday$'
      24
0015  4D 6F 6E 64 61 79    MON   DB    'Monday$'
      24
001C  54 75 65 73 64 61    TUE   DB    'Tuesday$'
      79 24
0024  57 65 64 6E 65 73    WED   DB    'Wednesday$'
      64 61 79 24
002E  54 68 75 72 73 64    THU   DB    'Thursday$'
      61 79 24
0037  46 72 69 64 61 79    FRI   DB    'Friday$'
      24
003E  53 61 74 75 72 64    SAT   DB    'Saturday$'
      61 79 24
0000                             .CODE                    ;start CODE segment
                                 .STARTUP                 ;start program
0017  B4 2A                      MOV   AH,2AH             ;get day of week
0019  CD 21                      INT   21H                ;access DOS
001B  E8 0004                    CALL  DAYS               ;display day of week
                                 .EXIT                    ;exit to DOS

0022                       DAYS  PROC  NEAR

0022  52                         PUSH  DX                 ;save DX and SI
0023  56                         PUSH  SI
0024  BE 0000 R                  MOV   SI,OFFSET DTAB     ;address lookup table
0027  B4 00                      MOV   AH,0               ;find day of week
0029  03 C0                      ADD   AX,AX
002B  03 F0                      ADD   SI,AX
002D  8B 14                      MOV   DX,[SI]            ;get address of day of week
002F  B4 09                      MOV   AH,9               ;display character string
0031  CD 21                      INT   21H
0033  5E                         POP   SI                 ;restore registers
0034  5A                         POP   DX
0035  C3                         RET

0036                       DAYS  ENDP
                                 END                      ;end of file
```

Before the INT 21H DOS function is called, the DS register is placed on the stack and loaded with the segment address of CS. This allows DOS function number 09H (display a string) to be used to display the day of the week. This procedure converts the numbers 0 to 6 to the days of the week.

The program that uses this procedure accesses the system clock by using a DOS function call to read the day of the week. Refer to Appendix A for the operation of DOS INT 21H function call number 2AH.

An Example Program Using Data Conversions

A program example is required to combine some of the DOS data conversion functions discussed thus far. Suppose you must display the time and date on the video screen. An example program (see Example 9–29) displays the time as 10:45 P.M., and the date as Tuesday, July 4, 1995. The program is short because it calls one procedure to display the time, and a second to display the date.

EXAMPLE 9–29

```
                              ;a program that displays the time and date in the following form:
                              ;10:34 A.M., Tuesday July 4, 1995
                              ;
                                      .MODEL SMALL                 ;select SMALL model
                                      .NOLISTMACRO                 ;don't expand macros
0000                                  .DATA                        ;start CODE segment
0000  0026 R 002F R           DTAB    DW     SUN,MON,TUE,WED,THU,FRI,SAT
        0038 R 0042 R
        004E R 0059 R
        0062 R
000E  006D R 0076 R           MTAB    DW     JAN,FEB,MAR,APR,MAY,JUN
        0080 R 0087 R
        008E R 0093 R
001A  0099 R 009F R                   DW     JUL,AUG,SEP,OCT,NOV,DCE
        00A7 R 00B2 R
        00BB R 00C5 R
0026  53 75 6E 64 61 79       SUN     DB     'Sunday, $'
      2C 20 24
002F  4D 6F 6E 64 61 79       MON     DB     'Monday, $'
      2C 20 24
0038  54 75 65 73 64 61       TUE     DB     'Tuesday, $'
      79 2C 20 24
0042  57 65 64 6E 65 73       WED     DB     'Wednesday, $'
      64 61 79 2C 20 24
004E  54 68 75 72 73 64       THU     DB     'Thursday, $'
      61 79 2C 20 24
0059  46 72 69 64 61 79       FRI     DB     'Friday, $'
      2C 20 24
0062  53 61 74 75 72 64       SAT     DB     'Saturday, $'
      61 79 2C 20 24
006D  4A 61 6E 75 61 72       JAN     DB     'January $'
      79 20 24
0076  46 65 62 72 75 61       FEB     DB     'February $'
      72 79 20 24
0080  4D 61 72 63 68 20       MAR     DB     'March $'
      24
0087  41 70 72 69 6C 20       APR     DB     'April $'
      24
008E  4D 61 79 20 24          MAY     DB     'May $'
0093  4A 75 6E 65 20 24       JUN     DB     'June $'
0099  4A 75 6C 79 20 24       JUL     DB     'July $'
009F  41 75 67 75 73 74       AUG     DB     'August $'
      20 24
00A7  53 65 70 74 65 6D       SEP     DB     'September $'
      62 65 72 20 24
00B2  4F 63 74 6F 62 65       OCT     DB     'October $'
      72 20 24
00BB  4E 6F 76 65 6D 62       NOV     DB     'November $'
      65 72 20 24
00C5  44 65 63 65 6D 62       DCE     DB     'December $'
      65 72 20 24
0000                                  .CODE                        ;start CODE segment
                              DISP    MACRO CHAR
                                      PUSH  AX                     ;;save AX and DX
                                      PUSH  DX
                                      MOV   DL,CHAR                ;;display character
                                      MOV   AH,2
                                      INT   21H
                                      POP   DX                     ;;restore AX and DX
                                      POP   AX
                                      ENDM
                                      .STARTUP                     ;start program
0017  E8 0007                         CALL  TIMES                  ;display time
```

```
001A  E8 00A3                CALL  DATES             ;display date
                             .EXIT                   ;exit to DOS

0021                   TIMES PROC  NEAR

0021  B4 2C                  MOV   AH,2CH            ;get time from DOS
0023  CD 21                  INT   21H
0025  B7 41                  MOV   BH,'A'            ;set 'A' for AM
0027  80 FD 0C               CMP   CH,12
002A  72 05                  JB    TIMES1            ;if below 12:00 noon
002C  B7 50                  MOV   BH,'P'            ;set 'P' for PM
002E  80 ED 0C               SUB   CH,12             ;adjust to 12 hour clock
0031                   TIMES1:
0031  0A ED                  OR    CH,CH             ;test for 0 hour
0033  75 02                  JNE   TIMES2            ;if not 0 hour
0035  B5 0C                  MOV   CH,12             ;change 0 hour to 12 hour
0037                   TIMES2:
0037  8A C5                  MOV   AL,CH
0039  B4 00                  MOV   AH,0
003B  D4 0A                  AAM                     ;convert hours to BCD
003D  0A E4                  OR    AH,AH
003F  74 0D                  JZ    TIMES3            ;if no tens of hours
0041  80 C4 30               ADD   AH,'0'            ;convert tens of hours to ASCII
                             DISP  AH                ;display tens of hours
004E                   TIMES3:
004E  04 30                  ADD   AL,'0'            ;convert units of hours to ASCII
                             DISP  AL                ;display units of hours
                             DISP  ':'               ;display colon
0064  8A C1                  MOV   AL,CL
0066  B4 00                  MOV   AH,0
0068  D4 0A                  AAM                     ;convert minutes to BCD
006A  05 3030                ADD   AX,3030H          ;convert minutes to ASCII
006D  50                     PUSH  AX
                             DISP  AH                ;display tens of minutes
0078  58                     POP   AX
                             DISP  AL                ;display units of minutes
                             DISP  ' '               ;display space
                             DISP  BH                ;display 'A' or 'P'
                             DISP  '.'               ;display .
                             DISP  'M'               ;display M
                             DISP  '.'               ;display .
                             DISP  ' '               ;display space
00BF  C3                     RET

00C0                   TIMES ENDP

00C0                   DATES PROC  NEAR

00C0  B4 2A                  MOV   AH,2AH            ;get date from DOS
00C2  CD 21                  INT   21H
00C4  52                     PUSH  DX
00C5  B4 00                  MOV   AH,0              ;get day of week
00C7  03 C0                  ADD   AX,AX
00C9  BE 0000 R              MOV   SI,OFFSET DTAB    ;address day lookup table
00CC  03 F0                  ADD   SI,AX
00CE  8B 14                  MOV   DX,[SI]           ;address day of week
00D0  B4 09                  MOV   AH,9              ;display day of week
00D2  CD 21                  INT   21H
00D4  5A                     POP   DX
00D5  52                     PUSH  DX
00D6  8A C6                  MOV   AL,DH             ;get month
00D8  FE C8                  DEC   AL
00DA  B4 00                  MOV   AH,0
00DC  03 C0                  ADD   AX,AX
00DE  BE 000E R              MOV   SI,OFFSET MTAB    ;address month lookup table
```

```
00E1  03 F0                 ADD   SI,AX
00E3  8B 14                 MOV   DX,[SI]          ;address month
00E5  B4 09                 MOV   AH,9             ;display month
00E7  CD 21                 INT   21H
00E9  5A                    POP   DX
00EA  8A C2                 MOV   AL,DL            ;get day of month
00EC  B4 00                 MOV   AH,0
00EE  D4 0A                 AAM                    ;convert to BCD
00F0  0A E4                 OR    AH,AH
00F2  74 0D                 JZ    DATES1           ;if tens of day of month is 0
00F4  80 C4 30              ADD   AH,30H           ;convert tens of day to ASCII
                            DISP  AH               ;display tens of day
0101               DATES1:
0101  04 30                 ADD   AL,30H           ;convert units of day to ASCII
                            DISP  AL               ;display units of day of month
                            DISP  ','              ;display comma
                            DISP  ' '              ;display space
0121  81 F9 07D0            CMP   CX,2000          ;test for year 2000
0125  72 19                 JB    DATES2           ;if below year 2000
0127  83 E9 64              SUB   CX,100           ;scale to 1900 - 1999
                            DISP  '2'              ;display 2
                            DISP  '0'              ;display 0
013E  EB 14                 JMP   DATES3
0140               DATES2:
                            DISP  '1'              ;display 1
                            DISP  '9'              ;display 9
0154               DATES3:
0154  81 E9 076C            SUB   CX,1900          ;scale to 00 - 99
0158  8B C1                 MOV   AX,CX
015A  D4 0A                 AAM                    ;convert to BCD
015C  05 3030               ADD   AX,3030H         ;convert to ASCII
                            DISP  AH               ;display tens of year
                            DISP  AL               ;display units of year
0173   C3                   RET

0174               DATES    ENDP
                            END                    ;end of file
```

The time is available from DOS using and INT 21H function call number 2CH. This function returns with the hours in CH and minutes in CL. Seconds are also available in DH, and hundredths of seconds are available in DL. The date is available using INT 21H function call number 2AH. This leaves the day of the week in AL, the year in CX, the day of the month in DH, and the month in DL.

The DATES procedure uses two ASCII lookup tables that convert the day and month to ASCII character strings. It also uses the AAM instruction to convert from binary to BCD for the time and date. Data display is handled in two ways: by character string (function 09H) and by single character (function 02H) located in a macro called DISP.

9–4 SUMMARY

1. The assembler program assembles modules that contain PUBLIC variables and segments plus EXTRN (external) variables. The linker program links modules and library files to create a run-time program executed from the DOS command line. The run-time program usually has the extension EXE.

2. The MACRO and ENDM directives create a new opcode for use in programs. These macros are similar to procedures, except there is no call or return. In place of them, the assembler inserts the code of the macro sequence into a program each time it is invoked. Macros can include variables that pass information and data to the macro sequence.
3. The DOS INT 21H function call provides a method of using the keyboard and video display. Function number 06H, placed into register AH, provides an interface to the keyboard and display. If DL = 0FFH, this function tests the keyboard for a keystroke. If no keystroke is detected, it returns equal. If a keystroke is detected, the standard ASCII character returns in AL. If an extended ASCII character is typed, it returns with AL = 00H, where the function must again be called to return with the extended ASCII character in AL. To display a character, DL is loaded with the character and AH with 06H, before INT 21H is used in a program.
4. Character strings are displayed using function number 09H. The DS:DX register combination addresses the character string, which must end with a $.
5. The INT 10H instruction accesses BIOS (basic I/O system) procedures that control the video display and keyboard. The BIOS functions are independent of DOS and function with any operating system.
6. Data conversion from binary to BCD is accomplished with the AAM instruction for numbers less than 100, or by repeated division by 10 for larger numbers. Once converted to BCD, a 30H is added to convert each digit to ASCII code for the video display.
7. When converting from an ASCII number to BCD, a 30H is subtracted from each digit. To obtain the binary equivalent, multiply by 10.
8. Lookup tables are used for code conversion with the XLAT instruction if the code is an 8-bit code. If the code is wider than 8 bits, a short procedure that accesses a lookup table provides the conversion. Lookup tables are also used to hold addresses, so that different parts of a program or different procedures can be selected.

9–5 QUESTIONS AND PROBLEMS

1. The assembler converts a source file into an _______________ file.
2. What files are generated from the source file TEST.ASM, as it is processed by MASM?
3. The linker program links object files and _______________ files to create an execution file.
4. What is the difference between an .EXE and a .COM file?
5. What does the PUBLIC directive indicate when placed in a program module?
6. What does the EXTRN directive indicate when placed in a program module?
7. What directives appear with labels defined external?
8. Describe how a library file works when it is linked to other object files by the linker program.
9. What assembler language directives delineate a macro sequence?

10. What is a macro sequence?
11. How are parameters transferred to a macro sequence?
12. Develop a macro called ADD32 that adds the 32-bit contents of DX–CX to the 32-bit contents of BX–AX.
13. Develop a macro called STUB that sign extends the 8-bit number in AL into a 64-bit sign-extended number in ECX–EBX. (Note that ECX is the most significant part of the 64-bit result.)
14. How is the LOCAL directive used within a macro sequence?
15. Develop a macro called ADDLIST PARA1,PARA2 that adds the contents of PARA1 to PARA2. Each of these parameters represents an area of memory. The number of bytes added are indicated by register CX before the macro is invoked.
16. Develop a macro that sums a list of byte-sized data invoked by the macro ADDM LIST,LENGTH. The label, LIST, is the starting address of the data block and LENGTH is the number of data added. The result must be a 16-bit sum found in AX at the end of the macro sequence.
17. What is the purpose of the INCLUDE directive?
18. Develop a procedure called RANDOM. This procedure must return an 8-bit random number in register CL at the end of the subroutine. (One way to generate a random number is to increment CL each time the DOS function 06H tests the keyboard and finds *no* keystroke. In this way, a random number is generated.)
19. Modify the procedure of Question 18 so the random number ranges in value from 1 through (and including) 6.
20. Develop a procedure that displays a character string that ends with a 00H. Your procedure must use the DS:DX register to address the start of the character string.
21. Develop a procedure that reads a key and displays the hexadecimal value of an extended ASCII-coded keyboard character if it is typed. If a normal character is typed, ignore it.
22. Use BIOS INT 10H to develop a procedure that positions the cursor at line 3, column 6.
23. When a number is converted from binary to BCD, the ______________ instruction accomplishes the conversion, provided the number is less than 100 decimal.
24. How is a large number (over 100 decimal) converted from binary to BCD?
25. A BCD digit is converted to ASCII code by adding a ______________.
26. An ASCII-coded number is converted to BCD by subtracting ______________.
27. Develop a procedure that reads an ASCII number from the keyboard and stores it as a BCD number into memory array DATA. The number ends when anything other than a number is typed.
28. Explain how a 3-digit ASCII-coded number is converted to binary.
29. Develop a procedure that converts all lowercase ASCII-coded letters into uppercase ASCII-coded letters. Your procedure may not change any other character except the letters a–z.
30. Develop a lookup table that converts hexadecimal data 00H–0FH into the ASCII-coded characters that represent the hexadecimal digits. Show the lookup table and any software required for the conversion.
31. Develop a program sequence that jumps to memory location ONE, if AL = 6, TWO if AL = 7, and THREE if AL = 8.

32. Show how to use the XLAT instruction to access a lookup table called LOOK located in the stack segment.
33. Write a program that reads any decimal number between 0 and 65,535 and displays the 16-bit binary version on the video display.
34. Write a program that displays the binary powers of two in decimal on the video screen for the powers 0 through 7. Your display should show 2^n = {value} for each power of 2.
35. Develop a program that accepts a four-digit number in hexadecimal code and displays it in base 7.
36. Using the technique learned in Question 18, develop a program that displays random numbers between 1 and 47 for a state lottery.
37. Develop a program that displays the hexadecimal contents of a block of 256 bytes of memory. Your software must be able to accept the starting address as a hexadecimal number between 00000H and FFF00H.
38. Develop a program that displays the hexadecimal number 123AH as an octal number on the video display. Use the technique learned for displaying decimal data, but divide by 8 instead of 10.

CHAPTER 10

Conditional Assembly and Video Displays

INTRODUCTION

Conditional assembly language statements relieve the programmer of many of the repetitive tasks normally associated with programming. They are an important part of any study of programming using the assembler. This chapter presents the many conditional statements available to the assembly language programmer. Note that MASM version 6.X is required to perform many of these conditional assembly functions.

Also presented is a study of some of the VGA video display modes that are accessible in all video display adapters. The chapter concentrates on the 320 × 200, 256-color and 640 × 480, 16-color modes. Although higher-resolution modes are available, no efficient standard yet exists for programming the many video display cards in existence. (This is changing. A VESA standard is currently evolving, but is not yet perfected.)

OBJECTIVES

Upon completion of this chapter, you will be able to:

1. Use conditional assembly language statements when writing assembly language programs.
2. Use some of the high-level language constructs available with conditional assembly language statements.
3. Program using the 256-color, 320 × 200 video display.
4. Program using the 16-color, 640 × 480 video display.

10–1 CONDITIONAL ASSEMBLY

Conditional assembly language statements are available for use in the assembly process and in macro sequences. Conditional statements for assembly flow control create instructions that control the flow of the program. They are variations of constructs used in high-

TABLE 10–1 Conditional assembly language IF statements

Statement	*Function*
IF	If the expression is true
IFB	If argument is blank
IFE	If the expression is not true
IFDEF	If the label has been defined
IFNB	If argument is not blank
IFNDEF	If the label has not been defined
IFIDN	If argument 1 equals argument 2
IFDIF	If argument 1 does not equal argument 2

level language programming languages (IF–THEN, IF–THEN–ELSE, DO-WHILE, and REPEAT-UNTIL). Conditional statements for macro sequence control are also available, but create instructions only at time of assembly.

Conditional Assembly

As mentioned, conditional assembly is implemented with the IF–THEN or IF–THEN–ELSE construct found in high-level languages. Table 10–1 shows the forms used for the IF statement in the conditional assembly process.

The IF and ENDIF statements allow portions of the program to assemble if some condition is met. Otherwise, the statements between IF and ENDIF do not assemble.

Example 10–1 shows how the IF, ELSE, and ENDIF statements are used to conditionally assemble values for the width and length of paper in a program. Note that TRUE and FALSE are defined as 1 and 0. This is important, because these values are not predefined by the assembler. Next, the width and length of the paper are adjusted by using TRUE and FALSE statements. This can be expanded to ask an entire series of questions about a program so custom versions can be created. Example 10–1(a) shows the original source code and Example 10–1(b) shows how the program assembles for TRUE answers for both the width and length. Example 10–1(c) shows the assembled output for a false width and a true length.

When Example 10–1(a) is assembled, TRUE and FALSE are equated to WIDT and LENGT to modify the way that the assembler forms the program. In Example 10–1(b), both WIDT and LENGT are defined as TRUE, which causes the assembler to modify the way the program is assembled, so a page is 72 columns wide and the length is continuous. Example 10–1(c) is another example where WIDT is FALSE and LENGT is TRUE, causing the assembler to form instructions that make the page width 80 columns and the length continuous. The only form not shown is where the page length is 66 lines.

Examples of some of the other forms listed in Table 10–1 appear later in the text. When one of these new conditional statements appears, it is explained and shown with an example.

EXAMPLE 10-1(a)

```
                    ;source program
                    ;
                    TRUE    EQU   1              ;define true
                    FALSE   EQU   0              ;define false

                    WIDT    EQU   FALSE          ;set to true for 72 columns and
                                                 ;false for 80 columns
                    LENGT   EQU   TRUE           ;set to true for continuous forms
                                                 ;false for 66 line paper

                            IF    WIDT           ;72 columns
                    WIDE    DB    72
                            ELSE
                    WIDE    DB    80             ;80 columns
                            ENDIF

                            IF    LENGT          ;if continuous
                    LONG    DB    -1
                            ELSE
                    LONG    DB    66             ;if 66 lines
                            ENDIF
```

EXAMPLE 10-1(b)

```
                    ;assembled portion of source with WIDT = TRUE and LENGT = TRUE
                    ;
                            IF    WIDT           ;72 columns
0000  48            WIDE    DB    72
                            ELSE
                            ENDIF

                            IF    LENGT          ;if continuous
0001  FF            LONG    DB    -1
                            ELSE
                            ENDIF
```

EXAMPLE 10-1(c)

```
                    ;assembled portion of source with WIDT = FALSE and LENGT = TRUE
                    ;
                            IF    WIDT           ;72 columns
                            ELSE
0000  50            WIDE    DB    80             ;80 columns
                            ENDIF
                            IF    LENGT          ;if continuous
0001  FF            LONG    DB    -1
                            ELSE
                            ENDIF
```

Controlling the Flow of an Assembly Language Program

The assembly language statements .IF, .ELSE, .ELSEIF, and .ENDIF are useful directives that control *the flow of the program,* rather than *how it assembles.* These statements always begin with a period. Control flow assembly language statements beginning with a period are only available to MASM version 6.X, not earlier versions. Note that these statements do not function within a macro block.

Example 10–2 shows how these statements are used to control the flow of a program by testing the system for the version of DOS. In this example, DOS INT 21H function number 30H is used to read the DOS version. The version is tested to determine whether it is above or below version 3.3. If below version 3.3, the program terminates using DOS INT 21H function number 4CH.

EXAMPLE 10–2(a)

```
                    ;source program sequence
                    ;
                            MOV     AH,30H
                            INT     21H             ;get DOS version

                            .IF     AL<3 && AH<30

                            MOV     AH,4CH          ;terminate program
                            INT     21H

                            .ENDIF
```

EXAMPLE 10–2(b)

```
                    ;assembled listing file of Example 10-2(a)
                    ;
0000  B4 30                 MOV     AH,30H
0002  CD 21                 INT     21H             ;get DOS version

                            .IF     AL<3 && AH<30
0004  3C 03         *       cmp     al,003h
0006  73 09         *       jae     @C0001
0008  80 FC 1E      *       cmp     ah,01Eh
000B  73 04         *       jae     @C0001

000D  B4 4C                 MOV     AH,4CH          ;terminate program
000F  CD 21                 INT     21H

                            .ENDIF
0011                *@C0001:
```

Example 10–2(a) shows the source program sequence as it was typed, and Example 10–2(b) shows the fully expanded output generated by the assembler program. Notice that assembler generated and inserted statements begin with an asterisk (*) in the listing. The .IF AL<3 && AH<30 statement tests for DOS version 3.30. If the major version number (AL) is less than 3 AND the minor version number (AH) is less than 30, the MOV AH,0 and INT 21H instructions execute. The && symbol represents the word AND in the IF statement. Refer to Table 10–2 for a complete list of relation operators used with the .IF statement. Note that many of these conditions (such as &&) are also used by high-level languages like C/C++.

Example 10–3 shows another example of the conditional .IF directive, which converts all ASCII-coded letters to uppercase. First the keyboard is read without echo using DOS INT 21H function 06H, then the .IF statement converts the character into upper case if necessary. In this example, the logical AND function (&&) is used to determine whether the character was lowercase. This program reads a key from the keyboard and converts it to uppercase before displaying it. Notice how the program terminates when the control C

TABLE 10–2 Relational operators used with .IF

Operator	*Function*
==	Equal or the same as
!=	Not equal
>	Greater than
<=	Greater than or equal
<	Less than
<=	Less than or equal
&	Bit test
!	Logical inversion
&&	Logical AND
||	Logical OR

key (ASCII = 03H) is typed. The .LISTALL directive causes all assembler-generated statements to be listed, including the label @Startup generated by the .STARTUP directive. The .EXIT directive is also expanded by .LISTALL, to show the use of the DOS INT 21H function 4CH, which returns control to DOS.

EXAMPLE 10–3

```
                   ;a program that reads the keyboard and converts all lowercase
                   ;data to uppercase before displaying it.
                   ;
                   ;this program uses a control C for termination
                   ;
                              .MODEL TINY          ;select TINY model
                              .LISTALL             ;list all assembler generated statements
0000                          .CODE                ;start CODE segment
                              .STARTUP             ;start program
0100            *  @Startup:
0100               MAIN1:
0100  B4 06                   MOV   AH,6           ;read key without echo
0102  B2 FF                   MOV   DL,0FFH
0104  CD 21                   INT   21H
0106  74 F8                   JE    MAIN1          ;if no key typed
0108  3C 03                   CMP   AL,3           ;test for control C key
010A  74 10                   JE    MAIN2          ;if control C key

                              .IF   AL>='a' && AL<='z'
010C  3C 61     *             cmp   al, 'a'
010E  72 06     *             jb    @C0001
0110  3C 7A     *             cmp   al, 'z'
0112  77 02     *             ja    @C0001
0114  2C 20                      SUB AL,20H
                              .ENDIF
0116            *  @C0001:
0116  8A D0                   MOV   DL,AL          ;echo character to display
0118  CD 21                   INT   21H
011A  EB E4                   JMP   MAIN1          ;repeat
011C               MAIN2:
                              .EXIT                ;exit to DOS on control C
```

```
011C  B4 4C    *           mov   ah, 04Ch
011E  CD 21    *           int   021h
                           END                ;end of file
```

In this program, a lowercase letter is converted to uppercase by use of the .IF AL >= 'a' && AL <= 'z' statement. If AL contains a value greater than or equal to a lowercase a, and less than or equal to a lowercase z, he statement between .IF and .ENDIF executes. This statement (SUB AL,20H) subtracts 20H from the lowercase letter to change it to uppercase letter. Notice how the assembler program implements the .IF statement (see lines that begin with *). The label @C0001 is an assembler-generated label used by the conditional jump statements placed in the program by the .IF statement.

Another example using conditional .IF statements appears in Example 10–4. This program reads a key from the keyboard and then converts it to hexadecimal code. This program is not listed in expanded form.

In this example, the .IF AL >='a' && AL<= 'f' statement causes the next instruction (SUB AL,57H) to execute if AL contains letters a through f, converting them to hexadecimal. If AL does not contain letters between a and f, the next .ELSEIF statement tests for the letters between A and F. If letters between A and F are found, 37H is subtracted from AL. If neither of these conditions are true, 30H is subtracted from AL before AL is stored at data segment memory location TEMP.

EXAMPLE 10–4

```
                        ;program that reads a key and stores its hexadecimal value
                        ;in memory location TEMP
                        ;
                                .MODEL SMALL                ;select SMALL model
0000                            .DATA                       ;start DATA segment
0000  00                TEMP    DB      ?                   ;define TEMP
0000                            .CODE                       ;start CODE segment
                                .STARTUP                    ;start program
0017  B4 01                     MOV     AH,1                ;read key
0019  CD 21                     INT     21H

                                .IF     AL>='a' && AL<='f'  ;if lowercase, subtract 57H
0023  2C 57                             SUB AL,57H

                                .ELSEIF AL>='A' && AL<='F'  ;if uppercase, subtract 37H
002F  2C 37                             SUB AL,37H

                                .ELSE                       ;otherwise subtract 30H
0033  2C 30                             SUB AL,30H

                                .ENDIF

0035  A2 0000 R                 MOV     TEMP,AL
                                .EXIT                       ;exit to DOS
                                END                         ;end of file
```

DO-WHILE Loops

Like most high-level languages, the assembler provides a DO-WHILE loop construct, available to MASM version 6.X. The .WHILE statement is used with a condition to begin the loop, and the .ENDW statement ends the loop.

Example 10–5 shows how the .WHILE statement is used to read data from the keyboard and store it in an array called BUF until the enter key (0DH) is typed. This program assumes that BUF is stored in the extra segment, because the STOSB instruction is used to store the keyboard data in memory. The .WHILE loop portion of the program is shown in expanded form so statements inserted by the assembler (beginning with a *) can be studied. After the enter key (0DH) is typed, the string is appended with a $ so it can be displayed with DOS INT 21H function number 9.

EXAMPLE 10–5

```
                              ;a program that reads a character string from the keyboard
                              ;and, after enter is typed, displays it again.
                              ;
                              .MODEL SMALL                    ;select small model
0000                          .DATA                           ;indicate start of DATA segment
0000  0D 0A           MES     DB      13,10                   ;store carriage return line feed
0002  0100 [          BUF     DB      256 DUP (?)             ;character string buffer
            00
              ]
0000                          .CODE                           ;indicate start of CODE segment
                              .STARTUP                        ;indicate start of program
0017  8C D8                   MOV     AX,DS                   ;make ES overlap DS
0019  8E C0                   MOV     ES,AX

001B  FC                      CLD                             ;select increment
001C  BF 0002 R               MOV     DI,OFFSET BUF           ;address buffer

                              .WHILE AL != 0DH                ;loop while AL is not an enter

001F  EB 05           *       jmp     @C0001
0021                  *  @C0002:

0021  B4 01                   MOV     AH,1                    ;read key with echo
0023  CD 21                   INT     21H
0025  AA                      STOSB                           ;store key code

                              .ENDW                           ;end while loop

0026                  *  @C0001:
0026  3C 0D           *       cmp     al, 00Dh
0028  75 F7           *       jne     @C0002

002A  C6 45 FF 24             MOV     BYTE PTR [DI-1],'$'     ;make it $ string
002E  BA 0000 R               MOV     DX,OFFSET MES           ;address MES
0031  B4 09                   MOV     AH,9                    ;display MES
0033  D 21                    INT     21H
                              .EXIT                           ;exit to DOS
                              END
```

The program in Example 10–5 functions perfectly, as long as we arrive at the .WHILE statement with AL containing some value other than 0DH. This can be accomplished by adding a MOV AL,0DH instruction before the .WHILE statement in Example 10–5. A better way to handle this problem is illustrated in Example 10–6. In that example, the .BREAK statement is used to break out of the .WHILE loop. A .WHILE 1 statement creates an infinite loop, and the .BREAK statement tests for a value of 0DH (enter) in AL. If AL = 0DH, the program breaks out of the infinite loop, correcting the problem exhibited in Example 10–5. Note that the .BREAK statement causes the break to occur at the point where it appears in the program. This is important, because it allows the point of the break to be selected by the programmer.

The .CONTINUE statement can be used to allow the DO-WHILE loop to continue if a certain condition is met. For example, a .CONTINUE .IF AL == 15 allows the loop to continue if AL equals 15. The .BREAK and .CONTINUE commands function in the same manner in a C-language program.

EXAMPLE 10–6

```
                                .MODEL SMALL
0000                            .DATA
0000  0D 0A          MES        DB     13,10               ;define string
0002  0100 [         BUF        DB     256 DUP (?)         ;memory for string
            00
              ]
0000                            .CODE
                                .STARTUP
0017  8C D8                     MOV    AX,DS               ;make ES overlap DS
0019  8E C0                     MOV    ES,AX

001B  FC                        CLD                        ;select increment
001C  BF 0002 R                 MOV    DI,OFFSET BUF       ;address BUF

                                .WHILE 1                   ;create an infinite loop
001F           *     @C0001:

001F  B4 01                     MOV    AH,1                ;read key
0021  CD 21                     INT    21H
0023  AA                        STOSB                      ;store key code in BUF

                                .BREAK .IF AL == 0DH       ;breaks the loop for a 0DH
0024  3C 0D    *                cmp    al, 00Dh
0026  74 02    *                je     @C0002

                                .ENDW
0028  EB F5    *                jmp    @C0001
002A           *     @C0002:

002A  C6 45 FF 24               MOV    BYTE PTR [DI-1],'$' ;make it a $ string
002E  BA 0000 R                 MOV    DX,OFFSET MES       ;display string
0031  B4 09                     MOV    AH,9
0033  CD 21                     INT    21H
                                .EXIT
                                END
```

Example 10–7 shows a practical way to use the DO-WHILE construct to display the contents of EAX in decimal on the video display. Note that the EAX register is initialized with a number (123455) to test this program. Two infinite loops are used to convert EAX to decimal. The first loop divides EAX by 10 until the quotient is zero. After each division, the remainder is saved on the stack as a significant digit in the result. Also located within the first infinite loop is a comma counter, stored in CL. Each time the quotient is *not* zero, the comma counter increments. If the comma counter reaches 3, a comma is pushed onto the stack for later display, and the comma counter is reset to zero. The second infinite loop displays the result. After each POP DX instruction, the break statement checks DX to find whether a 10 was pushed on the stack to indicate the end of the number. If DX contains a 10, the loop breaks. If it doesn't, a decimal digit or a comma is displayed. This procedure can be added to any program where a decimal number up to 4 billion must be displayed with commas at the correct places.

EXAMPLE 10–7

```
                    ;A program that displays the contents of EAX in decimal.
                    ;This program inserts commas between thousands, millions,
                    ;and billions.
                    ;
                                    .MODEL TINY
                                    .386                ;select 80386
0000                                .CODE
                                    .STARTUP

0100  66| B8 0001E23F       MOV     EAX,123455          ;load test data
0106  E8 0004               CALL    DISPE               ;display EAX in decimal
                            .EXIT
                    ;
                    ;the DISPE procedure displays EAX in decimal format.
                    ;

010D                DISPE   PROC    NEAR

010D  66| BB 0000000A       MOV     EBX,10              ;load 10 for decimal
0113  53                    PUSH    BX                  ;save end of number indicator (10)
0114  B1 00                 MOV     CL,0                ;load comma counter

                            .WHILE  1                   ;first infinite loop

0116  66| BA 00000000       MOV     EDX,0               ;clear EDX
011C  66| F7 F3             DIV     EBX                 ;divide EDX:EAX by 10
011F  80 C2 30              ADD     DL,30H              ;convert to ASCII
0122  52                    PUSH    DX                  ;save remainder

                            .BREAK  .IF EAX == 0        ;break if quotient is zero

0128  FE C1                 INC     CL                  ;increment comma counter

                            .IF     CL == 3             ;if comma count is 3
012F  6A 2C                 PUSH    ','                 ;save comma
0131  B1 00                 MOV     CL,0                ;clear comma counter
                            .ENDIF

                            .ENDW                       ;end first loop

                            .WHILE  1                   ;second infinite loop

0135  5A                    POP     DX                  ;get remainder
                            .BREAK  .IF DL == 10        ;break if remainder is 10

013B  B4 02                 MOV     AH,2                ;display decimal digit
013D  CD 21                 INT     21H

                            .ENDW

0141  C3                    RET

0142                DISPE   ENDP
                            END
```

REPEAT-UNTIL Loops

The REPEAT-UNTIL construct is another kind of loop available to the assembler. A series of instructions is repeated until some condition occurs. The .REPEAT statement defines the start of the loop, the .UNTIL statement, which contains a condition, defines the end of the loop. Note that .REPEAT and .UNTIL are available only to version 6.X of MASM.

Example 10–5 is reworked using the REPEAT-UNTIL construct. Refer to Example 10–8 for the program that reads keys from the keyboard and stores keyboard data into extra segment array BUF until the enter key is pressed. This program also fills the buffer with keyboard data until the enter key (0DH) is typed. Once enter is typed, the program displays the character string using DOS INT 21H function number 9, after appending the buffer data with the required dollar sign. Notice how the .UNTIL AL == 0DH statement generates a code (a statement beginning with *) to test for the enter key.

EXAMPLE 10–8

```
                               .MODEL  SMALL
0000                           .DATA
0000  0D 0A             MES    DB      13,10              ;define MES
0002  0100 [            BUF    DB      256 DUP (?)        ;reserve memory for BUF
               00
                 ]
0000                           .CODE
                               .STARTUP
0017  8C D8                    MOV     AX,DS              ;overlap DS and ES
0019  8E C0                    MOV     ES,AX
001B  FC                       CLD                        ;select increment
001C  BF 0002 R                MOV     DI,OFFSET BUF      ;address BUF

                               .REPEAT
001F               *    @C0001:

001F  B4 01                    MOV     AH,1               ;read key with echo
0021  CD 21                    INT     21H
0023  AA                       STOSB                      ;save key code in BUF

                               .UNTIL  AL == 0DH
0024  3C 0D        *           cmp     al, 00Dh
0026  75 F7        *           jne     @C0001

0028  C6 45 FF 24              MOV     BYTE PTR [DI-1],'$'  ;make $ string
002C  B4 09                    MOV     AH,9               ;display MES and BUF
002E  BA 0000 R                MOV     DX,OFFSET MES
0031  CD 21                    INT     21H
                               .EXIT
                               END
```

An .UNTILCXZ instruction uses a LOOP construction to check for the until condition. The .UNTILCXZ may have a condition or may just use the CX register as a counter to repeat a loop a specified number of times. Example 10–9 shows a sequence of instructions that use the .UNTILCXZ instruction to add the contents of byte-sized array ONE to byte sized array TWO. The sums are stored in array THREE. Each array contains 100 bytes of data, so the loop is repeated 100 times. This example assumes that array THREE is in the extra segment, and that arrays ONE and TWO are in the data segment.

EXAMPLE 10–9

```
012C  B9 0064                    MOV   CX,100            ;set count
012F  BF 00C8 R                  MOV   DI,OFFSET THREE   ;address arrays
0132  BE 0000 R                  MOV   SI,OFFSET ONE
0135  BB 0064 R                  MOV   BX,OFFSET TWO

                                 .REPEAT

0138                  *@C0001:

0138  AC                         LODSB
0139  02 07                      ADD   AL,[BX]
013B  AA                         STOSB
013C  43                         INC   BX

                                 .UNTILCXZ

013D  E2 F9           *          loop  @C0001
```

Using Conditional Statements in Macros

Macro sequences have their own set of conditional instructions that differ somewhat from the ones used with the assembler. For example, macros can use REPEAT and WHILE, but they do so without a period in front of the keyword. In a macro REPEAT has no corresponding UNTIL and WHILE has no corresponding ENDW. These statements are available to all versions of the assembler.

The macro WHILE and REPEAT commands are not preceded by a period and they use a different set of relational operators. Table 10-3 lists the relational operators used with WHILE and REPEAT in macro sequences. These operators are also used with any of the statements listed in Table 10–1. These operators are different than those specified in Table 10–2 for .WHILE and .REPEAT statements.

Repeat in a Macro. The REPEAT statement has a parameter associated with it to repeat the macro sequence a fixed number of times. The repeat sequence must end with an

TABLE 10–3 Relational operators used with WHILE and REPEAT in macro sequences

Operator	*Function*
EQ	Equal
NE	Not equal
LE	Less than or equal to
LT	Less than
GE	Greater than or equal to
GT	Greater than
NOT	Invert
AND	Logical AND
OR	Logical OR
XOR	Logical exclusive-OR

ENDM statement, just like any other macro. The repeat sequence inserts the instructions that appear between the REPEAT statement and the ENDM statement into the program the number of times indicated by the REPEAT statement.

Example 10–10 shows a macro called TESTS and its calling program, which sends the ten ASCII characters from 0 through 9 to the video screen. Notice how this macro is formed using the MACRO statement to name the macro TESTS, and how the REPEAT statement appears within macro TESTS with its own ENDM statement. The macro starts by placing a 6 into AH and the ASCII code for 0 in DL. This sets up the DOS INT 21H function call, so a zero is displayed on the video screen. Next, the REPEAT statement appears (note that it does not contain a period). This repeat statement is used only in macro sequences, and is available to all versions of MASM.

The repeated statements in this example are INT 21H (to display the ASCII contents of DL) and INC DL (to modify the ASCII code displayed). REPEAT 10 causes the statements between REPEAT 10 and the first ENDM to be repeated ten times, as illustrated. The 1 and 2 to the left of the instructions are listed to show that these statements are assembler-generated, not entered as part of the source program.

EXAMPLE 10–10

```
                              TESTS    MACRO

                                       MOV     AH,6
                                       MOV     DL,'0'

                                       REPEAT  10
                                       INT     21H
                                       INC     DL
                                       ENDM

                                       ENDM

0000                          MAIN     PROC    FAR

                                       TESTS                   ;display 0 through 9
0000  B4 06                   1        MOV     AH,6
0002  B2 30                   1        MOV     DL,'0'
0004  CD 21                   2        INT     21H
0006  FE C2                   2        INC     DL
0008  CD 21                   2        INT     21H
000A  FE C2                   2        INC     DL
000C  CD 21                   2        INT     21H
000E  FE C2                   2        INC     DL
0010  CD 21                   2        INT     21H
0012  FE C2                   2        INC     DL
0014  CD 21                   2        INT     21H
0016  FE C2                   2        INC     DL
0018  CD 21                   2        INT     21H
001A  FE C2                   2        INC     DL
001C  CD 21                   2        INT     21H
001E  FE C2                   2        INC     DL
0020  CD 21                   2        INT     21H
0022  FE C2                   2        INC     DL
0024  CD 21                   2        INT     21H
0026  FE C2                   2        INC     DL
0028  CD 21                   2        INT     21H
```

```
002A  FE C2          2        INC     DL

002C  B8 4C00                 MOV     AX,4C00H
002F  CD 21                   INT     21H             ;exit to DOS

0031                 MAIN     ENDP
```

While in a Macro. The WHILE statement is used in macro sequences in much the same way as REPEAT. That is, the WHILE loop is terminated with the ENDM statement. The expression associated with WHILE determines how many times the loop is repeated. Again, note that WHILE in the macro is different from the .WHILE statement described earlier. The WHILE statement is available to all versions of MASM.

Example 10–11 shows how the WHILE statement can be used to generate a table of squares, from 1^2 to whatever value fits into an array of byte-sized memory called SQUARE. The first statement of the sequence defines the label SQUARE for the first byte of data generated. WHILE RES LE 255 repeats the calculation (SEED*SEED) while the result is less than or equal to 255. Notice that the table generated contains the square of all numbers from 1^2 to 15^2 (225 or E1H) If you look closely at Example 10–11, the value of SEED + 1 and SEED * SEED show the numbers and their square.

EXAMPLE 10–11

```
                  ;table of byte-sized squares

                     ;
0000                 SQUARE  LABEL   BYTE               ;;define label
 = 0001              SEED    =       1
 = 0001              RES     =       SEED*SEED          ;;compute square
                     WHILE   RES LE 255
                             DB      RES
                             SEED    = SEED+1
                             RES     = SEED*SEED
                             ENDM
0000  01             1       DB      RES
 = 0002              1       SEED    = SEED+1
 = 0004              1       RES     = SEED*SEED
0001  04             1       DB      RES
 = 0003              1       SEED    = SEED+1
 = 0009              1       RES     = SEED*SEED
0002  09             1       DB      RES
 = 0004              1       SEED    = SEED+1
 = 0010              1       RES     = SEED*SEED
0003  10             1       DB      RES
 = 0005              1       SEED    = SEED+1
 = 0019              1       RES     = SEED*SEED
0004  19             1       DB      RES
 = 0006              1       SEED    = SEED+1
 = 0024              1       RES     = SEED*SEED
0005  24             1       DB      RES
 = 0007              1       SEED    = SEED+1
 = 0031              1       RES     = SEED*SEED
0006  31             1       DB      RES
 = 0008              1       SEED    = SEED+1
 = 0040              1       RES     = SEED*SEED
0007  40             1       DB      RES
 = 0009              1       SEED    = SEED+1
```

```
 = 0051          1      RES      = SEED*SEED
0008  51         1      DB       RES
 = 000A          1      SEED     = SEED+1
 = 0064          1      RES      = SEED*SEED
0009  64         1      DB       RES
 = 000B          1      SEED     = SEED+1
 = 0079          1      RES      = SEED*SEED
000A  79         1      DB       RES
 = 000C          1      SEED     = SEED+1
 = 0090          1      RES      = SEED*SEED
000B  90         1      DB       RES
 = 000D          1      SEED     = SEED+1
 = 00A9          1      RES      = SEED*SEED
000C  A9         1      DB       RES
 = 000E          1      SEED     = SEED+1
 = 00C4          1      RES       = SEED*SEED
000D  C4         1      DB       RES
 = 000F          1      SEED     = SEED+1
 = 00E1          1      RES      = SEED*SEED
000E  E1         1      DB       RES
 = 0010          1      SEED     = SEED+1
 = 0100          1      RES      = SEED*SEED
```

FOR Statement in a Macro. The FOR statement iterates a list of data. If you are familiar with BASIC, the FOR statement functions like the READ statement, and the list of data associated with it functions like the DATA statement. Example 10–12 shows how the FOR statement is used to display a series of characters on the video display. Notice that the CHR:VARARG indicates that variable name CHR is of variable size. The first use of the DISP macro generates the code required to display BARRY. The second use of DISP generates the code required to display BREY. The FOR statement counts the variable used after display and repeats the commands between FOR and ENDM for each variable.

EXAMPLE 10–12

```
                        DISP    MACRO CHR:VARARG
                                MOV   AH,2
                                FOR   ARG,<CHR>
                                MOV   DL,ARG
                                INT   21H
                                ENDM
                                ENDM

                                DISP  'B','A','R','R','Y',' '
0000  B4 02       1             MOV   AH,2
0002  B2 42       2             MOV   DL,'B'
0004  CD 21       2             INT  21H
0006  B2 41       2             MOV  DL,'A'
0008  CD 21       2             INT  21H
000A  B2 52       2             MOV  DL,'R'
000C  CD 21       2             INT  21H
000E  B2 52       2             MOV  DL,'R'
0010  CD 21       2             INT  21H
0012  B2 59       2             MOV  DL,'Y'
0014  CD 21       2             INT  21H
0016  B2 20       2             MOV  DL,' '
0018  CD 21       2             INT  21H
```

```
                                DISP  'B','R','E','Y'
001A  B4 02          1          MOV   AH,2
001C  B2 42          2          MOV   DL,'B'
001E  CD 21          2          INT   21H
0020  B2 52          2          MOV   DL,'R'
0022  CD 21          2          INT   21H
0024  B2 45          2          MOV   DL,'E'
0026  CD 21          2          INT   21H
0028  B2 59          2          MOV   DL,'Y'
002A  CD 21          2          INT   21H
```

IF, ELSE, and ENDIF. The IF statement is used in a macro to make decisions based on the parameters sent to the macro. Note that IF is used in a macro and .IF is used in a program. The IF statement is available to all versions of the assembler, but .IF is available only to version 6.X.

In Example 10–13, a macro is developed that uses a number of conditional assembly statements to read a key, display a character, or display a carriage return and line feed combination. This example illustrates the use of IF, IFB, INB, ENDIF, and ELSE. The macro is called IO. If IO is used on a line by itself, the assembler generates the code to read a key. If IO –1 appears as a statement, the assembler generates the code required to display a carriage return and line feed. If IO 'B' appears as a statement, the assembler generates the code required to display the letter B. This example is listed in expanded form, so the code generated by the assembler can be viewed and studied. Lines that contain a number between the hexadecimal code and the statement are assembler generated and are not included in the original source program.

EXAMPLE 10–13

```
                              .MODEL TINY
0000                          .CODE
                ;the IO macro functions in 3 ways
                ;
                ;(1) IO      read a key with echo
                ;(2) IO -1   display a carriage return & line feed
                ;(3) IO 'B'  display the letter 'B'
                ;or IO AL    display contents of AL
                ;
                IO      MACRO   CHAR
                        IFB     <CHAR>              ;;if CHAR is blank
                                MOV AH,1            ;;read key function number
                        ENDIF

                        IFNB    <CHAR>              ;;if CHAR is not blank
                                MOV AH,2            ;;display character function number

                                IF CHAR EQ -1       ;;if CHAR equals -1
                                   MOV DL,13        ;display carriage return
                                   INT 21H
                                   MOV DL,10        ;display line feed

                                ELSE                ;;if CHAR not equal to -1

                                  MOV DL,CHAR       ;;load CHAR to DL and display
                                ENDIF

                        ENDIF
```

```
                                INT     21H
                                ENDM
                                .STARTUP
                        ;
                        ;this program does a carriage return, line feed then displays
                        ;the letters BE on the video screen. Next it waits for a key
                        ;to be typed. Following the key, a carriage return, line feed
                        ;is displayed.
                        ;
                                IO      -1              ;carriage return & line feed
0100  B4 02     1               MOV     AH,2
0102  B2 0D     1               MOV     DL,13
0104  CD 21     1               INT     21H
0106  B2 0A     1               MOV     DL,10
0108  CD 21     1               INT     21H
                                IO      'B'             ;display 'B'
010A  B4 02     1               MOV     AH,2
010C  B2 42     1               MOV     DL,'B'
010E  CD 21     1               INT     21H
                                IO      'E'             ;display 'E'
0110  B4 02     1               MOV     AH,2
0112  B2 45     1               MOV     DL,'E'
0114  CD 21     1               INT     21H
                                IO                      ;read key
0116  B4 01     1               MOV     AH,1
0118  CD 21     1               INT     21H
                                IO      -1              ;carriage return & line feed
011A  B4 02     1               MOV     AH,2
011C  B2 0D     1               MOV     DL,13
011E  CD 21     1               INT     21H
0120  B2 0A     1               MOV     DL,10
0122  CD 21     1               INT     21H
                                .EXIT
                                END
```

The first part of the macro uses the IFB <CHAR> statement to test CHAR for a blank condition. If CHAR is blank, the assembler generates the MOV AH,1 instruction followed by the very last instruction in the macro, INT 21H, to read a key with echo. This is used in the program with the IO statement.

The second part of the macro contains the IFNB <CHAR> statement to test CHAR for a not-blank condition. If CHAR is *not* blank, another IF–ELSE–ENDIF sequence appears to test the contents of CHAR. If CHAR is a –1, the assembler generates the code required to display a carriage return and line feed combination. If CHAR is not a –1, the ELSE statement places CHAR into DL for display. This is a very powerful macro that can handle most keyboard and single-character display functions. It illustrates the power of conditional assembly statements used in a macro.

10–2 LOW-RESOLUTION GRAPHIC DISPLAYS (VGA)

Because most modern systems contain some form of a VGA graphic display, this topic is included as an introduction to graphic display systems. The study of graphic display could consume an entire textbook, so only the basics are explained here, to spark your interest in graphic displays.

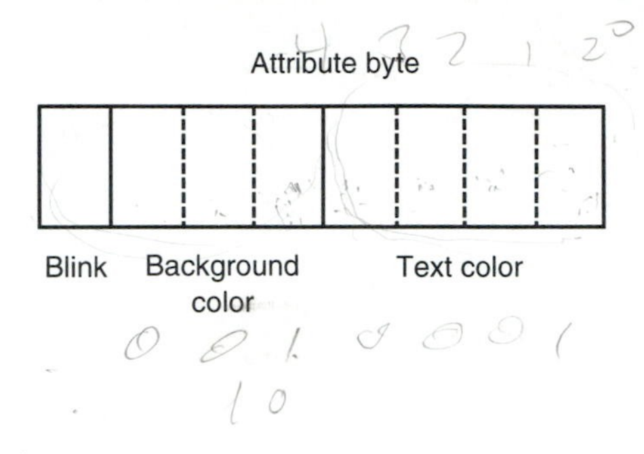

FIGURE 10–1 The attribute byte for text mode. Note that the text color is any color from Table 10–4 and the background color is 000–111.

The Basic VGA Display System

The basic VGA display system operates in several modes. In this text, we will concentrate on the modes most commonly used with DOS: the 16-color, 640 × 480 display and the 256-color, 320 × 200 display. These two basic graphic modes are used for most applications except the ones that require the display of extremely high-resolution video images. To display a high-resolution video image, use the 256-color, 800 × 600 or 1024 × 768 display modes, which are not available to all VGA displays. The most common resolutions available on more advanced VGA display adaptors are: 256-color, 640 × 480; 16-color, 800 × 600; 256-color, 800 × 600; 16-color, 1024 × 768; and 256-color, 1024 × 768. These displays are bit-mapped or graphics displays, instead of character-mode displays. This means that the display data are sent to the video adaptor as a series of bits or bit combinations, rather than ASCII characters with the DOS function calls as described earlier in this chapter.

Video memory exists at memory locations A0000H through AFFFFH for VGA graphics modes. (The text modes use video memory at locations B0000H–B7FFFH or B8000H–BFFFFH. If mode 7 is selected, the CGA black and white mode is activated.)

When the 256-color, 320 x 200 mode is selected, a byte in the video memory selects 1 color out of 256 for a single PEL or picture element. This type of display is often called a *bit-mapped display*. This graphics display mode (13H) uses 64,000 bytes of memory, slightly less than 64K bytes, which are directly addressed at locations A0000H–AF9FFH. This display mode displays 256 colors and is used by many video games for the personal computer. This area of memory is organized differently than the 16-color, 640 × 480 VGA display mode is described in the next section. The first byte of graphics memory (A000:0000) holds the upper-left PEL, and the last byte (A000:F9FF) holds the lower-right PEL.

The most common text mode, mode 3, uses memory beginning at location B8000H to store ASCII-coded text and attributes for each character. The first byte, at location B8000H, holds the ASCII character displayed in the upper-left corner of the video display. Location B8001H holds the attribute that describes this character. The second character and its attribute are stored at locations B8002H and B8003H. This continues until the last character and its attribute are stored at memory locations B8F9EH and B8F9FH for this 25-line, 80-character-per-line mode. Refer to Figure 10–1 for the contents of the attribute byte of the text display mode. Notice that it contains two sections: one for the text color and brightness, and the other for the background color and a bit that blinks the line. The 4-bit text color corresponds to the codes listed in Table 10–4. The background color corresponds to the first 8 entries in Table 10–4.

TABLE 10–4 Default color codes for all text and the first 16 colors for all graphics modes

Code	*Color*
0000	Black
0001	Blue
0010	Green
0011	Cyan
0100	Red
0101	Magenta
0110	Brown
0111	White
1000	Dark Gray
1001	Light blue
1010	Light green
1011	Light cyan
1100	Light red
1101	Light magenta
1110	Yellow
1111	Bright white

Example 10–14 lists a short program that addresses the text memory. This program first changes the attributes to black letters on a red video screen and stores blank spaces (20H) in all character display positions. Finally the program moves the cursor (using BIOS INT 10H) to the upper-left corner and displays the message "This is black on red." using DOS INT 21H function number 9. In most systems, type MODE co80 to return to the normal white-on-black display.

EXAMPLE 10–14

```
                                 ;program that blanks the test mode screen and makes it red.
                                 ;It then displays the message This is a test line, before
                                 ;returning to DOS.
                                 ;
                                       .MODEL SMALL
0000                                   .DATA
0000  54 68 69 73 20 69   MES          DB     'This is a test line.$'
         73 20 61 20 74 65
         73 74 20 6C 69 6E
         65 2E 24
0000                                   .CODE
                                       .STARTUP
0017  B8 B800                          MOV    AX,0B800H       ;address text segment
001A  8E C0                            MOV    ES,AX
001C  FC                               CLD                    ;select increment
001D  BF 0000                          MOV    DI,0            ;address text offset
0020  B4 40                            MOV    AH,40H          ;attribute black on red
0022  B0 20                            MOV    AL,20H          ;character is space
0024  B9 07D0                          MOV    CX,25*80        ;set count
0027  F3/ AB                           REP    STOSW           ;clear screen and change attributes

0029  B4 02                            MOV    AH,2            ;home cursor
```

```
002B  B7 00                 MOV   BH,0            ;page 0
002D  BA 0000               MOV   DX,0            ;row 0, char 0
0030  CD 10                 INT   10H

0032  BA 0000 R             MOV   DX,OFFSET MES   ;display "This is a test line."
0035  B4 09                 MOV   AH,9
0037  CD 21                 INT   21H
                            .EXIT
                            END
```

Programming the 256-Color, 320 × 200 Mode

Software for this bit-mapped 256-color, 320 × 200 display mode is fairly easy to write, because each byte represents a single PEL (picture element) on the video display. Before any information is displayed, the display adapter (assuming VGA is present) must be switched to mode 13H, the 256-color, 320 × 200 mode. This is accomplished with video BIOS INT 10H, function 00H. To select a new video mode place a 00H in AH and the new mode number in AL. Next, use video BIOS function call INT 10H to select the new mode. Note that when a new mode is selected, the video display is cleared. The normal DOS video mode is 03H.

Once in the new mode, a program can begin to display graphics information. Example 10–15 shows a short program that switches the display to the 256-color, 320 × 200 VGA mode and then displays four bands of a vertical color bar pattern on the screen to show the 256 colors programmed by default into the VGA adapter. Here, 5 PELs are used for each color across the entire width of the display. The colors displayed on the screen are the default colors programmed into the VGA display adapter. The first band contains colors 0 through 63; the second, colors 64 to 127; and so forth. After displaying the four bands, the program waits for any key to be typed on the keyboard, then returns to DOS with mode 3 selected. The first band contains the default colors, as illustrated in Table 10–4. The next sixteen colors (16–31) are shades of gray, and the remaining colors are selected by the video display adapter.

The BAND procedure displays a color bar that displays 64 colors across the screen. Each color in the band is displayed, 5 PELs wide and 40 PELs high. At the end of the BAND procedure a return is made, allowing 10 PELs between bands.

EXAMPLE 10–15

```
                  ;a program that displays all 256 colors available to the
                  ;320 x 200 video display mode (13H)
                  ;***use***
                  ;the BAND procedure to display 64 colors at a time in a band
                  ;on the display.
                  ;
                          .MODEL TINY
0000                      .CODE
                          .STARTUP
0100  B8 0013             MOV   AX,13H          ;select mode 13H
0103  CD 10               INT   10H

0105  B8 A000             MOV   AX,0A000H       ;address segment A000 with ES
0108  8E C0               MOV   ES,AX
010A  FC                  CLD                   ;select increment
010B  BF 0000             MOV   DI,0            ;address offset 0000
```

```
010E  B0 00                 MOV   AL,0              ;load starting test color of 00H
0110  E8 001C               CALL  BAND              ;display one band of 64 colors

0113  B0 40                 MOV   AL,64             ;load starting color of 40H
0115  E8 0017               CALL  BAND              ;display one band of 64 colors

0118  B0 80                 MOV   AL,128            ;load starting color of 80H
011A  E8 0012               CALL  BAND              ;display one band of 64 colors

011D  B0 C0                 MOV   AL,192            ;load starting color of C0H
011F  E8 000D               CALL  BAND              ;display one band of 64 colors

0122  B4 01                 MOV   AH,1              ;wait for any key
0124  CD 21                 INT   21H

0126  B8 0003               MOV   AX,3              ;switch back to DOS video mode
0129  CD 10                 INT   10H
                            .EXIT
                  ;
                  ;the BAND procedure displays a color band of 64 colors
                  ;***input parameter***
                  ;AL = starting color number
                  ;ES = A000H
                  ;DI = starting offset address for display
                  ;
012F              BAND      PROC  NEAR

012F  B7 28                 MOV   BH,40             ;load line count
0131              BAND1:
0131  50                    PUSH  AX                ;save starting color
0132  B9 0040               MOV   CX,64             ;load color across line count
0135              BAND2:
0135  B3 05                 MOV   BL,5              ;load times color is displayed
0137              BAND3:
0137  AA                    STOSB                   ;store color
0138  FE CB                 DEC   BL
013A  75 FB                 JNZ   BAND3             ;repeat 5 times
013C  FE C0                 INC   AL                ;change to next color
013E  E2 F5                 LOOP  BAND2             ;repeat for 64 colors
0140  58                    POP   AX                ;restore starting color
0141  FE CF                 DEC   BH
0143  75 EC                 JNZ   BAND1             ;repeat for 40 lines
0145  81 C7 0C80            ADD   DI,320*10         ;skip 10 lines
0149  C3                    RET

014A              BAND      ENDP
                            END
```

To change the default colors, you can reprogram the series of palette registers located on the video display card. There are 256 different locations in the palette; each represents a video display color. Each picture element is composed of the three basic video colors: red, green, and blue. These are the primary colors of light. To display a color, the monitor uses a combination of these three primary colors. The palette memory contains three 6-bit numbers (one for each primary color) that change the brightness of each primary color. The program in Example 10–15 uses the default palette, which has colors set so that the first 16 colors are compatible with other display modes, the second 16 colors are shades of gray, and the remaining colors vary.

Video BIOS INT 10H functions AH = 10H, AL = 10H, and BX = color number (00H–FFH) allow a palette location (color number) to be selected and changed to CH = green, CL = blue, and DH = red. The color values in CH, CL, and DH are 6-bit numbers ranging from 00H to 3FH. A value of 00H is off and 3FH is maximum brightness. For example, a very bright cyan (blue-green) is CH = 3FH, CL = 3FH, and DH = 00H. Varying the combinations of these three colors and their degreee of brightness produces all 256 colors.

To illustrate how the palette is changed, Example 10–16 contains a program that alters color numbers 80H–BFH to every possible brightness of red. It then displays a vertical color bar pattern, containing 64 bars, to illustrate each intensity of red. This program can be changed to display other colors by modifying the values loaded into the palette registers. Note that the BAND procedure is used to display the band of colors in this example.

EXAMPLE 10–16

```
                        ;a program that displays all possible brightness levels of the
                        ;color red for the 320 x 200, 256-color mode (13H)
                        ;
                              .MODEL TINY
0000                          .CODE
                              .STARTUP
0100  B8 0013                 MOV    AX,13H           ;switch to mode 13H
0103  CD 10                   INT    10H

0105  B8 A000                 MOV    AX,0A000H        ;address segment A000 with ES
0108  8E C0                   MOV    ES,AX
010A  FC                      CLD                     ;select increment

010B  B5 00                   MOV    CH,0             ;green value
010D  B1 00                   MOV    CL,0             ;blue value
010F  B6 00                   MOV    DH,0             ;red value
0111  BB 0080                 MOV    BX,80H           ;color register number 80H
0114  B8 1010                 MOV    AX,1010H         ;change palette color function
0117  B2 40                   MOV    DL,64            ;count to change colors 80H to BFH
0119                    PROG1:
0119  CD 10                   INT    10H              ;change a color value
011B  FE C6                   INC    DH               ;next color of red
011D  43                      INC    BX               ;next color palette register
011E  FE CA                   DEC    DL
0120  75 F7                   JNZ    PROG1            ;repeat for 64 colors

0122  BF 0000                 MOV    DI,0             ;address offset 0000
0125  B0 80                   MOV    AL,80H           ;starting color number
0127  E8 000D                 CALL   BAND             ;display 64 colors

012A  B4 01                   MOV    AH,1             ;wait for any key
012C  CD 21                   INT    21H

012E  B8 0003                 MOV    AX,3             ;switch back to DOS video mode
0131  CD 10                   INT    10H
                              .EXIT
                           ;
                           ;the BAND procedure displays a color band of 64 colors
                           ;***input parameter***
                           ;AL = starting color number
                           ;ES = A000H
                           ;DI = starting offset address for display
                           ;
```

```
0137                    BAND  PROC   NEAR

0137  B7 28                   MOV    BH,40           ;line count of 40
0139                    BAND1:
0139  50                      PUSH   AX              ;save starting color number
013A  B9 0040                 MOV    CX,64           ;color count of 64
013D                    BAND2:
013D  B3 05                   MOV    BL,5            ;load times color is displayed
013F                    BAND3:
013F  AA                      STOSB                  ;store color
0140  FE CB                   DEC    BL
0142  75 FB                   JNZ    BAND3           ;repeat 5 times
0144  FE C0                   INC    AL              ;get next color number
0146  E2 F5                   LOOP   BAND2           ;repeat for all 64 colors
0148  58                      POP    AX              ;restore original color number
0149  FE CF                   DEC    BH
014B  75 EC                   JNZ    BAND1           ;repeat for 40 raster lines
014D  81 C7 0C80              ADD    DI,320*10       ;skip 10 raster lines
0151  C3                      RET

0152                    BAND  ENDP
                              END
```

A graphics display, that can draw shapes and forms on the screen requires many procedures. This text cannot show all variations of these programs, but will show one. Example 10–17 shows a procedure that draws a box on the 256-color, 320 × 200 VGA display. The box is any size and can be placed at any location on the screen. Parameters are used to transfer size, color, and location, through registers, to this procedure. To illustrate the operation of the BOX procedure, a program is also included that draws one box on the video display screen.

EXAMPLE 10–17

```
                   ;a program that displays a green box on the video screen using
                   ;video mode 13H.
                   ;
                           .MODEL TINY
0000                       .CODE
                           .STARTUP
0100  FC                   CLD                       ;select auto-increment

0101  B8 0013              MOV     AX,13H            ;select mode 13H
0104  CD 10                INT 10H                   ;this also clears the screen

0106  B0 02                MOV     AL,2              ;use color 02H (green)
0108  B9 0064              MOV     CX,100            ;starting column number
010B  BE 000A              MOV     SI,10             ;starting row number
010E  BD 004B              MOV     BP,75             ;size
0111  E8 000D              CALL    BOX               ;display box

0114  B4 01                MOV     AH,1              ;wait for any key
0116  CD 21                INT     21H

0118  B8 0003              MOV AX,3                  ;switch to DOS video mode
011B  CD 10                INT 10H
                           .EXIT
                   ;
                   ;the BOX procedure displays a box on the mode 13H display.
```

```
                       ;***input parameters***
                       ;AL = color number (0-255)
                       ;CX = starting column number (0-319)
                       ;SI = starting row number (0-199)
                       ;BP = size of box
                       ;
0121                   BOX     PROC    NEAR

0121  BB A000                  MOV     BX,0A000H         ;address segment A000 with ES
0124  8E C3                    MOV     ES,BX
0126  50                       PUSH    AX                ;save color
0127  B8 0140                  MOV     AX,320            ;find starting PEL
012A  F7 E6                    MUL     SI
012C  8B F8                    MOV     DI,AX             ;address start of BOX
012E  03 F9                    ADD     DI,CX
0130  58                       POP     AX
0131  57                       PUSH    DI                ;save starting offset address
0132  8B CD                    MOV     CX,BP             ;save size in BP
0134                   BOX1:
0134  F3/ AA                   REP     STOSB             ;draw top line
0136  8B CD                    MOV     CX,BP
0138  83 E9 02                 SUB     CX,2              ;adjust CX
013B                   BOX2:
013B  5F                       POP     DI
013C  81 C7 0140               ADD     DI,320            ;address next row
0140  57                       PUSH    DI
0141  AA                       STOSB                     ;draw PEL
0142  03 FD                    ADD     DI,BP
0144  83 EF 02                 SUB     DI,2
0147  AA                       STOSB                     ;draw PEL
0148  E2 F1                    LOOP    BOX2

014A  5F                       POP     DI
014B  81 C7 0140               ADD     DI,320            ;address last row
014F  8B CD                    MOV     CX,BP
0151  F3/ AA                   REP     STOSB
0153  C3                       RET

0154                   BOX     ENDP
                               END
```

10–3 HIGH-RESOLUTION GRAPHICS DISPLAYS (VGA)

This section of the chapter explains the use and operation of the 16-color, 640 × 480 high-resolution graphics display. Although displays are available with higher resolutions and more colors, they are not standard. Additional information for higher resolution displays is available with each video card.

Organization of the Video Memory

The VGA graphics memory is 64K bytes in size and requires bit planes to address the 153,600 bytes of memory required for a 16-color, 640 × 480 display. Many video cards contain at least 256K bytes of memory, addressed in sections (pages) through the 64K byte memory window at locations A0000H–AFFFFH or B0000H–BFFFFH, depending on the mode and video adapter present. Each time a new bit plane is selected, the

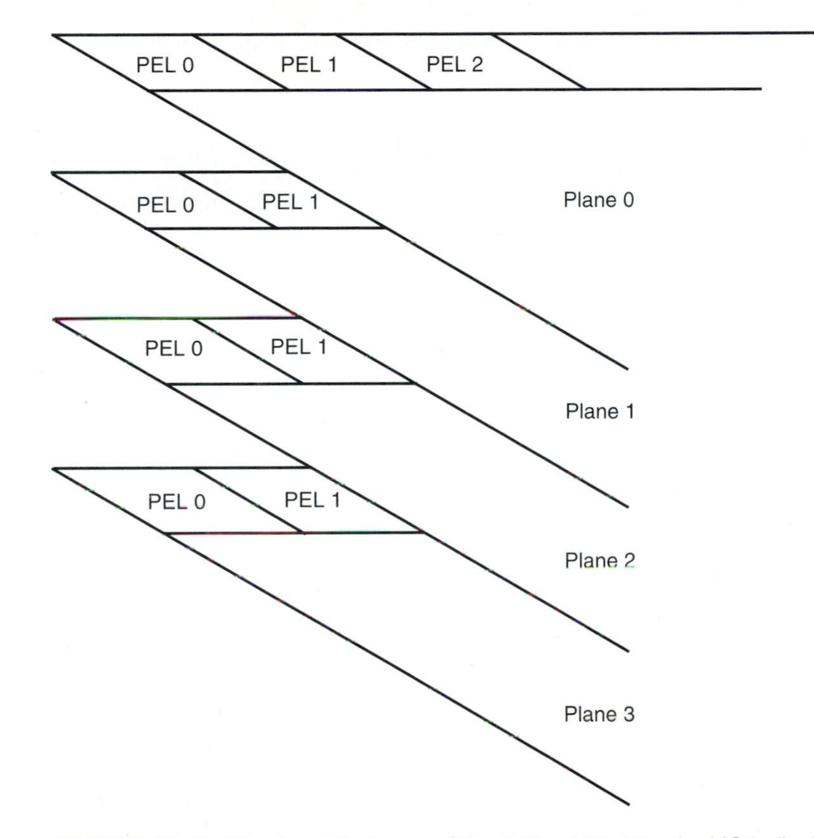

FIGURE 10–2 The four bit-planes of the 640 × 480, 16-color VGA display.

adapter internally addresses a separate 64K byte section of memory that appears at location A0000H–AFFFFH.

The 16-color, 640 × 480 VGA graphics mode is mode 12H when using the video BIOS. The memory organization for this mode is illustrated in Figure 10–2. The memory is organized in four bit-planes. Each byte in a bit-plane represents 8 picture elements (PELs) on the video screen. This means that a 640 × 480 display uses 38,400 bytes of memory in each bit-plane to address the 307,200 PELs found on this display. The first scanning line of 640 bits is stored in the first 80 bytes of video memory plane, the second scanning line is stored in the next 80 bytes of video memory plane, and so forth. A combination of the 4 bit-planes is used to specify one of the 16 colors for each of the PELs. To change the color of one PEL on the video display, four bits are changed, one in each bit-plane. Table 10–4 lists the color codes used for a standard VGA display using 4 bit-planes. If all 4 bit-planes are cleared, black is displayed for the PEL.

Programming the 16-Color, 640 × 480 Mode

Programming the 256-color, 320 × 200 mode was fairly easy. The 16-color mode requires more effort, because of the way the memory is organized in bit-planes. To plot a single dot:

1. Read the memory locations to be changed to load the bit-plane information into the video card.

2. Select and address a single bit (PEL) through the graphics address register (GAR) and bit-mask register (BMR).
3. Address and set the map-mask register (MMR) to 0FH and write 0 (black) to the address containing the PEL, to clear the old color from the PEL.
4. Set the desired PEL color through the MMR.
5. Write the PEL by changing the memory location that contains the video information.

The Bit-Mask Register. The bit-mask register (BMR) selects the bit or bits that are modified when a byte is written to the display adapter memory. Each bit position of the display adapter memory represents a PEL in the 16-color, 640 × 480 mode. Memory from locations A0000H through A95FFH store 38,400 bytes that represent the 307,200 PELs. (It is interesting to note that commercial television transmits 211,000 PELs to form a video image, but the number of colors displayed are infinite instead of 16.) Memory location A0000H holds the first 8 PELs, memory location A0001H holds the next 8 PELs, and so forth. If only the left-most PEL is changed, the bit-mask register is programmed with an 80H before any information is written to memory location A0000H. Likewise, other bits or multiple bits can be modified by changing the bit-mask register contents.

The BMR is accessed by programming the graphics-address register (GAR) with an index of 8 to select the BMR. The bit-mask is then programmed. The I/O address of the GAR is 03CEH and the I/O address of the BMR is 03CFH. Example 10–18 shows how the BMR is programmed so all eight bits (eight PELs) change together. The first set of three instructions selects the bit-mask register and the second set of three instructions selects all 8 bits. This setting stays in effect until it is changed by the same sequence of instructions.

EXAMPLE 10–18

```
0000  BA 03CE          MOV   DX,3CEH          ;graphics-address register
0003  B0 08            MOV   AL,8             ;select bit-mask register
0005  EE               OUT   DX,AL

0006  BA 03CF          MOV   DX,03CFH         ;bit-mask register
0009  B0 FF            MOV   AL,0FFH          ;select all 8 bits
000B  EE               OUT   DX,AL
```

The Map Mask Register. The map-mask register (MMR) selects the bit-planes enabled for a write operation. If all bit-planes are enabled, the color F (bright white) is written by default. Note that enabling a bit only allows it to be set. The only color that cannot be written in this manner is black (color 0). To write a new color to a PEL, if the bit-mask is FFH, set all four bit-planes (0FH) and write a 00H to memory. This sets the color to black. Next, select the color to be written by placing the binary color number (0–F) into the MMR, and write the new PEL to memory. This procedure works for up to 8 PELs.

Access to the MMR is provided by using index 2 for the sequence-address register (port 3C4H). Once the MMR is accessed, select bit-planes by writing to the MMR at I/O port 3C5H. Example 10–19 shows how to write the first (left-most) PEL on the video screen and set it to color 2 (green). The video memory location must be read before it is changed. If the location is not read, the result is unpredictable. The read operation loads the video byte into an internal latch so it can be changed before it is rewritten to the video memory by the display adapter.

EXAMPLE 10–19

```
0000  B4 00              MOV   AH,0                  ;set mode to 12H
0002  B0 12              MOV   AL,12H
0004  CD 10              INT   10H

0006  BA 03CE            MOV   DX,3CEH               ;graphics-address register
0009  B0 08              MOV   AL,8                  ;select bit-mask register
000B  EE                 OUT   DX,AL

000C  BA 03CF            MOV   DX,3CFH               ;bit-mask register
000F  B0 80              MOV   AL,080H               ;select left-most bit
0011  EE                 OUT   DX,AL

0012  BA 03C4            MOV   DX,3C4H               ;sequence-address register
0015  B0 02              MOV   AL,2                  ;select map-mask register
0017  EE                 OUT   DX,AL

0018  BA 03C5            MOV   DX,3C5H               ;map-mask register
001B  B0 0F              MOV   AL,0FH                ;enable all planes
001D  EE                 OUT   DX,AL

001E  B8 A000            MOV   AX,0A000H             ;address video memory
0021  8E D8              MOV   DS,AX
0023  BF 0000            MOV   DI,0
0026  8A 05              MOV   AL,[DI]               ;must read first
0028  C6 05 00           MOV   BYTE PTR [DI],0       ;clear old color

002B  B0 02              MOV   AL,02H                ;select color 2
002D  EE                 OUT   DX,AL

002E  C6 05 FF           MOV   BYTE PTR [DI],0FFH    ;write memory
```

Suppose a procedure to plot any PEL on the video display is required. Example 10–20 shows such a procedure and a program that displays some dots at various points, and in various colors, on the video screen. Look very closely for the single red dot below and to the right of the short cyan line when you execute this program.

The heart of this program is the DOT procedure, which displays one dot of any color at any location on the video display. Parameters are passed to DOT through BX (row address), SI (column address), and DL (color). The row address is multiplied by 80 to find the memory byte that corresponds to the row on the video display. Remember, each byte contains 8 PELs. Next, the column address is divided by 8. This provides a quotient, which is added to the row-address byte to locate the memory byte that corresponds to the desired PEL and a remainder. The remainder is used as a shift count, to shift an 80H right to locate the correct bit in the bit-mask register. Once the proper bit and memory location have been calculated, the remainder of the procedure sets the bit-mask register and displays the correct color.

EXAMPLE 10–20

```
                    ;a program that displays a short cyan line 10 PELs wide
                    ;with a red dot below and to the right of the cyan line.
                    ;
                           .MODEL TINY
0000                       .CODE
                           .STARTUP
```

```
0100  B8 A000             MOV   AX,0A000H        ;address video RAM at segment A000
0103  8E D8               MOV   DS,AX
0105  FC                  CLD                    ;select increment

0106  B8 0012             MOV   AX,12H           ;set mode to 12H
0109  CD 10               INT   10H              ;and clear screen

010B  B9 000A             MOV   CX,10            ;set dot count to 10
010E  BB 000A             MOV   BX,10            ;row address
0111  BE 0064             MOV   SI,100           ;column address
0114  B2 03               MOV   DL,3             ;color 3 (cyan)
0116              MAIN1:                         ;plot 10 dots
0116  E8 001B             CALL  DOT              ;display one dot
0119  46                  INC   SI
011A  E2 FA               LOOP  MAIN1            ;repeat 10 times

011C  BB 0028             MOV   BX,40            ;row address
011F  BE 00C8             MOV   SI,200           ;column address
0122  B2 04               MOV   DL,4             ;color 4 (red)
0124  E8 000D             CALL  DOT              ;display one red dot

0127  B4 01               MOV   AH,1             ;wait for key
0129  CD 21               INT   21H

012B  B8 0003             MOV   AX,3
012E  CD 10               INT   10H              ;return to DOS video mode
                          .EXIT
                  ;
                  ;the DOT procedure displays one dot or PEL on the video display.
                  ;BX = row address (0 to 479)
                  ;SI = column address (0 to 639)
                  ;DL = color (0 to 15)
                  ;
0134              DOT     PROC  NEAR

0134  51                  PUSH  CX
0135  52                  PUSH  DX               ;save color

0136  B8 0050             MOV   AX,80            ;find row address byte
0139  F7 E3               MUL   BX
013B  8B F8               MOV   DI,AX            ;save it
013D  8B C6               MOV   AX,SI            ;find column address byte
013F  B6 08               MOV   DH,8
0141  F6 F6               DIV   DH
0143  8A CC               MOV   CL,AH            ;get shift count
0145  B4 00               MOV   AH,0
0147  03 F8               ADD   DI,AX            ;form address of PEL byte
0149  B0 80               MOV   AL,80H
014B  D2 E8               SHR   AL,CL            ;find bit in bit-mask register
014D  50                  PUSH  AX               ;save bit mask

014E  BA 03CE             MOV   DX,3CEH          ;graphics-address register
0151  B0 08               MOV   AL,8             ;select bit-mask register
0153  EE                  OUT   DX,AL

0154  BA 03CF             MOV   DX,3CFH          ;bit-mask register
0157  58                  POP   AX               ;get bit mask
0158  EE                  OUT   DX,AL

0159  BA 03C4             MOV   DX,3C4H          ;sequence-address register
015C  B0 02               MOV   AL,2             ;select map-mask register
```

```
015E  EE                   OUT    DX,AL

015F  BA 03C5              MOV    DX,3C5H             ;map-mask register
0162  B0 0F                MOV    AL,0FH              ;enable all planes
0164  EE                   OUT    DX,AL

0165  8A 05                MOV    AL,[DI]             ;must read first
0167  C6 05 00             MOV    BYTE PTR [DI],0     ;clear old color
016A  58                   POP    AX                  ;get color from stack
016B  50                   PUSH   AX
016C  EE                   OUT    DX,AL
016D  C6 05 FF             MOV    BYTE PTR [DI],0FFH  ;write memory

0170  5A                   POP    DX                  ;restore registers
0171  59                   POP    CX
0172  C3                   RET

0173              DOT      ENDP
                           END
```

Because video displays are often organized into text areas, another procedure (BLOCK) plots a block that corresponds to text. This procedure, shown in Example 10–21, breaks the 640 × 480 display into a series of blocks 8 PELs wide and 9 PELs high, for a display that is 80 × 53. (3 raster lines are lost at the bottom of the display.) The BLOCK procedure uses BX for column number, SI for row number, and DL for the color of the block. The BLOCK procedure is useful for filling large areas of the video display at a high speed.

The program in Example 10–21 displays a cyan line across the top of the screen with a white background on the remainder of the display. This is accomplished by using the BLOCK procedure for filling the display.

EXAMPLE 10–21

```
                  ;a program that displays a cyan bar across the top of a white
                  ;screen.
                  ;
                           .MODEL TINY
0000                       .CODE
                           .STARTUP
0100  B8 A000              MOV    AX,0A000H           ;address video RAM at segment A000
0103  8E D8                MOV    DS,AX
0105  FC                   CLD                        ;select increment

0106  B8 0012              MOV    AX,12H              ;set mode to 12H
0109  CD 10                INT    10H                 ;and clear screen

010B  B9 0050              MOV    CX,80               ;block count
010E  BB 0000              MOV    BX,0                ;row address
0111  BE 0000              MOV    SI,0                ;column address
0114  B2 03                MOV    DL,3                ;color 3 (cyan)
0116              MAIN1:                              ;plot 80 blocks
0116  E8 0028              CALL   BLOCK               ;display a block
0119  46                   INC    SI                  ;address next column
011A  E2 FA                LOOP   MAIN1               ;repeat 80 times

011C  BB 0001              MOV    BX,1                ;row address
011F  B2 07                MOV    DL,7                ;color 7 (white)
0121  B6 34                MOV    DH,52               ;row count
```

```
0123                MAIN2:
0123  BE 0000               MOV     SI,0                ;column address
0126  B9 0050               MOV     CX,80               ;column count
0129                MAIN3:
0129  E8 0015               CALL    BLOCK               ;display a block
012C  46                    INC     SI                  ;address next column
012D  E2 FA                 LOOP    MAIN3               ;repeat 80 times
012F  43                    INC     BX                  ;increment row address
0130  FE CE                 DEC     DH
0132  75 EF                 JNZ     MAIN2               ;repeat 52 times

0134  B4 01                 MOV     AH,1                ;wait for key
0136  CD 21                 INT     21H

0138  B8 0003               MOV     AX,3
013B  CD 10                 INT     10H                 ;return to DOS video mode
                            .EXIT
                    ;
                    ;The BLOCK procedure displays one block that is 8 PELs
                    ;wide by 9 PELs high.
                    ;BX = row address (0 to 52)
                    ;SI = column address (0 to 79)
                    ;DL = block color (0 to 15)
                    ;
0141                BLOCK   PROC    NEAR

0141  51                    PUSH    CX
0142  52                    PUSH    DX                  ;save color

0143  BA 03CE               MOV     DX,3CEH             ;graphics-address register
0146  B0 08                 MOV     AL,8                ;select bit-mask register
0148  EE                    OUT     DX,AL
0149  BA 03CF               MOV     DX,3CFH             ;bit-mask register
014C  B0 FF                 MOV     AL,0FFH             ;enable all 8 bits
014E  EE                    OUT     DX,AL

014F  BA 03C4               MOV     DX,3C4H             ;sequence-address register
0152  B0 02                 MOV     AL,2                ;select map-mask register
0154  EE                    OUT     DX,AL

0155  B8 02D0               MOV     AX,80*9             ;find row address byte
0158  F7 E3                 MUL     BX
015A  8B F8                 MOV     DI,AX               ;save it
015C  03 FE                 ADD     DI,SI               ;form address of PEL byte

015E  B9 0009               MOV     CX,9                ;byte count
0161  BA 03C5               MOV     DX,3C5H             ;map-mask register
0164  58                    POP     AX                  ;get color
0165  50                    PUSH    AX
0166  8A E0                 MOV     AH,AL
0168                BLOCK1:
0168  B0 0F                 MOV     AL,0FH              ;enable all planes
016A  EE                    OUT     DX,AL
016B  8A 05                 MOV     AL,[DI]             ;must read first
016D  C6 05 00              MOV     BYTE PTR [DI],0     ;clear old color
0170  8A C4                 MOV     AL,AH
0172  EE                    OUT     DX,AL
0173  C6 05 FF              MOV     BYTE PTR [DI],0FFH  ;write memory
0176  83 C7 50              ADD     DI,80
0179  E2 ED                 LOOP    BLOCK1

017B  5A                    POP     DX
```

```
017C  59              POP    CX
017D  C3              RET

017E           BLOCK  ENDP
                      END
```

Text with Graphics. Displaying text in a graphics display mode, such as mode 12H, is also difficult. Sometimes, DOS function call 21H is used to place text on the screen, but only if the video attribute byte below the text can change both the text and its background color. A better way to display text in a graphics mode is to use a lookup table that contains the graphic display characters, so the text appears without a different background color.

The video BIOS ROM contains 8 × 8, 8 × 14, 9 × 14, 8 × 16, and 9 × 16 character sets. These character sets, or your own, may be used as a basis for displaying text in a graphic display mode. Figure 10–3 shows the format of the 8 × 8 character and a few example characters. The 8 × 8 character set is most often used for the 640 × 480 display. The character set is obtained from the video BIOS ROM via the INT 10H function, which returns the address of the character set in ES:BP.

EXAMPLE 10–22

```
0000                 GETC   PROC   FAR

0000  FC                    CLD                   ;select increment

0001  B4 11                 MOV    AH,11H         ;get ROM character set
0003  B0 30                 MOV    AL,30H
0005  B7 03                 MOV    BH,03H         ;select 8 x 8
0007  CD 10                 INT    10H

0009  8C C0                 MOV    AX,ES          ;address memory
000B  8E D8                 MOV    DS,AX
000D  B8 ---- R             MOV    AX,DATA
0010  8E C0                 MOV    ES,AX
0012  BF 0000 R             MOV    DI,OFFSET CHAR
0015  8B F5                 MOV    SI,BP
0017  B9 0400               MOV    CX,1024        ;load count
001A  F3/ A4                REP MOVSB             ;copy character set

001C  CB                    RET

001D                 GETC   ENDP
```

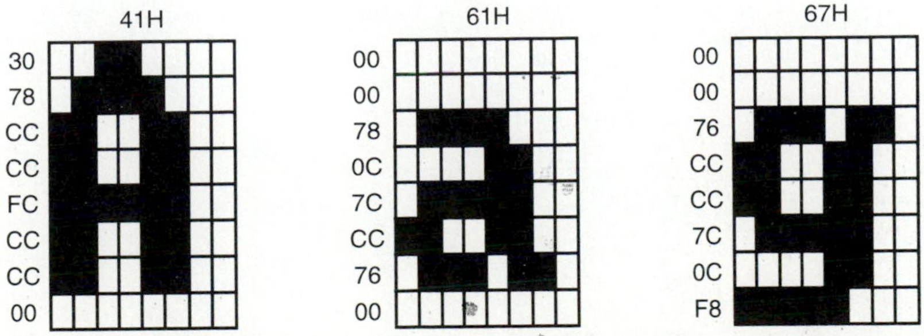

41H	61H	67H
30	00	00
78	00	00
CC	78	76
CC	0C	CC
FC	7C	CC
CC	CC	7C
CC	76	0C
00	00	F8

FIGURE 10–3 Some 8 × 8 characters for the VGA graphics display.

TABLE 10–5 Contents of BH for ROM character sets

BH	*Character set*
02H	8 (wide) x 14 (high)
03H	8 x 8 (standard)
04H	8 x 8 extended characters (ASCII 80H–FFH)
05H	9 x 14 (alternate)
06H	8 x 16
07H	8 x 16 (alternate)

Example 10–22 lists a procedure that copies the 8 × 8 standard character set from the video BIOS ROM into the data segment at memory location CHAR. Because each ASCII character requires 8 bytes of memory, and because the ASCII code contains 128 characters, the size of the character table is 1,024 bytes. Note that BH = 03H to obtain the 8 × 8 standard character set. The extended character set (code 128–255) is obtained in the same fashion, except BH = 04H. Table 10–5 lists the character sets available in most video card ROMs.

Once the character set is fetched from the video BIOS, it is used to display text on the graphics display. The procedure listed in Example 10–23 uses the techniques learned in Example 10–22 to obtain the character table used to display ASCII text on a graphics display. This procedure assumes the display uses 9 scanning lines for each character line and 80 characters appear across the screen on a character line. This allows 53 text lines, each containing 80 characters, to be displayed on the graphics mode screen. AL = ASCII to be displayed, DH = character line (0–52), DL = character column (0–79), and BL = text color (0–F). This procedure does not display a background color for the text. Instead, it superimposes the text on top of whatever graphics presentation appears on the video display. This procedure can display any character set, as long as it is an 8 × 8.

The program listed in Example 10–23 uses the CHAR procedure to display a cyan 'A' at row 5, column 0 and a bright red 'B' at row 0, column 0. The character table in the ROM is accessed each time the CHAR procedure is called. The CHAR procedure uses the bit-mask register to change only those bits that correspond to the ASCII character obtained from the video BIOS ROM.

EXAMPLE 10–23

```
                        ;program that displays a bright red B at row 0, column 0, and a
                        ;cyan A at row 5, column 0.
                                .MODEL TINY
0000                            .CODE
                                .STARTUP
0100  B8 A000                   MOV     AX,0A000H           ;address video RAM at segment A000
0103  8E D8                     MOV     DS,AX
0105  FC                        CLD                         ;select increment

0106  B8 0012                   MOV     AX,12H              ;set mode to 12H
0109  CD 10                     INT     10H                 ;and clear screen

010B  B0 41                     MOV     AL,'A'              ;display 'A'
```

```
010D  B2 03                  MOV     DL,3              ;cyan
010F  BB 0005                MOV     BX,5              ;row 5
0112  BE 0000                MOV     SI,0              ;column 0
0115  E8 001A                CALL    CHAR              ;display cyan 'A'

0118  B0 42                  MOV     AL,'B'            ;display 'B'
011A  B2 0C                  MOV     DL,12             ;bright red
011C  BB 0000                MOV     BX,0              ;row 0
011F  BE 0000                MOV     SI,0              ;column 0
0122  E8 000D                CALL    CHAR              ;display bright red 'B'

0125  B4 01                  MOV     AH,1              ;wait for key
0127  CD 21                  INT     21H

0129  B8 0003                MOV     AX,3
012C  CD 10                  INT     10H               ;return to DOS video mode
                             .EXIT
                       ;
                       ;The CHAR procedure displays a character (8 x 8) on the
                       ;mode 12H display without changing the background color.
                       ;AL = ASCII code
                       ;DL = color (0 to 15)
                       ;BX = row (0 to 52)
                       ;SI = column (0 to 79)
                       ;
0132                   CHAR  PROC    NEAR

0132  51                     PUSH    CX
0133  52                     PUSH    DX
0134  53                     PUSH    BX                ;save row address
0135  50                     PUSH    AX                ;save ASCII
0136  B8 1130                MOV     AX,1130H          ;get 8 x 8 set
0139  B7 03                  MOV     BH,3
013B  CD 10                  INT     10H               ;segment is in ES
013D  58                     POP     AX                ;get ASCII code
013E  B4 00                  MOV     AH,0
0140  D1 E0                  SHL     AX,1              ;multiply by 8
0142  D1 E0                  SHL     AX,1
0144  D1 E0                  SHL     AX,1
0146  03 E8                  ADD     BP,AX             ;index character in ROM
0148  5B                     POP     BX                ;get row address
0149  B8 02D0                MOV     AX,80*9           ;find row address
014C  F7 E3                  MUL     BX
014E  8B F8                  MOV     DI,AX
0150  03 FE                  ADD     DI,SI             ;add in column address
0152  B9 0008                MOV     CX,8              ;set count to 8 rows

0155  BA 03CE          C1:   MOV     DX,3CEH           ;address bit-mask register
0158  B0 08                  MOV     AL,8              ;load index 8
015A  26: 8A 66 00           MOV     AH,ES:[BP]        ;get character row
015E  45                     INC     BP                ;point to next row
015F  EF                     OUT     DX,AX             ;modify bit-mask register
0160  BA 03C4                MOV     DX,3C4H           ;address map-mask register
0163  B8 0F02                MOV     AX,0F02H
0166  EF                     OUT     DX,AX             ;select all planes
0167  42                     INC     DX
0168  8A 05                  MOV     AL,[DI]           ;read data
016A  C6 05 00               MOV     BYTE PTR [DI],0   ;write black
016D  58                     POP     AX                ;get color
016E  50                     PUSH    AX
016F  EE                     OUT     DX,AL             ;write color
```

```
0170  C6 05 FF             MOV    BYTE PTR [DI],0FFH
0173  83 C7 50             ADD    DI,80              ;address next raster row
0176  E2 DD                LOOP   C1                 ;repeat 8 times

0178  5A                   POP    DX
0179  59                   POP    CX
017A  C3                   RET

017B              CHAR     ENDP
                           END
```

Example 10–24 illustrates how to display a graphics screen and place text on top of the graphics. This program selects VGA mode 12H and then displays a cyan screen. Next, it places two lines of text in two different colors on top of the cyan screen. This program uses the BLOCK procedure to display a cyan screen by filling all the blocks with cyan. It then uses the CHAR procedure to place text on top of the cyan background. Note that a new procedure (LINE) is used to display a null string that represents test lines of text displayed by this program.

EXAMPLE 10–24 (pp. 362–365)

```
                          ;a program that displays two test lines of text on a cyan graphics
                          ;background screen.
                          ;
                                 .MODEL SMALL
0000                             .DATA
0000  54 68 69 73 20      MES1   DB              'This is test line 1.',0
      69 73 20 74 65
      73 74 20 6C 69
      6E 65 20 31 2E
      00
0015  54 68 69 73 20      MES2   D               'This is test line 2.',0
      69 73 20 74 65 73
      74 20 6C 69 6E 65
      20 32 2E 00
0000                             .CODE
                                 .STARTUP
0017  B8 A000                    MOV    AX,0A000H          ;address video RAM
001A  8E D8                      MOV    DS,AX
001C  FC                         CLD                       ;select increment

001D  B8 0012                    MOV    AX,12H             ;set mode to 12H
0020  CD 10                      INT    10H                ;and clear screen

0022  B2 03                      MOV    DL,3               ;color cyan
0024  B6 35                      MOV    DH,53              ;row counter
0026  BB 0000                    MOV    BX,0               ;row 0
0029                      MAIN1:
0029  B9 0050                    MOV    CX,80              ;column counter
002C  BE 0000                    MOV    SI,0               ;column 0
002F                      MAIN2:
002F  E8 0094                    CALL   BLOCK              ;display a cyan block
0032  46                         INC    SI                 ;address next column
0033  E2 FA                      LOOP   MAIN2              ;repeat 80 times
0035  43                         INC    BX                 ;address next row
0036  FE CE                      DEC    DH                 ;decrement row counter
0038  75 EF                      JNZ    MAIN1              ;repeat for 53 rows
```

```
003A  B8 ---- R         MOV   AX,@DATA          ;address data segment
003D  8E C0             MOV   ES,AX             ;with ES

003F  B2 09             MOV   DL,9              ;bright blue text
0041  BB 0005           MOV   BX,5              ;row 5
0044  BE 0000           MOV   SI,0              ;column 0
0047  BF 0000 R         MOV   DI,OFFSET MES1    ;address MES1
004A  E8 001B           CALL  LINE              ;display bright blue MES1

004D  B2 0C             MOV   DL,12             ;bright red
004F  BB 000F           MOV   BX,15             ;row 15
0052  BE 0000           MOV   SI,0              ;column 0
0055  BF 0015 R         MOV   DI,OFFSET MES2    ;address MES2
0058  E8 000D           CALL  LINE              ;display bright red MES2

005B  B4 01             MOV   AH,1              ;wait for key
005D  CD 21             INT   21H

005F  B8 0003           MOV   AX,3
0062  CD 10             INT   10H               ;return to DOS video mode
                        .EXIT
                  ;
                  ;The line procedure displays the line of text addressed by ES:DI
                  ;DL = color of text (0 to 15).
                  ;The text must be stored as a null string
                  ;BX = row
                  ;SI = column
                  ;
0068              LINE  PROC  NEAR

0068  26: 8A 05         MOV   AL,ES:[DI]        ;get character
006B  0A C0             OR    AL,AL             ;test for null
006D  74 0D             JZ    LINE1             ;if null
006F  06                PUSH  ES                ;save registers
0070  57                PUSH  DI
0071  56                PUSH  SI
0072  E8 0008           CALL  CHAR              ;display characters
0075  5E                POP   SI                ;restore registers
0076  5F                POP   DI
0077  07                POP   ES
0078  46                INC   SI                ;address next column
0079  47                INC   DI                ;address next character
007A  EB EC             JMP   LINE              ;repeat until null
007C              LINE1:
007C  C3                RET

007D              LINE  ENDP
                  ;
                  ;The CHAR procedure displays a character (8 x 8) on the
                  ;mode 12H display without changing the background color.
                  ;AL = ASCII code
                  ;DL = color (0 to 15)
                  ;BX = row (0 to 52)
                  ;SI = column (0 to 79)
                  ;
007D              CHAR  PROC  NEAR

007D  51                PUSH  CX
007E  52                PUSH  DX
007F  53                PUSH  BX                ;save row address
0080  50                PUSH  AX                ;save ASCII
```

```
0081  B8 1130                   MOV   AX,1130H            ;get 8 x 8 set
0084  B7 03                     MOV   BH,3
0086  CD 10                     INT   10H
0088  58                        POP   AX                  ;get ASCII code
0089  B4 00                     MOV   AH,0
008B  D1 E0                     SHL   AX,1                ;multiply by 8
008D  D1 E0                     SHL   AX,1
008F  D1 E0                     SHL   AX,1
0091  03 E8                     ADD   BP,AX               ;index character in ROM
0093  5B                        POP   BX                  ;get row address
0094  B8 02D0                   MOV   AX,80*9             ;find row address
0097  F7 E3                     MUL   BX
0099  8B F8                     MOV   DI,AX
009B  03 FE                     ADD   DI,SI               ;add in column address
009D  B9 0008                   MOV   CX,8                ;set count to 8 rows
00A0  BA 03CE         C1:       MOV   DX,3CEH             ;address bit mask register
00A3  B0 08                     MOV   AL,8                ;load index 8
00A5  26: 8A 66 00              MOV   AH,ES:[BP]          ;get character row
00A9  45                        INC   BP                  ;point to next row
00AA  EF                        OUT   DX,AX
00AB  BA 03C4                   MOV   DX,3C4H             ;address map-mask register
00AE  B8 0F02                   MOV   AX,0F02H
00B1  EF                        OUT   DX,AX               ;select all planes
00B2  42                        INC   DX
00B3  8A 05                     MOV   AL,[DI]             ;read data
00B5  C6 05 00                  MOV   BYTE PTR [DI],0     ;write black
00B8  58                        POP   AX                  ;get color
00B9  50                        PUSH  AX
00BA  EE                        OUT   DX,AL               ;write color
00BB  C6 05 FF                  MOV   BYTE PTR [DI],0FFH
00BE  83 C7 50                  ADD   DI,80               ;address next raster row
00C1  E2 DD                     LOOP  C1                  ;repeat 8 times
00C3  5A                        POP   DX
00C4  59                        POP   CX
00C5  C3                        RET

00C6                  CHAR      ENDP
                      ;
                      ;The BLOCK procedure displays one block that is 8 PELs
                      ;wide by 9 PELs high.
                      ;BX = row address (0 to 52)
                      ;SI = column address (0 to 79)
                      ;DL = block color (0 to 15)
                      ;
00C6                  BLOCK     PROC  NEAR

00C6  51                        PUSH  CX
00C7  52                        PUSH  DX                  ;save color

00C8  BA 03CE                   MOV   DX,3CEH             ;graphics-address register
00CB  B0 08                     MOV   AL,8                ;select bit-mask register
00CD  EE                        OUT   DX,AL
00CE  BA 03CF                   MOV   DX,3CFH             ;bit-mask register
00D1  B0 FF                     MOV   AL,0FFH             ;enable all 8 bits
00D3  EE                        OUT   DX,AL

00D4  BA 03C4                   MOV   DX,3C4H             ;sequence-address register
00D7  B0 02                     MOV   AL,2                ;select map-mask register
00D9  EE                        OUT   DX,AL

00DA  B8 02D0                   MOV   AX,80*9             ;find row address byte
```

```
00DD  F7 E3                  MUL    BX
00DF  8B F8                  MOV    DI,AX                 ;save it
00E1  03 FE                  ADD    DI,SI                 ;form address of PEL byte

00E3  B9 0009                MOV    CX,9                  ;byte count
00E6  BA 03C5                MOV    DX,3C5H               ;map-mask register
00E9  58                     POP    AX                    ;get color
00EA  50                     PUSH   AX
00EB  8A E0                  MOV    AH,AL
00ED                 BLOCK1:
00ED  B0 0F                  MOV    AL,0FH                ;enable all planes
00EF  EE                     OUT    DX,AL
00F0  8A 05                  MOV    AL,[DI]               ;must read first
00F2  C6 05 00               MOV    BYTE PTR [DI],0       ;clear old color
00F5  8A C4                  MOV    AL,AH
00F7  EE                     OUT    DX,AL
00F8  C6 05 FF               MOV    BYTE PTR [DI],0FFH    ;write memory
00FB  83 C7 50               ADD    DI,80
00FE  E2 ED                  LOOP   BLOCK1

0100  5A                     POP    DX
0101  59                     POP    CX
0102  C3                     RET

0103                 BLOCK   ENDP
                             END
```

Although this has not been a complete discussion of graphics displays, it provides a foundation in video display modes 12H and 13H. It also shows how text and graphics can be mixed in the graphics display modes.

10–4 SUMMARY

1. Conditional .IF/.ELSE/.ENDIF statements aid in assembling software. These statements generate instructions that perform the indicated conditional task in program.
2. The DO-WHILE construct is implemented in assembly language by using the .WHILE and .ENDW statements. Associated with the .WHILE statement is a condition that is tested during the executing of the DO-WHILE construct.
3. The REPEAT-UNTIL construct is implemented in assembly language by using the .REPEAT and .UNTIL or UNTILCXZ statements. .UNTIL usually places a conditional LOOP instruction in the program, while .UNTILCXZ uses the unconditional LOOP instruction.
4. Macro sequences also use conditional assembly statements, but they are not preceded by a period. For example, the IF/ELSE/ENDIF statements are used in macros, rather than .IF/.ELSE/.ENDIF.
5. The REPEAT-UNTIL construct in a macro uses REPEAT with a condition or count and the ENDM statement to end the construct. This is different from the .REPEAT and .UNTIL constructs for the assembler. The function is also different because the macro REPEAT-UNTIL construct generates copies of the code between the REPEAT and ENDM statements, while the assembler versions do not.

6. The FOR statement is used in macro sequences to function as a READ and DATA statement. The FOR statement is used to read a string of parameters and use them in the macro to generate results.
7. The video display is accessed through the video BIOS or directly through the video display memory. For a VGA display adapter, video display memory exists at location A0000H–AFFFFH for the graphics modes 12H and 13H. Mode 12H is a 16-color, 640 × 480 display mode. Mode 13H is a 256-color, 320 × 200 display mode.
8. The 256-color mode stores one picture element per byte of memory, and the 16-color mode is a bit-mapped mode, storing 8 PELs per memory byte.
9. The bit-mask register selects the PEL to be written in the 16-color, 640 × 480 video display. The map-mask register selects the bit-planes (color) to be written in the 16-color, 640 × 480 video display.
10. The video BIOS ROM contains character sets that are copied to an application program to display text on the graphics mode VGA screen. This allows the display of standard characters, custom characters or character sets, or even multiple character sets.

10–5 QUESTIONS AND PROBLEMS

1. If the .IF AL = 7 statement appears in a program, the assembler will generate what instruction to accomplish this task?
2. What is the purpose of the .ENDIF statement?
3. Develop a procedure that uses the .IF, .ELSE, and .ENDIF statements to convert from a binary number in AL, between 00H and 0FH, to ASCII–coded data in AL, between 30H and 39H, for the values 00H–09H, and to ASCII-coded data in AL between 41H and 46H for the values 0AH–0FH.
4. Develop a procedure that uses the .IF and .ENDIF statements to convert from the binary value in AL to the decimal value displayed on the video screen.
5. What is used with the .WHILE statement to create an infinite loop?
6. What is the purpose of the .BREAK statement?
7. Develop a procedure that uses the .WHILE and .ENDW statement to search through the character string stored at memory location DATA:TEMP for the $ (24H) code. The return must carry the address of the $ to the calling sequence in the BP register.
8. Where is the condition placed when the .REPEAT and .UNTIL statements are used in a program?
9. Develop a procedure that uses .REPEAT and .UNTIL to search through data beginning at the memory address pointed to by DS:SI, which are transferred to the procedure from the calling sequence. The search code is also transferred to the procedure through the AL register.
10. Develop a procedure that uses .REPEAT and .UNTILCXZ to clear a memory buffer to 00H. The address of the buffer and length are passed to the procedure through DS:SI and CX.
11. Write a macro sequence that uses REPEAT and ENDM to store 23 NOP instructions in memory.
12. Is it possible to use the .BREAK statement in a REPEAT/UNTIL loop?

13. Write a macro called PRINT that has one parameter: The macro must display an ASCII-coded value on the video display. To use the macro, type PRINT 'A' to display an A, or PRINT with no parameter to display a carriage return (0DH) and a line feed (0AH). Use the IFB or IFNB statement to test for a blank parameter.
14. What is meant by the statement .IF AL < 3 && AL >5?
15. Where is the video data stored when mode 13H is selected for the VGA video display adapter?
16. Where is the video text data stored in the personal computer?
17. How many bytes of memory are required to store a screen of information for the 256-color, 320×200 video display mode?
18. How is the INT 10H instruction used to change to a new video display mode?
19. Write a short program that displays color number 19 at each PEL on the mode 13H video display.
20. Develop a program that places the video adapter into mode 13H and then displays a red bar (color number 4) at the left margin of the screen, 50 PELs high and 30 PELs wide. Place this bar 20 PELs down from the top of the screen.
21. What is a bit-plane?
22. How is the bit-mask register used in the 16-color, 640×480 VGA video display mode?
23. How is the map-mask register used in the 16-color, 640×480 VGA video display mode?
24. Develop a program that places the video adapter into mode 12H and displays a thin bright blue line, 1 PEL wide, from the top center of the screen to the bottom center of the screen.
25. Develop a program that places the video adapter into mode 12H and then draws a green line, 2 PELs wide, across the bottom of the screen.
26. Develop a program that places the video adapter into mode 12H and then displays a vertical color bar pattern that shows the 16 colors available in this mode in 16 bars, each 40 PELs wide.
27. Develop a program that places the video adapter into mode 12H and displays a white screen (color 7), with a cyan horizontal bar across the top 10 scanning lines high.
28. Develop your own character set for mode 12H of the VGA display. Your character set must contain characters that are 16 bits wide and 16 bits high. Luckily you only need the characters N, a, m, e, space, and =.
29. Using the characters developed in Question 28, develop a program using mode 12H that displays a blue video screen that contains the red character line "Name = " at the upper-left margin of the screen.

CHAPTER 11

Disk Memory Functions and the Mouse

INTRODUCTION

Programming requires a familiarity with disk memory systems and the disk operations performed by DOS. This chapter details the organization of disk data and directories, and explains how to use the DOS INT 21H function call to access disk memory data and program files via assembly language.

The DOS INT 21H function call is used to create, delete, read, write, and close files on the disk. Complete coverage of this function is provided with examples of its use. Also presented is a description of software that allows access to both random and sequential access disk data files.

Because the mouse is an important I/O device, perhaps as important as the keyboard, this chapter also introduces the mouse and the mouse interrupt (INT 33H). The mouse is also used in example programs to show how a TSR can be used to control screen items with the DOS command processor.

OBJECTIVES

Upon completion of this chapter, you will be ablt to:

1. Use DOS INT 21H function to create, open, read, write, delete, rename, and close disk files.
2. Create and use random and sequential access files.
3. Append data to a file.
4. Insert data in the middle of a file.
5. Manipulate directory information and entries.
6. Retrieve data from the DOS command line.
7. Use INT 33H to read the mouse position.
8. Use INT 33H to enable and disable the mouse.

11–1 SEQUENTIAL ACCESS DISK FILES

Files are normally accessed through the DOS INT 21H function call. Although files may also be accessed through BIOS INT 13H, it is fairly difficult, because that function does not use the DOS directory structure. There are two ways to access a file using an INT 21H. One method uses a file control block and the other uses a file handle. Today, all software file accesses are via a file handle, so this text uses file handles for file access. Note that even though this chapter presents access to disk files using the assembler, such access is often accomplished with high-level programming languages. File control blocks are a carryover from an earlier operating system called CP/M[1] (control program/microprocessor), which was designed for 8-bit computer systems.

All DOS files are sequential access files. A sequential access file is stored and addressed from its beginning toward its end. This means that the first byte, and all bytes between it and the last, must be accessed to read the last byte. Fortunately, files are read and written with the DOS INT 21H function calls (refer to Appendix A), which simplifies file access and data manipulation. This section of the text describes how to create, read, write, delete, and rename a sequential access file using the DOS INT 21H function calls.

File Creation

Before a file is used, it must exist on the disk. A file is *created* with INT 21H function call number 3CH. To create a file, the name of the file and its extension are stored in memory as an ASCII-Z string, at a location addressed by DS:DX. The CX register must also contain the attribute of the file created by function 3CH. DOS function 5BH also creates a file and uses the same parameters as function 3CH. The difference is that function 3CH erases the file if it already exists, and 5BH signals an error if the file already exists. Which function to use depends on the program and whether or not the file that has the same name must be detected or erased.

EXAMPLE 11–1

```
0000  44 4F 47 2E 54 58    FILE1 DB    'DOG.TXT',0;file name DOG.TXT
      54 00
0008  43 3A 44 41 54 41    FILE2 DB    'C:DATA.DOC',0          ;file C:DATA.DOC
      2E 44 4F 43 00
0013  43 3A 5C 44 52 45    FILE3 DB    'C:\DREAD\ERROR.FIL',0  ;file C:\DREAD\ERROR.FIL
      41 44 5C 45 52 52
      4F 52 2E 46 49 4C
      00
```

A **file name** is always stored as an ASCII-Z string and may contain the disk drive letter and directory path(s) if needed. Example 11–1 shows several ASCII-Z string file names stored in memory for access by file utilities. An **ASCII-Z string** is a character string that ends with a null character (00H).

---

[1]CP/M is a registered trademark of Digital Research Corporation.

Suppose that a 256-byte memory buffer area is filled with data that must be stored in a new file called DATA.NEW on the default disk drive and current directory path. Before data is written to this new file, it must first be created. Example 11–2 lists a short program that creates this new file on the disk.

EXAMPLE 11–2

```
                         ;a program that creates file DATA.NEW
                         ;DO NOT RUN this program because it does not close the file
                         ;
                                 .MODEL SMALL
0000                             .DATA
0000  44 41 54 4         FILEN   DB 'DATA.NEW',0          ;file name
      2E 4E 45 57
      00
0000                             .CODE
                                 .STARTUP
0017  B4 3C                      MOV AH,3CH               ;create file function
0019  B9 0000                    MOV CX,0                 ;normal file attribute
001C  BA 0000 R                  MOV DX,OFFSET FILEN      ;address file name
001F  CD 21                      INT 21H                  ;access DOS
                                 .EXIT
                                 END
```

Whenever a file is created, the CX register must contain the attributes or characteristics of the file. In Example 11–2, a file is created without any attributes. Table 11–1 lists the attribute bit positions and defines them. A logic one in a bit selects the attribute, while a logic zero does not. For example, to create a hidden archive file, CX is loaded with a 0022H. The hidden and archive bits are set by 0022H or 02H plus 20H.

After returning from the INT 21H, the carry flag indicates whether an error occurred (C = 1) during the creation of the file. Some errors that occur during file creation (which are obtained, if needed, by INT 21H function call number 59H) are: path not found, no file handles available, or media error. If carry is cleared (C = 0), no error occurred during file creation and the AX register contains a file handle. The **file handle** is a number that refers to the file after it is created or opened. The file handle allows a file to be accessed without

TABLE 11–1 File attribute definitions

Bit Position	*Value*	*Attribute*	*Function*
0	01H	Read-only	A read-only file or subdirectory
1	02H	Hidden	Prevents the file or subdirectory name from appearing in the directory when a DIR is used from the DOS command line
2	04H	System	Specifies a file as a system file
3	08H	Volume	Specifies the disk volume label
4	10H	Subdirectory	Specifies a subdirectory name
5	20H	Archive	Indicates that a file has been changed and should be archived

TABLE 11–2 The first five file handles

Handle	*Function*
0000H	Reads the keyboard (CON:)
0001H	Display on the video display (CON:)
0002H	Error display (CON:)
0003H	Uses the COM1 port (AUX:)
0004H	Uses the printer port (PRN:)

using the ASCII-Z string name of the file, speeding the operation for subsequent file accesses.

In DOS, you may use as many file handles as necessary, but the number is normally restricted to 50 or fewer. File handles specify newly created files or opened files. When a file is closed, the file handle is released for the next creation or open. File handles also specify the common I/O devices connected to the personal computer. The first five file handles, and the devices they address, are listed in Table 11–2. If necessary, data can be sent to the video display through file handle 0001H, or the keyboard can be read using file handle 0000H. File handles are not normally used for the keyboard and display, but they could be.

Writing to a File

Now that a new file, called FILE.NEW, has been created in Example 11–2, data can be written to it. Before writing to a file, you must create or open one. When a file is created or opened, the file handle returns in the AX register. The file handle is used to refer to the file whenever data are written.

Function number 40H writes data to an opened or newly created file. In addition to loading a 40H into AH, BX is loaded with the file handle, CX is loaded with the number of bytes to be written, and DS:DX is loaded with the address of the memory buffer data area to be written to the disk.

EXAMPLE 11–3

```
                              .        .
                              .        .
                              .        .
0010  8B D8                   MOV      BX,AX               ;move handle to BX
0012  B4 40                   MOV      AH,40H              ;load write function
0014  B9 0100                 MOV      CX,256              ;load count
0017  BA 0009 R               MOV      DX,OFFSET BUFFER    ;address BUFFER
001A  CD 21                   INT      21H                 ;write 256 bytes from BUFFER

001C  72 32                   JC       ERROR1              ;on write error
                              .        .
                              .        .
                              .        .
```

Suppose that 256 bytes are written from data segment memory area BUFFER to a file. This is accomplished, as illustrated in Example 11–3, using function 40H. If an error occurs during a write operation, the carry flag is set. If no error occurs, the carry flag is cleared and the number of bytes written to the file are returned in the AX register. Errors that occur during write operations usually indicate that the disk is full or that there is some type of media error. Media errors occur for a bad disk (floppy) or if there is a problem with the disk electronics. They may also occur if the disk drive door is left open or no disk is placed into the drive.

Opening, Reading, and Closing a File

To read an existing file, it first must be opened. When a file is opened, DOS checks the directory to determine whether the file exists and returns the DOS file handle in register AX. The DOS file handle must be used for reading, writing, and closing a file.

EXAMPLE 11–4

```
                                ;a program that opens the file TEMP.ASM and reads the first
                                ;256 bytes into an area of memory called BUF
                                ;
                                        .MODEL SMALL
0000                                    .DATA
0000  54 45 4D 50       FILEN   DB      'TEMP.ASM',0          ;file name
      2E 41 53 4D
      00
0009  0100 [            BUF     DB      256 DUP (?)           ;buffer area
                 00
                ]
0000                                    .CODE
                                        .STARTUP
0017  B8 3D02                           MOV   AX,3D02H              ;open file function
001A  BA 0000 R                         MOV   DX,OFFSET FILEN       ;address file name
001D  CD 21                             INT   21H                   ;access DOS
001F  8B D8                             MOV   BX,AX                 ;file handle to BX

0021  B4 3F                             MOV   AH,3FH                ;read file function
0023  B9 0100                           MOV   CX,256                ;read 256 bytes
0026  BA 0009 R                         MOV   DX,OFFSET BUF         ;store data at BUF
0029  CD 21                             INT   21H                   ;access DOS

002B  B4 3E                             MOV   AH,3EH                ;close file function
002D  CD 21                             INT   21H                   ;access DOS
                                        .EXIT
                                        END
```

Example 11–4 shows a program that opens a file, reads 256 bytes from the file into memory area BUF, and then closes the file. When a file is opened (AH = 3DH), the AL register specifies the type of operation allowed for the opened file. If AL = 00H, the file is opened for a read operation; if AL = 01H, the file is opened for a write operation; and if AL = 02H, the file is opened for a read or a write operation. In this example it is opened for a read/write operation.

Function number 3FH causes a file to be read. BX contains the file handle; CX contains the number of bytes to be read; and DS:DX contains the location of a memory area where the data are stored. As with all disk functions, the carry flag indicates an error when C = 1. If C = 0, the AX register indicates the number of bytes read from the file. In this example, the errors are ignored by not testing the carry flag after the INT 21H instructions. That's what is done in this example, in order to shorten it.

Closing a file is very important. If a file is left open, some serious problems can occur that may actually destroy the disk and all its data. If a file is written and not closed, the FAT can become corrupted, making it difficult or impossible to retrieve data from the disk. Always be certain to close a file after it is read or written. If you suspect that a file has been written but not closed, execute the DOS utility program CHKDSK (check disk) before any subsequent write to the disk. The CHKDSK program tests the disk and looks for lost file chains. If it detects a lost file chain (usually caused by forgetting to close a file), it can correct the problem. Run the CHKDSK /F command at the DOS prompt to fix any lost file chains. You can also use the SCANDISK program with DOS version 6.X. Be careful to never use CHKDSK from Windows.

The File Pointer

Whenever a file is opened, written, or read, the file pointer addresses the current location in the sequential file. When a file is opened, the file pointer always addresses the first byte of the file. If a file is 1,024 bytes in length, and a read function (3FH) reads 1,023 bytes, the file pointer addresses the last byte of the file, but not the end of the file.

The **file pointer** is a 32-bit number that addresses any byte in a file. Once a file is opened, the file pointer can be changed with the move file pointer function number 42H to access any location in the file. A file pointer can be moved from the start of the file (AL = 00H), from the current location (AL = 01H), or from the end of the file (AL = 02H). In practice, all three directions are used to access different parts of the file. The distance moved by the file pointer is specified by registers CX and DX. The DX register holds the least significant part of the distance, and the CX register holds the most significant part of the distance. Register BX must contain the file handle before using function 42H to move the file pointer.

Suppose a file exists on the disk and you must append the file with 256 bytes of new information. When the file is opened, the file pointer addresses the first byte of the file. If you attempt to write new data without moving the file pointer to the end of the file, the new data will overwrite the first 256 bytes of the file. Example 11–5 shows a program that opens a file, moves the file pointer to the end of the file, writes 256 bytes of data, and then closes the file. This *appends* the file with 256 new bytes of data.

EXAMPLE 11–5

```
                                ;a program that opens FILE.NEW and appends it with 256 bytes
                                ;of data from BUF
                                ;
                                        .MODEL SMALL
0000                                    .DATA
0000 46 49 4C 45        FILEN   DB      'FILE.NEW',0            ;file name
     2E 4E 45 57
     00
0009 0100 [             BUF     DB      256 DUP (?)             ;buffer
               00
                   ]
0000                                    .CODE
                                        .STARTUP
0017 B8 3D02                            MOV     AX,3D02H        ;open FILE.NEW
001A BA 0000 R                          MOV     DX,OFFSET FILEN
001D CD 21                              INT     21H
```

```
001F 8B D8               MOV   BX,AX

0021 B8 4202             MOV   AX,4202H          ;move file pointer to end of file
0024 BA 0000             MOV   DX,0
0027 B9 0000             MOV   CX,0
002A CD 21               INT   21H

002C B4 40               MOV   AH,40H            ;write BUF to end of file
002E B9 0100             MOV   CX,256
0031 BA 0009 R           MOV   DX,OFFSET BUF
0034 CD 21               INT   21H

0036 B4 3E               MOV   AH,3EH            ;close file
0038 CD 21               INT   21H
                         .EXIT
                         END
```

One of the more difficult file maneuvers is inserting new data into the middle of a file. Figure 11–1 shows how this is accomplished by creating a second temporary file. The part of the old file *before* the insertion point is copied into the new file. This is followed by the new information, then the remainder of the old file is appended to the end of the new file. Once the new file is complete, the old file is deleted and the new file is renamed to the old file name.

Example 11–6 shows a program that inserts new data into an old file. The new data comes from the DATA.NEW file. This information, 256 bytes in length, is added between the first 256 bytes of file DATA.OLD and the remainder of the file. This is accomplished by creating a new file on the disk called DATA.TMP. The first 256 bytes of DATA.OLD are read and stored in DATA.TMP. Next, the 256 bytes from DATA.NEW are read and stored in DATA.TMP. Then the remainder of DATA.OLD is read and stored in DATA.TMP in 256 byte sections until no more data is available. Finally, the DATA.OLD file is deleted and the DATA.TMP file is renamed DATA.OLD. This is a fairly long program, but it is clearly listed and commented so it can be understood.

EXAMPLE 11–6

```
                         ;a program that adds the 256 byte contents of the file DATA.NEW
                         ;to DATA.OLD at a point between the first 256 bytes of DATA.OLD
                         ;and the remainder of the file.
                         ;
                                 .MODEL SMALL
0000                             .DATA
0000  0000               HAN1    DW     ?              ;file handle for DATA.TMP
0002  0000               HAN2    DW     ?              ;file handle for DATA.OLD
0004  44 41 54 41        FILE1   DB     'DATA.TMP',0
      2E 54 4D 50
      00
000D  44 41 54 4         FILE2   DB     'DATA.OLD',0
      2E 4F 4C 44
      00
0016  44 41 54 4         FILE3   DB     'DATA.NEW',0
      2E 4E 45 57
      00
001F  0100 [             BUF     DB     256 DUP (?)    ;data buffer area
                  00
```

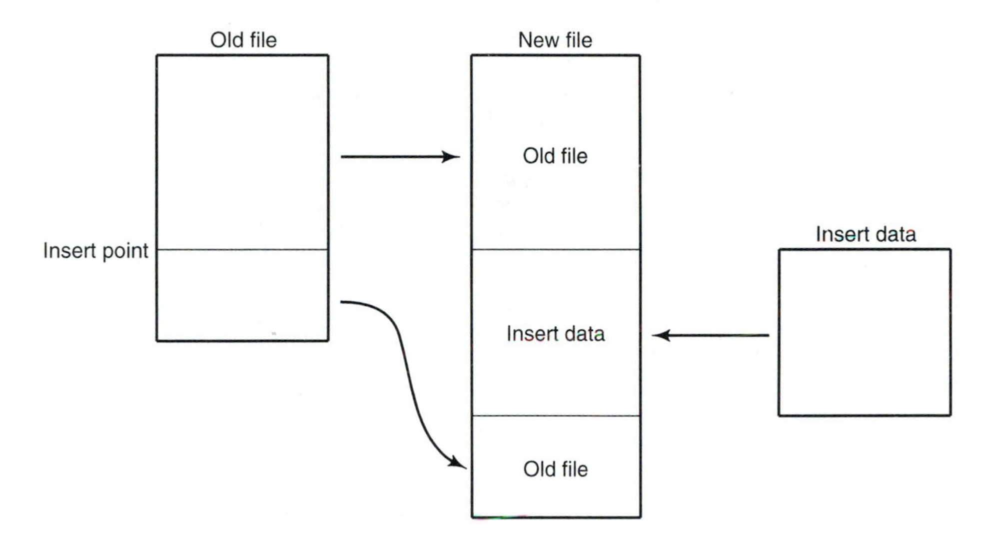

FIGURE 11–1 Inserting new data within an old file.

```
0000                     .CODE
                         .STARTUP
0017  B4 3C              MOV    AH,3CH          ;create DATA.TMP
0019  B9 0000            MOV    CX,0
001C  BA 0004 R          MOV    DX,OFFSET FILE1
001F  CD 21              INT    21H
0021  A3 0000 R          MOV    HAN1,AX         ;save handle at HAN1

0024  B8 3D02            MOV    AX,3D02H        ;open DATA.OLD
0027  BA 000D R          MOV    DX,OFFSET FILE2
002A  CD 21              INT    21H
002C  8B D8              MOV    BX,AX
002E  A3 0002 R          MOV    HAN2,AX         ;save handle at HAN2

0031  B4 3F              MOV    AH,3FH          ;read 256 bytes of DATA.OLD into BUF
0033  B9 0100            MOV    CX,256
0036  BA 001F R          MOV    DX,OFFSET BUF
0039  CD 21              INT    21H

003B  B4 40              MOV    AH,40H          ;write BUF to DATA.TMP
003D  8B 1E 0000 R       MOV    BX,HAN1         ;get handle
0041  B9 0100            MOV    CX,256
0044  BA 001F R          MOV    DX,OFFSET BUF
0047  CD 21              INT    21H

0049  B8 3D02            MOV    AX,3D02H        ;open DATA.NEW
004C  BA 0016 R          MOV    DX,OFFSET FILE3
004F  CD 21              INT    21H
0051  8B D8              MOV    BX,AX

0053  B4 3F              MOV    AH,3FH          ;read 256 bytes from DATA.NEW to BUF
0055  B9 0100            MOV    CX,256
```

```
0058  BA 001F R              MOV   DX,OFFSET BUF
005B  CD 21                  INT   21H

005D  B4 3E                  MOV   AH,3EH          ;close DATA.NEW
005F  CD 21                  INT   21H

0061  B4 40                  MOV   AH,40H          ;write BUF to DATA.TMP
0063  8B 1E 0000 R           MOV   BX,HAN1         ;get handle
0067  B9 0100                MOV   CX,256
006A  BA 001F R              MOV   DX,OFFSET BUF
006D  CD 21                  INT   21H
006F                  MAIN1:
006F  B4 3F                  MOV   AH,3FH          ;read 256 bytes from DATA.OLD to BUF
0071  8B 1E 0002 R           MOV   BX,HAN2
0075  B9 0100                MOV   CX,256
0078  BA 001F R              MOV   DX,OFFSET BUF
007B  CD 21                  INT   21H
007D  0B C0                  OR    AX,AX           ;test for zero byte read
007F  74 10                  JZ    MAIN2           ;if file empty

0081  B4 40                  MOV   AH,40H          ;write BUF to DATA.TMP
0083  8B 1E 0000 R           MOV   BX,HAN1
0087  B9 0100                MOV   CX,256
008A  BA 001F R              MOV   DX,OFFSET BUF
008D  CD 21                  INT   21H
008F  EB DE                  JMP   MAIN1
0091                  MAIN2:
0091  B4 3E                  MOV   AH,3EH          ;close DATA.OLD
0093  CD 21                  INT   21H

0095  B4 41                  MOV   AH,41H          ;delete DATA.OLD
0097  BA 000D R              MOV   DX,OFFSET FILE2
009A  CD 21                  INT   21H

009C  B4 3E                  MOV   AH,3EH          ;close DATA.TMP
009E  8B 1E 0000 R           MOV   BX,HAN1
00A2  CD 21                  INT   21H

00A4  8C D8                  MOV   AX,DS
00A6  8E C0                  MOV   ES,AX           ;overlap DS and ES

00A8  B4 56                  MOV   AH,56H          ;rename DATA.TMP to DATA.OLD
00AA  BA 0004 R              MOV   DX,OFFSET FILE1 ;old name
00AD  BF 000D R              MOV   DI,OFFSET FILE2 ;new name
00B0  CD 21                  INT   21H
                             .EXIT
                             END
```

This program uses two new INT 21H function calls. The delete (41H) and rename (56H) function calls are used to delete the old file before the temporary file is renamed to the old file name. These functions, and all other DOS file functions, are listed in Appendix A for reference.

Another method exists for creating a temporary file for this program. DOS function 5AH creates a unique file name that can be used as the name of a temporary file. Upon entry, CX = attribute code and DS:DX is an ASCII-Z string that contains the path name ending with a backslash (\). Upon exit, DOS returns a unique file name appended to the end of the path. It also creates and opens this file.

11–2 RANDOM ACCESS FILES

Random access files are developed through software. A random access file is a DOS sequential access file, indexed by record numbers to represent a random access file. A random access file is addressed by a record number, rather than by passing through the file searching for data. The move pointer function is very important when random access files are created and a record is addressed. Random access files are much easier to use for large volumes of data than sequential access files. The only disadvantage to a random access file is that it requires more room on the disk, because it often contains many empty records.

Creating a Random Access File

Planning is paramount when creating a random access file system. Suppose a random access file is required for storing the names of customers and other information about customers. Each customer record requires 16 bytes for the last name, 16 bytes for the first name, and 1 byte for the middle initial. Each customer record contains two street address lines of 32 bytes each, a city line of 16 bytes, 2 bytes for the state code, and 9 bytes for the ZIPCode. So, the basic customer information requires 105 bytes. Additional information expands the record to 256 bytes in length. Because the business is growing, provisions are made for 5,000 customers. This means that the total required random access file is 1,280,000 bytes in length.

EXAMPLE 11–7

```
                        ;a program that creates CUST.FIL and then fills 5,000 records of
                        ;256 bytes each with zeros.
                        ;
                                .MODEL SMALL
0000                            .DATA
0000  43 55 53 5      FILE1     DB      'CUST.FIL',0  ;file name
      2E 46 49 4C
      00
0009  0100 [          BUF       DB      256 DUP (0)   ;buffer
          00
            ]
0000                            .CODE
                                .STARTUP
0017  B4 3C                     MOV  AH,3CH          ;create CUST.FIL
0019  B9 0000                   MOV  CX,0
001C  BA 0000 R                 MOV  DX,OFFSET FILE1
001F  CD 21                     INT  21H
0021  8B D8                     MOV  BX,AX           ;handle to BX

0023  BD 1388                   MOV  BP,5000         ;record counter
0026                  MAIN1:
0026  B4 40                     MOV  AH,40H          ;write record
0028  B9 0100                   MOV  CX,256
002B  BA 0009 R                 MOV  DX,OFFSET BUF
002E  CD 21                     INT  21H
0030  4D                        DEC  BP              ;decrement record count
0031  75 F3                     JNZ  MAIN1           ;if 5000 records have not been written
```

```
0033  B4 3E                MOV  AH,3EH          ;close file
0035  CD 21                INT  21H
                           .EXIT
                           END
```

Example 11–7 illustrates a short program that creates a file called CUST.FIL and inserts 5,000 blank records of 256 bytes each. A blank record contains 00H in each of its bytes. This appears to be a large file, but it fits on a single high-density $5^1/4''$ or $3^1/2''$ floppy disk drive. This program takes a considerable amount of time to execute, because it writes data to virtually every byte on the floppy disk. If executed to create a file on a hard disk drive, the program takes much less time.

Reading and Writing a Record

Whenever a record is read, the record number is loaded into the BP register before the procedure listed in Example 11–8 is called. This procedure (READ) assumes that FIL contains the file handle number and that CUST.FIL remains open at all times. An example might be the database in a video tape rental store. As long as the store is open, the CUST.FIL stays open for access to customer records.

EXAMPLE 11–8

```
                    ;the READ procedure reads one record from the opened CUST.FIL
                    ;Input parameters are:
                    ;FIL (word) = CUST.FIL handle
                    ;BP = record number
                    ;Output parameters are:
                    ;BUFFER (256 bytes) = customer record
                    ;
0000                READ   PROC   FAR

0000 8B 1E 0100 R   MOV    BX,FIL                   ;get handle
0004 B8 0100               MOV    AX,256            ;multiply by 256
0007 F7 E5                 MUL    BP
0009 8B CA                 MOV    CX,DX
000B 8B D0                 MOV    DX,AX
000D B8 4200               MOV    AX,4200H          ;move pointer
0010 CD 21                 INT    21H

0012 B4 3F                 MOV    AH,3FH            ;read record
0014 B9 0100               MOV    CX,256
0017 BA 0000 R             MOV    DX,OFFSET BUFFER
001A CD 21                 INT    21H
001C CB                    RET

001D                READ   ENDP
```

The record number is multiplied by 256 to obtain a count for the move pointer function. In each case, the file pointer is moved from the start of the file to the desired record before it is read into memory area BUFFER. Writing a record is performed in the same manner as reading. Note that DOS does not read the disk information for the move file pointer. The move file pointer causes the disk read\write head to be positioned over the area that contains the information. This operation is called a *seek*.

Cataloging Random Access Files

Obviously, it takes considerable time to search each record for a customer name or address. Suppose a second file called CATNAME is created to hold all the customer names in the random access file and the record number where the data for each customer name appears. This short file can be searched quickly, improving search times when a customer must be found in the database. The same is true for the address or any other field in the database. This type of file is often called an *index file*.

Example 11–7 used a database with 5,000 entries of 256 bytes each. The database is stored in a huge file that is 5000 × 256, or 1,280,000, bytes in length. A catalog file called CATNAME requires a second file of 5000 × 34, or 170,000, bytes of disk space, but requires much less time to load and search than the entire database. The entire CATNAME file could be loaded into memory and be searched in one operation, which takes little time. Each entry contains 32 bytes for the customer name and 2 bytes for the record number. Another advantage of the CATNAME catalog is that is can be sorted in alphabetical order without moving any of the records in the database, which saves a tremendous amount of time if an alphabetic list is needed.

Example 11–9 illustrates a procedure that opens, reads, and searches the CATNAME file. The name to be located in CATNAME is transferred to the procedure in the DS:SI register, and the name of the CATNAME file is stored at address DS:DX, which is also transferred to the procedure. A buffer area for file storage must be 8,192 bytes, whose address is passed to the procedure in DS:BP. Each entry in CATNAME and the name to be searched for are padded with 00H, if the name is not 32 bytes in length. The last two bytes of each entry in CATNAME contain the record number, which is returned to the calling program in the BP register with carry cleared if the name is found. If the name is not found, a return with carry set occurs.

EXAMPLE 11–9

```
                        ;CALLING PARAMETERS: DS:BP = buffer address (8192 bytes)
                        ;DS:SI = search name address, and DS:DX = file name address
                        ;RETURN PARAMETER: BP = record number
                        ;
                                PUBLIC SEARCH              ;declare search public

0000                    SEARCH  PROC   FAR

0000  B8 3D02                   MOV    AX,3D02H            ;open file
0003  CD 21                     INT    21H
0005  72 43                     JC     SEARCH5             ;if error
0007  8B D8                     MOV    BX,AX               ;get handle
0009  06                        PUSH   ES                  ;make ES = DS
000A  8C D8                     MOV    AX,DS
000C  8E C0                     MOV    ES,AX
000E                    SEARCH1:
000E  B4 3F                     MOV    AH,3FH
0010  B9 2000                   MOV    CX,8192
0013  8B D5                     MOV    DX,BP
0015  CD 21                     INT    21H                 ;read file (8192 bytes)
0017  72 2B                     JC     SEARCH4             ;if error
0019  83 F8 00                  CMP    AX,0
001C  74 26                     JE     SEARCH4             ;if no more data
```

```
001E  B1 22                 MOV    CL,34
0020  F6 F1                 DIV    CL                    ;how many records are read?
0022  8A D0                 MOV    DL,AL                 ;save it as count
0024  8B FD                 MOV    DI,BP
0026               SEARCH2:
0026  56                    PUSH   SI
0027  57                    PUSH   DI
0028  B9 0020               MOV    CX,32                 ;record length
002B  F3/ A6                REPE   CMPSB                 ;compare for name
002D  E3 0B                 JCXZ   SEARCH3               ;if found
002F  5F                    POP    DI
0030  5E                    POP    SI
0031  83 C7 22              ADD    DI,34                 ;go to next record
0034  FE CA                 DEC    DL                    ;decrement count
0036  75 EE                 JNZ    SEARCH2               ;look again
0038  EB D4                 JMP    SEARCH1               ;read more of file
003A               SEARCH3:
003A  8B 2D                 MOV    BP,[DI]               ;get record number
003C  5F                    POP    DI
003D  5E                    POP    SI                    ;clear stack
003E  B4 3E                 MOV    AH,3EH
0040  CD 21                 INT    21H                   ;close file
0042  07                    POP    ES                    ;restore ES
0043  CB                    RET
0044               SEARCH4:
0044  B4 3E                 MOV    AH,3EH                ;close file
0046  CD 21                 INT    21H
0048  F9                    STC                          ;show error
0049  07                    POP    ES                    ;restore ES
004A               SEARCH5:
004A  CB                    RET

004B               SEARCH   ENDP
```

11–3 THE DISK DIRECTORY

So far, files and the manipulation of file data have been discussed. Another feature of the disk memory system is the directory, which is also controlled using DOS function INT 21H calls. Functions used for directory control are: create subdirectory (39H), erase subdirectory (3AH), change subdirectory (3BH), read current directory (47H), select default disk drive (0EH), find first matching file (4EH), and find next matching file (4FH). Although the last two functions do not change a directory, they are used to search a directory for a file name.

Using the Directory Commands

To illustrate how some of the directory functions are used, this part of the text develops a program that changes to the root directory of drive C and then creates the subdirectory name typed at the command line. This requires the use of the change directory function and also the create subdirectory function. The program allows the use of the command line to place parameters into a program via the PSP (program segment prefix). This is important because many programs use command line parameters.

EXAMPLE 11–10

```
                    ;
                    ;The EX procedure copies the first parameter from the command line
                    ;into the location addressed by ES:DI.
                    ;***Input parameters***
                    ;ES:DI = address for storage of command line parameter
                    ;DS = segment of PSP
                    ;***Output parameters***
                    ;Carry = 0 for normal return
                    ;Carry = 1 if no parameter is found
                    ;
0000                EX    PROC    NEAR

0000  BE 0081             MOV     SI,81H              ;address command line

                          .REPEAT                     ;skip leading spaces and tabs
0003  AC                          LODSB
                          .UNTIL  AL > ' ' || AL == 13

000C  3C 0D               CMP     AL,13               ;test for enter
000E  F9                  STC                         ;set no parameter error
000F  74 1A               JE      EX1                 ;if enter
0011  4E                  DEC     SI                  ;adjust pointer

                          .REPEAT                     ;store uppercase parameter
0012  AC                          LODSB
                                  .IF AL >= 'a' && AL <= 'z'
001B  2C 20                         SUB AL,20H
                                  .ENDIF
001DAA                              STOSB
                          .UNTIL    AL == 13 || AL <= ' ' || AL == 2CH

0026  4F                  DEC     DI                  ;adjust pointer
0027  B0 00               MOV     AL,0                ;make it an ASCII-Z string
0029  AA                  STOSB
002A  F8                  CLC                         ;indicate parameter
002B                EX1:
002B  C3                  RET

002C                EX    ENDP
```

Example 11–10 shows a procedure that extracts a parameter from the command line. This example procedure uses two .REPEAT-.UNTIL loops. The first loop skips over any ASCII character that is one space or less, except for enter. The second loop stores uppercase characters from the command line in memory addressed by ES:DI and is terminated by an enter (13), comma (2CH), or any character equal to or below a space. Fortunately, DOS provides the command line information in the PSP. Refer to Appendix A, Figure A–5 for the contents of the PSP. The PSP appends the front of an execute file and is 256 bytes in length. When a program is started, the DS and ES segment registers both address the PSP. The command line is stored as it was typed, beginning at offset address 81H, with room for 127 characters. The command line is partially parsed at offsets 5CH and 6CH. The first word on the command line is placed in memory beginning at 5CH and the second, at 6CH. These areas were designed for use before DOS used subdirectories, and are only 16 bytes in length. Because there is not enough room to store any possible file name with subdirectories in these areas, they are seldom used.

FIGURE 11–2 The loader program (a part of DOS) places the program and its segment in memory following the program segment prefix.

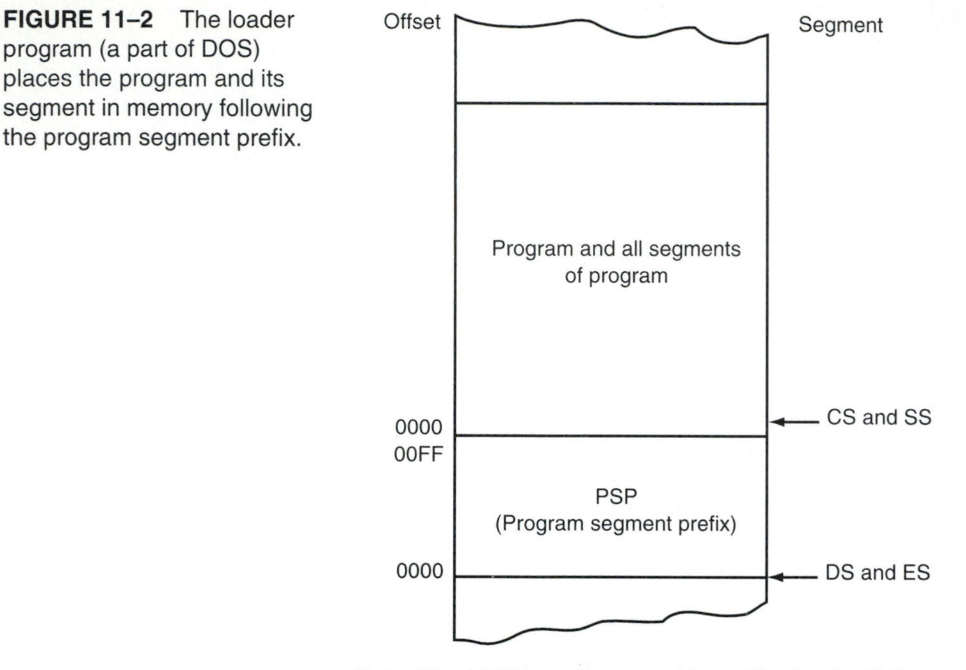

Refer to Figure 11–2 for the organization of the memory and the locations addressed by various segment and pointer registers when DOS executes a program. Notice that the CS and SS registers both address the start of the code segment that is placed in memory by the linker immediately following the PSP. This assumes that no stack segment has been designated in the program; otherwise the stack segment appears between the program and the PSP. Also note that the contents of SP are 0000H, and the contents of IP are loaded with the address specified by the END directive in the program.

The linker program stores the segments in memory exactly as they appear in the assembler file. In our examples, this is (1) stack, (2) data and extra, and (3) code. This can be changed, but it is usually not necessary. An option Microsoft uses with many assembly language programs is called the Microsoft DOS segment order. If you need to generate a program that follows the DOS segment order, place the .DOSSEG directive as the first line of the program. This causes segments of the program to be stored in the memory in the following order: (1) code, (2) data and extra segments, and (3) the stack segment.

The PSP contains whatever information was entered on the command line, beginning at offset address 81H. (Note that 81H almost always contains a 20H or space, but to make sure, start at offset 81H.) For example, if a program named FOO is executed at the DOS prompt as c:\>FOO BETTER memory location 82H contains the letter B (42H), location 83H contains the E (45H), and so forth. The last entry, following the letter R (52H), is a carriage return or 0DH. The case of the command prompt is preserved in the memory beginning at 81H.

Because the DS and ES registers both address the PSP when a program is started, it is fairly simple to extract the command line information beginning at offset address

81H. Example 11–10 extracts the command line parameter and stores it into the data segment location addressed by DI. Leading spaces are skipped. Note the end of the command line data at any space (20H), carriage return (0DH), comma (2CH), or tab (09H). If more than a single parameter exists, the procedure must be modified to extract additional parameters.

Example 11–11 illustrates a program that changes to drive C, selects the root directory, and creates a new directory. The program name is NEWSUB, and it is followed by the name of the directory created in the root. An example is NEWSUB FROG, which creates the FROG directory in the root directory.

EXAMPLE 11–11

```
                        ;a program that creates a new directory in the root of drive C
                        ;the name of the directory is a parameter
                        ;
                               .MODEL SMALL
0000                           .DATA
0000  0080 [            FILE1  DB      128 DUP (0)        ;space for parameter
               00
                   ]
0080  43 3A 5C 00       ROOTS  DB      ' C:\',0           ;root directory
0000                           .CODE
                               .STARTUP
0017  1E                       PUSH    DS                 ;exchange ES with DS
0018  06                       PUSH    ES
0019  1F                       POP     DS
001A  07                       POP     ES
001B  BF 0000 R                MOV     DI,OFFSET FILE1    ;address FILE1
001E  E8 001E                  CALL    EX                 ;get command line parameter
0021  72 18                    JC      ERR                ;if no parameter
0023  8C C0                    MOV     AX,ES
0025  8E D8                    MOV     DS,AX              ;make ES and DS overlap

0027  B4 0E                    MOV     AH,0EH             ;log to drive C
0029  B2 02                    MOV     DL,2               ;select drive C (2)
002B  CD 21                    INT     21H

002D  B4 3B                    MOV     AH,3BH             ;log to root
002F  BA 0080 R                MOV     DX,OFFSET ROOTS
0032  CD 21                    INT     21H

0034  B4 39                    MOV     AH,39H             ;make new directory
0036  BA 0000 R                MOV     DX,OFFSET FILE1
0039  CD 21                    INT     21H
003B                    ERR:
                               .EXIT
                        ;
                        ;The EX procedure copies the first parameter from the command line
                        ;into the location addressed by ES:DI.
                        ;***Input parameters***
                        ;ES:DI = address for storage of command line parameter
                        ;DS = segment of PSP
                        ;***Output parameters***
                        ;Carry = 0 for normal return
                        ;Carry = 1 if no parameter is found
                        ;
003F                    EX     PROC    NEAR
003F  BE 0081                  MOV     SI,81H             ;address command line
```

```
                              .REPEAT
0042  AC                               LODSB
                              .UNTIL   AL > ' ' || AL == 13

004B  3C 0D                   CMP      AL,13                ;test for enter
004D  F9                      STC
004E  74 1E                   JE       EX1                  ;if enter
0050  4E                      DEC      SI                   ;adjust pointer

                              .REPEAT

0051  AC                               LODSB
                                       .IF AL >= 'a' && AL <= 'z'
005A  2C 20                               SUB AL,20H
                                       .ENDIF
005C  AA                               STOSB
                              .UNTIL   AL == 13 || AL <= ' ' || AL == 2CH

0069  4F                      DEC      DI                   ;adjust pointer
006A  B0 00                   MOV      AL,0                 ;make it an ASCII-Z string
006C  AA                      STOSB
006D  F8                      CLC
006E                   EX1:
006E  C3                      RET

006F                   EX     ENDP
                              END
```

11–4 USING THE MOUSE

The mouse pointing device is controlled with INT 33H. Refer to Appendix A for a list of the Microsoft compatible mouse functions associated with INT 33H. With INT 33H, the function number is selected through the AL register and AH is usually set to 00H. There are a total of fifty mouse functions available; only the main functions are described in this section of the text.

Testing for a Mouse

To determine whether a mouse driver is installed in the system, test the contents of interrupt vector 33H. If interrupt vector 33H contains a 0000:0000, a mouse driver is not installed in the system. In some systems, a vector exists even though no mouse driver is present. In this instance, the INT 33H vector address points to an IRET instruction (CFH). The interrupt vector address is retrieved by using the DOS INT 21H function 35H. The address is then tested for 0000:0000 and if it contains another value, the contents of the address pointed to by interrupt vector 33H are tested for CFH. Refer to Example 11–12 for a procedure that tests for the existence of a mouse driver.

Once it is determined that a mouse driver exists, the mouse is reset to make certain it is connected to the system and functioning. The mouse reset is accomplished using mouse function 00H. The return from function 00H is AX = 0000H if no mouse is present. The CHKM procedure returns with carry cleared if the mouse exists and with carry set if no mouse exists.

EXAMPLE 11–12

```
                        ;procedure that tests for the presence of a mouse driver
                        ;***Output parameters***
                        ;Carry = 1, if no mouse present
                        ;Carry = 0, if mouse is present
                        ;
0000                    CHKM  PROC   NEAR

0000 B8 3533                  MOV    AX,3533H             ;get INT 33H vector
0003 CD 21                    INT    21H                  ;returns vector in ES:BX

0005 8C C0                    MOV    AX,ES
0007 0B C3                    OR     AX,BX                ;test for 0000:0000
0009 F9                       STC                         ;indicate no mouse
000A 74 13                    JZ     CHKM1                ;if no mouse driver

000C 26: 80 3F CF             CMP    BYTE PTR ES:[BX],0CFH
0010 F9                       STC
0011 74 0C                    JE     CHKM1                ;if no mouse driver

0013 B8 0000                  MOV    AX,0
0016 CD 33                    INT    33H                  ;reset mouse
0018 83 F8 00                 CMP    AX,0
001B F9                       STC
001C 74 01                    JZ     CHKM1                ;if no mouse
001E F8                       CLC
001F                    CHKM1:
001F C3                       RET

0020                    CHKM  ENDP
```

Which Mouse and Driver?

The mouse function interrupt determines the type of mouse connected to the system and also the driver version number. Example 11–13 lists a program that displays the mouse type and driver version number after it is determined that a mouse is present. Mouse INT 33H function 24H locates the mouse driver version number and mouse driver type. The return from function 24H leaves the mouse driver number in BX (BH=major and BL=minor) and the mouse type in CH. If the mouse driver version is 8.00, then BH = 08H and BL = 00H. The mouse types that are returned in register CH are currently: bus = 1, serial = 2, InPort = 3, PS/2 = 4, and Hewlett-Packard = 5. As time passes, this list of mouse types may grow.

EXAMPLE 11–13

```
                                   ;a program that displays the mouse driver version number and
                                   ;the type of mouse installed.
                                   ;
                                             .MODEL SMALL
0000                                         .DATA
0000   0D 0A 4E 6F 20 4D          MES1      DB     13,10,'No MOUSE or MOUSE DRIVER found.$'
       4F 55 53 45 20 6F
       72 20 4D 4F 55 53
       45 20 44 52 49 56
       45 52 20 66 6F 75
       6E 64 2E 24
```

```
0022  0D 0A 4D 6F 75 73    MES2   DB     13,10,'Mouse driver version'
      65 20 64 72 69 76
      65 72 20 76 65 72
      73 69 6F 6E 20
0039  20 20 20 20 20 20    M1     DB     '      ',13,10,'$'
      20 0D 0A 24
0043  004D R 0051 R        TYPES  DW     T1,T2,T3,T4,T5
      0058 R 005F R
      0064 R
004D  42 75 73 24          T1     DB     'Bus$'
0051  53 65 72 69 61 6C    T2     DB     'Serial$'
      24
0058  49 6E 50 6F 72 74    T3     DB     'InPort$'
      24
005F  50 53 2F 32 24       T4     DB     'PS/2$'
0064  48 50 24             T5     DB     'HP$'
0067  20 6D 6F 75 73 65    MES3   DB     ' mouse installed.',13,10,'$'
      20 69 6E 73 74 61
      6C 6C 65 64 2E 0D
      0A 24
0000                              .CODE
                                  .STARTUP
0017  E8 0041                     CALL   CHKM                 ;test for mouse
001A  73 05                       JNC    MAIN1                ;if mouse present
001C  BA 0000 R                   MOV    DX,OFFSET MES1
001F  EB 32                       JMP    MAIN2                ;if no mouse
0021                       MAIN1:
0021  B8 0024                     MOV    AX,24H
0024  CD 33                       INT    33H                  ;get driver version and type
0026  BF 0039 R                   MOV    DI,OFFSET M1
0029  8A C7                       MOV    AL,BH                ;save ASCII major version
002B  E8 004D                     CALL   DISP
002E  C6 05 2E                    MOV    BYTE PTR [DI],'.'    ;save period
0031  47                          INC    DI

0032  8A C3                       MOV    AL,BL                ;save ASCII minor version
0034  E8 0044                     CALL   DISP

0037  BA 0022 R                   MOV    DX,OFFSET MES2       ;display version number
003A  B4 09                       MOV    AH,9
003C  CD 21                       INT    21H

003E  BE 0043 R                   MOV    SI,OFFSET TYPES      ;index type
0041  B4 00                       MOV    AH,0
0043  8A C5                       MOV    AL,CH
0045  48                          DEC    AX
0046  03 F0                       ADD    SI,AX
0048  03 F0                       ADD    SI,AX
004A  8B 14                       MOV    DX,[SI]              ;display type
004C  B4 09                       MOV    AH,9
004E  CD 21                       INT    21H
0050  BA 0067 R                   MOV    DX,OFFSET MES3
0053                       MAIN2:
0053  B4 09                       MOV    AH,9
0055  CD 21                       INT    21H
                                  .EXIT
                                  ;procedure that tests for the presence of a mouse driver
                                  ;***Output parameters***
                                  ;Carry = 1, if no mouse present
                                  ;Carry = 0, if mouse is present
```

```
                             ;
005B                         CHKM    PROC    NEAR

005B  B8 3533                        MOV     AX,3533H           ;get INT 33H vector
005E  CD 21                          INT     21H                ;returns vector in ES:BX

0060  8C C0                          MOV     AX,ES
0062  0B C3                          OR      AX,BX              ;test for 0000:0000
0064  F9                             STC
0065  74 13                          JZ      CHKM1              ;if no mouse driver
0067  26: 80 3F CF                   CMP     BYTE PTR ES:[BX],0CFH
006B  F9                             STC
006C  74 0C                          JE      CHKM1              ;if no mouse driver
006E  B8 0000                        MOV     AX,0
0071  CD 33                          INT     33H                ;reset mouse
0073  83 F8 00                       CMP     AX,0
0076  F9                             STC
0077  74 01                          JZ      CHKM1              ;if no mouse
0079  F8                             CLC
007A                         CHKM1:
007A  C3                             RET

007B                         CHKM    ENDP
                             ;
                             ;save the ASCII coded version number
                             ;***input parameters***
                             ;AL = version
                             ;DS:DI = address where stored
                             ;***output parameters***
                             ;ASCII version number stored at DS:DI
                             ;
007B                         DISP    PROC    NEAR

007B  B4 00                          MOV     AH,0
007D  D4 0A                          AAM                        ;convert to BCD
007F  05 3030                        ADD     AX,3030H
0082  80 FC 30                       CMP     AH,30H             ;save ASCII version
0085  74 03                          JE      DISP1              ;suppress leading zero
0087  88 25                          MOV     [DI],AH
0089  47                             INC     DI
008A                         DISP1:
008A  88 05                          MOV     [DI],AL
008C  47                             INC     DI
008D  C3                             RET

008E                         DISP    ENDP
                                     END
```

Using the Mouse

The mouse functions in either the text or graphics mode. This section illustrates how to enable the mouse for use with a text mode program. It also functions in a graphic mode, but, in the graphics mode, an arrow is displayed as the cursor, rather than a block or mouse pointer. The first step is to check for the presence of a mouse driver. Example 11–14 uses the CHKM procedure to test for the presence of a mouse. If no mouse is present, a return from TM_ON occurs with the carry flag set. If a mouse is present, a return from TM_ON occurs with the carry flag cleared.

EXAMPLE 11–14

```
                        ;the TM_ON procedure tests for the presence of a mouse
                        ;and enables the mouse pointer.
                        ;uses the CHKM (check for mouse) procedure
                        ;
                        ;***output parameters***
                        ;Carry = 0, if mouse is present pointer enabled
                        ;Carry = 1, if no mouse present
                        ;
0000                    TM_ON  PROC   NEAR

0000  E8 FFDD                  CALL   CHKM            ;test for mouse
0003  72 06                    JC     TM_ON1          ;if no mouse
0005  B8 0001                  MOV    AX,1            ;show mouse pointer
0008  CD 33                    INT    33H
000A  F8                       CLC                    ;show mouse present
000B                    TM_ON1:
000B  C3                       RET

000C                    TM_ON  ENDP
```

Example 11–14 only enables the mouse and displays the mouse cursor. To *use* the mouse, a program must be written that tracks the mouse and its position. Such a program appears in Example 11–15.

EXAMPLE 11–15

```
                             ;a program that displays the mouse pointer and its X and Y
                             ;position.
                             ;
                                    .MODEL SMALL
0000                                .DATA
0000  0D 58 20 50 6F 73      MES    DB     13,'X Position= '
      69 74 69 6F 6E 3D
      20
000D  20 20 20 20 20 20      MX     DB     '  '
0013  59 20 50 6F 73 69             DB     'Y Position= '
      74 69 6F 6E 3D 20
001F  20 20 20 20 20 20      MY     DB     ' $'
      24
0026  0000                   X      DW     ?                ;X position
0028  0000                   Y      DW     ?                ;Y position
0000                                .CODE
                                    .STARTUP
0017  E8 006D                       CALL   TM_ON            ;enable mouse
001A  72 47                         JC     MAIN4            ;if no mouse
001C                         MAIN1:
001C  B8 0003                       MOV    AX,3             ;get mouse status
001F  CD 33                         INT    33H
0021  83 FB 01                      CMP    BX,1
0024  74 38                         JE     MAIN3            ;if left button pressed

0026  3B 0E 0026 R                  CMP    CX,X
002A  75 06                         JNE    MAIN2            ;if X position changed
002C  3B 16 0028 R                  CMP    DX,Y
0030  74 EA                         JE     MAIN1            ;if Y position did not change
0032                         MAIN2:
0032  89 0E 0026 R                  MOV    X,CX             ;save new position
```

```
0036  89 16 0028 R                  MOV    Y,DX
003A  BF 000D R                     MOV    DI,OFFSET MX
003D  8B C1                         MOV    AX,CX
003F  E8 0051                       CALL   PLACE            ;store ASCII X
0042  BF 001F R                     MOV    DI,OFFSET MY
0045  A1 0028 R                     MOV    AX,Y
0048  E8 0048                       CALL   PLACE            ;store ASCII Y

004B  B8 0002                       MOV    AX,2
004E  CD 33                         INT    33H              ;hide mouse pointer

0050  B4 09                         MOV    AH,9
0052  BA 0000 R                     MOV    DX,OFFSET MES
0055  CD 21                         INT    21H              ;display position

0057  B8 0001                       MOV    AX,1
005A  CD 33                         INT    33H              ;show mouse pointer

005C  EB BE                         JMP    MAIN1            ;do again
005E                         MAIN3:
005E  B8 0000                       MOV    AX,0             ;reset mouse
0061  CD 33                         INT    33H

0063                         MAIN4:
                                    .EXIT
                                    ;
                                    ;procedure that tests for the presence of a mouse driver
                                    ;***Output parameters***
                                    ;Carry = 1, if no mouse present
                                    ;Carry = 0, if mouse is present
                                    ;
0067                         CHKM   PROC   NEAR

0067  B8 3533                       MOV    AX,3533H         ;get INT 33H vector
006A  CD 21                         INT    21H              ;returns vector in ES:BX

006C  8C C0                         MOV    AX,ES
006E  0B C3                         OR     AX,BX            ;test for 0000:0000
0070  F9                            STC
0071  74 13                         JZ     CHKM1            ;if no mouse driver
0073  26: 80 3F CF                  CMP    BYTE PTR ES:[BX],0CFH
0077  F9                            STC
0078  74 0C                         JE     CHKM1            ;if no mouse driver
007A  B8 0000                       MOV    AX,0
007D  CD 33                         INT    33H              ;reset mouse
007F  83 F8 00                      CMP    AX,0
0082  F9                            STC
0083  74 01                         JZ     CHKM1            ;if no mouse
0085  F8                            CLC
0086                         CHKM1:
0086  C3                            RET

0087                         CHKM   ENDP
                             ;
                             ;the TM_ON procedure tests for the presence of a mouse
                             ;and enables mouse pointer.
                             ;uses the CHKM (check for mouse) procedure
                             ;
                             ;***output parameters***
                             ;Carry = 0, if mouse is present pointer enabled
                             ;Carry = 1, if no mouse present
```

```
                              ;
0087                          TM_ON   PROC    NEAR

0087  E8 FFDD                         CALL    CHKM               ;test for mouse
008A  72 06                           JC      TM_ON1
008C  B8 0001                         MOV     AX,1               ;show mouse pointer
008F  CD 33                           INT     33H
0091  F8                              CLC
0092                          TM_ON1:
0092  C3                              RET

0093                          TM_ON   ENDP
                              ;
                              ;The PLACE procedure converts the contents of AX into a
                              ;decimal ASCII coded number stored at the memory location
                              ;addressed by DS:DI
                              ;***input parameters***
                              ;AX = number to be converted to decimal ASCII code
                              ;DS:DI = address where number is stored
                              ;
0093                          PLACE   PROC    NEAR

0093  B9 0000                         MOV     CX,0               ;clear count
0096  BB 000A                         MOV     BX,10              ;set divisor
0099                          PLACE1:
0099  BA 0000                         MOV     DX,0               ;clear DX
009C  F7 F3                           DIV     BX                 ;divide by 10
009E  52                              PUSH    DX
009F  41                              INC     CX
00A0  83 F8 00                        CMP     AX,0
00A3  75 F4                           JNE     PLACE1             ;repeat until quotient 0
00A5                          PLACE2:
00A5  BB 0005                         MOV     BX,5
00A8  2B D9                           SUB     BX,CX
00AA                          PLACE3:
00AA  5A                              POP     DX
00AB  80 C2 30                        ADD     DL,30H             ;convert to ASCII
00AE  88 15                           MOV     [DI],DL            ;store digit
00B0  47                              INC     DI
00B1  E2 F7                           LOOP    PLACE3
00B3  83 FB 00                        CMP     BX,0
00B6  74 08                           JE      PLACE5
00B8  8B CB                           MOV     CX,BX
00BA                          PLACE4:
00BA  C6 05 20                        MOV     BYTE PTR [DI],20H
00BD  47                              INC     DI
00BE  E2 FA                           LOOP    PLACE4
00C0                          PLACE5:
00C0  C3                              RET

00C1                          PLACE   ENDP
                                      END
```

The program in Example 11–15 displays the mouse cursor by placing a 0001H into AX, followed by the INT 33H instruction. Next, the status of the mouse is read with function AX = 0003H. The status function returns with the status of the mouse buttons in BX, the X coordinate of the mouse pointer in CX, and the Y coordinate of the mouse in DX. (Refer to Appendix A for more complete information on the status of the mouse.) In this example, if the left mouse button is pressed, the program terminates; otherwise, the coordinates are compared with the values saved in X and Y. If a change has occurred in these coordinates, the new coordinates are calculated and displayed. Notice that before the video display is accessed, the mouse pointer is hidden, using INT 33H with AX = 0002H. This is very important. If you don't hide the mouse pointer, the display will become unstable, and the computer may even reboot. In most cases, a copy of the mouse pointer remains on the screen if data are displayed without turning the mouse off. Additional mouse applications are explained in the next chapter.

11–5 SUMMARY

1. Files are accessed by, first, opening or creating them using DOS function 21H calls. Once a file is opened or created, it can be read or written using other DOS function 21H calls. Before a program that uses a file ends, the file must be closed, or problems can result. Files are accessed by the use of a file name stored in an ASCII-Z string. An ASCII-Z string is an ASCII character string that ends with a null (00H).
2. Once a file is opened via an ASCII-Z string, it is referred to by a file handle. The file handle allows DOS to process file references more efficiently. File handles can also be used to access the keyboard and video display.
3. The file pointer is a 32-bit number used to refer to any byte in a file. A DOS function 21H call is provided (42H) that allows the file pointer to be moved any number of bytes from the start of the file, the current location, or the end of the file.
4. Random access files are created by using sequential access files and the file pointer. The file pointer can be used to randomly access any portion of the sequential file.
5. The directory entry is accessed in much the same manner as a file, except that the 3BH function is used to change to a new directory.
6. The command line begins at offset address 81H in the program segment prefix (PSP). This information is often used within a DOS program to ask for help or enable some feature of the DOS program.
7. The mouse is accessed via mouse interrupt INT 33H. Through this interrupt, the position, type, and various other information about the mouse is obtained. Refer to Appendix A for a complete listing of the INT 33H mouse functions.

11–6 QUESTIONS AND PROBLEMS

1. How many characters can be used in a file extension?
2. How many bytes are used to store a directory name?
3. Develop a short sequence of instructions that opens a file whose name is stored at data offset address FILE1 as a read-only file.
4. Develop a short sequence of instructions that create a hidden file whose name is stored at data segment offset address FILES.
5. Develop a program that creates a file whose name is FROG.LST and fills it with 512 bytes of 00H. (Don't forget to close the file.)

6. Write a macro instruction sequence called READ. The syntax for this macro is READ BUFFER,COUNT,HANDLE, where BUFFER is the offset address of a file buffer, COUNT is the number of bytes read, and HANDLE is the address of the file handle.
7. Develop a macro sequence called APPEND. The syntax for this macro is APPEND BUFFER,COUNT,HANDLE, where BUFFER is the offset address of the file buffer, COUNT is the number of bytes to append to the file, and HANDLE is the address of the file handle. (Don't forget to use the move file pointer function to append the end of the file.)
8. Create a random access file (RAN.FIL) containing 200 records with a record length of 100 bytes. To initialize this random access file, place a 00H in every byte.
9. Using the random access file developed in Question 8, develop a macro sequence that accesses and reads any record. The syntax for this macro is ACCESS BUFFER,RECORD,HANDLE, where BUFFER is the offset address of the record buffer, RECORD is the record number, and HANDLE is the address of the file handle.
10. Modify the macro of Question 9 so it can read or write a record. The syntax is ACCESS T,BUFFER,RECORD,HANDLE, where T is added to specify a read ('R') or a write ('W') using the ASCII characters for the letters R and W.
11. Develop a short program that displays the first 256 bytes of a file on the video screen. Your display must appear like that in Example 11–16, showing the actual data in any file. It must ask for the name of the file to be displayed and then display the first 256 bytes. (Use text mode.)

EXAMPLE 11–16

```
FILENAME = TEST.TXT
00: 00 00 00 00 00 00 00 00 00 00 00 00 00 00 00 00
10: 00 00 00 00 00 00 00 00 00 00 00 00 00 00 00 00
20: 00 00 00 00 00 00 00 00 00 00 00 00 00 00 00 00
30: 00 00 00 00 00 00 00 00 00 00 00 00 00 00 00 00
40: 00 00 00 00 00 00 00 00 00 00 00 00 00 00 00 00
50: 00 00 00 00 00 00 00 00 00 00 00 00 00 00 00 00
60: 00 00 00 00 00 00 00 00 00 00 00 00 00 00 00 00
70: 00 00 00 00 00 00 00 00 00 00 00 00 00 00 00 00
80: 00 00 00 00 00 00 00 00 00 00 00 00 00 00 00 00
90: 00 00 00 00 00 00 00 00 00 00 00 00 00 00 00 00
A0: 00 00 00 00 00 00 00 00 00 00 00 00 00 00 00 00
B0: 00 00 00 00 00 00 00 00 00 00 00 00 00 00 00 00
C0: 00 00 00 00 00 00 00 00 00 00 00 00 00 00 00 00
D0: 00 00 00 00 00 00 00 00 00 00 00 00 00 00 00 00
E0: 00 00 00 00 00 00 00 00 00 00 00 00 00 00 00 00
F0: 00 00 00 00 00 00 00 00 00 00 00 00 00 00 00 00
```

12. Repeat Question 11 using the graphics mode to display black numerals on a cyan video screen.
13. Develop a program that uses DOS INT 21H function number 0EH to log on to disk drive B.
14. Develop a program that displays only the execute files (.EXE) found in the root directory of drive C. You must refer to Appendix A for information on the find first matching file function (4EH), and find subsequent matching file function (4FH). You may use wildcards to represent the file name for these functions.
15. Develop a program that displays the directory entries found in the root directory of drive C.

16. When the find functions are used in DOS, where is the file or directory name found by these functions located?
17. What is the DTA?
18. Use DOS function number 5AH in a program to create a unique file in the root directory of disk drive C.
19. What INT instruction is used to access the mouse?
20. Describe how to test for the existence of the mouse in a computer system.
21. How is it determined whether the mouse is a serial or a bus mouse?
22. How is it determined whether the right mouse button is pressed?
23. Why must the mouse be disabled when data are displayed in the video display?
24. Develop a program that enables the mouse in graphics mode 13H. Describe the appearance of the mouse pointer.
25. How is the velocity of the mouse pointer adjusted? (Refer to Appendix A.)

CHAPTER 12

Interrupt Hooks and the TSR

INTRODUCTION

In order to use most personal computer applications, a few additional software techniques are required. This chapter presents a technique called an interrupt hook that allows a program to hook into the interrupt structure of the computer system. Interrupts occur for almost every I/O event in a computer. For example, typing a key causes an interrupt. An interrupt hook allows assembly language access to these systems events. Most interrupt hooks are written with assembly language rather than a high-level language such as C/C++.

A TSR is a terminate and stay resident program that remains in the memory, dormant, until activated by some event. A TSR is often activated by an interrupt or a special key called a hot key. A hot key might be a shift F2, or an alternate C, or any other nonstandard ASCII keyboard code. A program accessed via a hot key is often called a pop-up program.

OBJECTIVES

Upon completion of this chapter, you will be able to:

1. Understand the function and application of the interrupt structure of the personal computer.
2. Hook into the interrupts for the keyboard and clock-tick to perform tasks on the computer.
3. Develop terminate and stay resident software, using a hot key to access a program.
4. Control the timer and speaker in the personal computer.
5. Use the mouse in an interrupt hook.
6. Install and remove TSR programs.

12–1 INTERRUPT HOOKS

Interrupt hooks are used to tap into the interrupt structure of the microprocessor. For example, keyboard interrupt can be hooked, or intercepted, so that a special keystroke, called a hot key, can be detected. Whenever the hot key is typed, a terminate and stay resident (TSR) program is activated to perform a special task. Examples of hot key software are pop-up calculators and pop-up clocks. The clock-tick interrupt is also intercepted to accurately time events.

Figure 12–1 shows a time line and some interrupt-activated events in the personal computer. The interrupt steals slices of time from an application to perform some task. The interrupts in Figure 12–1 include the clock-tick interrupt, which keeps the time of day and occurs about 18.2 times per second; the keyboard interrupt, which occurs whenever a key is typed on the keyboard; and the printer interrupt, which occurs whenever the printer is ready for another character. Without interrupts, it would be almost impossible to integrate system hardware into software.

Linking to an Interrupt

Interrupts are used to control almost all I/O devices connected to the personal computer system. At times, these controlling interrupts must be intercepted to perform some additional tasks for an I/O device or personal computer feature. For example, the personal computer contains a keyboard interrupt that is intercepted to use a hot key. Figure 12–2 illustrates the interrupt vector and normal keyboard interrupt service procedure before and after an interrupt hook is installed. Notice how the interrupt hook gains access prior to the normal interrupt service procedure. This *intercept* allows the installed interrupt hook to process keyboard data before the normal interrupt service procedure. Processing often includes the detection of hot keys or macro sequences.

In order to intercept or hook into an interrupt, use a DOS INT 21H function call that reads the current address from the interrupt vector. DOS INT 21H function call number 35H reads the current interrupt vector, and DOS INT 21H function call number 25H changes the address of the current vector. With both DOS function calls, AL indicates the vector type number (00H–FFH) intercepted, and AH indicates the DOS function call number. Never access the interrupt vectors directly. Always use the DOS INT 21H function to access the interrupt, to prevent any possible conflict with DOS.

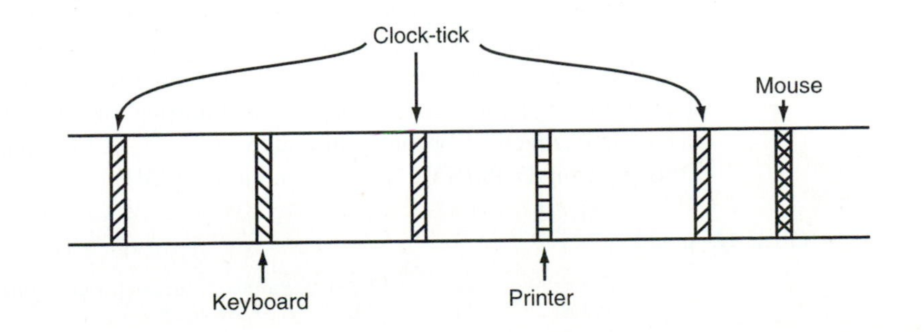

FIGURE 12–1 A time line showing how the clock, keyboard, mouse, and printer steal slices of time from the system through interrupts.

FIGURE 12–2 The normal keyboard interrupt (a) and the keyboard interrupt, showing how a hook (b) has intercepted the keystroke.

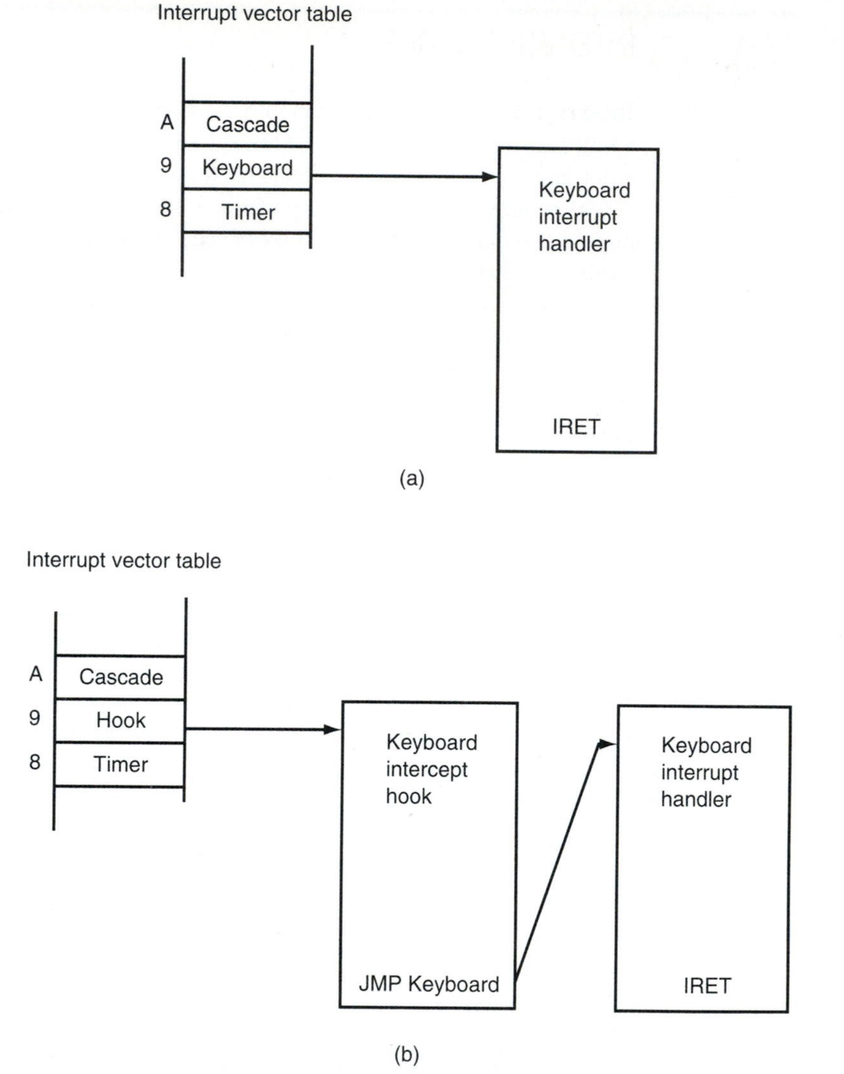

When the vector is read using function 35H, the offset address is returned in register BX and the segment address is returned in register ES. These two registers contain the interrupt service procedure address. The address is saved so it can be restored when the interrupt hook is removed from memory and normal interrupt service is restored. When the vector is changed, it becomes the address stored at the memory location addressed by DS:DX, using DOS INT 21H function number 25H.

EXAMPLE 12–1

```
;a sequence of instructions that shows the installation of a new
;interrupt for vector 0 (divide error).
;Note this is not a complete program.
```

```
                  ;
                         .MODEL TINY
0000                     .CODE
                         .STARTUP
0100  EB 05              JMP    MAIN              ;skip data and interrupt procedures

0102  00000000     ADDR  DD     ?                 ;old interrupt vector

0106               NEW   PROC   FAR               ;new interrupt procedure

0106  CF                 IRET                     ;do nothing interrupt

0107               NEW   ENDP

0107               MAIN:

0107  8C C8              MOV    AX,CS             ;address CS with DS
0109  8E D8              MOV    DS,AX

                   ;get vector 0 address

010B  B8 3500            MOV    AX,3500H
010E  CD 21              INT    21H

                   ;save vector address at ADDR

0110  89 1E 0102 R       MOV    WORD PTR ADDRESS,BX
0114  8C 06 0104 R       MOV    WORD PTR ADDRESS+2,ES

                   ;install new interrupt vector 0 address

0118  B8 2500            MOV    AX,2500H
011B  BA 0106 R          MOV    DX,OFFSET NEW
011E  CD 21              INT    21H

                   ;other installation software continues here
```

The process of installing an interrupt handler through a hook is illustrated in Example 12–1. This procedure reads the current interrupt vector's address (DOS INT 21H function 35H) and stores it in a doubleword memory location for access by the new interrupt service procedure. This doubleword is addressed by using the WORD PTR to force a word-sized transfer. Next, the address of the new interrupt service procedure (NEW), located in DS:DX, is placed into the vector by DOS INT 21H function 25H. This interrupt procedure replaces the normal divide error message with an IRET instruction, so when a divide error occurs, it is ignored. This sequence of instructions is organized as a .COM file, which must have an origin of 100H. When models are used, the TINY model and .STARTUP directive set the offset address to 0100H. As mentioned in an earlier chapter, a .COM file is created from an .EXE file by using the EXE2BIN program with the MASM 5.1 assembler, or by using an option selected in the programmer's workbench program with MASM 6.X. Notice the placement of the data and the interrupt service procedure before the section of the program that installs it.

Some words of caution about interrupt handlers and interrupt hooks. An interrupt invades a program whenever a hardware event occurs. If any register is changed in the interrupt handler, it causes catastrophic results in the interrupted program. To prevent this, make sure that all registers used in the interrupt handler are saved before they are used and restored after they are used. The only register that need not be saved is the flag register, which is saved by the interrupt.

The interrupt hook intercepts the normal interrupt. The normal interrupt must continue or the result can be a clock that doesn't keep time or a keyboard that doesn't function. If the interrupt handler requires more than a few microseconds, check to see that the old interrupt is called. This is accomplished by pushing the flags on the stack, then calling the old interrupt with a CALL CS:old_vector command, as in the examples in this chapter. Once the old interrupt is called, the hook returns with an IRET. If the software only takes a few microseconds, jump to the old interrupt with a JMP CD:old_vector. Don't forget to enable future interrupts in the interrupt hook with STI. When an interrupt is processed, it disables all future interrupts. The STI instruction prevents the lockup that would occur if interrupts remained disabled.

Hooking into the Clock-Tick Interrupt

The clock-tick interrupt is a constant on all personal computer systems and is an excellent way to time events and programs. The clock-tick interrupt occurs about 18.2 times a second (the exact time is closer to 18.206 Hz). The clock-tick interrupt uses vector 8 in the personal computer. When any interrupt occurs, it pauses the program currently executing in order to execute the interrupt service procedure addressed by the interrupt vector. After the interrupt service procedure executes, an interrupt return instruction (IRET) resumes the paused program at its point of interruption.

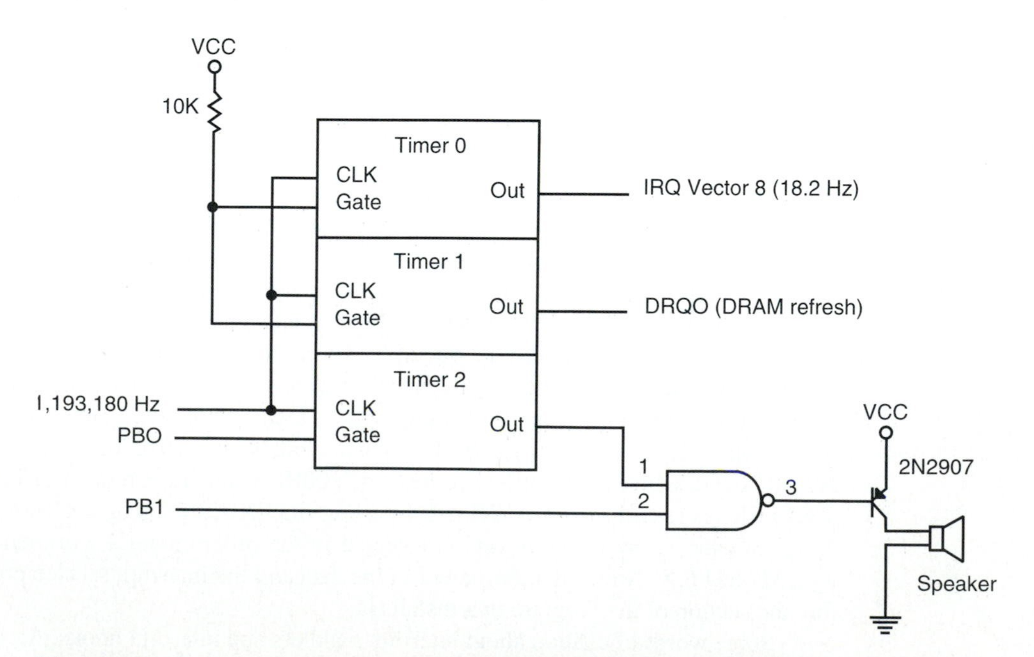

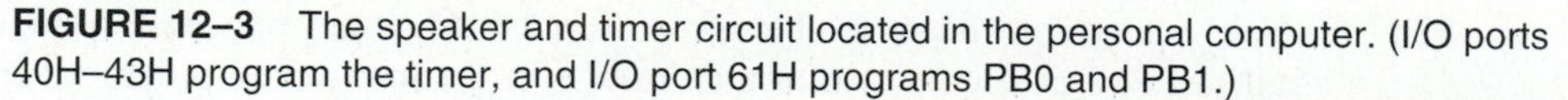

FIGURE 12–3 The speaker and timer circuit located in the personal computer. (I/O ports 40H–43H program the timer, and I/O port 61H programs PB0 and PB1.)

Suppose a program is needed that beeps the speaker once on each half-hour and twice on the hour. This is accomplished by hooking into interrupt vector 8, which occurs 18.2 times per second. As each clock-tick interrupt occurs, the time is tested to see whether it is at the half-hour or hour mark. If it is, the interrupt service procedure beeps the speaker the correct number of times.

Before this program is written, the control and operation of the speaker in the personal computer must be understood. Figure 12–3 illustrates the internal organization of the timer (an 8253 triple-timer integrated circuit) and speaker in the personal computer. The 8253's timer 0 provides the clock-tick interrupt, timer 1 provides an interrupt that controls DRAM refresh in some early personal computers, and timer 2 provides a signal to the speaker through a NAND gate. Never change or modify timer 0 or 1, or the computer will crash. The NAND gate selects the speaker, and was required in early personal computers, which used timer 2 to generate a signal for storing data on an audiocassette recorder. Today, the audiocassette recorder is replaced by floppy disks, but the circuit is still there for the purpose of compatibility with early personal computers.

Before hooking into the interrupt and programming the half-hour/hour chime, a program is written to control the speaker and play a few notes. Example 12–2 shows a short program that programs timer 2 and also generates a few audio tones on the speaker. The example also programs the PB0 and PB1 bits that control the NAND gate.

To program the timer, first calculate the count required to produce an audio tone. The input to timer 2 is 1,193,180 Hz in all personal computers. If 1,193,180 Hz is divided by the desired audio frequency for the speaker, a count is calculated that generates the desired audio frequency. This is the count that is programmed into timer 2. For example, to generate an 800 Hz tone, divide 1,193,180 Hz by 800 Hz. This produces a count of 1491.475. The count programmed into timer 2 must be an integer. If timer 2 is programmed to divide by 1491, it generates an output frequency of 800.255 Hz, which is very close to 800 Hz. This process can allow the timer to generate virtually any output frequency desired. The count is sent to the timer through I/O port number 42H, the count control register for timer 2, which starts the timer.

Once the timer is programmed, bits PB0 and PB1 (port B) of the 8255 peripheral controller are both set to allow an audio tone to pass through to the speaker via the NAND gate. The length of time bits PB0 and PB1 remain a logic 1 determines the duration of the audio tone. The I/O port used for port B is port number 61H. It is important that these bits are set and reset without disturbing the remaining bits of port B. The OR and XOR instructions are used with IN and OUT to accomplish this task.

One way to control the duration of the tone is by timing the length of the tone with the contents of a few personal computer memory locations that contain a count. Doubleword memory location 0000:046C contains a number that changes 18.2 times per second. This count does not reflect the time of day or any other information. The memory resident counter is used to time the duration of the beep. If 18 counts elapse, the beep duration is about 1 second. Any number of counts can elapse to produce a time delay of any length that is fairly accurate and not dependent on the operating frequency of the computer. This technique is a good way to time long events in the personal computer.

EXAMPLE 12–2

```
                        ;a program that beeps the speaker with some sample audio tones
                        ;that each have a duration of 1/3 second.
                        ;
                                .MODEL TINY
0000                            .CODE
                                .STARTUP
0100  B8 0000                   MOV     AX,0
0103  8E D8                     MOV     DS,AX               ;address segment 0000H

0105  B9 0004                   MOV     CX,4                ;set count to 4
0108  E4 61                     IN      AL,61H              ;enable timer and speaker
010A  0C 03                     OR      AL,3                ;set PB0 and PB1
010C  E6 61                     OUT     61H,AL
010E                    MAIN1:
010E  BB 03E8                   MOV     BX,1000             ;select 1000 Hz tone
0111  E8 0018                   CALL    BEEP
0114  BB 04B0                   MOV     BX,1200             ;select 1200 Hz tone
0117  E8 0012                   CALL    BEEP
011A  BB 0578                   MOV     BX,1400             ;select 1400 Hz tone
011D  E8 000C                   CALL    BEEP
0120  E2 EC                     LOOP    MAIN1               ;repeat 4 times

0122  E4 61                     IN      AL,61H              ;turn speaker off
0124  34 03                     XOR     AL,3                ;clear PB0 and PB1
0126  E6 61                     OUT     61H,AL
                                .EXIT
                        ;
                        ;the BEEP procedure programs timer 2 to beep the speaker for
                        ;1/3 of a second with the frequency of the audio tone in BX.
                        ;***input parameters***
                        ;BX = desired audio tone
                        ;***uses***
                        ;WAITS procedure to wait for 1/2 second
                        ;
012C                    BEEP    PROC    NEAR                ;beep speaker for 1/3 second

012C  B8 34DC                   MOV     AX,34DCH            ;load AX with 1,193,180
012F  BA 0012                   MOV     DX,12H
0132  F7 F3                     DIV     BX                  ;find count
0134  E6 42                     OUT     42H,AL              ;program timer 2 with count
0136  8A C4                     MOV     AL,AH
0138  E6 42                     OUT     42H,AL
013A  E8 0001                   CALL    WAITS               ;wait 1/3 second
013D  C3                        RET

013E                    BEEP    ENDP

                        ;
                        ;the WAITS procedure waits 1/3 of a second
                        ;***uses***
                        ;memory doubleword location 0000:46CH to time the wait
                        ;
013E                    WAITS   PROC    NEAR

013E  BA 0006                   MOV     DX,6                ;number of clock ticks
0141  BB 0000                   MOV     BX,0
0144  03 16 046C                ADD     DX,DS:[46CH]        ;get tick count plus time
0148  13 1E 046E                ADC     BX,DS:[46EH]
014C                    WAIT1:
```

```
014C  8B 2E 046C            MOV    BP,DS:[46CH]          ;test for elapsed time
0150  A1 046E               MOV    AX,DS:[46EH]
0153  2B EA                 SUB    BP,DX
0155  1B C3                 SBB    AX,BX
0157  72 F3                 JC     WAIT1                 ;keep testing

0159  C3                    RET

015A              WAITS     ENDP
                            END
```

Example 12–2 lists a short program that beeps the speaker for 1/3 second (6 counts) for each of three audio tones. It repeats each beep 4 times. Notice how the timer is programmed and how the duration of the beeps are timed. Also note how the 8255 peripheral controller is programmed to enable the speaker and timer. Try changing the count to vary the beep length and the tone (in BX) to vary the frequency. Notice how the count at doubleword memory location 0000:046C is used to time the WAITS procedure for a time delay. The number in this location is read from memory and the delay count (6 in this example) is added to form the end of the delay time. The procedure then waits until doubleword memory location 0000:046C counts up to this new value before returning.

Now that the speaker's operation is understood, a **terminate and stay resident** (TSR) program is written to monitor the clock-tick interrupt (vector 8). A program or procedure is made to terminate and stay resident in the memory by using DOS INT 21H function 31H. This function causes DOS to keep the program in the memory until the machine is turned off. It is used to terminate the program in place of DOS INT 21H function 4CH (embodied in .EXIT), which removes the program from memory. Both functions 4CH and 31H return control or exit to DOS. To place a program in memory as a TSR, the TSR function (31H) is called with the number of hexadecimal paragraphs of memory to set aside for the TSR software. This is accomplished by placing the number of paragraphs (16-byte divisions) into register DX before invoking function 31H. For example, if the program is 39 bytes in length, a 3 is passed to function 31H in register DX to set aside 3×16 (48) bytes of memory for the TSR. Generally, interrupt service procedures must remain in memory so they can intercept the interrupt. A program or procedure that is made resident stays in memory until the computer is turned off.

Example 12–3 shows the CHIME program. Notice that this software is written as a .COM program to reduce the amount of memory required to store it. A .COM program contains one segment and is originated at location 100H. To form a .COM program, always use the TINY model and always assemble it as a .COM file. Because a .COM file assembles in one segment, and only the CS register is initialized by .STARTUP, it is important that either (a) all data addressed in the code segment is addressed using the segment override prefix (CS:) or that (b) the DS register is loaded with the contents of CS. In most TSR programs, the interrupt service procedures address data using the segment override prefix.

The CHIME program hooks into interrupt vector 8 and beeps the speaker once on the half-hour and twice on the hour. This program is a TSR and remains active until the computer is turned off. Note how the TSR is installed and how the interrupt vector is hooked. Also notice that the normal interrupt vector 8 procedure continues to execute even as the beeper is activated.

EXAMPLE 12–3

```
                        ;a terminate and stay resident program that hooks into interrupt
                        ;vector 8 to beep the speaker one time per half-hour and two
                        ;times per hour.
                        ;***must be assembled as a .COM file***
                        ;
                                      .MODEL TINY
0000                                  .CODE
                                      .STARTUP
0100  E9 00CE                  JMP    INSTALL           ;install interrupt service procedure
= 03E8                  TONE   EQU    1000              ;set tone at 1000 Hz
0103  00                COUNT  DB     0                 ;elapsed time counter
0104  00000000          ADD8   DD     ?                 ;old vector address
0108  00                PASS   DB     0                 ;1 or 2 beeps
0109  00                BEEP   DB     0                 ;beep or silent
010A  00                FLAG   DB     0                 ;busy flag

010B                    VEC8   PROC   FAR               ;interrupt service procedure

010B  2E: 80 3E 010A R         CMP    CS:FLAG,          ;test busy flag
      00
0111  74 05                    JE     VEC81             ;if not busy
0113  2E: FF 2E 0104 R         JMP    CS:ADD8           ;if busy do normal INT 8
0118                    VEC81:
0118  9C                       PUSHF                    ;do normal INT 8
0119  2E: FF 1E 0104 R         CALL   CS:ADD8
011E  2E: C6 06 010A R         MOV    CS:FLAG,1         ;show busy
      01
0124  FB                       STI                      ;allow other interrupts
0125  2E: 80 3E 0108 R         CMP    CS:PASS,0
      00
012B  75 2C                    JNE    VEC83             ;if beep counter active
012D  50                       PUSH   AX                ;save registers
012E  51                       PUSH   CX
012F  52                       PUSH   DX
0130  B4 02                    MOV    AH,2              ;get time from BIOS
0132  CD 1A                    INT    1AH
0134  80 FE 00                 CMP    DH,0              ;is it 00 seconds
0137  75 68                    JNE    VEC86             ;not time yet, so return
0139  80 F9 00                 CMP    CL,0              ;test for hour
013C  74 10                    JE     VEC82             ;if hour beep 2 times
013E  80 F9 30                 CMP    CL,30H            ;test for half-hour
0141  75 5E                    JNE    VEC86             ;if not half-hour
0143  E8 0065                  CALL   BEEPS             ;start speaker beep
0146  2E: C6 06 0108 R         MOV    CS:PASS,1         ;set number of beeps to 1
      01
014C  EB 53                    JMP    VEC86             ;end it
014E                    VEC82:
014E  E8 005A                  CALL   BEEPS             ;start speaker beep
0151  2E: C6 06 0108 R         MOV    CS:PASS,2         ;set number of beeps to 2
      02
0157  EB 48                    JMP    VEC86             ;end it
0159                    VEC83:
0159  2E: 80 3E 0103 R         CMP    CS:COUNT,0        ;test for end of delay
      00
015F  74 07                    JE     VEC84             ;if time delay has elapsed
0161  2E: FE 0E 0103 R         DEC    CS:COUNT
0166  EB 3C                    JMP    VEC88             ;end it
0168                    VEC84:
0168  2E: 80 3E 0109 R         CMP    CS:BEEP,0         ;test beep on
```

```
       00
016E  75 1C                  JNE   VEC85           ;if beep is on
0170  2E: FE 0E 0108 R       DEC   CS:PASS         ;test for 2 beeps
0175  74 2D                  JZ    VEC88           ;if second beep not needed end
0177  2E: C6 06 0103 R       MOV   CS:COUNT,9      ;reset count
       09
017D  2E: C6 06 0109 R       MOV   CS:BEEP,1       ;beep on for second beep
       01
0183  50                     PUSH  AX
0184  E4 61                  IN    AL,61H          ;enable speaker for beep
0186  0C 03                  OR    AL,3
0188  E6 61                  OUT   61H,AL
018A  EB 17                  JMP   VEC87           ;end it
018C                   VEC85:
018C  2E: C6 06 0103 R       MOV   CS:COUNT,9      ;reset count
       09
0192  2E: C6 06 0109 R       MOV   CS:BEEP,0       ;show beep is off
       00
0198  50                     PUSH  AX
0199  E4 61                  IN    AL,61H          ;disable speaker for beep off
019B  34 03                  XOR   AL,3
019D  E6 61                  OUT   61H,AL
019F  EB 02                  JMP   VEC87           ;end it
01A1                   VEC86:
01A1  5A                     POP   DX              ;restore registers
01A2  59                     POP   CX
01A3                   VEC87:
01A3  58                     POP   AX
01A4                   VEC88:
01A4  2E: C6 06 010A         MOV   CS:FLAG,0       ;show not busy
       00
01AA  CF                     IRET  ;interrupt return

01AB                   VEC8  ENDP
                       ;
                       ;the BEEPS procedure programs the speaker for the frequency stored
                       ;as TONE using an equate at assembly time. The duration of the
                       ;beep is 1/2 second.
                       ;***uses registers AX, CX, and DX***
                       ;
01AB                   BEEPS PROC  NEAR            ;beep speaker for 1/2 second

01AB  2E: 8B 0E 03E8         MOV   CX,CS:TONE      ;set tone
01B0  B8 34DC                MOV   AX,34DCH        ;load AX with 1,193,180
01B3  BA 0012                MOV   DX,12H
01B6  F7 F1                  DIV   CX              ;calculate count
01B8  E6 42                  OUT   42H,AL          ;program timer 2 with count
01BA  8A C4                  MOV   AL,AH
01BC  E6 42                  OUT   42H,AL

01BE  E4 61                  IN    AL,61H          ;speaker on
01C0  0C 03                  OR    AL,3
01C2  E6 61                  OUT   61H,AL
01C4  2E: C6 06 0103 R       MOV   CS:COUNT,9      ;set count for 1/2 second
       09
01CA  2E: C6 06 0109         RMOV  CS:BEEP,1       ;indicate beep is on
       01
01D0 C3                      RET

01D1                   BEEPS ENDP
```

```
01D1                      INSTALL:                            ;install interrupt procedure VEC8
01D1   8C C8                      MOV    AX,CS                ;overlap CS and DS
01D3   8E D8                      MOV    DS,AX

01D5   B8 3508                    MOV    AX,3508H             ;get current vector 8
01D8   CD 21                      INT    21H                  ;and save it
01DA   89 1E 0104 R               MOV    WORD PTR ADD8,BX
01DE   8C 06 0106 R               MOV    WORD PTR ADD8+2,ES

01E2   B8 2508                    MOV    AX,2508H
01E5   BA 010B R                  MOV    DX,OFFSET VEC8       ;address interrupt procedure
01E8   CD 21                      INT    21H                  ;install vector 8

01EA   BA 01D1 R                  MOV    DX,OFFSET INSTALL ;find paragraphs
01ED   B1 04                      MOV    CL,4
01EF   D3 EA                      SHR    DX,CL
01F1   42                         INC    DX

01F2   B8 3100                    MOV    AX,3100H             ;exit to DOS, but make VEC 8 a TSR
01F5   CD 21                      INT    21H
                                  END
```

The CHIME program uses several memory locations as flags to signal the operation of the interrupt service procedure. The first flag tested by CHIME is the busy flag (FLAG), which indicates that a part of the interrupt service procedure is active. If FLAG = 1 (busy condition), the procedure jumps to the normal vector 8 interrupt (JMP CS:ADD8), which ends VEC8's execution. If FLAG = 0 (not busy), the interrupt service procedure continues at VEC81. The default address for all direct memory data is the data segment. In the TSR software used in this example and others, it is important to use the segment override prefix (CS:) to ensure that the program addresses data in the code segment, where it appears.

At VEC81, the normal vector 8 interrupt is executed with a forced interrupt call (PUSHF followed by a CALL CS:ADD8). Upon return from the normal vector 8 interrupt (required to keep accurate time), the busy flag is set to show a busy condition (FLAG = 1) and other interrupts are enabled with the STI instruction.

The PASS flag is now tested to see whether the VEC8 procedure is currently beeping the speaker. If PASS = 0 (not beeping speaker), the time of day is retrieved from BIOS using the INT 1AH instruction. It is important not to access DOS from within a TSR or interrupt service procedure. If DOS is accessed at this time, and it is in the process of executing an operation that affects the interrupt, the program will crash. The INT 1AH instruction returns the number of seconds (DH), minutes (CL), and hours (CH) in BCD form. After obtaining the current time, the number of seconds is tested for zero. If it is not zero seconds, the interrupt procedure ends. If it is zero seconds, then CL is tested for 00 minute (hour) and 30 minutes (half-hour). If either case is true, the speaker is enabled and TONE is programmed in the timer by a call to BEEPS. If neither case is true, the interrupt ends. Notice that the BEEPS procedure programs timer 2, enables the speaker, and sets the count to 9.

The time delay counter (COUNT) is decremented each time the interrupt occurs. If the count reaches zero, the procedure tests BEEP to control the speaker. If the speaker is beeping, the procedure turns it off and resets the time delay count to 9. If the speaker is not beeping, the procedure tests PASS to determine whether another beep is required on the hour. The time delay is $^1/_2$ second (COUNT = 9) in this program and cannot be less. If a delay of less than 1/2 second is chosen, the speaker will beep twice for both the hour and half-hour. The reason is that the clock (INT 1AH) is checked for the zero second. If a time delay of less than 1/2 second is used, the half-hour will be picked up twice.

The TSR program is loaded into memory at the DOS command line by typing the name of the program; in this case, CHIME. If DOS version 5.0 or 6.X is in use, CHIME can be loaded into the upper memory or high memory area by typing LOADHIGH CHIME. Once this program loads into memory, it continues to beep the time until the power to the computer is disconnected or until it is rebooted. This is an excellent—and not too annoying—way to keep track of time. The next section of the text describes hot keys. If desired, a hot key could be used to enable and disable CHIME.

12–2 HOT KEYS

Another important programming technique is the installation of a hot key. A **hot key** is a special keystroke, or combination of keystrokes, that invokes a terminate and stay resident program. For example, the ALT C key can be defined as a hot key that causes a clock program to execute. To install a hot key, hook into both interrupts 8 and 9. Interrupt 8 is the clock-tick interrupt and interrupt 9 is the keyboard interrupt. The keyboard interrupt occurs every time a key is pressed on the keyboard and allows access to almost any combination of keys.

Keyboard Operation

The keyboard connected to the personal computer generates a type 9 interrupt whenever a key is typed. This interrupt is intercepted to perform tasks in the computer system. When interrupt 9 is intercepted by a TSR interrupt handler, the interrupt handler reads the keyboard code directly from I/O port 60H (port A of the 8255 PIA). The I/O port returns the keyboard scan code, which is different from the ASCII-coded characters normally obtained from the keyboard. Refer to Table 9–1 for the keyboard scan codes of each of the key on the keyboard. Notice that a few keys, such as print screen, return more than one code.

If the desired keyboard scan code appears at I/O port 60H, it is thrown away and processing for the hot key begins. When the hot key is processed, a TSR is executed, which processes the hot key and performs some task. In addition to the scan code, a hot key also uses the alternate\shift keyboard data located at word-sized memory location 0000:0417. Figure 12–4 shows the binary bit pattern and function of each bit in this word-sized memory location. Access to the alternate, shift, control, and other keys not returned as scan codes are provided through this word-sized location in the memory. For example, a hot key could be an alternate + shift + F, or any other key sequence.

This text assumes that a PC/XT/AT system is in use. The IBM PS/2 series keyboard functions differently. The PS/2 system must use INT 15H (miscellaneous services interrupt) to read the keyboard scanning code. The INT 15H instruction returns the key scan code when called with AH = 4FH. In the PS/2, an INT 15H must be used in place of the IN AL,60H used in the PC/XT/AT system. The shift/alternate keys are still returned at word-sized memory location 0000:0417 in the PS/2. If the PS/2 is in use, the software for testing the scan code must be changed or included.

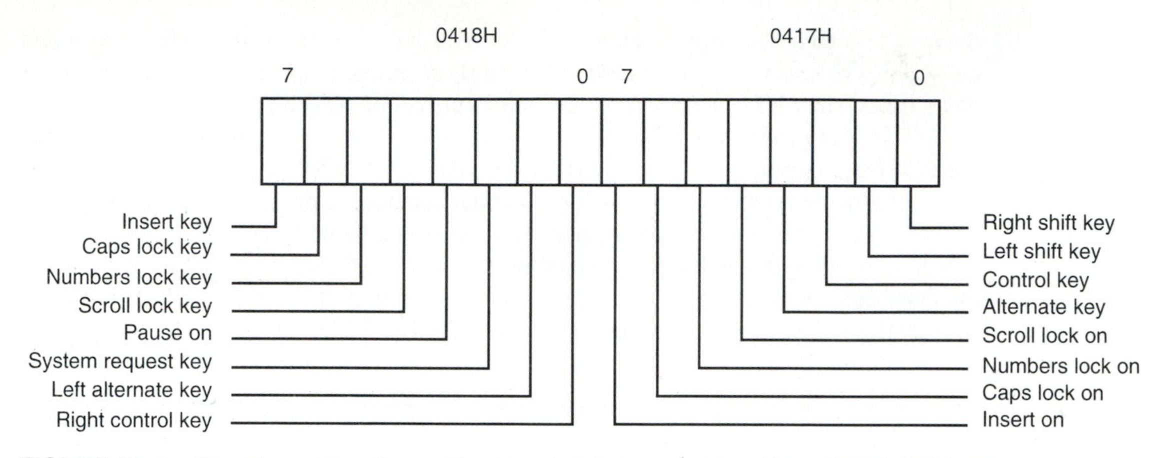

FIGURE 12–4 The binary bit pattern of keyboard status word at locations 0000:0417H and 0000:0418H.

An example that illustrates the use of the keyboard interrupt and the clock-tick interrupt appears in Example 12–4. This sample TSR program dims the video display after five minutes of keyboard inactivity. A video display can become permanently etched if a video image is displayed at full intensity too long. This TSR monitors the keyboard interrupt and, if activity is absent for more than 5 minutes, the palette colors are reduced in intensity to dim the display. As soon as any key is typed, the display rebounds to its normal intensity until another five minutes of inactivity occurs. The amount of delay time is determined by COUNT. In this example COUNT is 1,092 ($18.2 \times 60 \times 5$ for a five-minute delay). The COUNT can be changed for any delay time up to an hour (COUNT = 64K). If the five-minute delay seems too short or too long, change the DELAY EQU statement and reassemble the program. Note that this software only functions with a 256-color VGA display system. It does not function if more colors are present and in use. This program will not function correctly in the Windows environment unless it is listed as a TSR in the WIN.INI file.

Because all video BIOS INT 10H function calls are notoriously slow, the video I/O ports are accessed directly by this program. The palette registers are accessed directly through I/O ports 3C7H, 3C8H, and 3C9H. Port 3C7H selects the read mode and starting PEL register (color) number located in AL; port 3C8H selects the write mode and starting PEL number located in AL; and port 3C9H reads/writes the colors from/to the PEL register number. Note that the PEL number automatically increments after the three colors (red, green, and blue) are read or written. Also note that interrupts must be disabled as the colors are read or written, and at least 240 ns must elapse between the reads or writes. If additional time is required because of the clock speed of your computer you must add NOP instructions (instructions that perform no operation, except to require execution time) between the reads and writes.

Example 12–4 detects mouse pointer movement or buttons as well as the striking of any key on the keyboard. Notice that the INT 33H is added to the clock-tick interrupt to read the status (AX = 3) of the mouse. If a button is pressed (for a two- or three-button mouse), a jump to MICKEY1 occurs, where VEC85 resets the count and brings the display to normal brightness if dimmed. The software also compares the X and Y coordinates of the mouse pointer with the coordinates previously stored in memory. If the mouse is moved, a jump to MICKEY occurs where, again, the count is reset and the display is brought to normal brightness if needed. The mouse is tested 18.2 times per second in the clock-tick interrupt.

EXAMPLE 12–4 (pp. 407–410)

```
                              ;a TSR interrupt hook that monitors the keyboard and the mouse.
                              ;if the keyboard and mouse are idle for more than 5 minutes, the
                              ;screen dims by a factor of four. If the mouse or keyboard
                              ;is used, the display rebounds to normal brightness.
                              ;
                                     .MODEL TINY
                                     .386                        ;uses comes 80386 instructions
0000                                 .CODE
                                     .STARTUP
0100  E9 0623                        JMP    INSTALL              ;install VEC8 and VEC9
= 1554                        DELAY  EQU    1092*5               ;delay time (5 minutes)
0103  00                      BUSY   DB     0                    ;busy flag
0104  00000000                ADD8   DD     ?                    ;old vector 8 address
0108  00000000                ADD9   DD     ?                    ;old vector 9 address
010C  00000000                SADD   DD     ?                    ;old stack pointer
0110  1554                    COUNT  DW     DELAY                ;count
0112  0000                    X      DW     ?                    ;X mouse position
0114  0000                    Y      DW     ?                    ;Y mouse position
0116  0100 [                         DW     256 DUP (?)          ;stack area
                0000
                              ]
0316  = 0316                  STT    EQU    THIS WORD
0316  00                      KFLAG  DB     0                    ;key detected
0317  00                      PFLAG  DB     0                    ;palette flag
0318  0300 [                  PAL    DB     256*3 DUP (?)        ;palette memory
                  00
                             ]

0618                          VEC9   PROC   FAR                  ;keyboard intercept

0618  2E: C6 06 0316 R               MOV    CS:KFLAG,1           ;indicate any key typed
        01
061E  2E: FF 2E 0108 R               JMP    CS:ADD9              ;do keyboard interrupt

0623                          VEC9   ENDP

0623                          VEC8   PROC   FAR                  ;clock tick interrupt procedure

0623  2E: 80 3E 0103 R               CMP    CS:BUSY,0            ;test busy flag
        00
0629  74 05                          JZ     VEC81                ;if not busy
062B  2E: FF 2E 0104 R               JMP    CS:ADD8              ;do old INT 8
0630                     VEC81:
0630  2E: C6 06 0103 R               MOV    CS:BUSY,1            ;indicate busy
        01
0636  9C                             PUSHF                       ;do normal INT 8 interrupt
0637  2E: FF 1E 0104 R               CALL   CS:ADD8
063C  FA                             CLI                         ;interrupts off
063D  2E: 89 26 010C R               MOV    WORD PTR CS:SADD,SP
0642  8C D4                          MOV    SP,SS                ;save old SS:SP
0644  2E: 89 26 010E R               MOV    WORD PTR CS:SADD+2,SP
```

```
0649  8C CC                MOV   SP,CS
064B  8E D4                MOV   SS,SP              ;load new SS:SP
064D  BC 0316 R            MOV   SP,OFFSET STT
0650  FB                   STI                      ;interrupts on
0651  2E: 80 3E 0316 R     CMP   CS:KFLAG,0         ;test key active
        00
0657  0F 85 0083           JNZ   VEC85              ;if key active
065B  60                   PUSHA                    ;test mouse
065C  B8 0003              MOV   AX,3
065F  CD 33                INT   33H                ;get mouse status
0661  83 E3 07             AND   BX,7               ;2- or 3-button mouse
0664  75 77                JNZ   MICKEY1            ;for any button
0666  2E: 39 0E 0112 R     CMP   CS:X,CX
066B  75 66                JNE   MICKEY             ;if mouse moved
066D  2E: 39 16 0114 R     CMP   CS:Y,DX
0672  75 5F                JNE   MICKEY             ;if mouse moved
0674  61                   POPA
0675  2E: 80 3E 0317 R     CMP   CS:PFLAG,0         ;test for dim
        00
067B  0F 85 009A           JNE   VEC88              ;if already dim, end
067F  2E: FF 0E 0110 R     DEC   CS:COUNT           ;decrement count
0684  0F 85 0091           JNZ   VEC88              ;if time delay not done, end
0688                 VEC82:                         ;dim screen
0688  60                   PUSHA                    ;save registers
0689  06                   PUSH  ES
068A  1E                   PUSH  DS
068B  FC                   CLD
068C  8C C8                MOV   AX,CS              ;load ES and DS
068E  8E C0                MOV   ES,AX
0690  8E D8                MOV   DS,AX
0692  BF 0318 R            MOV   DI,OFFSET PAL      ;address palette storage area
0695  8B F7                MOV   SI,DI
0697  B9 0100              MOV   CX,256             ;load count
069A  BA 03C7              MOV   DX,3C7H            ;select read, set color to 00H
069D  32 C0                XOR   AL,AL
069F  EE                   OUT   DX,AL
06A0  BA 03C9              MOV   DX,3C9H            ;address PEL register
06A3                 VEC83:                         ;save old palette and dim it
06A3  FA                   CLI                      ;disable interrupts
06A4  EC                   IN    AL,DX              ;get red and save it
06A5  AA                   STOSB
06A6  EC                   IN    AL,DX              ;get green and save it
06A7  AA                   STOSB
06A8  EC                   IN    AL,DX              ;get blue and save it
06A9  AA                   STOSB
06AA  FB                   STI                      ;enable interrupts
06AB  E2 F6                LOOP  VEC83              ;read all 256 colors
06AD  B9 0100              MOV   CX,256             ;load count
06B0  4A                   DEC   DX                 ;select write port
06B1  32 C0                XOR   AL,AL              ;color number 00H
06B3  EE                   OUT   DX,AL
06B4  42                   INC   DX                 ;address PEL register
06B5                 VEC84:
06B5  FA                   CLI                      ;disable interrupts
06B6  AC                   LODSB                    ;get red and dim it
06B7  C0 E8 02             SHR   AL,2
06BA  EE                   OUT   DX,AL
06BB  AC                   LODSB                    ;get green and dim it
06BC  C0 E8 02             SHR   AL,2
06BF  EE                   OUT   DX,AL
06C0  AC                   LODSB                    ;get blue and dim it
```

```
06C1  C0 E8 02                SHR   AL,2
06C4  EE                      OUT   DX,AL
06C5  FB                      STI                       ;enable interrupts
06C6  E2 ED                   LOOP  VEC84               ;write all colors
06C8  2E: C6 06 0317 R        MOV   CS:PFLAG,1          ;indicate dim
      01
06CE  1F                      POP   DS                  ;restore registers
06CF  07                      POP   ES
06D0  61                      POPA
06D1  EB 46                   JMP   VEC88
06D3                   MICKEY:
06D3  2E: 89 0E 0112 R        MOV   CS:X,CX             ;save new position
06D8  2E: 89 16 0114 R        MOV   CS:Y,DX
06DD                   MICKEY1:
06DD  61                      POPA
06DE                   VEC85:
06DE  2E: 80 3E 0317 R        CMP   CS:PFLAG,0
      00
06E4  74 26                   JZ    VEC87               ;if display normal
06E6  60                      PUSHA                     ;save registers
06E7  1E                      PUSH  DS
06E8  8C C8                   MOV   AX,CS               ;load DS
06EA  8E D8                   MOV   DS,AX
06EC  BE 0318 R               MOV   SI,OFFSET PAL
06EF  FC                      CLD
06F0  B9 0100                 MOV   CX,256              ;load count
06F3  BA 03C8                 MOV   DX,3C8H             ;select write, color 00H
06F6  32 C0                   XOR   AL,AL
06F8  EE                      OUT   DX,AL
06F9  42                      INC   DX                  ;address PEL register
06FA                   VEC86:                           ;reload palette
06FA  FA                      CLI                       ;disable interrupts
06FB  AC                      LODSB                     ;reload red
06FC  EE                      OUT   DX,AL
06FD  AC                      LODSB
06FE  EE                      OUT   DX,AL               ;reload green
06FF  AC                      LODSB
0700  EE                      OUT   DX,AL               ;reload blue
0701  FB                      STI                       ;enable interrupts
0702  E2 F6                   LOOP  VEC86               ;repeat for all colors
0704  2E: C6 06 0317 R        MOV   CS:PFLAG,           ;indicate normal palette
      00
070A  1F                      POP   DS                  ;restore registers
070B  61                      POPA
070C                   VEC87:
070C  2E: C7 06 0110 R        MOV   CS:COUNT,DELAY      ;reset count to DELAY
      1554
0713  2E: C6 06 0316 R        MOV   CS:KFLAG,0          ;clear key active
      00
0719                   VEC88:
0719  2E: C6 06 0103 R        MOV   CS:BUSY,0           ;indicate not busy
      00
071F  2E: 0F B2 26 010C R     LSS   SP,CS:SADD
0725  CF                      IRET

0726                   VEC8   ENDP

0726                   INSTALL:

0726  8C C8                   MOV   AX,CS               ;load DS
0728  8E D8                   MOV   DS,AX
```

```
072A  B8 3508           MOV   AX,3508H            ;get current vector 8
072D  CD 21             INT   21H                 ;and save it
072F  89 1E 0104 R      MOV   WORD PTR ADD8,BX
0733  8C 06 0106 R      MOV   WORD PTR ADD8+2,ES

0737  B8 3509           MOV   AX,3509H            ;get current vector 9
073A  CD 21             INT   21H                 ;and save it
073C  89 1E 0108 R      MOV   WORD PTR ADD9,BX
0740  8C 06 010A R      MOV   WORD PTR ADD9+2,ES

0744  B8 2508           MOV   AX,2508H            ;install new interrupt 8
0747  BA 0623 R         MOV   DX,OFFSET VEC8
074A  CD 21             INT   21H

074C  B8 2509           MOV   AX,2509H            ;install new interrupt 9
074F  BA 0618 R         MOV   DX,OFFSET VEC9
0752  CD 21             INT   21H

0754  BA 0726 R         MOV   DX,OFFSET INSTALL
0757  C1 EA 04          SHR   DX,4                ;determine paragraphs
075A  42                INC   DX

075B  B8 3100           MOV   AX,3100H            ;install as TSR
075E  CD 21             INT   21H
                        END
```

The program in Example 12–4 uses a fair amount of stack space because of the PUSHA instruction. If more than one or two pushes appear in a TSR interrupt service procedure, it is important that a local stack appear. In this example, the address of the old stack is stored in SADD and a new stack area is obtained. Before returning from the VEC8 interrupt service procedure, the LSS SP,CS:SADD instruction loads SP and SS with the old stack area address.

Installing a Hot Key

A hot key is installed with a TSR program and interrupt hooks. To illustrate a hot key program, let us develop a program that counts keystrokes. The keystroke counter program permits productivity to be assessed in a business that uses computers for data entry or other tasks. The keystroke counter program counts each keystroke and displays the count when the ALT + K key is pressed. It is important to note that this program spies on workers; it is the duty of any company using the program to notify workers. It may even be necessary for the company to obtain permission from workers before such a program is placed into service.

This program can be modified to keep track of keystrokes by the hour or any other time unit. In this example, the keystroke count (up to 4 billion) accumulates keystrokes as long as power is applied to the computer. The program also stores the installation time for security purposes. This is important because if a machine is reset, the start time for this TSR will be reset.

This program hooks into interrupts 8 and 9 to count keys. The interrupt 9 hook detects the hot key (ALT K) and counts key strokes. When the hot key is detected, the 18.2 Hz interrupt 8 activates the hot key program, which displays the keystroke count and time of installation. This type of TSR is often called a **pop-up program,** because it pops up when the hot key is typed. Notice that this program uses INT 16H to test the keyboard.

Never use a DOS INT 21H function call within a TSR or interrupt hook because serious problems can arise. This program also uses direct manipulation of the video text memory that begins at location B8000H. This memory is organized with 2 bytes per ASCII character. The first byte contains the ASCII code and the second byte contains the background and character color.

EXAMPLE 12–5 (pp. 411–415)

```
                              ;a TSR program that counts keystrokes and reports the time of
                              ;installation and number of accumulated keystrokes whenever
                              ;the ALT K key combination is activated.
                              ;***requires an 80386 or newer microprocessor***
                              ;
                                     .MODEL TINY
                                     .386
0000                                 .CODE
                                     .STARTUP
0100   E9 0241                       JMP    INSTALL                   ;install VEC8 and VEC9

0103  00                      HFLAG  DB     0                         ;Hot-key detected
0104  00000000                ADD8   DD     ?                         ;old vector 8 address
0108  00000000                ADD9   DD     ?                         ;old vector 9 address
010C  00000000                COUNT  DD     0                         ;Keystroke counter
0110  00                      HOUR   DB     ?                         ;start-up time
0111  00                      MIN    DB     ?
0112  00                      SFLAG  DB     0                         ;start-up flag
0113  00                      FLAG8  DB     0                         ;interrupt 8 busy
0114  25                      KEY    DB     25H                       ;scan code for K
0115  08                      HMASK  DB     8                         ;alternate key mask
0116  08                      MKEY   DB     8                         ;alternate key
0117  00A0 [                  SCRN   DB     160 DUP (?)               ;screen buffer
               00
                  ]
01B7  54 69 6D 65             MES1   DB     'Time = '
         20 3D 20
01BE  20 20 20 4B             MES2   DB     ' KeyStrokes = '
         65 79 53 74
         72 6F 6B 65
         73 20 3D 20

01CE                          VEC9   PROC   FAR                       ;keyboard intercept

01CE  FB                             STI                              ;enable interrupts
01CF  66| 50                         PUSH   EAX                       ;save EAX
01D1  E4 60                          IN     AL,60H                    ;get scan code
01D3  2E: 3A 06 0114 R               CMP    AL,CS:KEY                 ;test for K
01D8  75 16                          JNE    VEC91                     ;no hot-key
01DA  B8 0000                        MOV    AX,0                      ;address segment 0000
01DD  1E                             PUSH   DS                        ;save DS
01DE  8E D8                          MOV    DS,AX
01E0  A0 0417                        MOV    AL,DS:[417H]              ;get shift/alternate data
01E3  1F                             POP    DS
01E4  2E: 22 06 0115 R               AND    AL,CS:HMASK               ;isolate alternate key
01E9  2E: 3A 06 0116 R               CMP    AL,CS:MKEY                ;test for alternate key
01EE  74 2A                          JE     VEC93                     ;if hot key found
01F0                          VEC91:
01F0  51                             PUSH   CX                        ;add one to BCD COUNT
01F1  B9 0003                        MOV    CX,3
01F4  66| 2E: A1 010C R              MOV    EAX,CS:COUNT
```

```
01F9  66| 83 C0 01          ADD    EAX,1
01FD  27                    DAA                      ;make result BCD
01FE              VEC92:
01FE  9C                    PUSHF
01FF  66| C1 C8 08          ROR    EAX,8
0203  9D                    POPF
0204  14 00                 ADC    AL,0              propagate carry
0206  27                    DAA
0207  E2 F5                 LOOP   VEC92
0209  66| C1 C8 08          ROR    EAX,8
020D  66| 2E: A3 010C R     MOV    CS:COUNT,EAX
0212  59                    POP    CX
0213  66| 58                POP    EAX
0215  2E: FF 2E 0108 R      JMP    CS:ADD9           ;do normal interrupt
021A              VEC93:                             ;if hot key pressed
021A  FA                    CLI                      ;interrupts off
021B  E4 61                 IN     AL,61H            ;clear keyboard and
021D  0C 80                 OR     AL,80H            ;throw away hot key
021F  E6 61                 OUT    61H,AL
0221  24 7F                 AND    AL,7FH
0223  E6 61                 OUT    61H,AL
0225  B0 20                 MOV    AL,20H            ;reset keyboard interrupt
0227  E6 20                 OUT    20H,AL
0229  FB                    STI                      ;enable interrupts
022A  2E: C6 06 0103 R      MOV    CS:HFLAG,1        ;indicate hot key pressed
         01
0230  66| 58                POP    EAX
0232  CF                    IRET

0233              VEC9      ENDP

0233              VEC8      PROC   FAR               ;clock-tick interrupt procedure

0233  2E: 80 3E 0113 R      CMP    CS:FLAG8,0
         00
0239  74 05                 JZ     VEC81             ;if not busy
023B  2E: FF 2E 0104 R      JMP    CS:ADD8           ;if busy
0240              VEC81:
0240  2E: 80 3E 0103 R      CMP    CS:HFLAG,0
         00
0246  75 37                 JNZ    VEC83             ;if hot key detected
0248  2E: 80 3E 0112 R      CMP    CS:SFLAG,0
         00
024E  74 05                 JZ     VEC82             ;if start-up
0250  2E: FF 2E 0104 R      JMP    CS:ADD8           ;if not hot key or start-up
0255              VEC82:
0255  9C                    PUSHF                    ;do old interrupt 8
0256  2E: FF 1E 0104 R      CALL   CS:ADD8
025B  2E: C6 06 0113 R      MOV    CS:FLAG8,1        ;indicate busy
         01
0261  FB                    STI                      ;enable interrupts
0262  50                    PUSH   AX
0263  51                    PUSH   CX
0264  52                    PUSH   DX
0265  B4 02                 MOV    AH,2              ;get start-up time
0267  CD 1A                 INT    1AH
0269  2E: 88 2E 0110 R      MOV    CS:HOUR,CH        ;save hour
026E  2E: 88 0E 0111 R      MOV    CS:MIN,CL         ;save minute
0273  5A                    POP    DX                ;restore registers
0274  59                    POP    CX
0275  58                    POP    AX
```

```
0276  2E: C6 06 0112 R       MOV    CS:SFLAG,1           ;indicate started
        01
027C  E9 00A5                JMP    VEC89                ;end it
027F                 VEC83:                              ;do hot key display
027F  9C                     PUSHF                       ;do old interrupt 8
0280  2E: FF 1E 0104 R       CALL   CS:ADD8
0285  2E: C6 06 0113 R       MOV    CS:FLAG8,1           ;indicate busy
        01
028B  FB                     STI                         ;enable interrupts
028C  50                     PUSH   AX                   ;save registers
028D  53                     PUSH   BX
028E  B4 0F                  MOV    AH,0FH               ;get video mode
0290  CD 10                  INT    10H
0292  3C 03                  CMP    AL,3
0294  76 05                  JBE    VEC84                ;if DOS text mode
0296  5B                     POP    BX                   ;ignore if graphics mode
0297  58                     POP    AX
0298  E9 0083                JMP    VEC88
029B                 VEC84:                              ;for text mode
029B  51                     PUSH   CX
029C  66| 52                 PUSH   EDX
029E  57                     PUSH   DI
029F  56                     PUSH   SI
02A0  1E                     PUSH   DS
02A1  06                     PUSH   ES
02A2  FC                     CLD
02A3  8C C8                  MOV    AX,CS                ;address this segment
02A5  8E C0                  MOV    ES,AX
02A7  B8 B800                MOV    AX,0B800H            ;address text memory
02AA  8E D8                  MOV    DS,AX
02AC  B9 00A0                MOV    CX,160               ;save top screen line
02AF  BF 0117 R              MOV    DI,OFFSET SCRN
02B2  BE 0000                MOV    SI,0
02B5  F3/ A4                 REP    MOVSB
02B7  1E                     PUSH   DS                   ;swap segments
02B8  06                     PUSH   ES
02B9  1F                     POP    DS
02BA  07                     POP    ES
02BB  BF 0050                MOV    DI,80                ;start display at center
02BE  BE 01B7 R              MOV    SI,OFFSET MES1
02C1  B4 0F                  MOV    AH,0FH               ;load white on black
02C3  B9 0007                MOV    CX,7
02C6                 VEC85:
02C6  AC                     LODSB                       ;display Time =
02C7  AB                     STOSW
02C8  E2 FC                  LOOP   VEC85
02CA  2E: 8A 16 0111 R       MOV    DL,CS:MIN
02CF  2E: 8A 36 0110 R       MOV    DH,CS:HOUR
02D4  66| C1 E2 10           SHL    EDX,16
02D8  B9 0002                MOV    CX,2
02DB  B3 30                  MOV    BL,30H
02DD  E8 004B                CALL   DISP                 ;display hours
02E0  B0 3A                  MOV    AL,':'
02E2  AB                     STOSW                       ;display colon
02E3  B9 0002                MOV    CX,2
02E6  B3 80                  MOV    BL,80H
02E8  E8 0040                CALL   DISP                 ;display minutes
02EB  BE 01BE R              MOV    SI,OFFSET MES2       ;display  KeyStrokes =
02EE  B9 0010                MOV    CX,16
02F1                 VEC86:
02F1  AC                     LODSB
```

```
02F2  AB                          STOSW
02F3  E2 FC                       LOOP    VEC86
02F5  66| 2E: 8B 16 010C R        MOV     EDX,CS:COUNT          ;get count
02FB  B9 0008                     MOV     CX,8
02FE  B3 30                       MOV     BL,30H
0300  E8 0028                     CALL    DISP                  ;display count
0303                       VEC87:
0303  B4 01                       MOV     AH,1                  ;wait for any key (BIOS)
0305  CD 16                       INT     16H
0307  74 FA                       JZ      VEC87
0309  FC                          CLD
030A  BE 0117 R                   MOV     SI,OFFSET SCRN        ;restore text
030D  BF 0000                     MOV     DI,0
0310  B9 00A0                     MOV     CX,160
0313  F3/ A4                      REP     MOVSB
0315  07                          POP     ES
0316  1F                          POP     DS
0317  5E                          POP     SI
0318  5F                          POP     DI
0319  66| 5A                      POP     EDX
031B  59                          POP     CX
031C  5B                          POP     BX
031D  58                          POP     AX
031E                       VEC88:
031E  2E: C6 06 0103 R            MOV     CS:HFLAG,0            ;kill hot_key
         00
0324                       VEC89:
0324  2E: C6 06 0113 R            MOV     CS:FLAG8,0            ;indicate not busy
         00
032A  CF                          IRET

032B                       VEC8   ENDP
                           ;
                           ;the DISP procedure displays the BCD contents of EDX.
                           ;***input parameters***
                           ;CX = number of digits
                           ;BL = 30H for blank leading zeros or 80H for no blanking
                           ;ES = segment address of text mode display
                           ;DI = offset address of text mode display
                           ;
032B                       DISP   PROC    NEAR                  ;display

032B  66| C1 C2 04                ROL     EDX,4                 ;position number
032F  8A C2                       MOV     AL,DL
0331  24 0F                       AND     AL,0FH
0333  04 30                       ADD     AL,30H                ;convert to ASCII
0335  AB                          STOSW                         ;store in text display
0336  3A C3                       CMP     AL,BL                 ;test for blanking
0338  74 04                       JE      DISP1                 ;if blanking needed
033A  B3 80                       MOV     BL,80H                ;turn off blanking
033C  EB 03                       JMP     DISP2                 ;continue
033E                       DISP1:
033E  83 EF 02                    SUB     DI,2                  ;blank digit
0341                       DISP2:
0341  E2 E8                       LOOP    DISP
0343  C3                          RET

0344                       DISP   ENDP

0344                       INSTALL:                             ;install VEC8 and VEC9

0344  8C C8                       MOV     AX,CS                 ;load DS
```

```
0346  8E D8            MOV   DS,AX

0348  B8 3508          MOV   AX,3508H            ;get current vector 8
034B  CD 21            INT   21H                 ;and save it
034D  89 1E 0104 R     MOV   WORD PTR ADD8,BX
0351  8C 06 0106 R     MOV   WORD PTR ADD8+2,ES

0355  B8 3509          MOV   AX,3509H            ;get current vector 9
0358  CD 21            INT   21H                 ;and save it
035A  89 1E 0108 R     MOV   WORD PTR ADD9,BX
035E  8C 06 010A R     MOV   WORD PTR ADD9+2,ES

0362  B8 2508          MOV   AX,2508H
0365  BA 0233 R        MOV   DX,OFFSET VEC8      ;address interrupt procedure
0368  CD 21            INT   21H                 ;install vector 8

036A  B8 2509          MOV   AX,2509H
036D  BA 01CE R        MOV   DX,OFFSET VEC9      ;address interrupt procedure
0370  CD 21            INT   21H                 ;install vector 9

0372  BA 0344 R        MOV   DX,OFFSET INSTALL   ;find paragraphs
0375  C1 EA 04         SHR   DX,4
0378  42               INC   DX

0379  B8 3100          MOV   AX,3100H            ;set as a TSR
037C  CD 21            INT   21H
                       END
```

The program that counts keystrokes appears in Example 12–5. Note that the pop-up portion of this program only functions in the text mode, and will count any unseen keystrokes DOS generates. It also counts shift, alternate, and other keys as they are pressed and released. For example, the capital A would be counted as 2 or 3 keystrokes. The count will be inflated, but this program *is* useful for counting keystrokes by a given operator. If the operator reboots the system, the new reboot time is displayed and the count cleared to zero.

The VEC9 interrupt service procedure intercepts all keystrokes. The IN AL,60H instruction reads the scan code from the keyboard interface within the personal computer. This is then tested for the K scan code. If the K scan code is not found, the procedure increments the BCD count stored at location COUNT and returns to the normal keyboard interrupt handler. If the K scan code is detected, the contents of memory location 0000:0417 are tested for the alternate key. If an alternate key is detected, the program sets the HFLAG to 1, tosses away the hot key, and returns. Notice how the hot key is discarded by strobing I/O port number 61H. The keyboard is cleared by sending a logic 1 in bit position 7 of port 61H, followed by a logic 0 in bit position 7. The interrupt controller in the computer must also be cleared, by sending a 20H to I/O port number 20H.

The VEC8 interrupt service procedure tests the HFLAG for the hot key and the SFLAG for system start-up. If the SFLAG = 0, the system has just been installed and the time is stored in HOUR and MIN. If the HFLAG = 1, a hot key was detected by VEC9. The VEC8 procedure responds to the hot key by storing the contents of the top line of the text display at memory array SCRN. Once the *to* line of the text display is stored, the message *Time* = is displayed, followed by the installation time. Next, the message *KeyStrokes* = is displayed, followed by the BCD number stored in COUNT. Recall that the count is incremented each time VEC9 detects a key typed on the keyboard.

Password Protection TSR

Example 12–6 shows another hot key program that adds password protection in the form of a TSR interrupt hook. The keyboard interrupt is intercepted to test for the hot key combination ALT L, which can be changed. Once the hot key is detected, the keyboard interrupt is locked out, and LOCKS is set to a logic 1. The LOCKS byte shows that the hot key was detected by the keyboard interrupt.

EXAMPLE 12–6

```
                        ;a TSR program that locks the keyboard when an ALT + L key
                        ;is typed. To unlock the keyboard, type ESC (escape) followed by the
                        ;password, in this case SESAME.
                        ;***uses the 80386 or newer microprocessor***
                        ;
                                .MODEL TINY
                                .386
0000                            .CODE
                                .STARTUP
0100  E9 00B4                   JMP     INSTALL             ;install VEC8 and VEC9
0103  00                LOCKS   DB      0                   ;lock flag
0104  00000000          ADD8    DD      ?                   ;old vector 8 address
0108  00000000          ADD9    DD      ?                   ;old vector 9 address
010C  26                KEY     DB      26H                 ;for L hot key scan code
010D  08                BMASK   DB      8                   ;alternate key mask
010E  08                AKEY    DB      8                   ;alternate key
010F  53 45 53 41       PASS    DB      'SESAME',0          ;password
      4D 45 00
0116  00                FLAG8   DB      0                   ;busy flag for vector 8

0117                    VEC8    PROC    FAR                 ;clock-tick handler

0117  2E: 80 3E 0116 R          CMP     CS:FLAG8,0
      00
011D  75 08                     JNE     VEC81               ;if busy
011F  2E: 80 3E 0103 R          CMP     CS:LOCKS,0
      00
0125  75 05                     JNE     VEC82               ;if locked
0127                    VEC81:
0127  2E: FF 2E 0104 R          JMP     CS:ADD8             ;do old INT 8
012C                    VEC82:
012C  2E: C6 06 0116 R          MOV     CS:FLAG8,1          ;show busy
      01
0132  9C                        PUSHF                       ;force old INT 8
0133  2E: FF 1E 0104 R          CALL    CS:ADD8
0138  FB                        STI                         ;enable interrupts
0139  60                        PUSHA                       ;save registers
013A  06                        PUSH    ES
013B  8C C8                     MOV     AX,CS
013D  8E C0                     MOV     ES,AX               ;address this segment
013F  FC                        CLD
0140                    VEC83:
0140  BF 010F R                 MOV     DI,OFFSET PASS      ;address password
0143                    VEC84:
0143  B4 10                     MOV     AH,10H
0145  CD 16                     INT     16H                 ;read key
0147  3C 1B                     CMP     AL,1BH              ;escape key resets
0149  74 F5                     JE      VEC83
014B  AE                        SCASB                       ;compare
```

```
014C  75 08                  JNE   VEC85                  ;if wrong do nothing
014E  26: 80 3D 00           CMP   BYTE PTR ES:[DI],0
0152  74 0C                  JE    VEC86                  ;if correct
0154  EB ED                  JMP   VEC84                  ;get next
0156               VEC85:
0156  B4 10                  MOV   AH,10H
0158  CD 16                  INT   16H                    ;read key
015A  3C 1B                  CMP   AL,1BH
015C  74 E2                  JE    VEC83                  ;exit on escape
015E  EB F6                  JMP   VEC85
0160               VEC86:
0160  07                     POP   ES                     ;restore registers
0161  61                     POPA
0162  2E: C6 06 0103 R       MOV   CS:LOCKS,0             ;unlock keyboard
          00
0168  2E: C6 06 0116 R       MOV   CS:FLAG8,0             ;show not busy
          00
016E  FA                     CLI
016F  CF                     IRET

0170               VEC8      ENDP

0170               VEC9      PROC  FAR                    ;keyboard handler

0170  50                     PUSH  AX
0171  FB                     STI                          ;enable interrupts
0172  2E: 80 3E 0103 R       CMP   CS:LOCKS,0
         00
0178  75 1F                  JNE   VEC91                  ;if locked
017A  E4 60                  IN    AL,60H                 ;get scan code
017C  2E: 3A 06 010C R       CMP   AL,CS:KEY              ;test for L
0181  75 16                  JNE   VEC91                  ;not an L
0183  06                     PUSH  ES
0184  2B C0                  SUB   AX,AX
0186  8E C0                  MOV   ES,AX                  ;address segment 0000
0188  26: A0 0417            MOV   AL,ES:[417H]
018C  2E: 22 06 010D R       AND   AL,CS:BMASK            ;mask alternate
0191  2E: 3A 06 010E R       CMP   AL,CS:AKEY
0196  07                     POP   ES
0197  74 06                  JE    VEC92                  ;if an ALT + L
0199               VEC91:
0199  58                     POP   AX
019A  2E: FF 2E 0108 R       JMP   CS:ADD9                ;do normal keyboard interrupt
019F               VEC92:
019F  FA                     CLI                          ;interrupts off
01A0  E4 61                  IN    AL,61H                 ;throw away ALT + L key away
01A2  0C 80                  OR    AL,80H
01A4  E6 61                  OUT   61H,AL
01A6  24 7F                  AND   AL,7FH
01A8  E6 61                  OUT   61H,AL
01AA  B0 20                  MOV   AL,20H                 ;reset keyboard interrupt
01AC  E6 20                  OUT   20H,AL
01AE  FB                     STI                          ;enable interrupts
01AF  2E: C6 06 0103 R       MOV CS:LOCKS,1               ;lock keyboard
          01
01B5  58                     POP   AX
01B6  CF                     IRET

01B7               VEC9      ENDP

01B7               INSTALL:                               ;install VEC8 and VEC9
```

```
01B7  8C C8              MOV    AX,CS                ;load DS
01B9  8E D8              MOV    DS,AX

01BB  B8 3508            MOV    AX,3508H             ;get current vector 8
01BE  CD 21              INT    21H
01C0  89 1E 0104 R       MOV    WORD PTR ADD8,BX
01C4  8C 06 0106 R       MOV    WORD PTR ADD8+2,ES

01C8  B8 3509            MOV    AX,3509H             ;get current vector 9
01CB  CD 21              INT    21H
01CD  89 1E 0108 R       MOV    WORD PTR ADD9,BX
01D1  8C 06 010A R       MOV    WORD PTR ADD9+2,ES

01D5  B8 2508            MOV    AX,2508H             ;install new interrupt 8
01D8  BA 0117 R          MOV    DX,OFFSET VEC8
01DB  CD 21              INT    21H

01DD  B8 2509            MOV    AX,2509H             ;install new interrupt 9
01E0  BA 0170 R          MOV    DX,OFFSET VEC9
01E3  CD 21              INT    21H

01E5  BA 01B7 R          MOV    DX,OFFSET INSTALL
01E8  C1 EA 04           SHR    DX,4                 ;find paragraphs
01EB  42                 INC    DX

01EC  B8 3100            MOV    AX,3100H             ;install as TSR
01EF  CD 21              INT    21H
                         END
```

The clock-tick interrupt detects a logic 1 in LOCKS and begins comparing keyboard data, obtained with BIOS INT 16H, with the contents of the PASSWORD character string. Again, recall that the INT 21H DOS function does not work correctly within a TSR interrupt hook. In this example, the password is set to SESAME, upper case. If an error occurs as the password is typed, the escape key resets to the start of the password string. To use this TSR, type ALT + L to lock the keyboard. The keyboard remains locked until the correct password is typed. The password can only be changed by reassembling this TSR program. This can be modified to include a password change feature.

12–3 REMOVABLE TSRs

All the TSR software discussed thus far has been permanent. That is, once installed, the TSR software remains in the memory until the system is either rebooted or turned off and restarted. In some cases, it is desirable to remove a TSR at some point. DOS does not directly support the removal of a TSR. A removable TSR can be installed if the INT 2FH (DOS multiplex interrupt) is used during the installation process. The DOS multiplex interrupt allows TSR software to communicate with other TSR software and it also allows, although indirectly, the removal of a TSR.

A Removable TSR

The DOS INT 2FH function, the multiplex interrupt, is called to install and remove a TSR. The AH register contains the ID number of the TSR, ranging in value from 192 to 255. Any ID number below 192 is reserved for DOS. Each installed TSR also contains a character string that identifies the TSR so it can be removed and detected. The example of a pop-up clock is used to illustrate the installation and removal of a TSR. To make this program as flexible as possible, parameters are passed to the TSR from the command line. If the program is named CLOCK, it is installed with CLOCK and removed with CLOCK /R, where the /R switch removes the TSR. To activate the pop-up clock, type the alternate/left shift/C key combination on the keyboard. This key combination can be changed by modifying the HOT_K memory location to the scan code of a different hot key.

EXAMPLE 12–7 (pp. 419–426)

```
                        ;a TSR program that displays the time in the upper right corner
                        ;of any text mode display.
                        ;Installation is accomplished by typing the program name.
                        ;Removal is accomplished by typing the program name followed by /R.
                        ;***uses the 80386 or newer microprocessor***
                        ;
                               .MODEL TINY
                               .386
0000                           .CODE
                               .STARTUP
0100  E9 01F7                  JMP    START                 ;start installation/removal

0103  00                BUSY   DB     0                     ;busy flag
0104  00000000          ADD8   DD     ?                     ;old address of vector 8
0108  00000000          ADD9   DD     ?                     ;old address of vector 9
010C  00000000          ADD2F  DD     ?                     ;old address of vector 2F
0110  00000000          ADD15  DD     ?                     ;old address of vector 15
0114  000C              ID_L   DW     12                    ;TSR ID string length
0116  0000              PSPAD  DW     ?                     ;PSP address for TSR
0118  0186 R 020E R     VECS   DW     VEC8,VEC9,VEC2F,VEC15
      0284 R 024E R
0120  08 09 2F 15       LIST   DB     8,9,2FH,15H
0124  00                HFLAG  DB     0                     ;hot key flag
0125  2E                HKEY   DB     2EH                   ;hot key scan code for C
0126  0A                HMASK  DB     0AH                   ;hot mask alternate/left shift
0127  0A                HCODE  DB     0AH                   ;hot code alternate/left shift
0128  0010 [           VIDBUF DB     16 DUP (?)
               00
                 ]
0138  00                PSFLAG DB     0                     ;PS/2 flag
0139  00                ID_N   DB     ?                     ;TSR ID number
013A  50 4F 50 2D 55    ID_S   DB     'POP-UP CLOCK'        ;TSR ID string
      50 20 43 4C 4F
      43 4B
0146  0D 0A 41 6C 72    ABORT  DB     13,10,'Already Installed or Cannot Install.',13,10,'$'
      65 61 64 79 20
      49 6E 73 74 61
      6C 6C 65 64 20
      6F 72 20 43 61
      6E 6E 6F 74 20
      49 6E 73 74 61
      6C 6C 2E 0D 0A
      24
016F  0D 0A 54 53 52    NOTS   DB   13,10,'TSR not Installed.',13,10,'$'
      20 6E 6F 74 20
      49 6E 73 74 61
      6C 6C 65 64 2E
      0D 0A 24
```

```
0186                    VEC8   PROC   FAR               ;clock-tick interrupt

0186  2E: 80 3E 0103 R         CMP    CS:BUSY,0         ;check for busy
          00
018C  75 08                    JNE    VEC81             ;if busy
018E  2E: 80 3E 0124 R         CMP    CS:HFLAG,0        ;check for hot key
          00
0194  75 05                    JNE    VEC82             ;if hot key active
0196                    VEC81:
0196  2E: FF 2E 0104 R         JMP    CS:ADD8           ;do old interrupt
019B                    VEC82:
019B  FB                       STI                      ;enable interrupts
019C  9C                       PUSHF                    ;simulate an interrupt
019D  2E: FF 1E 0104 R         CALL   CS:ADD8
01A2  2E: C6 06 0103 R         MOV    CS:BUSY,1         ;indicate busy
          01
01A8  60                       PUSHA                    ;save registers
01A9  1E                       PUSH   DS
01AA  06                       PUSH   ES
01AB  8C C8                    MOV    AX,CS             ;address video buffer
01AD  8E C0                    MOV    ES,AX
01AF  BF 0128 R                MOV    DI,OFFSET VIDBUF
01B2  B8 B800                  MOV    AX,0B800H         ;address video memory
01B5  8E D8                    MOV    DS,AX
01B7  BE 0090                  MOV    SI,2*72           ;address column 72
01BA  B9 0010                  MOV    CX,16             ;load count
01BD  F3/ A4                   REP    MOVSB             ;copy from video memory
01BF  1E                       PUSH   DS                ;exchange segments
01C0  06                       PUSH   ES
01C1  1F                       POP    DS
01C2  07                       POP    ES
01C3  BF 0090                  MOV    DI,2*72           ;address video memory
01C6  B4 02                    MOV    AH,02H            ;get time from BIOS
01C8  CD 1A                    INT    1AH
01CA  B4 0F                    MOV    AH,0FH            ;set video attribute
01CC  8A C5                    MOV    AL,CH             ;display hours
01CE  E8 002D                  CALL   DISP
01D1  B0 3A                    MOV    AL,':'            ;display colon
01D3  AB                       STOSW
01D4  8A C1                    MOV    AL,CL             ;display minutes
01D6  E8 0025                  CALL   DISP
01D9                    VEC83:
01D9  B4 11                    MOV    AH,11H            ;test keyboard
01DB  CD 16                    INT    16H
01DD  74 FA                    JE     VEC83             ;wait for any key
01DF  B4 10                    MOV    AH,10H            ;dump keystroke
01E1  CD 16                    INT    16H
01E3  BF 0090                  MOV    DI,2*72           ;restore video
01E6  BE 0128 R                MOV    SI,OFFSET VIDBUF
01E9  B9 0010                  MOV    CX,16
01EC  F3/ A4                   REP    MOVSB
01EE  07                       POP    ES                ;restore register
01EF  1F                       POP    DS
01F0  61                       POPA
01F1  2E: C6 06 0124 R         MOV    CS:HFLAG,0        ;clear hot key
          00
01F7  2E: C6 06 0103 R         MOV    CS:BUSY,0         ;clear busy
          00
01FD  CF                       IRET

01FE                    VEC8   ENDP
```

```
                        ;the DISP procedure displays the BCD contents of AL on
                        ;the text mode video displays at the offset address DI.
                        ;
01FE                    DISP    PROC    NEAR
01FE  50                        PUSH    AX
01FF  C0 E8 04                  SHR     AL,4
0202  24 0F                     AND     AL,0FH
0204  04 30                     ADD     AL,30H             ;mask ASCII
0206  AB                        STOSW
0207  58                        POP     AX
0208  24 0F                     AND     AL,0FH
020A  04 30                     ADD     AL,30H
020C  AB                        STOSW
020D  C3                        RET

020E                    DISP    ENDP

020E                    VEC9    PROC    FAR                ;keyboard vector (PC/XT/AT)

020E  FB                        STI                        ;enable interrupts
020F  50                        PUSH    AX
0210  E4 60                     IN      AL,60H             ;get scan code
0212  2E: 3A 06 0125 R          CMP     AL,CS:HKEY         ;test scan code
0217  75 2F                     JNE     VEC91              ;if no hot key
0219  06                        PUSH    ES
021A  B8 0000                   MOV     AX,0               ;address segment 0000
021D  8E C0                     MOV     ES,AX
021F  26: A0 0417               MOV     AL,BYTE PTR ES:[417H]
0223  07                        POP     ES
0224  2E: 22 06 0126 R          AND     AL,CS:HMASK
0229  2E: 3A 06 0127 R          CMP     AL,CS:HCODE
022E  75 18                     JNE     VEC91              ;if no hot key
0230  FA                        CLI                        ;disable interrupt
0231  E4 61                     IN      AL,61H
0233  0C 80                     OR      AL,80H             ;clear keyboard
0235  E6 61                     OUT     61H,AL
0237  24 7F                     AND     AL,7FH
0239  E6 61                     OUT     61H,AL
023B  B0 20                     MOV     AL,20H             ;clear keyboard interrupt
023D  E6 20                     OUT     20H,AL
023F  FB                        STI                        ;enable interrupts
0240  2E: C6 06 0124 R          MOV     CS:HFLAG,1         ;signal hot key
          01
0246  58                        POP     AX
0247  CF                        IRET
0248                    VEC91:
0248  58                        POP     AX
0249  2E: FF 2E 0108 R          JMP     CS:ADD9            ;do normal keyboard interrupt

024E                    VEC9    ENDP

024E                    VEC15   PROC    FAR                ;interrupt 15H for PS/2

024E  FB                        STI                        ;enable interrupt
024F  80 FC 4F                  CMP     AH,4FH             ;keyboard service?
0252  75 2A                     JNE     VEC151             ;if not for keyboard
0254  50                        PUSH    AX
0255  2E: 3A 06 0125 R          CMP     AL,CS:HKEY         ;test hot key code
025A  75 22                     JNE     VEC151             ;if not hot key
025C  06                        PUSH    ES
025D  B8 0000                   MOV     AX,0
```

```
0260  8E C0                    MOV   ES,AX                 ;address segment 0000
0262  26: A0 0417              MOV   AL,BYTE PTR ES:[417H]
0266  07                       POP   ES
0267  2E: 22 06 0126 R         AND   AL,CS:HMASK
026C  2E: 3A 06 0127 R         CMP   AL,CS:HCODE
0271  75 0B                    JNE   VEC151                ;if not hot key
0273  58                       POP   AX
0274  2E: C6 06 0124 R         MOV   CS:HFLAG,1            ;signal hot key
      01
027A  F8                       CLC                         ;signal BIOS no key
027B  CA 0002                  RET   2
027E                     VEC151:
027E  FA                       CLI                         ;disable interrupts
027F  2E: FF 2E 0110 R         JMP   CS:ADD15              ;do old interrupt

0284                     VEC15  ENDP
                         ;
                         ;the DOS multiplex interrupt provides a way of passing
                         ;parameters to the TSR.
                         ;This handler accepts two commands: AL = 0 and AL = 1.
                         ;***input parameters***
                         ;AH = TSR ID number
                         ;AL = function number
                         ;***functions provided***
                         ;AL = 0; returns the address of the TSR ID in ES:DI
                         ;AL = 1; returns the address of the PSP in ES:DI
                         ;
0284                     VEC2F  PROC  FAR                  ;multiplex interrupt

0284  2E: 3A 26 0139 R         CMP   AH,CS:ID_N            ;test this TSR ID number
0289  74 05                    JE    VEC2F1                ;if this ID
028B  2E: FF 2E 010C R         JMP   CS:ADD2F              ;if not, do old interrupt
0290                     VEC2F1:
0290  3C 00                    CMP   AL,0                  ;function 0? (verify presence)
0292  75 09                    JNE   VEC2F2                ;if not function 0
0294  B0 FF                    MOV   AL,0FFH               ;indicate ID good
0296  0E                       PUSH  CS
0297  07                       POP   ES                    ;address this segment
0298  BF 013A R                MOV   DI,OFFSET ID_S        ;address TSR ID string
029B  EB 0C                    JMP   VEC2F3
029D                     VEC2F2:
029D  3C 01                    CMP   AL,1                  ;function 1 (get PSP)
029F  75 08                    JNE   VEC2F3
02A1  2E: 8E 06 0116 R         MOV   ES,CS:PSPAD           ;get PSP address
02A6  BF 0000                  MOV   DI,0                  ;clear DI
02A9                     VEC2F3::                          ;make global for use in MULT
02A9  CF                       IRET

02AA                     VEC2F  ENDP
                         ;
                         ;the MULT procedure interrogates the multiplex interrupt
                         ;to get an ID number and indicate whether or not POP-UP CLOCK
                         ;is installed.
                         ;***return codes***
                         ;AX = 0; cannot install (no ID code available)
                         ;AX = 1; POP-UP CLOCK is installed
                         ;AX = 2; POP-UP CLOCK is not installed
                         ;
02AA                     MULT   PROC  NEAR                 ;get ID and functions

02AA  50                       PUSH  AX
```

```
02AB  B8 352F                         MOV   AX,352FH             ;test for DOS version 2.X
02AE  CD 21                           INT   21H
02B0  8C C0                           MOV   AX,ES
02B2  0B C3                           OR    AX,BX
02B4  75 08                           JNE   MULT1                ;if vector loaded
02B6  BA 02A9 R                       MOV   DX,OFFSET VEC2F3     ;IRET vector address
02B9  B8 252F                         MOV   AX,252FH
02BC  CD 21                           INT   21H
02BE                       MULT1:
02BE  B6 C0                           MOV   DH,192               ;initial test ID number
02C0                       MULT2:
02C0  8A E6                           MOV   AH,DH                ;load test ID
02C2  32 C0                           XOR   AL,AL                ;set function 00H
02C4  52                              PUSH  DX
02C5  1E                              PUSH  DS
02C6  CD 2F                           INT   2FH                  ;do multiplex interrupt
02C8  1F                              POP   DS
02C9  5A                              POP   DX
02CA  0A C0                           OR    AL,AL
02CC  74 22                           JZ    MULT5                ;if ID number available
02CE  3C FF                           CMP   AL,0FFH
02D0  75 0C                           JNE   MULT3                ;if not ready
02D2  BE 013A R                       MOV   SI,OFFSET ID_S       ;check ID string
02D5  2E: 8B 0E 0114 R                MOV   CX,CS:ID_L
02DA  F3/ A6                          REPE  CMPSB
02DC  74 09                           JE    MULT4                ;if this TSR installed
02DE                       MULT3:
02DE  FE C6                           INC   DH
02E0  75 DE                           JNZ   MULT2                ;keep looking for free ID
02E2  B8 0000                         MOV   AX,0                 ;no ID available code
02E5  EB 11                           JMP   MULT6                ;end it
02E7                       MULT4:
02E7  8A E6                           MOV   AH,DH                ;get ID number
02E9  CD 2F                           INT   2FH
02EB  B8 0001                         MOV   AX,1                 ;is installed code
02EE  EB 08                           JMP   MULT6
02F0                       MULT5:
02F0  2E: 88 36 0139 R                MOV   CS:ID_N,DH           ;save ID number
02F5  B8 0002                         MOV   AX,2                 ;not installed
02F8                       MULT6:
02F8  5B                              POP   BX
02F9  C3                              RET

02FA                       MULT       ENDP

02FA                       START:                                ;begin installation/deinstallation
02FA  8C C8                           MOV   AX,CS
02FC  8E D8                           MOV   DS,AX                ;address data segment
02FE  FC                              CLD
02FF  80 3E 0080 03                   CMP   BYTE PTR DS:[80H],3
0304  75 09                           JNE   ST1
0306  80 3E 0082 2F                   CMP   BYTE PTR DS:[82H],'/'
030B  0F 84 0080                      JE    ST6                  ;to deinstall
030F                       ST1:
030F  B0 00                           MOV   AL,0
0311  E8 FF96                         CALL  MULT                 ;request multiplex interrupt 2F
0314  83 F8 02                        CMP   AX,2
0317  0F 85 0110                      JNE   ST12                 ;if already installed
031B  C6 06 0138 R 00                 MOV   PSFLAG,0             ;indicate PC/XT/AT
0320  8C 0E 0116 R                    MOV   PSPAD,CS             ;save PSP address
0324  B4 30                           MOV   AH,30H
```

```
0326  CD 21                        INT   21H                  ;check DOS version
0328  86 C4                        XCHG  AL,AH
032A  3D 031E                      CMP   AX,031EH
032D  72 18                        JB    ST2                  ;if DOS version below 3.3
032F  B8 0C00                      MOV   AX,0C00H             ;test for PS/2
0332  CD 15                        INT   15H
0334  FB                           STI                        ;enable interrupts
0335  74 10                        JZ    ST2                  ;if not PS/2
0337  0A E4                        OR    AH,AH
0339  75 0C                        JNZ   ST2                  ;if not PS/2
033B  26: F6 47 05 10              TEST  BYTE PTR ES:[BX+5],10H
0340  74 05                        JZ    ST2                  ;if not PS/2
0342  C6 06 0138 R 01              MOV   PSFLAG,1             ;if PS/2
0347                         ST2:
0347  B9 0004                      MOV   CX,4
034A  BF 0104 R                    MOV   DI,OFFSET ADD8
034D  BE 0120 R                    MOV   SI,OFFSET LIST
0350                         ST3:
0350  B4 35                        MOV   AH,35H               ;get and save vectors
0352  AC                           LODSB
0353  CD 21                        INT   21H
0355  89 1D                        MOV   [DI],BX
0357  8C 45 02                     MOV   [DI+2],ES
035A  83 C7 04                     ADD   DI,4
035D  E2 F1                        LOOP  ST3
035F  B9 0003                      MOV   CX,3
0362  BF 0118 R                    MOV   DI,OFFSET VECS       ;load new vectors
0365  BE 0120 R                    MOV   SI,OFFSET LIST
0368                         ST4:
0368  8B 15                        MOV   DX,[DI]
036A  B4 25                        MOV   AH,25H
036C  AC                           LODSB
036D  CD 21                        INT   21H
036F  83 C7 02                     ADD   DI,2
0372  E2 F4                        LOOP  ST4
0374  80 3E 0138 R 01              CMP   PSFLAG,1
0379  75 08                        JNE   ST5
037B  B8 2515                      MOV   AX,2515H
037E  BA 024E R                    MOV   DX,OFFSET VEC15
0381  CD 21                        INT   21H
0383                         ST5:
0383  BA 02AA R                    MOV   DX,OFFSET MULT
0386  C1 EA 04                     SHR   DX,4
0389  42                           INC   DX
038A  B8 3100                      MOV   AX,3100H             ;make resident
038D  CD 21                        INT   21H
038F                         ST6:
038F  80 26 0083 DF                AND   BYTE PTR DS:[83H],0DFH
0394  80 3E 0083 52                CMP   BYTE PTR DS:[83H],'R'
0399  0F 85 008E                   JNE   ST12                 ;abort
039D  32 C0                        XOR   AL,AL
039F  E8 FF08                      CALL  MULT                 ;multiplex
03A2  83 F8 01                     CMP   AX,1
03A5  74 0A                        JE    ST7                  ;if installed
03A7  B4 09                        MOV   AH,9
03A9  BA 016F R                    MOV   DX,OFFSET NOTS
03AC  CD 21                        INT   21H
03AE  E9 0085                      JMP   ST13                 ;exit to DOS
03B1                         ST7:
03B1  06                           PUSH  ES
03B2  1F                           POP   DS
```

```
03B3  B9 0003                    MOV   CX,3
03B6  BE 0120 R                  MOV   SI,OFFSET LIST
03B9  BF 0118 R                  MOV   DI,OFFSET VECS
03BC                       ST8:
03BC  B4 35                      MOV   AH,35H            ;test vectors
03BE  AC                         LODSB
03BF  CD 21                      INT   21H
03C1  3B 1D                      CMP   BX,[DI]
03C3  75 66                      JNE   ST12              ;if wrong vector
03C5  8C C0                      MOV   AX,ES
03C7  8C DB                      MOV   BX,DS
03C9  3B C3                      CMP   AX,BX
03CB  75 5E                      JNE   ST12              ;if wrong vector
03CD  83 C7 02                   ADD   DI,2
03D0  E2 EA                      LOOP  ST8
03D2  80 3E 0138 R 01            CMP   PSFLAG,1
03D7  75 11                      JNE   ST9
03D9  B4 35                      MOV   AH,35H
03DB  AC                         LODSB
03DC  CD 21                      INT   21H
03DE  3B 1D                      CMP   BX,[DI]
03E0  75 49                      JNE   ST12              ;if wrong vector
03E2  8C C0                      MOV   AX,ES
03E4  8C DB                      MOV   BX,DS
03E6  3B C3                      CMP   AX,BX
03E8  75 41                      JNE   ST12              ;if wrong vector
03EA                       ST9:
03EA  B0 01                      MOV   AL,1
03EC  E8 FEBB                    CALL  MULT              ;get PSP segment
03EF  06                         PUSH  ES
03F0  B9 0003                    MOV   CX,3
03F3  BE 0120 R                  MOV   SI,OFFSET LIST
03F6  BF 0104 R                  MOV   DI,OFFSET ADD8
03F9                       ST10:
03F9  AC                         LODSB
03FA  B4 25                      MOV   AH,25H
03FC  1E                         PUSH  DS
03FD  C5 15                      LDS   DX,[DI]
03FF  CD 21                      INT   21H
0401  1F                         POP   DS
0402  83 C7 04                   ADD   DI,4
0405  E2 F2                      LOOP  ST10
0407  80 3E 0138 R 01            CMP   PSFLAG,1
040C  75 09                      JNE   ST11              ;if not PS/2
040E  AC                         LODSB
040F  B4 25                      MOV   AH,25H
0411  1E                         PUSH  DS
0412  C5 15                      LDS   DX,[DI]
0414  CD 21                      INT   21H
0416  1F                         POP   DS
0417                       ST11:
0417  1F                         POP   DS                ;get PSP address
0418  A1 002C                    MOV   AX,DS:[2CH]       ;release environment
041B  8E C0                      MOV   ES,AX
041D  B4 49                      MOV   AH,49H
041F  CD 21                      INT   21H
0421  72 08                      JC    ST12              ;if error
0423  1E                         PUSH  DS
0424  07                         POP   ES
0425  B4 49                      MOV   AH,49H
0427  CD 21                      INT   21H               ;release TSR memory
```

```
0429  73 0B              JNC   ST13              ;if no error, exit to DOS
042B              ST12:
042B  8C C8              MOV   AX,CS
042D  8E D8              MOV   DS,AX
042F  B4 09              MOV   AH,9              ;indicate already installed
0431  BA 0146 R          MOV   DX,OFFSET ABORT
0434  CD 21              INT   21H
0436              ST13:
                         .EXIT
                         END
```

Example 12–7 lists the interrupt service procedure for interrupt vectors 8, 9, and 2FH. This software is long, because it illustrates just about every function required for most TSR software, including installation and removal as well as type of computer system and DOS version. Vector 8 detects the hot key trapped by interrupt 9 and interrupt 2FH (the multiplex interrupt) allows the TSR to be removed once it is installed. Once the hot key is activated, the clock-tick interrupt causes the time of day to appear in the upper right corner of the screen. The time is updated each minute and remains on the screen until any key is typed. When any key is typed, the time disappears and is replaced by previous screen's information. This function only works in text modes 0–3 for DOS versions 2.0 or higher. It also functions on a PS/2 model because the required INT 15H intercept is included for detecting the scan code from the PS/2 keyboard.

The VEC8 and VEC9 procedures are very similar to others presented earlier in this chapter. The VEC9 interrupt handler detects the hot key (in this case, alternate/left shift/C) by setting the HFLAG. The VEC8 handler monitors the HFLAG to determine whether a hot key was typed. If the HFLAG = 1, the VEC8 handler accesses the system clock through the BIOS, and displays the time through the use of the DISP procedure. Note that in an IBM PS/2 system, the VEC15 handler places a 1 into HFLAG, instead of into VEC9.

In order to remove a TSR, an additional interrupt, the DOS multiplex interrupt, must be monitored. The DOS multiplex interrupt (VEC2F) indicates whether the TSR is installed and also provides the address of the ID string (ES:DI) and the address of the PSP (ES:DI). The address of the PSP is needed so the TSR can be removed from the memory.

Upon execution of this program, location START is accessed, to begin installation or deinstallation. The command line is accessed to determine whether a /R was typed for removal or just the program name. If the /R appears after the program name on the DOS command line, the START portion of the program jumps to location ST6. The MULT procedure is called to determine whether the TSR is installed. If it is not installed, the message NOTS is displayed ("TSR not installed"), and an exit to DOS occurs. If MULT is displayed, that indicates the TSR is installed, and execution continues at ST7.

The TSR is deinstalled by first restoring the interrupt vectors using DOS INT 21H function call 25H. Once the interrupt vectors are restored, the environment block, set up by DOS upon execution of the TSR, is located by the contents of address 2CH in the program segment prefix. This address is moved to ES and DOS INT 21H function 49H (release memory) is called to release the memory allocated to the environment block. Finally, the address of the PSP is placed in ES and function 49H is again called to release the memory used by the TSR.

If the /R switch is missing at location START, the program continues at ST1. At ST1, the MULT procedure is called to determine whether the TSR is already installed. If it is installed, the program continues at location ST12, where a message is displayed indicating that the TSR is already installed. This is followed by an exit to DOS. If, at ST1, the MULT procedure does not detect that POP-UP CLOCK is installed, the program continues by saving the PSP address (needed for removal), and installing the interrupt vectors for interrupts, 8, 9, 15H, and 2FH.

12–4 SUMMARY

1. An interrupt hook allows a program to intercept the interrupt structure of the computer system. Interrupts that are commonly hooked into are the clock-tick interrupt (vector 8) and the keyboard interrupt (vector 9). Although these are the interrupts most often used with hooks, any interrupt can be hooked for any purpose.
2. An interrupt is intercepted by first removing the current interrupt service procedure address from the vector location by DOS INT 21H, function call 35H. Next, a new interrupt vector is installed using DOE INT 21H, function call 25H. The new vector installs the user-provided interrupt service procedure.
3. The clock-tick interrupt (vector 8) occurs about 18.2 times a second (the actual time is 18.206). The clock-tick allows a program to time events independent of the system clock frequency, because it is a constant on all personal computer systems.
4. The 8253 timer generates the clock-tick interrupt (timer 0), refreshes the dynamic RAM in the system (timer 1), and can provide a tone at the speaker (timer 2).
5. A terminate and stay resident (TSR) program is executed and then placed in the memory until the computer is turned off. The TSR is accessed by the interrupt structure of the computer and can be activated by a special key sequence called a hot key. Most TSR programs are assembled as .COM programs instead of .EXE programs.
6. The keyboard can be tested through the number 9 interrupt vector. Access to the keyboard is via I/O port 60H unless the personal computer is a series PS/2. The PS/2 accesses the keyboard via interrupt vector 15H. The key scan code is returned from INT 15H or IN AL,60H. The key scan code, unlike an ASCII code, indicates which physical key is activated on the keyboard. Memory location 0000:0417 indicates which shift keys, alternate keys, control keys, and so forth, are active at the time of the vector 9 interrupt.
7. A hot key is a special key sequence that accesses a TSR program. A hot key is composed of a key scan code and a code found in memory location 0000:0417. For example, the C/left alternate/right shift combination could be a hot key, as could any other key and control combination. A program accessed by a hot key is often called a pop-up program.
8. The multiplex interrupt (vector 2FH) is used by DOS to keep track of TSR programs. Through this interrupt, it is possible for TSR programs to communicate with each other and it is also possible to remove the TSR from the memory. Functions provided by the multiplex interrupt often return the address of the name of the TSR and its PSP for removal.

12–5 QUESTIONS AND PROBLEMS

1. What is an interrupt hook?
2. What DOS function is used to copy the current interrupt vector address from the interrupt vector table?
3. Develop a sequence of instructions that copy interrupt vectors 10H, 11H, and 12H into doubleword-sized memory locations ADD10, ADD11, and ADD12.
4. What DOS function is used to install a new interrupt service procedure address into an interrupt vector?
5. Develop a sequence of instructions that install VEC11 as interrupt vector 11H, VEC12 as interrupt vector 12H, and VEC13 as interrupt vector 13H.
6. How is a program organized so it becomes a .COM program?
7. The clock-tick interrupt occurs _______________ times per second and is found at interrupt vector _______________.
8. Timer 2 can provide an audio tone to a personal computer's speaker. Develop a program (not a TSR) that beeps the speaker with a 3,000 Hz tone for 2 seconds.
9. What constants are changed in the CHIME program of Example 12–3 to change the duration of the audio tone?
10. What is changed in Example 12–3 to change the frequency of the audio tone?
11. What DOS function causes a program to become a TSR?
12. How are the number of bytes required for a TSR indicated to the DOS system?
13. Develop a TSR that beeps the speaker with a 2,000 Hz tone for 1 second, ten minutes after the installation of the TSR. Your TSR may beep only one time and then deactivate itself, but it must remain resident.
14. What is a hot key?
15. How is the scan code for a hot key retrieved from the keyboard interface in an AT-style computer system?
16. How is the scan code for a hot key retrieved in a PS/2 series personal computer system?
17. What is the scan code for the F11 key?
18. If the right-shift key is activated, what bit of word-sized memory location 0000:0417 indicates this event?
19. What is the purpose of the alternate/shift mask?
20. What can be stated about the access time to the PEL color register through the video BIOS and vs. direct access using the OUT instruction?
21. Develop a hot key TSR that beeps the speaker with a 250 Hz tone whenever alternate/F12 key is pressed.
22. Develop a hot key TSR that clears the screen whenever alternate/F11 is pressed. This TSR only functions for text modes 0–3.
23. Develop a hot key TSR that clears the screen whenever an alternate/left shift/F10 is pressed. This TSR only functions for graphics modes 12H and 13H.
24. Develop a TSR, which can be installed and removed, that displays the date in the upper right corner of the screen when activated by the hot key sequence alternate/D.
25. Develop a TSR that enables the mouse when an ALT O is typed and disables the mouse when an ALT F is typed.

CHAPTER 13

The Arithmetic Coprocessor

INTRODUCTION

The Intel family of arithmetic coprocessors for use with the 80486SX microprocessor includes the 8087, 80287, 80387SX, 80387DX, and the 80487SX. The 80486DX and Pentium microprocessors both contain their own built-in arithmetic coprocessors. Some of the cloned 80486 microprocessors (from IBM and Cyrix) do not contain arithmetic coprocessors. The instruction sets and programming for all devices are almost identical, but each coprocessor is designed to function with a different Intel microprocessor. Because the 80486DX and Pentium microprocessors, which are so commonplace, have built-in arithmetic coprocessors, many programs now require, or at least benefit from, a coprocessor. This chapter provides detail on the entire family of arithmetic coprocessors.

The 80X87 family of coprocessors is able to multiply, divide, add, subtract, find the square root, partial tangent, partial arctangent, and logarithms. Data types include: 16-, 32-, and 64-bit signed integers; 18-digit BCD data; and 32-, 64-, and 80-bit floating-point numbers. The operations performed by the 80X87 generally execute many times faster than equivalent operations written with the most efficient programs using the microprocessor's normal instruction set. With the improved Pentium coprocessor, operations execute about 5 times faster than those performed by the 80486 microprocessor with an equal clock frequency. The Pentium can often execute a coprocessor instruction and two integer instructions simultaneously.

OBJECTIVES

Upon completion of this chapter, you will be able to:

1. Convert between decimal data and signed integer, BCD, and floating-point data for use by the arithmetic coprocessor.
2. Explain the operation of the 80X87 arithmetic coprocessor.

3. Explain the operation and addressing modes of each arithmetic coprocessor instruction.
4. Develop programs that solve complex arithmetic problems using the arithmetic coprocessor.

13–1 DATA FORMATS FOR THE ARITHMETIC COPROCESSOR

This section of the text presents the types of data used by all arithmetic coprocessor family members. (Refer to Table 13–1 for a listing of all Intel microprocessors and their companion coprocessors.) These data types include signed-integer, BCD, and floating-point. Each has a specific use in a system, and many systems require all three data types. Note that assembly language programming with the coprocessor is often limited to modifying the coding generated by a high-level language such as C/C++. In order to accomplish any such modification, the instruction set and some basic programming concepts are required.

Signed Integers

The signed integers used with the coprocessor are basically the same as those described in Chapter 1. When used with the arithmetic coprocessor, signed integers are 16 (*word*) , 32 (*short integer*), or 64 bits (*long integer*) in width. Conversion between decimal and signed-integer format is handled in exactly the same manner as described in Chapter 1. As you will recall, positive numbers are stored in true form with a left-most sign bit of 0, and negative numbers are stored in two's complement form with a left-most sign bit of 1.

The word integers range in value from –32,768 to +32,767. The short integer range is $\pm 2 \times 10^{+9}$, and the long integer range is $\pm 9 \times 10^{+18}$. Integer data types are found in some applications that use the arithmetic coprocessor. Refer to Figure 13–1, which shows these three forms of signed-integer data.

Data are stored in memory using the same assembler directives described and used in earlier chapters: DW defines words, DD defines short integers, and DQ defines long-integers. Example 13–1 shows how several different sizes of signed integers are defined for use by the assembler and arithmetic coprocessor.

TABLE 13–1 Microprocessor and Intel coprocessor compatibility

Microprocessor	*Coprocessor*
8086	8087
8088	8087
80286	80287
80386SX	80387SX
80386DX	80387DX
80486SX	80487
80486DX	Built into microprocessor
Pentium	Built into microprocessor

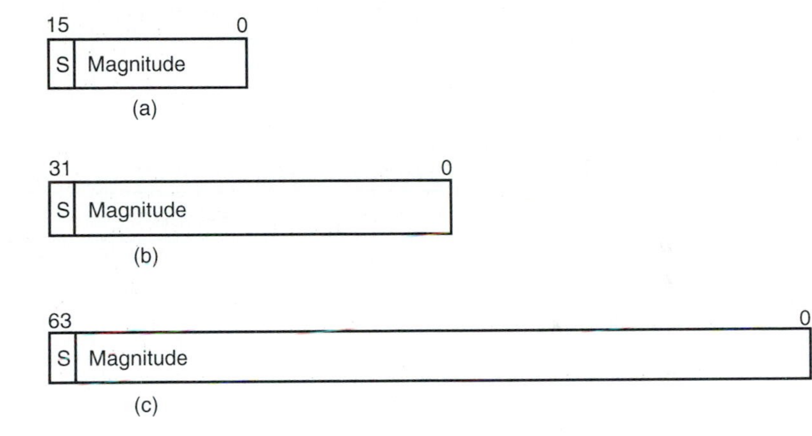

FIGURE 13–1 Integer formats for the 80X87 family of arithmetic coprocessors: (a) word, (b) short, and (c) long.

EXAMPLE 13–1

```
0000  0002                 DATA1  DW   +2        ;16-bit integer
0002  FFDE                 DATA2  DW   -34       ;16-bit integer
0004  000004D2             DATA3  DD   +1234     ;short integer
0008  FFFFFF9C             DATA4  DD   -100      ;short integer
000C  0000000000005BA0     DATA5  DQ   +23456    ;long integer
0014  FFFFFFFFFFFFFF86     DATA6  DQ   -122      ;long integer
```

Binary-Coded Decimal (BCD) Data

The binary-coded decimal (BCD) form requires 80 bits of memory. Each number is stored as an 18-digit packed integer in 9 bytes of memory, two digits per byte. The tenth byte contains only a sign bit for the 18-digit signed BCD number. Figure 13–2 shows the format of the BCD number used with the arithmetic coprocessor. Note that both positive and negative numbers are stored in true form, never in ten's complement form. The DT directive stores BCD data in the memory as illustrated in Example 13–2.

EXAMPLE 13–2

```
0000                    DATA1   DT   200       ;200 decimal stored as BCD
   00000000000000000200

000A                    DATA2   DT   -10       ;-10 decimal stored as BCD
   80000000000000000010

0014                    DATA3   DT   10020     ;10,020 decimal stored as BCD
   00000000000000010020
```

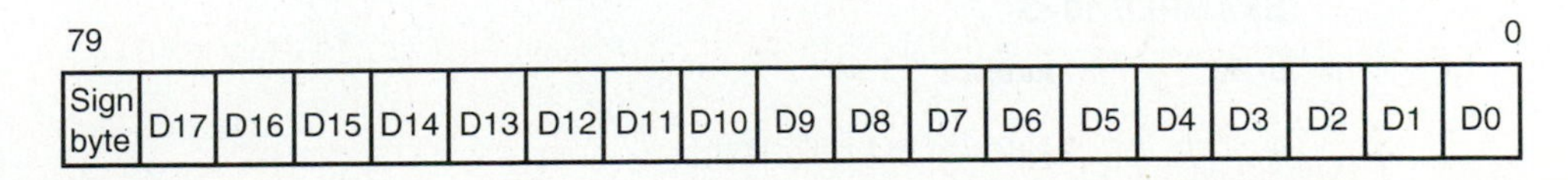

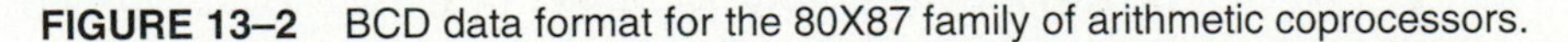

FIGURE 13–2 BCD data format for the 80X87 family of arithmetic coprocessors.

FIGURE 13–3 Floating-point (real) format for the 80X87 family of arithmetic coprocessors. (a) Short (single-precision) with a bias of 7FH, (b) long (double-precision) with a bias of 3FFH, and (c) temporary (extended-precision) with a bias of 3FFFH.

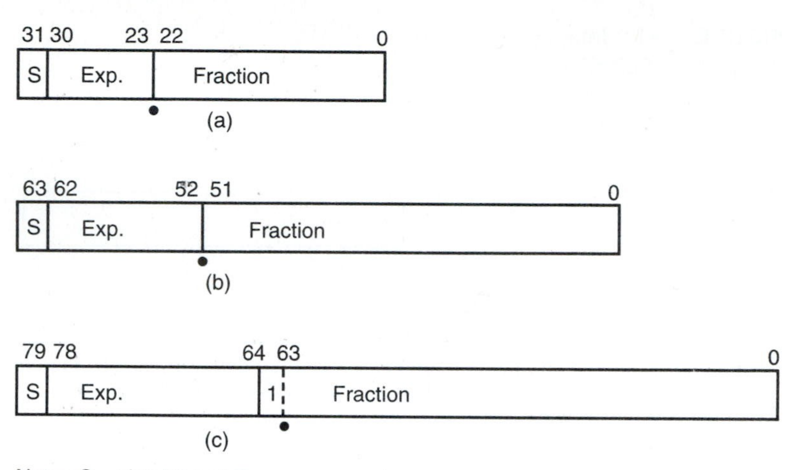

Floating-Point Numbers

Floating-point numbers are often called **real numbers** because they hold signed integers, fractions, and mixed numbers. A floating-point number has three parts: a **sign bit, a biased exponent,** and a **significand.** Floating-point numbers are written in **scientific binary notation.** The Intel family of arithmetic coprocessors supports three types of floating-point numbers: short (32 bits), long (64 bits), and temporary (80 bits). Refer to Figure 13–3 for the three forms of the floating-point number. The short form is also called a **single precision number** and the long form is called a **double precision number.** Sometimes the 80-bit temporary form is called an **extended precision number.** The floating-point numbers and the operations performed by the arithmetic coprocessor conform to the IEEE-754 standard adopted by all major personal computer software producers. This includes Microsoft, which has recently stopped supporting the Microsoft floating-point format and the ANSI floating-point standard popular in mainframe computer systems.

Converting to Floating-Point Form. Converting from decimal to floating-point form is a simple task that is accomplished through the following steps:

1. Convert the decimal number into binary.
2. Normalize the binary number.
3. Calculate the biased exponent.
4. Store the number in floating-point format.

These four steps are illustrated for the decimal number 100.25 in Example 13–3. Here, the decimal number is converted to a single precision (32-bit) floating-point number.

EXAMPLE 13–3

Step	**Result**
1	100.25 = 1100100.01
2	1100100.01 = 1.10010001 x 2^6

```
3           110 + 01111111 = 10000101

4           Sign = 0
            Exponent = 10000101
            Significand = 10010001000000000000000
```

In Step 3, the biased exponent is the exponent, a +6 (110), plus a bias of 01111111 (7FH), or 10000101 (85H). All single precision numbers use a bias of 7FH, all double precision numbers use a bias of 3FFH, and all extended precision numbers use a bias of 3FFFH.

In Step 4, the prior information is combined to form the floating-point number. The left-most bit is the sign bit of the number. In this case, the sign bit is 0, because the number is +100.25. The biased exponent follows the sign bit. The significand is a 23-bit number with an implied one-bit. Note that the significand of 1.XXXX is the XXXX portion. The 1. is an implied one-bit that is only stored in the extended precision form of the floating-point number as an explicit one-bit.

Some special rules apply to a few numbers. The number 0, for example, is stored as all zeros except for the sign bit, which can be a logic 1 to represent a negative zero. Plus and minus infinity are stored as logic 1's in the exponent, with a significand of all zeros and a sign bit that represents plus or minus. A NAN (*not-a-number*) is an invalid floating-point result that has all ones in the exponent with a significand that is *not* all zeros.

Converting from Floating-Point Form. Conversion to a decimal number from a floating-point number is summarized in the following steps:

1. Separate the sign bit, biased exponent, and significand.
2. Convert the biased exponent into a true exponent by subtracting the bias.
3. Write the number as a normalized binary number.
4. Convert it to a denormalized binary number.
5. Convert the denormalized binary number to decimal.

These five steps convert a single precision floating-point number to decimal in Example 13–4. The sign bit of 1 makes the decimal result negative. The implied one-bit is added to the normalized binary result in Step 3.

EXAMPLE 13–4

```
Step        Result

1           Sign = 1
            Exponent = 10000011
            Significand = 10010010000000000000000

2           100 = 10000011 - 01111111

3           1.1001001 x 2^4

4           11001.001

5           -25.125
```

Storing Floating-Point Data in Memory. Floating-point numbers are stored using the DD directive for single precision, DQ for double precision, and DT for extended precision. Some examples of floating-point data storage are shown in Example 13–5. The Microsoft version 6.0 macro assembler contains an error that does not allow a plus sign to be used with positive floating-point numbers. Thus, a +92.45 must be defined as 92.45 for the assembler to function correctly. Microsoft has assured this author that the error has been corrected in version 6.11 of MASM if the REAL4, REAL8, or REAL10 directives are used in place of DD, DQ, and DT to specify floating-point data. Note that the assembler provides access 8087 emulator if your system does not contain a Pentium, 80486, or any other microprocessor with a coprocessor. The emulator comes with all Microsoft high-level languages, or with shareware programs such as EM87. The emulator is accessed by including the OPTION EMULATOR statement immediately following the .MODEL statement in a program. Be aware that the emulator does not emulate some of the coprocessor instructions. Do not use this option if your system contains a coprocessor.

EXAMPLE 13–5

```
0000  C377999A          DATA7   DD     -247.6        ;define single precision
0004  40000000          DATA8   DD     2.0           ;define single precision
0008  486F4200          DATA9   REAL4  2.45E+5       ;define single precision
000C                    DATA10  DQ     100.25        ;define double precision
        4059100000000000
0014                    DATA11  REAL8  0.001235      ;define double precision
        3F543BF727136A40
001C                    DATA12  REAL10 33.9876       ;define extended precision
        400487F34D6A161E4F76
```

13–2 80X87 ARCHITECTURE

The 80X87 is designed to operate concurrently with the microprocessor. Note that the 80486DX and Pentium microprocessors contain their own *internal* and fully compatible versions of the 80387 coprocessor. With other family members, the coprocessor is an external integrated circuit that parallels most of the connections on the microprocessor. The 80X87 executes 68 different instructions. The microprocessor executes all normal instructions and the 80X87 executes arithmetic coprocessor instructions. The microprocessor and coprocessor can execute their respective instructions simultaneously or concurrently. The numeric or **arithmetic coprocessor** is a special-purpose microprocessor specially designed to efficiently execute arithmetic and transcendental operations.

The microprocessor intercepts and executes the normal instruction set, and the coprocessor intercepts and executes only the coprocessor instructions. Recall that the coprocessor instructions are actually escape (ESC) instructions. These instructions are used by the microprocessor to generate a memory address for the coprocessor, so the coprocessor can execute a coprocessor instruction.

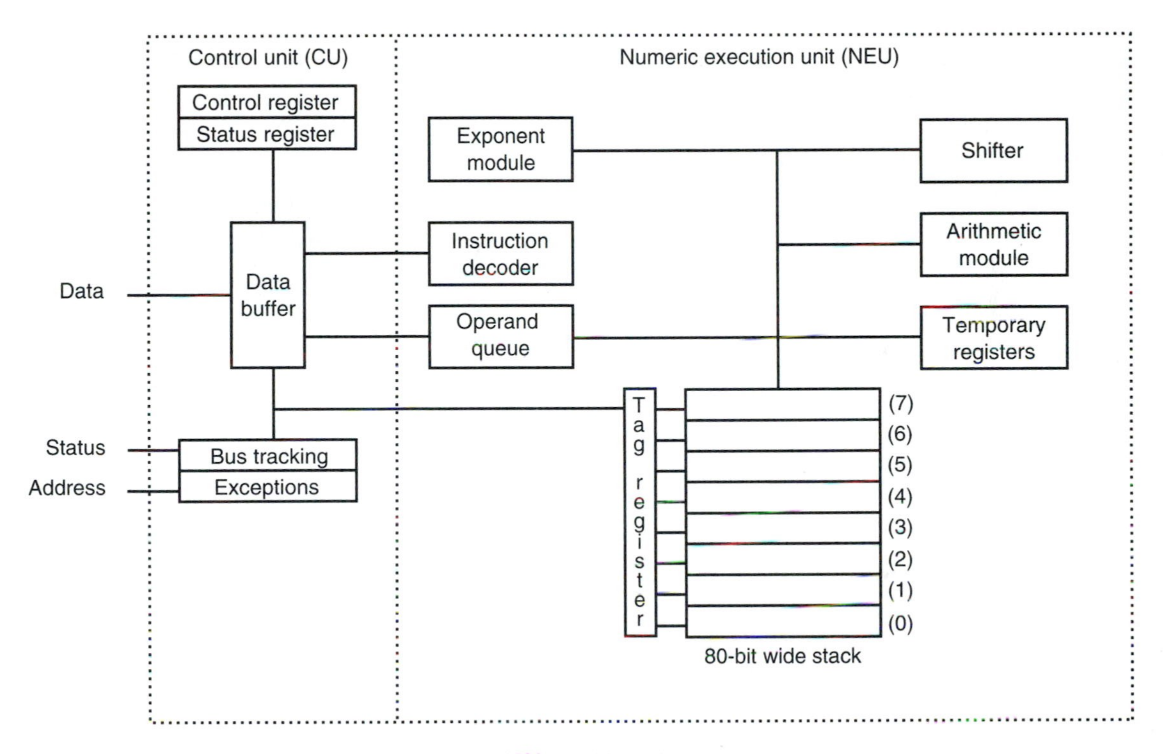

FIGURE 13–4 The internal structure of the 80X87 arithmetic coprocessor.

Internal Structure of the 80X87

Figure 13–4 shows the internal structure of the arithmetic coprocessor. Notice that this device is divided into two major sections: the **control unit** and the **numeric execution unit.**

The control unit interfaces the coprocessor to the microprocessor system data bus. Both the devices monitor the instruction stream. If the instruction is an ESCape (*coprocessor*) instruction, the coprocessor executes it, and if it is not, the microprocessor executes it.

The numeric execution unit (NEU) is responsible for executing all coprocessor instructions. The NEU has an eight-register stack that holds operands for arithmetic instructions and the results of arithmetic instructions. Instructions either address data in specific stack data registers or use a push and pop mechanism to store and retrieve data on the top of the stack. Other registers in the NEU are status, control, tag, and exception pointers. A few instructions transfer data between the coprocessor and the AX register in the microprocessor. The FSTSW AX instruction is the only instruction available to the coprocessor that allows direct communications to the microprocessor through the AX register. Note that the 8087 does not contain the FSTSW AX instruction.

The stack within the coprocessor contains eight registers that are each 80 bits in width. These stack registers *always* contain an 80-bit extended precision floating-point number. The only time data appear in any other form is when they reside in the memory system. The

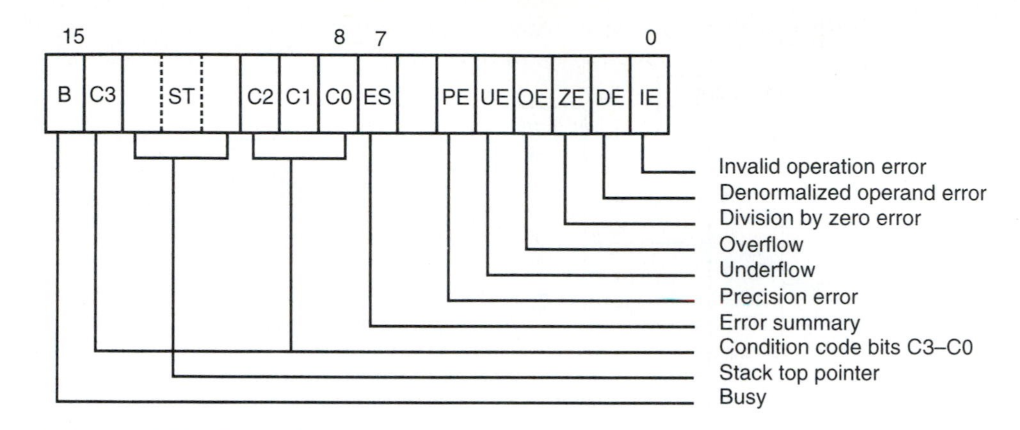

FIGURE 13–5 The 80X87 arithmetic coprocessor status register.

coprocessor converts from signed integer, BCD, single precision, or double precision form as the data are moved between the memory and the coprocessor register stack.

Status Register. The status register (see Figure 13–5) reflects the overall operation of the coprocessor. The status register is accessed by executing the instruction FSTSW, which stores the contents of the status register into a word of memory. The FSTSW AX instruction copies the status register directly to the microprocessor's AX register on the 80287 or above coprocessor. Once status is stored in memory or the AX register, the bit positions of the status register can be examined by normal software. Coprocessor/microprocessor communications are carried out through I/O ports 00FAH–00FFH on the 80287 and I/O ports 800000FAH–800000FFH on the 80386 through the Pentium.

The newer coprocessors (80287 and above) use status bit position 6 (SF) to indicate a stack overflow or underflow error. Following is a list of the status bits, except for SF, and their application:

B (Busy) Indicates that the coprocessor is busy executing a task. Busy can be tested by examining the status register or by using the FWAIT instruction. Newer coprocessors automatically synchronize with the microprocessor, so the busy flag need not be tested before performing additional coprocessor tasks.

C3–C0 (Condition Code Bits) Refer to Table 13–2 for a complete listing of each combination of these bits and their function. These bits have different meanings for different instructions, as indicated in the table. The top of the stack is denoted as ST.

TOP (Top-of-Stack) Indicates the current register addressed as the top-of-the-stack (ST). This is normally register 0.

ES (Error Summary) Set if any unmasked error bit (PE, UE, OE, ZE, DE, or IE) is set. In the 8087 coprocessor, the error summary also causes a coprocessor interrupt. Since

TABLE 13–2 The 80X87 status register condition code bits

Instruction	*C3*	*C2*	*C1*	*C0*	*Function*
FTST,FCOM	0	0	X	0	ST > Operand or (0 FTST)
	0	0	X	1	ST < Operand or (0 FTST)
	1	0	X	0	ST = Operand or (0 FTST)
	1	1	X	1	ST is not comparable
FPREM	Q1	0	Q0	Q2	right-most 3 bits of quotient
	?	1	?	?	incomplete
FXAM	0	0	0	0	+ unnormal
	0	0	0	1	+ NAN
	0	0	1	0	– unnormal
	0	0	1	1	– NAN
	0	1	0	0	+ normal
	0	1	0	1	+ ∞
	0	1	1	0	– normal
	0	1	1	1	– ∞
	1	0	0	0	+ 0
	1	0	0	1	empty
	1	0	1	0	– 0
	1	0	1	1	empty
	1	1	0	0	+ denormal
	1	1	0	1	empty
	1	1	1	0	– denormal
	1	1	1	1	empty

Notes: Unnormal = leading bits of the significand are zero
Denormal = exponent at its most negative value
Normal = standard floating-point form
NAN (not-a-number) = an exponent of all ones and a significand not equal to zero

the 80287, the coprocessor interrupt has been absent from the family.

PE (Precision Error) Result or operands exceed selected precision.

UE (Underflow Error) Indicates a nonzero result too small to represent with the current precision selected by the control word.

OE (Overflow Error) Indicates a result too large to be represented. If this error is masked, the coprocessor generates infinity for an overflow error.

ZE (Zero Error) Indicates the divisor was zero while the dividend was a noninfinity or nonzero number.

DE (Denormalized Error) Indicates at least one of the operands is denormalized.

IE (Invalid Error) Indicates a stack overflow or underflow, indeterminate form (0/0, ∞, –∞, etc.) or the use of a NAN as an operand. This flag indicates errors such as those produced by taking the square root of a negative number, etc.

EXAMPLE 13–6

```
                         ;using TEST to isolate the divide by zero error bit

0000  67& D8 35 00000000 R  FDIV   DATA1
0007  9B DF E               FSTSW  AX                    ;copy status register to AX
000A  A9 000                TEST   AX,4                  ;test bit position 2
000D  75 18                 JNZ    DIVIDE_ERROR

                         ;using TEST to isolate the invalid operation error bit
                         ;after a FSQRT instruction

000F  D9 FA                 FSQRT
0011  9B DF E0              FSTSW  AX                    ;copy status register to AX
0014  A9 0001               TEST   AX,1                  ;test bit position 1
0017  75 0E                 JNZ    FSQRT_ERROR

                         ;using the SAHF instruction and conditional jumps to
                         ;test for the conditions in Table 13-3 after an FCOM

0019  67& D8 15 00000000 R  FCOM   DATA1
0020  9B DF E0              FSTSW  AX                    ;copy status register to AX
0023  9E                    SAHF                         ;copy status bits to flags
0024  74 04                 JE     ST_EQUAL
0026  72 02                 JB     ST_BELOW
0028  77 00                 JA     ST_ABOVE
```

There are two ways to test the bits of the status register once they are moved into the AX register with the FSTSW AX instruction. One method uses the TEST instruction to test individual bits of the status register. The other uses the SAHF instruction to transfer the leftmost 8 bits of the status register into the microprocessor's flag register. Both methods are illustrated in Example 13–6. This example uses the DIV instruction to divide the top of the stack by the contents of DATA1 and the FSQRT instruction to find the square root of the top of the stack. The example also uses the FCOM instruction to compare the contents of the stack top with DATA1. Note that the conditional jump instructions are used with the SAHF instruction to test for the condition listed in Table 13–3. Although SAHF and conditional jumps cannot test all possible operating conditions of the coprocessor, they can help to reduce the complexity of certain tested conditions. Note that SAHF places C0 into the carry flag, C2 into the parity flag, and C3 into the zero flag.

TABLE 13–3 Coprocessor conditions tested with the conditional jump instructions and SAHF after FCOM or FTST, as illustrated in Example 13–6

C3	*C2*	*C0*	*Condition*	*Jump Instruction*
0	0	0	ST > Operand	JA (jump if ST above)
0	0	1	ST < Operand	JB (jump if ST below)
1	0	0	ST = Operand	JE (jump if ST equal)

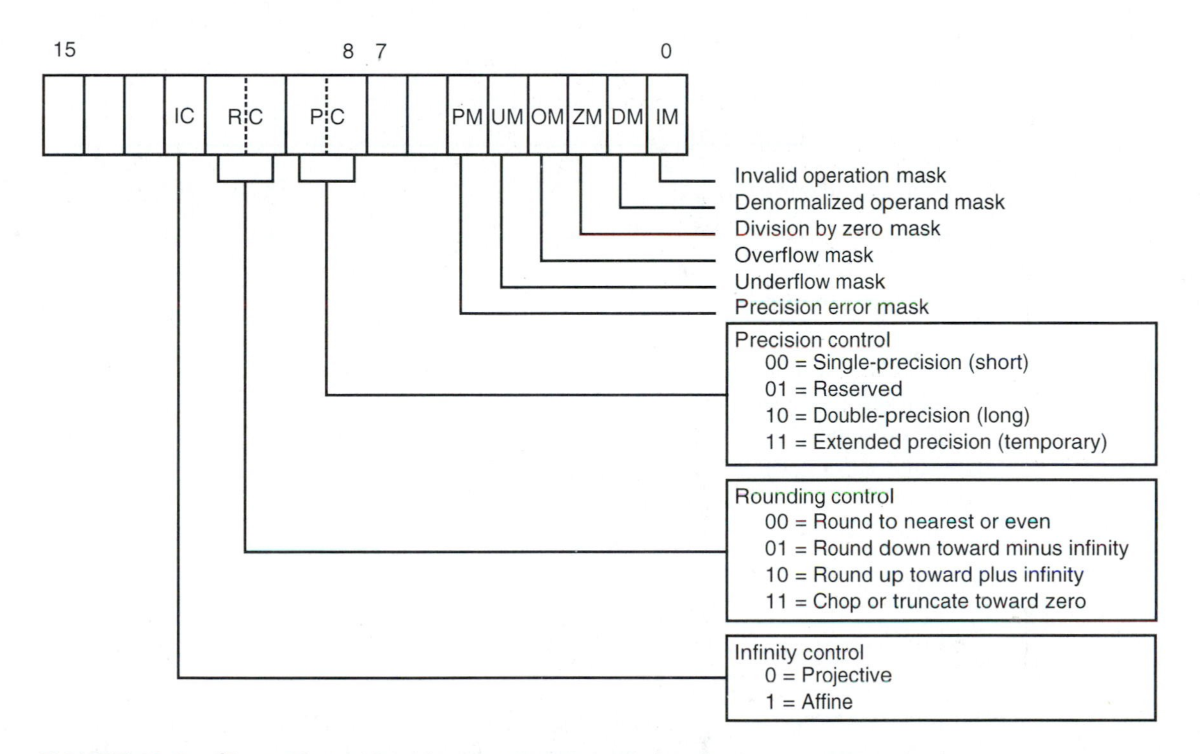

FIGURE 13–6 The 80X87 arithmetic coprocessor control register.

When the FXAM instruction and FSTSW AX are executed and followed by the SAHF instruction, the zero flag will contain C3. Notice from Table 13–2 that C3 indicates a ± 0 when set along with other errors. The FXAM instruction could be used to test a divisor before a division for a zero value by using the JZ instruction following FXAM, FSTSW AX, and SAHF.

Control Register. The control register is pictured in Figure 13–6. The control register selects precision, rounding control, and infinity control. It also masks and unmasks the exception bits that correspond to the right-most six bits of the status register. The FLDCW instruction is used to load a value into the control register.

Following is a description of each bit or grouping of bits found in the control register:

IC (Infinity Control) Selects either affine or projective infinity. Affine allows positive and negative infinity, while projective assumes infinity is unsigned.

RC (Rounding Control) Selects the type of rounding, as defined in Figure 13–6.

PC (Precision Control) Selects the precision of the result, as defined in Figure 13–6.

Exception Masks Determine whether the error indicated by the exception affects the error bit in the status register. If a logic 1 is placed in one of the exception control bits, the corresponding status register bit is masked off.

FIGURE 13–7 The 80X87 arithmetic coprocessor tag register.

15			8	7			0
TAG (7)	TAG (6)	TAG (5)	TAG (4)	TAG (3)	TAG (2)	TAG (1)	TAG (0)

TAG VALUES:
00 = VALID
01 = ZERO
10 = INVALID or INFINITY
11 = EMPTY

Tag Register. The tag register indicates the contents of each location in the coprocessor stack. Figure 13–7 illustrates the tag register and the status indicated by each tag. The tag indicates whether a register is valid, zero, invalid or infinity, or empty. The only way a program can view the tag register is by storing the coprocessor environment, using the FSTENV, FSAVE, or FRSTOR instructions. Each of these instructions stores the tag register along with other coprocessor data.

13–3 INSTRUCTION SET

The arithmetic coprocessor executes 68 different instructions. Whenever a coprocessor instruction references memory, the microprocessor automatically generates the memory address for the instruction. The coprocessor uses the data bus for data transfers during coprocessor instructions and the microprocessor uses the data bus during normal instructions. The 80287 uses the Intel-reserved I/O ports 00F8H–00FFH for communications between the coprocessor and the microprocessor. These ports are used mainly for the FSTSW AX instruction. The 80387–Pentium use I/O ports 800000F8H–800000FFH for these communications.

This section of the text describes the function of each instruction and lists its assembly language form. Because the coprocessor uses the microprocessor memory addressing modes, not all possible forms of each instruction are illustrated. Each time the assembler encounters one of the coprocessor mnemonic opcodes, it converts it into a machine language ESC instruction. The ESC instruction represents an opcode to the coprocessor.

Data Transfer Instructions

There are three basic data transfers: floating-point, signed-integer, and BCD. The only time that data ever appears in signed-integer or BCD form is in the memory. Inside the coprocessor, data are always stored as an 80-bit extended precision, floating-point number.

Floating-Point Data Transfers. There are four floating-point data transfer instructions in the coprocessor instruction set: FLD (load real), FST (store real), FSTP (store real and pop), and FXCH (exchange).

The FLD instruction loads floating-point memory data to the top of the internal stack, referred to as ST (stack top). This instruction stores the data on the top of the stack and then decrements the stack pointer by one. Data loaded to the top of the stack are from any memory location or from another coprocessor register. For example, an FLD ST(2) instruction copies the contents of register 2 to the stack top, which is ST. The top of the stack is register 0 when the coprocessor is reset or initialized. Another example is the FLD DATA7 instruction, which copies the contents of memory location DATA7 to the top of the stack. The size of the transfer is automatically determined by the assembler through the directives DD or REAL4 for single precision numbers, DQ or REAL 8 for double precision numbers, and DT or REAL10 for extended precision numbers.

The FST instruction stores a copy of the top of the stack into the memory location or coprocessor register indicated by the operand. At the time of storage, the internal, extended precision floating-point number is rounded to the size of the floating-point number indicated by the control register.

The FSTP (floating-point store and pop) instruction stores a copy of the top of the stack into memory or any coprocessor register and then pops the data from the top of the stack. You might think of FST as a *copy instruction* and FSTP as a *removal instruction*.

The FXCH instruction exchanges the register indicated by the operand with the top of the stack. For example, the FXCH ST(2) instruction exchanges the top of the stack with register 2.

Integer Data Transfer Instructions. The coprocessor supports three integer data transfer instructions: FILD (load integer), FIST (store integer), and FISTP (store integer and pop). These three instructions function just like FLD, FST, and FSTP, except the data transferred are integer data. The coprocessor automatically converts the internal, extended precision, floating-point data to integer data. The size of the data is determined by the way the label is defined in the assembly language program (with DW, DD, or DQ).

BCD Data Transfer Instructions. Two instructions load or store BCD signed integer data. The FBLD instruction loads the top of the stack with BCD memory data, and the FBSTP stores the top of the stack and does a pop.

Example 13–7 shows how the assembler automatically adjusts the FLD, FILD, and FBLD instructions for different sized operands. (Look closely at the machine coded forms of the instructions.) Note, this example begins with the .386 and .387 directives, which identify the microprocessor as an 80386 and the coprocessor as an 80387. If the 80286 microprocessor is in use with its own coprocessor, the directives .286 and .287 appear. The assembler assumes by default that the software is assembled for an 8086/8088 with an 8087 coprocessor. The .486, .487, .586, and .587 switches are also available for use with the 80486 and Pentium microprocessors. Even though the program in Example 13–7 executes, CodeView, or some other debugging tool, must be used to view any changes to the coprocessor stack.

EXAMPLE 13–7

```
                                .MODEL   SMALL
                                .386                    ;select 80386 microprocessor
                                .387                    ;select 80387 coprocessor
0000                            .DATA
0000  41F00000      DATA1       DD     30.0             ;single precision
0004                DATA2       DQ     100.25           ;double precision
        4059100000000000
```

```
000C                   DATA3   DT     33.9876      ;extended precision
     400487F34D6A161E4F76
0016  001E             DATA4   DW     30           ;16-bit integer
0018  0000001E         DATA5   DD     30           ;32-bit integer
001C                   DATA6   DQ     30           ;64-bit integer
     000000000000001E
0024                   DATA7   DT     30H          ;BCD 30
     00000000000000000030
0000                           .CODE
                               .STARTUP

0010  D9 06 0000 R             FLD    DATA1
0014  DD 06 0004 R             FLD    DATA2
0018  DB 2E 000C R             FLD    DATA3

001C  DF 06 0016 R             FILD   DATA4
0020  DB 06 0018 R             FILD   DATA5
0024  DF 2E 001C R             FILD   DATA6

0028  DF 26 0024 R             FBLD   DATA7
                               .EXIT
                               END
```

Arithmetic Instructions

Arithmetic instructions for the coprocessor include addition, subtraction, multiplication, division, and square root. The arithmetic-related instructions are scaling, rounding, absolute value, and changing the sign.

Table 13–4 shows the basic addressing modes allowed for the arithmetic operations. Each addressing mode is shown with an example using the FADD (real addition) instruction. All arithmetic operations are floating-point, except when memory data are referenced as an operand.

The **classic stack form** of addressing operand data (stack addressing) uses the top of the stack as the source operand and the next-to-the-top of the stack as the destination operand. Afterwards, a pop removes the source datum from the stack, and only the result in the destination register remains at the top of the stack. To use this addressing mode, the instruction is placed in the program without any operands.

TABLE 13–4 Arithmetic addressing modes

Mode	*Form*	*Examples*
Stack	ST(1),ST	FADD
Register	ST,ST(n)	FADD ST,ST(2)
	ST(n),ST	FADD ST(6),ST
Register pop	ST(n),ST	FADDP ST(3),ST
Memory	Operand	FADD DATA

Note: Stack addressing is fixed as ST(1),ST and includes a pop, so only the result remains at the top of the stack. n = register number 0–7. Register addressing for any instruction can use a destination of ST or ST(n), as illustrated.

The FADD instruction adds ST to ST(1) and stores the answer at the top of the stack. It also removes the original two datum from the stack by popping. FSUB subtracts ST from ST(1) and leaves the difference at ST. Therefore, a reverse subtraction (FSUBR) subtracts ST(1) from ST and leaves the difference at ST. (An error exists in Intel documentation, in describing the operation of some reverse instructions.) Another use for reverse operations is finding a reciprocal ($^1/_x$). This is accomplished, if x is at the top of the stack, by loading a 1.0 to ST (FLD1), followed by the FDIVR instruction. The FDIVR instruction divides ST(1) into ST, or x into 1, and leaves the reciprocal ($^1/_x$) at ST.

The register-addressing mode uses ST for the top of the stack and ST(n) for another location, where n is the register number. With this form, one operand is ST and the other is ST(n). To double the top of the stack, use the FADD ST,ST(0) instruction, where ST(0) also addresses the top of the stack. One of the two operands in the register-addressing mode must be ST, while the other must be in the form ST(n), where n is a stack register 0–7. For many instructions, either ST or ST(n) can be the destination. It is fairly important that the top of the stack is ST(0). This is accomplished by resetting or initializing the coprocessor before using it in a program. Another example of register-addressing is FADD ST(1),ST, where the contents of ST is added to ST(1) and the result is placed into ST(1).

The top of the stack is always used as the destination for the memory-addressing mode because the coprocessor is a *stack-oriented machine*. For example, the FADD DATA instruction adds the real-number contents of memory location DATA to the top of the stack.

Arithmetic Operations. The letter P in an opcode specifies a register pop after the operation (FADDP compared to FADD). The letter R in an opcode (subtraction and division only) indicates reverse mode. The reverse mode is useful for memory data, because normally memory data subtracts from the top of the stack. A reversed subtract instruction subtracts the top of the stack from memory and stores the result in the top of the stack. For example, if the top of the stack contains a 10 and memory location DATA1 contains a 1, an FSUB DATA1 instruction results in a +9 on the stack top, and an FSUBR instruction results in a –9. Another example is FSUBR ST,ST(1), which will subtract ST from ST(1) and store the result on ST. FSUBR ST(1),ST will subtract ST(1) from ST and store the result on ST(1).

The letter I as a second letter in an opcode indicates that the memory operand is an integer. For example, an FADD DATA instruction is a floating-point addition, while an FIADD DATA is an integer addition that adds the integer at memory location DATA to the floating-point number at the top of the stack. The same rules apply to FADD, FSUB, FMUL, and FDIV instructions.

Arithmetic-Related Operations. Other operations that are arithmetic in nature include FSQRT (*square root*), FSCALE (*scale a number*), FPREM/FPREM1 (*find partial remainder*), FRNDINT (*round to integer*), FXTRACT (*extract exponent and significand*), FABS (*find absolute value*), and FCHG (*change sign*). These instructions and the functions they perform follow.

FSQRT Finds the square root of the top of the stack and leaves the resultant square root at the top of the stack. An invalid error occurs for the square root of a negative number. For this reason, the IE bit of the status register should be tested whenever an invalid result can occur. The IE bit can be tested by loading the status register to AX with the FSTSW AX instruction, followed by TEST AX,1 to test the IE status bit.

FSCALE Adds the contents of ST(1) (interpreted as an integer) to the exponent at the top of the stack. FSCALE multiplies or divides rapidly by powers of two. The value in ST(1) must be between 2^{-15} and 2^{+15}.

FPREM/FPREM1 Performs modulo division of ST by ST(1). The resultant remainder is found in the top of the stack and has the same sign as the original dividend. Note that a modulo division results in a remainder without a quotient. Note that FPREM is supported for the 8086 and 80287 and FPREM1 should be used in newer coprocessors.

FRNDINT Rounds the top of the stack to an integer.

FXTRACT Decomposes the number at the top of the stack into two separate parts that represent the value of the unbiased exponent and the value of the significand. The extracted significand is found at the top of the stack and the unbiased exponent at ST(1). This instruction is often used to convert a floating-point number into a form that can be printed as a mixed number.

FABS Changes the sign of the top of the stack to positive.

FCHS Changes the sign of a floating-point number from positive to negative or negative to positive.

Comparison Instructions

The comparison instructions examine data at the top of the stack in relation to another element, and return the result of the comparison in the status register condition code bits C3–C0. Comparisons that are allowed by the coprocessor are: FCOM (*floating-point compare*), FCOMP (*floating-point compare with a pop*), FCOMPP (*floating-point compare with 2 pops*), FICOM (*integer compare*), FICOMP (*integer compare and pop*), FSTS (*test*), and FXAM (*examine*). Following is a list of these instructions with a description of their function.

FCOM Compares the floating-point data at the top of the stack with an operand, which may be any register or any memory operand. If the operand is not coded with the instruction, the next stack element ST(1) is compared with the stack top, ST.

FCOMP and **FCOMPP** Both instructions perform like FCOM, but they also pop one or two datum from the stack.

FICOM and **FICOMP** The top of the stack is compared with the integer stored at a memory operand. In addition to the compare, FICOMP also pops the top of the stack.

FTST Tests the contents of the top of the stack against a zero. The result of the comparison is coded in the status register condition-code bits as illustrated in Table 13–2. See Table 13–3 for a way of using SAHF and the conditional jump instruction with FTST.

FXAM Examines the stack top and modifies the condition-code bits to indicate whether the contents are positive, negative, normalized, etc. Refer to the status register in Table 13–2.

Transcendental Operations

The transcendental instructions include FPTAN (*partial tangent*), FPATAN (*partial arctangent*), FSIN, (*sine*), FCOS (*cosine*), FSINCOS (*sine and cosine*)), F2XM1 ($2^x - 1$), FYL2X (Y $\log_2$X), and FYL2XP1 (Y $\log_2$(X + 1)). A list of these operations follows, with a description of each transcendental operation.

FPTAN Finds the partial tangent of Y/X = tan θ. The value of θ is at the top of the stack, and must be between 0 and π/4 radians for the 8087 and 80287, and less than 2^{63} for the 80387, 80486/7 and Pentium microprocessors. The result is a ratio is found as ST = X and ST(1) = Y. If the value is outside the allowable range, an invalid error occurs as indicated by the status register IE bit. ST(7) must be empty for this instruction to function properly.

FPATAN Finds the partial arctangent as θ = ARCTAN X/Y. The value of X is at the top of the stack and Y is at ST(1). The values of X and Y must be as follows: $0 \leq Y < X < \infty$. The instruction pops the stack and leaves θ at the top of the stack.

F2XM1 Finds the function $2^x - 1$. The value of X is taken from the top of the stack and the result is returned to the top of the stack. To obtain 2^x, add 1 to the result at the top of the stack. The value of X must be between –1 and +1. The F2XM1 instruction is used to derive the functions listed in Table 13–5. Note that the constants $\log_2 10$ and $\log_2 e$ are built in as standard values for the coprocessor.

FSIN/FCOS Find the sine or cosine of the argument located in ST, expressed in radians (360° = 2π radians), with the result found in ST. The values of ST must be less than 2^{63}.

TABLE 13–5 Exponential functions

Function	*Equation*
10^y	$2^y \times \log_2 10$
e^y	$2^y \times \log_2 \varepsilon$
X^y	$2^y \times \log_2 X$

TABLE 13–6 Constant operations

Instruction	Constant Pushed to ST
FLDZ	+ 0.0
FLD1	+ 1.0
FLDPI	π
FLDL2T	$\log_2 10$
FLDL2E	$\log_2 \varepsilon$
FLDLG2	$\log_{10} 2$
FLDLN2	$\log_\varepsilon 2$

FSINCOS Finds the sine and cosine of ST, expressed in radians, and leaves the results as ST = sine and ST(1) = cosine. As with FSIN or FCOS, the initial value of ST must be less than 2^{63}.

FYL2X Finds Y $\log_2$X. X is taken from the stack top and Y is taken from ST(1). The result is found at the top of the stack after a pop. The value of X must be between 0 and ∞, and the value of Y must be between –∞ and +∞. A logarithm with any positive base (b) is found by the equation $LOG_b\ X = (LOG_2\ b)^{-1} \times LOG_2\ X$

FYL2XP1 Finds Y $\log_2$ (X + 1). The value of X is taken from the stack top and Y is taken from ST(1). The result is found at the top of the stack after a pop. The value of X must range between 0 and $1 - \sqrt{2}/2$, and the value of Y must be between –∞ and +∞.

Constant Operations

The coprocessor instruction set includes opcodes that return constants to the top of the stack. A list of these instructions appears in Table 13–6.

Coprocessor Control Instructions

The coprocessor has control instructions for initialization, exception handling, and task switching. The control instructions have two forms. For example, FINIT and FNINIT initialize the coprocessor. The difference is that FNINIT does not cause any wait states, while FINIT does. The microprocessor waits for the FINIT instruction by testing the $\overline{\text{BUSY}}$ pin on the coprocessor. All control instructions have these two forms. Following is a list of each control instruction with its function.

FINIT/FNINIT This instruction performs a reset operation on the arithmetic coprocessor. (Refer to Table 13–7 for the reset conditions.) The coprocessor operates with a closure of projective (unsigned infinity), rounds to the nearest or even, and uses extended precision numbers when reset or initialized. It also sets register 0 as the top of the stack.

FSETPM Changes the addressing mode of the coprocessor to the protected addressing mode when the microprocessor is operated in the protected mode. Protected mode can only

TABLE 13–7 Coprocessor state after a reset or initialization

Field	*Value*	*Condition*
Infinity	0	Projective
Rounding	00	Round to nearest number
Precision	11	Extended precision
Error masks	11111	Error bits disabled
Busy	0	Not busy
C3–C0	????	Unknown
TOP	000	Register 000
ES	0	No error
Error bits	00000	No errors
All tags	11	Empty
Registers	–	Not changed

be exited by a hardware reset or, in the case of the 80386 through the Pentium, with a change to the control register.

FLDCW Loads the control register with the word addressed by the operand.

FSTCW/FNSTCW Stores the control register into the word-sized memory operand.

FSTSW AX/FNSTSW AX Copies the contents of the control register to the AX register. This instruction is not available to the 8087 coprocessor.

FCLEX/FNCLEX Clears the error flags in the status register and also the busy flag.

FSAVE/FNSAVE Writes the entire state of the machine to memory. Figure 13–8 shows the memory layout for this instruction.

FRSTOR Restores the state of the machine from memory. This instruction is used to restore the information saved by FSAVE/FNSAVE.

FSTENV/FNSTENV Stores the environment of the coprocessor as shown in Figure 13–9.

FLDENV Reloads the environment saved by FSTENV/FNSTENV.

FINCST Increments the stack pointer.

FDECSTP Decrements the stack pointer.

FFREE Frees a register by changing the destination register's tag to empty. It does not affect the contents of the register.

FNOP Floating-point coprocessor NOP.

FWAIT Causes the microprocessor to wait for the coprocessor to finish an operation. FWAIT should be used before the microprocessor accesses memory data affected by the coprocessor.

FIGURE 13–8 Memory format when the 80X87 registers are saved with the FSAVE instruction.

Offset	15 … 0	
5CH	S | Exponent 0–14	Last stack element ST(7)
5AH	Fraction 48–63	
58H	Fraction 32–47	
56H	Fraction 16–31	
54H	Fraction 0–15	
20H	S | Exponent 0–14	Next stack element ST(1)
1EH	Fraction 48–63	
1CH	Fraction 32–47	
1AH	Fraction 16–31	
18H	Fraction 0–15	
16H	S | Exponent 0–14	Stack top element ST(0)
14H	Fraction 48–63	
12H	Fraction 32–47	
10H	Fraction 16–31	
0EH	Fraction 0–15	
0CH	OP 16–19 | 0	
0AH	Operand pointer (OP) 0–15	
08H	IP 16–19 | Opcode	
06H	IP 0–15	
04H	Tag register	
02H	Status register	
00H	Control register	

Coprocessor Instructions

Newer coprocessors contain the same basic instructions provided by the earlier versions with a few additional instructions. The 80387, 80486, 80487SX, and Pentium contain the following additional instructions: FCOS (cosine), FPREM1 (partial remainder), FSIN (sine), FSINCOS (sine and cosine), and FUCOM/FUCOMP/FUCOMPP (unordered

FIGURE 13–9 Memory format for the FSTENV instruction: (a) real mode and (b) protected mode.

(a)

Offset	
0CH	OP 16–19 | 0
0AH	Operand pointer 0–15
08H	IP 16–19 | Opcode
06H	Instruction pointer 0–15
04H	Tag register
02H	Status register
00H	Control register

(b)

Offset	
0CH	Operand selector
0AH	Operand offset
08H	CS selector
06H	IP offset
04H	Tag register
02H	Status register
00H	Control register

compare). The sine and cosine instructions are the most significant addition to the instruction set. In earlier versions of the coprocessor, sine and cosine were calculated from the tangent.

Table 13–8 lists the instruction sets for all versions of the coprocessor. It also lists the number of clocking periods required to execute each instruction. Execution times are

TABLE 13–8 The instruction set of the arithmetic coprocessor (pp. 449–468)

F2XM1 $2^{ST} - 1$

11011001 11110000

Example		Clocks
F2XM1	8087	310–630
	80287	310–630
	80387	211–476
	80486/7	140–279
	Pentium	13–57

FABS Absolute value of ST

11011001 11100001

Example		Clocks
FABS	8087	10–17
	80287	10–17
	80387	22
	80486/7	3
	Pentium	1

FADD/FADDP/FIADD Addition

11011000 oo000mmm disp	32-bit memory (FADD)
11011100 oo000mmm disp	64-bit memory (FADD)
11011d00 11000rrr	FADD ST,ST(rrr)
11011110 11000rrr	FADDP ST,ST(rrr)
11011110 oo000mmm disp	16-bit memory (FIADD)
11011010 oo000mmm disp	32-bit memory (FIADD)

Format	Examples		Clocks
FADD FADDP FIADD	FADD DATA FADD ST,ST(1) FADDP FIADD NUMBER FADD ST,ST(3) FADDP ST,ST(2) FADD ST(2),ST	8087	70–143
		80287	70–143
		80387	23–72
		80486/7	8–20
		Pentium	1–7

FCLEX/FNCLEX Clear errors

11011011 11100010

Example		Clocks
FCLEX FNCLEX	8087	2–8
	80287	2–8
	80387	11
	80486/7	7
	Pentium	9

FCOM/FCOMP/FCOMPP/FICOM/FICOMP Compare

11011000 oo010mmm disp	32-bit memory (FCOM)
11011100 oo010mmm disp	64-bit memory (FCOM)
11011000 11010rrr	FCOM ST(rrr)
11011000 oo011mmm disp	32-bit memory (FCOMP)
11011100 oo011mmm disp	64-bit memory (FCOMP)
11011000 11011rrr	FCOMP ST(rrr)
11011110 11011001	FCOMPP
11011110 oo010mmm disp	16-bit memory (FICOM)
11011010 oo010mmm disp	32-bit memory (FICOM)
11011110 oo011mmm disp	16-bit memory (FICOMP)
11011010 oo011mmm disp	32-bit memory (FICOMP)

Format	Examples		Clocks
FCOM FCOMP FCOMPP FICOM FICOMP	FCOM ST(2) FCOMP DATA FCOMPP FICOM NUMBER FICOMP DATA3	8087	40–93
		80287	40–93
		80387	24–63
		80486/7	15–20
		Pentium	1–8

FCOS Cosine of ST

11011001 11111111

Example		Clocks
FCOS	8087	—
	80287	—
	80387	123–772
	80486/7	193–279
	Pentium	18–124

FDECSTP Decrement stack pointer

11011001 11110110

Example		Clocks
FDECSTP	8087	6–12
	80287	6–12
	80387	22
	80486/7	3
	Pentium	1

FDISI/FNDISI Disable interrupts

11011011 11100001

(ignored on the 80287, 80387, 80486/7, and Pentium)

Example	Clocks	
FDISI FNDISI	8087	2–8
	80287	—
	80387	—
	80486/7	—
	Pentium	—

FDIV/FDIVP/FIDIV Division

```
11011000 oo110mmm disp     32-bit memory (FDIV)
11011100 oo100mmm disp     64-bit memory (FDIV)
11011d00 11111rrr          FDIV ST,ST(rrr)
11011110 11111rrr          FDIVP ST,ST(rrr)
11011110 oo110mmm disp     16-bit memory (FIDIV)
11011010 oo110mmm disp     32-bit memory (FIDIV)
```

Format	Examples	Clocks	
FDIV FDIVP FIDIV	FDIV DATA FDIV ST,ST(3) FDIVP FIDIV NUMBER FDIV ST,ST(5) FDIVP ST,ST(2) FDIV ST(2),ST	8087	191–243
		80287	191–243
		80387	88–140
		80486/7	8–89
		Pentium	39–42

FDIVR/FDIVRP/FIDIVR Division reversed

11011000 oo111mmm disp	32-bit memory (FDIVR)
11011100 oo111mmm disp	64-bit memory (FDIVR)
11011d00 11110rrr	FDIVR ST,ST(rrr)
11011110 11110rrr	FDIVRP ST,ST(rrr)
11011110 oo111mmm disp	16-bit memory (FIDIVR)
11011010 oo111mmm disp	32-bit memory (FIDIVR)

Format	Examples	Clocks	
FDIVR FDIVRP FIDIVR	FDIVR DATA FDIVR ST,ST(3) FDIVRP FIDIVR NUMBER FDIVR ST,ST(5) FDIVRP ST,ST(2) FDIVR ST(2),ST	8087	191–243
		80287	191–243
		80387	88–140
		80486/7	8–89
		Pentium	39–42

FENI/FNENI Disable interrupts

11011011 11100000

(ignored on the 80287, 80387, 80486/7 and Pentium)

Example	Clocks	
FENI FNENI	8087	2–8
	80287	—
	80387	—
	80486/7	—
	Pentium	—

FFREE Free register

11011101 11000rrr

Format	Examples	Clocks	
FFREE	FFREE FFREE ST(1) FFREE ST(2)	8087	9–16
		80287	9–16
		80387	18
		80486/7	3
		Pentium	1

FINCSTP Increment stack pointer

11011001 11110111

Example	Clocks	
FINCSTP	8087	6–12
	80287	6–12
	80387	21
	80486/7	3
	Pentium	1

FINIT/FNINIT Initialize coprocessor

11011001 11110110

Example	Clocks	
FINIT FNINIT	8087	2–8
	80287	2–8
	80387	33
	80486/7	17
	Pentium	12–16

FLD/FILD/FBLD Load data to ST(0)

```
11011001 oo000mmm disp    32-bit memory (FLD)
11011101 oo000mmm disp    64-bit memory (FLD)
11011011 oo101mmm disp    80-bit memory (FLD)
11011111 oo000mmm disp    16-bit memory (FILD)
11011011 oo000mmm disp    32-bit memory (FILD)
11011111 oo101mmm disp    64-bit memory (FILD)
11011111 oo100mmm disp    80-bit memory (FBLD)
```

Format	Examples	Clocks	
FLD FILD FBLD	FLD DATA FILD DATA1 FBLD DEC_DATA	8087	17–310
		80287	17–310
		80387	14–275
		80486/7	3–103
		Pentium	1–3

FLD1 Load +1.0 to ST(0)

```
11011001 11101000
```

Example	Clocks	
FLD1	8087	15–21
	80287	15–21
	80387	24
	80486/7	4
	Pentium	2

FLDZ Load +0.0 to ST(0)

11011001 11101110

Example		Clocks
FLDZ	8087	11–17
	80287	11–17
	80387	20
	80486/7	4
	Pentium	2

FLDPI Load π to ST(0)

11011001 11101011

Example		Clocks
FLDPI	8087	16–22
	80287	16–22
	80387	40
	80486/7	8
	Pentium	3–5

FLDL2E Load $\log_2 e$ to ST(0)

11011001 11101010

Example		Clocks
FLDL2E	8087	15–21
	80287	15–21
	80387	40
	80486/7	8
	Pentium	3–5

FLDL2T Load $\log_2 10$ to ST(0)

11011001 11101001

Example		Clocks
FLDL2T	8087	16–22
	80287	16–22
	80387	40
	80486/7	8
	Pentium	3–5

FLDLG2 Load $\log_{10} 2$ to ST(0)

11011001 11101000

Example		Clocks
FLDLG2	8087	18–24
	80287	18–24
	80387	41
	80486/7	8
	Pentium	3–5

FLDLN2 Load $\log_e 2$ to ST(0)

11011001 11101101

Example		Clocks
FLDLN2	8087	17–23
	80287	17–23
	80387	41
	80486/7	8
	Pentium	3–5

FLDCW Load control register

11011001 oo101mmm disp

Format	Examples		Clocks
FLDCW	FLDCW DATA FLDCW STATUS	8087	7–14
		80287	7–14
		80387	19
		80486/7	4
		Pentium	7

FLDENV Load environment

11011001 oo100mmm disp

Format	Examples		Clocks
FLDENV	FLDENV ENVIRON FLDENV DATA	8087	35–45
		80287	25–45
		80387	71
		80486/7	34–44
		Pentium	32–37

FMUL/FMULP/FIMUL Multiplication

11011000 oo001mmm disp	32-bit memory (FMUL)
11011100 oo001mmm disp	64-bit memory (FMUL)
11011d00 11001rrr	FMUL ST,ST(rrr)
11011110 11001rrr	FMULP ST,ST(rrr)
11011110 oo001mmm disp	16-bit memory (FIMUL)
11011010 oo001mmm disp	32-bit memory (FIMUL)

Format	Examples		Clocks
FMUL FMULP FIMUL	FMUL DATA FMUL ST,ST(2) FMUL ST(2),ST FMULP FIMUL DATA3	8087	110–168
		80287	110–168
		80387	29–82
		80486/7	11–27
		Pentium	1–7

FNOP No operation

11011001 11010000

Example		Clocks
FNOP	8087	10–16
	80287	10–16
	80387	12
	80486/7	3
	Pentium	1

FPATAN Partial arctangent of ST(0)

11011001 11110011

Example		Clocks
FPATAN	8087	250–800
	80287	250–800
	80387	314–487
	80486/7	218–303
	Pentium	17–173

FPREM Partial remainder

11011001 11111000

Example	Clocks	
FPREM	8087	15–190
	80287	15–190
	80387	74–155
	80486/7	70–138
	Pentium	16–64

FPREM1 Partial remainder (IEEE)

11011001 11110101

Example	Clocks	
FPREM1	8087	—
	80287	—
	80387	95–185
	80486/7	72–167
	Pentium	20–70

FPTAN Partial tangent of ST(0)

11011001 11110010

Example	Clocks	
FPTAN	8087	30–450
	80287	30–450
	80387	191–497
	80486/7	200–273
	Pentium	17–173

FRNDINT Round ST(0) to an integer

11011001 11111100

Example		Clocks
FRNDINT	8087	16–50
	80287	16–50
	80387	66–80
	80486/7	21–30
	Pentium	9–20

FRSTOR Restore state

11011101 oo110mmm disp

Format	Examples		Clocks
FRSTOR	FRSTOR DATA FRSTOR STATE FRSTOR MACHINE	8087	197–207
		80287	197–207
		80387	308
		80486/7	120–131
		Pentium	70–95

FSAVE/FNSAVE Save machine state

11011101 oo110mmm disp

Format	Examples		Clocks
FSAVE FNSAVE	FSAVE STATE FNSAVE STATUS FSAVE MACHINE	8087	197–207
		80287	197–207
		80387	375
		80486/7	143–154
		Pentium	124–151

FSCALE Scale ST(0) by ST(1)

11011001 11111101

Example		Clocks
FSCALE	8087	32–38
	80287	32–38
	80387	67–86
	80486/7	30–32
	Pentium	20–31

FSETPM Set protected mode

11011011 11100100

Example		Clocks
FSETPM	8087	—
	80287	2–18
	80387	12
	80486/7	—
	Pentium	—

FSIN Sine of ST(0)

11011001 11111110

Example		Clocks
FSIN	8087	—
	80287	—
	80387	122–771
	80486/7	193–279
	Pentium	16–126

FSINCOS Find sine and cosine of ST(0)

11011001 11111011

Example		Clocks
FSINCOS	8087	—
	80287	—
	80387	194–809
	80486/7	243–329
	Pentium	17–137

FSQRT Square root of ST(0)

11011001 11111010

Example		Clocks
FSQRT	8087	180–186
	80287	180–186
	80387	122–129
	80486/7	83–87
	Pentium	70

FST/FSTP/FIST/FISTP/FBSTP Store

11011001 oo010mmm disp	32-bit memory (FST)
11011101 oo010mmm disp	64-bit memory (FST)
11011101 11010rrr	FST ST(rrr)
11011011 oo011mmm disp	32-bit memory (FSTP)
11011101 oo011mmm disp	64-bit memory (FSTP)
11011011 oo111mmm disp	80-bit memory (FSTP)
11011101 11001rrr	FSTP ST(rrr)
11011111 oo010mmm disp	16-bit memory (FIST)
11011011 oo010mmm disp	32-bit memory (FIST)
11011111 oo011mmm disp	16-bit memory (FISTP)
11011011 oo011mmm disp	32-bit memory (FISTP)
11011111 oo111mmm disp	64-bit memory (FISTP)
11011111 oo110mmm disp	80-bit memory (FBSTP)

Format	Examples	Clocks	
FST FSTP FIST FISTP FBSTP	FST DATA FST ST(3) FST FSTP FIST DATA2 FBSTP DATA6 FISTP DATA9	8087	15–540
		80287	15–540
		80387	11–534
		80486/7	3–176
		Pentium	1–3

FSTCW/FNSTCW Store control register

11011001 oo111mmm disp

Format	Examples	Clocks	
FSTCW FNSTCW	FSTCW CONTROL FNSTCW STATUS FSTCW MACHINE	8087	12–18
		80287	12–18
		80387	15
		80486/7	3
		Pentium	2

FSTENV/FNSTENV Store environment

11011001 oo110mmm disp

Format	Examples	Clocks	
FSTENV FNSTENV	FSTENV CONTROL FNSTENV STATUS FSTENV MACHINE	8087	40–50
		80287	40–50
		80387	103–104
		80486/7	58–67
		Pentium	48–50

FSTSW/FNSTSW Store status register

11011101 oo111mmm disp

Format	Examples	Clocks	
FSTSW FNSTSW	FSTSW CONTROL FNSTSW STATUS FSTSW MACHINE FSTSW AX	8087	12–18
		80287	12–18
		80387	15
		80486/7	3
		Pentium	2–5

FSUB/FSUBP/FISUB Subtraction

11011000 oo100mmm disp	32-bit memory (FSUB)
11011100 oo100mmm disp	64-bit memory (FSUB)
11011d00 11101rrr	FSUB ST,ST(rrr)
11011110 11101rrr	FSUBP ST,ST(rrr)
11011110 oo100mmm disp	16-bit memory (FISUB)
11011010 oo100mmm disp	32-bit memory (FISUB)

Format	Examples	Clocks	
FSUB FSUBP FISUB	FSUB DATA FSUB ST,ST(2) FSUB ST(2),ST FSUBP FISUB DATA3	8087	70–143
		80287	70–143
		80387	29–82
		80486/7	8–35
		Pentium	1–7

FSUBR/FSUBRP/FISUBR Reverse subtraction

11011000 oo101mmm disp	32-bit memory (FSUBR)
11011100 oo101mmm disp	64-bit memory (FSUBR)
11011d00 11100rrr	FSUBR ST,ST(rrr)
11011110 11100rrr	FSUBRP ST,ST(rrr)
11011110 oo101mmm disp	16-bit memory (FISUBR)
11011010 oo101mmm disp	32-bit memory (FISUBR)

Format	Examples		Clocks
FSUBR FSUBRP FISUBR	FSUBR DATA FSUBR ST,ST(2) FSUBR ST(2),ST FSUBRP FISUBR DATA3	8087	70–143
		80287	70–143
		80387	29–82
		80486/7	8–35
		Pentium	1–7

FTST Compare ST(0) with + 0.0

11011001 11100100

Example		Clocks
FTST	8087	38–48
	80287	38–48
	80387	28
	80486/7	4
	Pentium	1–4

FUCOM/FUCOMP/FUCOMPP Unordered compare

11011101 11100rrr FUCOM ST,ST(rrr)
11011101 11101rrr FUCOMP ST,ST(rrr)
11011101 11101001 FUCOMPP

Format	Examples		Clocks
FUCOM FUCOMP FUCOMPP	FUCOM ST,ST(2) FUCOM FUCOMP ST,ST(3) FUCOMP FUCOMPP	8087	—
		80287	—
		80387	24–26
		80486/7	4–5
		Pentium	1–4

FWAIT Wait

10011011

Example		Clocks
FWAIT	8087	4
	80287	3
	80387	6
	80486/7	1–3
	Pentium	1–3

FXAM Examine ST(0)

11011001 11100101

Example		Clocks
FXAM	8087	12–23
	80287	12–23
	80387	30–38
	80486/7	8
	Pentium	21

FXCH Exchange ST(0) with another register

11011001 11001rrr FXCH ST,ST(rrr)

Format	Examples		Clocks
FXCH	FXCH ST,ST(1) FXCH FXCH ST,ST(4)	8087	10–15
		80287	10–15
		80387	18
		80486/7	4
		Pentium	1

FXTRACT Extract components of ST(0)

11011001 11110100

Example		Clocks
FXTRACT	8087	27–55
	80287	27–55
	80387	70–76
	80486/7	16–20
	Pentium	13

FYL2X ST(1) x $\log_2$ ST(0)

11011001 11110001

Example		Clocks
FYL2X	8087	900–1100
	80287	900–1100
	80387	120–538
	80486/7	196–329
	Pentium	22–111

FXL2XP1 ST(1) x $\log_2$ [ST(0) + 1.0]

11011001 11111001

Example		Clocks
FXL2XP1	8087	700–1000
	80287	700–1000
	80387	257–547
	80486/7	171–326
	Pentium	22–103

Notes: d = direction, where d = 0 for ST as the destination, and d = 1 for ST as the source; rrr = floating-point register number; oo = mode; mmm = r/m field; and disp = displacement.

listed for the 8087, 80287, 80387, 80486, 80487, and Pentium. To determine the execution time of an instruction, the clock time is multiplied by the execution time. The FADD instruction requires 70–143 clocks for the 80287. If an 8 MHz clock is used, the clocking period would be $^1/_8$ MHz, or 125 ns. The FADD instruction requires between 8.75 μs and 17.875 μs to execute. Using a 33 MHz (33 ns) 80486DX2, this instruction requires between 0.264 μs and 0.66 μs to execute. On the Pentium operated at 66 MHz (15.2 ns), the FADD requires between 0.0152 μs and 0.106 μs.

Table 13–8 uses some shorthand notations to represent the displacement that may or may not be required for an instruction that uses a memory-addressing mode. It also uses the abbreviations 00 to represent the mode, mmm to represent a register/memory addressing mode, and rrr to represent one of the floating-point coprocessor registers ST(0)–ST(7). The d (destination) bit that appears in some instruction opcodes defines the direction of the data flow, as in FADD ST,ST(2) or FADD ST(2),ST. The d bit is a logic 0 for flow *towards ST,* as in FADD ST,ST(2) where ST holds the sum after the addition, and a logic 1 flow *away from ST,* as in FADD ST(2),ST where ST(2) holds the sum.

Also note that some instructions allow a choice of whether a wait is inserted. For example, the FSTSW AX instruction copies the status register into AX. The FNSTSW AX instruction also copies the status register to AX, but without a wait.

13–4 PROGRAMMING WITH THE ARITHMETIC COPROCESSOR

This section provides programming examples for the arithmetic coprocessor. Each example is chosen to illustrate a programming technique for the coprocessor.

Calculating the Area of a Circle

This first programming example illustrates a simple method of addressing the coprocessor stack. First recall that the equation for calculating the area of a circle is $A = \pi R^2$. A program that performs this calculation is shown in Example 13–8. This program takes test data from array RAD, which contains five sample radii. The five areas are stored in a second array called AREA. No attempt is made in this program to use the data from the AREA array.

EXAMPLE 13–8

```
                    ;a short program that finds the area of five circles whose
                    ;radii are stored in array RAD.
                    ;
                            .MODEL SMALL
                            .386                              ;select 80386
                            .387                              ;select 80387
0000                        .DATA
0000  4015C28F      RAD     DD      2.34,5.66,9.33,234.5,23.4
      40B51EB8
      411547AE
      436A8000
      41BB3333
0014  0005 [        AREA    DD      5 DUP (?)
          00000000
```

```
                              ]
0000                               .CODE
                                   .STARTUP
0010  BE 0000                      MOV SI,0                  ;source element 0
0013  BF 0000                      MOV DI,0                  ;destination element 0
0016  B9 0005                      MOV CX,5                  ;count of 5
0019                    MAIN1:
0019  D9 84 0000 R                 FLD    RAD [SI]           ;radius to ST
001D  D8 C8                        FMUL   ST,ST(0)           ;square radius
001F  D9 EB                        FLDPI                     ;π to ST
0021  DE C9                        FMUL                      ;multiply ST = ST x ST(1)
0023  D9 9D 0014 R                 FSTP   AREA [DI]          ;save area
0027  46                           INC    SI
0028  47                           INC    DI
0029  E2 EE                        LOOP   MAIN1
                                   .EXIT
                                   END
```

Although this is a simple program, it illustrates the operation of the stack. To provide a better understanding of the operation of the stack, Figure 13–10 shows the contents of the stack after each instruction of Example 13–8 executes. Only one pass through the loop is illustrated, though the program calculates five areas.

The first instruction loads the contents of memory location RAD [SI], one of the elements of the array, to the top of the stack. Next, the FMUL ST,ST(0) instruction squares the radius on the top of the stack. The FLDPI instruction loads π to the stack top. The

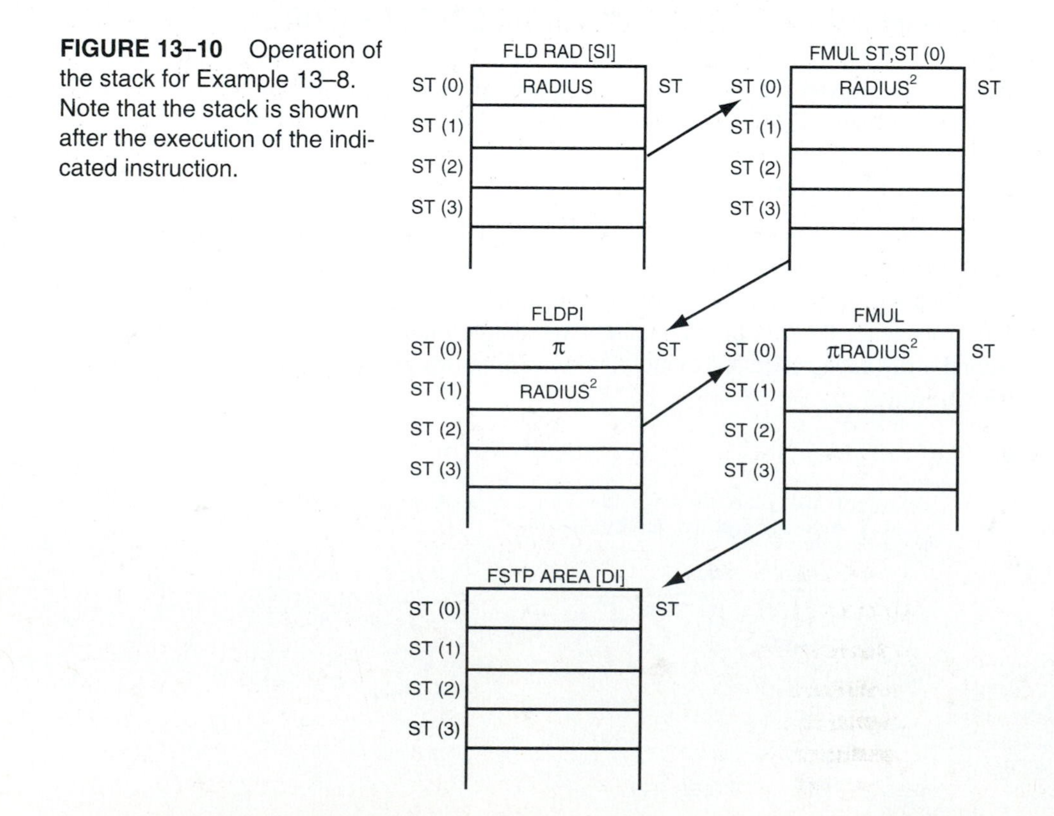

FIGURE 13–10 Operation of the stack for Example 13–8. Note that the stack is shown after the execution of the indicated instruction.

FMUL instruction uses the classic stack addressing mode to multiply ST by ST(1). After the multiplication, the prior values of ST and ST(1) are removed from the stack and the product replaces them at the top of the stack. Finally, the FSTP [DI] instruction copies the top of the stack, the area, to an array memory location AREA and clears the stack.

Notice how care is taken to always remove all stack data. This is important, because if data remain on the stack at the end of the procedure, the stack top will no longer be register 0. This could cause problems, because software assumes that the top of the stack is register 0. Another way of insuring that the coprocessor is initialized is to place the FINIT (initialization) instruction at the start of the program.

Finding the Resonant Frequency

An equation commonly used in electronics is the formula for determining the resonant frequency of an LC circuit. The equation solved by the program illustrated in Example 13–9 is $Fr = 1/(2\pi\sqrt{LC})$. This example uses L1 for the inductance L, C1 for the capacitor C, and RESO for the resultant resonant frequency.

EXAMPLE 13–9

```
                        ;a sample program that finds the resonant frequency of an LC
                        ;tank circuit.
                        ;
                               .MODEL SMALL
                               .386
                               .387
0000                           .DATA
0000  00000000          RESO   DD     ?                      ;resonant frequency
0004  358637BD          L1     DD     0.000001               ;inductance
0008  358637BD          C1     DD     0.000001               ;capacitance
000C  40000000          TWO    DD     2.0                    ;constant
0000                           .CODE
                               .STARTUP
0010  D9 06 0004 R             FLD    L1                     ;get L
0014  D8 0E 0008 R             FMUL   C1                     ;find LC

0018  D9 FA                    FSQRT                         ;find √LC

001A  D8 0E 000C R             FMUL   TWO                    ;find 2√LC

001E  D9 EB                    FLDPI                         ;get π
0020  DE C9                    FMUL                          ;get 2π√LC

0022  D9 E8                    FLD1                          ;get 1
0024  DE F1                    FDIVR                         ;form 1/(2π√LC)

0026  D9 1E 0000 R             FSTP   RESO                   ;save frequency
                               .EXIT
                               END
```

Notice the straightforward manner in which the program solves this equation. Very little data manipulation is required, because of the stack inside the coprocessor. Also notice how the constant TWO is defined for the program and how the DIVRP, using classic stack addressing, is used to form the reciprocal. If you own a reverse-polish entry calculator, such as those produced by Hewlett-Packard, you are familiar with stack addressing. If not, using the coprocessor will increase your experience with this type of entry.

Finding the Roots Using the Quadratic Equation

This example illustrates how to find the roots of a polynomial expression ($ax^2 + bx + c = 0$) using the quadratic equation. The quadratic equation is $\frac{(b \pm \sqrt{b^2 - 4ac})}{2a}$. Example 13–10 illustrates a program that finds the roots (R1 and R2) for the quadratic equation. The constants are stored in memory locations A1, B1, and C1. No attempt is made to determine the roots if they are imaginary. This example tests for imaginary roots and exits to DOS with a zero in the roots (R1 and R2) if it finds them. In practice, imaginary roots could be solved for and stored in a separate set of memory locations.

EXAMPLE 13–10

```
                        ;a program that finds the roots of a polynomial equation using
                        ;the quadratic equation. Note imaginary roots are indicated if
                        ;both root 1 (R1) and root 2 (R2) are zero.
                        ;
                              .MODEL SMALL
                              .386
                              .387
0000                          .DATA
0000 40000000           TWO   DD     2.0
0004 40800000           FOUR  DD     4.0
0008 3F800000           A1    DD     1.0
000C 00000000           B1    DD     0.0
0010 C1100000           C1    DD     -9.0
0014 00000000           R1    DD     ?
0018 00000000           R2    DD     ?
0000                          .CODE
                              .STARTUP
0010  D9 EE                   FLDZ
0012  D9 16 0014 R            FST    R1                 ;clear roots
0016  D9 1E 0018 R            FSTP   R2
001A  D9 06 0000 R            FLD    TWO
001E  D8 0E 0008 R            FMUL   A1                 ;form 2a
0022  D9 06 0004 R            FLD    FOUR
0026  D8 0E 0008 R            FMUL   A1
002A  D8 0E 0010 R            FMUL   C1                 ;form 4ac
002E  D9 06 000C R            FLD    B1
0032  D8 0E 000C R            FMUL   B1                 ;form b²
0036  DE E1                   FSUBR                     ;form b²-4ac
0038  D9 E4                   FTST                      ;test b²-4ac for zero
003A  9B DF E0                FSTSW  AX                 ;copy status register to AX
003D  9E                      SAHF                      ;move to flags
003E  74 0E                   JZ     ROOTS1             ;if b²-4ac is zero
0040  D9 FA                   FSQRT                     ;find square root of b²-4ac
0042  9B DF E0                FSTSW  AX
0045  A9 0001                 TEST   AX,1               ;test for invalid error (negative)
0048  74 04                   JZ     ROOTS1
004A  DE D9                   FCOMPP                    ;clear stack
004C  EB 18                   JMP    ROOTS2             ;end
004E                    ROOTS1:
004ED9 06 000C R              FLD    B1
0052D8 E1                     FSUB   ST,ST(1)
0054D8 F2                     FDIV   ST,ST(2)
0056D9 1E 0014 R              FSTP   R1                 ;save root 1
005A  D9 06 000C R            FLD    B1
005E  DE C1                   FADD
0060  DE F1                   FDIVR
```

```
0062  D9 1E 0018 R        FSTP   R2                 ;save root 2
0066                  ROOTS2:
                            .EXIT
                            END
```

Using a Memory Array to Store Results

The next programming example illustrates the use of a memory array and the scaled-indexed addressing mode to access the array. Example 13–11 shows a program that calculates 100 values of inductive reactance. The equation for inductive reactance is XL = 2πFL. In this example, the frequency range is between 10Hz and 1000Hz for F, and the inductance is 4H. The instruction FSTP DWORD PTR CS:[EDI+4*ECX] is used to store the reactance for each frequency, beginning with the last at 1000Hz and ending with the first at 10Hz. The FCOMP instruction is used to clear the stack just before the RET instruction.

EXAMPLE 13–11

```
                      ;a program that calculates the inductive reactance of L
                      ;at a frequency range of 10Hz to 1000Hz and stores it
                      ;in array XL. Note the increment is 10Hz.
                             .MODEL SMALL
                             .386
                             .387
0000                         .DATA
0000  40800000     L         DD     4.0                 ;4.0H test value
0004  0064 [       XL        DD     100 DUP (?)
              00000000
                         ]
0194  447A0000     F         DD     1000.0              ;start at 1000Hz
0198  41200000     TEN       DD     10.0                ;increment of 10Hz
0000                         .CODE
                             .STARTUP
0010  66| B9 00000064        MOV    ECX,100             ;load count
0016  66| BF 00000000 R      MOV    EDI,OFFSET XL-4     ;address result
001C  D9 EB                  FLDPI                      ;get π
001E  D8 C0                  FADD   ST,ST(0)            ;form 2π
0020  D8 0E 0000 R           FMUL   L                   ;form 2πL
0024               L1:
0024  D9 06 0194 R           FLD    F                   ;get F
0028  D8 C9                  FMUL   ST,ST(1)
002A  67& D9 1C 8F           FSTP   DWORD PTR [EDI+4*ECX]
002E  D9 06 0194 R           FLD    F
0032  D8 26 0198 R           FSUB   TEN                 ;change frequency
0036  D9 1E 0194 R           FSTP   F
003A  E2 E8                  LOOP   L1
003C  D8 D9                  FCOMP
                             .EXIT
                             END
```

Displaying a Single Precision Floating-Point Number

This section of the text shows how to take the floating-point contents of a 32-bit single precision floating-point number and display it on the video display. The procedure displays the floating-point number as a mixed number, with an integer part and a fractional part separated by a decimal point. In order to simplify the procedure, a limit is placed on the display size of the mixed number, so the integer portion is a 32-bit binary number and the fraction is a 24-bit binary number. The procedure will not function properly for larger or smaller numbers.

EXAMPLE 13–12

```
                        ;a program that displays the floating-point contents of NUMB
                        ;as a mixed decimal number.
                               .MODEL SMALL
                               .386
                               .387
0000                           .DATA
0000  C50B0200          NUMB   DD      -2224.125                    ;test data
0004  0000              TEMP   DW      ?
0006  00000000          WHOLE  DD      ?
000A  00000000          FRACT  DD      ?
0000                           .CODE
                               .STARTUP
0010  E8 000B                  CALL    DISP                         ;display NUMB
                               .EXIT
                        ;
                        ;procedure that displays the ASCII code from AL
                        ;
0017                    DISPS  PROC    NEAR

0017  B4 06                    MOV     AH,6                         ;display AL
0019  8A D0                    MOV     DL,AL
001B  CD 21                    INT     21H
001D  C3                       RET

001E                    DISPS  ENDP
                        ;
                        ;procedure that displays the floating-point contents of NUMB
                        ;in decimal form.
                        ;
001E                    DISP   PROC    NEAR

001E  9B D9 3E 0004 R          FSTCW   TEMP                         ;save current control word
0023  81 0E 0004 R 0C00        OR      TEMP,0C00H                   ;set rounding to chop
0029  D9 2E 0004 R             FLDCW   TEMP
002D  D9 06 0000 R             FLD     NUMB                         ;get NUMB
0031  D9 E4                    FTST                                 ;test NUMB
0033  9B DF E0                 FSTSW   AX                           ;status to AX
0036  25 4500                  AND     AX,4500H                     ;get C3, C2, and C0
                               .IF AX == 0100H
003E  B0 2D                            MOV  AL,'-'
0040  E8 FFD4                          CALL  DISPS
0043  D9 E1                            FABS
                               .ENDIF
0045  D9 C0                    FLD     ST
0047  D9 FC                    FRNDINT                              ;get integer part
0049  DB 16 0006 R             FIST    WHOLE
004D  DE E1                    FSUBR
004F  D9 E1                    FABS
0051  D9 1E 000A R             FSTP    FRACT                        ;save fraction
0055  66| A1 0006 R            MOV     EAX,WHOLE
0059  66| BB 0000000A          MOV     EBX,10
005F  B9 0000                  MOV     CX,0
0062  53                       PUSH    BX
                               .WHILE 1                             ;divide until quotient = 0
```

```
0063  66| BA 00000000               MOV    EDX,0
0069  66| F7 F3                     DIV    EBX
006C  80 C2 30                      ADD    DL,30H
006F  52                            PUSH   DX
                            .BREAK .IF EAX == 0
0075  41                            INC    CX
                                   .IF    CX == 3
007B  6A 2C                                PUSH   ','
007D  B9 0000                              MOV    CX,0
                                   .ENDIF
                            .ENDW
                            .WHILE 1                           ;display whole number part
0082  5A                            POP   DX
                            .BREAK .IF DX == BX
0087  8A C2                         MOV    AL,DL
0089  E8 FF8B                       CALL   DISPS
                            .ENDW
008E  B0 2E                 MOV    AL,'.'                      ;display decimal point
0090  E8 FF84               CALL   DISPS
0093  66| A1 000A R         MOV    EAX,FRACT
0097  9B D9 3E 0004 R       FSTCW  TEMP                        ;save current control word
009C  81 36 0004 R 0C00     XOR    TEMP,0C00H                  ;set rounding to nearest
00A2  D9 2E 0004 R          FLDCW  TEMP
00A6  D9 06 000A R          FLD    FRACT
00AA  D9 F4                 FXTRACT
00AC  D9 1E 000A R          FSTP   FRACT
00B0  D9 E1                 FABS
00B2  DB 1E 0006 R          FISTP  WHOLE
00B6  66| 8B 0E 0006 R      MOV    ECX,WHOLE
00BB  66| A1 000A R         MOV    EAX,FRACT
00BF  66| C1 E0 09          SHL    EAX,9
00C3  66| D3 D8             RCR    EAX,CL
                            .REPEAT
00C6  66| F7 E3                     MUL EBX
00C9  66| 50                        PUSH EAX
00CB  66| 92                        XCHG EAX,EDX
00CD  04 30                         ADD AL,30H
00CF  E8 FF45                       CALL DISPS
00D2  66| 58                        POP EAX
                            .UNTIL EAX == 0
00D9  C3                    RET

00DA                DISP    ENDP
                    END
```

Example 13–12 lists a program that calls a procedure for displaying the contents of memory location NUMB on the video display at the current cursor position. The procedure first tests the sign of the number and displays a minus sign for a negative number. Then, if necessary, the number is made positive by the FABS instruction. Next, it is divided into an integer and fractional part and stored at WHOLE and FRACT. Notice how the FRNDINT instruction is used to round the top of the stack (using the chop mode) to form the whole number part of NUMB. The whole number part is then subtracted from the original number to generate the fractional part. This is accomplished with the FSUB instruction, which subtracts the contents of ST(1) from ST.

The last part of the procedure displays the whole number part followed by the fractional part. The techniques are the same as introduced earlier in the text: divide a number by ten and display the remainder in reverse order to convert and display an integer. Multiplication by

10 converts a fraction to decimal for displaying. Note that the fractional part may contain a rounding error for certain values. This occurs because the number has not been adjusted to remove the rounding error inherent in floating-point factional numbers.

Reading a Mixed-Number from the Keyboard

If floating-point arithmetic is used in a program, a method of reading the number from the keyboard and converting it to floating-point must be developed. The procedure listed in Example 13–13 reads a signed mixed number from the keyboard and converts it to a floating-point number located at memory location NUMB.

EXAMPLE 13-13

```
                        ;a program that reads a mixed number from the keyboard.
                        ;The result is stored at memory location NUMB as a
                        ;double precision floating-point number.
                        ;
                              .MODEL SMALL
                              .386
                              .387
0000                          .DATA
0000  00                SIGN   DB     ?                   ;sign indicator
0001  0000              TEMP1  DW     ?                   ;temporary storage
0003  41200000          TEN    DD     10.0                ;10.0
0007  00000000          NUMB   DD     ?                   ;result
0000                          .CODE
                        GET    MACRO                      ;;read key macro
                               MOV   AH,1
                               INT   21H
                               ENDM
                               .STARTUP
0010  D9 EE                    FLDZ                       ;clear ST
                               GET                        ;read a character
                               .IF AL == '+'              ;test for +
001A  C6 06 0000 R 00              MOV SIGN,0             ;clear sign indicator
                               GET
                               .ENDIF
                               .IF AL == '-'              ;test for -
0027  C6 06 0000 R 01              MOV SIGN,1             ;set sign indicator
                                   GET
                               .ENDIF
                               .REPEAT
0030  D8 0E 0003 R                 FMUL TEN               ;multiply result by 10
0034  B4 00                        MOV AH,0
0036  2C 30                        SUB AL,30H             ;convert from ASCII
0038  A3 0001 R                    MOV TEMP1,AX
003B  DE 06 0001 R                 FIADD TEMP1            ;add it to result
                                   GET                    ;get next character
                               .UNTIL AL < '0' || AL > '9'
                               .IF AL == '.'              ;do if -
004F  D9 E8                        FLD1                   ;get one
                                   .WHILE 1
0051  D8 36 0003   R                 FDIV TEN
                                     GET
                                     .BREAK .IF AL < '0' || AL > '9'
0061  B4 00                          MOV AH,0
0063  2C 30                          SUB AL,30H           ;convert from ASCII
0065  A3 0001 R                      MOV TEMP1,AX
```

```
0068  DF 06 0001 R          FILD TEMP1
006C  D8 C9                 FMUL ST,ST(1)
006E  DC C2                 FADD ST(2),ST
0070  D8 D9                 FCOMP
                          .ENDW
0074  D8 D9               FCOMP                   ;clear stack
                      .ENDIF
                      .IF SIGN == 1
007D  D9 E0               FCHS                    ;make negative
                      .ENDIF
007F  D9 1E 0007 R    FSTP   NUMB                 ;save result
                      .EXIT
                      END
```

Unlike other examples in this chapter, this example uses some of the high-level language constructs presented in earlier chapters to reduce its size. Here, the sign is first read from the keyboard and, if present, saved for later use as a 0 for positive and a one for negative. Next, the integer portion of the number is read. The .REPEAT-.UNTIL loop is used to read the number until something other than a number (0–9) is typed. This portion terminates with a period, space, or carriage return. If a period is typed, the procedure continues and reads a fractional part by using an .IF–.ENDIF construct. If a space or carriage return is entered, the number is converted to floating-point form and stored at NUMB. Notice how a .WHILE–.ENDW loop is used to convert the fractional part of the number. The whole number portion is converted with a multiplication by ten, and the fractional portion is converted with a division by ten.

13–5 SUMMARY

1. The arithmetic coprocessor functions in parallel with the microprocessor. This means that the microprocessor and coprocessor can execute their instructions simultaneously.
2. The data types manipulated by the coprocessor include signed-integer, floating-point, and binary-coded decimal (BCD).
3. Three forms of integers are used with the coprocessor: word (16 bits), short (32 bits), and long (64 bits). Each integer contains a signed number in true magnitude form for positive numbers and two's complement form for negative numbers.
4. A BCD number is stored as an 18-digit number in 10 bytes of memory. The most significant byte contains the sign bit, and the remaining 9 bytes contain an 18-digit packed BCD number.
5. The coprocessor supports three types of floating-point numbers: single precision (32 bits), double precision (64 bits), and extended precision (80 bits). A floating-point number has three parts: the sign, the biased exponent, and the significand. In the coprocessor, the exponent is biased with a constant; the integer bit of the normalized number is not stored in the significand except in extended precision form.
6. Decimal numbers are converted to floating-point numbers by: (a) converting the number to binary, (b) normalizing the binary number, (c) adding the bias to the exponent, and (d) storing the number in floating-point form.

7. Floating-point numbers are converted to decimal by: (a) subtracting the bias from the exponent, (b) unnormalizing the number, and (c) converting it to decimal.
8. The 80287 uses I/O space for the execution of some of its instructions. This space is invisible to the program and is used internally by the 80286/80287 system. These 16-bit I/O addresses—00F8H, 00FAH, and 00FCH—must not be used for I/O data transfers in a system that contains an 80287. The 80387, 80486/7, and Pentium use I/O addresses 800000F8H–800000FFH.
9. The coprocessor contains a status register that indicates busy, what conditions follow a compare or test, the location of the top of the stack, and the state of the error bits. The FSTSW AX instruction followed by SAHF is often used with conditional jump instructions to test for some coprocessor conditions.
10. The control register of the coprocessor contains control bits that select infinity, rounding, precision, and error masks.
11. The following directives are often used with the coprocessor for storing data: DW (define word), DD (define doubleword), DQ (define quadword), and DT (define 10 bytes).
12. The coprocessor uses a stack to transfer data between itself and the memory system. Generally, data are loaded to the top of the stack or removed from the top of the stack for storage.
13. All internal coprocessor data are in the 80-bit extended precision form. The only time data is in any other form is when it is stored or loaded from the memory.
14. The coprocessor addressing modes include: the classic stack mode, register, register with a pop, and memory. Stack addressing is implied: the data at ST becomes the source, ST(1) becomes the destination, and the result is found in ST.
15. The coprocessor's arithmetic operations include: addition, subtraction, multiplication, division, and square root.
16. There are transcendental functions in the coprocessor's instruction set. These functions find the partial tangent or arctangent, 2x – 1, Y $\log_2$ X, and Y $\log_2$ (X + 1). The 80387, 80486/7, and Pentium also include sine and cosine functions.
17. Constants stored inside the coprocessor provide +0.0, +1.0, π, $\log_2 10$, $\log_2 e$, $\log_{10} 2$, and $\log_e 2$.
18. The 80387 functions with the 80386 microprocessor, and the 80487SX functions with the 80486SX microprocessor, but the 80486DX and Pentium contain their own internal arithmetic coprocessor. The instructions performed by the earlier versions are available on these coprocessors. In addition to these instructions, the 80387, 80486/7, and Pentium also can find sine and cosine.

13–6 QUESTIONS AND PROBLEMS

1. List the three types of data that are loaded or stored in memory by the coprocessor.
2. List the three integer data types, the range of the integers stored in them, and the number of bits allotted to each.
3. Explain how a BCD number is stored in memory by the coprocessor.

4. List the three types of floating-point numbers used with the coprocessor and the number of binary bits assigned to each.
5. Convert the following decimal numbers into single precision floating-point numbers:
   (a) 28.75
   (b) 624
   (c) –0.615
   (d) +0.0
   (e) –1000.5
6. Convert the following single precision floating-point numbers into decimal:
   (a) 11000000 11110000 00000000 00000000
   (b) 00111111 00010000 00000000 00000000
   (c) 01000011 10011001 00000000 00000000
   (d) 01000000 00000000 00000000 00000000
   (e) 01000001 00100000 00000000 00000000
   (f) 00000000 00000000 00000000 00000000
7. Explain what the coprocessor does when a normal microprocessor instruction executes.
8. Explain what the microprocessor does when a coprocessor instruction executes.
9. What is the purpose of the C3–C0 bits in the status register?
10. What operation is accomplished with the FSTSW AX instruction?
11. What is the purpose of the IE bit in the status register?
12. How can SAHF and a conditional jump instruction be used to determine whether the top of the stack (ST) is equal to register ST(2)?
13. How is the rounding mode selected in the 80X87?
14. What coprocessor instruction uses the microprocessor's AX register?
15. What I/O ports are reserved for coprocessor use with the 80287?
16. How are data stored inside the coprocessor?
17. What is a NAN?
18. Whenever the coprocessor is reset, the top of the stack register is register number ______________.
19. What does the term *chop* mean in the rounding control bits of the control register?
20. What is the difference between *affine* and *projective* infinity control?
21. What microprocessor instruction forms the opcodes for the coprocessor?
22. The FINIT instruction selects ______________ precision for all coprocessor operations.
23. Using assembler pseudo-opcodes, form a statement that accomplishes the following:
    (a) Store a 23.44 into a double precision floating-point memory location named FROG.
    (b) Store a –123 into a 32-bit signed integer location named DATA3.
    (c) Store a –23.8 into a single precision floating-point memory location named DATA1.
    (d) Reserve a double-precision memory location named DATA2.
24. Describe how the FST DATA instruction functions. Assume that DATA is defined as a 64-bit memory location.
25. What does the FILD DATA instruction accomplish?
26. Form an instruction that adds the contents of register 3 to the top of the stack.
27. Describe the operation of the FADD instruction.
28. Choose an instruction that subtracts the contents of register 2 from the top of the stack and stores the result in register 2.
29. What is the function of the FBSTP DATA instruction?

30. What is the difference between forward and reverse division?
31. Develop a procedure that finds the reciprocal of a single precision floating-point number. The number is passed to the procedure in EAX and must be returned as a reciprocal in EAX.
32. What is the difference between the FTST instruction and the FXAM instruction?
33. Explain what the F2XM1 instruction calculates.
34. Which coprocessor status register bit should be tested after the FSQRT instruction executes?
35. Which coprocessor instruction pushes π onto the top of the stack?
36. Which coprocessor instruction places a 1.0 at the top of the stack?
37. What will FFREE ST(2) accomplish when executed?
38. Which instruction stores the environment?
39. What does the FSAVE instruction save?
40. Develop a procedure that finds the area of a rectangle ($A = L \times W$). Memory locations for this procedure are single precision floating-point locations A, L, and W.
41. Write a procedure that finds the capacitive reactance ($XC = 1/2\pi FC$). Memory locations for this procedure are single precision floating-point locations XL, F, and C.
42. Develop a procedure that generates a table of square roots for the integers 2 through 10. The results must be stored as single precision floating-point numbers in an array called ROOTS.
43. When is the FWAIT instruction used in a program?
44. What is the difference between the FSTSW instruction and the FNSTSW instruction?
45. Given the series/parallel circuit equation illustrated in Figure 13–11, develop a program, using single precision values for R1, R2, R3, and R4, that finds the total resistance and stores the result at single precision location RT.
46. Develop a procedure that finds the cosine of a single precision floating-point number. The angle, in degrees, is passed to the procedure in EAX and the cosine is returned in EAX. Recall that FCOS finds the cosine of an angle expressed in radians.
47. Given two arrays of double precision floating-point data that each contain 100 elements (ARRAY1 and ARRAY2), develop a procedure that finds the product of ARRAY1 times ARRAY2 and stores the double precision floating-point result in a third array (ARRAY3).
48. Develop a procedure that takes the single precision contents of register EBX times π, and stores the result in register EBX as a single precision floating-point number. You must use memory to accomplish this task.

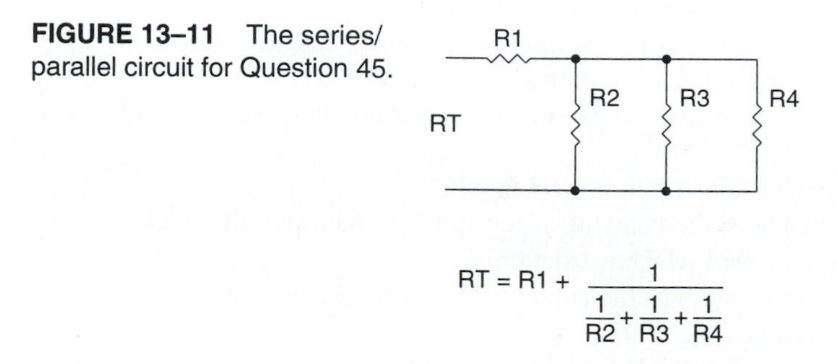

FIGURE 13–11 The series/parallel circuit for Question 45.

49. Write a procedure that raises a single precision floating-point number X to the power Y. Parameters are passed to the procedure with EAX = X and EBX = Y. The result is passed back to the calling sequence in ECX.
50. Given that $LOG_{10} X = (LOG_2 10)^{-1} \times LOG_2 X$, write a procedure called LOG10 that finds the LOG_{10} of the value at the stack top. Return the LOG_{10} at the stack top at the end of the procedure.
51. Use the procedure developed in Question 50 to solve the equation Gain in decibels = 20 LOG_{10} (Vout/Vin). The program should take arrays of single precision values for VOUT and VIN and store the decibel gains in a third array called DBG. There are 100 values for VOUT and VIN.

CHAPTER 14

Advanced Assembly Language Techniques

INTRODUCTION

Some techniques not developed in Chapters 1–13 appear in this chapter, including the DOS EXEC function, which executes a program from within a program. DOS EXEC is useful for chaining applications when required. Also described in this chapter are the EMS (expanded memory system emulator), XMS (extended memory system emulator), VCPI (virtual control program interface), and DPMI (DOS protected mode interface) standards for access to expanded and extended memory in the personal computer. Some DOS applications require more than 640K bytes of memory, which means that they must access the expanded or extended memory system. Also explained is IOCTL and the construction and application of a device driver.

One of the last areas explained is protected-mode programming from DOS. This includes the technique required to switch from the real mode to the protected mode and back. Although this requires an 80386 or newer system for effective access (80286 is too slow), it is becoming important because of the proliferation of these newer machines and large memory systems.

The last section discusses the use of assembly language procedures with high-level languages such as C/C++, BASIC, PASCAL, and FORTRAN. None of these advanced programming languages are adept at handling video display with the speed or efficiency needed. This opens them to the application of assembly language procedures to speed the execution of software.

OBJECTIVES

Upon completion of this chapter, you will be able to:

1. Develop software that accesses another application through the DOS EXEC function.
2. Explain the operation of and access to expanded and extended memory using EMS, XMS, VCPI, and DPMI.

3. Develop an application that uses extended memory.
4. Detail the purpose of IOCTL.
5. Develop a device driver for a DOS application.
6. Explain how a program can make the switch to protected-mode operation.
7. Develop a protected-mode application that accesses extended memory.
8. Show how to interface assembly language programs to C/C++, BASIC, PASCAL, and FORTRAN programs.

14–1 USING DOS EXEC TO EXECUTE A PROGRAM

The DOS EXEC (execute) function is accessed via DOS INT 21H function number 4BH. The EXEC function allows one assembly language program to execute another program, load another program without executing it, and load a file as a program overlay. A program that uses the EXEC function to load or execute another program is called the **parent.** The parent remains in memory while the **child,** the program loaded into memory by the parent, executes. The process of creating a child program in memory is often called **spawning.**

The EXEC Function

To execute another program with function 4BH, load AH with a 4BH and AL with the type of operation. The type of operation loaded to AL is 00H = load and execute a child program; 01H = load a child; 03H = load a program overlay; and 05H = enter the execute state. The DS:DX register combination addresses the child name or state pointer, and the ES:BX register combination is a pointer to a parameter block that contains information used to spawn (load and execute) the child. The state pointer is also used to specify the address of a program (overlay) that is loaded into memory with function 05H.

The Child File Name. The name of the child process is stored as an ASCII-Z string in the memory location addressed by the DS:DX register for AL = 00H, 01H, or 03H. For example, to execute a program called MYFILE.COM in the current directory, the ASCII-Z string shown in Example 14–1 appears at the location addressed by DS:DX. Note that the program file name must include any necessary extension (COM or EXE) and may not contain any wild cards.

EXAMPLE 14–1

```
0000  4D 59 46 49 4      EHAN    DB     'MYFILE.COM',0     ;child program name
         45 2E 43 4F 4D
         00
```

The Parameter Block. In addition to naming the program through the ASCII-Z string stored at the location address by DS:DX, the EXEC function also requires a parameter block. The parameter block is addressed by the ES:BX register combination and contains 14 bytes of information used by the EXEC function to build the environment for the child. Refer to Table A–6 for the contents of the parameter block.

The first entry in the parameter block contains the segment address of the environment block. The contents of the environment block are built by the CONFIG.SYS file, using the SET, PROMPT, and PATH commands in the AUTOEXEC.BAT file and or other places. Normally, the environment block is addressed by placing a 0000H into the first entry of the parameter block. This can be changed to address a different environment block, but that is rare. Most systems contain one environment block.

The next entry in the parameter block (offset 02H) contains the segment and offset address of the command line associated with the child. This is the command line parameter appended to the end of a command to select an option. For example, DIR *.BAK lists all files with the extension .BAK. The command line parameter is *.BAK in this example. The first byte of the command line contains the length of the command line, the next byte contains the command line, and the last byte contains a 13 or 0DH to indicate the end. Example 14–2 shows the command line structure for the DIR *.BAK command. The first byte holds the length or count of the number of characters in '*.BAK'. This count never includes the carriage return code (13).

EXAMPLE 14–2

```
000B  05               CLINE   DB     CR-CLINE-1         ;child command line
000C  2A 2E 42 41 4B           DB     '*.BAK'
0011  0D               CR      DB     13
```

The remaining two parameters in the parameter block are file control block pointers, which are not normally used with modern software. These are holdovers from DOS version 2.0 and are not recommended for newer programs.

Executing a Program with EXEC

Now that the basics of the EXEC function are known, let's see an example showing how to apply them with DOS INT 21H function 4BH. Example 14–3 shows an assembly language program that uses the DOS memory command (MEM) as a child process. The command line parameter in Example 14–3 asks whether a directory is needed and then asks for which area of the disk (e.g., drive). This program assumes that the MEM command is stored on hard disk drive C:\DOS, as indicated by CHILD. If MEM is stored elsewhere, this character string must be changed.

EXAMPLE 14–3

```
                        ;a program that accesses the DOS MEM.EXE program to display either
                        ;a summary of memory or a complete listing.
                        ;
                                .MODEL SMALL
                                .386
                                .STACK 1024              ;1K stack
0000                            .DATA
0000  0067 R 008A R     TAB     DW      SUM,COMP,BYE     ;jump table
         00AD R
0006  0000              OLDSS   DW      ?                ;stack pointer storage
0008  0000              OLDSP   DW      ?

000A  0000              PARA1   DW      0                ;standard environment
000C  0035 R                    DW      OFFSET C1        ;address blank command line
```

```
000E  ---- R                  DW     SEG C1
0010  0000 0000 0000          DW     0,0,0,0              ;FCBs
        0000

0018  0000            PARA2   DW     0                    ;standard environment
001A  0037 R                  DW     OFFSET C2            ;address /C /P command line
001C  ---- R                  DW     SEG C2
001E  0000 0000 0000          DW     0,0,0,0;FCBs
        0000

0026  43 3A 5C 44 4F  CHILD   DB     'C:\DOS\MEM.EXE',0   ;name of MEM program
      53 5C 4D 45 4D
      2E 45 58 45 00
0035  00 0D           C1      DB     0,13                 ;blank command line
0037  05 2F 43 20 2F  C2      DB     5,'/C /P',13         ;/C /P command line
      50 0D

003E  0D 0A 0A 4D 45  MES1    DB     13,10,10,'MEMORY DISPLAY',13,10,10
      4D 4F 52 59 20
      44 49 53 50 4C
      41 59 0D 0A 0A
0052  31 20 2D 20 44          DB     '1 - Display memory summary',13,10
      69 73 70 6C 61
      79 20 6D 65 6D
      6F 72 79 20 73
      75 6D 6D 61 72
      79 0D 0A
006E  32 20 2D 20 44          DB     '2 - Display memory complete',13,10
      69 73 70 6C 61
      79 20 6D 65 6D
      6F 72 79 20 63
      6F 6D 70 6C 65
      74 65 0D 0A
008B  33 20 2D 20 45          DB     '3 - Exit to DOS',13,10,10
      78 69 74 20 74
      6F 20 44 4F 53
      0D 0A 0A
009D  45 6E 74 65 72          DB     'Enter choice $'
      20 63 68 6F 69
      63 65 20 24
00AB  54 79 70 65 20  MES2    DB     'Type any key to continue: $'
      61 6E 79 20 6B
      65 79 20 74 6F
      20 63 6F 6E 74
      69 6E 75 65 3A
      20 24
0000                          .CODE
                      KEY     MACRO                       ;;read key, no echo
                              LOCAL  K1,K2
                      K1:
                              MOV    AH,6
                              MOV    DL,0FFH
                              INT    21H
                              JZ     K1                   ;;if no key typed
                              CMP    AL,0
                              JNZ    K2                   ;;if good ASCII code
                              INT    21H                  ;if extended ASCII
                              JMP    K1                   ;;ignore extended ASCII
                      K2:
                              ENDM

                              .STARTUP
0010  8C DB                   MOV    BX,DS                ;get DS
0012  8C C0                   MOV    AX,ES                ;load ES with segment address
0014  2B D8                   SUB    BX,AX                ;calculate length of data paragraphs
```

```
0016  8B C4                    MOV   AX,SP              ;find length of stack
0018  C1 E8 04                 SHR   AX,4
001B  40                       INC   AX
001C  03 D8                    ADD   BX,AX              ;find length of all segments
001E  B4 4A                    MOV   AH,4AH             ;reserve memory for this program
0020  CD 21                    INT   21H                ;its data, stack, and PSP
0022                    MAIN1:                          ;display menu
0022  B4 09                    MOV   AH,9
0024  BA 003E R                MOV   DX,OFFSET MES1
0027  CD 21                    INT   21H
0029                    MAIN2:
                               KEY                      ;get key, no echo
0039  2C 31                    SUB   AL,31H             ;convert from ASCII
003B  72 EC                    JB    MAIN2              ;if below 1 key
003D  3C 02                    CMP   AL,2
003F  77 E8                    JA    MAIN2              ;if above 3 key
0041  B4 00                    MOV   AH,0
0043  03 C0                    ADD   AX,AX              ;double choice
0045  BE 0000 R                MOV   SI,OFFSET TAB      ;address lookup table
0048  03 F0                    ADD   SI,AX
004A  8B 1C                    MOV   BX,[SI]            ;get address of procedure
004C  FF D3                    CALL  BX                 ;execute choice
004E  BA 00AB R                MOV   DX,OFFSET MES2     ;display any key
0051  B4 09                    MOV   AH,9
0053  CD 21                    INT   21H
                               KEY                      ;get any key
0065  EB BB                    JMP   MAIN1              ;repeat menu display

0067                    SUM    PROC  NEAR               ;display memory summary (1)

0067  B4 4B                    MOV   AH,4BH             ;execute MEM
0069  B0 00                    MOV   AL,0               ;operation load end execute
006B  BA 0026 R                MOV   DX,OFFSET CHILD    ;address program name
006E  1E                       PUSH  DS                 ;get DS to ES
006F  07                       POP   ES
0070  BB 000A R                MOV   BX,OFFSET PARA1    ;address command line C1
0073  FA                       CLI                      ;disable interrupts
0074  8C 16 0006 R             MOV   OLDSS,SS           ;save stack at OLD
0078  89 26 0008 R             MOV   OLDSP,SP
007C  FB                       STI                      ;enable interrupts
007D  CD 21                    INT   21H                ;execute MEM child process
007F  FA                       CLI                      ;disable interrupts
0080  8E 16 0006 R             MOV   SS,OLDSS
0084  8B 26 0008 R             MOV   SP,OLDSP           ;reload stack
0088  FB                       STI                      ;enable interrupts
0089  C3                       RET

008A                    SUM    ENDP

008A                    COMP   PROC  NEAR               ;display memory complete (2)

008A  B4 4B                    MOV   AH,4BH             ;execute MEM /C /P
008C  B0 00                    MOV   AL,0
008E  BA 0026 R                MOV   DX,OFFSET CHILD    ;address program name
0091  1E                       PUSH  DS                 ;DS to ES
0092  07                       POP   ES
0093  BB 0018 R                MOV   BX,OFFSET PARA2    ;address command line C2
0096  FA                       CLI                      ;disable interrupts
0097  8C 16 0006 R             MOV   OLDSS,SS           ;save old stack area
009B  89 26 0008 R             MOV   OLDSP,SP
009F  FB                       STI                      ;enable interrupts
00A0  CD 21                    INT   21H                ;execute MEM child process
00A2  FA                       CLI                      ;disable interrupts
00A3  8E 16 0006 R             MOV   SS,OLDSS
00A7  8B 26 0008 R             MOV   SP,OLDSP           ;get old stack area
```

```
00AB   FB                   STI                    ;enable interrupts
00AC   C3                   RET

00AD                COMP    ENDP

00AD                BYE     PROC    NEAR           ;for exit to DOS (3)

                            .EXIT                  ;exit to DOS

00B1                BYE     ENDP
                            END
```

The first part of this program is important because, before the EXEC function is used, the amount of space required for the current program and all segments associated with it must be reserved. This is accomplished with DOS INT 21H function number 4AH, which reserves memory in paragraphs. By subtracting the ES from the DS, the size of the program and code segment is determined in 16-byte paragraphs. The stack, which is placed at the end of the code and data segments, is converted to a paragraph length and added to the length of the code and data segment. This is then used by the allocate memory function, 4AH, to reserve memory for the program listed in Example 14–3. This is important, because when a child is executed, the parent must remain in memory.

Another important part of this example is saving the contents of SP and SS before the child is spawned and executed. The EXEC command saves all registers except the SP and SS registers when a child is executed. It is very important to save these registers before the child executes. It is also important to restore their contents after the child terminates.

This example program uses a lookup table (TAB) to locate the name of a procedure that corresponds to the numbers 1–3. The CALL BX instruction accesses the procedures to execute MEM with one of two parameters (C1 or C2) or exit to DOS.

Overlay Files

The EXEC function is also used with overlay files. An overlay file is a file that is loaded into memory only when needed. This allows a section of the memory to be shared with many different overlays, making more effective use of the memory system. The name of the overlay file is placed into an area of memory as an ASCII-Z string and addressed by DS:DX. This is identical to the load and execute child function.

The main difference between the load and execute child function and the overlay file is the parameter block. The parameter block for the overlay contains two words. The word at offset 00H defines the segment address where the overlay file is stored. The next word, at offset 02H, defines the relocation factor for the overlay file. Once the overlay is loaded into memory, it is accessed as data or software used with the parent.

14–2 MEMORY MANAGEMENT DRIVERS

The expanded and extended memory systems cannot be accessed from DOS without the installation of memory management drivers or programs. There are quite a few of these available; one of the more common ones is described. The **EMS (expanded memory**

system) driver was the first made available with the addition of expanded memory systems. Later, the **XMS (extended memory system)** driver was added, with the 80286 microprocessor, to control its extended memory system. Recently the **VCPI (virtual control program interface)** has been added to control the memory system in the 80386 and newer microprocessors. VCPI services are provided by DOS version 6.X as part of the EMM386.EXE program. Finally, **DPMI (DOS protected mode interface)** provides the best control if Windows is available. Windows includes the DPMI interface if the application is executed from the DOS applications box inside Windows.

EMS

The EMS (expanded memory system) memory manager divides the upper memory area or systems area into page frames that begin at any 4K byte boundary and are 64K bytes in length. A page frame is divided into four pages, each 16K bytes in length. This area of memory is known as expanded memory, because it also is used in early 8088-based personal computers to expand the memory system. This memory expansion standard, called *LIM 4.0* (Lotus/Intel/Microsoft), uses INT 67H for accessing and controlling the pages of memory. The EMS manager found in most systems is EMM386.EXE, designed to provide memory management for 80386 or newer microprocessors. Although EMS seems dated, it is still used by many personal computer video games. Eventually, this memory manager will disappear, and it is strongly recommended that the XMS manager be used in its place.

Figure 14–1 shows the area of memory managed by EMM386.EXE. The page frame is any 64K byte contiguous section of system memory. In the illustration, location

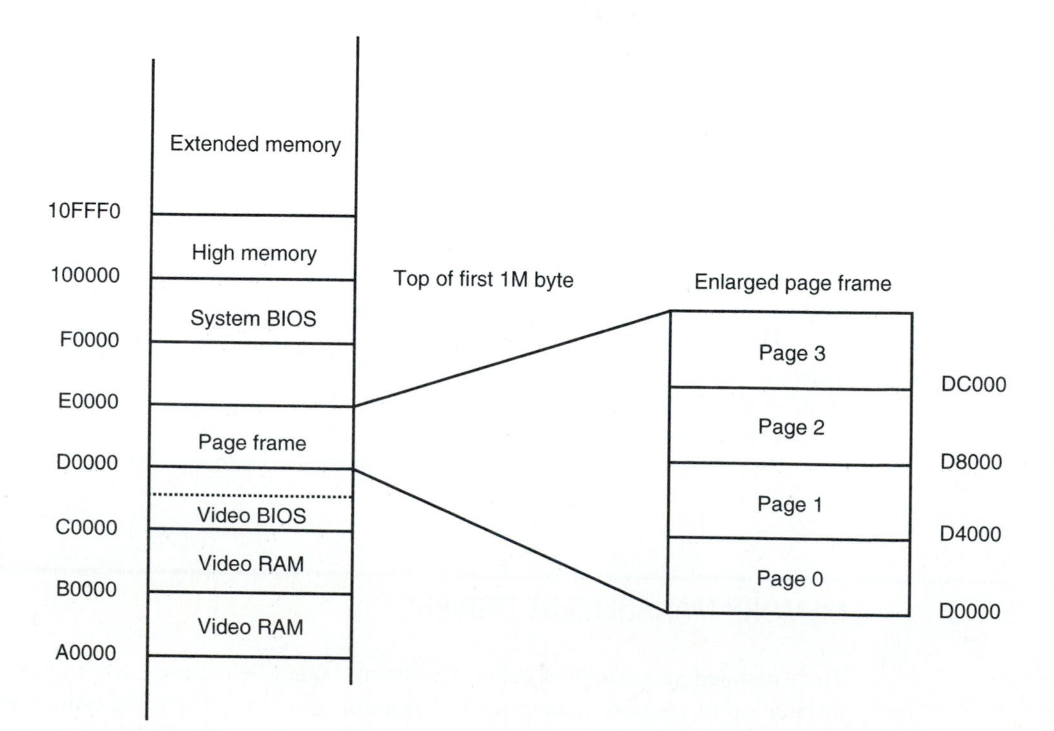

FIGURE 14–1 The page frame and expanded memory pages as specified by LIM 4.0.

D0000H–DFFFFH is chosen for the page frame. The 64K byte page frame is divided into four 16K byte pages. The EMM386.EXE program uses the microprocessor's internal hardware paging mechanism to assign extended memory to each page located in the page frame. In older systems, the memory manager activated hardware memory pages on a plug-in memory board using a bank swapping technique.

Is EMM386.EXE Installed? Before expanded memory is used by a program, it must be known whether EMM386.EXE or some other expanded memory manager is installed. To find out if the expanded memory manager is installed:

1. Open a file with the namc EMMXXXX0 using DOS INT 21H function 3DH (the open function). If there is a carry upon return, the manager is not installed.
2. If an EMS manager *is* installed, see if it is ready to accept output information by using IOCTL (I/O control) function 07H. If the return value is not a –1, the memory manager is not present.

If both tests are passed, use INT 67H to access the manager and expanded memory. If not, indicate that the memory manager is not present.

EXAMPLE 14–4

```
                              ;a program that tests for the presence of an extended memory
                              ;manager and expanded memory. If present, the program displays
                              ;the version number, page frame address, total expanded memory,
                              ;and available expanded memory.
                              ;
                                      .MODEL SMALL
                                      .386
0000                                  .DATA
0000  45 4D 4D 58 58          EMM     DB      'EMMXXXX0',0          ;EMM driver name
      58 58 30 00
0009  0D 0A 0A 4E 6F          ERR     DB      13,10,10,'No EMM driver present.$'
      20 45 4D 4D 20
      64 72 69 76 65
      72 20 70 72 65
      73 65 6E 74 2E
      24
0023  0D 0A 0A 45 4D          MES1    DB      13,10,10,'EMM version number $'
      4D 20 76 65 72
      73 69 6F 6E 20
      6E 75 6D 62 65
      72 20 24
003A  0D 0A 50 61 67          MES2    DB      13,10,'Page frame address = $'
      65 20 66 72 61
      6D 65 20 61 64
      64 72 65 73 73
      20 3D 20 24
0052  0D 0A 54 6F 74          MES3    DB      13,10,'Total expanded memory = $'
      61 6C 20 65 78
      70 61 6E 64 65
      64 20 6D 65 6D
      6F 72 79 20 3D
      20 24
006D  0D 0A 41 76 61          MES4    DB      13,10,'Available expanded memory = $'
      69 6C 61 62 6C
      65 20 65 78 70
      61 6E 64 65 64
      20 6D 65 6D 6F
```

```
        72 79 20 3D 20
        24
0000                            .CODE
                                .STARTUP
0010  E8 0069                   CALL    ISEMS           ;see if EMS driver present
0013  73 09                     JNC     MAIN1           ;if EMM driver present
0015  B4 09                     MOV     AH,9
0017  BA 0009 R                 MOV     DX,OFFSET ERR   ;display no EMM driver
001A  CD 21                     INT     21H
001C  EB 5A                     JMP     MAIN2           ;end it
001E                    MAIN1:
001E  B4 09                     MOV     AH,9
0020  BA 0023 R                 MOV     DX,OFFSET MES1
0023  CD 21                     INT     21H             ;display EMM version number

0025  B4 46                     MOV     AH,46H
0027  CD 67                     INT     67H             ;get version number

0029  8A E0                     MOV     AH,AL           ;get minor version
002B  80 E4 0F                  AND     AH,0FH
002E  C0 E8 04                  SHR     AL,4            ;get major version
0031  05 3030                   ADD     AX,3030H        ;make ASCII
0034  50                        PUSH    AX
0035  8A D0                     MOV     DL,AL           ;display version number
0037  B4 06                     MOV     AH,6
0039  CD 21                     INT     21H
003B  B2 2E                     MOV     DL,'.'
003D  CD 21                     INT     21H
003F  5A                        POP     DX
0040  8A D6                     MOV     DL,DH
0042  CD 21                     INT     21H
0044  B4 09                     MOV     AH,9
0046  BA 003A R                 MOV     DX,OFFSET MES2
0049  CD 21                     INT     21H             ;page frame address
004B  B4 41                     MOV     AH,41H
004D  CD 67                     INT     67H             ;get page frame address
004F  E8 0049                   CALL    DISPH           ;display hex address
0052  B4 09                     MOV     AH,9
0054  BA 0052 R                 MOV     DX,OFFSET MES3
0057  CD 21                     INT     21H             ;total expanded memory
0059  B4 42                     MOV     AH,42H
005B  CD 67                     INT     67H             ;get number of pages
005D  8B C2                     MOV     AX,DX
005F  B9 4000                   MOV     CX,4000H
0062  F7 E1                     MUL     CX              ;find total
0064  E8 0052                   CALL    DISPD           ;display decimal
0067  B4 09                     MOV     AH,9
0069  BA 006D R                 MOV     DX,OFFSET MES4
006C  CD 21                     INT     21H             ;available expanded memory
006E  8B C3                     MOV     AX,BX
0070  B9 4000                   MOV     CX,4000H
0073  F7 E1                     MUL     CX
0075  E8 0041                   CALL    DISPD
0078                    MAIN2:
                                .EXIT
                        ;
                        ;the ISEMS procedure tests to determine if EMM is loaded.
                        ;***return parameter***
                        ;if carry = 1, EMM is not installed
                        ;if carry = 0, EMM is installed
                        ;
007C                    ISEMS   PROC    NEAR
```

```
007C  B8 3D00               MOV    AX,3D00H
007F  BA 0000 R             MOV    DX,OFFSET EMM
0082  CD 21                 INT    21H              ;open file
0084  72 0E                 JC     ISEMS1           ;if EMM not there
0086  8B D8                 MOV    BX,AX            ;save handle
0088  B8 4407               MOV    AX,4407H         ;access IOCTL function 7
008B  CD 21                 INT    21H
008D  72 05                 JC     ISEMS1           ;if error
008F  3C FF                 CMP    AL,-1
0091  74 01                 JE     ISEMS1           ;if present
0093  F9                    STC
0094                 ISEMS1:
0094  9C                    PUSHF                   ;save carry
0095  B4 3E                 MOV    AH,3EH           ;close file
0097  CD 21                 INT    21H
0099  9D                    POPF
009A  C3                    RET

009B                 ISEMS  ENDP
                     ;
                     ;the DISPH is places the contents of BX as a 4 digit
                     ;hexadecimal number.
                     ;
009B                 DISPH  PROC   NEAR

009B  B5 04                 MOV    CH,4
009D                 DISPH1:
009D  C1 C3 04              ROL    BX,4             ;get digit
00A0  8A D3                 MOV    DL,BL
00A2  80 E2 0F              AND    DL,0FH
00A5  80 C2 30              ADD    DL,30H           ;convert to ASCII
                            .IF DL > 39H
00AD  80 C2 07                ADD DL,7              ;if a letter, add 7 more
                            .ENDIF
00B0  B4 06                 MOV    AH,6             ;display contents of DL
00B2  CD 21                 INT    21H
00B4  FE CD                 DEC    CH               ;count off 4
00B6  75 E5                 JNZ    DISPH1           ;repeat 4 times
00B8  C3                    RET

00B9                 DISPH  ENDP
                     ;
                     ;the DISPD procedure displays the contents of EAX in decimal
                     ;with commas.
                     ;
00B9                 DISPD  PROC   NEAR

00B9  53                    PUSH   BX               ;save BX
00BA  BD 002C               MOV    BP,','
00BD  66| C1 E2 10          SHL    EDX,16
00C1  66| 25 FFFF0000 AND   EAX,0FFFF0000H
00C7  66| 03 C2             ADD    EAX,EDX
00CA  B9 0003               MOV    CX,3             ;comma counter
00CD  BB FFFF               MOV    BX,-1
00D0  53                    PUSH   BX               ;indicate end
00D1  66| BB 0000000A MOV   EBX,10                  ;division factor
00D7                 DISPD1:
00D7  66| 33 D2             XOR    EDX,EDX
00DA  66| F7 F3             DIV    EBX
00DD  80 C2 30              ADD    DL,30H           ;make ASCII
00E0  52                    PUSH   DX
00E1  66| 83 F8 00          CMP    EAX,0
```

```
00E5  74 08                 JE     DISPD2               ;if done
00E7  E2 EE                 LOOP   DISPD1               ;do three times
00E9  55                    PUSH   BP                   ;save comma
00EA  B9 0003               MOV    CX,3
00ED  EB E8                 JMP    DISPD1
00EF                DISPD2:
00EF  B4 06                 MOV    AH,6                 ;display decimal digit
00F1  5A                    POP    DX
00F2  80 FA FF              CMP    DL,-1
00F5  74 04                 JE     DISPD3
00F7  CD 21                 INT    21H
00F9  EB F4                 JMP    DISPD2
00FB                DISPD3:
00FB  5B                    POP    BX                   ;restore BX
00FC  C3                    RET

00FD                DISPD  ENDP
                           END
```

To illustrate this sequence, Example 14–4 tests for the presence of EMS management and then uses INT 67H to get the version number, page frame address, total EMS memory, and available EMS memory. For a list of INT 67H calls, refer to Appendix A. You must install EMM386.EXE with expanded memory for this example to display the presence of expanded memory. An example is DEVICE=EMM386.EXE 1024, which installs 1M byte of expanded memory from the CONFIG.SYS file. ISEMS tests for the presence of EMM and DISPH and DISPD display hexadecimal or decimal data.

Using Expanded Memory in a Program. Now that a check for expanded memory has been made, let's investigate the use of expanded memory. Example 14–5 shows how 512K bytes of expanded memory is accessed and filled with 00H. Although this is not a practical program, it does illustrate the access to a considerable amount of expanded memory for an application.

EXAMPLE 14–5

```
                        ;a program that accesses expanded memory and stores a logic 00H into
                        ;512K bytes. Although not practical, this program does illustrate
                        ;how expanded memory is allocated, accessed and deallocated.
                        ;
                                .MODEL SMALL
                                .386
0000                            .DATA
0000  0000              HAN     DW     ?                ;EMM handle
0002  0000              PAGES   DW     ?                ;logical page number
0004  45 4D 4D 58 58    EMM     DB     'EMMXXXX0',0     ;EMM driver name
      58 58 30 00
000D  0D 0A 0A 45 4D    ERR1    DB     13,10,10,'EMM driver error.$'
      4D 20 64 72 69
      76 65 72 20 65
      72 72 6F 72 2E
      24
0000                            .CODE
                                .STARTUP
0010  E8 0064                   CALL   ISEMS            ;see if EMS driver present
0013  73 09                     JNC    MAIN1            ;if EMM driver present
0015                    ERR:
0015  B4 09                     MOV    AH,9
0017  BA 000D R                 MOV    DX,OFFSET ERR1   ;display no EMM driver
```

```
001A  CD 21                          INT    21H
001C  EB 55                          JMP    MAIN3             ;end it
001E                      MAIN1:
001E  B4 43                          MOV    AH,43H            ;allocate 32 pages
0020  BB 0020                        MOV    BX,32             ;or 512K bytes
0023  CD 67                          INT    67H
0025  80 FC 00                       CMP    AH,0              ;test for error
0028  75 EB                          JNZ    ERR               ;if error
002A  89 16 0000 R                   MOV    HAN,DX            ;save handle

002E  C7 06 0002 R 0000              MOV    PAGES,0           ;address of first page
0034  BD 0020                        MOV    BP,32             ;page counter
0037  B4 41                          MOV    AH,41H            ;get page frame address
0039  CD 67                          INT    67H
003B  80 FC 00                       CMP    AH,0
003E  75 D5                          JNE    ERR               ;if error
0040  8E C3                          MOV    ES,BX             ;load ES
0042                      MAIN2:
0042  B4 44                          MOV    AH,44H            ;select page
0044  B0 00                          MOV    AL,0              ;page 0 in frame
0046  8B 1E 0002 R                   MOV    BX,PAGES          ;first page in frame
004A  8B 16 0000 R                   MOV    DX,HAN            ;load handle
004E  CD 67                          INT    67H
0050  80 FC 00                       CMP    AH,0
0053  75 C0                          JNE    ERR               ;if error

0055  B9 4000                        MOV    CX,4000H          ;fill a page with 00H
0058  B0 00                          MOV    AL,0
005A  BF 0000                        MOV    DI,0
005D  F3/ AA                         REP    STOSB

005F  FF 06 0002 R                   INC    PAGES             ;get next page
0063  4D                             DEC    BP
0064  75 DC                          JNZ    MAIN2             ;do 32 times for 32 pages

0066  B4 45                          MOV    AH,45H
0068  8B 16 0000 R                   MOV    DX,HAN
006C  CD 67                          INT    67H               ;release memory
006E  80 FC 00                       CMP    AH,0
0071  75 A2                          JNE    ERR
0073                      MAIN3:
                                     .EXIT
                          ;
                          ;the ISEMS procedure tests for the presence of an EMS manager.
                          ;***return parameter***
                          ;if carry = 1, no EMS manager is present
                          ;if carry = 0, EMS manager is present
                          ;
0077                      ISEMS      PROC   NEAR

0077  B8 3D00                        MOV    AX,3D00H
007A  BA 0004 R                      MOV    DX,OFFSET EMM
007D  CD 21                          INT    21H               ;open file
007F  72 0E                          JC     ISEMS1            ;if not there
0081  8B D8                          MOV    BX,AX             ;save handle
0083  B8 4407                        MOV    AX,4407H          ;access IOCTL function 7
0086  CD 21                          INT    21H
0088  72 05                          JC     ISEMS1            ;if error
008A  3C FF                          CMP    AL,-1
008C  74 01                          JE     ISEMS1            ;if present
008E  F9                             STC
008F                      ISEMS1:
008F  9C                             PUSHF                    ;save carry
0090  B4 3E                          MOV    AH,3EH            ;close file
0092  CD 21                          INT    21H
```

```
0094  9D                  POPF
0095  C3                  RET

0096              ISEMS   ENDP
                          END
```

INT 67H functions 41H, 43H, 44H, and 45H are important parts of this program. The page frame address is obtained using function 41H. Function 43H allocates 32 pages, 16K bytes each (512K bytes), of expanded memory for this program. If less than 512K bytes of expanded memory exists, the INT 67H returns with AH not equal to zero. The page frame handle must be saved to refer to the allocated memory. All other functions require the page frame handle in register DX to access the allocated memory. The page frame address is then loaded into the ES register for accessing the page frame, to store 00H in all 512K bytes of the expanded memory. Function 44H maps a page into the page frame. In this example, a 16K-byte page is mapped into page frame page 0. This is repeated 32 times, to access a total of 512K bytes of expanded memory. Each time a page is mapped, the page number in BX is changed. Note that there are 64K pages available, for a total of 16K × 64K, or 1G bytes, of expanded memory available per handle. Of course, if you have less than 1G byte of memory, less is available. Note that AL is the page number in the frame (0, 1, 2, or 3) and BX is the logical page number (0, 1, . . . FFFFH). The last part of the program releases expanded memory by using function 45H. It is very important to release any expanded memory allocated by an application. If the expanded memory is not released, it is not available to other applications and will become available only if the system is rebooted. Some video games fail to release expanded memory when terminated.

XMS

The XMS (extended memory system) manager is installed with the HIMEM.SYS driver in the CONFIG.SYS file. The XMS driver allows access to the extended memory system in the personal computer. The HIMEM.SYS manages extended memory and provides high memory to DOS programs. The high memory area begins at memory address 100000H and extends to 10FFEFH. Data located at address 110000H through whatever is available are managed as extended memory by HIMEM.SYS. In most cases, high memory is used to load DOS and is unavailable for any other purpose. Many newer video games and other DOS applications access extended memory through the XMS services provided by HIMEM.SYS.

The XMS management software is accessed at two places. One is at the DOS multiplex interrupt (INT 2FH) and the other is through an entry point provided by the multiplex interrupt.

Is HIMEM.SYS Installed? Like EMS, the XMS system must be tested to see whether it is installed and operational before it is used. Example 14–6 lists a program that tests for the presence of XMS and reports the driver version number, and the amount of extended memory available.

EXAMPLE 14–6

```
;a program that tests for the presence of an XMS driver.
;If the driver is present, it displays the driver version
;number and the size of available and free extended memory
;in bytes.
;
```

```
                                .MODEL SMALL
                                .386
0000                            .DATA
0000 00000000          XMS      DD      ?                  ;address of XMS driver
0004 0D 0A 0A 58 4D    ERR1     DB      13,10,10,'XMS driver not present.$'
     53 20 64 72 69
     76 65 72 20 6E
     6F 74 20 70 72
     65 73 65 6E 74
     2E 24
001F 0D 0A 0A 58 4D    MES1     DB      13,10,10,'XMS version number = $'
     53 20 76 65 72
     73 69 6F 6E 20
     6E 75 6D 62 65
     72 20 3D 20 24
0038 20 2E 20 24       MES2     DB      ' . $'
003C 0D 0A 54 6F 74    MES3     DB      13,10,'Total extended memory = $'
     61 6C 20 65 78
     74 65 6E 64 65
     64 20 6D 65 6D
     6F 72 79 20 3D
     20 24
0057 0D 0A 46 72 65    MES4     DB      13,10,'Free extended memory = $'
     65 20 65 78 74
     65 6E 64 65 64
     20 6D 65 6D 6F
     72 79 20 3D 20
     24
0000                   .CODE
                       .STARTUP
0010 E8 0049                    CALL    ISXMS              ;is XMS driver present
0013 73 09                      JNC     MAIN1              ;if installed
0015 B4 09                      MOV     AH,9
0017 BA 0004 R                  MOV     DX,OFFSET ERR1
001A CD 21                      INT     21H                ;not installed
001C EB 3A                      JMP     MAIN2
001E                   MAIN1:
001E B4 09                      MOV AH,9
0020 BA 001F R                  MOV DX,OFFSET MES1
0023 CD 21                      INT 21H
0025 B4 00                      MOV AH,0                   ;get XMS version
0027 FF 1E 0000 R               CALL XMS                   ;access XMS driver
002B 05 3030                    ADD AX,3030H               ;make ASCII
002E 88 26 0038 R               MOV MES2,AH                ;save version
0032 A2 003A R                  MOV MES2+2,AL
0035 B4 09                      MOV AH,9
0037 BA 0038 R                  MOV DX,OFFSET MES2
003A CD 21                      INT     21H                ;display version
003C B4 09                      MOV     AH,9
003E BA 003C R                  MOV     DX,OFFSET MES3
0041 CD 21                      INT     21H
0043 B4 08                      MOV     AH,8               ;get memory info
0045 FF 1E 0000 R               CALL    XMS
0049 50                         PUSH    AX                 ;save free memory
004A E8 0028                    CALL    DISPD              ;display total memory
004D B4 09                      MOV     AH,9
004F BA 0057 R                  MOV     DX,OFFSET MES4
0052 CD 21                      INT     21H
0054 5A                         POP     DX
0055 E8 001D                    CALL    DISPD              ;display free memory
0058                   MAIN2:
                                .EXIT
```

```
                         ;
                         ;the ISXMS procedure tests for the presence of the XMS driver.
                         ;If the driver is not present, a return with carry = 1 occurs
                         ;If the driver is present, the address of the driver entry point
                         ;is stored at location XMS for later access and a return with
                         ;carry = 0 occurs.
                         ;
005C                     ISXMS  PROC   NEAR

005C  B8 4300                   MOV    AX,4300H             ;check XMS
005F  CD 2F                     INT    2FH                  ;do multiplex interrupt
0061  3C 00                     CMP    AL,0
0063  74 0E                     JE     ISXMS1               ;if not installed
0065  B8 4310                   MOV    AX,4310H             ;get driver address
0068  CD 2F                     INT    2FH
006A  89 1E 0000 R              MOV    WORD PTR XMS,BX      ;save driver address
006E  8C 06 0002 R              MOV    WORD PTR XMS+2,ES
0072  F9                        STC
0073                     ISXMS1:
0073  F5                        CMC                         ;return carry = not installed
0074  C3                        RET

0075                     ISXMS  ENDP
                         ;
                         ;the DISPD procedure displays the contents of DX after it is
                         ;multiplied by 1,024. This displays the size of memory in
                         ;bytes.
                         ;
0075                     DISPD  PROC   NEAR
0075  66| B8 00000000           MOV    EAX,0
007B  66| 81 E2 0000FFFF        AND    EDX,0FFFFH
0082  B8 0400                   MOV    AX,400H              ;find 1K times blocks
0085  F7 E2                     MUL    DX
0087  66| C1 E2 10              SHL    EDX,16
008B  66| 03 C2                 ADD    EAX,EDX
008E  66| BB 0000000A           MOV    EBX,10
0094  53                        PUSH   BX                   ;mark end
0095  B9 0003                   MOV    CX,3                 ;load comma count
0098                     DISPD1:
0098  66| BA 00000000           MOV EDX,0
009E  66| F7 F3                 DIV    EBX                  ;divide by 10
00A1  80 C2 30                  ADD    DL,30H               ;make ASCII
00A4  52                        PUSH   DX
00A5  66| 83 F8 00              CMP    EAX,0
00A9  74 0A                     JE     DISPD2
00AB  E2 EB                     LOOP   DISPD1
00AD  B2 2C                     MOV    DL,','
00AF  52                        PUSH   DX                   ;save comma
00B0  B9 0003                   MOV    CX,3
00B3  EB E3                     JMP    DISPD1
00B5                     DISPD2:
00B5  B4 06                     MOV    AH,6
00B7  5A                        POP    DX                   ;get data
00B8  3A D3                     CMP    DL,BL
00BA  74 04                     JE     DISPD3               ;if done
00BC  CD 21                     INT    21H                  ;display it
00BE  EB F5                     JMP    DISPD2
00C0                     DISPD3:
00C0  C3                        RET

00C1                     DISPD  ENDP
                                END
```

This example shows how to test for the presence of XMS using INT 2FH (the DOS multiplex interrupt) function 4300H, which returns 00H in AL if XMS is not present. If XMS is present, INT 2FH function 4310H is used to obtain the address of the XMS handler. The handler address is stored at memory location XMS for access to the extended memory system later in the program.

To obtain the version number, a 00H is loaded into AH and a CALL XMS is executed to access the XMS driver. The driver version number is returned in AX. Next, an 8 is loaded into AH and a second call to XMS obtains the number of 1K-byte blocks of extended memory free in AX and the number in DX of available 1K-byte blocks. These are displayed by the DISPD procedure.

Using Extended Memory with XMS Access. Extended memory functions in much the same manner as EMS, at least for allocation and deallocation. Access to extended memory requires the use of the move function (0BH) and a buffer area placed in the first 1M of memory. The data from extended memory is moved between extended memory and the buffer designated in real memory. A better method of accessing extended memory is with the VCPI interface provided by the current version of EMM386.EXE, or by the DPMI interface provided by Windows. For this reason, we do not show access to extended memory using the technique presented in Example 14–6.

VCPI

The VCPI (virtual control program interface) allows access to both extended and expanded memory if DOS version 6.X is installed, using the current version of the EMM386.EXE driver. The main advantage of VCPI is that expanded memory installed by EMM386.EXE can be used as either extended memory or expanded memory. With VCPI, memory is defined as 4M for expanded or extended memory if EMM386.EXE 4096 is installed. Before VCPI, expanded memory could only be allocated as expanded memory, which meant that it could not be used as extended memory. With VCPI, memory can be used as either expanded or extended memory. Like the EMS manager, VCPI uses INT 67H to access the driver.

Is VCPI Installed as Part of EMM386.EXE? To determine whether VCPI is installed, the existence of EMM386.EXE must be determined. The ISEMS procedure is called to determine whether EMM386.EXE is installed. If it is installed, the presence of VCPI is queried with INT 67H function AX = DE00H.

EXAMPLE 14–7

```
                               ;a program that tests to determine whether VCPI is installed.
                               ;
                                      .MODEL SMALL
                                      .386
0000                                  .DATA
0000  45 4D 4D 58 58     EMS    DB    'EMMXXXX0',0          ;EMS name
      58 58 30 00
0009  0D 0A 0A 56 43     ERR1   DB    13,10,10,'VCPI not installed.$'
      50 49 20 6E 6F
      74 20 69 6E 73
      74 61 6C 6C 65
      64 2E 24
```

```
0020  0D 0A 0A 56 43     MES1   DB     13,10,10,'VCPI version'
      50 49 20 76 65
      72 73 69 6F 6E
      20
0030  20 2E 20 24        VER    DB     '. $'
0034  0D 0A 46 72 65     MES2   DB     13,10,'Free VCPI memory = $'
      65 20 56 43 50
      49 20 6D 65 6D
      6F 72 79 20 3D
      20 24
0000                            .CODE
                                .STARTUP
0010  E8 0036                   CALL   ISVCPI
0013  73 09                     JNC    MAIN1             ;if EMS $ VCPI is installed
0015  B4 09              ERR:   MOV    AH,9
0017  BA 0009 R                 MOV    DX,OFFSET ERR1
001A  CD 21                     INT    21H               ;display error
001C  EB 27                     JMP    MAIN2
001E  81 C3 3030         MAIN1: ADD    BX,3030H          ;make version ASCII
0022  88 3E 0030 R              MOV    VER,BH            ;save major
0026  88 1E 0032 R              MOV    VER+2,BL          ;save minor
002A  B4 09                     MOV    AH,9
002C  BA 0020 R                 MOV    DX,OFFSET MES1
002F  CD 21                     INT    21H               ;display version
0031  B8 DE03                   MOV    AX,0DE03H         ;get memory info
0034  CD 67                     INT    67H
0036  80 FC 00                  CMP    AH,0
0039  75 DA                     JNE    ERR               ;if error
003B  66| C1 E2 0C              SHL    EDX,12            ;multiply EDX by 4K
003F  66| 8B C2                 MOV    EAX,EDX
0042  E8 0030                   CALL   DISPD             ;display free VCPI memory
0045                     MAIN2:
                                .EXIT
                         ;
                         ;the ISVCPI procedure detects the VCPI driver.
                         ;***return parameter***
                         ;carry = 1, if VCPI is not installed
                         ;carry = 0, if VCPI is installed
                         ;
0049                     ISVCPI PROC   NEAR

0049  E8 000D                   CALL   ISEMS             ;test for EMS
004C  72 0A                     JC     ISVCPI1           ;if no EMS
004E  B8 DE00                   MOV    AX,0DE00H         ;test for VCPI
0051  CD 67                     INT    67H
0053  3C 00                     CMP    AL,0
0055  74 01                     JE     ISVCPI1           ;if VCPI is found
0057  F9                        STC
0058                     ISVCPI1:
0058  C3                        RET

0059                            ISVCPI ENDP
                         ;
                         ;the ISEMS procedure determines whether EMS is installed.
                         ;***return parameter***
                         ;carry = 1, if no EMS installed
                         ;carry = 0, if EMS is installed
                         ;
0059                     ISEMS  PROC   NEAR

0059  B8 3D00                   MOV    AX,3D00H
005C  BA 0000 R                 MOV    DX,OFFSET EMS
```

```
005F  CD 21                   INT     21H                 ;open EMS driver
0061  72 0B                   JC      ISEMS1              ;if not found
0063  8B D8                   MOV     BX,AX               ;save handle
0065  B8 4407                 MOV     AX,4407H            ;check IOCTL
0068  CD 21                   INT     21H
006A  72 02                   JC      ISEMS1              ;if not found
006C  3C FF                   CMP     AL,0FFH
006E                  ISEMS1:
006E  9C                      PUSHF
006F  B4 3E                   MOV     AH,3EH              ;close file
0071  CD 21                   INT     21H
0073  9D                      POPF
0074  C3                      RET

0075                  ISEMS   ENDP
                      ;
                      ;the DISPD procedure displays the amount of free memory
                      ;available to VCPI.
                      ;
0075                  DISPD   PROC    NEAR

0075  66| 50                  PUSH    EAX                 ;save number
0077  B4 09                   MOV     AH,9
0079  BA 0034 R               MOV     DX,OFFSET MES2      ;display free memory string
007C  CD 21                   INT     21H
007E  66| 58                  POP     EAX
0080  66| BB 0000000A         MOV     EBX,10
0086  B9 FFFF                 MOV     CX,-1
0089  51                      PUSH    CX                  ;save end
008A  B9 0003                 MOV     CX,3
008D                  DISPD1:
008D  66| BA 00000000         MOV     EDX,0               ;clear EDX
0093  66| F7 F3               DIV     EBX                 ;divide by 10
0096  80 C2 30                ADD     DL,30H              ;convert to ASCII
0099  52                      PUSH    DX
009A  66| 83 F8 00            CMP     EAX,0
009E  74 0A                   JE      DISPD2              ;if done
00A0  E2 EB                   LOOP    DISPD1
00A2  B2 2C                   MOV     DL,','              ;save comma
00A4  52                      PUSH    DX
00A5  B9 0003                 MOV     CX,3
00A8  EB E3                   JMP     DISPD1
00AA                  DISPD2:
00AA  B4 06                   MOV     AH,6
00AC  5A                      POP     DX
00AD  80 FA FF                CMP     DL,-1
00B0  74 04                   JE      DISPD3              ;if done
00B2  CD 21                   INT     21H                 ;display digit
00B4  EB F4                   JMP     DISPD2
00B6                  DISPD3:
00B6  C3                      RET

00B7                  DISPD   ENDP
                              END
```

Example 14–7 lists a program that locates VCPI and then displays the version number and memory attributes. This is very similar to the example that tested the system to determine whether EMS was installed. This examples tests for EMS and then tests to see if VCPI is installed. Although VCPI has application in some video game programs, newer software uses the DPMI (DOS protected-mode interface).

DPMI

The most extensive control of the personal computer is obtained with the DOS protected-mode interface, or DPMI. This interface is provided by WINDOWS and is accessed by shelling to DOS from within Windows. It is also provided by other sources, such as Phar Lap for use with DOS applications not using Windows. A total of 77 functions are provided by DPMI. These include control of the real and extended memory systems, interrupt management, page table management, local descriptor management, debugging, and switching between real- and protected-mode operation. Refer to Appendix A for a complete list of the DPMI functions. The next version of Windows and the next version of DOS should both include this memory manager. The DPMI is often referred to as the **host program** and is said to provide DPMI services to its **clients** (usually Windows and DOS, if a DOS extender such as Phar Lap is included).

Is DPMI Installed? The DPMI interface must be tested before it can be used. When an application is run from Windows, DPMI is usually present. If a Windows application is executed, it must be executed from the DOS icon under Windows. The program that appears in Example 14–8 detects the DPMI interface and displays information about the system using the get-DPMI entry point function (1687H). Although DOS is still the main operating system in many applications, it may eventually disappear. When it does, the DPMI functions will be used as the main interface to DOS applications through the Windows environment.

EXAMPLE 14–8

```
                                 ;a program that tests for DPMI and displays information
                                 ;about the installation of DPMI. Note that this program must
                                 ;be executed from Windows or with a DOS extender.
                                 ;
                                        .MODEL SMALL
                                        .386
0000                                    .DATA
0000  0000                       SUP    DW     ?                    ;support flag
0002  00                         PROCS  DB     ?                    ;processor type
0003  0000                       VERS   DW     ?                    ;version number
0005  00000000                   ENTR   DD     ?                    ;entry point
0009  0D 0A 0A 4D 75             ERR    DB     13,10,10,'Must run from WINDOWS.$'
      73 74 20 72 75
      6E 20 66 72 6F
      6D 20 57 49 4E
      44 4F 57 53 2E
      24
0023  0D 0A 0A 44 50             MES1   DB     13,10,10,'DPMI version = $'
      4D 49 20 76 65
      72 73 69 6F 6E
      20 3D 20 24
0036  0D 0A 43 50 55             MES2   DB     13,10,'CPU version = $'
      20 76 65 72 73
      69 6F 6E 20 3D
      20 24
0047  0D 0A 50 72 6F             MES3   DB     13,10,'Protected mode entry point = $'
      74 65 63 74 65
      64 20 6D 6F 64
      65 20 65 6E 74
```

```
     72 79 20 70 6F
     69 6E 74 20 3D
     20 24
0067 006D R 0074 R      CPUT    DW     C1,C2,C3
     007B R
006D 38 30 32 38 36     C1      DB     '80286.$'
     2E 24
0074 38 30 33 38 36     C2      DB     '80386.$'
     2E 24
007B 38 30 34 38 36     C3      DB     '80486.$'
     2E 24
0082 75 6E 6B 6E 6F     C4      DB     'unknown.$'
     77 6E 2E 24
008B 0D 0A 44 50 4D     C5      DB     13,10,'DPMI supports 16-bit mode.$'
     49 20 73 75 70
     70 6F 72 74 73
     20 31 36 2D 62
     69 74 20 6D 6F
     64 65 2E 24
00A8 0D 0A 44 50 4D     C6      DB     13,10,'DPMI supports 32-bit mode.$'
     49 20 73 75 70
     70 6F 72 74 73
     20 33 32 2D 62
     69 74 20 6D 6F
     64 65 2E 24
0000                            .CODE
                                .STARTUP
0010 E8 007A                    CALL   ISDPMI              ;is DPMI loaded?
0013 72 09                      JC     MAIN1               ;if it is loaded
0015 B4 09                      MOV    AH,9
0017 BA 0009 R                  MOV    DX,OFFSET ERR
001A CD 21                      INT    21H                 ;display not loaded
001C EB 6B                      JMP    MAIN4               ;end program
001E                    MAIN1:
001E B4 09                      MOV    AH,9                ;display version
0020 BA 0023 R                  MOV    DX,OFFSET MES1
0023 CD 21                      INT    21H
0025 8B 1E 0003 R               MOV    BX,VERS
0029 8A C7                      MOV    AL,BH
002B E8 0081                    CALL   DISPB
002E B4 06                      MOV    AH,6
0030 B2 2E                      MOV    DL,'.'
0032 CD 21                      INT    21H
0034 A1 0003 R                  MOV    AX,VERS
0037 E8 0075                    CALL   DISPB               ;display CPU type
003A B4 09                      MOV    AH,9
003C BA 0036 R                  MOV    DX,OFFSET MES2
003F CD 21                      INT    21H
0041 B4 00                      MOV    AH,0
0043 A0 0002 R                  MOV    AL,PROCS
0046 83 E8 02                   SUB    AX,2
0049 BA 0082 R                  MOV    DX,OFFSET C4        ;unknown CPU
004C 72 0D                      JB     MAIN2
004E 3C 02                      CMP    AL,2
0050 77 09                      JA     MAIN2
0052 03 C0                      ADD    AX,AX
0054 BE 0067 R                  MOV    SI,OFFSET CPUT
0057 03 F0                      ADD    SI,AX
0059 8B 14                      MOV    DX,[SI]
005B                    MAIN2:
005B B4 09                      MOV    AH,9
```

```
005D  CD 21                       INT    21H
005F  BA 008B R                   MOV    DX,OFFSET C5
0062  83 3E 0000 R 00             CMP    SUP,0
0067  74 03                       JE     MAIN3
0069  BA 00A8 R                   MOV    DX,OFFSET C6
006C                      MAIN3:
006C  B4 09                       MOV    AH,9
006E  CD 21                       INT    21H                 ;support
0070  B4 09                       MOV    AH,9
0072  BA 0047 R                   MOV    DX,OFFSET MES3
0075  CD 21                       INT    21H                 ;entry point
0077  A1 0007 R                   MOV    AX,WORD PTR ENTR+2
007A  E8 004D                     CALL   DISPH
007D  B4 06                       MOV    AH,6
007F  B2 3A                       MOV    DL,':'
0081  CD 21                       INT    21H
0083  A1 0005 R                   MOV    AX,WORD PTR ENTR
0086  E8 0041                     CALL   DISPH
0089                      MAIN4:
                                  .EXIT
                          ;
                          ;the ISDPMI procedure tests for the presence of DPMI.
                          ;***return parameter***
                          ;if carry = 1, DPMI is installed
                          ;if carry = 0, DPMI is not installed
                          ;
008D                      ISDPMI PROC   NEAR

008D  B8 1687                     MOV    AX,1687H
0090  CD 2F                       INT    2FH                 ;get entry point
0092  0B C0                       OR     AX,AX
0094  75 18                       JNE    ISDPMI1             ;if not loaded
0096  83 E3 01                    AND    BX,1
0099  89 1E 0000 R                MOV    SUP,BX              ;save support
009D  88 0E 0002 R                MOV    PROCS,CL            ;save processor type
00A1  89 16 0003 R                MOV    VERS,DX             ;save version
00A5  89 3E 0005 R                MOV    WORD PTR ENTR,DI    ;save entry point
00A9  8C 06 0007 R                MOV    WORD PTR ENTR+2,ES
00AD  F9                          STC                        ;set carry for loaded
00AE                      ISDPMI1:
00AE  C3                          RET

00AF                      ISDPMI ENDP



                          ;
                          ;the DISPB procedure displays contents of AL in
                          ;decimal format.
                          ;
00AF                      DISPB  PROC   NEAR

00AF  B4 00                       MOV    AH,0
00B1  D4 0A                       AAM                        ;convert to BCD
00B3  05 3030                     ADD    AX,3030H            ;convert to ASCII
00B6  80 FC 30                    CMP    AH,30H              ;test for leading zero
00B9  74 08                       JE     DISPB1              ;blank leading zero
00BB  50                          PUSH   AX
00BC  8A D4                       MOV    DL,AH               ;display non blank
00BE  B4 06                       MOV    AH,6
00C0  CD 21                       INT    21H
```

```
00C2  58                  POP     AX
00C3                DISPB1:
00C3  8A D0               MOV     DL,AL             ;display second digit
00C5  B4 06               MOV     AH,6
00C7  CD 21               INT     21H
00C9  C3                  RET

00CA                DISPB   ENDP
                    ;
                    ;the DISPH procedure displays the contents of BX as a
                    ;4 digit hexadecimal value.
                    ;
00CA                DISPH   PROC    NEAR

00CA  B9 0004             MOV     CX,4              ;set digit count
00CD                DISPH1:
00CD  C1 C3 04            ROL     BX,4              ;position digit
00D0  8A D3               MOV     DL,BL
00D2  80 E2 0F            AND     DL,0FH
                          .IF DL > 9                ;convert to ASCII
00DA  80 C2 37                ADD DL,37H
                          .ELSE
00DF  80 C2 30                ADD DL,30H
                          .ENDIF
00E2  B4 06               MOV     AH,6              ;display digit
00E4  CD 21               INT     21H
00E6  E2 E5               LOOP    DISPH1            ;repeat 4 times
00E8  C3                  RET

00E9                DISPH   ENDP
                          END
```

The ISDPMI procedure tests the DOS multiplex interrupt (2FH) to determine whether the DPMI driver is installed. If it is installed, the address of the entry point is stored at ENTR, the version number is stored at VERS, the processor type is stored at PROCS, and the type of support (16- or 32-bit) is stored at SUP.

Once it is known that DPMI is installed, the program displays all of the information obtained from ISDPMI. The DPMI is used to switch to protected-mode operation from DOS. It also provides memory management for the extended memory system. Protected-mode operation is described and used with some of the DPMI functions in Section 14–4.

14–3 IOCTL AND DEVICE DRIVERS

The IOCTL (I/O control) functions are provided by DOS INT 21H function number 44H (refer to Appendix A). The **IOCTL functions** provide control of device drivers, which in turn control physical devices attached to the system. The IOCTL functions are most commonly used in the development of device drivers. In the last section, IOCTL interrogated the EMM386.EXE driver to determine whether it was ready to accept output data by using IOCTL function AX = 4407H. Without IOCTL, the presence of EMM386.EXE could be tested with the DOS open file function. The purpose of IOCTL is to determine whether EMM386.EXE is ready to accept output information.

IOCTL Devices

Two types of devices are controlled by the IOCTL functions: character devices and block devices. The **character devices** process a single byte at a time, such as the keyboard, mouse, and printer. The **block devices** are devices that process a block of bytes simultaneously, such as a disk drive or CDROM. The purpose of IOCTL function 44H is to control these devices, not necessarily to transfer data to or from them. Data transfer is handled by communicating to the I/O device though DOS or other functions provided by BIOS, or by drivers provided with the I/O device.

Character Devices. Character devices are controlled in either binary or ASCII mode by the IOCTL functions. They are named using DOS conventions, as listed in Table 14–1. EMMXXXX0, for example, was the name used in the last section for the EMM386.EXE driver. The names in Table 14–1 apply to many other character I/O devices in the system, but not all.

To access a character-mode device, it is opened with file open function 3DH, just as if it were a disk file. But instead of using a file name, a name listed in Table 14–1 is used. Example 14–9 tests for the presence of the LPT1 port and also checks the device-related information using IOCTL function 00H. If the name of any device is changed, this program displays the IOCTL status for that device. If the contents of DEV contain a file name, the IOCTL status is returned with information about the file. Try using some of the other device names by changing DEV and reassembling, linking, and executing the file.

TABLE 14–1 IOCTL character device names

Name	*Device*
AUX	The COM1 device in most systems
CLOCK$	The real-time system clock
COM1	Serial communications port number 1
COM2	Serial communications port number 2
COM3	Serial communications port number 3
COM4	Serial communications port number 4
CON	Console, the keyboard for input and the video display for output
EMMXXXX0	The expanded memory manager EMM386.EXE
LPT1	Parallel port number 1
LPT2	Parallel port number 2
LPT3	Parallel port number 3
LPT4	Parallel port number 4
NUL	A nonexistent device
PRN	LPT1 in most systems
XMSXXXX0	The extended memory manager HIMEM.SYS

Note: Not all devices may be present, and additional devices may be found.

EXAMPLE 14–9

```
                              ;a program that tests for an IOCTL device and lists its
                              ;status information.
                                     .MODEL SMALL
                                     .386
0000                                 .DATA
0000   0015 R 002E R          NTAB   DW      M0,M1,M2,M3,M4,M5,M6
       0048 R 0056 R
       0066 R 007A R
       008B R
000E   4C 50 54 31 00         DEV    DB      'LPT1',0            ;name for LPT1
0013   53 74 61 6E 64         M0     DB      'Standard input device.'
       61 72 64 20 69
       6E 70 75 74 20
       64 65 76 69 63
       65 2E
0029   0D 0A 24               CRLF   DB      13,10,'$'
002C   53 74 61 6E 64         M1     DB      'Standard output device.',13,10,'$'
       61 72 64 20 6F
       75 74 70 75 74
       20 64 65 76 69
       63 65 2E 0D 0A
       24
0046   4E 55 4C 20 64         M2     DB      'NUL device.',13,10,'$'
       65 76 69 63 65
       2E 0D 0A 24
0054   43 6C 6F 63 6B         M3     DB      'Clock device.',13,10,'$'
       20 64 65 76 69
       63 65 2E 0D 0A
       24
0064   53 75 70 70 6F         M4     DB      'Supports INT 28H.',13,10,'$'
       72 74 73 20 49
       4E 54 20 32 38
       48 2E 0D 0A 24
0078   42 69 6E 61 72         M5     DB      'Binary device.',13,10,'$'
       79 20 64 65 76
       69 63 65 2E 0D
       0A 24
0089   44 6F 65 73 20         M6     DB      'Does not signal EOF.',13,10,'$'
       6E 6F 74 20 73
       69 67 6E 61 6C
       20 45 4F 46 2E
       0D 0A 24
00A0   49 4F 43 54 4C         ME     DB      'IOCTL functions 02H and 03H supported.',13,10,'$'
       20 66 75 6E 63
       74 69 6F 6E 73
       20 30 32 48 20
       61 6E 64 20 30
       33 48 20 73 75
       70 70 6F 72 74
       65 64 2E 0D 0A
       24
00C9   0D 0A 45 52 52         ERR1   DB      13,10,'ERROR$'
       4F 52 24
0000                                 .CODE
                                     .STARTUP
0010   B8 3D02                       MOV     AX,3D02H            ;open device
0013   BA 0010 R                     MOV     DX,OFFSET DEV
0016   CD 21                         INT     21H
```

```
0018  72 3E                JC     ERR              ;if not found
001A  8B D8                MOV    BX,AX            ;save handle in BX
001C  B8 4400              MOV    AX,4400H         ;get IOCTL status
001F  CD 21                INT    21H
0021  72 35                JC     ERR              ;if error
0023  F6 C2 80             TEST   DL,80H           ;is it a file?
0026  74 30                JZ     ERR              ;if file
0028  52                   PUSH   DX               ;save status
0029  B4 09                          MOV AH,9
002B  BA 002B R            MOV    DX,OFFSET CRLF   ;get new line
002E  5A                   POP    DX               ;get status
002F  52                   PUSH   DX
0030  BE 0000 R            MOV    SI,OFFSET NTAB   ;address message table
0033  B9 0007              MOV    CX,7             ;load count
0036              MAIN1:
0036  D0 EA                SHR    DL,1             ;test feature
0038  72 14                JC     MAIN3            ;display message if 1
003A              MAIN2:
003A  83 C6 02             ADD SI,2                ;address next message
003D  E2 F7                LOOP   MAIN1            ;keep checking 7 bits
003F  5A                   POP    DX
0040  F6 C6 40             TEST   DH,40H           ;test bit position 14
0043  74 1A                JZ     MAIN4
0045  B4 09                MOV    AH,9
0047  BA 00A2 R            MOV    DX,OFFSET ME
004A  CD 21                INT    21H
004C  EB 11                JMP    MAIN4            ;if done
004E              MAIN3:
004E  52                   PUSH   DX
004F  B4 09                MOV    AH,9
0051  8B 14                MOV    DX,[SI]
0053  CD 21                INT 21H                 ;display message
0055  5A                   POP    DX
0056  EB E2                JMP    MAIN2            ;repeat
0058              ERR:                             ;if error
0058  B4 09                MOV    AH,9             ;display ERROR
005A  BA 00CB R            MOV    DX,OFFSET ERR1
005D  CD 21                INT    21H
005F              MAIN4:
                           .EXIT
                           END
```

Figure 14–2 shows the contents of register DX after the IOCTL status read operation. Bit position 7 indicates a file (1) or a device (0). If a device is present, bit position 14 indicates a character device (1) or a block device (0). Bit positions 0 through 6 indicate various information about the device.

The IOCTL functions are used primarily in developing device drivers, as will be explained later. Normally, the DOS INT 21H functions are used to access devices such as CON, LPT, and so forth.

Block Devices. Block devices transfer information a block at a time. The most common block device is a disk drive. The block device, whether a disk drive or other device, is referenced through drive letters. For the block functions, BL is loaded with a 00H for the default block device, 01H for device A, 02H for device B, and so forth.

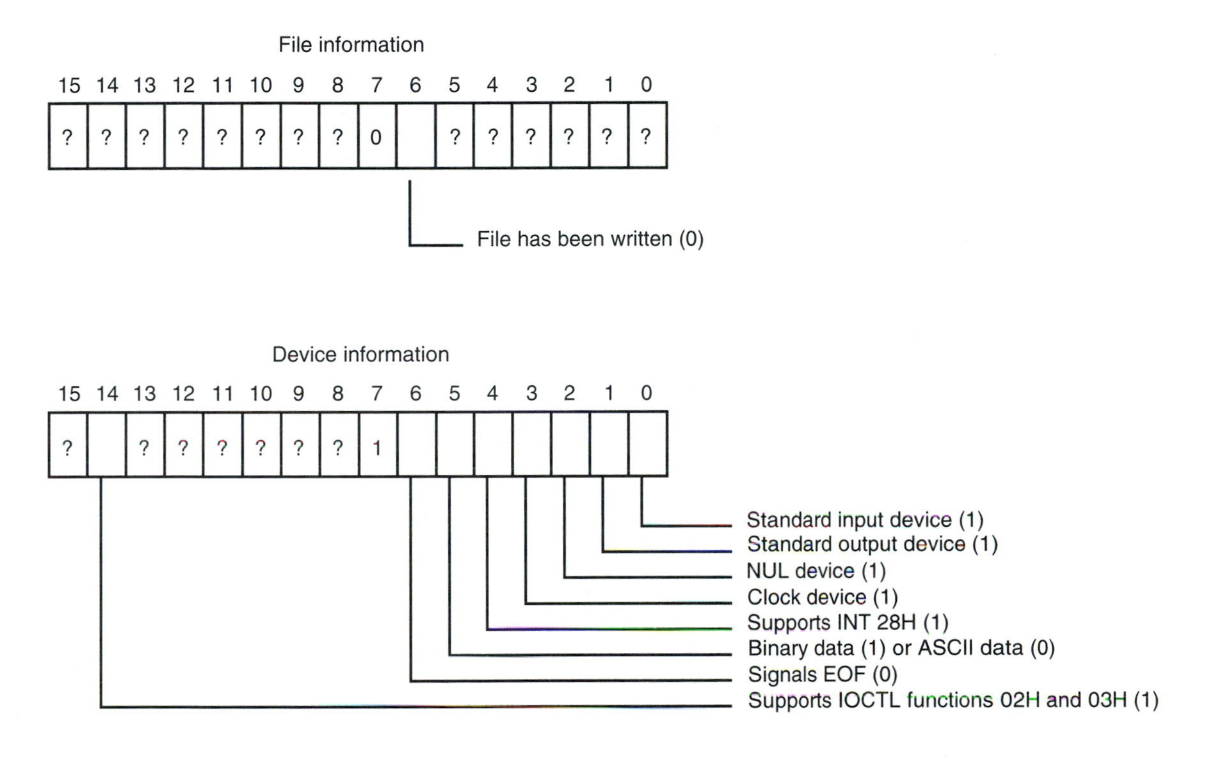

Note: (1) = a logic 1 is true, or (0) = a logic 0 is true. For example, a 1 in bit 0 of DX indicates standard in-out device.

FIGURE 14–2 The contents of register DX after an IOCTL status-read operation.

EXAMPLE 14–10

```
                        ;a program that lists the information from the parameter
                        ;block for a block device.
                               .MODEL SMALL
                               .386
0000                           .DATA
0000  01                FUNC   DB     1              ;get disk parameter block
0001  00                DTYPE  DB     ?              ;disk type
0002  0000              DATT   DW     ?              ;disk attribute
0004  0000              CYL    DW     ?              ;number of cylinders
0006  00                MTYPE  DB     ?              ;media type
0007  0000              SEC    DW     ?              ;bytes per sector
0009  00                CLUST  DB     ?              ;sectors per cluster
000A  0000              RSECT  DW     ?              ;reserved sectors
000C  00                NFAT   DB     ?              ;number of FATs
000D  0000              ROOT   DW     ?              ;maximum root entries
000F  0000              NSEC   DW     ?              ;total number of sectors
0011  00                MEDIA  DB     ?              ;media ID
0012  0000              SFAT   DW     ?              ;sectors per FAT
0014  0000              TRACK  DW     ?              ;number of sectors per track
0016  0000              NHEAD  DW     ?              ;number of heads
0018  00000000          NHID   DD     ?              ;number of hidden sectors
```

```
001C  00000000          HUGE    DD      ?                   ;total sectors if NSEC = 0
0020  0006 [                    DB      6 DUP (?)           ;reserved
            00
                  ]
0026  003A R 0052 R     DTAB    DW      T0,T1,T2,T3,T4,T5,T6,T7,T8,T9
      0065 R 0078 R
      0091 R 00AA R
      00B6 R 00C2 R
      00D6 R 00EA R
003A  33 32 30 4B 2F    T0      DB      '320K/360K, 5-1/4" disk.$'
      33 36 30 4B 2C
      20 35 2D 31 2F
      34 22 20 64 69
      73 6B 2E 24
0052  31 2E 32 4D 2C    T1      DB      '1.2M, 5-1/4" disk.$'
      20 35 2D 31 2F
      34 22 20 64 69
      73 6B 2E 24
0065  37 32 30 4B 2C    T2      DB      '720K, 3-1/3" disk.$'
      20 33 2D 31 2F
      33 22 20 64 69
      73 6B 2E 24
0078  53 69 6E 67 6C    T3      DB      'Single-density, 8" disk.$'
      65 2D 64 65 6E
      73 69 74 79 2C
      20 38 22 20 64
      69 73 6B 2E 24
0091  44 6F 75 62 6C    T4      DB      'Double-density, 8" disk.$'
      65 2D 64 65 6E
      73 69 74 79 2C
      20 38 22 20 64
      69 73 6B 2E 24
00AA  46 69 78 65 64    T5      DB      'Fixed disk.$'
      20 64 69 73 6B
      2E 24
00B6  54 61 70 65 20    T6      DB      'Tape drive.$'
      64 72 69 76 65
      2E 24
00C2  31 2E 34 34 4D    T7      DB      '1.44M, 3-1/2" disk.$'
      2C 20 33 2D 31
      2F 32 22 20 64
      69 73 6B 2E 24
00D6  32 2E 38 38 4D    T8      DB      '2.88M, 3-1/2" disk.$'
      2C 20 33 2D 31
      2F 32 22 20 64
      69 73 6B 2E 24
00EA  4F 74 68 65 72    T9      DB      'Other block device.$'
      20 62 6C 6F 63
      6B 20 64 65 76
      69 63 65 2E 24
00FE  0106 R 0118 R     DATAB   DW      DA0,DA1,DA2,DA3
      012D R 0146 R
0106  52 65 6D 6F 76    DA0     DB      'Removable medium.$'
      61 62 6C 65 20
      6D 65 64 69 75
      6D 2E 24
0118  4E 6F 6E 72 65    DA1     DB      'Nonremovable medium.$'
      6D 6F 76 61 62
      6C 65 20 6D 65
      64 69 75 6D 2E
      24
```

```
012D  44 6F 6F 72 20   DA2     DB      'Door lock not supported.$'
      6C 6F 63 6B 20
      6E 6F 74 20 73
      75 70 70 6F 72
      74 65 64 2E 24
0146  44 6F 6F 72 20   DA3     DB      'Door lock supported.$'
      6C 6F 63 6B 20
      73 75 70 70 6F
      72 74 65 64 2E
      24
015B  42 79 74 65 73   BPS     DB      'Bytes per sector = $'
      20 70 65 72 20
      73 65 63 74 6F
      72 20 3D 20 24
016F  4E 75 6D 62 65   NCYL    DB      'Number of cylinders = $'
      72 20 6F 66 20
      63 79 6C 69 6E
      64 65 72 73 20
      3D 20 24
0186  53 65 63 74 6F   AU      DB      'Sectors per cluster = $'
      72 73 20 70 65
      72 20 63 6C 75
      73 74 65 72 20
      3D 20 24
019D  4E 75 6D 62 65   CYLS    DB      'Number of reserved sectors from sector 0 = $'
      72 20 6F 66 20
      72 65 73 65 72
      76 65 64 20 73
      65 63 74 6F 72
      73 20 66 72 6F
      6D 20 73 65 63
      74 6F 72 20 30
      20 3D 20 24
01C9  4E 75 6D 62 65   FAT     DB      'Number of FATs = $'
      72 20 6F 66 20
      46 41 54 73 20
      3D 20 24
01DB  4E 75 6D 62 65   ROOTS   DB      'Number of root directory entries = $'
      72 20 6F 66 20
      72 6F 6F 74 20
      64 69 72 65 63
      74 6F 72 79 20
      65 6E 74 72 69
      65 73 20 3D 20
      24
01FF  54 6F 74 61 6C   SECS    DB      'Total number of sectors = $'
      20 6E 75 6D 62
      65 72 20 6F 66
      20 73 65 63 74
      6F 72 73 20 3D
      20 24
021A  4E 75 6D 62 65   SECF    DB      'Number of sectors per FAT = $'
      72 20 6F 66 20
      73 65 63 74 6F
      72 73 20 70 65
      72 20 46 41 54
      20 3D 20 24
0237  4E 75 6D 62 65   STRA    DB      'Number of sectors per track = $'
      72 20 6F 66 20
      73 65 63 74 6F
      72 73 20 70 65
```

```
     72 20 74 72 61
     63 6B 20 3D 20
     24
0256 4E 75 6D 62 65   HEADS  DB     'Number of heads = $'
     72 20 6F 66 20
     68 65 61 64 73
     20 3D 20 24
0269 4E 75 6D 62 65   HID    DB     'Number of hidden sectors = $'
     72 20 6F 66 20
     68 69 64 64 65
     6E 20 73 65 63
     74 6F 72 73 20
     3D 20 24
0285 45 52 52 4F 52   ERR1   DB     'ERROR$'
     24
0000                         .CODE
                      ;;the CRLF macro displays a carriage return
                      ;;and line feed.
                      ;;
                      CRLF   MACRO
                             MOV    AH,6
                             MOV    DL,13
                             INT    21H
                             MOV    DL,10
                             INT    21H
                             ENDM
                      ;;
                      ;;the STRING macro display a character string
                      ;;
                      STRING MACRO  WHAT
                             CRLF
                             MOV    AH,9
                             MOV    DX,OFFSET WHAT
                             INT    21H
                             ENDM
                             .STARTUP
0010 B8 440D                 MOV    AX,440DH          ;generic block control
0013 B3 03                   MOV    BL,3              ;drive C
0015 B9 0860                 MOV    CX,860H           ;get device parameters
0018 BA 0000 R               MOV    DX,OFFSET FUNC    ;address parameter block
001B CD 21                   INT    21H
001D 73 14                   JNC    MAIN1             ;if no error
                             STRING ERR1              ;display error
0030 E9 014B                 JMP    MAIN6             ;exit to DOS
0033                  MAIN1:                          ;if no error detected
0033 BE 0001 R               MOV    SI,OFFSET FUNC+1
0036 AC                      LODSB                    ;get device type
0037 B4 00                   MOV    AH,0
0039 03 C0                   ADD    AX,AX
003B BF 0026 R               MOV    DI,OFFSET DTAB    ;address device type table
003E 03 F8                   ADD    DI,AX
                             CRLF
004A 8B 15                   MOV    DX,[DI]
004C B4 09                   MOV    AH,9              ;display device type
004E CD 21                   INT    21H
0050 AD                      LODSW                    ;get attribute
0051 BF 00FE R               MOV    DI,OFFSET DATAB   ;address DATAB table
0054 A8 01                   TEST   AL,1              ;test bit 0
0056 74 03                   JE     MAIN2             ;if bit 0 = 0
0058 83 C7 02                ADD    DI,2              ;else
005B                  MAIN2:
```

```
005B  50                    PUSH    AX                  ;save attribute
                            CRLF
0066  8B 15                 MOV     DX,[DI]
0068  B4 09                 MOV     AH,9                ;display removable/
006A  CD 21                 INT     21H                 ;nonremovable
006C  58                    POP     AX                  ;get attribute
006D  BF 0102 R             MOV     DI,OFFSET DATAB+4   ;address entry 2
0070  A8 02                 TEST AL,2                   ;test door lock support
0072  74 03                 JE      MAIN3               ;if not supported
0074  83 C7 02              ADD     DI,2                ;if supported
0077                  MAIN3:
                            CRLF
0081  8B 15                 MOV     DX,[DI]
0083  B4 09                 MOV     AH,9
0085  CD 21                 INT     21H                 ;display door lock status
                            STRING NCYL
0098  AD                    LODSW
0099  E8 00F0               CALL    DISPW               ;display cylinders
009C  AC                    LODSB                       ;media type
                            STRING BPS
00AE  AD                    LODSW                       ;bytes per sector
00AF  E8 00DA               CALL    DISPW
                            STRING AU                   ;sectors per cluster
00C3  AC                    LODSB
00C4  E8 00BB               CALL    DISPB
                            STRING CYLS                 ;number of cylinders
00D8  AD                    LODSW                       ;sectors per cylinder
00D9  E8 00B0               CALL    DISPW
                            STRING FAT
00ED  AC                    LODSB
00EE  E8 0091               CALL    DISPB               ;number of FATs
                            STRING ROOTS
0102  AD                    LODSW
0103  E8 0086               CALL    DISPW               ;number of root entries
                            STRING SECS
0117  AD                    LODSW
0118  0B C0                 OR      AX,AX
011A  74 05                 JE      MAIN4               ;if huge
011C  E8 006D               CALL    DISPW               ;total sectors
011F  EB 07                 JMP     MAIN5
0121                  MAIN4:
0121  66| A1 001C R         MOV     EAX,HUGE
0125  E8 006E               CALL    DISPD
0128                  MAIN5:
0128  AC                    LODSB
                            STRING SECF
013A  AD                    LODSW
013B  E8 004E               CALL    DISPW               ;sectors per FAT
                            STRING STRA
014F  AD                    LODSW
0150  E8 0039               CALL    DISPW               ;sectors per track
                            STRING HEADS
0164  AD                    LODSW
0165  E8 0024               CALL    DISPW               ;number of heads
                            STRING HID
0179  66| AD                LODSD
017B  E8 0018               CALL    DISPD               ;number of hidden sectors
017E                  MAIN6:
                            .EXIT
                      ;the DISPB procedure displays the contents of AL as a
                      ;decimal number.
```

```
                                ;***uses DISPD***
                                ;
0182                            DISPB   PROC    NEAR

0182  66| 25 000000FF                   AND     EAX,0FFH
0188  E8 000B                           CALL    DISPD
018B  C3                                RET
018C                            DISPB   ENDP
                                ;
                                ;the DISPW procedure displays the contents of AX as a
                                ;decimal number.
                                ;***uses DISPD***
                                ;
018C                            DISPW   PROC    NEAR

018C  66| 25 0000FFFF                   AND     EAX,0FFFFH
0192  E8 0001                           CALL    DISPD
0195  C3                                RET

0196                                                    DISPW ENDP
                                ;the DISPD procedure displays the contents of EAX as a
                                ;decimal number with leading zeros suppressed, including
                                ;commas.
                                ;
0196                            DISPD   PROC    NEAR

0196  66| BB 0000000A                   MOV     EBX,10          ;load 10 for decimal
019C  53                                PUSH    BX              ;save end marker
019D  B9 0003                           MOV     CX,3            ;load comma counter
01A0                            DISPD1:
01A0  66| 33 D2                         XOR EDX,EDX             ;clear EDX to zero
01A3  66| F7 F3                         DIV EBX                 ;divide by 10
01A6  80 C2 30                          ADD DL,30H              ;convert remainder to ASCII
01A9  52                                PUSH DX                 ;save remainder
01AA  66| 83 F8 00                      CMP EAX,0               ;test quotient for zero
01AE  74 0A                             JE DISPD2               ;if zero display data
01B0  E2 EE                             LOOP DISPD1             ;test for comma
01B2  B9 0003                           MOV     CX,3            ;reload comma counter
01B5  B2 2C                             MOV     DL,','          ;get comma
01B7  52                                PUSH    DX              ;save comma
01B8  EB E6                             JMP     DISPD1          ;repeat divisions
01BA                            DISPD2:
01BA  B4 06                             MOV     AH,6            ;select display function
01BC  5A                                POP     DX              ;get digit from stack
01BD  3B D3                             CMP     DX,BX           ;test for end marker
01BF  74 04                             JE      DISPD3          ;if end marker encountered
01C1  CD 21                             INT     21H             ;display digit
01C3  EB F5                             JMP     DISPD2          ;repeat until end marker
01C5                            DISPD3:
01C5  C3                                RET

01C6                            DISPD   ENDP
                                        END
```

To illustrate access to a block device, Example 14–10 shows how to read the status information for disk drive C. This program uses generic block device control function AX = 440DH to read the parameters from the disk drive interface. Notice that BL = 03H for disk drive C, CH = 08H for access to a disk drive, CL = 60H to select the read device parameters, and DS:DX = the address of a disk parameter table where the parameters are stored by this function. Refer to Figure 14–3 for the format of the disk parameter block.

FIGURE 14–3 The format of the disk parameter block.

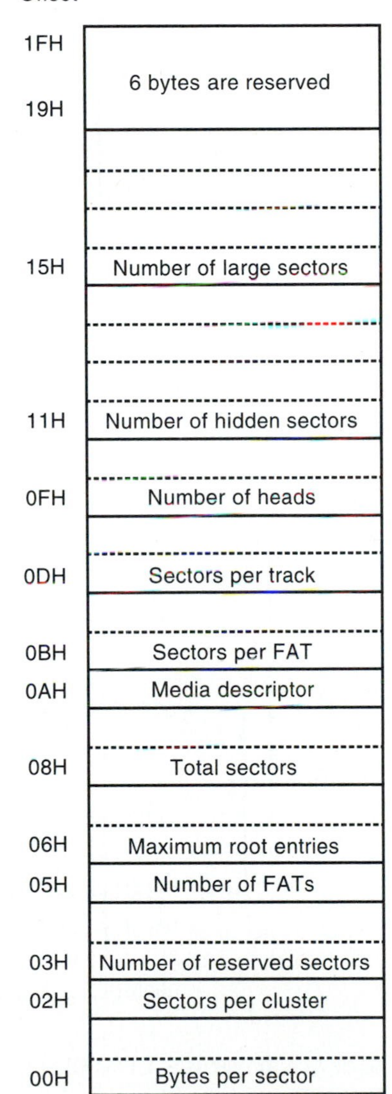

Device Drivers

Device drivers are special programs installed by the CONFIG.SYS file to control installable devices. Thus, personal computers can be expanded at some future time by the installation of new devices. The newest device is the CDROM drive, which is controlled by a device driver added to the system in the CONFIG.SYS file. This device was not available when DOS or the personal computer was created, but the device driver allows it to be installed without any modification to the system software except inclusion in the CONFIG.SYS file.

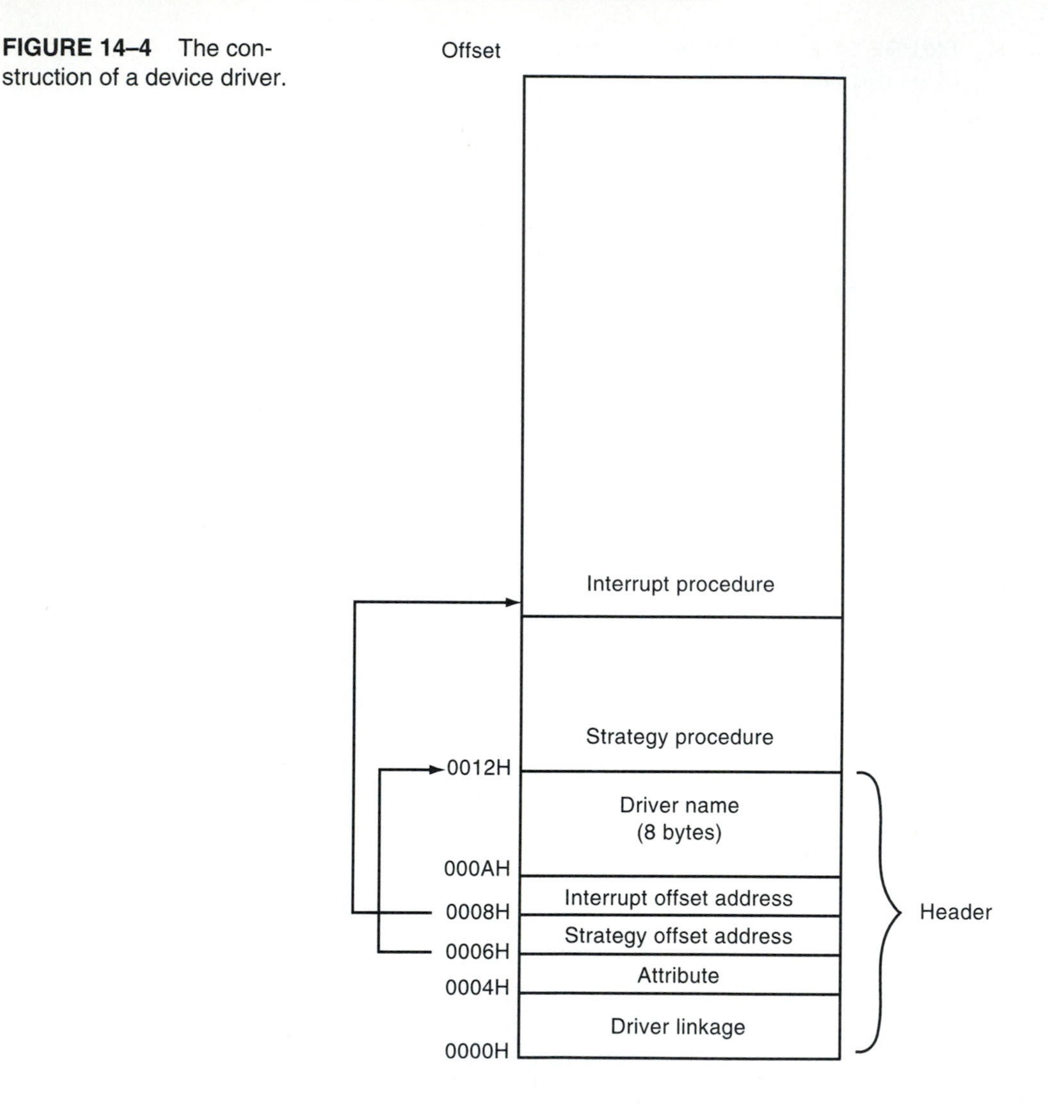

FIGURE 14–4 The construction of a device driver.

The **device driver** is a .COM file (TINY model) organized in three parts (see Figure 14–4): the header, the strategy procedure, and the interrupt procedure. The driver has either the .SYS or .EXE extension and is originated at offset address 0000H. Note that .COM files normally originate at location 0100H. The header contains information that allows DOS to identify the driver. It also contains pointers that allow it to chain to other drivers loaded into the system. The strategy procedure is called when loaded into memory by DOS or whenever the controlled device requests service. The interrupt procedure is not a true interrupt handler because it is called by DOS immediately after the strategy call during an I/O request.

The Header. The **header** section of a device driver is 18 bytes in length and contains pointers and the name of the driver. Example 14–11 shows the assembly language struc-

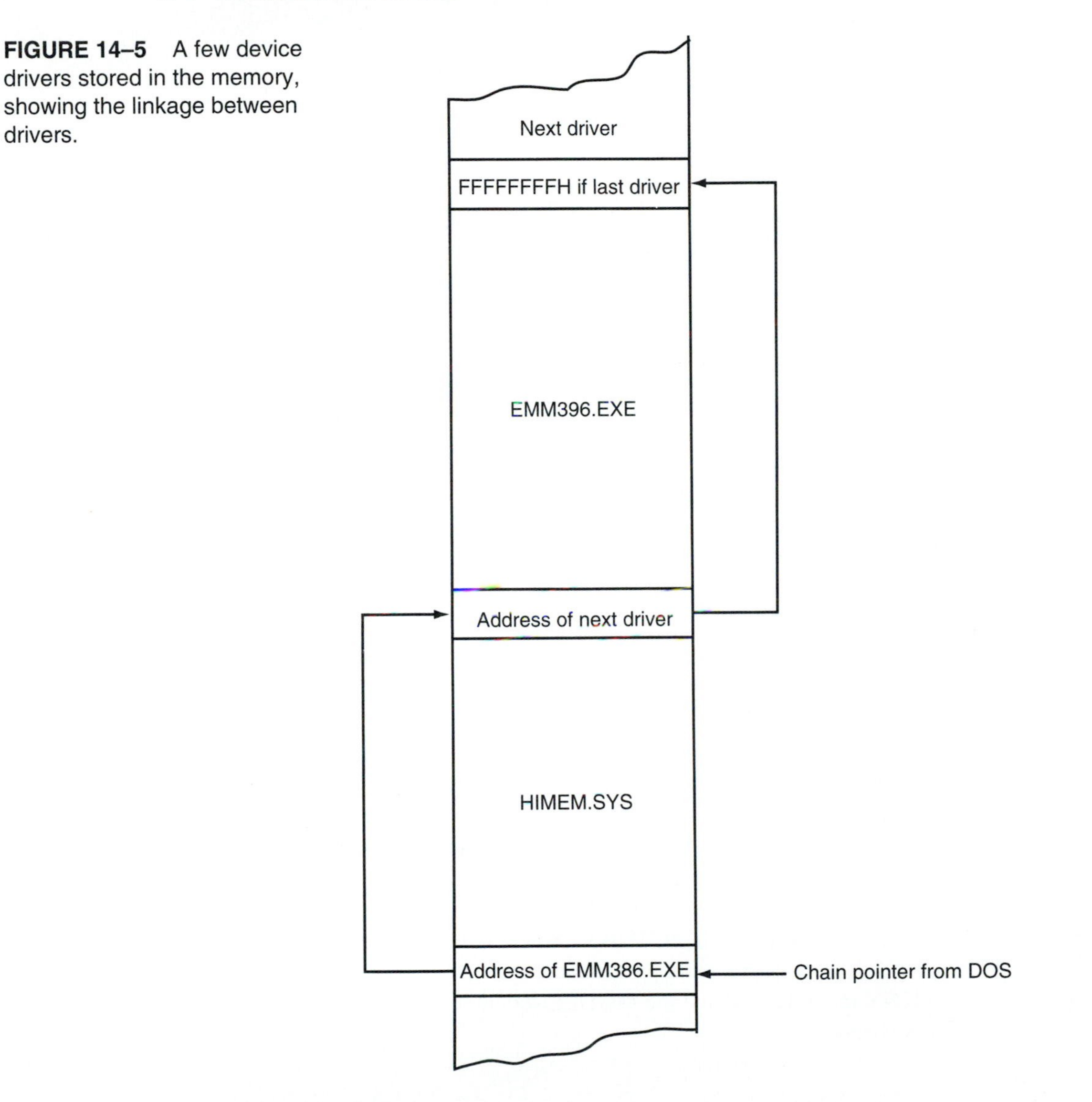

FIGURE 14–5 A few device drivers stored in the memory, showing the linkage between drivers.

ture of the header. The first doubleword contains a –1 that informs DOS this is the last driver in a chain. If additional drivers are added, DOS inserts a chain address in this doubleword as the segment and offset address. The chain address points to the next driver in the chain. This allows additional drivers to be installed at any time. Figure 14–5 shows a series of device drivers in memory and the contents of the chain pointer for each.

EXAMPLE 14–11

```
0000  FFFFFFFF        CHAIN  DD    -1             ;link to next driver in chain
0004  0000            ATTR   DW    0              ;driver attribute
0006  0012 R          STRT   DW    STRAT          ;address of strategy
```

```
0008  0013 R          INTR   DW    INTT          ;address of interrupt
000A  4D 59 44 52 49  DNAME  DB    'MYDRIVER'    ;driver name
      56 45 52

0012                  STRAT  PROC  FAR

                      ;strategy is placed here

0013                  STRAT  ENDP

0013                  INTT   PROC  FAR

                      ;interrupt is placed here

0014                  INTT   ENDP
```

Following the chain pointer in the header is the attribute word. Figure 14–6 shows the structure of the attribute word. The attribute word indicates the type of header included for the driver and the type of device the driver installs. It also indicates whether the driver controls a character device or a block device.

The next two words contain the offset addresses of the strategy and the interrupt. Finally, the last eight bytes of the header contain the name of the driver. In Example 14–11, the driver name is MYDRIVER. Any name can be used as long as it fills all eight bytes. If the name is shorter than eight bytes, it is padded with blank spaces (20H).

The Strategy. The **strategy procedure** is called whenever the device driver is loaded into memory or whenever any I/O requests are issued. The main purpose of the strategy is to save the request header and its address (see Table 14–2 for contents of the request header) for use by the interrupt procedure. The **request header** is used by DOS to communicate commands and other information to the interrupt procedure in the device driver. In most cases, the strategy procedure merely transfers the address of the request header to memory by saving the contents of ES:BX, which addresses the request header. An example of a typical strategy procedure is included with the header in Example 14–12. Notice that the TINY mode is used and the MODEL statement is followed by .CODE instead of .DATA, as is customary.

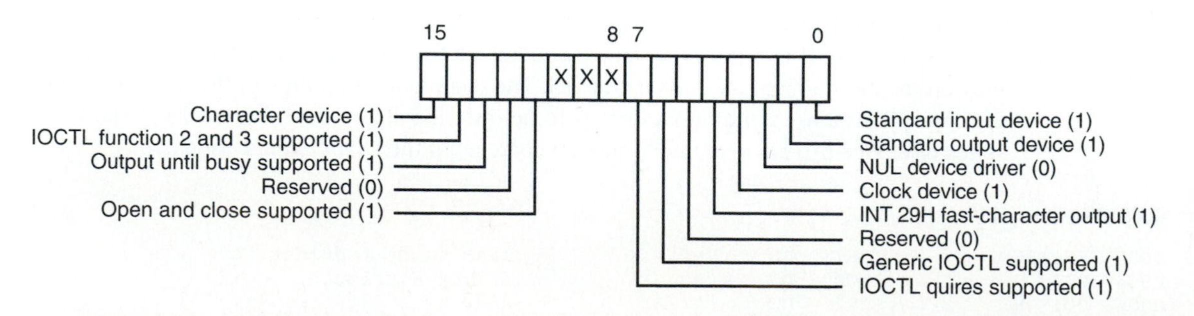

FIGURE 14–6 The structure of the header attribute word.

EXAMPLE 14–12

```
                          ;basic template for developing a device driver.
                          ;
                                .MODEL TINY
0000                            .CODE
0000  FFFFFFFF            CHAIN DD    -1            ;link to next driver in chain
0004  0000                ATTR  DW    0             ;driver attribute
0006  0016 R              STRT  DW    STRAT         ;address of strategy
0008  001F R              INTR  DW    INTT          ;address of interrupt
000A  4D 59 44 52 49      DNAME DB    'MYDRIVER'    ;driver name
         56 45 52
0012  00000000            REQ   DD    ?             ;request header address

0016                      STRAT PROC  FAR           ;strategy procedure (must be FAR)

0016  89 1E 0012 R              MOV   WORD PTR REQ,BX   ;save request header address
001A  8C 06 0014 R              MOV   WORD PTR REQ+2,ES
001E  CB                        RET

001F                      STRAT ENDP

001F                      INTT  PROC  FAR           ;interrupt procedure (must be FAR)

001F  CB                        RET                 ;procedure placed here

0020                      INTT  ENDP
                                END
```

The request header contains the length of the request header as its first byte. This is necessary because the length of the request header varies from command to command. The unit number must be provided for block control devices and is ignored for character devices. The *return status word* communicates information back to DOS from the device driver. Figure 14–7 illustrates the binary bit pattern and the meaning of each bit in the return status word portion of the request header. The low-order part of the return status word contains error codes returned to DOS from the device driver.

The device driver responds to the commands listed in Table 14–3. DOS uses these commands to request an action from the device driver. The commands themselves are described in detail in this section of the text.

The *initialize driver command* (00H) is always executed when DOS initializes the device driver. The initialization command passes messages to the video display indicating

TABLE 14–2 The request header for a device driver

Offset	*Size*	*Purpose*
00H	Byte	Length of request header
01H	Byte	Unit number
02H	Byte	Command
03H	Word	Return status word
05H	Quadword	Reserved by DOS
0DH	Variable	Meaning depends on command

Note: The unit number is used for block devices to number the unit.

TABLE 14–3 Device driver commands as issued in the request header by DOS

Command	*Purpose*
00H	Initialize driver
01H	Media check
02H	Build BPB (block device parameter block)
03H	Control read
04H	Read
05H	Nondestructive read
06H	Input status
07H	Input flush
08H	Write
09H	Write with verify
0AH	Output status
0BH	Output flush
0CH	Control write
0DH	Open device
0EH	Close device
0FH	Removable media
10H	Output until busy
13H	Generic IOCTL
17H	Get logical device
18H	Set logical device
19H	IOCTL query

that the driver is loaded into the system, and returns to DOS the amount of memory used by the driver. You may only use DOS INT 21H functions 00H–0CH and 30H in the initialize driver command procedure. If a block device is initialized, it must provide the number of units and a pointer to the BIOS parameter block. Table 14–4 lists the contents of the request header as provided by DOS to the device driver for initialization. Notice that some information is provided by DOS and some must be returned to DOS, as illustrated in the table.

The *media check command* (01H) is used with block devices to determine whether the disk drive has been changed. When the driver is called by DOS, it passes the media ID to the driver at offset 0DH. This should be checked to determine that the media is in the requested drive. The return code, at offset 0EH, is returned to DOS as a –1 to indicate the disk has been changed, a 0 if it cannot determine whether a disk has been changed, and a +1 if the disk has not been changed. If the disk has been changed, the address stored at offset 0FH must point to the previous ASCII-Z volume label. Refer to Table 14–5 for the contents of the request header for the media check command.

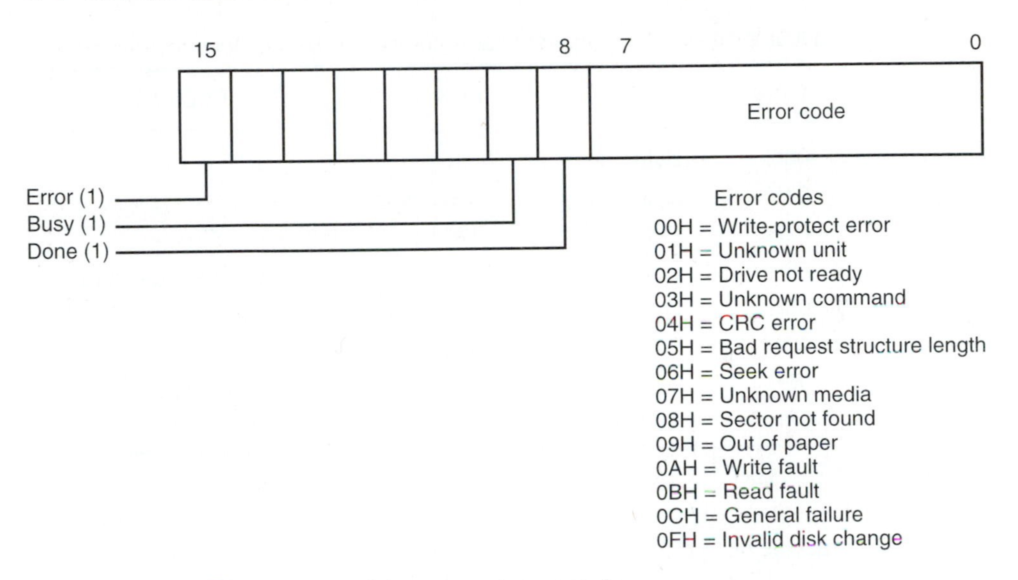

FIGURE 14–7 The contents of the return status word.

The *build BPB command* (02H) is used with block devices to build the block device parameter block. The build BPB command is issued when DOS detects, or suspects, that the media has been changed to obtain a new BPB. Table 14–6 lists the contents of the request header for the build BPB command.

The *control read command* (03H) functions with both block and character devices and provides a means of direct control for an application or device controlled by the device driver. Table 14–7 lists the contents of the request header for the control read command. Note that this command is seldom used with drivers. The request header shown in Table 14–7 also applies to the *control write command* (0CH), except for the byte at offset 02H,

TABLE 14–4 The request header for the initialize driver command (00H)

Offset	*Size*	*From DOS*	*To DOS*
00H	Byte	19H	19H
01H	Byte	—	—
02H	Byte	00H	—
03H	Word	—	Return status word
05H	Quadword	Reserved	Reserved
0DH	Byte	—	Number of units
0EH	Doubleword	—	Ending address of driver
12H	Doubleword	—	BPB pointer
16H	Byte	Next drive unit	—
17H	Word	—	Error message flag

TABLE 14–5 The request header for the media check command (01H)

Offset	*Size*	*From DOS*	*To DOS*
00H	Byte	13H	13H
01H	Byte	Unit code	—
02H	Byte	01H	—
03H	Word	—	Return status word
05H	Quadword	Reserved	Reserved
0DH	Byte	Media ID	—
0EH	Byte	—	Change code
0FH	Doubleword	—	Volume label address

which is a 0CH. The control write command allows control information to be passed to the device operated by the device driver.

The *output until busy command* (10H) is used with character devices to send a steady stream of data to an output device. The transfer continues until the device requests a stop, or until the data are entirely transferred to the device.

The *read command* (04H) functions with both block and character devices and is used to transfer information from the controlled device to DOS through a memory buffer. Refer to Table 14–8 for the contents of the request header for the read command.

The *nondestructive read command* (05H) is used with character devices such as keyboards and serial ports to read a character from the device before it is needed. This does not remove the character from the buffer, it only copies it to DOS through the device driver. Table 14–9 lists the contents of the request header for the nondestructive read command.

The *input status command* (06H) is used with character devices to determine whether data are available. The busy bit of the return status register is cleared by the driver to indicate that a character is available for reading. Table 14–10 shows the configuration of the request header for the input status register command. The *output status command*

TABLE 14–6 The request header for the build BPB command (02H)

Offset	*Size*	*From DOS*	*To DOS*
00H	Byte	16H	16H
01H	Byte	Unit code	—
02H	Byte	02H	—
03H	Word	—	Return status word
05H	Quadword	Reserved	Reserved
0DH	Byte	Media ID	—
0EH	Doubleword	Buffer address	—
12H	Doubleword	—	BPB address

TABLE 14–7 The request header for the control read (03H), control write (0CH), and output until busy (10H) commands

Offset	*Size*	*From DOS*	*To DOS*
00H	Byte	14H	14H
01H	Byte	Unit code	—
02H	Byte	03H*	—
03H	Word	—	Return status word
05H	Quadword	Reserved	Reserved
0DH	Media ID	—	—
0EH	Doubleword	Buffer address	—
12H	Word	Byte count	Actual byte count

Note: *For a control write command, the byte at offset 02H becomes 0CH and for an output until busy, it becomes 10H.

TABLE 14–8 The request header for the read command (04H)

Offset	*Size*	*From DOS*	*To DOS*
00H	Byte	1EH	1EH
01H	Byte	Unit code	—
02H	Byte	04H	04H
03H	Word	—	Return status word
05H	Quadword	Reserved	Reserved
0DH	Byte	Media ID	—
0EH	Doubleword	Buffer address	—
12H	Word	Byte/sector count	Actual transfer count
14H	Word	Starting sector	—
16H	Doubleword	—	Volume identifier
1AH	Doubleword	32-bit sector	Volume address if error

TABLE 14–9 The request header for the nondestructive read command (05H)

Offset	*Size*	*From DOS*	*To DOS*
00H	Byte	0EH	0EH
01H	Byte	—	—
02H	Byte	05H	—
03H	Word	—	Return status word
05H	Quadword	Reserved	Reserved
0DH	Byte	—	Character

TABLE 14–10 The request header for the input status command (06H) or the output status command (0AH)

Offset	*Size*	*From DOS*	*To DOS*
00H	Byte	0DH	0DH
01H	Byte	—	—
02H	Byte	06H*	—
03H	Word	—	Return status word
05H	Quadword	Reserved	Reserved

Note: *Byte 02H is 0AH for the output status command.

(0AH) uses the same request status format as the input status command, except the byte at offset 02H is 0AH instead of 06H. The output status command functions with character devices and clears the busy bit if the output device is ready to receive data.

The *input flush command* (07H) is used with character devices to clear or flush the input buffer. Any information in the buffer is lost when it is flushed. Refer to Table 14–11 for the configuration of the request header for the input flush command. The *output flush command* (0BH) uses the same request header format as the input flush command, except the byte at offset 02H is a 0BH instead of 07H. The output flush command clears data from an output device.

The *open device command* (0DH) and *close device command* (0EH) also use the request header in Table 14–11. The open and close device commands function with block or character devices and are used to open and close block devices. With a character device, these commands may be used to pass a control string to the device.

The *removable media command* (0FH) functions with a block device to determine whether a block device supports removable media. This function is used only when the device is accessed via DOS.

TABLE 14–11 The request header for input flush (07H), output flush (0BH), open device (0DH), close device (0EH), removable media (0FH), get logical device (17H), and set logical device (18H) commands

Offset	*Size*	*From DOS*	*To DOS*
00H	Byte	0DH	0DH
01H	Byte	Unit code	—
02H	Byte	07H*	—
03H	Word	—	Return status word
05H	Quadword	Reserved	Reserved

Note: *The byte at offset 02H is a 0BH for an output flush, 0DH for an open device, 0EH for a close device, 0FH for the removable media, 17H for a get logical device, and 18H for a set logical device command.

TABLE 14–12 The request header for the write command (08H) and write with verify command (09H)

Offset	*Size*	*From DOS*	*To DOS*
00H	Byte	1EH	1EH
01H	Byte	Unit code	—
02H	Byte	08H*	—
03H	Word	—	Return status word
05H	Quadword	Reserved	Reserved
0DH	Byte	Media ID	—
0EH	Doubleword	Buffer address	—
12H	Word	Byte/sector count	Actual transfer count
14H	Word	Starting sector	—
16H	Doubleword	—	Volume identifier
1AH	Doubleword	32-bit sector	Volume label address if error

Note: *09H for the write with verify command.

The *get logical device command* (17H) and *set logical device command* (18H) apply to block devices to get and set logical devices. These functions are often used in a system that contains a single floppy disk drive that is referred to as both A and B for functions such as disk copy.

The *write command* (08H) is used with both block and character devices to write data from a memory buffer to the device controlled by the device driver. Table 14–12 illustrates the contents of the request header for the write command. The *write with verify command* (09H) uses exactly the same request header, except it verifies the write and the command at offset address 02H is 09H instead of 08H.

The *generic IOCTL command* (13H) functions with both block and character devices and should be made to function like DOS INT 21H function 44H. Refer to Table 14–13 for the contents of the request header for this command.

The last device driver command is the *IOCTL query command* (19H) used with both block and character devices in DOS version 5.0 or newer. DOS uses this device driver function when the INT 21H instruction executes with AX = 4410H or 4411H. Table 14–14 shows the contents of the request header for this command.

The Interrupt Procedure. The **interrupt procedure** uses the request header to determine the function requested by DOS. It also performs all functions for the device driver. The interrupt procedure must respond to at least the initialize driver command (00H) and any other commands required to control the device operated by the device driver. If any of the commands are not implemented, the device driver should place a 8003H into the return status word of the request header to indicate that the command is not present. Note that the interrupt procedure is different from an interrupt handler. The interrupt handler services

TABLE 14–13 The request header for the generic IOCTL command (13H)

Offset	*Size*	*From DOS*	*To DOS*
00H	Byte	17H	17H
01H	Byte	Unit number	—
02H	Byte	13H	—
03H	Word	—	Return status word
05H	Quadword	Reserved	Reserved
0DH	Byte	Major code	—
0EH	Byte	Minor code	—
0FH	Word	Contents of SI	—
11H	Word	Contents of DI	—
13H	Doubleword	Buffer address	—

hardware interrupts as they are requested by I/O devices, while the interrupt procedure included in the driver is just a procedure called by DOS.

The Device Driver. This section presents a device driver called DUMP that allows a file to be displayed in hexadecimal format. Although this could be written as a program, it is interesting to see how a device driver can be built to control a character device. To use this driver, type COPY TARGET.FIL DUMP. This causes the file named TARGET.FIL to be copied to the video screen in hexadecimal format by the DUMP driver. Or, type TYPE TARGET.FIL > DUMP to redirect the TARGET.FIL to the DUMP driver. Another example is DIR > DUMP, which will display the directory of the current drive in hexadecimal format. Like any driver, DUMP.EXE must be installed in the CONFIG.SYS file using either a DEVICE or DEVICEHIGH command.

TABLE 14–14 The request header for the generic IOCTL query command (19H)

Offset	*Size*	*From DOS*	*To DOS*
00H	Byte	17H	17H
01H	Byte	Unit number	—
02H	Byte	19H	—
03H	Word	—	Return status word
05H	Quadword	Reserved	Reserved
0DH	Byte	Command	—
0EH	Byte	Minor code	—
0FH	Doubleword	Reserved	—
13H	Doubleword	Buffer address	—

EXAMPLE 14–13

```
                         ;a device driver called DUMP that allows hexadecimal data
                         ;to be displayed from the DOS pipe or by using the DOS
                         ;COPY command. This file must be loaded as a DEVICE in the
                         ;CONFIG.SYS file.
                                .MODEL TINY
                                .386
0000                            .CODE
                         ;HEADER
                         ;
0000  FFFFFFFF           CHAIN  DD    -1                ;link to next driver in chain
0004  A800               ATTR   DW    0A800H            ;driver attribute
0006  006F R             STRT   DW    STRAT             ;address of strategy
0008  007A R             INTR   DW    INTT              ;address of interrupt
000A  44 55 4D 50 20     DNAME  DB    'DUMP '           ;driver name
      20 20 20
0012  00000000           REQ    DD    ?                 ;request header address
0016  01A6 R 00C1 R      CT     DW    C0,CZ,CZ,CZ,CZ,C5,CZ,CZ ;command lookup table
      00C1 R 00C1 R
      00C1 R 00C5 R
      00C1 R 00C1 R
0026  00C9 R 00C1 R             DW    C8,CZ,CZ,CZ,CZ,CD,CZ,CZ
      00C1 R 00C1 R
      00C1 R 0196 R
      00C1 R 00C1 R
0036  00C9 R 00BD R             DW    C8,CU,CU,CZ,CU,CU,CU
      00BD R 00C1 R
      00BD R 00BD R
      00BD R
0044  00C1 R 00C1 R             DW    CZ,CZ,CZ
      00C1 R
004A  0000               COUNT  DW    ?                 ;transfer counter
004C  00000000           ADDR1  DD    ?                 ;address
0050  00                 LINES  DB    ?                 ;line counter
0051  50 72 65 73 73     ANY    DB    'Press any key to continue ...',0
      20 61 6E 79 20
      6B 65 79 20 74
      6F 20 63 6F 6E
      74 69 6E 75 65
      20 2E 2E 2E 00
                         ;STRATEGY
                         ;
006F                     STRAT  PROC  FAR               ;strategy procedure (must be FAR)

006F  2E: 89 1E 0012 R          MOV   WORD PTR CS:REQ,BX  ;save request header address
0074  2E: 8C 06 0014 R          MOV   WORD PTR CS:REQ+2,ES
0079  CB                        RET

007A                     STRAT  ENDP
                         ;
                         ;INTERRUPT procedure processes commands
                         ;
007A                     INTT   PROC  FAR               ;interrupt procedure (must be FAR)

007A  9C                        PUSHF                   ;save all registers
007B  66| 60                    PUSHAD
007D  1E                        PUSH  DS
007E  06                        PUSH  ES
007F  8C C8                     MOV   AX,CS
0081  8E D8                     MOV   DS,AX             ;load DS
```

```
0083  8B 36 0012 R            MOV   SI,WORD PTR REQ  ;get request header address
0087  8E 06 0014 R            MOV   ES,WORD PTR REQ+2

008B  26: 8A 44 02            MOV   AL,ES:[SI+2]     ;get command
008F  3C 19                   CMP   AL,19H           ;check range
0091  76 05                   JBE   INTT1            ;if possible command
0093  B8 9003                 MOV   AX,9003H         ;get bad command code
0096  EB 0D                   JMP   INTT_END         ;end program
0098                    INTT1:
0098  B4 00                   MOV   AH,0
009A  03 C0                   ADD   AX,AX            ;double code
009C  05 0016 R               ADD   AX,OFFSET CT     ;address command table
009F  8B F8                   MOV   DI,AX
00A1  8B 05                   MOV   AX,[DI]          ;get address from table
00A3  FF D0                   CALL  AX               ;call command
00A5                    INTT_END:
00A5  26: 89 44 03            MOV ES:[SI+3],AX;save return status word
00A9  07                      POP   ES               ;restore registers
00AA  1F                      POP   DS
00AB  66| 61                  POPAD
00AD  9D                      POPF
00AE  CB                      RET

00AF                    INTT  ENDP
                        ;
                        ;the DISP procedure displays the null character string
                        ;stored at the location addressed by DS:SI.
                        ;
00AF                    DISP  PROC  NEAR

00AF  B7 00                   MOV   BH,0             ;select video page 0
00B1                    DISP1:
00B1  AC                      LODSB                  ;get byte
00B2  0A C0                   OR    AL,AL            ;test for null (0)
00B4  74 06                   JZ    DISP2            ;if null
00B6  B4 0E                   MOV   AH,0EH           ;display AL
00B8  CD 10                   INT   10H              ;call BIOS display function
00BA  EB F5                   JMP   DISP1            ;repeat until null
00BC                    DISP2:
00BC  C3                      RET

00BD                    DISP  ENDP
                        ;
                        ;process any unknown command
                        ;
00BD                    CU    PROC  NEAR

00BD  B8 9003                 MOV   AX,9003H         ;indicate error
00C0  C3                      RET

00C1                    CU    ENDP
                        ;
                        ;process an unimplemented command
                        ;
00C1                    CZ    PROC  NEAR

00C1  B8 0100                 MOV   AX,100H          ;indicate not implemented
00C4  C3                      RET

00C5                    CZ    ENDP
                        ;
```

```
                     ;process the nondestructive read command
                     ;
00C5                 C5      PROC   NEAR            ;nondestructive read

00C5  B8 0300                MOV    AX,300H
00C8  C3                     RET

00C9                 C5      ENDP
                     ;
                     ;process the write command
                     ;
                     ;this command displays hexadecimal data on the video display.
                     ;
00C9                 C8      PROC   NEAR            ;write command (08H)

00C9  06                     PUSH   ES              ;save data segment
00CA  56                     PUSH   SI              ;save request header offset
00CB  26: 8B 44 0E           MOV    AX,ES:[SI+0EH]  ;get buffer address
00CF  26: 8B 5C 10           MOV    BX,ES:[SI+10H]
00D3  26: 8B 4C 12           MOV    CX,ES:[SI+12H]  ;get transfer count
00D7  89 0E 004A R           MOV    COUNT,CX
00DB  E3 0E                  JCXZ   C8E             ;if no data to transfer
00DD  8E C3                  MOV    ES,BX
00DF  8B F0                  MOV    SI,AX
00E1                 C81:
00E1  51                     PUSH   CX              ;save count
00E2  E8 0014                CALL   HADDR           ;display address
00E5  E8 0075                CALL   HBYTE           ;display byte
00E8  59                     POP    CX              ;get count
00E9  E2 F6                  LOOP   C81             ;repeat until count = 0
00EB                 C8E:                           ;finish write
00EB  5E                     POP    SI              ;restore registers
00EC  07                     POP    ES
00ED  8B 0E 004A R           MOV    CX,COUNT        ;save transfer count
00F1  26: 89 4C 12           MOV    ES:[SI+12H],CX
00F5  B8 0100                MOV    AX,100H
00F8  C3                     RET

00F9                 C8      ENDP
                     ;
                     ;the HADDR procedure displays the contents of ADDR1 as a
                     ;hexadecimal address whenever it represents a paragraph
                     ;boundary. Otherwise, a return occurs without displaying
                     ;any address.
                     ;
00F9                 HADDR   PROC   NEAR            ;display address

00F9  66| A1 004C R          MOV    EAX,ADDR1       ;get address
00FD  A8 0F                  TEST   AL,0FH          ;test for paragraph
00FF  75 4C                  JNZ    HADDR4          ;if not XXXXXXX0
0101  FE 0E 0050 R           DEC    LINES           ;decrement line counter
0105  75 1E                  JNZ    HADDR1          ;if not last line of screen
0107  66| 50                 PUSH   EAX
0109  56                     PUSH   SI
010A  BA 1800                MOV    DX,1800H
010D  B7 00                  MOV    BH,0
010F  B4 02                  MOV    AH,2
0111  CD 10                  INT    10H             ;cursor to bottom line
0113  BE 0051 R              MOV    SI,OFFSET ANY
0116  E8 FF96                CALL   DISP            ;display "type any key"
0119  5E                     POP    SI
```

```
011A  B4 00                   MOV   AH,0
011C  CD 16                   INT   16H              ;read any key
011E  E8 0056                 CALL  CLS              ;clear screen and home cursor
0121  66| 58                  POP   EAX
0123  EB 0E                   JMP   HADDR2
0125                  HADDR1:
0125  66| 50                  PUSH  EAX
0127  B8 0E0D                 MOV   AX,0E0DH
012A  CD 10                   INT   10H              ;display carriage return
012C  B8 0E0A                 MOV   AX,0E0AH
012F  CD 10                   INT   10H              ;display line feed
0131  66| 58                  POP   EAX
0133                  HADDR2:
0133  B9 0008                 MOV   CX,8             ;set count for 8 digit address
0136                  HADDR3:
0136  66| C1 C0 04            ROL   EAX,4            ;position digit
013A  66| 50                  PUSH  EAX
013C  E8 000F                 CALL  DIP              ;display one hex digit
013F  66| 58                  POP   EAX
0141  E2 F3                   LOOP  HADDR3           ;eight times
0143  B8 0E3A                 MOV   AX,0E3AH
0146  CD 10                   INT   10H              ;display colon
0148  B8 0E20                 MOV   AX,0E20H
014B  CD 10                   INT   10H              ;display space
014D                  HADDR4:
014D  C3                      RET

014E                  HADDR ENDP
                      ;
                      ;the DIP procedure displays a single hexadecimal digit
                      ;from the AL register bit 3 through bit 0.
                      ;
014E                  DIP     PROC  NEAR

014E  24 0F                   AND   AL,0FH           ;get digit
0150  04 30                   ADD   AL,30H           ;make it ASCII
                              .IF AL > 39H
0156  04 07                      ADD AL,7            ;if A through F
                              .ENDIF
0158  B4 0E                   MOV   AH,0EH
015A  CD 10                   INT   10H              ;display digit
015C  C3                      RET

015D                  DIP   ENDP
                      ;
                      ;the HBYTE procedure displays the contents of ES:SI as a
                      ;2-digit hexadecimal number.
                      ;
015D                  HBYTE  PROC  NEAR

015D  26: 8A 04               MOV   AL,ES:[SI]       ;get byte
0160  46                      INC   SI               ;address next byte
0161  50                      PUSH  AX
0162  C0 C0 04                ROL   AL,4
0165  E8 FFE6                 CALL  DIP              ;display first digit
0168  58                      POP   AX
0169  E8 FFE2                 CALL  DIP              ;display second digit
016C  B8 0E20                 MOV   AX,0E20H
016F  CD 10                   INT   10H              ;display space
0171  66| FF 06 004C R        INC   ADDR1            ;increment address
0176  C3                      RET
```

```
0177                         HBYTE  ENDP
                             ;
                             ;the CLS procedure clears the screen and homes the cursor
                             ;
0177                         CLS    PROC  NEAR

0177  51                            PUSH  CX
0178  B8 0600                       MOV   AX,600H          ;clear screen
017B  B7 07                         MOV   BH,7
017D  B9 0000                       MOV   CX,0
0180  B6 18                         MOV   DH,24
0182  B2 4F                         MOV   DL,79
0184  CD 10                         INT   10H
0186  BA 0000                       MOV   DX,0             ;home cursor
0189  B7 00                         MOV   BH,0
018B  B4 02                         MOV   AH,2
018D  CD 10                         INT   10H
018F  59                            POP   CX
0190  C6 06 0050 R 17               MOV   LINES,23         ;reset line counter
0195  C3                            RET

0196                         CLS    ENDP
                             ;
                             ;process the open device command
                             ;
0196                         CD     PROC  NEAR             ;open device (0DH)

0196  E8 FFDE                       CALL  CLS              ;clear screen
0199  66| C7 06 004C R              MOV   ADDR1,0          ;set address to 00000000
         00000000
01A2  B8 0100                       MOV   AX,100H          ;show done
01A5  C3                            RET

01A6                         CD     ENDP
                             ;
                             ;process the initialize driver command
                             ;
01A6                         C0     PROC  NEAR             ;initialize driver

01A6  56                            PUSH  SI               ;save request header address
01A7  BE 01BC R                     MOV   SI,OFFSET HELLO
01AA  E8 FF02                       CALL  DISP             ;display hello
01AD  5E                            POP   SI
                             ;save end address
01AE  26: C7 44 0E 0000             MOV   ES:[SI+0EH],OFFSET C0
01B4  26: 8C 4C 10                  MOV   ES:[SI+10H],CS
01B8  B8 0100                       MOV   AX,100H
01BB  C3                            RET

01BC                         C0     ENDP

01BC  0D 0A 0A 2E 2E         HELLO  DB    13,10,10,'...initializing hexadecimal DUMP',13,10,0
      2E 69 6E 69 74
      69 61 6C 69 7A
      69 6E 67 20 68
      65 78 61 64 65
      63 69 6D 61 6C
      20 44 55 4D 50
      0D 0A 00
                                    END
```

Example 14–13 shows the DUMP device driver program. This program is assembled as an ordinary execute file (.EXE) with one segment. Do not include a .STACK or .DATA segment, or an error ensues. When this program is built with the workbench or linked with the linker, two warnings will appear that should be ignored. The warnings are "no stack" and "no starting address". Notice that the first section of the driver contains the header, which is where it must be placed. This is followed by data used by the program, the strategy procedure, and interrupt procedure. Also notice that the device driver uses INT 10H for any video display information and INT 16H to read the keyboard. It is very important that INT 21H does not appear in the device driver for any reason. If DOS INT 21H is accessed by a device driver, DOS may be busy at the time and the system will stop.

This device driver uses commands 00H (initialize driver), 08H (write), 0DH (open device), and 10H (output until busy). The 00H command must be placed at the end of the program because once it executes at initialization, it is discarded from the memory. Recall that the initialize driver function only executes when the device driver is loaded into memory. The initialize driver part of the program is discarded when procedure C0 (command 00H) executes at the end of the driver. Please notice how a command lookup table accesses the addresses of procedures used to respond to each command.

14–4 PROTECTED-MODE OPERATION

Chapter 4 introduced the basics of protected-mode operation. In this chapter, the basics are built upon to allow an assembly language program access to the protected mode and all the additional memory that provides. As described in Chapter 4, memory above the first 1M byte is accessed through protected-mode operation. Another way to access memory above the first 1M byte is by using either VCPI or DPMI, as described in this chapter.

This section uses the DPMI interface provided with Windows to enter and leave the protected mode and to access extended memory. Why DPMI? Soon, personal computers will include Windows, not as an option, but as an essential part of the operating system. Using DPMI to accomplish the switch between real and protected mode allows this essential component of the personal computer to be utilized with assembly language programs. Recall that to use DPMI to execute a DOS application, you must shell out of Windows to DOS.

The real and protected modes use and execute instructions in the same manner. That is, if a MOV AL,[DI] instruction executes, the real-mode operation uses the combination of DS and DI to generate the memory address. The same thing is true in protected mode. The difference is in the way that the system uses the data segment register to generate the memory address. In the real mode, DS is loaded with the starting paragraph number of the data segment, which is appended with a 0H and added to DI to generate the address. In the protected mode, DS contains a 13-bit selector that accesses an 8-byte long descriptor. The descriptor contains the starting address of the memory segment, its length or limit, and an access rights byte that defines how the segment behaves.

EXAMPLE 14–14

```
                                ;a program that uses DPMI (provided by Windows) to access
                                ;protected-mode operation and display a message from
                                ;protected mode.
                                ;
                                        .MODEL SMALL
                                        .386
                                        .STACK 1024             ;make stack 1,024 bytes
0000                                    .DATA
0000  00000000                  ENTRY   DD      ?               ;DPMI entry point
0004  0000                      MSIZE   DW      ?               ;memory needed for DPMI
0006  0D 0A 0A 44 50            ERR1    DB      13,10,10,'DPMI not present.$'
      4D 49 20 6E 6F
      74 20 70 72 65
      73 65 6E 74 2E
      24
001B  0D 0A 0A 4E 6F            ERR2    DB      13,10,10,'Not enough real memory.$'
      74 20 65 6E 6F
      75 67 68 20 72
      65 61 6C 20 6D
      65 6D 6F 72 79
      2E 24
0036  0D 0A 0A 43 6F            ERR3    DB   13,10,10,'Could not move to protected mode.$'
      75 6C 64 20 6E
      6F 74 20 6D 6F
      76 65 20 74 6F
      20 70 72 6F 74
      65 63 74 65 64
      20 6D 6F 64 65
      2E 24
005B  0D 0A 0A 49 20            MES1    DB      13,10,10,'I am displayed from protected mode.$'
      61 6D 20 64 69
      73 70 6C 61 79
      65 64 20 66 72
      6F 6D 20 70 72
      6F 74 65 63 74
      65 64 20 6D 6F
      64 65 2E 24
                                ;
                                ;register array storage for DPMI function 0300H
                                ;
0082 = 0082                     ARRAY   EQU     THIS BYTE
0082  00000000                  REDI    DD      0               ;EDI
0086  00000000                  RESI    DD      0               ;ESI
008A  00000000                  REBP    DD      0               ;EBP
008E  00000000                          DD      0               ;reserved
0092  00000000                  REBX    DD      0               ;EBX
0096  00000000                  REDX    DD      0               ;EDX
009A  00000000                  RECX    DD      0               ;ECX
009E  00000000                  REAX    DD      0               ;EAX
00A2  0000                      RFLAG   DW      0               ;flags
00A4  0000                      RES     DW      0               ;ES
00A6  0000                      RDS     DW      0               ;DS
00A8  0000                      RFS     DW      0               ;FS
00AA  0000                      RGS     DW      0               ;GS
00AC  0000                      RIP     DW      0               ;IP
00AE  0000                      RCS     DW      0               ;CS
00B0  0000                      RSP     DW      0               ;SP
00B2  0000                      RSS     DW      0               ;SS
```

```
0000                            .CODE
                                .STARTUP
0010  8C C0                     MOV   AX,ES
0012  8C DB                     MOV   BX,DS          ;find size of program and data
0014  2B D8                     SUB   BX,AX
0016  8B C4                     MOV   AX,SP          ;find stack length
0018  C1 E8 04                  SHR   AX,4
001B  40                        INC   AX
001C  03 D8                     ADD   BX,AX          ;BX = length in paragraphs
001E  B4 4A                     MOV   AH,4AH
0020  CD 21                     INT   21H            ;modify memory allocation

0022  E8 0051                   CALL  ISDPMI         ;is DPMI loaded?
0025  72 09                     JC    MAIN1          ;if DPMI present
0027  B4 09                     MOV   AH,9
0029  BA 0006 R                 MOV   DX,OFFSET ERR1
002C  CD 21                     INT   21H            ;display DPMI not present
002E  EB 42                     JMP   MAIN4          ;end
0030                       MAIN1:
0030  B8 0000                   MOV   AX,0           ;indicate 0 memory needed
0033  83 3E 0004 R 00           CMP   MSIZE,0
0038  74 13                     JE    MAIN2          ;if DPMI needs no memory
003A  8B 1E 0004 R              MOV   BX,MSIZE       ;get amount
003E  B4 48                     MOV   AH,48H
0040  CD 21                     INT   21H            ;allocate memory for DPMI
0042  73 09                     JNC   MAIN2          ;if enough memory exists
0044  B4 09                     MOV   AH,9           ;if not enough real memory
0046  BA 001B R                 MOV   DX,OFFSET ERR2
0049  CD 21                     INT   21H            ;display not enough memory
004B  EB 25                     JMP   MAIN4          ;end
004D                       MAIN2:
004D  8E C0                     MOV   ES,AX
004F  B8 0000                   MOV   AX,0           ;16-bit application
0052  FF 1E 0000 R              CALL  ENTRY          ;switch to protected mode
0056  73 09                     JNC   MAIN3          ;if in protected mode
0058  B4 09                     MOV   AH,9           ;if switch failed
005A  BA 0036 R                 MOV   DX,OFFSET ERR3
005D  CD 21                     INT   21H            ;display switch failed
005F  EB 11                     JMP   MAIN4          ;end
                           ;
                           ;PROTECTED MODE
                           ;
0061                       MAIN3:
0061  BE 005B R                 MOV SI,OFFSET MES1   ;address MES1
0064                       MAIN3A:                   ;display MES1
0064  AC                        LODSB
0065  8A D0                     MOV   DL,AL
0067  3C 24                     CMP   AL,'$'         ;test for end of string
0069  74 07                     JE    MAIN4          ;if end of string
006B  B4 02                     MOV   AH,2           ;display characters
006D  E8 001D                   CALL  INT21H         ;emulate INT 21H
0070  EB F2                     JMP   MAIN3A
0072                       MAIN4:
                                .EXIT
                           ;
                           ;the ISDPMI procedure tests for the presence of the DPMI
                           ;manager.
                           ;***return parameters***
                           ;carry = 1; if DPMI is present
                           ;carry = 0; if DPMI is not present
                           ;
```

```
0076                        ISDPMI PROC   NEAR

0076  B8 1687                      MOV    AX,1687H           ;get DPMI status
0079  CD 2F                        INT    2FH                ;DOS multiplex
007B  0B C0                        OR     AX,AX              ;test AX
007D  75 0D                        JNZ    ISDPMI1            ;if no DPMI
007F  89 36 0004 R                 MOV    MSIZE,SI           ;save amount of memory needed
0083  89 3E 0000 R                 MOV    WORD PTR ENTRY,DI
0087  8C 06 0002 R                 MOV    WORD PTR ENTRY+2,ES
008B  F9                           STC
008C                        ISDPMI1:
008C  C3                           RET

008D                        ISDPMI ENDP
                            ;
                            ;the INT21H procedure is required to emulate a DOS INT 21H
                            ;function from protected-mode operation.
                            ;***entry parameters***
                            ;AH = DOS INT 21H function number
                            ;other registers as required by the function
                            ;
008D                        INT21H PROC   NEAR

008D  66| A3 009E R                MOV    REAX,EAX           ;save registers
0091  66| 89 1E 0092 R             MOV    REBX,EBX
0096  66| 89 0E 009A R             MOV    RECX,ECX
009B  66| 89 16 0096 R             MOV    REDX,EDX
00A0  66| 89 36 0086 R             MOV    RESI,ESI
00A5  66| 89 3E 0082 R             MOV    REDI,EDI
00AA  66| 89 2E 008A R             MOV    REBP,EBP
00AF  9C                           PUSHF
00B0  58                           POP    AX
00B1  A3 00A2 R                    MOV    RFLAG,AX
00B4  06                           PUSH   ES                 ;do DOS interrupt
00B5  B8 0300                      MOV    AX,0300H
00B8  BB 0021                      MOV    BX,21H
00BB  B9 0000                      MOV    CX,0
00BE  1E                           PUSH   DS
00BF  07                           POP    ES
00C0  BF 0082 R                    MOV    DI,OFFSET ARRAY
00C3  CD 31                        INT    31H
00C5  07                           POP    ES
00C6  A1 00A2 R                    MOV    AX,RFLAG           ;restore registers
00C9  50                           PUSH   AX
00CA  9D                           POPF
00CB  66| 8B 3E 0082 R             MOV    EDI,REDI
00D0  66| 8B 36 0086 R             MOV    ESI,RESI
00D5  66| 8B 2E 008A R             MOV    EBP,REBP
00DA  66| A1 009E R                MOV    EAX,REAX
00DE  66| 8B 1E 0092 R             MOV    EBX,REBX
00E3  66| 8B 0E 009A R             MOV    ECX,RECX
00E8  66| 8B 16 0096 R             MOV    EDX,REDX
00ED  C3                           RET

00EE                        INT21H ENDP
                                   END
```

When the switch is made to protected mode, the descriptors must be defined so memory can be accessed. This is accomplished by the tools provided by DPMI or VCPI. To illustrate the switch to protected mode through the DPMI interface, a simple program is developed in Example 14–14. Notice that this program displays a message from protected mode and then exits to DOS.

At the start of this example program—after allocating memory for the program, data, and stack areas—the program queries whether DPMI is present. If DPMI is present, the switch to protected mode is made with the CALL ENTRY instruction. The CALL ENTRY command uses the address stored by the INT 2FH in the ISDPMI procedure as the entry point to protected-mode operation provided by DPMI. After the return from CALL ENTRY, the segment descriptors are adjusted so CS, SS, DS contain selectors that access segments with limits of 64K bytes. The FS and GS segments are set to 0000H, and ES addresses the PSP with a limit of 256 bytes.

This program also accesses the real-mode DOS INT 21H function through a procedure called INT21H. This procedure stores the contents of the register into an array used by INT 31H to emulate the DOS real-mode INT 21H function. This program does not use function 09H to display the character string for a good reason: the real-mode interrupt (INT 21H) expects DS to address a real-mode segment. The DS register, instead, contains a selector which will not work in the real mode to address a character string in the data segment. Function 02H, used in place of function 09H, displays one character at a time because no real-mode address is needed with this function. Function 02H merely transfers the ASCII contents of register DL to the video screen.

The program ends in real or protected mode with AX = 4C00H and an INT 21H to exit to DOS. The DPMI handler intercepts this function and switches back to the real mode for the exit to DOS.

Displaying Extended Memory from Protected Mode

A more extensive example using DPMI appears to display the contents of memory in hexadecimal format. To access extended memory, the microprocessor is placed into protected mode. This allows extended memory to be accessed through the DPMI. The program illustrated in Example 14–15 uses the command line to pass the starting and ending location of the area of extended memory to be displayed in hexadecimal format. The syntax for this program, called EDUMP, is EDUMP 100000,1000FF to display the contents of extended memory locations 100000H through 1000FFH. Any range of addresses can be displayed with this function, including real memory below location 100000H.

EXAMPLE 14–15

```
                        ;a program that displays the contents of any area of memory
                        ;including extended memory.
                        ;***command line syntax***
                        ;EDUMP XXXX,YYYY where XXXX is the start address and YYYY is
                        ;the end address.
                        ;Note: this program must be executed from Windows.
                        ;
                               .MODEL SMALL
                               .386
                               .STACK 1024                ;stack area of 1,024 bytes
0000                           .DATA
0000  00000000          ENTRY  DD     ?                   ;DPMI entry point
0004  00000000          EXIT   DD     ?                   ;DPMI exit point
0008  00000000          FIRST  DD     ?                   ;first address
```

```
000C  00000000          LAST1   DD    ?                   ;last address
0010  0000              MSIZE   DW    ?                   ;memory needed for DPMI
0012  0D 0A 0A 50 61    ERR1    DB    13,10,10,'Parameter error.$'
      72 61 6D 65 74
      65 72 20 65 72
      72 6F 72 2E 24
0026  0D 0A 0A 44 50    ERR2    DB    13,10,10,'DPMI not present.$'
      4D 49 20 6E 6F
      74 20 70 72 65
      73 65 6E 74 2E
      24
003B  0D 0A 0A 4E 6F    ERR3    DB    13,10,10,'Not enough real memory.$'
      74 20 65 6E 6F
      75 67 68 20 72
      65 61 6C 20 6D
      65 6D 6F 72 79
      2E 24
0056  0D 0A 0A 43 6F    ERR4    DB    13,10,10,'Could not move to protected mode.$'
      75 6C 64 20 6E
      6F 74 20 6D 6F
      76 65 20 74 6F
      20 70 72 6F 74
      65 63 74 65 64
      20 6D 6F 64 65
      2E 24
007B  0D 0A 0A 43 61    ERR5    DB    13,10,10,'Cannot allocate selector.$'
      6E 6E 6F 74 20
      61 6C 6C 6F 63
      61 74 65 20 73
      65 6C 65 63 74
      6F 72 2E 24
0098  0D 0A 0A 43 61    ERR6    DB    13,10,10,'Cannot use base address.$'
      6E 6E 6F 74 20
      75 73 65 20 62
      61 73 65 20 61
      64 64 72 65 73
      73 2E 24
00B4  0D 0A 0A 43 61    ERR7    DB    13,10,10,'Cannot allocate 64K to limit.$'
      6E 6E 6F 74 20
      61 6C 6C 6F 63
      61 74 65 20 36
      34 4B 20 74 6F
      20 6C 69 6D 69
      74 2E 24
00D5  0D 0A 24          CRLF    DB    13,10,'$'
00D8  50 72 65 73 73    MES1    DB    'Press any key...$'
      20 61 6E 79 20
      6B 65 79 2E 2E
      2E 24
                        ;
                        ;register array storage for DPMI function 0300H
                        ;
00E9 = 00E9             ARRAY   EQU   THIS BYTE
00E9  00000000          REDI    DD    0                   ;EDI
00ED  00000000          RESI    DD    0                   ;ESI
00F1  00000000          REBP    DD    0                   ;EBP
00F5  00000000                  DD    0                   ;reserved
00F9  00000000          REBX    DD    0                   ;EBX
00FD  00000000          REDX    DD    0                   ;EDX
0101  00000000          RECX    DD    0                   ;ECX
0105  00000000          REAX    DD    0                   ;EAX
```

```
0109  0000              RFLAG  DW     0                ;flags
010B  0000              RES    DW     0                ;ES
010D  0000              RDS    DW     0                ;DS
010F  0000              RFS    DW     0                ;FS
0111  0000              RGS    DW     0                ;GS
0113  0000              RIP    DW     0                ;IP
0115  0000              RCS    DW     0                ;CS
0117  0000              RSP    DW     0                ;SP
0119  0000              RSS    DW     0                ;SS
0000                           .CODE
                               .STARTUP
0010  8C C0                    MOV    AX,ES
0012  8C DB                    MOV    BX,DS            ;find size of program and data
0014  2B D8                    SUB    BX,AX
0016  8B C4                    MOV    AX,SP            ;find stack size
0018  C1 E8 04                 SHR    AX,4
001B  40                       INC    AX
001C  03 D8                    ADD    BX,AX            ;BX = length in paragraphs
001E  B4 4A                    MOV    AH,4AH
0020  CD 21                    INT    21H              ;modify memory allocation
0022  E8 00D1                  CALL   GETDA            ;get command line information
0025  73 0A                    JNC    MAIN1            ;if parameters are good
0027  B4 09                    MOV    AH,9             ;parameter error
0029  BA 0012 R                MOV    DX,OFFSET ERR1
002C  CD 21                    INT    21H
002E  E9 00AA                  JMP    MAINE            ;exit to DOS
0031                    MAIN1:
0031  E8 00AB                  CALL   ISDPMI           ;is DPMI loaded?
0034  72 0A                    JC     MAIN2            ;if DPMI present
0036  B4 09                    MOV    AH,9
0038  BA 0026 R                MOV    DX,OFFSET ERR2
003B  CD 21                    INT    21H              ;display DPMI not present
003D  E9 009B                  JMP    MAINE            ;exit to DOS
0040                    MAIN2:
0040  B8 0000                  MOV    AX,0             ;indicate 0 memory needed
0043  83 3E 0010 R 00          CMP    MSIZE,0
0048  74 F6                    JE     MAIN2            ;if DPMI needs no memory
004A  8B 1E 0010 R             MOV    BX,MSIZE         ;get amount
004E  B4 48                    MOV    AH,48H
0050  CD 21                    INT    21H              ;allocate memory for DPMI
0052  73 09                    JNC    MAIN3
0054  B4 09                    MOV    AH,9             ;if not enough real memory
0056  BA 003B R                MOV    DX,OFFSET ERR3
0059  CD 21                    INT    21H
005B  EB 7E                    JMP    MAINE            ;exit to DOS
005D                    MAIN3:
005D  8E C0                    MOV    ES,AX
005F  B8 0000                  MOV    AX,0             ;16-bit application
0062  FF 1E 0000 R             CALL   DS:ENTRY         ;switch to protected mode
0066  73 09                    JNC    MAIN4
0068  B4 09                    MOV    AH,9             ;if switch failed
006A  BA 0056 R                MOV    DX,OFFSET ERR4
006D  CD 21                    INT    21H
006F  EB 6A                    JMP    MAINE            ;exit to DOS
                        ;
                        ;PROTECTED MODE
                        ;
0071                    MAIN4:
0071  B8 0000                  MOV    AX,0000H         ;get local selector
0074  B9 0001                  MOV    CX,1             ;only one is needed
0077  CD 31                    INT    31H
```

```
0079  72 48                  JC      MAIN7              ;if error
007B  8B D8                  MOV     BX,AX              ;save selector
007D  8E C0                  MOV     ES,AX              ;load ES with selector
007F  B8 0007                MOV     AX,0007H           ;set base address
0082  8B 0E 000A R           MOV     CX,WORD PTR FIRST+2
0086  8B 16 0008 R           MOV     DX,WORD PTR FIRST
008A  CD 31                  INT     31H
008C  72 3D                  JC      MAIN8              ;if error
008E  B8 0008                MOV     AX,0008H
0091  B9 0000                MOV     CX,0
0094  BA FFFF                MOV     DX,0FFFFH          ;set limit to 64K
0097  CD 31                  INT     31H
0099  72 38                  JC      MAIN9              ;if error
009B  B9 0018                MOV     CX,24              ;load line count
009E  BE 0000                MOV     SI,0               ;load offset
00A1                 MAIN5:
00A1  E8 00F4                CALL    DADDR              ;display address, if needed
00A4  E8 00CE                CALL    DDATA              ;display data
00A7  46                     INC     SI                 ;point to next data
00A8  66| A1 0008 R          MOV     EAX,FIRST          ;test for end
00AC  66| 3B 06 000C R       CMP     EAX,LAST1
00B1  74 07                  JE      MAIN6              ;if done
00B3  66| FF 06 0008 R       INC     FIRST
00B8  EB E7                  JMP     MAIN5
00BA                 MAIN6:
00BA  B8 0001                MOV     AX,0001H           ;release descriptor
00BD  8C C3                  MOV     BX,ES
00BF  CD 31                  INT     31H
00C1  EB 18                  JMP     MAINE              ;exit to DOS
00C3                 MAIN7:
00C3  BA 007B R              MOV     DX,OFFSET ERR5
00C6  E8 0096                CALL    DISPS              ;display cannot allocate selector
00C9  EB 10                  JMP     MAINE              ;exit to DOS
00CB                 MAIN8:
00CB  BA 0098 R              MOV     DX,OFFSET ERR6
00CE  E8 008E                CALL    DISPS              ;display cannot use base address
00D1  EB E7                  JMP     MAIN6              ;release descriptor
00D3                 MAIN9:
00D3  BA 00B4 R              MOV     DX,OFFSET ERR7
00D6  E8 0086                CALL    DISPS              ;display cannot allocate 64K limit
00D9  EB DF                  JMP     MAIN6              ;release descriptor
00DB                 MAINE:
                             .EXIT
                     ;
                     ;the ISDPMI procedure tests for the presence of DPMI
                     ;***exit parameters***
                     ;carry = 1; if DPMI is present
                     ;carry = 0; if DPMI is not present
                     ;
00DF                 ISDPMI PROC   NEAR

00DF  B8 1687                MOV     AX,1687H           ;get DPMI status
00E2  CD 2F                  INT     2FH                ;DOS multiplex
00E4  0B C0                  OR      AX,AX
00E6  75 0D                  JNZ     ISDPMI1            ;if no DPMI
00E8  89 36 0010 R           MOV     MSIZE,SI           ;save amount of memory needed
00EC  89 3E 0000 R           MOV     WORD PTR ENTRY,DI
00F0  8C 06 0002 R           MOV     WORD PTR ENTRY+2,ES
00F4  F9                     STC
00F5                 ISDPMI1:
00F5  C3                     RET
```

```
00F6                        ISDPMI ENDP
                            ;
                            ;the GETDA procedure retrieves the command line parameters
                            ;for memory display in hexadecimal.
                            ;FIRST = the first address from the command line
                            ;LAST1 = the last address from the command line
                            ;***return parameters***
                            ;carry = 1; if error
                            ;carry = 0; for no error
                            ;
00F6                        GETDA  PROC   NEAR

00F6  1E                           PUSH   DS
00F7  06                           PUSH   ES
00F8  1F                           POP    DS
00F9  07                           POP    ES                ;exchange ES with DS
00FA  BE 0081                      MOV    SI,81H            ;address command line
00FD                        GETDA1:
00FD  AC                           LODSB                    ;skip spaces
00FE  3C 20                        CMP    AL,' '
0100  74 FB                        JE     GETDA1            ;if space
0102  3C 0D                        CMP    AL,13
0104  74 1E                        JE     GETDA3            ;if enter = error
0106  4E                           DEC    SI                ;adjust SI
0107                        GETDA2:
0107  E8 0020                      CALL   GETNU             ;get first number
010A  3C 2C                        CMP    AL,','
010C  75 16                        JNE    GETDA3            ;if no comma = error
010E  66| 26: 89 16 0008 R         MOV    ES:FIRST,EDX
0114  E8 0013                      CALL   GETNU             ;get second number
0117  3C 0D                        CMP    AL,13
0119  75 09                        JNE    GETDA3            ;if error
011B  66| 26: 89 16 000C R         MOV    ES:LAST1,EDX
0121  F8                           CLC                      ;indicate no error
0122  EB 01                        JMP    GETDA4            ;return no error
0124                        GETDA3:
0124  F9                           STC                      ;indicate error
0125                        GETDA4:
0125  1E                           PUSH   DS                ;exchange ES with DS
0126  06                           PUSH   ES
0127  1F                           POP    DS
0128  07                           POP    ES
0129  C3                           RET

012A                        GETDA  ENDP
                            ;
                            ;the GETNU procedure extracts a number from the command line
                            ;and returns with it in EDX and last command line character in
                            ;AL as a delimiter.
                            ;
012A                        GETNU  PROC   NEAR

012A  66| BA 00000000              MOV    EDX,0             ;clear result
0130                        GETNU1:
0130  AC                           LODSB                    ;get digit from command line
                                   .IF AL >= 'a' && AL <= 'z'
0139  2C 20                           SUB AL,20H            ;make uppercase
                                   .ENDIF
013B  2C 30                        SUB    AL,'0'            ;convert from ASCII
013D  72 12                        JB     GETNU2            ;if not a number
                                   .IF AL > 9               ;convert A-F from ASCII
0143  2C 07                           SUB AL,7
```

```
                                   .ENDIF
0145  3C 0F                        CMP    AL,0FH
0147  77 08                        JA     GETNU2             ;if not 0--F
0149  66| C1 E2 04                 SHL    EDX,4
014D  02 D0                        ADD    DL,AL              ;add digit to EDX
014F  EB DF                        JMP    GETNU1             ;get next digit
0151                        GETNU2:
0151  8A 44 FF                     MOV    AL,[SI-1]          ;get delimiter
0154  C3                           RET

0155                        GETNU  ENDP
                            ;
                            ;the DISPC procedure displays the ASCII character found
                            ;in register AL.
                            ;***uses***
                            ;INT21H
                            ;
0155                        DISPC  PROC   NEAR

0155  52                           PUSH   DX
0156  8A D0                        MOV    DL,AL
0158  B4 06                        MOV    AH,6
015A  E8 0084                      CALL   INT21H             ;do real INT 21H
015D  5A                           POP    DX
015E  C3                           RET

015F                        DISPC  ENDP
                            ;
                            ;the DISPS procedure displays a character string from
                            ;protected mode addressed by DS:EDX.
                            ;***uses***
                            ;DISPC
                            ;
015F                        DISPS  PROC   NEAR

015F  66| 81 E2 0000FFFF           AND    EDX,0FFFFH
0166  67& 8A 02                    MOV    AL,[EDX]           ;get character
0169  3C 24                        CMP    AL,'$'             ;test for end
016B  74 07                        JE     DISP1              ;if end
016D  66| 42                       INC    EDX                ;address next character
016F  E8 FFE3                      CALL   DISPC              ;display character
0172  EB EB                        JMP    DISPS              ;repeat until $
0174                        DISP1:
0174  C3                           RET

0175                        DISPS  ENDP
                            ;
                            ;the DDATA procedure displays a byte of data at the location
                            ;addressed by ES:SI. The byte is followed by one space.
                            ;***uses***
                            ;DIP and DISPC
                            ;
0175                        DDATA  PROC   NEAR

0175  26: 8A 04                    MOV    AL,ES:[SI]         ;get byte
0178  C0 E8 04                     SHR    AL,4
017B  E8 000C                      CALL   DIP                ;display first digit
017E  26: 8A 04                    MOV    AL,ES:[SI]         ;get byte
0181  E8 0006                      CALL   DIP                ;display second digit
0184  B0 20                        MOV    AL,' '             ;display space
0186  E8 FFCC                      CALL   DISPC
0189  C3                           RET
```

```
018A                      DDATA   ENDP
                          ;
                          ;the DIP procedure displays the right nibble found in AL as a
                          ;hexadecimal digit.
                          ;***uses***
                          ;DISPC
                          ;
018A                      DIP     PROC    NEAR

018A  24 0F                       AND     AL,0FH          ;get right nibble
018C  04 30                       ADD     AL,30H          ;convert to ASCII
                                  .IF AL > 39H            ;if A-F
0192  04 07                          ADD AL,7
                                  .ENDIF
0194  E8 FFBE                     CALL    DISPC           ;display digit
0197  C3                          RET

0198                      DIP   ENDP
                          ;
                          ;the DADDR procedure displays the hexadecimal address found
                          ;in DS:FIRST if it is a paragraph boundary.
                          ;***uses***
                          ;DIP, DISPS, DISPC, and INT21H
                          ;
0198                      DADDR  PROC   NEAR

0198  66| A1 0008 R               MOV     EAX,FIRST       ;get address
019C  A8 0F                       TEST    AL,0FH          ;test for XXXXXXX0
019E  75 40                       JNZ     DADDR4          ;if not, don't display address
01A0  BA 00D5 R                   MOV     DX,OFFSET CRLF
01A3  E8 FFB9                     CALL    DISPS           ;display CR and LF
01A6  49                          DEC     CX              ;decrement line count
01A7  75 18                       JNZ     DADDR2          ;if not end of page
01A9  BA 00D8 R                   MOV     DX,OFFSET MES1  ;if end of page
01AC  E8 FFB0                     CALL    DISPS           ;display press any key
01AF                      DADDR1:
01AF  B4 06                       MOV     AH,6            ;get any key, no echo
01B1  B2 FF                       MOV     DL,0FFH
01B3  E8 002B                     CALL    INT21H          ;do real INT 21H
01B6  74 F7                       JZ      DADDR1          ;if nothing typed
01B8  BA 00D5 R                   MOV     DX,OFFSET CRLF
01BB  E8 FFA1                     CALL    DISPS           ;display CRLF
01BE  B9 0018                     MOV     CX,24           ;reset line count
01C1                      DADDR2:
01C1  51                          PUSH    CX              ;save line count
01C2  B9 0008                     MOV     CX,8            ;load digit count
01C5  66| 8B 16 0008 R            MOV     EDX,FIRST       ;get address
01CA                      DADDR3:
01CA  66| C1 C2 04                ROL     EDX,4
01CE  8A C2                       MOV     AL,DL
01D0  E8 FFB7                     CALL    DIP             ;display digit
01D3  E2 F5                       LOOP    DADDR3          ;repeat 8 times
01D5  59                          POP     CX              ;retrieve line count
01D6  B0 3A                       MOV     AL,':'
01D8  E8 FF7A                     CALL    DISPC           ;display colon
01DB  B0 20                       MOV     AL,' '
01DD  E8 FF75                     CALL    DISPC           ;display space
01E0                      DADDR4:
01E0  C3                          RET

01E1                      DADDR  ENDP
                          ;
```

```
                          ;the INT21H procedure gains access to the real-mode DOS
                          ;INT 21H instruction with the parameters intact.
                          ;
01E1                      INT21H PROC   NEAR

01E1  66| A3 0105 R              MOV    REAX,EAX          ;save registers
01E5  66| 89 1E 00F9 R           MOV    REBX,EBX
01EA  66| 89 0E 0101 R           MOV    RECX,ECX
01EF  66| 89 16 00FD R           MOV    REDX,EDX
01F4  66| 89 36 00ED R           MOV    RESI,ESI
01F9  66| 89 3E 00E9 R           MOV    REDI,EDI
01FE  66| 89 2E 00F1 R           MOV    REBP,EBP
0203  9C                         PUSHF
0204  58                         POP    AX
0205  A3 0109 R                  MOV    RFLAG,AX
0208  06                         PUSH   ES                ;do DOS interrupt
0209  B8 0300                    MOV    AX,0300H
020C  BB 0021                    MOV    BX,21H
020F  B9 0000                    MOV    CX,0
0212  1E                         PUSH   DS
0213  07                         POP    ES
0214  BF 00E9 R                  MOV    DI,OFFSET ARRAY
0217  CD 31                      INT    31H
0219  07                         POP    ES
021A  A1 0109 R                  MOV    AX,RFLAG          ;restore registers
021D  50                         PUSH   AX
021E  9D                         POPF
021F  66| 8B 3E 00E9 R           MOV    EDI,REDI
0224  66| 8B 36 00ED R           MOV    ESI,RESI
0229  66| 8B 2E 00F1 R           MOV    EBP,REBP
022E  66| A1 0105 R              MOV    EAX,REAX
0232  66| 8B 1E 00F9 R           MOV    EBX,REBX
0237  66| 8B 0E 0101 R           MOV    ECX,RECX
023C  66| 8B 16 00FD R           MOV    EDX,REDX
0241  C3                         RET

0242                      INT21H ENDP
                                 END
```

The memory locations passed to the program in this example are accessed by allocating a local descriptor. Next, the first address from the command line is placed into the local descriptor as the base address. The limit is set to 64K bytes, so the software displays up to 64K bytes of memory at a time. If the DPMI supports the 32-bit mode, this can be changed to a limit of up to 4G bytes. Notice how the selector, returned by the allocate local descriptor, is positioned in the ES register to access the section of memory to be displayed in hexadecimal format. Like Example 14–14, the real mode INT 21H is accessed via a procedure called INT21H, which saves the contents of the register before using DPMI to access INT 21H.

14–5 INTERFACING ASSEMBLY LANGUAGE TO HIGH-LEVEL LANGUAGES

The assembler and assembly language are often used with high-level languages such as C/C++, BASIC, PASCAL, and FORTRAN in the form of in-line assembly language code or through a procedure. The in-line assembly language code varies from one high-level language to another and is not explained in this text. The procedure, or function as it is often called, is described for interface to high-level language programs. Although this text concentrates on assembly language procedures for use with Microsoft languages, most other languages can also use them, provided the calling conventions are compatible.

The Calling Conventions

The assembler uses two calling conventions when interfaced to high-level languages: the C-language calling convention for C/C++ language and the PASCAL-language calling convention for BASIC, PASCAL, and FORTRAN applications. The calling convention language is specified in the MODEL statement or with the OPTION directive associated with the PROC statement. It can also be specified without the use of these statements using full-segment definitions.

The C-Language Calling Convention. The C-language calling convention pushes arguments from the right to the left as they are placed in the caller's argument list. The return is made from the assembly language procedure with the arguments in place on the stack. A single result is found in AX or DX:AX for most languages.

The values found in BP, DI, SI, DS, and SS and the direction flag must be saved by the assembly language procedure if they are changed. These registers are used, or are assumed to be used, by the high-level language. Figure 14–8 shows the stack frames for both near and far calls to functions from C-language calling conventions.

Access to the parameters stored in the stack frame is provided by the C-language calling convention, as illustrated in Example 14–16. The statements that contain an asterisk (*) are added by the assembler to adjust or conform to the C-language calling convention. Notice how the contents of BP are pushed to the stack and how BP is loaded with SP to allow access to the parameters placed on the stack by the C-language program.

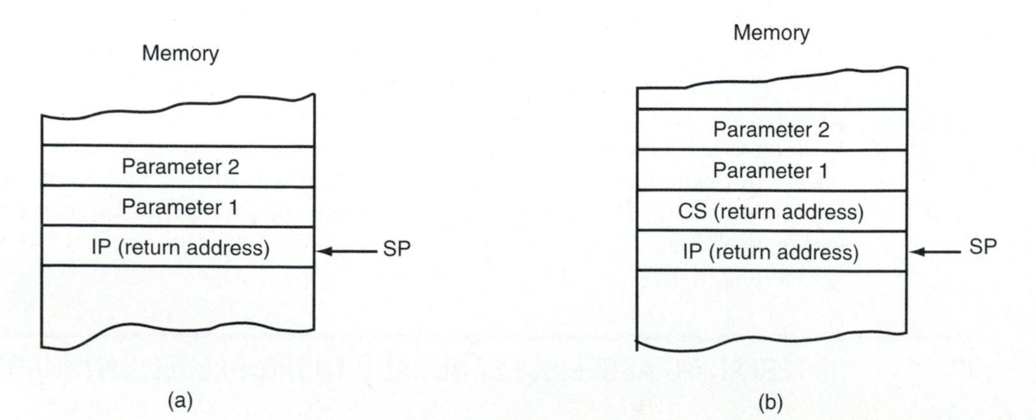

FIGURE 14–8 The stack frames for the C-language calling convention: (a) the near function call and (b) the far function call.

EXAMPLE 14–16

```
                            .listall
                            .MODEL SMALL,C
                            sss     PROTO  C, a:SWORD, a:SWORD

0000                        .CODE

0000                        sss     PROC   C a:SWORD,b:SWORD

0000  55            *               push   bp           ;save BP
0001  8B EC         *               mov    bp, sp       ;address stack frame with BP
0003  8B 46 04                      MOV    AX,a         ;access parameter a
0006  03 46 06                      ADD    AX,b         ;access parameter b
                                    RET
0009  5D            *               pop    bp           ;retrieve BP
000A  C3            *               ret    00000h

000B                        sss     ENDP

                                    END
```

Assembly language procedures that are written with the C-language calling convention are called from assembly language with the INVOKE directive. INVOKE is used in place of a standard CALL. INVOKE follows the C-language calling convention, allowing assembly language programs access to C-language programs and functions. The INVOKE directive is used to access the procedure in Example 14–16 as INVOKE sss, 20 , 30 to add 20 to 30 and leave the result in AX on the return. INVOKE differs from CALL because it removes the parameters from the stack as required by the calling convention. The syntax for the INVOKE directive requires the name of the procedure followed by any parameters, all separated by a comma as a delimiter.

The PASCAL-Language Calling Convention. The PASCAL-language calling convention is used with the BASIC, PASCAL, and FORTRAN programming languages. It is also used to access functions in the Windows library (WINH.LIB) for use in Windows. The PASCAL-language calling convention differs from the C-language calling convention in that the arguments are pushed onto the stack from the left to the right. The PASCAL-language calling convention also requires that registers SI, DI, DS, and SS are saved if used in the assembly language procedure. The direction flag is cleared upon entry and must also be returned in the cleared state. The INVOKE statement is used to access high-level language procedures from assembly language, just as it is in the C-language calling convention. This is particularly important when using assembly language with Windows. An example is the INVOKE MessageBeep, –1, which produces a beep from the speaker if the WINH.LIB library is loaded with the program. Other parameters can be used with MessageBeep to play any of the default sounds from Windows. In a C-language program, access to the Windows applications interface (API) is obtained by using the #include <WINDOWS.H> directive at the start of the C-language source file. Note that this only applies to Microsoft C/C++ languages.

Figure 14–9 shows the stack frame for function calls from BASIC, FORTRAN, or PASCAL. Note that all functions called from these languages are considered far and BP addresses the stack frame where the parameters are stored. The right-most parameter is addressed by BP + 6, in the opposite order of storage as the C-language calling convention.

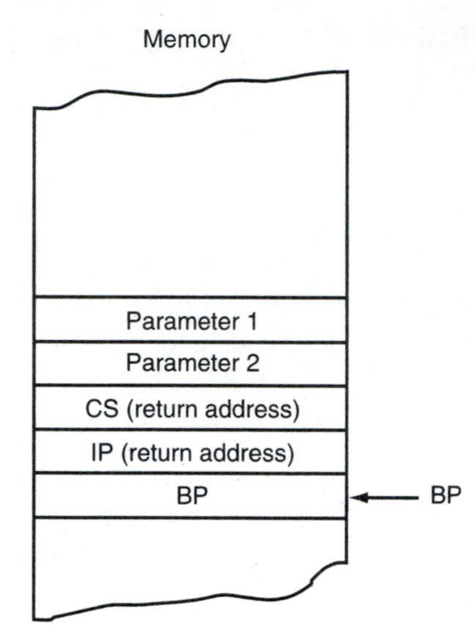

FIGURE 14–9 The stack frame for the PASCAL-language convention.

The hidden instructions push BP to the stack, and then load BP with SP, just as they did in the C-language calling convention.

Data Types

The data types used by MASM are different than those used by high-level languages. Table 14–15 compares the MASM data types with the equivalent data types used by C/C++, BASIC, PASCAL, and FORTRAN.

Language Interface

A procedure for a C-language program is illustrated in Example 14–17. The procedure uses the MODEL statement to specify the size of the memory model (small in this example) and the calling convention language (C). The PROTO statement declares the name of the procedure as external and also specifies the sizes and names of any parameters associated with the procedure. The procedure itself is formed as a near procedure.

EXAMPLE 14–17

```
                                .MODEL SMALL,C

                                X1      PROTO  C, l:SWORD, f:SWORD

0000                            .CODE

0000                            X1      PROC   C  l:SWORD, f:SWORD

0003  9B D9 EB                          FLDPI                  ;get pi
0006  9B D8 C0                          FADD   ST,ST(0)        ;form 2*pi
0009  8B 46 04                          MOV    AX,l            ;get l
000C  2E: A3 0035 R                     MOV    TEMP,AX
0010  9B 2E: DF 06 0035 R               FILD   TEMP            ;get l
0016  9B DE C9                          FMUL                   ;form 2*pi*l
0019  8B 46 06                          MOV    AX,f            ;get f
```

TABLE 14–15 Comparison of data types

MASM Type	*C-Type*	*BASIC Type*	*FORTRAN Type*
BYTE	Unsigned char	—	CHARACTER*1
SBYTE	Char	—	INTEGER*1
WORD	Unsigned short	STRING*1	—
SWORD	Short int	INTEGER (%X)	INTEGER*2
DWORD	Unsigned long	—	—
SDWORD	Long	LONG (&X)	INTEGER*4
REAL4	Float	SINGLE (X!)	REAL*4
REAL8	Double	DOUBLE (X#)	REAL*8
REAL10	Long double	—	—

Note: PASCAL data types are the same as BASIC data types except for case, i.e., INTEGER (BASIC) versus integer (PASCAL).

```
001C  2E: A3 0035 R              MOV    TEMP,AX
0020  9B 2E: DF 06 0035 R        FILD   TEMP          ;get f
0026  9B DE C9                   FMUL                 ;form 2*pi*l*f
0029  9B 2E: DF 1E 0035 R        FISTP  TEMP          ;save result
002F  2E: A1 0035 R              MOV    AX,TEMP       ;get result for return to C
                                 RET

0035                      X      ENDP

0035  0000                TEMP   DW     ?             ;temporary storage
                                 END
```

This example uses the arithmetic coprocessor to calculate the inductive reactance (Xl = 2πfl) of the values *l* and *f* that are passed from C-language to this assembly language procedure. The return value (Xl) is passed to the C-language program in AX as a word-sized integer. The data for *l* and *f* are retrieved from the stack using the [BP+4] addressing mode to get *l* and the [BP+6] address mode to get *f* from the caller's stack. Examine the instruction 8B 46 04 (MOV AX,l) in the example. If the instruction is converted from machine language to symbolic code, it is a MOV AX,[BP+4] instruction that retrieves the value of *l* from the stack.

Example 14–18 shows the C-language calling sequence that is the main module. This sequence prints the value of inductive reactance for an inductance of 4H and a frequency of 1000 Hz. Note that the result is truncated to an integer in this example.

EXAMPLE 14–18

```
#include <stdio.h>

extern int Xl( int l, int f);

void main()
{
  printf("The inductive reactance of 4 H at 1,000 Hz is %d\n", Xl (4, 1000) );
}
```

The same procedure can be rewritten for use with other programming languages. Example 14–19 shows the same procedure rewritten for use with a BASIC language program such as QBASIC. Notice how the MODEL is medium for the QBASIC assembly language module. Also notice how pointers (PTR) are used to obtain the data from the BASIC calling stack. The use of the word PASCAL is found in all assembly language procedures except for C-language.

EXAMPLE 14–19

```
                              .MODEL MEDIUM

                        Xl    PROTO  PASCAL, l:PTR WORD, f:PTR WORD

0000                          .CODE

0000                    Xl    PROC   PASCAL l:PTR WORD, f:PTR WORD

0003  9B D9 EB                FLDPI                    ;get pi
0006  9B D8 C0                FADD   ST,ST(0)          ;form 2*pi
0009  8B 5E 08                MOV    BX,WORD PTR l
000C  8B 07                   MOV    AX,[BX]           ;get l
000E  2E: A3 003B R           MOV    TEMP,AX
0012  9B 2E: DF 06 003B R     FIL    D       TEMP      ;get l
0018  9B DE C9                FMUL                     ;form 2*pi*l
001B  8B 5E 06                MOV    BX,WORD PTR f
001E  8B 07                   MOV    AX,[BX]           ;get f
0020  2E: A3 003B R           MOV    TEMP,AX
0024  9B 2E: DF 06 003B R     FIL    D       TEMP      ;get f
002A  9B DE C9                FMUL                     ;form 2*pi*l*f
002D  9B 2E: DF 1E 003B R     FISTP  TEMP              ;save result
0033  2E: A1 003B R           MOV    AX,TEMP           ;get result for return to C
                              RET

003B                    Xl    ENDP

003B 0000               TEMP  DW     ?                 ;temporary storage

                              END
```

This example stores temporary data in the code segment. The difference in the calling convention is clear if variables *l* and *f* are examined. Here, variable *l* is at memory offset [SP+8] and variable *f* is at memory offset [SP+6]. In the C-language example, variable *l* is at offset [SP+4] and *f* is at offset [SP+6]. Notice how that stack data are reversed by the C-language calling convention as compared to the PASCAL calling convention.

Example 14–20 shows the BASIC language program that calculates the inductive reactance via the assembly language procedure in Example 14–19.

EXAMPLE 14–20

```
DEFINT A-Z

DECLARE FUNCTION Xl (A AS INTEGER, B AS INTEGER)
PRINT "The inductive reactance of 4 H and 1000 Hz is ";
PRINT Xl(4,1000)

END
```

Using Full Segments with C-Language

The prior example of a C-language procedure used the PROTO statement and also noted the C-language calling convention in the PROC statement. The same effect can be accomplished using standard assembly language, as illustrated in Example 14–21. This procedure receives two parameters from the caller, multiplies them, and returns the result in AX. Note how the PUBLIC statement defines the procedure as a public procedure for a C-language program. The EXTERN statement is placed in the C-language program to declare this subroutine as external. The PUBLIC statement is generated by the PROTO statement as explained earlier.

EXAMPLE 14–21

```
0000            CODE    SEGMENT 'code'
                        ASSUME CS:CODE

                        PUBLIC C Subr

0000            Subr    PROC    NEAR

0000 55                 PUSH    BP                  ;save BP
0001 8B EC              MOV     BP,SP               ;address parameter space

0003 8B 46 04           MOV     AX,[BP+4]           ;get leftmost parameter
0006 F7 66 06           MUL     WORD PTR [BP+6]     ;get rightmost parameter

0009 5D                 POP     BP                  ;cleanup stack
000A C3                 RET

000B            Subr    ENDP

000B            CODE    ENDS
                        END
```

14–6 SUMMARY

1. The DOS EXEC (execute) function (4BH) allows a program to be loaded into memory and executed, or allows an overlay file to be loaded for access. The program that loads another program or overlay into memory is called the parent. The program loaded into memory by the parent is called the child. The process of creating a child program is often called spawning.
2. The child name associated with the EXEC function is an ASCII-Z string that contains the name of a program or an overlay file. It must also include any applicable file extension. The parameter block contains the environment segment address and the address of the command line if needed for the child. If the function loads an overlay, the first word of the parameter block contains the segment address for storing the overlay file.
3. The EMS (expanded memory system) memory manager uses a 64K-byte page frame located in the systems area at any 4K byte boundary. This is usually located at

memory D0000H–DFFFFH. The page frame is divided into four 16K-byte expanded memory pages. Most systems use EMM386.EXE for the EMS memory manager.

4. The presence of EMM386.EXE is determined by opening a file with the name EMSXXXX0 using DOS INT 21H function 3DH. This opens the driver, instead of a file. Once the driver is opened, it is tested with IOCTL function 07H to determine whether it is available for operation. The EMS system is accessed through the INT 67H instruction.
5. The XMS (extended memory system) extended memory manager is installed by the HIMEN.SYS device driver in most systems from the CONFIG.SYS file. The HIMEM.SYS driver provides access to high memory at locations 100000H–10FFFEH and to any extended memory above location 110000H.
6. The presence of XMS is determined with DOS multiplex interrupt INT 2FH function AX = 4300H. Once XMS is detected, INT 2FH function AX = 4310H obtains the address of the XMS driver. To access the XMS driver for control of extended memory, a far CALL to the XMS driver address is made to access the XMS functions.
7. The VCPI (virtual control program interface) is a part of EMM386.EXE that allows expanded and extended memory to be accessed. The VCPI functions are obtained by executing the INT 67H instruction to allocate and use expanded and extended memory. Note that this interface dynamically allocates the memory.
8. The DPMI (DOS protected-mode interface) provides access to extended memory and protected mode if Windows is present in the system. The DPMI server is called the host and any applications that use it are called its clients. Access to DPMI is provided only if the DOS application is run from Windows or the shell to DOS part of Windows.
9. The IOCTL (I/O control) functions provide control of the device driver, which in turn controls the I/O device. The IOCTL functions are not normally used for data transfer. Devices controlled by IOCTL are character devices, which process one byte at a time, and block devices, which process a block of bytes.
10. The device driver is a special program that controls installable devices. The device can be either a hardware component or a software function. The device driver is a .COM file that contains three parts: header, strategy, and interrupt. The device driver header provides the system with a chain address, attribute word, address of the strategy and interrupt procedures, and name of the driver. The strategy procedure stores the address of the request header, which is used to transfer commands and data to the interrupt procedure. Finally, the interrupt procedure decodes the commands that are sent from DOS and controls and installs the device.
11. Protected-mode operation allows access to all memory locations in the system: real, upper, high, expanded, and extended. The DPMI services are used to switch between protected and real mode, to allocate memory, to specify descriptors and selectors, and to access real-mode procedures and interrupts from protected mode and vice versa.
12. When the switch to protected mode is made with DPMI, the DS, SS, and CS registers contain selectors that access segments with limits of 64K bytes each. The FS and GS segments are cleared to 0000H, and ES is loaded with a selector that addresses a segment that points to the PSP, with a limit of 256 bytes.

13. Assembly-language procedures used with high-level languages use either the C-language calling convention (for C-language programs) or the PASCAL-language calling convention (for BASIC, PASCAL, or FORTRAN).
14. The PROTO statement defines the language type and variable as external for a procedure called from a high-level language.

14–7 QUESTIONS AND PROBLEMS

1. Define the terms *parent, child,* and *spawn* as they apply to the EXEC function provided by DOS INT 21H function 4BH.
2. What is the *child process name* and how is it stored and accessed?
3. What is the purpose of the parameter block associated with the EXEC function?
4. Describe the information found in the parameter block for the EXEC function as it applies to a child process.
5. Write an assembly-language program that executes the ATTRIB *.* program. Your program must display the attributes of all files in the current directory when executed.
6. List the four memory managers available to DOS.
7. Write an assembly language program that uses the EXEC function to format the $3^1/_2''$ disk in drive A as either a 1.44M or 720K byte disk. Your program must display a menu that allows the user to select either format.
8. The EMS expanded memory manager is provided by what DOS device driver program?
9. Briefly describe expanded memory, highlighting its location in the memory system, the size of the page frame, and the size of the page.
10. Which INT instruction accesses the EMS device driver?
11. Explain how the presence of EMM386.EXE is detected.
12. The XMS extended memory manager is provided by what DOS device driver?
13. What DOS device driver provides access to memory locations 100000H–10FFFEH from the real mode?
14. Explain how to obtain the entry point (address of) the XMS device driver program.
15. What program installs the VCPI device driver?
16. Which area(s) of memory is(are) accessed by the VCPI?
17. Which DOS INT instruction provides access to VCPI?
18. What is DPMI and which program provides it for DOS?
19. *Host* is another name for what?
20. *Client* is another name for what?
21. What are the IOCTL functions and how are they accessed?
22. The character device processes binary or ASCII characters a ______________ at a time.
23. Define the term *block device.*
24. What is the purpose of each of the following character devices?
    (a) CON
    (b) LPT1

   (c) COM2
   (d) PRN

25. What is a device driver?
26. List the three parts of a device driver.
27. The device header contains what information for the device driver?
28. The strategy procedure provides what information to the interrupt procedure in a device driver?
29. The request header provides a location for the return status word. What is the return status word and what information does it contain?
30. The device driver command 00H (initialize driver) must accomplish what tasks to initialize the driver?
31. What is the BPB as it applies to a device driver?
32. Explain the difference between the read command and the nondestructive read command in relation to device drivers.
33. The device driver interrupt procedure is responsible for what device driver task?
34. Describe how to determine whether DPMI is installed.
35. Write a device driver called DISP that displays the contents of any file on the video screen in ASCII code. Your driver must display one screen at a time. It must also convert any control codes (00H-1FH and 80H–FFH) into hexadecimal format and display them surrounded by parentheses. For example, control code 7 should display as (07H).
36. Explain how the switch to protected mode is made with the DPMI.
37. What INT instruction accesses the DPMI services?
38. Describe how DPMI functions with DOS INT 21H function 4CH.
39. Write a program that uses the DPMI services to search through the entire memory system for any word or phrase typed on the command line. The output must display the linear address of each occurrence.
40. Develop a procedure and C-language calling program that displays the value of the variable "Number" passed to the procedure from C. The assembly language procedure will display the unsigned word-sized integer at the current cursor position when function Display() is called.

APPENDIX A

The Assembler, Disk Operating System, Basic I/O System, Mouse, and Memory Managers

This appendix explains how the assembler is used and also how the DOS (disk operating system), BIOS (basic I/O system) function calls, and mouse function calls are used by assembly language to control the personal computer. The function calls control everything from reading and writing disk data, to managing the keyboard and displays, to controlling the mouse. The assembler represented in this text is the Microsoft ML (version 6.X) and MASM (version 5.10) macro assembler programs. It is fairly important that version 6.X be used, instead of the dated 5.10 version. Also presented are the memory managers: EMS, XMS, VCPI, and DPMI and their function calls.

ASSEMBLER USAGE

The assembler program requires that a symbolic program first be written using a word processor, text editor, or the workbench program provided with the assembler package. The editor provided with version 5.10, M.EXE, is strictly a full-screen editor. The editor provided with version 6.X, PWB.EXE, is a fully integrated development system that contains extensive help. Refer to the documentation that accompanies your assembler package for details on the operation of the editor program. If at all possible, use version 6.X, because it contains a detailed help file that guides the user through assembly language statements, directives, and even the DOS and BIOS interrupt function calls.

If you are using a word processor to develop your software, make sure it is initialized to generate a pure ASCII file. The source file you generate must use the extension .ASM, which is required for the assembler to properly identify your source program.

Once your source file is prepared, it must be assembled. If you are using the workbench provided with version 6.X, this is accomplished by selecting the compile feature with your mouse. If you are using a word processor and DOS command lines with version 5.10, see Example A–1 for the dialog for version 5.10 to assemble a file called FROG.ASM. This example shows the portions typed by the user in italics.

EXAMPLE A–1

```
A>MASM

Microsoft (R) Macro Assembler Version 5.10
Copyright (C) Microsoft Corp 1981, 1989. All rights reserved.

Source filename [.ASM]:FROG
Object filename [FROG.OBJ]:FROG
List filename [NUL.LST]:FROG
Cross reference [NUL.CRF]:FROG
```

Once a program is assembled, it must be linked before it can be executed. The linker converts the object file into an executable file (.EXE). Example A–2 shows the dialog required for the linker using an MASM version 5.10 object file. If the ML version 6.X assembler is in use, it automatically assembles and links a program using the COMPILE or BUILD command from workbench. After compiling with ML, workbench allows the program to be debugged with a debugging tool called Codeview. Codeview is also available with MASM, but CV must be typed at the DOS command line to access it.

EXAMPLE A–2

```
A:\>LINK

Microsoft (R) Overlay Linker Version 3.64
Copyright (C) Microsoft Corp 1983-1988. All rights reserved.

Object modules [.OBJ]:FROG
Run file [FROG.EXE]:FROG
List file [NUL.MAP]:FROG
Libraries [.LIB]:SUBR
```

If MASM version 6.X is in use, the comand line syntax differs a little. Example A–3 shows the command line syntax for ML, the assembler and linker for MASM version 6.X.

EXAMPLE A–3

```
C:\>ML /FITEST.LST TEST.ASM

Microsoft (R) Macro Assembler Version 6.11
Copyright (C) Microsoft Corp 1981-1993. All rights reserved.

 Assembling: TEST.ASM

Microsoft (R) Segmented-Executable Linker Version 5.13
Copyright (C) Microsoft Corp 1984-1993. All rights reserved.

Object Modules [.OBJ]: TEST.obj/t
Run File [TEST.com]: "TEST.com"
List File [NUL.MAP]: NUL
Libraries [.LIB]:
Definitions File [NUL.DEF]: ;
```

Version 6.X of the Microsoft MASM program contains the programmer's workbench program. Programmer's workbench allows an assembly language program to be developed with its full-screen editor and tool bar. Figure A–1 illustrates the display found

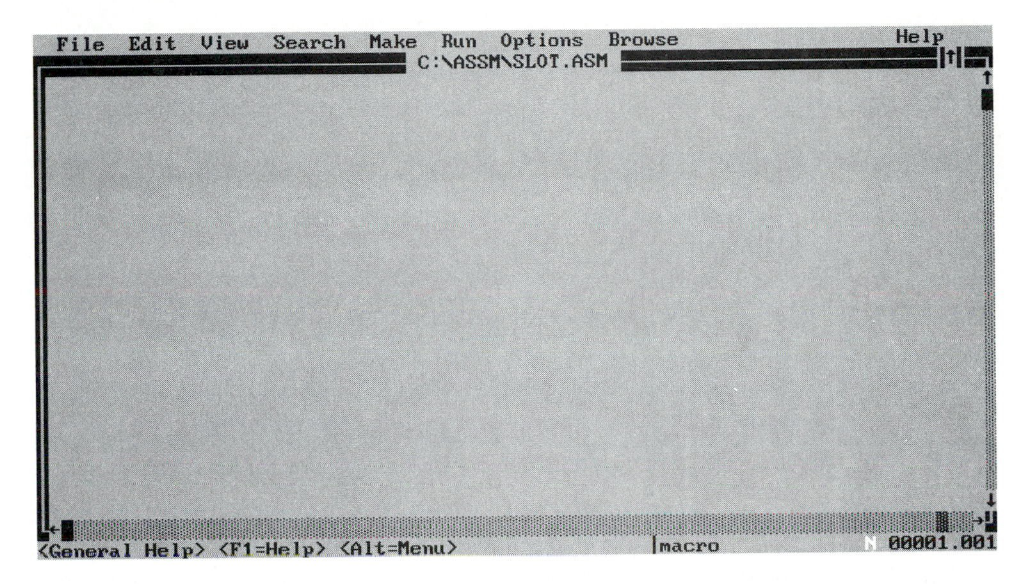

FIGURE A–1 The edit screen of Programmer's WorkBench Version 2.0, as provided with MASM 6.11

with programmer's workbench. To access this program, type PWB at the DOS prompt. The make option allows a program to be automatically assembled and linked, making these tasks simple in comparison to version 5.10 of the assembler.

ASSEMBLER MEMORY MODELS

Memory models and the .MODEL statement are introduced in Chapter 5 and used exclusively throughout the text. Here we completely define the memory models available for software development. Each model defines the way a program is stored in the memory system. Table A–1 describes the different models available with MASM and ML.

The tiny model is used to create a .COM file instead of an execute file. The .COM file is different because all data and codes fit into one code segment. A .COM file must use an origin of offset address 0100H as the start of the program. A .COM file loads from the disk and executes faster than the normal execute (.EXE) file. For most applications, use the execute file (.EXE) and the small memory model.

When models are used to create a program, certain defaults apply, as illustrated in Table A–2. In this table, the *directive* is used to start a particular type of segment for the models listed in the table. If the .CODE directive is placed in a program, it indicates the beginning of the code segment. Likewise, .DATA indicates the start of a data segment. The *name* indicates the name of the segment. *Align* indicates whether the segment is aligned on a word, doubleword, or 16-byte paragraph. *Combine* indicates the type of segment created. *Class* indicates the class of the segment, such as 'CODE' or 'DATA'. *Group* indicates the group type of the segment.

TABLE A–1 Memory models for the assembler

Type	*Description*
Tiny	All data and code fit into one segment, the code segment. Tiny model programs are written in the .COM file format, which means that the program must be originated at memory location 0100H. This model is most often used with small programs.
Small	All data fit into a single 64K-byte data segment and all codes fit into another single 64K-byte code segment. This allows all codes to be accessed with near jumps and calls.
Medium	All data fit into a single 64K-byte data segment, and codes fit into more than one code segment. This allows codes to exist in multiple segments.
Compact	All codes fit into a single 64K-byte code segment, and data fit into more than one data segment.
Large	Both code and data fit into multiple code and data segments.
Huge	Same as large, but allows data segments that are larger than 64K bytes.
Flat	Not available in MASM version 5.10. The flat memory model uses one segment, with a maximum length of 512M bytes, to store data and codes.

Example A-4 shows a program that uses the small model. The small model is used for programs that contain one DATA and one CODE segment, which is common. Notice that not only is the program listed, but so is all the information generated by the assembler. Here, the .DATA directive and .CODE directive indicate the start of each segment. Notice how the DS register is loaded in this program. As presented throughout the text, the .STARTUP directive can be used to load the data segment register, set up the stack, and define the starting address of a program. In this example, an alternate method is illustrated for loading the data segment register and defining the starting address of the program (END BEGIN).

EXAMPLE A-4

```
Microsoft (R) Macro Assembler Version 6.11

                          .MODEL SMALL
                          .STACK 100H
0000                      .DATA

0000 0A                   FROG      DB        10
0001 0064 [               DATA1     DB        100 DUP (2)
              02
                ]

0000                      .CODE

0000 B8 ---- R            BEGIN:    MOV       AX,DGROUP          ;set up DS
```

TABLE A–2 Defaults for the .MODEL directive

Model	*Directive*	*Name*	*Align*	*Combine*	*Class*	*Group*
Tiny	.CODE	_TEXT	Word	PUBLIC	'CODE'	DGROUP
	.FARDATA	FAR_DATA	Para	Private	'FAR_DATA'	
	.FARDATA?	FAR_BSS	Para	Private	'FAR_BSS'	
	.DATA	_DATA	Word	PUBLIC	'DATA'	DGROUP
	.CONST	CONST	Word	PUBLIC	'CONST'	DGROUP
	.DATA?	_BSS	Word	PUBLIC	'BSS'	DGROUP
Small	.CODE	_TEXT	Word	PUBLIC	'CODE'	
	.FARDATA	FAR_DATA	Para	Private	'FAR_DATA'	
	.FARDATA?	FAR_BSS	Para	Private	'FAR_BSS'	
	.DATA	_DATA	Word	PUBLIC	'DATA'	DGROUP
	.CONST	CONST	Word	PUBLIC	'CONST'	DGROUP
	.DATA?	_BSS	Word	PUBLIC	'BSS'	DGROUP
	.STACK	STACK	Para	STACK	'STACK'	DGROUP
Medium	.CODE	name_TEXT	Word	PUBLIC	'CODE'	
	.FARDATA	FAR_DATA	Para	Private	'FAR_DATA'	
	.FARDATA?	FAR_BSS	Para	Private	'FAR_BSS'	
	.DATA	_DATA	Word	PUBLIC	'DATA'	DGROUP
	.CONST	CONST	Word	PUBLIC	'CONST'	DGROUP
	.DATA?	_BSS	Word	PUBLIC	'BSS'	DGROUP
	.STACK	STACK	Para	STACK	'STACK'	DGROUP
Compact	.CODE	_TEXT	Word	PUBLIC	'CODE'	
	.FARDATA	FAR_DATA	Para	Private	'FAR_DATA'	
	.FARDATA?	FAR_BSS	Para	Private	'FAR_BSS'	
	.DATA	_DATA	Word	PUBLIC	'DATA'	DGROUP
	.CONST	CONST	Word	PUBLIC	'CONST'	DGROUP
	.DATA?	_BSS	Word	PUBLIC	'BSS'	DGROUP
	.STACK	STACK	Para	STACK	'STACK'	DGROUP
Large or huge	.CODE	name_TEXT	Word	PUBLIC	'CODE'	
	.FARDATA	FAR_DATA	Para	Private	'FAR_DATA'	
	.FARDATA?	FAR_BSS	Para	Private	'FAR_BSS'	
	.DATA	_DATA	Word	PUBLIC	'DATA'	DGROUP
	.CONST	CONST	Word	PUBLIC	'CONST'	DGROUP
	.DATA?	_BSS	Word	PUBLIC	'BSS'	DGROUP
	.STACK	STACK	Para	STACK	'STACK'	DGROUP
Flat	.CODE	_TEXT	Dword	PUBLIC	'CODE'	
	.FARDATA	_DATA	Dword	PUBLIC	'DATA'	
	.FARDATA?	_BSS	Dword	PUBLIC	'BSS'	
	.DATA	_DATA	Dword	PUBLIC	'DATA'	
	.CONST	CONST	Dword	PUBLIC	'CONST'	
	.DATA?	_BSS	Dword	PUBLIC	'BSS'	
	.STACK	STACK	Dword	PUBLIC	'STACK'	

```
0003 8E D8                           MOV       DS,AX
                                     .
                                     .
                                     .
                                     END       BEGIN

Segments and Groups:

        N a m e          Size        Length     Align     Combine    Class

DGROUP..........         GROUP
_DATA...........         16 Bit      0065       Word      Public     'DATA'
STACK...........         16 Bit      0100       Para      Stack      'STACK'
_TEXT...........         16 Bit      0005       Word      Public     'CODE'


Symbols:

        N a m e          Type        Value      Attr

@CodeSize.......         Number      0000h
@DataSize.......         Number      0000h
@Interface......         Number      0000h
@Model..........         Number      0002h
@code...........         Text                   _TEXT
@data...........         Text                   DGROUP
@fardata?.......         Text                   FAR_BSS
@fardata........         Text                   FAR_DATA
@stack..........         Text                   DGROUP
BEGIN...........         L Near      0000       _TEXT
DATA1...........         Byte        0001       _DATA
FROG............         Byte        0000       _DATA

                0 Warnings
                0 Errors
```

Example A–5 lists a program that uses the large model. Notice how it differs from the small model program in Example A–4. Models can be very useful in developing software, but often we use full-segment descriptions, as depicted in most examples in the text.

EXAMPLE A–5

```
Microsoft (R) Macro Assembler Version 6.11

                               .MODEL LARGE
                               .STACK 1000H
0000                           .FARDATA?

0000  00                       FROG      DB        ?
0001  0064 [                   DATA1     DW        100 DUP (?)
            0000
                ]

0000                           .CONST

0000 54 68 69 73 20 69         MES1      DB        'This is a character string'
     73 20 61 20 63 68
     61 72 61 63 74 65
```

```
     72 20 73 74 72 69
     6E 67
001A 53 6F 20 69 73 20     MES2      DB        'So is this!'
     74 68 69 73 21

0000                       .DATA

0000 000C                  DATA2     DW        12
0002 00C8 [                DATA3     DB        200 DUP (1)
            01
               ]

0000                       .CODE

0000                       FUNC      PROC      FAR
                                     .
                                     .
                                     .
                                     .
0000 CB                              RET

0001                       FUNC      ENDP

                                     END       FUNC

Segments and Groups:

       N a m e             Size      Length    Align    Combine    Class

DGROUP.................    GROUP
_DATA..................    16 Bit    00CA      Word     Public     'DATA'
STACK..................    16 Bit    1000      Para     Stack      'STACK'
CONST..................    16 Bit    0025      Word     Public     'CONST' ReadOnly
EXA_TEXT...............    16 Bit    0001      Word     Public     'CODE'
FAR_BSS................    16 Bit    00C9      Para     Private    'FAR_BSS'
_TEXT..................    16 Bit    0000      Word     Public     'CODE'

Procedures, parameters and locals:

       N a m e             Type      Value     Attr

FUNC...................    P Far     0000      EXA_TEXT            Length= 0001 Public


Symbols:

       N a m e             Type      Value     Attr

@CodeSize..............    Number    0001h
@DataSize..............    Number    0001h
@Interface.............    Number    0000h
@Model.................    Number    0005h
@code..................    Text                EXA_TEXT
@data..................    Text                DGROUP
@fardata?..............    Text                FAR_BSS
@fardata...............    Text                FAR_DATA
@stack.................    Text                DGROUP
DATA1..................    Word      0001      FAR_BSS
DATA2..................    Word      0000      _DATA
DATA3..................    Byte      0002      _DATA
```

```
FROG..................  Byte   0000   FAR_BSS
MES1..................  Byte   0000   CONST
MES2..................  Byte   001A   CONST

            0 Warnings
            0 Errors
```

DOS FUNCTION CALLS

In order to use DOS function calls, always place the function number into register AH and load all other pertinent information into registers as described in Table A–3. Once this is accomplished, follow with an INT 21H to execute the DOS function. Example A–6 shows how to display an ASCII A on the CRT screen at the current cursor position with a DOS function call.

EXAMPLE A–6

```
0000 B4 06                  MOV   AH,6      ;load function 06H
0002 B2 41                  MOV   DL,'A'    ;select letter 'A'
0004 CD 21                  INT   21H       ;call DOS function
```

Table A–3 is a complete listing of the DOS function calls. Some function calls require a segment and offset address, for example, DS:DI. This means the data segment is the segment address and DI is the offset address. All of the function calls use INT 21H and AH contains the function call number. Note that functions marked with an @ should not be used unless DOS version 2.XX is in use. Also note that not all function numbers are implemented. As a rule, DOS function calls save all registers not used as exit data, but in certain cases, some registers may change. In order to prevent problems, it is advisable to save registers where problems occur. (Note: Figures referenced in Table A–3 appear on pp. 588–590.)

BIOS FUNCTION CALLS

In addition to DOS function call INT 21H, some BIOS function calls prove useful in controlling the I/O environment of the computer. Unlike INT 21H, which exists in the DOS program, the BIOS function calls are found stored in the system and video BIOS ROMs. These BIOS ROM functions directly control the I/O devices, with or without DOS loaded into a system.

INT 10H

The INT 10H BIOS interrupt is often called the video services interrupt, because it directly controls the video display in a system. The INT 10H instruction uses register AH to select the video service provided by this interrupt. The video BIOS ROM is located on the video board and varies from one video card to another.

TABLE A–3 DOS function calls (pp. 559–588)

00H	TERMINATE A PROGRAM
Entry	AH = 00H CS = program segment prefix address
Exit	DOS is entered
01H	READ THE KEYBOARD
Entry	AH = 01H
Exit	AL = ASCII character
Notes	If AL = 00H, the function call must be invoked again to read an extended ASCII character. Refer to Table 9–1 for a listing of the extended ASCII keyboard codes. This function call automatically echoes whatever is typed to the video screen.
02H	WRITE TO STANDARD OUTPUT DEVICE
Entry	AH = 02H DL = ASCII character to be displayed
Notes	This function call normally displays data on the video display.
03H	READ CHARACTER FROM COM1
Entry	AH = 03H
Exit	AL = ASCII character read from the communications port
Notes	This function call reads data from the serial communications port.

04H	WRITE TO COM1
Entry	AH = 04H DL = character to be sent out of COM1
Notes	This function transmits data through the serial communications port. The COM port assignments can be changed to use other COM ports with functions 03H and 04H by using the DOS MODE command to reassign COM1 to another COM port.

05H	WRITE TO LPT1
Entry	AH = 05H DL = ASCII character to be printed
Notes	Prints DL on the line printer attached to LPT1. Note that the line printer port can be changed with the DOS MODE command.

06H	DIRECT CONSOLE READ/WRITE
Entry	AH = 06H DL = 0FFH or DL = ASCII character
Exit	AL = ASCII character
Notes	If DL = 0FFH on entry, then this function reads the console. If DL = ASCII character, then this function displays the ASCII character on the console video screen. If a character is read from the console keyboard, the zero flag (ZF) indicates whether a character was typed. A zero condition indicates no key is typed, and a not-zero condition indicates that AL contains the ASCII code of the key or a 00H. If AL = 00H, the function must again be invoked to read an extended ASCII character from the keyboard. Note that the key does not echo to the video screen.

07H	DIRECT CONSOLE INPUT WITHOUT ECHO
Entry	AH = 07H
Exit	AL = ASCII character
Notes	This functions exactly like function number 06H with DL = 0FFH, but it will not return from the function until the key is typed.

08H	READ STANDARD INPUT WITHOUT ECHO
Entry	AH = 08H
Exit	AL = ASCII character
Notes	Performs like function 07H, except it reads the standard input device. The standard input device can be assigned as either the keyboard or the COM port. This function also responds to a control-break, where function 06H and 07H do not. A control-break causes INT 23H to execute. By default this function as does function 07H.

09H	DISPLAY A CHARACTER STRING
Entry	AH = 09H DS:DX = address of the character string
Notes	The character string must end with an ASCII $ (24H). The character string can be of any length and may contain control characters, such as carriage return (0DH) and line feed (0AH).

0AH	BUFFERED KEYBOARD INPUT
Entry	AH = 0AH DS:DX = address of keyboard input buffer
Notes	The first byte of the buffer contains the size of the buffer (up to 255). The second byte is filled with the number of characters typed upon return. The third byte, through the end of the buffer, contains the character string typed, followed by a carriage return (0DH). This function continues to read the keyboard (displaying data as typed) until either the specified number of characters are typed or a carriage return (enter) key is typed.

0BH	TEST STATUS OF THE STANDARD INPUT DEVICE
Entry	AH = 0BH
Exit	AL = status of the input device
Notes	This function tests the standard input device to determine if data are available. If AL = 00, no data are available. If AL = 0FFH, then data are available that must be input using function number 08H.

0CH	CLEAR KEYBOARD BUFFER AND INVOKE KEYBOARD FUNCTION
Entry	AH = 0CH AL = 01H, 06H, 07H, or 0AH
Exit	See exit for functions 01H, 06H, 07H, or 0AH.
Notes	The keyboard buffer holds keystrokes while programs execute other tasks. This function empties or clears the buffer and then invokes the keyboard function located in register AL.

0DH	FLUSH DISK BUFFERS
Entry	AH = 0DH
Notes	Erases all file names stored in disk buffers. This function does not close the files specified by the disk buffers, so care must be exercised in its use.

0EH	SELECT DEFAULT DISK DRIVE
Entry	AH = 0EH DL = desired default disk drive number
Exit	AL = the total number of drives present in the system
Notes	Drive A = 00H, drive B = 01H, drive C = 02H, and so forth.

0FH	@OPEN FILE WITH FCB
Entry	AH = 0FH DS:DX = address of the unopened file control block (FCB)
Exit	AL = 00H if file found AL = 0FFH if file not found
Notes	The file control block (FCB) is only used with early DOS software and should never be used with new programs. File control blocks do not allow path names, as do the newer file access function codes presented later. Figure A–2 illustrates the structure of the FCB. To open a file, the file must either be present on the disk or be created with function call 16H.

10H	@CLOSE FILE WITH FCB
Entry	AH = 10H DS:DX = address of the opened file control block (FCB)
Exit	AL = 00H if file closed AL = 0FFH if error found
Notes	Errors that occur usually indicate either that the disk is full or the media is bad.

11H	@SEARCH FOR FIRST MATCH (FCB)
Entry	AH = 11H DS:DX = address of the file control block to be searched
Exit	AL = 00H if file found AL = 0FFH if file not found
Notes	Wild card characters (? or *) may be used to search for a file name. The ? wild card character matches any character, and the * matches any name or extension.

12H	@SEARCH FOR NEXT MATCH (FCB)
Entry	AH = 12H DS:DX = address of the file control block to be searched
Exit	AL = 00H if file found AL = 0FFH if file not found
Notes	This function is used after function 11H finds the first matching file name.

13H	@DELETE FILE USING FCB
Entry	AH = 13H DS:DX = address of the file control block to be deleted
Exit	AL = 00H if file deleted AL = 0FFH if error occurred
Notes	Errors that most often occur are defective media errors.

14H	@SEQUENTIAL READ (FCB)
Entry	AH = 14H DS:DX = address of the file control block to be read
Exit	AL = 00H if read successful AL = 01H if end of file reached AL = 02H if DTA had a segment wrap AL = 03H if less than 128 bytes were read

15H	@SEQUENTIAL WRITE (FCB)
Entry	AH = 15H DS:DX = address of the file control block to be written
Exit	AL = 00H if write successful AL = 01H if disk is full AL = 02H if DTA had a segment wrap

16H	@CREATE A FILE (FCB)
Entry	AH = 16H DS:DX = address of an unopened file control block
Exit	AL = 00H if file created AL = 01H if disk is full

17H	@RENAME A FILE (FCB)
Entry	AH = 17H DS:DX = address of a modified file control block
Exit	AL = 00H if file renamed AL = 01H if error occurred
Notes	Refer to Figure A–3 for the modified FCB used to rename a file.

19H	RETURN CURRENT DRIVE
Entry	AH = 19H
Exit	AL = current drive
Notes	AL = 00H for drive A, 01H for drive B, and so forth.

1AH	SET DISK TRANSFER AREA
Entry	AH = 1AH DS: DX = address of new DTA
Notes	The disk transfer area is normally located within the program segment prefix at offset address 80H. The DTA is used by DOS for all disk data transfers using file control blocks.

1BH	GET DEFAULT DRIVE FILE ALLOCATION TABLE (FAT)
Entry	AH = 1BH
Exit	AL = number of sectors per cluster DS:BX = address of the media descriptor CX = size of a sector in bytes DX = number of clusters on drive
Notes	Refer to Figure A–4 for the format of the media descriptor byte. The DS register is changed by this function, so make sure to save it before using this function.

1CH	GET ANY DRIVE FILE ALLOCATION TABLE (FAT)
Entry	AH = 1CH DL = disk drive number
Exit	AL = number of sectors per cluster DS:BX = address of the media descriptor CX = size of a sector in bytes DX = number of clusters on drive

21H	@RANDOM READ USING FCB
Entry	AH = 21H DS:DX = address of opened FCB
Exit	AL = 00H if read successful AL = 01H if end of file reached AL = 02H if the segment wrapped AL = 03H if less than 128 bytes read

22H	@RANDOM WRITE USING FCB
Entry	AH = 22H DS:DX = address of opened FCB
Exit	AL = 00H if write successful AL = 01H if disk full AL = 02H if the segment wrapped

23H	@RETURN NUMBER OF RECORDS (FCB)
Entry	AH = 23H DS:DX = address of FCB
Exit	AL = 00H number of records AL = 0FFH if file not found

24H	@SET RELATIVE RECORD SIZE (FCB)
Entry	AH = 24H DS:DX = address of FCB
Notes	Sets the record field to the value contained in the FCB.

25H	SET INTERRUPT VECTOR
Entry	AH = 25H AL = interrupt vector number DS:DX = address of new interrupt procedure
Notes	Before changing the interrupt vector, it is suggested that the current interrupt vector is saved using DOS function 35H. This allows a back-link, so the original vector can later be restored.

26H	CREATE NEW PROGRAM SEGMENT PREFIX
Entry	AH = 26H DX = segment address of new PSP
Notes	Figure A–5 illustrates the structure of the program segment prefix.

27H	@RANDOM FILE BLOCK READ (FCB)
Entry	AH = 27H CX = the number of records DS:DX = address of opened FCB
Exit	AL = 00H if read successful AL = 01H if end of file reached AL = 02H if the segment wrapped AL = 03H if less than 128 bytes read CX = the number of records read

28H	@RANDOM FILE BLOCK WRITE (FCB)
Entry	AH = 28H CX = the number of records DS:DX = address of opened FCB
Exit	AL = 00H if write successful AL = 01H if disk full AL = 02H if the segment wrapped CX = the number of records written

29H	@PARSE COMMAND LINE (FCB)
Entry	AH = 29H AL = parse mask DS:SI = address of FCB DS:DI = address of command line
Exit	AL = 00H if no file name characters found AL = 01H if file name characters found AL = 0FFH if drive specifier incorrect DS:SI = address of character after name DS:DI = address first byte of FCB

2AH	READ SYSTEM DATE
Entry	AH = 2AH
Exit	AL = day of the week CX = the year (1980–2099) DH = the month DL = day of the month
Notes	The day of the week is encoded as Sunday = 00H through Saturday = 06H. The year is a binary number equal to 1980 through 2099.

2BH	SET SYSTEM DATE
Entry	AH = 2BH CX = the year (1980–2099) DH = the month DL = day of the month

2CH	READ SYSTEM TIME
Entry	AH = 2CH
Exit	CH = hours (0–23) CL = minutes DH = seconds DL = hundredths of seconds
Notes	All times are returned in binary form.
2DH	**SET SYSTEM TIME**
Entry	AH = 2DH CH = hours CL = minutes DH = seconds DL = hundredths of seconds
2EH	**DISK VERIFY WRITE**
Entry	AH = 2EH AL = 00H to disable verify on write AL = 01H to enable verify on write
Notes	By default disk verify is disabled.
2FH	**READ DISK TRANSFER AREA**
Entry	AH = 2FH
Exit	ES:BX = contains DTA address
30H	**READ DOS VERSION NUMBER**
Entry	AH = 30H
Exit	AH = fractional version number AL = whole number version number
Notes	For example, DOS version number 3.2 is returned as a 3 in AL and a 14H in AH.

31H	TERMINATE AND STAY RESIDENT (TSR)
Entry	AH = 31H AL = the DOS return code DX = number of paragraphs to reserve
Notes	A paragraph is 16 bytes, and the DOS return code is read at the batch file level with ERRORCODE.
33H	TEST CONTROL-BREAK
Entry	AH = 33H AL = 00H to request current control-break AL = 01H to change control-break DL = 00H to disable control-break DL = 01H to enable control-break
Exit	DL = current control-break state
34H	GET ADDRESS OF InDOS FLAG
Entry	AH = 34H
Exit	ES:BX = address of InDOS flag
Notes	The InDOS flag is available in DOS versions 3.2 or newer and indicates DOS activity. If InDOS = 00H, DOS is inactive or 0FFH if DOS is active.
35H	READ INTERRUPT VECTOR
Entry	AH = 35H AL = interrupt vector number
Exit	ES:BX = address stored at vector
Notes	This DOS function is used with function 25H to install/remove interrupt handlers.

36H	DETERMINE FREE DISK SPACE
Entry	AH = 36H DL = drive number
Exit	AX = FFFFH if drive invalid AX = number of sectors per cluster BX = number of free clusters CX = bytes per sector DX = number of clusters on drive
Notes	The default disk drive is DL = 00H, drive A = 01H, drive B = 02H, and so forth.

38H	RETURN COUNTRY CODE
Entry	AH = 38H AL = 00H for current country code BX = 16-bit country code DS:DX = data buffer address
Exit	AX = error code if carry set BX = counter code DS:DX = data buffer address

39H	CREATE SUBDIRECTORY
Entry	AH = 39H DS:DX = address of ASCII-Z string subdirectory name
Exit	AX = error code if carry set
Notes	The ASCII-Z string is the name of the subdirectory in ASCII code ended with a 00H instead of a carriage return/line feed.

3AH	ERASE SUBDIRECTORY
Entry	AH = 3AH DS:DX = address of ASCII-Z string subdirectory name
Exit	AX = error code if carry set

3BH	CHANGE SUBDIRECTORY
Entry	AH = 3BH DS:DX = address of new ASCII-Z string subdirectory name
Exit	AX = error code if carry set

3CH	CREATE A NEW FILE
Entry	AH = 3CH CX = attribute word DS:DX = address of ASCII-Z string file name
Exit	AX = error code if carry set AX = file handle if carry cleared
Notes	The attribute word can contain any of the following (added together): 01H = read-only access, 02H = hidden file or directory,04H = system file, 08H = volume label, 10H = subdirectory, and 20H = archive bit. In most cases, a file is created with 0000H.

3DH	OPEN A FILE
Entry	AH = 3DH AL = access code DS:DX = address of ASCII-Z string file name
Exit	AX = error code if carry set AX = file handle if carry cleared
Notes	The access code is AL = 00H for a read-only access, AL = 01H for a write-only access, and AL = 02H for a read/write access. For shared files in a network environment, bit 4 of AL = 1 will deny read/write access, bit 5 of AL = 1 will deny a write access, bits 4 and 5 of AL = 1 will deny read access, bit 6 of AL = 1 denies none, bit 7 of AL = 0 causes the file to be inherited by child, and bit 7 of AL = 1 file is restricted to current process.

3EH	CLOSE A FILE
Entry	AH = 3EH BX = file handle
Exit	AX = error code if carry set

3FH	READ A FILE
Entry	AH = 3FH BX = file handle CX = number of bytes to be read DS:DX = address of file buffer to hold data read
Exit	AX = error code if carry set AX = number of bytes read if carry cleared

40H	WRITE A FILE
Entry	AH = 40H BX = file handle CX = number of bytes to write DS:DX = address of file buffer that holds write data
Exit	AX = error code if carry set AX = number of bytes written if carry cleared

41H	DELETE A FILE
Entry	AH = 41H DS:DX = address of ASCII-Z string file name
Exit	AX = error code if carry set

42H	MOVE FILE POINTER
Entry	AH = 42H AL = move technique BX = file handle CX:DX = number of bytes pointer moved
Exit	AX = error code if carry set AX:DX = bytes pointer moved
Notes	The move technique causes the pointer to move from the start of the file if AL = 00H, from the current location if AL = 01H, and from the end of the file if AL = 02H. The count is stored so DX contains the least-significant 16 bits, and either CX or AX contains the most-significant 16 bits.

43H	READ/WRITE FILE ATTRIBUTES
Entry	AH = 43H AL = 00H to read attributes AL = 01H to write attributes CX = attribute word (see function 3CH) DS:DX = address of ASCII-Z string file name
Exit	AX = error code if carry set CX = attribute word of carry cleared

44H	I/O DEVICE CONTROL (IOTCL)
Entry	AH = 44H AL = sub function code (see notes)
Exit	AX = error code (see function 59H) if carry set
Notes	The sub function codes found in AL are as follows: 00H = read device status (status is illustrated in Figure 14–2) Entry: BX = file handle Exit: DX = status 01H = write device status Entry: BX = file handle, DH = 0, DL = device information Exit: AX = error code if carry set

	02H = read control data from character device Entry: BX = file handle, CX = number of bytes, DS:DX = I/O buffer address Exit: AX = number of bytes read 03H = write control data to character device Entry: BX = file handle, CX = number of bytes, DS: DX = I/O buffer address Exit: AX = number of bytes written 04H = read control data from block device Entry: BL = drive number (0 = default, 1 = A, 2 = B, etc.), CX = number of bytes, DS:DX = I/O buffer address Exit: AX = number of bytes read 05H = write control data to block device Entry: BL = drive number, CX = number of bytes, DS:DX = I/O buffer address Exit: AX = number of bytes written 06H = check input status Entry: BX = file handle Exit: AL = 00H ready or FFH not ready 07H = check input status Entry: BX = file handle Exit: AL = 00H ready or FFH not ready 08H = removable media? Entry: BL = drive number Exit: AL = 00H removable, 01H fixed 09H = network block device? Entry: BL = drive number Exit: bit 12 of DX set for network block device 0AH = local or network character device? Entry: BX = file handle Exit: bit 15 of DX set for network character device 0BH = change entry count (must have SHARE.EXE loaded) Entry: CX = delay loop count, DX = retry count Exit: AX = error code if carry set

<table>
<tr><td></td><td>0CH = genericl/O control for character devices
Entry: BX = file handle, CH = category, CL = function
Categories: 00H = unknown, 01H = COM port,
02H = CON, 05H = LPT ports
Function:
CL = 45H; set iteration count
CL = 4AH; select code page
CL = 4CH; start code page preparation
CL = 4DH; end code page preparation
CL = 5FH; set display information
CL = 65H; get iteration count
CL = 6AH; query selected code page
CL = 6BH; query preparation list
CL = 7FH; get display information
0DH = generic I/O control for block devices
Entry: BL = drive number, CH = category, CL = function, DS: DX = address of parameter block
Category: 08H = disk drive
Function:
CL = 40H; set device parameters
CL = 41H; write track
CL = 42H; format and verify track
CL = 46H; set media ID code
CL = 47H; set access flag
CL = 60H; get device parameters
CL = 61H; read track
CL = 62H; verify track
CL = 66H; get media ID code
CL = 67H; get access code
0EH = return logical device map
Entry: BL = drive number
Exit: AL = number of last device
0FH = change logical device map
Entry: BL = drive number
Exit: Al = number of last device</td></tr>
<tr><td>45H</td><td>DUPLICATE FILE HANDLE</td></tr>
<tr><td>Entry</td><td>AH = 45H
BX = current file handle</td></tr>
<tr><td>Exit</td><td>AX = error code if carry set
AX = duplicate file handle</td></tr>
</table>

46H	FORCE DUPLICATE FILE HANDLE
Entry	AH = 46H BX = current file handle CX = new file handle
Exit	AX = error code if carry set
Notes	This function works like function 45H, except function 45H allows DOS to select the new handle while this function allows the user to select the new handle.

47H	READ CURRENT DIRECTORY
Entry	AH = 47H DL = drive number DS:SI = address of a 64-byte buffer for directory name
Exit	DS:SI addresses current directory name if carry cleared
Notes	Drive A = 00, Drive B = 01, and so forth

48H	ALLOCATE MEMORY BLOCK
Entry	AH = 48H BX = number of paragraphs to allocate CX = new file handle
Exit	BX = largest block available if carry cleared

49H	RELEASE ALLOCATED MEMORY BLOCK
Entry	AH = 49H ES = segment address of block to be released CX = new file handle
Exit	Carry indicates an error if set

4AH	MODIFY ALLOCATED MEMORY BLOCK
Entry	AH = 4AH BX = new block size in paragraphs ES = segment address of block to be modified
Exit	BX = largest block available if carry cleared

4BH	LOAD OR EXECUTE A PROGRAM
Entry	AH = 4BH AL = function code ES:BX = address of parameter block DS:DX = address of ASCII-Z string command
Exit	Carry indicates an error if set
Notes	The function codes are: AL = 00H to load and execute a program, AL = 01H to load a program, but not execute it, AL = 03H to load a program overlay, and AL = 05H to enter the EXEC state. Figure A–6 shows the parameter block used with this function.

4CH	TERMINATE A PROCESS
Entry	AH = 4CH AL = error code
Exit	Returns control to DOS
Notes	This function returns control to DOS with the error code saved so it can be obtained using DOS ERROR LEVEL batch processing system. We normally use this function with an error code of 00H to return to DOS.

4DH	READ RETURN CODE
Entry	AH = 4DH
Exit	AX = return error code
Notes	This function is used to obtain the return status code created by executing a program with DOS function 4BH. The return codes are: AX = 0000H for a normal—no error—termination, AX = 0001H for a control-break termination, AX = 0002H for a critical device error, and AX = 0003H for a termination by an INT 31H.

4EH	FIND FIRST MATCHING FILE
Entry	AH = 4EH CX = file attributes DS:DX = address ASCII-Z string file name
Exit	Carry is set for file not found
Notes	This function searches the current or named directory for the first matching file. Upon exit, the DTA contains the file information. See Figure A–7 for the disk transfer area (DTA).

4FH	FIND NEXT MATCHING FILE
Entry	AH = 4FH
Exit	Carry is set for file not found
Notes	This function is used after the first file is found with function 4EH.

50H	SET PROGRAM SEGMENT PREFIX (PSP) ADDRESS
Entry	AH = 50H BX = offset address of the new PSP
Notes	Extreme care must be used with this function because no error recovery is possible.

51H	GET PSP ADDRESS
Entry	AH = 51H
Exit	BX = current PSP segment address

54H	READ DISK VERIFY STATUS
Entry	AH = 54H
Exit	AL = 00H if verify off AL = 01H if verify on

56H	RENAME FILE
Entry	AH = 56H ES:DI = address of ASCII-Z string containing new file name DS:DX = address of ASCII-Z string containing file to be renamed
Exit	Carry is set for error condition

57H	READ FILE'S DATE AND TIME STAMP
Entry	AH = 57H AL = function code BX = file handle CX = new time DX = new date
Exit	Carry is set for error condition CX = time if carry cleared DX = date if carry cleared
Notes	AL = 00H to read date and time or 01H to write date and time.

59H	GET EXTENDED ERROR INFORMATION
Entry	AH = 59H BX = 0000H for DOS version 3.X
Exit	AX = extended error code BH = error class BL = recommended action CH = locus
Notes	Following are the error codes found in AX: 0001H = invalid function number 0002H = file not found 0003H = path not found 0004H = no file handles available 0005H = access denied 0006H = file handle invalid 0007H = memory control block failure 0008H = insufficient memory 0009H = memory block address invalid 000AH = environment failure 000BH = format invalid 000CH = access code invalid 000DH = data invalid 000EH = unknown unit 000FH = disk drive invalid 0010H = attempted to remove current directory 0011H = not same device 0012H = no more files 0013H = disk write-protected 0014H = unknown unit 0015H = drive not ready 0016H = unknown command 0017H = data error (CRC check error) 0018H = bad request structure length 0019H = seek error 001AH = unknown media type 001BH = sector not found 001CH = printer out of paper 001DH = write fault 001EH = read fault 001FH = general failure 0020H = sharing violation 0021H = lock violation

	0022H = disk change invalid 0023H = FCB unavailable 0024H = sharing buffer exceeded 0025H = code page mismatch 0026H = handle end of file operation not completed 0027H = disk full 0028H–0031H reserved 0032H = unsupported network request 0033H = remote machine not listed 0034H = duplicate name on network 0035H = network name not found 0036H = network busy 0037H = device no longer exists on network 0038H = netBIOS command limit exceeded 0039H = error in network adapter hardware 003AH = incorrect response from network 003BH = unexpected network error 003CH = remote adapter is incompatible 003DH = print queue is full 003EH = not enough room for print file 003FH = print file was deleted 0040H = network name deleted 0041H = network access denied 0042H = incorrect network device type 0043H = network name not found 0044H = network name exceeded limit 0045H = netBIOS session limit exceeded 0046H = temporary pause 0047H = network request not accepted 0048H = print or disk redirection pause 0049H–004FH reserved 0050H = file already exists 0051H = duplicate FCB 0052H = cannot make directory 0053H = failure in INT 24H (critical error) 0054H = too many redirections 0055H = duplicate redirection 0056H = invalid password 0057H = invalid parameter 0058H = network write failure 0059H = function not supported by network 005AH = required system component not installed 0065H = device not selected

	Following are the error class codes found in BH: 01H = no resources available 02H = temporary error 03H = authorization error 04H = internal software error 05H = hardware error 06H = system failure 07H = application software error 08H = item not found 09H = invalid format 0AH = item blocked 0BH = media error 0CH = item already exists 0DH = unknown error Following is the recommended action found in BL: 01H = retry operation 02H = delay and retry operation 03H = user retry 04H = abort processing 05H = immediate exit 06H = ignore error 07H = retry with user intervention Following is a list of locus in CH: 01H = unknown source 02H = block device error 03H = network area 04H = serial device error 05H = memory error

5AH	CREATE UNIQUE FILE NAME
Entry	AH = 5AH CX = attribute code DS:DX = address of the ASCII-Z string directory path
Exit	Carry is set for error condition AX = file handle if carry cleared DS:DX = address of the appended directory name
Notes	The ASCII-Z file directory path must end with a backslash (\). On exit the directory name is appended with a unique file name.

5BH	CREATE A DOS FILE
Entry	AH = 5BH CX = attribute code DS:DX = address of the ASCII-Z string containing the file name
Exit	Carry is set for error condition AX = file handle if carry cleared
Notes	The function only works in DOS version 3.X or higher. It is almost identical to function 3CH, except function 3CH erases the file, if it already exists, while function 5BH reports that the file exists with erasing it.

5CH	LOCK/UNLOCK FILE CONTENTS
Entry	AH = 5CH BX = file handle CX:DX = offset address of locked/unlocked area SI:DI = number of bytes to lock or unlock beginning at offset
Exit	Carry is set for error condition

5DH	SET EXTENDED ERROR INFORMATION
Entry	AH = 5DH AL = 0AH DS:DX = address of the extended error data structure
Notes	This function is used by DOS version 3.1 or higher to store extended error information.
5EH	**NETWORK/PRINTER**
Entry	AH = 5EH AL = 00H (get network name) DS:DX = address of the ASCII-Z string containing network name
Exit	Carry is set for error condition CL = netBIOS number if carry cleared
Entry	AH = 5EH AL = 02H (define network printer) BX = redirection list CX = length of setup string DS:DX = address of printer setup buffer
Exit	Carry is set for error condition
Entry	AH = 5EH AL = 03H (read network printer setup string) BX = redirection list DS:DX = address of printer setup buffer
Exit	Carry is set for error condition CX = length of setup string if carry cleared ES:DI = address of printer setup buffer
62H	**GET PSP ADDRESS**
Entry	AH = 62H
Exit	BX = segment address of the current program
Notes	The function only works in DOS version 3.0 or higher.

65H	GET EXTENDED COUNTRY INFORMATION
Entry	AH = 65H AL = function code ES:DI = address of buffer to receive information
Exit	Carry is set for error condition CX = length of country information
Notes	The function only works in DOS version 3.3 or higher.

66H	GET/SET CODE PAGE
Entry	AH = 66H AL = function code BX = code page number
Exit	Carry is set for error condition BX = active code page number DX = default code page number
Notes	A function code in AL of 01H gets the code page number, and a code of 02H sets the code page number.

67H	SET HANDLE COUNT
Entry	AH = 67H BX = number of handles desired
Exit	Carry is set for error condition
Notes	This function is available for DOS version 3.3 or higher.

68H	COMMIT FILE
Entry	AH = 68H BX = handle number
Exit	Carry is set for error condition Else, the date and time stamp is written to directory
Notes	This function is available for DOS version 3.3 or higher.

6CH	EXTENDED OPEN FILE
Entry	AH = 6CH AL = 00H BX = open mode CX = attributes DX = open flag DS:SI = address of ASCII-Z string file name
Exit	AX = error code if carry is set AX = handle if carry is cleared CX = 0001H file existed and was opened CX = 0002H file did not exist and was created
Notes	This function is available for DOS version 4.0 or higher.

FIGURE A–2 The contents of the file control block (FCB)

Offset	Contents
21H	Relative record number
20H	Current record number
16H	Reserved
14H	Creation date
10H	File size
0EH	Record size
0CH	Current block number
09H	3-character filename extension
01H	8-character filename
00H	Drive

FIGURE A–3 The contents of the modified file control block (FCB)

Offset	Contents
16H	Second filename
14H	Creation date
10H	File size
0EH	Record size
0CH	Current block number
09H	3-character filename extension
01H	8-character filename
00H	Drive

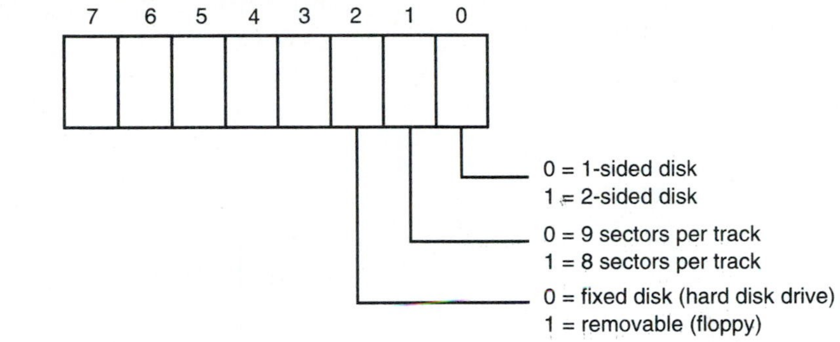

FIGURE A–4 The media descriptor byte

FIGURE A–5 Contents of the program segment prefix (PSP)

Offset	Contents
81H	Command line
80H	Command line length
6CH	File control block 2
5CH	File control block 1
52H	Reserved
50H	DOS call
2EH	Reserved
2CH	Environment address (segment)
16H	Reserved
14H	Critical error address (segment)
12H	Critical error address (offset)
10H	Control break address (segment)
0EH	Control break address (offset)
0CH	Terminate address (segment)
0AH	Terminate address (offset)
06H	Number of bytes in segment
05H	Opcode
04H	Reserved
02H	Top of memory
00H	INT 20H

FIGURE A–6 The parameter blocks used with function 4BH (EXEC): (a) function code 00H and (b) function code 03H

Offset	Contents
0CH	File control block 2 (segment)
0AH	File control block 2 (offset)
08H	File control block 1 (segment)
06H	File control block 1 (offset)
04H	Command line address (segment)
02H	Command line address (offset)
00H	Environment address (segment)

(a)

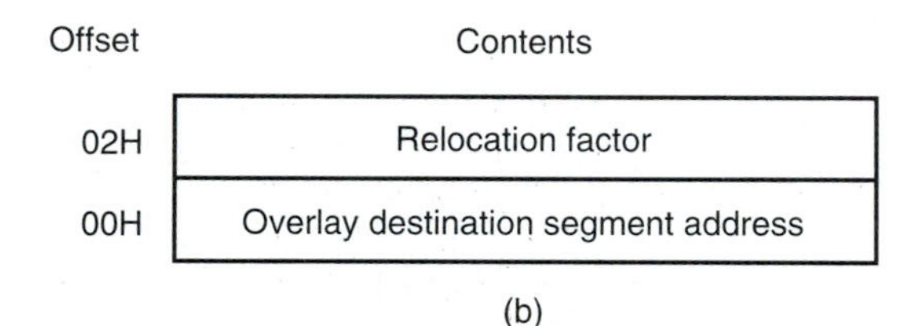

(b)

FIGURE A–7 Data transfer area (DTA) used to find a file

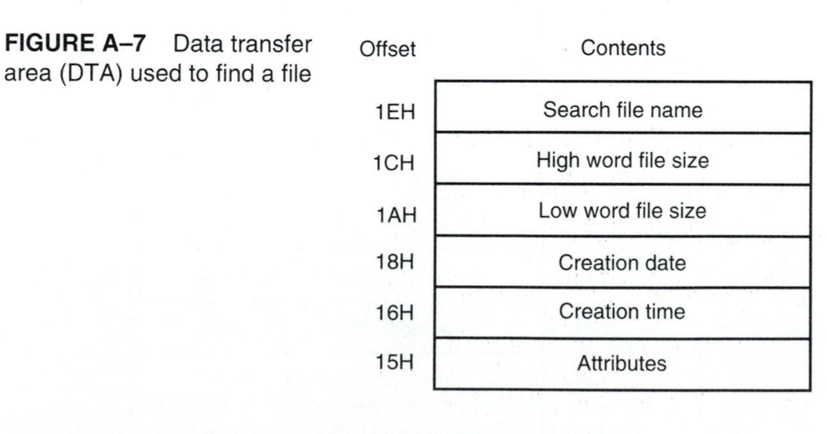

Offset
Contents
1EH
Search file name
1CH
High word file size
1AH
Low word file size
18H
Creation date
16H
Creation time
15H
Attributes

Video Mode Selection. The mode of operation for the video display is accomplished by placing a 00H into AH, followed by one of many mode numbers in AL. Table A-4 lists the modes of operation found in video display systems using standard video modes. The VGA can use any mode listed, while the other displays are more restrictive in use. Additional, higher-resolution modes are explained later in this section.

Example A–7 lists a short sequence of instructions that place the video display into mode 03H operation. This mode is available on CGA, EGA, and VGA displays. This mode allows the display to draw test data with 16 colors at various resolutions, dependent upon the display adapter.

TABLE A–4 Video display modes

Mode	*Type*	*Columns*	*Rows*	*Resolution*	*Standard*	*Colors*
00H	Text	40	25	320 × 200	CGA	2
00H	Text	40	25	320 × 350	EGA	2
00H	Text	40	25	360 × 400	VGA	2
01H	Text	40	25	320 × 200	CGA	16
01H	Text	40	25	320 × 350	EGA	16
01H	Text	40	25	360 × 400	VGA	16
02H	Text	80	25	640 × 200	CGA	2
02H	Text	80	25	640 × 350	EGA	2
02H	Text	80	25	720 × 400	VGA	2
03H	Text	80	25	640 × 200	CGA	16
03H	Text	80	25	640 × 350	EGA	16
03H	Text	80	25	720 × 400	VGA	16
04H	Graphics	40	25	320 × 200	CGA	4
05H	Graphics	40	25	320 × 200	CGA	2
06H	Graphics	80	25	640 × 200	CGA	2
07H	Text	80	25	720 × 350	EGA	4
07H	Text	80	25	720 × 400	VGA	4
0DH	Graphics	80	25	320 × 200	CGA	16
0EH	Graphics	80	25	640 × 200	CGA	16
0FH	Graphics	80	25	640 × 350	EGA	4
10H	Graphics	80	25	640 × 350	EGA	16
11H	Graphics	80	30	640 × 480	VGA	2
12H	Graphics	80	30	640 × 480	VGA	16
13H	Graphics	40	25	320 × 200	VGA	256

EXAMPLE A–7

```
0000 B4 00          MOV     AH,0        ;select mode
0002 B0 03          MOV     AL,3        ;mode is 03H
0004 CD 10          INT     10H
```

Cursor Control and other Standard Features. Table A-5 shows the function codes used to control the cursor on the video display. These cursor control functions will work on any video display, from the CGA display to the latest super VGA display. It also lists the functions used to display data and change to a different character set.

If an SVGA (super VGA), EVGA (extended VGA), or XVGA (also extended VGA) adapter is available, the super VGA mode is set by using INT 10H function call AX = 4F02H, with BX = to the VGA mode for these advanced display adapters. This conforms to the VESA standard for VGA adapters. Table A-6 shows the modes selected by register BX for this INT 10H function call. Most video cards are equipped with a driver called VVESA.COM or VVESA.SYS that conforms the card to the VESA standard functions.

TABLE A–5 Video BIOS (INT 10H) functions (pp. 592–595)

00H	**SELECT VIDEO MODE**
Entry	AH = 00H AL = mode number
Exit	Mode changed and screen cleared
01H	**SELECT CURSOR TYPE**
Entry	AH = 01H CH = starting line number CL = ending line number
Exit	Cursor size changed
02H	**SELECT CURSOR POSITION**
Entry	AH = 02H BH = page number (usually 0) DH = row number (beginning with 0) DL = column number (beginning with 0)
Exit	Changes cursor to new position
03H	**READ CURSOR POSITION**
Entry	AH = 03H BH = page number
Exit	CH = starting line (cursor size) CL = ending line (cursor size) DH = current row DL = current column
04H	**READ LIGHT PEN**
Entry	AH = 04H (not supported in VGA)
Exit	AH = 0, light pen triggered BX = pixel column CX = pixel row DH = character row DL = character column

05H	SELECT DISPLAY PAGE
Entry	AH = 05H AL = page number
Exit	Page number selected. Following are the valid page numbers. Modes 0 and 1 support pages 0–7 Modes 2 and 3 support pages 0–7 Modes 4, 5, and 6 support page 0 Modes 7 and D support pages 0–7 Mode E supports pages 0–3 Modes F and 10 support pages 0–1 Modes 11, 12, and 13 support page 0
Notes	Most modern displays use page 0 for most operations.

06H	SCROLL PAGE UP
Entry	AH = 06H AL = number of lines to scroll (0 clears window) BH = character attribute for new lines CH = top row of scroll window CL = left column of scroll window DH = bottom row of scroll window DL = right column of scroll window
Exit	Scrolls window from the bottom toward the top of the screen. Blank lines fill the bottom using the character attribute in BH.

07H	SCROLL PAGE DOWN
Entry	AH = 07H AL = number of lines to scroll (0 clears window) BH = character attribute for new lines CH = top row of scroll window CL = left column of scroll window DH = bottom row of scroll window DL = right column of scroll window
Exit	Scrolls window from the top toward the bottom of the screen. Blank lines fill from the top using the character attribute in BH.

08H	READ ATTRIBUTE/CHARACTER AT CURRENT CURSOR POSITION
Entry	AH = 08H BH = page number
Exit	AL = ASCII character code AH = character attribute Note: this function does not advance the cursor.
Notes	The attribute byte is illustrated in Figure 9–3.
09H	WRITE ATTRIBUTE/CHARACTER AT CURRENT CURSOR POSITION
Entry	AH = 09H AL = ASCII character code BH = page number BL = character attribute CX = number of characters to write
Exit	Note: this function does not advance the cursor.
0AH	WRITE CHARACTER AT CURRENT CURSOR POSITION
Entry	AH = 0AH AL = ASCII character code BH = page number CX = number of characters to write
Exit	Note: this function does not advance the cursor.
0FH	READ VIDEO MODE
Entry	AH = 0FH
Exit	AL = current video mode AH = number of character columns BH = page number

10H	SET VGA PALETTE REGISTER
Entry	AH = 10H AL = 10H BX = color number (0–255) CH = green (0–63) CL = blue (0–63) DH = red (0–63)
Exit	Palette register color is changed. Note: the first 16 colors (0–15) are used in the 16-color, VGA text mode and other modes.

10H	READ VGA PALETTE REGISTER
Entry	AH = 10H AL = 15H BX = color number (0–255)
Exit	CH = green CL = blue DH = red

11H	GET ROM CHARACTER SET
Entry	AH = 11H AL = 30H BH = 2 = ROM 8 × 14 character set BH = 3 = ROM 8 × 8 character set BH = 4 = ROM 8 × 8 extended character set BH = 5 = ROM 9 × 14 character set BH = 6 = ROM 8 × 16 character set BH = 7 = ROM 9 × 16 character set
Exit	CX = bytes per character DL = rows per character ES:BP = address of character set

TABLE A–6 Extended VGA functions

BX	*Function*
100H	640 × 400 with 256 colors
101H	640 × 480 with 256 colors
102H	800 × 600 with 16 colors
103H	800 × 600 with 256 colors
104H	1,024 × 768 with 16 colors
105H	1,024 × 768 with 256 colors
106H	1,280 × 1,024 with 16 colors
107H	1,280 × 1,024 with 256 colors
108H	80 × 60 in text mode
109H	132 × 25 in text mode
10AH	132 × 43 in text mode
10BH	132 × 50 in text mode
10CH	132 × 60 in text mode

INT 11H

This function is used to determine the type of equipment installed in the system. To use this call, the AX register is loaded with an FFFFH and then the INT 11H instruction is executed. On return, an INT 11H provides information as listed in Figure A–8.

INT 12H

The memory size is returned by the INT 12H instruction. After executing the INT 12H instruction, the AX register contains the number of 1K byte blocks of memory (conventional memory in the first 1M bytes of address space) installed in the computer.

INT 13H

This call controls the diskettes ($5^{1}/_{4}''$ or $3^{1}/_{2}''$) as well as the fixed or hard disk drives attached to the system. Table A–7 lists the functions available to this interrupt via register AH. The direct control of a floppy disk or hard disk can lead to problems. Therefore, we provide only a list of the functions without detail on their use. Before using these functions, refer to the BIOS literature available from the company that produced your version of the BIOS ROM. Never use these functions for normal disk operations.

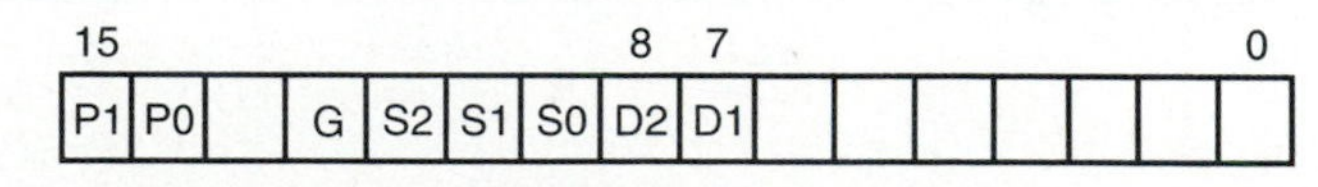

FIGURE A–8 The contents of AX after an INT 11H, which indicates what equipment is attached to the PC

P1, P0 = number of parallel ports
G = 1 if game port is attached
S2, S1, S0 = number of serial ports
D2, D1 = number of disk drives

TABLE A–7 Disk I/O function via INT 13H

AH	*Function*
00H	Reset disk system
01H	Get disk system status into AL
02H	Read sector
03H	Write sector
04H	Verify sector
05H	Format track
06H	Format bad track
07H	Format drive
08H	Get drive parameters
09H	Initialize fixed disk characteristics
0AH	Read long sector
0BH	Write long sector
0CH	Seek
0DH	Reset fixed disk system
0EH	Read sector buffer
0FH	Write sector buffer
10H	Get drive status
11H	Recalibrate drive
12H	Controller RAM diagnostics
13H	Controller drive diagnostics
14H	Controller internal diagnostics
15H	Get disk type
16H	Get disk change status
17H	Set disk type
18H	Set media type
19H	Park heads
1AH	Format ESDI drive

INT 14H

Interrupt 14H controls the serial COM (communications) ports attached to the computer. Most computer systems contain two COM ports, COM1 and COM2. Newer, AT-style machines add two more communications ports, COM3 and COM4. Communications ports are normally controlled with software packages that allow data transfer through a modem and the telephone lines. The INT 14H instruction controls these ports as illustrated in Table A–8.

TABLE A–8 COM port interrupt INT 14H

AH	*Function*
00H	Initialize communications port
01H	Send character
02H	Receive character
03H	Get COM port status
04H	Extended initialize communications port
05H	Extended communications port control

TABLE A–9 The I/O subsystem interrupt INT 15H

AH	*Function*
00H	Cassette motor on
01H	Cassette motor off
02H	Read cassette
03H	Write cassette
0FH	Format ESDI drive periodic interrupt
21H	Keyboard intercept
80H	Device open
81H	Device closed
82H	Process termination
83H	Event wait
84H	Read joystick
85H	System request key
86H	Delay
87H	Move extended block of memory
88H	Get extended memory size
89H	Enter protected mode
90H	Device wait
91H	Device power on self test (POST)
C0H	Get system environment
C1H	Get address of extended BIOS data area
C2H	Mouse pointer
C3H	Set watch-dog timer
C4H	Programmable option select

INT 15H

The INT 15H instruction controls many of the I/O devices interfaced to the computer. It also allows access to protected-mode operation and the extended memory system on an 80286, 80386 or 80486 system. Table A-9 lists the functions supported by INT 15H.

INT 16H

The INT 16H instruction is used as a keyboard interrupt. This interrupt is accessed by DOS interrupt INT 21H, but can be accessed directly. Table A–10 shows the functions performed by INT 16H.

TABLE A–10 Keyboard interrupt INT 16H

AH	*Function*
00H	Read keyboard character
01H	Get keyboard status
02H	Get keyboard flags
03H	Set repeat rate
04H	Set keyclick
05H	Push character and scan code

TABLE A–11 Parallel printer interrupt INT 17H

AH	*Function*
00H	Print character
01H	Initialize printer
02H	Get printer status

INT 17H

The INT 17H instruction accesses the parallel printer port, labeled LPT1 in most systems. Table A–11 lists the three functions available for the INT 17H instruction.

DOS SYSTEM MEMORY MAP

Figure A–9 illustrates the memory map used by a DOS computer system. The first 1M byte of memory is listed with all the areas containing different devices and programs used with DOS. The transient program area (TPA) is where DOS applications programs are loaded and executed. The size of the TPA is usually slightly over 500K bytes, unless a number of TSR programs and drivers have filled memory before the TPA.

DOS LOW-MEMORY ASSIGNMENTS

Table A–12 shows the low-memory assignments (00000H–005FFH) for the DOS-based microprocessor system. This area of memory contains the interrupt vectors, BIOS data area, and the DOS/BIOS data area illustrated in Figure A–9.

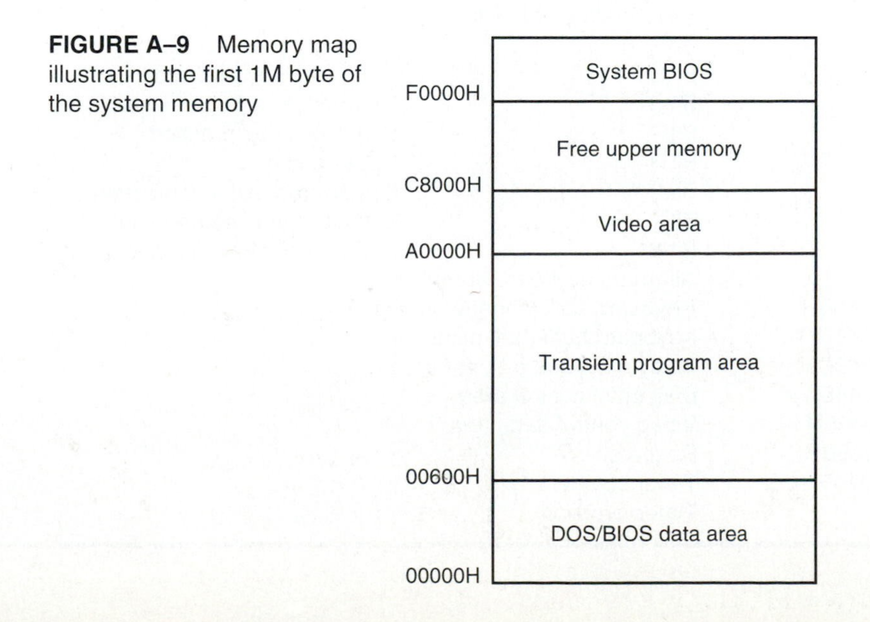

FIGURE A–9 Memory map illustrating the first 1M byte of the system memory

TABLE A–12 DOS low-memory assignments

Location	*Purpose*	
00000H–002FFH	System interrupt vectors	
00300H–003FFH	System interrupt vectors, power on, and bootstrap area	
00400H–00407H	COM1–COM4 I/O port base addresses	
00408H–0040FH	LPT1–LPT4 I/O port base addresses	
00410H–00411H	Equipment flag word, returned in AX by an INT 11H	
	Bits 15–14	Number of parallel printers (LPT1–LPT4)
	Bit 13	Internal MODEM installed
	Bit 12	Joystick installed
	Bits 11–9	Number of serial ports (COM1–COM4)
	Bit 8	Unused
	Bits 7–6	Number of disk drives
	Bits 5–4	Video mode
	Bits 3–2	Unused
	Bit 1	Math coprocessor installed
	Bit 0	Disk installed
00412H	Reserved	
00413H–00414H	Memory size in K bytes (0–640K)	
00415H–00416H	Reserved	
00417H	Keyboard control byte	
	Bit 7	Insert locked
	Bit 6	Caps locked
	Bit 5	Numbers locked
	Bit 4	Scroll locked
	Bit 3	Alternate key pressed
	Bit 2	Control key pressed
	Bit 1	Left shift key pressed
	Bit 0	Right shift key pressed
00418H	Keyboard control byte	
	Bit 7	Insert key pressed
	Bit 6	Caps lock key pressed
	Bit 5	Numbers lock key pressed
	Bit 4	Scroll lock key pressed
	Bit 3	Pause locked
	Bit 2	System request key pressed
	Bit 1	Left alternate key pressed
	Bit 0	Right control key pressed
00419H	Alternate keyboard entry	
0041AH–0041BH	Keyboard buffer header pointer	
0041CH–0041DH	Keyboard buffer tail pointer	
0041EH–0043DH	32-byte keyboard buffer area	
0043EH–00448H	Disk drive control area	
00449H–00466H	Video control data area	
00467H–0046BH	Reserved	
0046CH–0046FH	Timer counter	
00470H	Timer overflow	

TABLE A–12 (Continued)

Location	*Purpose*
00471H	Break key state
00472H–00473H	Reset flag
00474H–00477H	Hard disk drive data area
00478H–0047BH	LPT1–LPT4 timeout area
0047CH–0047FH	COM1–COM4 timeout area
00480H–00481H	Keyboard buffer start offset pointer
00482H–00483H	Keyboard buffer end offset pointer
00484H–0048AH	Video control data area
0048BH–00495H	Hard drive control area
00496H	Keyboard mode, state, and type flag
00497H	Keyboard LED flags
00498H–00499H	Offset address of user wait complete flag
0049AH–0049BH	Segment address of user wait complete flag
0049CH–0049DH	User wait count (low word)
0049EH–0049FH	User wait count (high word)
004A0H	Wait active flag
004A1H–004A7H	Reserved
004A8H–004ABH	Pointer to video parameters
004ACH–004EFH	Reserved
004F0H–004FFH	Applications program communications area
00500H	Print screen status
00501H–00503H	Reserved
00504H	Single drive mode status
00505H–0050FH	Reserved
00510H–00521H	Used by ROM BASIC
00522H–0052FH	Used by DOS for disk initialization
00530H–00533H	Used by MODE command
00534H–005FFH	Reserved

DOS VERSION 5/6 MEMORY MAP

Microsoft DOS version 5/6 has a slightly different memory map than earlier versions of DOS, because of its ability to load drivers and programs in the system area. If the microprocessor is an 80386 or 80486, ROM memory located between addresses A0000H and FFFFFH can be backfilled with extended memory for drivers and programs. In many systems, memory area D0000H–DFFFFH is unused, as is E0000H–EFFFFH. These areas can be filled with extended memory through memory paging, found in the 80386 and 80486 microprocessors. This new memory area can then be filled with and addressed by normal real-mode memory programs, extending the memory available to DOS applications.

The HIMEM.SYS and EMM386.SYS drivers are used to accomplish the backfilling. If you want to use memory area E0000H–EFFFFH, you must load EMM386.SYS as EMM386.SYS I=E000-EFFF. Using these drivers increases the DOS TPA to more than 600K bytes. A typical CONFIG.SYS file for DOS 5/6 appears in Example A–8. Notice that drivers after EMM386.SYS are loaded in high memory with the DEVICEHIGH directive instead of the DEVICE directive. Programs are loaded using the LOADHIGH or LH directive in front of the program name.

EXAMPLE A–8

```
(CONFIG.SYS file)

FILES=30
BUFFERS=30
STACKS=64,128
FCBS=48
SHELL=C:\DOS\COMMAND.COM C:\DOS\ /E:256 /P
DEVICE=C:\DOS\HIMEM.SYS
DOS=HIGH,UMB
DEVICE=C:\DOS\EMM386.EXE I=C800-EFFF NOEMS
DEVICEHIGH SIZE=1EB0 C:\LASERLIB\SONY_CDU.SYS /D:SONY_001 /B:340 /Q:* /T:* /M:H
DEVICEHIGH SIZE=0190 C:\DOS\SETVER.EXE
DEVICEHIGH SIZE=3150 C:\MOUSE1\MOUSE.SYS
LASTDRIVE = F

(AUTOEXEC.BAT file)
PATH C:\DOS;C:\;C:\MASM\BIN;C:\MASM\BINB\;C:\UTILITY;C:\WS;C:\LASERLIB
SET BLASTER=A220 I7 D1 T3
SET INCLUDE=C:\MASM\INCLUDE\
SET HELPFILES=C:\MASM\HELP\*.HLP
SET INIT=C:\MASM\INIT\
SET ASMEX=C:\MASM\SAMPLES\
SET TMP=C:\MASM\TMP
SET SOUND=C:\SB
LOADHIGH C:\LASERLIB\MSCDEX.EXE /D:SONY_001 /L:F /M:8
LOADHIGH C:\LASERLIB\LLTSR.EXE ALT-Q
LOADHIGH C:\DOS\FASTOPEN C:=256
LOADHIGH C:\DOS\DOSKEY /BUFSIZE=1024
LOADHIGH C:\LASERLIB\PRINTF.COM
DOSKEY GO=DOSSHELL
DOSSHELL
```

MOUSE FUNCTIONS

The mouse is controlled and adjusted with INT 33H function call instructions. These functions provide complete control over the mouse and information provided by the mouse driver program. Table A–13 lists the mouse INT 33H functions by number, details the parameters required, and notes any comments needed to use them. Refer to Chapter 10 for a discussion of the mouse driver and some example programs that access the mouse functions through INT 33H.

TABLE A–13 Mouse (INT 33H) funtions (pp. 603–617)

00H	RESET MOUSE
Entry	AL = 00H
Exit	BX = number of mouse buttons Both software and hardware are reset to their default values
Notes	The default values are: CRT Page = 0 Cursor = off Current cursor position = center of screen Minimum horizontal position = 0 Minimum vertical position = 0 Maximum horizontal position = maximum for display mode Maximum vertical position = maximum for display mode Horizontal mickey-to-pixel ratio = 1 to 1 Vertical mickey-to-pixel ratio = 2 to 1 Double-speed threshold = 64 per second Graphics cursor = arrow Text cursor = reverse block Light-pen emulation = on Interrupt call mask = 0

01H	SHOW MOUSE CURSOR
Entry	AL = 01H
Exit	Displays the mouse cursor

02H	HIDE MOUSE CURSOR
Entry	AL = 02H
Exit	Hides the mouse cursor
Notes	When displaying data on the screen, it is important to hide the mouse cursor. If you don't, problems with the display will occur and the computer may reset and reboot.

03H	READ MOUSE STATUS
Entry	AL = 03H
Exit	BX = button status CX = horizontal cursor position DX = vertical cursor position
Notes	The right bit (bit 0) of BX contains the status of the left mouse button and bit 1 contains the status of the right mouse button. A 1 indicates the button is active.
04H	**SET MOUSE CURSOR POSITION**
Entry	AL = 04H CX = horizontal position DX = vertical position
05H	**GET BUTTON PRESS INFORMATION**
Entry	AL = 05H BX = desired button (0 for left and 1 for right)
Exit	AX = button status BX = number of presses CX = horizontal position of last press DX = vertical position of last press
06H	**GET BUTTON RELEASE INFORMATION**
Entry	AL = 06H BX = desired button
Exit	AX = button status BX = number of releases CX = horizontal position of last release DX = vertical position of last release

07H	SET HORIZONTAL BOUNDARY
Entry	AL = 07H CX = minimum horizontal position DX = maximum horizontal position
Exit	Horizontal boundary is changed.

08H	SET VERTICAL BOUNDARY
Entry	AL = 08H CX = minimum vertical position DX = maximum vertical position
Exit	Vertical boundary is changed.

09H	SET GRAPHICS CURSOR
Entry	AL = 09H BX = horizontal center CX = vertical center ES:DX = address of 16 x 16 bit map of cursor
Exit	New graphics cursor installed
Notes	The center is where the mouse pointer position is set. For example, a center of 0,0 is the upper left corner and 15,15 is the lower right corner. The pixel bit map mask is stored at the address passed through ES:DX and is a 16 x 16 array. Following the pixel bit mask is the cursor mask, also 16 x 16. The contents of the bit map mask are ANDed with a 16 x 16 portion of the video display, after which the contents of the cursor mask is Exclusive-ORed with a 16 x 16 portion of the video display to produce a cursor.

0AH	SET TEXT CURSOR
Entry	AL = 0AH BX = cursor type (0 = software, 1 = hardware) CX = pixel bit mask or beginning scan line DX = cursor mask or ending scan line
Exit	Changes the text cursor.

0BH	READ MOTION COUNTERS
Entry	AL = 0BH
Exit	CX = horizontal distance DX = vertical distance
Notes	Returns the distance traveled by the mouse since the last call to this function.

0CH	SET INTERRUPT SUBROUTINE
Entry	AL = 0CH CX = interrupt mask ES:DX = address of interrupt service procedure
Exit	New interrupt handler is installed.
Notes	The interrupt mask defines the actions that request the installed interrupt handler. Following is a list of the actions that cause the interrupt when placed in the interrupt mask. Note that these actions can appear singly or in combination. For example, an interrupt mask of 8 plus 2, or 10 (0AH), causes an interrupt for either the left or right mouse button. 1 = any change in cursor position 2 = left mouse button pressed 4 = left mouse button released 8 = right mouse button pressed 16 = right mouse button released After the installed interrupt is called by one of the selected changes, the following registers contain information about the mouse: AX = interrupt mask BX = button status CX = horizontal position DX = vertical position SI = horizontal change DI = vertical change

0DH	ENABLE LIGHT-PEN EMULATION
Entry	AL = 0DH
Exit	Light-pen emulation is enabled.
Notes	Used whenever the mouse must replace the action of a light-pen.

0EH	DISABLE LIGHT-PEN EMULATION
Entry	AL = 0EH
Exit	Light-pen emulation is disabled.
Notes	Used to disable light-pen emulation.

0FH	SET MICKEY-TO-PIXEL RATIO
Entry	AL = 0FH CX = horizontal ratio DX = vertical ratio
Exit	Mickey-to-pixel ratio changed.
Notes	This function allows the screen tracking speed of the mouse cursor to change. The default value is 1. If it is changed to 2, the speed will be reduced to half the normal.

10H	BLANK MOUSE CURSOR
Entry	AL = 10H CX = left corner DX = upper position SI = right corner DI = bottom position
Exit	Blanks the cursor in the window specified.
Notes	The mouse pointer is blanked in the area of the screen selected by registers CX, DX, SI, and DI.

13H	SET DOUBLE-SPEED
Entry	AL = 13H DX = threshold value
Exit	Changes the threshold for double-speed pointer movement.
Notes	The default threshold value for mouse pointer acceleration is 64. This default threshold is changed by this function to any other value.

14H	SWAP INTERRUPTS
Entry	AL = 14H CX = interrupt mask ES:DX = interrupt service procedure address
Exit	CX = old interrupt mask ES:DX = old interrupt service procedure address
Notes	Like mouse INT 33H function 0CH, function 14H installs a new interrupt handler. The difference is that function 14H replaces the handler already installed with function 0CH.

15H	GET MOUSE STATE SIZE
Entry	AL = 15H
Exit	BX = size of buffer required to store mouse state
Notes	Indicates the amount of memory required to store the state of the mouse with mouse function INT 33H number 16H.

16H	SAVE MOUSE STATE
Entry	AL = 16H ES:DX = address where state of mouse is stored
Exit	Saves the current state of the mouse.

17H	RELOAD MOUSE STATE
Entry	AL = 17H ES:DX = address where state of mouse is stored
Exit	Reloads the saved state of the mouse.

18H	SET ALTERNATE INTERRUPT SUBROUTINE
Entry	AL = 18H CX = alternate interrupt mask ES:DX = address of alternate interrupt service procedure Carry = 0
Exit	Alternate interrupt handler is installed.
Notes	The alternate interrupt handler is accessed after the primary handler. The alternate interrupt mask defines the actions that request the alternate interrupt handler. Following is a list of the actions that cause an alternate interrupt when placed in the alternate interrupt mask: 1 = any change in cursor position 2 = left mouse button pressed 4 = left mouse button released 8 = right mouse button pressed 16 = right mouse button released 32 = shift key with mouse button 64 = control key with mouse button 128 = alternate key with mouse button After the installed interrupt is called by one of the selected changes, the following registers contain information about the mouse: Carry = 1 AX = interrupt mask (see function 0CH) BX = button status CX = horizontal position DX = vertical position SI = horizontal change DI = vertical change

19H	GET ALTERNATE INTERRUPT ADDRESS
Entry	AL = 19H CX = alternate interrupt mask
Exit	AX = –1 if unsuccessful ES:DX = address of interrupt service procedure
Notes	Used to read the address of the alternate interrupt handler.

1AH	SET MOUSE SENSITIVITY
Entry	AL = 1AH BX = horizontal sensitivity CX = vertical sensitivity DX = double-speed threshold
Exit	Sensitivity is changed.
Notes	The default values for vertical and horizontal sensitivity are both 50. The default value for the double-speed threshold is 64. The values for sensitivity range between 1 and 100, which represent ratios of between 33% and 350%.

1BH	GET MOUSE SENSITIVITY
Entry	AL = 1BH
Exit	BX = horizontal sensitivity CX = vertical sensitivity DX = double-speed threshold

1CH	SET INTERRUPT RATE
Entry	AL = 1CH BX = number of interrupts per second
Exit	Changes the interrupt rate for the InPort mouse only.
Notes	The default value is 30 interrupts per second, but this can be changed to any value listed: 0 = interrupt off 1 = 30 interrupts per second 2 = 50 interrupts per second 3 = 100 interrupts per second 4 = 200 interrupts per second

1DH	SET CRT PAGE
Entry	AL = 1DH BX = CRT page number
Exit	Changes to a new CRT page for mouse cursor support.

1EH	GET CRT PAGE
Entry	AL = 1EH
Exit	BX = CRT page

1FH	DISABLE MOUSE DRIVER
Entry	AL = 1FH
Exit	AX = –1 if unsuccessful ES:BX = Address of mouse driver

20H	ENABLE MOUSE DRIVER
Entry	AL = 20H
Exit	The mouse driver is enabled.

21H	SOFTWARE RESET
Entry	AL = 21H
Exit	AX = 0FFFFH if successful BX = 2 if successful

22H	SET LANGUAGE
Entry	AL = 22H BX = language number
Exit	Language is changed only with the international version.
Notes	The languages selected by BX are as folllows: 0 = English 1 = French 2 = Dutch 3 = German 4 = Swedish 5 = Finnish 6 = Spanish 7 = Portuguese 8 = Italian

23H	GET LANGUAGE
Entry	AL = 23H
Exit	BX = language number

24H	GET DRIVER VERSION
Entry	AL = 24H
Exit	BH = major version number BL = minor version number CH = mouse type CL = interrupt request number

25H	GET DRIVER INFORMATION
Entry	AL = 25H
Exit	AX contains the following driver information in the bits indicated: Bits 13, 12 00 = software text cursor active 01 = hardware text cursor active 1X = graphics cursor active Bit 14 0 = non-integrated mouse display driver 1 = integrated mouse display driver Bit 15 0 = driver is a .SYS file 1 = driver is a .COM file

26H	GET MAXIMUM VIRTUAL COORDINATES
Entry	AL = 26H
Exit	BX = mouse driver status CX = maximum horizontal coordinate DX = maximum vertical coordinate

27H	GET CURSOR MASKS AND COUNTS
Entry	AL = 27H
Exit	AX = screen mask or beginning scanning line BX = cursor mask or ending scan line CX = horizontal mickey count DX = vertical mickey count

28H	SET VIDEO MODE
Entry	AL = 28H CX = video mode DX = font size
Exit	CX = 0 if successful
29H	GET SUPPORTED VIDEO MODES
Entry	AL = 29H CX = search flag
Exit	BX:DX = address of ASCII description of video mode CX = mode number of 0 if unsuccessful
Notes	The search flag is 0 to search for the first video mode and non-zero to search for the next video mode.
2AH	GET CURSOR HOT SPOT
Entry	AL = 2AH
Exit	AX = display flag BX = horizontal hot spot CX = vertical hot spot DX = mouse type
Notes	The hot spot is the position on the cursor that is returned when a mouse button is clicked. AX (display flag) indicates that the cursor is active (0) or inactive (1).

2BH	SET ACCELERATION CURVE
Entry	AL = 2BH BX = curve number ES:SI = address of the acceleration curve
Exit	AX = 0 if successful
Notes	Following are the contents of the acceleration curve table that contains curves 1 through 4. Offset Meaning 00H = curve 1 counts 01H = curve 2 counts 02H = curve 3 counts 03H = curve 4 counts 04H = curve 1 mouse count and threshold 24H = curve 2 mouse count and threshold 44H = curve 3 mouse count and threshold 64H = curve 4 mouse count and threshold 84H = curve 1 scale factor array A4H = curve 2 scale factor array C4H = curve 3 scale factor array E4H = curve 4 scale factor array 104H = curve 1 name 114H = curve 2 name 124H = curve 3 name 134H = curve 4 name

2CH	GET ACCELERATION CURVE
Entry	AL = 2CH BX = current curve ES:SI = address of current acceleration curves
Exit	AX = 0 if successful

2DH	GET ACTIVE ACCELERATION CURVE
Entry	AL = 2DH BX = curve number or –1 for current
Exit	AX = 0 if successful BX = curve number ES:SI = address of acceleration curves

2FH	MOUSE HARDWARE RESET
Entry	AL = 2FH
Exit	AX = 0 if unsuccessful
30H	**SET/GET BALLPOINT INFORMATION**
Entry	AL = 30H BX = rotation angle (±32K) CX = command (0 for read and 1 for write)
Exit	AX = FFFFH if unsuccessful or button state BX = rotation angle CX = button masks
31H	**GET VIRTUAL COORDINATES**
Entry	AL = 31H
Exit	AX = minimum horizontal BX = minimum vertical CX = maximum horizontal DX = maximum vertical
32H	**GET ACTIVE ADVANCED FUNCTIONS**
Entry	AL = 32H
Exit	AX = function flag
Notes	The function flags indicate which INT 33H advanced functions are available. The leftmost bit indicates function 25H; the rightmost bit indicates function 34H.

33H	GET SWITCH SETTING
Entry	AL = 33H CX = buffer length ES:DI = address of buffer
Exit	AX = 0 CX = byte in buffer ES:DI = address of buffer

34H	GET MOUSE.INI
Entry	AL = 34H
Exit	AX = 0 ES:DX = buffer address

EMS, THE EXPANDED MEMORY MANAGER

The expanded memory system is managed by the EMS manager program and accessed via INT 67H. The requested function number is placed in AH, along with any other required parameters in other registers before the INT 67H instruction executes. This interrupt allows access to the EMM386.EXE program, used by many systems for expanded memory management. Note that these functions are compatible with LIM version 4.0. Refer to Table A–14 for an abbreviated listing of the EMS functions. Only the nonhardware-specific functions are included in this table. Refer to Chapter 13 for a discussion and examples illustrating the use of EMS and the application of INT 67H.

TABLE A–14 EMS (INT 67H) functions (pp. 617–620)

40H	GET EMS STATUS
Entry	AH = 40H
Exit	AH = 00H if EMS is functional AL = <> 00H if EMS is not funtional

41H	GET PAGE FRAME SEGMENT ADDRESS
Entry	AH = 41H
Exit	AH = 00H if successful AH <> 00H if not successful BX = page frame segment address
42H	GET NUMBER OF PAGES
Entry	AH = 42H
Exit	AH = 00H if successful AH <> 00H if not successful BX = number of available pages DX = number of total pages
43H	ALLOCATE PAGES
Entry	AH = 43H BX = number of pages to allocate
Exit	AH = 00H if successful AH <> 00H if not successful DX = EMS handle number
44H	MAP MEMORY PAGES
Entry	AH = 44H AL = frame page location (0, 1, 2, or 3) BX = logical page location DX = EMS handle number
Exit	AH = 00H if successful AH <> 00H if not successful

45H	RELEASE ALLOCATED PAGES
Entry	AH = 45H DX = EMS handle number
Exit	AH = 00H if successful AH <> 00H if not successful
46H	GET EMS VERSION NUMBER
Entry	AH = 46H
Exit	AH = 00H if successful AH <> 00H if not successful AL = EMS version number
Notes	The version number is in AL in BCD code. For example, version 4.0 appears as 40H, while version 3.2 appears as 32H.
47H	SAVE PAGE MAP
Entry	AH = 47H DX = EMS handle number
Exit	AH = 00H if successful AH <> 00H if not successful
Notes	This function saves the state of the page mapping register for only the current 64K byte page frame. The page map is restored with function 48H.
48H	RESTORE PAGE MAP
Entry	AH = 48H DX = EMS handle number
Exit	AH = 00H if successful AH <> 00H if not successful

4BH	GET HANDLE COUNT
Entry	AH = 4BH
Exit	AH = 00H if successful AH <> 00H if not successful BX = number of allocated memory handles
4CH	GET PAGES FOR ONE HANDLE
Entry	AH = 4CH DX = EMS handle number
Exit	AH = 00H if successful AH <> 00H if not successful BX = number of pages allocated to handle
4DH	GET PAGES FOR ALL HANDLES
Entry	AH = 4DH ES:DI = buffer address
Exit	AH = 00H if successful AH <> 00H if not successful BX = active handle count
Notes	Each handle is stored in memory in a 4-byte area per handle. The first word contains the handle number and the second word contains the number of pages assigned to the handle. Note that the buffer requires up to 1,024 bytes of memory.

XMS, THE EXTENDED MEMORY MANAGER

The XMS (extended memory system) manager, HIMEM.SYS in DOS, allows access to the extended memory system. This function does not use a DOS interrupt vector for access; instead, it uses the address of the driver as provided by the multiplex interrupt (2FH). The XMS functions are listed in Table A–15 as obtained by a CALL to the address returned in ES:BX from function AX = 4310H of INT 2FH. Refer to Chapter 13 for the application and examples of the XMS manager.

TABLE A–15 XMS functions (pp. 621–625)

00H	GET XMS VERSION
Entry	AH = 00H
Exit	AX = version number BX = driver version DX = high memory flag
Notes	The version number is returned in AX as AH = major and AL = minor. For example, version 3.0 is returned as 0300H in AX. The driver version is provided by the developer and is not normally used. The high memory flag indicates whether high memory exists.
01H	REQUEST HIGH MEMORY
Entry	AH = 01H DX = space requested in bytes
Exit	AL = 01H if successful AL = 00H if not successful BL = error code
02H	RELEASE HIGH MEMORY
Entry	AH = 02H
Exit	AL = 01H if successful AL = 00H if not successful BL = error code

03H	GLOBAL ENABLE A20 LINE
Entry	AH = 03H
Exit	AL = 01H if successful AL = 00H if not successful BL = error code
Notes	The A20 line provides access to the high memory area. This function enables the A20 line to provide access to high memory.
04H	GLOBAL DISABLE A20 LINE
Entry	AH = 04H
Exit	AL = 01H if successful AL = 00H if not successful BL = error code
Notes	This function disables the A20 line. If DOS version 5.0 or 6.X is installed and DOS is loaded high, this function crashes the system.
05H	LOCAL ENABLE A20 LINE
Entry	AH = 05H
Exit	AL = 01H if successful AL = 00H if not successful BL = error code
Notes	The A20 line provides access to the high memory area. This function enables the A20 line in systems that are multitasked.
06H	DISABLE LOCAL A20 LINE
Entry	AH = 06H
Exit	AL = 01H if successful AL = 00H if not successful BL = error code

07H	QUERY A20 LINE
Entry	AH = 07H
Exit	AL = 01H line is enabled AL = 00H line is disabled if BL also 0 BL = error code
08H	QUERY EXTENDED MEMORY
Entry	AH = 08H
Exit	AL = largest block size free AX = 00H if not successful BL = error code DX = total installed extended memory
Notes	Both AX and DX return the amount of extended memory in 1K block sizes.
09H	ALLOCATE EXTENDED MEMORY
Entry	AH = 09H DX = number of blocks desired
Exit	AL = 01H if successful AL = 00H if not successful BL = error code DX = block handle number
0AH	FREE EXTENDED MEMORY
Entry	AH = 0AH DX = block handle number
Exit	AL = 01H if successful AL = 00H if not successful BL = error code

0BH	MOVE EXTENDED MEMORY
Entry	AH = 0BH DS:SI = parameter table address
Exit	AL = 01H if successful AL = 00H if not successful BL = error code
Notes	The parameter table contains 16 bytes of information passed to the XMS driver. Offset — Size — Function 00H — Doubleword — Number of bytes moved 04H — Word — Source block handle 06H — Doubleword — Source offset 0AH — Word — Destination block handle 0CH — Doubleword — Destination offset

Offset	Size	Function
00H	Doubleword	Number of bytes moved
04H	Word	Source block handle
06H	Doubleword	Source offset
0AH	Word	Destination block handle
0CH	Doubleword	Destination offset

0CH	LOCK BLOCK
Entry	AH = 0CH DX = block handle number
Exit	AL = 01H if successful AL = 00H if not successful DX:BX = linear block address
Notes	Anchors a block of extended memory at any linear address. Do not leave the block locked for extended periods.

0DH	UNLOCK BLOCK
Entry	AH = 0DH DX = block handle number
Exit	AL = 01H if successful AL = 00H if not successful BL = error code

0EH	GET HANDLE INFORMATION
Entry	AH = 0EH DX = block handle number
Exit	AL = 01H if successful AL = 00H if not successful BH = lock count BL = number of available handles DX = block size
Notes	If BH is greater than 00H, the block is locked.

0FH	REALLOCATE BLOCK
Entry	AH = 0FH BX = new size DX = block handle number
Exit	AL = 01H if successful AL = 00H if not successful BL = error code

10H	REQUEST UPPER MEMORY BLOCK
Entry	AH = 10H DX = desired block size
Exit	AL = 01H if successful AL = 00H if not successful BX = block segment address DX = block size delivered
Notes	Upper memory blocks are handled by DOS versions 5.0 and 6.X and should not normally be accessed via the XMS driver.

11H	RELEASE UPPER MEMORY BLOCK
Entry	AH = 11H DX = block segment address
Exit	AL = 01H if successful AL = 00H if not successful BL = error code

VCPI MEMORY CONTROL FUNCTIONS

The expanded and extended memory systems are controlled by the VCPI (virtual control program interface) provided by the DOS version 6.X EMM386.EXE program. Like EMS, VCPI is accessed through INT 67H. A list of VCPI functions provided by EMM386.EXE is given in Table A–16. Please note that all VCPI functions are accessed with AH = DEH and AL = function number. Refer to Chapter 13 for a discussion and applications depicting the VCPI interface.

TABLE A–16 XMS functions (pp. 626–630)

00H	**VCPI VERSION**
Entry	AH = DEH AL = 00H
Exit	AH = 00H if successful AH <> 00H if unsuccessful BX = version number
Notes	The version number is returned in BX as: BH = major and BL = minor.
01H	**GET PROTECTED MODE**
Entry	AH = DEH AL = 01H DS:SI = GDT address ES:DI = page table buffer
Exit	AH = 00H if successful AH <> 00H if unsuccessful ES:DI = address of first uninitialized entry in page table buffer
Notes	This prepares VCPI for a switch to the protected mode. The buffer is a 4K byte section of memory used to construct a page table for mapping the first 4M bytes of memory. The GDT descriptor table, the global descriptor table, is filled by the VCPI with the protected-mode entry point found in EBX as an offset, using the first descriptor as a code segment.

02H	GET MAXIMUM ADDRESS
Entry	AH = DEH AL = 02H
Exit	AH = 00H if successful AH <> 00H if unsuccessful EDX = maximum address
Notes	The maximum address is the highest page address assigned by VCPI.
03H	RETURN NUMBER OF FREE PAGES
Entry	AH = DEH AL = 03H
Exit	AH = 00H if successful AH <> 00H if unsuccessful EDX = number of free 4K byte pages
04H	ALLOCATE A PAGE
Entry	AH = DEH AL = 04H
Exit	AH = 00H if successful AH <> 00H if unsuccessful EDX = physical page address
Notes	The physical address is a 32-bit linear address
05H	RELEASE A PAGE
Entry	AH = DEH AL = 05H EDX = physical page address to release
Exit	AH = 00H if successful AH <> 00H if unsuccessful

06H	GET REAL PAGE ADDRESS
Entry	AH = DEH AL = 06H CX = page number
Exit	AH = 00H if successful AH <> 00H if unsuccessful EDX = physical page address
Notes	The page number requested in CX is the 32-bit linear address, shifted right 12 places to form the page number. The page number must be in real memory (00H–FFH). The EDX register contains the physical page address mapped into the real memory from the EMS memory.

07H	READ CONTROL REGISTER 0
Entry	AH = DEH AL = 07H
Exit	AH = 00H if successful AH <> 00H if unsuccessful EBX = control register 0 value

08H	READ DEBUG REGISTERS
Entry	AH = DEH AL = 08H ES:DI = address of storage array
Exit	AH = 00H if successful AH <> 00H if unsuccessful
Notes	This reads and stores the debug registers into a memory array addressed by ES:DI. The array contains 8 doublewords that contain the contents of the 8 debug registers. Debug register 0 is stored in the first doubleword, and debug register 7 in the last.

09H	LOAD DEBUG REGISTERS
Entry	AH = DEH AL = 09H ES:DI = address of register array
Exit	AH = 00H if successful AH <> 00H if unsuccessful
0AH	GET HARDWARE INTERRUPT VECTORS
Entry	AH = DEH AL = 0AH
Exit	AH = 00H if successful AH <> 00H if unsuccessful BX = starting vector for IRQ0 CX = starting vector for IRQ8
0BH	SET HARDWARE INTERRUPT VECTORS
Entry	AH = DEH AL = 0BH BX = IRQ0 starting vector number CX = IRQB8 starting vector number
Exit	AH = 00H if successful AH <> 00H if unsuccessful

0CH	SWITCH CPU MODE (CALLED FROM REAL MODE)
Entry	AH = DEH AL = 0CH ESI = linear address of system register table
Exit	unspecified
Notes	The linear address loaded into ESI before this function points to a table 21 bytes in length that contains data used to switch to protected-mode operation.

Offset	Size	Purpose
00H	Doubleword	Value loaded to CR3
04H	Doubleword	GDTR linear address in real memory
08H	Doubleword	IDTR linear address in real memory
0CH	Word	LDTR selector value
0EH	Word	TR selector value
10H	6 byte	CS:EIP protected-mode entry point

0CH	SWITCH CPU MODE (CALLED FROM PROTECTED MODE)
Entry	AH = DEH AL = 0CH ESP = linear address of system register table DS = data segment selector SS = stack segment selector
Exit	unspecified
Notes	The linear address loaded into ESP before this function points to a table that contains data loaded into a real register upon the from protected-mode operation.

Offset	Size	Contents
00H	8 bytes	Far return address (unused)
08H	Doubleword	EIP value
0CH	Doubleword	CS value
10H	Doubleword	EFLAGS
14H	Doubleword	ESP
18H	Doubleword	SS
1CH	Doubleword	ES
20H	Doubleword	DS
24H	Doubleword	FS
28H	Doubleword	GS

DPMI CONTROL FUNCTIONS

The DPMI (DOS protected-mode interface) provides DOS and Windows applications with access to the protected mode and to real and extended memory. Four functions are accessed through the DOS multiplex interrupt (INT 2FH) and the remaining functions are accessed via INT 31H. The functions accessed by the INT 2FH interrupt release a time slice (AX = 1680H), get the CPU mode (AX = 1686H), get the mode switch entry point (AX = 1687H), and get the API entry point (AX = 168AH). The INT 2FH functions are listed in Table A–17 and the INT 31H functions appear in Table A–18. Refer to Chapter 13 for a complete discussion and examples of the DPMI interface.

TABLE A–17 INT 2FH DPMI functions (pp. 631–632)

80H	**RELEASE TIME SLICE**
Entry	AX = 1680H
Exit	AX = 0000H if successful AX <> 1680H if unsuccessful
Notes	A time slice is a portion of time allowed to an application. If the application is idle, it should release the remaining portion of its time slice to other applications using this function.
86H	**GET CPU MODE**
Entry	AX = 1686H
Exit	AX = 0000H if CPU in protected mode AX <> 0000H if CPU in real or virtual mode

87H	GET MODE ENTRY POINT
Entry	AX = 1687H
Exit	AX = 0000H if successful AX <> 0000H if unsuccessful BX = support flag CL = processor type DX = DPMI version number SI = private paragraph count ES:DI = protected mode entry point
Notes	Used to allocate the entry point required to switch the CPU into the protected mode using DPMI. BX is a logic one if 32-bit support is provided by DPMI. CL indicates the microprocessor (2, 3, or 4 for the 80286, 80386, or 80486). DX = the DPMI version number returned in binary value in DH (major) and DL (minor). SI = the number of paragraphs required by DPMI for proper operation. ES:DI = the address used to switch to protected mode.

8AH	GET API ENTRY POINT
Entry	AX = 168AH DS:DI = address of vendor name (ASCII-Z)
Exit	AX = 0000H if successful AX <> 168AH if unsuccessful ES:DI = API extensions entry point
Notes	Used to interface a specific vendor's extensions to the DPMI interface.

TABLE A–18 INT 31H DPMI functions (pp. 633–657)

0000H	ALLOCATE LDT DESCRIPTOR
Entry	AX = 0000H CX = number of descriptors needed
Exit	AX = base selector, carry = 0 AX = error code, carry = 1
Notes	Allocates one or more local descriptors with the base or starting descriptor number returned in AX.

0001H	RELEASE LDT DESCRIPTOR
Entry	AX = 0001H BX = selector
Exit	Carry = 0 if successful Carry = 1 if unsuccessful, AX = error code
Notes	Release one descriptor selected by the selector placed in BX on the call to the INT 31H function.

0002H	MAP REAL SEGMENT TO DESCRIPTOR
Entry	AX = 0002H BX = segment address
Exit	Carry = 0 if successful, AX = selector Carry = 1 if unsuccessful, AX = error code
Notes	Maps a real-mode segment to a protected-mode descriptor. This cannot be released and should only be used to map the start of the descriptor table or other global protected-mode segments.

0003H	GET SEGMENT INCREMENT
Entry	AX = 0003H
Exit	Carry = 0 if successful, AX = increment
Notes	Used with function AX = 0000H to determine the increment of the selector returned as the base selector.

0006H	GET SEGMENT BASE ADDRESS
Entry	AX = 0006H BX = selector
Exit	Carry = 0 if successful Carry = 1 if unsuccessful, AX = error code CX:DX = base address of segment
Notes	Returns the segment address as selected by BX in CX:DX, where CX = the high-order word and DX = the low-order word expressed as a 32-bit linear address.

0007H	SET SEGMENT BASE ADDRESS
Entry	AX = 0007H BX = selector CX:DX = linear base address
Exit	Carry = 0 if successful Carry = 1 if unsuccessful, AX = error code
Notes	Sets the base address to the 32-bit linear address found in CX:DX, where CX = high-order word and DX = low-order word.

0008H	SET SEGMENT LIMIT
Entry	AX = 0008H BX = selector CX:DX = segment limit in bytes
Exit	Carry = 0 if successful Carry = 1 if unsuccessful, AX = error code

0009H	SET DESCRIPTOR ACCESS RIGHTS
Entry	AX = 0009H BX = selector CX = access rights
Exit	Carry = 0 if successful Carry = 1 if unsuccessful, AX = error code
Notes	Access rights determine how a segment is accessed in the protected mode. The definition of each bit of CX follows: Bit — Rights 0 — Descriptor access (1) 1 — Read/write (1) or Read-only (0) 2 — Expand segment up (0) or expand segment down (1) 3 — Code segment (1) or data segment (0) 4 — Must be 1 5, 6 — Descriptor desired privilege level 7 — Present (1) or not present (0) 8–13 — Must be 000000 14 — 32-bit instruction mode (1) or 16-bit mode (0) 15 — Granularity bit is 0 for 1X multiplier and 1 for 4K multiplier for limit field Bits 15–8 must be 0000 0000 for the 80286

000AH	CREATE ALIAS DESCRIPTOR
Entry	AX = 000AH BX = selector
Exit	Carry = 0 if successful, AX = alias selector Carry = 1 if unsuccessful, AX = error code
Notes	Creates a new "carbon copy" as an alias local descriptor.

000BH	GET DESCRIPTOR
Entry	AX = 000BH BX = selector ES:DI = address of buffer
Exit	Carry = 0 if successful Carry = 1 if unsuccessful, AX = error code
Notes	Copies the 8-byte descriptor into the buffer addressed by ES:DI

000CH	SET DESCRIPTOR
Entry	AX = 000CH BX = selector ES:DI = address of buffer
Exit	Carry = 0 if successful Carry = 1 if unsuccessful, AX = error code
Notes	Copies the 8-byte descriptor from the buffer to the descriptor table.

000DH	ALLOCATE SPECIFIC LDT DESCRIPTOR
Entry	AX = 000DH BX = selector
Exit	Carry = 0 if successful Carry = 1 if unsuccessful, AX = error code
Notes	Creates a descriptor based in the selector of your choice (4–7CH); as many as 16 descriptors can be assigned by this function.

000EH	GET MULTIPLE DESCRIPTORS
Entry	AX = 000EH BX = number to get ES:DI = buffer address
Exit	Carry = 0 if successful, CX = number copied to buffer Carry = 1 if unsuccessful, AX = error code
Note	Copies multiple descriptors to a buffer addressed by ES:DI

000FH	SET MULTIPLE DESCRIPTORS
Entry	AX = 000FH BX = number to set ES:DI = buffer address
Exit	Carry = 0 if successful, number copied from buffer Carry = 1 if unsuccessful, AX = error code
Notes	Creates descriptors from the buffer at the location addressed by ES:DI. Each entry is 10 bytes in length; the first 2 bytes contain the selector number, and the next 8 bytes contain descriptor data.

0100H	ALLOCATE DOS MEMORY BLOCK
Entry	AX = 0100H DX = paragraphs to allocate
Exit	Carry = 0 if successful AX = real-mode segment DX = selector of descriptor for memory block Carry = 1 if unsuccessful AX = error code BX = size of available block
Notes	Allocates a local descriptor for the DOS memory block. Release with function 0101H only.

0101H	RELEASE DOS MEMORY BLOCK
Entry	AX = 0101H DX = selector
Exit	Carry = 0 if successful Carry = 1 if unsuccessful, AX = error code

0200H	GET REAL-MODE INTERRUPT
Entry	AX = 0200H BX = interrupt number
Exit	CX = vector segment DX = vector offset
Notes	Release one descriptor selected by the selector placed in BX on the call to the INT 31H function.
0201H	SET REAL-MODE INTERRUPT VECTOR
Entry	AX = 0201H BX = interrupt number CX = vector segment DX = vector offset
0202H	GET EXCEPTION HANDLER VECTOR
Entry	AX = 0202H BX = exception number 0–1FH
Exit	Carry = 0 if successful CX = handler selector DX = handler offset Carry = 1 if unsuccessful AX = error code
0203H	SET EXCEPTION HANDLER VECTOR
Entry	AX = 0203H BX = exception number 0–1FH CX = handler sector DX = handler offset
Exit	Carry = 0 if successful Carry = 1 if unsuccessful, AX = error code

0204H	GET PROTECTED-MODE INTERRUPT VECTOR
Entry	AX = 0204H BX = interrupt number
Exit	Carry = 0 if successful CX = handler selector DX = handler offset Carry = 1 if unsuccessful AX = error code

0205H	SET PROTECTED-MODE INTERRUPT VECTOR
Entry	AX = 0205H BX = interrupt number CX = handler selector DX = handler offset
Exit	Carry = 0 if successful Carry = 1 if unsuccessful, AX = error code

0210H	GET EXTENDED REAL EXCEPTION HANDLER VECTOR
Entry	AX = 0210H BX = exception number 0–1FH
Exit	Carry = 0 if successful CX = handler selector DX = handler offset Carry = 1 if unsuccessful AX = error code

0211H	GET EXTENDED PROTECTED EXCEPTION HANDLER VECTOR
Entry	AX = 0211H BX = exception number 0–1FH
Exit	Carry = 0 if successful CX = handler selector DX = handler offset Carry = 1 if unsuccessful AX = error code
0212H	SET EXTENDED PROTECTED EXCEPTION HANDLER VECTOR
Entry	AX = 0212H BX = exception number 0–1FH CX = handler selector DX = handler offset
Exit	Carry = 0 if successful Carry = 1 if unsuccessful, AX = error code
0213H	SET EXTENDED REAL EXCEPTION HANDLER VECTOR
Entry	AX = 0213H BX = exception number 0–1FH CX = handler sector DX = handler offset
Exit	Carry = 0 if successful Carry = 1 if unsuccessful, AX = error code

0300H	SIMULATE REAL-MODE INTERRUPT
Entry	AX = 0300H BX = interrupt number CX = word copy count ES:DI = buffer address
Exit	Carry = 0 if successful, ES:DI = buffer address Carry = 1 if unsuccessful, AX = error code
Notes	Copies the number of words (CX) from the protected mode stack to the real mode stack. The ES:DI register combination addresses a memory buffer that specifies the contents of the register when the switch to the real-mode interrupt occurs. The register set is stored as: Offset / Size / Register 00H doubleword EDI 04H doubleword ESI 08H doubleword EBP 0CH doubleword must be 00000000H 10H doubleword EBX 14H doubleword EDX 18H doubleword ECX 1CH doubleword EAX 20H word flags 22H word ES 24H word DS 26H word FS 28H word GS 2AH word IP 2CH word CS 2EH word SP 30H word SS

Offset	Size	Register
00H	doubleword	EDI
04H	doubleword	ESI
08H	doubleword	EBP
0CH	doubleword	must be 00000000H
10H	doubleword	EBX
14H	doubleword	EDX
18H	doubleword	ECX
1CH	doubleword	EAX
20H	word	flags
22H	word	ES
24H	word	DS
26H	word	FS
28H	word	GS
2AH	word	IP
2CH	word	CS
2EH	word	SP
30H	word	SS

0301H	CALL FAR REAL-MODE PROCEDURE
Entry	AX = 0301H BH = 00H CX = word copy count ES:DI = buffer address
Exit	Carry = 0 if successful, ES:DI = buffer address Carry = 1 if unsuccessful, AX = error code
Notes	Calls a real-mode procedure from protected mode. The buffer contains the register set as defined for function 0300H, which is loaded on the switch to real mode.

0302H	CALL REAL-MODE INTERRUPT PROCEDURE
Entry	AX = 0302H BH = 00H CX = word copy count ES:DI = buffer address
Exit	Carry = 0 if successful, ES:DI = buffer address Carry = 1 if unsuccessful, AX = error code
Notes	Calls the real-mode interrupt procedure from protected mode. The buffer contains the register set as defined for function 0300H, which is loaded on the switch to real mode.
0303H	ALLOCATE REAL-MODE CALLBACK ADDRESS
Entry	AX = 0303H DS:SI = protected-mode procedure address ES:DI = buffer address
Exit	Carry = 0 if successful, CX:DX = callback address Carry = 1 if unsuccessful, AX = error code
Notes	Calls a protected-mode procedure from the real mode. The real mode address is found in CX (segment) and DX (offset).
0304H	RELEASE REAL-MODE CALLBACK ADDRESS
Entry	AX = 0304H CX:DX = callback address
Exit	Carry = 0 if successful Carry = 1 if unsuccessful, AX = error code
Note	Release the callback address allocated by function 0303H.

0305H	GET STATE SAVE AND RESTORE ADDRESS
Entry	AX = 0305H
Exit	Carry = 0 if successful AX = buffer size BX:CX = real-mode procedure address SI:DI = protected-mode procedure address Carry = 1 if unsuccessful, AX = error code
Notes	The procedure addresses returned by this function save the real- or protected-mode registers in memory before switching modes. In the real mode BX:CX contains the far address or a procedure that saves the protected-mode registers. Likewise, if operating in the protected mode, save the real-mode register with a call to the protected-mode address in SI:DI. The value of AL dictates how the procedure called by the procedure address functions. If AL = 0, the registers are saved. If AL = 1, the registers are restored.

0306H	GET RAW CPU MODE SWITCH ADDRESS
Entry	AX = 0306H
Exit	Carry = 0 if successful BX:CX = switch to protected-mode address SI:DI = switch to real-mode address Carry = 1 if unsuccessful, AX = error code
Notes	To switch to protected mode from real mode, use a far JMP to the address found in BX:CX. To switch to real mode from protected mode, use a JMP to the address found in SI:DI. The register must be preloaded as follows before jumping to the switch address: AX = new DS BX = new SP CX = new ES DX = new SS SI = new CS DI = new IP

0400H	GET DPMI VERSION
Entry	AX = 0400H
Exit	Carry = 0 if successful AX = DPMI version number BX = implementation flag CL = processor type DH = master interrupt controller base address DL = slave interrupt controller base address Carry = 1 if unsuccessful, AX = error code
Notes	See Chapter 14 for a description of the contents of the registers and an example that displays the version number and so forth. The interrupt controller vector numbers are usually returned as 08H for the master and 70H for the slave.

0401H	GET DPMI CAPABILITIES
Entry	AX = 0401H ES:DI = buffer address
Exit	Carry = 0 if successful, AX = capabilities Carry = 1 if unsuccessful, AX = error code
Notes	This function is supported under version 1.0 of DPMI and uses AX to return the following capabilities: Bit Purpose 6 write-protect host (DPMI) 5 write-protect client (DOS/WINDOWS) 4 demand zero fill 3 conventional memory mapping 2 device mapping 1 exceptions restartability 0 page dirty

0500H	GET FREE MEMORY
Entry	AX = 0500H ES:DI = buffer address
Exit	Carry = 0 if successful Carry = 1 if unsuccessful
Notes	Information about the system memory is returned by this function in the buffer addressed by the buffer addresses. The contents of the buffer are: Offset / Size / Purpose 00H doubleword largest free block in bytes 04H doubleword maximum unlocked pages 08H doubleword maximum located pages 0CH doubleword linear address in pages 10H doubleword total unlocked pages 14H doubleword total free pages 18H doubleword total physical pages 1CH doubleword free linear pages 20H doubleword size of paging file in pages 24H 12 bytes reserved

Offset	Size	Purpose
00H	doubleword	largest free block in bytes
04H	doubleword	maximum unlocked pages
08H	doubleword	maximum located pages
0CH	doubleword	linear address in pages
10H	doubleword	total unlocked pages
14H	doubleword	total free pages
18H	doubleword	total physical pages
1CH	doubleword	free linear pages
20H	doubleword	size of paging file in pages
24H	12 bytes	reserved

0501H	ALLOCATE MEMORY BLOCK
Entry	AX = 0501H BX:CX = memory block size in bytes
Exit	Carry = 0 if successful BX:CX = linear base address of block SI:DI = memory block handle Carry = 1 if unsuccessful, AX = error code
Notes	Allocates a block of memory whose size in bytes is in BX:CX on the call to this function. On the return BX:CX = the starting address of the memory block and SI:DI contains the block handle.

0502H	RELEASE MEMORY BLOCK
Entry	AX = 0502H SI:DI = memory block handle
Exit	Carry = 0 if successful Carry = 1 if unsuccessful, AX = error code

0503H	RESIZE MEMORY BLOCK
Entry	AX = 0503H BX:CX = new memory block size in bytes SI:DI = memory block handle
Exit	Carry = 0 if successful BX:CX = linear base address of block SI:DI = memory block handle Carry = 1 if unsuccessful, AX = error code

0504H	ALLOCATE LINEAR MEMORY BLOCK
Entry	AX = 0504H EBX = desired linear base address ECX = block size in bytes EDX = action code
Exit	Carry = 0 if successful EBX = linear base address of block ESI = memory block handle Carry = 1 if unsuccessful, AX = error code
Notes	Used with a 32-bit DPMI host to allocate a linear memory block. The action code is 0 to create an uncommitted block and 1 to create a committed block.

0505H	RESIZE LINEAR MEMORY BLOCK
Entry	AX = 0505H ES:EBX = buffer address ECX = block size in bytes EDX = action code (see function 0504H) ESI = memory block handle
Exit	Carry = 0 if successful BX:CX = linear base address of block SI:DI = memory block handle Carry = 1 if unsuccessful, AX = error code

0506H	GET PAGE ATTRIBUTES
Entry	AX = 0506H EBX = page offset within memory block ECX = page count ES:EDX = buffer address ESI = memory block handle
Exit	Carry = 0 if successful Carry = 1 if unsuccessful, AX = error code
Notes	The memory buffer contains a word of attribute information for each page in the memory block. The attribute word stored in the buffer indicates the following information: Bit Function 0 0 = uncommitted and 1 = committed 1 1 = mapped 3 0 = read-only and 1 = read/write 4 0 = dirty bit invalid and 1 = dirty bit valid 5 0 = page unaccessed and 1 = page accessed 6 0 = page unmodified and 1 = page modified

0507H	SET PAGE ATTRIBUTES
Entry	AX = 0507H EBX = page offset within memory block ECX = page count ES:EDX = buffer address ESI = memory block handle
Exit	Carry = 0 if successful Carry = 1 if unsuccessful, AX = error code
Notes	Like function 0506H, the memory buffer contains a word for each page that defines the attribute for each page. Following is the meaning of the bits in the page attribute word: Bit Function 0 0 = make uncommitted and 1 = make committed 1 1 = modify attributes, but not page type 3 0 = make read-only and 1 = make read/write 4 1 = modify dirty bit 5 1 = mark page accessed 6 1 = mark page modified

0508H	MAP DEVICE IN MEMORY BLOCK
Entry	AX = 0508H EBX = page offset within memory block ECX = page count EDX = device address ESI = memory block handle
Exit	Carry = 0 if successful Carry = 1 if unsuccessful, AX = error code
Note	Assigns the physical address of a device (EDX) to a linear address in a memory block.

0509H	MAP CONVENTIONAL MEMORY
Entry	AX = 0509H EBX = page offset within memory block ECX = page count EDX = linear address of conventional memory ESI = memory block handle
Exit	Carry = 0 if successful Carry = 1 if unsuccessful, AX = error code
Note	Allocates a linear (real) address to a memory block.

050AH	GET MEMORY BLOCK SIZE
Entry	AX = 050AH SI:DI = memory block handle
Exit	Carry = 0 if successful SI:DI = memory block size Carry = 1 if unsuccessful, AX = error code

050BH	GET MEMORY INFORMATION
Entry	AX = 050BH DI:SI = buffer address
Exit	Carry = 0 if successful Carry = 1 if unsuccessful, AX = error code
Notes	This function fills a 128-byte buffer addressed by DI:SI with information about the memory. Following is the contents of the buffer: Offset Size Function 00H doubleword allocated physical memory 04H doubleword allocated virtual memory (host) 08H doubleword available virtual memory (host) 0CH doubleword allocated virtual memory (machine) 10H doubleword available virtual memory (machine) 14H doubleword allocated virtual memory (client) 18H doubleword available virtual memory (client) 1CH doubleword locked memory (client) 20H doubleword maximum locked memory (client) 24H doubleword maximum linear address (client) 28H doubleword maximum free memory block size 2CH doubleword minimum allocation unit 30H doubleword allocation alignment unit 34H 76 bytes reserved

Offset	Size	Function
00H	doubleword	allocated physical memory
04H	doubleword	allocated virtual memory (host)
08H	doubleword	available virtual memory (host)
0CH	doubleword	allocated virtual memory (machine)
10H	doubleword	available virtual memory (machine)
14H	doubleword	allocated virtual memory (client)
18H	doubleword	available virtual memory (client)
1CH	doubleword	locked memory (client)
20H	doubleword	maximum locked memory (client)
24H	doubleword	maximum linear address (client)
28H	doubleword	maximum free memory block size
2CH	doubleword	minimum allocation unit
30H	doubleword	allocation alignment unit
34H	76 bytes	reserved

0600H	LOCK LINEAR REGION
Entry	AX = 0600H BX:CX = linear address of memory to lock SI:DI = number of bytes to lock
Exit	Carry = 0 if successful Carry = 1 if unsuccessful, AX = error code

0601H	UNLOCK LINEAR REGION
Entry	AX = 0601H BX:CX = linear address of memory to unlock SI:DI = number of bytes to unlock
Exit	Carry = 0 if successful Carry = 1 if unsuccessful, AX = error code

0602H	MARK REAL-MODE REGION PAGABLE
Entry	AX = 0602H BX:CX = linear address of memory to mark SI:DI = number of bytes in region
Exit	Carry = 0 if successful Carry = 1 if unsuccessful, AX = error code
0603H	RELOCK REAL-MODE REGION
Entry	AX = 0603H BX:CX = linear address to relock SI:DI = number of bytes to relock
Exit	Carry = 0 if successful Carry = 1 if unsuccessful, AX = error code
0604H	GET PAGE SIZE
Entry	AX = 0604H
Exit	Carry = 0 if successful BX:CX = page size in bytes Carry = 1 if unsuccessful, AX = error code
0702H	MARK PAGE AS DEMAND PAGING CANDIDATE
Entry	AX = 0702H BX:CX = linear address of memory region SI:DI = number of bytes in region
Exit	Carry = 0 if successful Carry = 1 if unsuccessful, AX = error code

0703H	DISCARD PAGE CONTENTS
Entry	AX = 0703H BX:CX = linear address of memory region SI:DI = number of bytes in region
Exit	Carry = 0 if successful Carry = 1 if unsuccessful, AX = error code
Notes	This releases the memory for other uses by DPMI. A discarded page still contains data, but it is undefined.

0800H	PHYSICAL ADDRESS MAPPING
Entry	AX = 0800H BX:CX = base physical address
Exit	Carry = 0 if successful BX:CX = base linear address Carry = 1 if unsuccessful, AX = error code
Note	Converts a physical address to a linear address.

0801H	RELEASE PHYSICAL ADDRESS MAPPING
Entry	AX = 0801H BX:CX = linear address
Exit	Carry = 0 if successful Carry = 1 if unsuccessful, AX = error code

0900H	GET AND DISABLE VIRTUAL INTERRUPT STATE
Entry	AX = 0900H
Exit	Carry = 0 if successful Carry = 1 if unsuccessful, AX = error code

0901H	GET AND ENABLE VIRTUAL INTERRUPT STATE
Entry	AX = 0901H
Exit	Carry = 0 if successful Carry = 1 if unsuccessful, AX = error code

0902H	GET VIRTUAL INTERRUPT STATE
Entry	AX = 0902H
Exit	Carry = 0 if successful Carry = 1 if unsuccessful, AX = error code

0A00H	GET API ENTRY POINT
Entry	AX = 0A00H DS:SI = vendor ASCII offset
Exit	Carry = 0 if successful, AL = 0 if disabled ES:DI = entry point address Carry = 1 if unsuccessful, AX = error code

0B00H	SET DEBUG WATCHPOINT
Entry	AX = 0B00H BX:CX = linear watchpoint address DH = watchpoint type DL = watchpoint size
Exit	Carry = 0 if successful BX = watchpoint handle Carry = 1 if unsuccessful, AX = error code
Notes	The BX:CX register provides the watchpoint address to the function. The DH register provides the type of watchpoint (0 = instruction executed at watchpoint address, 1 = memory write to watchpoint address, and 2 = a read or write to watchpoint address). The DL register holds the size in bytes of the watchpoint address for types 1 and 2. When the watchpoint is triggered, function AX = 0B02H is used to test for the trigger.

0B01H	CLEAR DEBUG WATCHPOINT
Entry	AX = 0B01H BX = watchpoint handle
Exit	Carry = 0 if successful Carry = 1 if unsuccessful, AX = error code
Note	Erase a watchpoint assigned with function 0B00H.
0B02H	GET STATE OF DEBUG WATCHPOINT
Entry	AX = 0B02H BX = watchpoint handle
Exit	Carry = 0 if successful AX = watchpoint status Carry = 1 if unsuccessful, AX = error code
Notes	Used to detect a watchpoint trigger. A status of 0 indicates that a watchpoint has not been detected and 1 indicates it has.
0B03H	RESET DEBUG WATCHPOINT
Entry	AX = 0B03H BX = watchpoint handle
Exit	Carry = 0 if successful Carry = 1 if unsuccessful, AX = error code
Note	Clears the watchpoint status, but does not erase the watchpoint.

0C00H	INSTALL RESIDENT SERVICE PROVIDER CALLBACK
Entry	AX = 0C00H ES:DI = buffer address
Exit	Carry = 0 if successful Carry = 1 if unsuccessful, AX = error code
Notes	Used for a protected-mode TSR to notify the DPMI to call the TSR whenever another machine is in the same virtual memory. The buffer contains the following information for DPMI:

Offset	Size	Function
00H	quadword	16-bit data segment descriptor
08H	quadword	16-bit code segment descriptor
10H	word	16-bit callback procedure offset
12H	word	reserved
14H	quadword	32-bit data segment descriptor
1CH	quadword	32-bit code segment descriptor
24H	doubleword	32-bit callback procedure offset

0C01H	TERMINATE AND STAY RESIDENT
Entry	AX = 0C01H BL = exit code DX = number of paragraphs to reserve
Exit	none

0D00H	ALLOCATE SHARED MEMORY
Entry	AX = 0D00H ES:DI = buffer address
Exit	Carry = 0 if successful Carry = 1 if unsuccessful, AX = error code
Notes	The buffer contains information for DPMI used to allocate shared memory. Its contents follow: Offset Size Function 00H doubleword desired block size in bytes 04H doubleword actual memory block size in bytes 08H doubleword memory block handle 0CH doubleword linear address of memory block 10H doubleword offset of ASCII-Z string name 14H word selector of ASCII-Z string name 16H word reserved 18H doubleword must be set to 00000000H

0D01H	RELEASE SHARED MEMORY
Entry	AX = 0D01H SI:DI = memory block handle
Exit	Carry = 0 if successful Carry = 1 if unsuccessful, AX = error code

0D02H	SERIALIZE ON SHARED MEMORY
Entry	AX = 0D02H DX = option flag SI:DI = memory block handle
Exit	Carry = 0 if successful Carry = 1 if unsuccessful, AX = error code

0D03H	RELEASE SERIALIZATION ON SHARED MEMORY
Entry	AX = 0D03H DX = option flag SI:DI = memory block handle
Exit	Carry = 0 if successful Carry = 1 if unsuccessful, AX = error code
Notes	The DX register contains the option code as follows: Bit Function 0 0 = suspend until serialization and 1 = return error code 1 0 = exclusive serialization and 1 = shared serialization

0E00H	GET COPROCESSOR STATUS
Entry	AX = 0E00H
Exit	Carry = 0 if successful AX = status code Carry = 1 if unsuccessful, AX = error code
Notes	The following is the coprocessor status: Bit Function 0 1 = coprocessor enabled 1 1 = emulation of coprocessor enabled for client 2 1 = coprocessor present 3 1 = emulation of coprocessor enabled for host 7–4 0000 = no coprocessor 80287 present 80287 present 80487/Pentium present

0E01H	SET COPROCESSOR EMULATION
Entry	AX = 0E01H BX = action code
Exit	Carry = 0 if successful Carry = 1 if unsuccessful, AX = error code
Notes	The contents of register BX follow: Bit Function 0 1 = enable coprocessor for client 1 1 = client will supply emulation

APPENDIX B

Instruction Set Summary

The instruction set summary contains a complete listing of all instructions for the 8086, 8088, 80286, 80386, 80486, and Pentium microprocessors. Note that numeric coprocessor instructions for the 8087 through the Pentium appear in Chapter 13.

Each instruction entry lists the mnemonic opcode plus a brief description of the purpose of the instruction. Also listed is the binary machine language coding for each instruction, plus any other data required to form the instruction, such as displacement or immediate data. Next to the binary machine language version of the instruction appear the flag register bits and any change that might occur for a given instruction. In this listing, a blank indicates no change, a ? indicates a change with an unpredictable outcome, a * indicates a predictable change, a 1 indicates the flag is set, and a 0 indicates the flag is cleared.

Before the instruction listing begins, some information about the bit settings in the binary machine language versions of the instructions is required. Table B–1 shows the modifier bits, coded as oo in the instruction listings, so instructions can be formed with a register, displacement, or no displacement.

Table B-2 lists the memory-addressing modes available with the register/memory field, coded as mmm. This table applies to all versions of the microprocessor.

Table B-3 lists the register options (rrr) when encoded for either an 8-bit or a 16-bit register. This table also lists the 32-bit registers used with the 80386, 80486, and Pentium microprocessors.

TABLE B–1 The modifier bits, coded as oo in the instruction listing

oo	*Function*
00	If mmm = 110, then a displacement follows the opcode; otherwise, no displacement is used
01	An 8-bit signed displacement follows the opcode
10	A 16-bit signed displacement follows the opcode
11	mmm specifies a register, instead of an addressing mode

TABLE B–2
Register/memory field (mmm) description

mmm	*Function*
000	DS:[BX+SI]
001	DS:[BX+DI]
010	SS:[BP+SI]
011	SS:[BP+DI]
100	DS:[SI]
101	DS:[DI]
110	SS:[BP]
111	DS:[BX]

Table B–4 lists the segment register bit assignments (rrr) for the MOV, PUSH, and POP instructions.

When the 80386, 80486, and Pentium microprocessors are used, some of the definitions provided in Tables B–1 through B–4 will change. Refer to Tables B–5 and B–6 for these changes as they apply to the 80386, 80486, and Pentium microprocessors.

The instruction set summary that follows lists all of the instructions, with examples, for the 8086, 8088, 80286, 80386, 80486, and Pentium microprocessors. Missing are the segment override prefixes: CS (2EH), SS (36H), DS (3EH), ES (26H), FS (64H), and GS (65H). These prefixes are one byte in length and placed in memory before the instruction that is prefixed.

Table B-7 lists the effective address calculations that apply only to the 8086 and 8088 microprocessors as ea or a in the instruction set summary. For example, the 8086 ADD DATA,AL instruction requires 16 + ea clocks. Because Table B–7 lists displacement addressing as requiring 6 clocks, this instruction requires 22 clocks to execute. All times listed are maximum times; in some cases, the microprocessor may execute the instruction in less time.

The D-bit, in the code segment descriptor, indicates the default size of the operand and the addresses for the 80386, 80486, and Pentium microprocessors. If D = 1, then all addresses and operands are 32 bits. If D = 0, all addresses and operands are 16 bits. In the real mode, the D-bit in the 80386, 80486, and Pentium microprocessors is set to zero, so operands and addresses are 16 bits.

TABLE B–3 Register field (rrr) options

rrr	*W = 0*	*W = 1*	*32-bit Register*
000	AL	AX	EAX
001	CL	CX	ECX
010	DL	DX	EDX
011	BL	BX	EBX
100	AH	SP	ESP
101	CH	BP	EBP
110	DH	SI	ESI
111	BH	DI	EDI

TABLE B–4 Register field assignments (rrr) that are used to represent the segment registers

rrr	*Register*
000	ES
001	CS
010	SS
011	DS
100	FS
101	GS

TABLE B–5 Index registers specified with rrr in the 80386, 80486, and Pentium microprocessors

rrr	*Index Register*
000	EAX
001	ECX
010	EDX
011	EBX
100	No index
101	EBP
110	ESI
111	EDI

TABLE B–6 Possible combinations of oo, mmm, and rrr for the 80386, 80486, and Pentium instruction set using the 32-bit addressing mode

oo	*mmm*	*rrr*	*Function*
00	000	—	DS:[EAX]
00	001	—	DS:[ECX]
00	010	—	DS:[EDX]
00	011	—	DS:[EBX]
00	100	000	DS:[EAX+scaled-index]
00	100	001	DS:[ECX+scaled-index]
00	100	010	DS:[EDX+scaled-index]
00	100	011	DS:[EBX+scaled-index]
00	100	100	SS:[ESP+scaled-index]
00	100	101	DS:[disp32+scaled-index]
00	100	110	DS:[ESI+scaled-index]
00	100	111	DS:[EDI+scaled-index]
00	101	—	DS:disp32
00	110	—	DS:[ESI]
00	111	—	DS:[EDI]
01	000	—	DS:[EAX+disp8]
01	001	—	DS:[ECX+disp8]
01	010	—	DS:[EDX+disp8]
01	011	—	DS:[EBX+disp8]
01	100	000	DS:[EAX+scaled-index+disp8]
01	100	001	DS:[ECX+scaled-index+disp8]
01	100	010	DS:[EDX+scaled-index+disp8]
01	100	011	DS:[EBX+scaled-index+disp8]

TABLE B–6 (Continued)

oo	*mmm*	*rrr*	*Function*
01	100	100	SS:[ESP+scaled-index+disp8]
01	100	101	SS:[EBP+scaled-index+disp8]
01	100	110	DS:[ESI+scaled-index+disp8]
01	100	111	DS:[EDI+scaled-index+disp8]
01	101	—	SS:[EBP+disp8]
01	110	—	DS:[ESI+disp8]
01	111	—	DS:[EDI+disp8]
10	000	—	DS:[EAX+disp32]
10	001	—	DS:[ECX+disp32]
10	010	—	DS:[EDX+disp32]
10	011	—	DS:[EBX+disp32]
10	100	000	DS:[EAX+scaled-index+disp32]
10	100	001	DS:[ECX+scaled-index+disp32]
10	100	010	DS:[EDX+scaled-index+disp32]
10	100	011	DS:[EBX+scaled-index+disp32]
10	100	100	SS:[ESP+scaled-index+disp32]
10	100	101	SS:[EBP+scaled-index+disp32]
10	100	110	DS:[ESI+scaled-index+disp32]
10	100	111	DS:[EDI+scaled-index+disp32]
01	101	—	SS:[EBP+disp32]
01	110	—	DS:[ESI+disp32]
01	111	—	DS:[EDI+disp32]

Note: disp8 = 8-bit displacement. disp32 = 32-bit displacement.

TABLE B–7 Effective address calculations for the 8086 and 8088

Type	*Clocks*	*Example*
Base or index	5	MOV CL,[DI]
Displacement	3	MOV AL,DATA
Base plus index [BP+DI] or [BX+SI]	7	MOV BL,[BP+DI]
Base plus index [BP+SI] or [BX+DI]	8	MOV CL,[BP+SI]
Displacement plus base or index	9	MOV DH,[DI+20H]
Base plus index plus displacement [BP+DI+disp] or [BX+SI+disp]	11	MOV CX,DATA[BX+SI]
Base plus index plus displacement [BP+SI+disp] or [BX+DI+disp]	12	MOV CX,[BX+DI+2]
Segment override prefix	ea+2	MOV AL,ES:DATA

To change the default size as selected by the D-bit, the address-size prefix (67H) must be placed before instructions in the 80386, 80486, and Pentium. For example, the MOV AX,[ECX] instruction must have the address-size prefix placed before it in machine code if the default size is 16 bits. If the default size is 32 bits, the address prefix is not needed with this instruction. The operand-override prefix (66H) functions in much the same way as the address-size prefix. In the previous example, the operand size is 16 bits. If the D-bit selects 32-bit operands and addresses, this instruction requires the operand-size prefix.

The Pentium microprocessor can often execute two instructions that are not dependent on each other simultaneously. For example, the ADD AL,6 instruction, listed as 1 clock, may execute with another instruction listed as 1 clock. The result is two instructions executed in one clocking period.

INSTRUCTION SET SUMMARY (pp. 662–744)

AAA	ASCII adjust AL after addition	
00110111	O D I T S Z A P C ? ? ? * ? *	
Example		Clocks
AAA	8086	8
	8088	8
	80286	3
	80386	4
	80486	3
	Pentium	3

AAD	ASCII adjust AX before division	
11010101 00001010	O D I T S Z A P C ? * * ? * ?	
Example		Clocks
AAD	8086	60
	8088	60
	80286	14
	80386	19
	80486	14
	Pentium	10

AAM — ASCII adjust AX after multiplication

11010100 00001010

O	D	I	T	S	Z	A	P	C
?				*	*	?	*	?

Example		Clocks
AAM	8086	83
	8088	83
	80286	16
	80386	17
	80486	15
	Pentium	18

AAS — ASCII adjust AL after subtraction

00111111

O	D	I	T	S	Z	A	P	C
?				?	?	*	?	*

Example		Clocks
AAS	8086	8
	8088	8
	80286	3
	80386	4
	80486	3
	Pentium	3

ADC — Addition with carry

000100dw oorrrmmm disp

O	D	I	T	S	Z	A	P	C
*				*	*	*	*	*

Format	Examples		Clocks
ADC reg,reg	ADC AX,BX ADC AL,BL ADC EAX,EBX ADC CX,SI ADC ESI,EDI	8086	3
		8088	3
		80286	2
		80386	2
		80486	1
		Pentium	1

ADC mem,reg	ADC DATAY,AL ADC LIST,SI ADC DATAX [DI],CL ADC [EAX],BL ADC [EBX+2*ECX],EDX	8086	16 + ea
		8088	24 + ea
		80286	7
		80386	7
		80486	3
		Pentium	3
ADC reg,mem	ADC BL,DATA1 ADC SI,LIST ADC CL,DATA2 [DI] ADC CL,[EAX] ADC EDX,[EBX+100H]	8086	9 + ea
		8088	13 + ea
		80286	7
		80386	6
		80486	2
		Pentium	2
100000sw oo010mmm disp data			
Format	Examples		Clocks
ADC reg,imm	ADC CX,3 ADC DI,1AH ADC DL,34H ADC EAX,12345 ADC CX,1234H	8086	4
		8088	4
		80286	3
		80386	2
		80486	1
		Pentium	1
ADC mem,imm	ADC DATA4,33 ADC LIST,'A' ADC DATA [DI],2 ADC BYTE PTR [EAX],3 ADC WORD PTR [DI],669H	8086	17 + ea
		8088	23 + ea
		80286	7
		80386	7
		80486	3
		Pentium	3

0001010w data Format	Examples		Clocks
ADC acc,imm	ADC AX,3 ADC AL,1AH ADC AH,34 ADC EAX,3 ADC AL,'Z'	8086	4
		8088	4
		80286	3
		80386	2
		80486	1
		Pentium	1

ADD Addition

000000dw oorrrmmm disp

O	D	I	T	S	Z	A	P	C
*				*	*	*	*	*

Format	Examples		Clocks

ADD reg,reg	ADD AX,BX
	ADD AL,BL
	ADD EAX,EBX
	ADD CX,SI
	ADD ESI,EDI
		8086	3

		8088	3
		80286	2
		80386	2
		80486	1
		Pentium	1

ADD mem,reg	ADD DATA5,AL
	ADD LIST,SI
	ADD DATA6 [DI],CL
	ADD [EAX],CL
	ADD [EBX+4*EDX],EBX
		8086	16 + ea

		8088	24 + ea
		80286	7
		80386	7
		80486	3
		Pentium	3

ADD reg,mem	ADD BL,DATA
	ADD SI,LIST
	ADD CL,DATA7 [DI]
	ADD CL,[EAX]
	ADD EDX,[EBX+200H]
		8086	9 + ea

		8088	13 + ea
		80286	7
		80386	6
		80486	2
		Pentium	2

100000sw oo000mmm disp data Format	Examples		Clocks

ADD reg,imm	ADD CX,3
	ADD DI,1AH
	ADD DL,34H
	ADD EAX,123456
	ADD CX,18AFH
		8086	4

		8088	4
		80286	3
		80386	2
		80486	1
		Pentium	1

ADD mem,imm	ADD DATAF,33
	ADD LIST,'A'
	ADD DATAM [DI],2
	ADD BYTE PTR [EAX],3
	ADD WORD PTR [DI],6A8H
		8086	17 + ea

		8088	23 + ea
		80286	7
		80386	7
		80486	3
		Pentium	3
0000010w data Format	Examples		Clocks

ADD acc,imm	ADD AX, 3
	ADD AL,1AH
	ADD AH,56
	ADD EAX,3
	ADD AL,'D'
		8086	4

		8088	4
		80286	3
		80386	2
		80486	1
		Pentium	1

AND Logical AND

001000dw oorrrmmm disp

O	D	I	T	S	Z	A	P	C
0				*	*	?	*	0

Format	Examples		Clocks
AND reg,reg	AND CX,BX AND DL,BL AND ECX,EBX AND BP,SI AND EDX,EDI	8086	3
		8088	3
		80286	2
		80386	2
		80486	1
		Pentium	1
AND mem,reg	AND BIT,CH AND LIST,DI AND DATAZ [BX],CL AND [ECX],AL AND [EDX+8*ECX],EDI	8086	16 + ea
		8088	24 + ea
		80286	7
		80386	7
		80486	3
		Pentium	3
AND reg,mem	AND BL,DATAW AND SI,LIST AND CL,DATAQ [DI] AND CL,[EAX] AND EDX,[EBX+34AH]	8086	9 + ea
		8088	13 + ea
		80286	7
		80386	6
		80486	2
		Pentium	2

100000sw oo100mmm disp data

Format	Examples		Clocks
AND reg,imm	AND BP,1 AND DI,10H AND DL,34H AND EBP,12345 AND SP,1234H	8086	4
		8088	4
		80286	3
		80386	2
		80486	1
		Pentium	1

AND mem,imm	AND DATAD,33 AND LIST,4 AND DATAS [SI],2 AND BYTE PTR [EAX],3 AND DWORD PTR [DI],32	8086	17 + ea
		8088	23 + ea
		80286	7
		80386	7
		80486	3
		Pentium	3

0010010w data

Format	Examples		Clocks
AND acc,imm	AND AX,15 AND AL,1FH AND AH,34 AND EAX,3 AND AL,'R'	8086	4
		8088	4
		80286	3
		80386	2
		80486	1
		Pentium	1

ARPL Adjust requested privilege level

01100011 oorrrmmm disp

O	D	I	T	S	Z	A	P	C
					*			

Format	Examples		Clocks
ARPL reg,reg	ARPL AX,BX ARPL BX,SI ARPL CX,DX ARPL BX,AX ARPL DI,SI	8086	—
		8088	—
		80286	10
		80386	20
		80486	9
		Pentium	7
ARPL mem,reg	ARPL NUMB,AX ARPL LIST,DI ARPL DATA2 [BX],CX ARPL [ECX],AX ARPL [EDX+4*ECX],DI	8086	—
		8088	—
		80286	11
		80386	21
		80486	9
		Pentium	7

BOUND Check array against bounds

01100010 oorrrmmm disp

O	D	I	T	S	Z	A	P	C

Format	Examples		Clocks
BOUND reg,mem	BOUND AX,BETS BOUND BX,SAID BOUND CX,DATA3 BOUND BX,[DI] BOUND DI,[BX+2]	8086	—
		8088	—
		80286	13
		80386	10
		80486	7
		Pentium	8

BSF Bit scan forward

00001111 10111100 oorrmmm disp

O	D	I	T	S	Z	A	P	C
?				?	*	?	?	?

Format	Examples		Clocks
BSF reg,reg	BSF AX,BX BSF BX,SI BSF ECX,EBX BSF EBX,EAX BSF DI,SI	8086	—
		8088	—
		80286	—
		80386	10 + 3n
		80486	6–42
		Pentium	6–42
BSF reg,mem	BSF AX,DATAD BSF BP,LISTG BSF ECX,MEMORY BSF EAX,DATA6 BSF DI,[ECX]	8086	—
		8088	—
		80286	—
		80386	10 + 3n
		80486	7–43
		Pentium	6–43

BSR Bit scan reverse

00001111 10111101 oorrmmm disp

O	D	I	T	S	Z	A	P	C
?				?	*	?	?	?

Format	Examples		Clocks
BSR reg,reg	BSR AX,BX BSR BX,SI BSR ECX,EBX BSR EBX,EAX BSR DI,SP	8086	—
		8088	—
		80286	—
		80386	10 + 3n
		80486	6–103
		Pentium	7–71
BSR reg,mem	BSR AX,DATAE BSR BP,LISTG BSR ECX,MEMORY BSR EAX,DATA6 BSR DI,[EBX]	8086	—
		8088	—
		80286	—
		80386	10 + 3n
		80486	7–104
		Pentium	7–72

BSWAP Byte swap

00001111 11001rrr

O	D	I	T	S	Z	A	P	C

Format	Examples		Clocks
BSWAP reg	BSWAP EAX BSWAP EBX BSWAP ECX BSWAP EDX BSWAP EDI	8086	—
		8088	—
		80286	—
		80386	—
		80486	1
		Pentium	1

BT Bit test

00001111 10111010 oo100mmm disp data

O	D	I	T	S	Z	A	P	C
								*

Format	Examples		Clocks
BT reg,imm8	BT AX,2 BT CX,4 BT BP,10H BT CX,8 BT BX,2	8086	—
		8088	—
		80286	—
		80386	3
		80486	3
		Pentium	4
BT mem,imm8	BT DATA1,2 BT LIST,2 BT DATA4 [DI],2 BT [BX],1 BT FROG,3	8086	—
		8088	—
		80286	—
		80386	6
		80486	3
		Pentium	4

00001111 10100011 disp

Format	Examples		Clocks
BT reg,reg	BT AX,CX BT CX,DX BT BP,AX BT SI,CX BT CX,BP	8086	—
		8088	—
		80286	—
		80386	3
		80486	3
		Pentium	4
BT mem,reg	BT DATA1,AX BT LIST,DX BT DATA3,CX BT DATA9,BX BT DATA3 [DI],AX	8086	—
		8088	—
		80286	—
		80386	12
		80486	8
		Pentium	9

BTC Bit test and complement

00001111 10111010 oo111mmm disp data

O	D	I	T	S	Z	A	P	C
								*

Format	Examples		Clocks
BTC reg,imm8	BTC AX,2 BTC CX,4 BTC BP,10H BTC CX,8 BTC BX,2	8086	—
		8088	—
		80286	—
		80386	6
		80486	6
		Pentium	7
BTC mem,imm8	DATA1,2 BTC LIST,2 BTC DATA [DI],3 BTC [BX],1 BTC TOAD,5	8086	—
		8088	—
		80286	—
		80386	8
		80486	8
		Pentium	8

00001111 10111011 disp

Format	Examples		Clocks
BTC reg,reg	BTC AX,CX BTC CX,DX BTC BP,AX BTC SI,CX BTC CX,BX	8086	—
		8088	—
		80286	—
		80386	6
		80486	6
		Pentium	7
BTC mem,reg	BTC DATA1,AX BTC LIST,DX BTC DATA3,CX BTC DATA9,BX BTC DATAS [DI],AX	8086	—
		8088	—
		80286	—
		80386	13
		80486	13
		Pentium	13

BTR Bit test and reset

00001111 10111010 oo110mmm disp data

O	D	I	T	S	Z	A	P	C
								*

Format	Examples	Clocks	
BTR reg,imm8	BTR AX,2 BTR CX,4 BTR BP,10H BTR CX,8 BTR BX,2	8086	—
		8088	—
		80286	—
		80386	6
		80486	6
		Pentium	7
BTR mem,imm8	BTR DATA1,2 BTR LIST,2 BTR DATAZ [DI],4 BTR [BX],1 BTR SLED,6	8086	—
		8088	—
		80286	—
		80386	8
		80486	8
		Pentium	8

00001111 10110011 disp

Format	Examples	Clocks	
BTR reg,reg	BTR AX,CX BTR CX,DX BTR BP,AX BTR SI,CX BTR BP,CX	8086	—
		8088	—
		80286	—
		80386	6
		80486	6
		Pentium	7
BTR mem,reg	BTR DATA1,AX BTR LIST,DX BTR DATA3,CX BTR DATA9,BX BTR DATAD [BX],AX	8086	—
		8088	—
		80286	—
		80386	13
		80486	13
		Pentium	13

BTS Bit test and set

00001111 10111010 oo101mmm disp data

O	D	I	T	S	Z	A	P	C
								*

Format	Examples		Clocks
BTS reg,imm8	BTS AX,2 BTS CX,4 BTS BP,10H BTS CX,8 BTS BX,3	8086	—
		8088	—
		80286	—
		80386	6
		80486	6
		Pentium	7
BTS mem,imm8	BTS DATA1,2 BTS LIST,2 BTS DATAT [BP],7 BTS [BX],1 BTS FROG,3	8086	—
		8088	—
		80286	—
		80386	8
		80486	8
		Pentium	8

00001111 10101011 disp

Format	Examples		Clocks
BTS reg,reg	BTS AX,CX BTS CX,DX BTS BP,AX BTS SI,CX BTS CX,BP	8086	—
		8088	—
		80286	—
		80386	6
		80486	6
		Pentium	7
BTS mem,reg	BTS DATA1,AX BTS LIST,DX BTS DATA3,CX BTS DATA9,BX BTS DATAL [BP],AX	8086	—
		8088	—
		80286	—
		80386	13
		80486	13
		Pentium	13

CALL   Call procedure (subroutine)

11101000 disp

O D I T   S Z A P C

Format	Examples	Clocks	
CALL label (near)	CALL FOR_FUN CALL HOME CALL ET CALL WAITING CALL SOMEONE	8086	19
		8088	23
		80286	7
		80386	3
		80486	3
		Pentium	1

10011010 disp

Format	Examples	Clocks	
CALL label (far)	CALL FAR PTR DATES CALL WHAT CALL WHERE CALL FARCE CALL WHOM	8086	28
		8088	36
		80286	13
		80386	17
		80486	18
		Pentium	4

11111111 oo010mmm

Format	Examples	Clocks	
CALL reg (near)	CALL AX CALL BX CALL CX CALL DI CALL SI	8086	16
		8088	20
		80286	7
		80386	7
		80486	5
		Pentium	2
CALL mem (near)	CALL ADDRESS CALL NEAR PTR [DI] CALL DATA CALL FROG CALL HERO	8086	21 + ea
		8088	29 + ea
		80286	11
		80386	10
		80486	5
		Pentium	2

11111111 oo011mmm Format	Examples		Clocks
CALL mem (far)	CALL FAR_LIST [SI] CALL FROM_HERE CALL TO_THERE CALL SIXX CALL OCT	8086	37 + ea
		8088	53 + ea
		80286	16
		80386	22
		80486	17
		Pentium	5

CBW Convert byte to word

10011000 Example	O D I T S Z A P C Clocks	
CBW	8086	2
	8088	2
	80286	2
	80386	3
	80486	3
	Pentium	3

CDQ Convert doubleword to quadword

10011001 Example	O D I T S Z A P C Clocks	
CDQ	8086	—
	8088	—
	80286	—
	80386	2
	80486	2
	Pentium	2

CLC Clear carry flag

11111000	O	D	I	T	S	Z	A	P	C
									0

Example		Clocks
CLC	8086	2
	8088	2
	80286	2
	80386	2
	80486	2
	Pentium	2

CLD Clear direction flag

11111100	O	D	I	T	S	Z	A	P	C
		0							

Example		Clocks
CLD	8086	2
	8088	2
	80286	2
	80386	2
	80486	2
	Pentium	2

CLI Clear interrupt flag

11111010	O	D	I	T	S	Z	A	P	C
			0						

Example		Clocks
CLI	8086	2
	8088	2
	80286	3
	80386	3
	80486	5
	Pentium	7

CLTS	Clear task switched flag (CR0)	
00001111 00000110	O D I T S Z A P C	
Example		Clocks
CLTS	8086	—
	8088	—
	80286	2
	80386	5
	80486	7
	Pentium	10

CMC	Complement carry flag	
10011000	O D I T S Z A P C	
	C: *	
Example		Clocks
CMC	8086	2
	8088	2
	80286	2
	80386	2
	80486	2
	Pentium	2

CMP	Compare operands		
001110dw oorrrmmm disp		O D I T S Z A P C	
		O: * S: * Z: * A: * P: * C: *	
Format	Examples		Clocks

CMP reg,reg	CMP AX,BX
	CMP AL,BL
	CMP EAX,EBX
	CMP CX,SI
	CMP ESI,EDI
		8086	3

		8088	3
		80286	2
		80386	2
		80486	1
		Pentium	1

CMP mem,reg	CMP DATAF,AL
	CMP LIST,SI
	CMP DATAE [BX],CL
	CMP [EAX],AL
	CMP [EBX+2*ECX],EBX
		8086	9 + ea

		8088	13 + ea
		80286	7
		80386	5
		80486	2
		Pentium	2

CMP reg,mem	CMP BL,DATAZ
	CMP SI,LIST
	CMP CL,DATAD [DI]
	CMP CL,[EAX]
	CMP EDX,[EBX+200H]
		8086	9 + ea

		8088	13 + ea
		80286	6
		80386	6
		80486	2
		Pentium	2
100000sw oo111mmm disp data			
Format	Examples		Clocks

CMP reg,imm	CMP CX,3
	CMP DI,1AH
	CMP DL,34H
	CMP EBX,12345
	CMP CX,123AH
		8086	4

		8088	4
		80286	3
		80386	2
		80486	1
		Pentium	1

CMP mem,imm	CMP DATAS,33
	CMP LIST,'A'
	CMP DATAF [DI],87H
	CMP BYTE PTR [EAX],3
	CMP WORD PTR [SI],7
		8086	10 + ea

		8088	14 + ea
		80286	6
		80386	5
		80486	2
		Pentium	2

0011110w data

Format	Examples		Clocks

CMP acc,imm	CMP AX,3
	CMP AL,1AH
	CMP AH,34
	CMP EAX,3
	CMP AL,'I'
		8086	4

		8088	4
		80286	3
		80386	2
		80486	1
		Pentium	1

CMPS Compare strings

1010011w

O D I T S Z A P C
* * * * * *

Format	Examples		Clocks
CMPSB CMPSW CMPSD	CMPSB CMPSW CMPSD CMPS DATA1 REPE CMPSB REPNE CMPSW	8086	22
		8088	30
		80286	8
		80386	10
		80486	8
		Pentium	5

CMPXCHG Compare and exchange

00001111 1011000w 11rrrrrr

O D I T S Z A P C
* * * * * *

Format	Examples		Clocks
CMPXCHG reg,reg	CMPXCHG EAX,EBX CMPXCHG ECX,EDX	8086	—
		8088	—
		80286	—
		80386	—
		80486	6
		Pentium	6

00001111 1011000w oorrrmmm

Format	Examples	Clocks	
CMPXCHG mem,reg	CMPXCHG DATAD,EAX CMPXCHG DATA2,EBX	8086	—
		8088	—
		80286	—
		80386	—
		80486	7
		Pentium	6

CMPXCHG8B Compare and exchange 8 bytes

00001111 11000111 11rrrrrr

O	D	I	T	S	Z	A	P	C
					*			

Format	Examples	Clocks	
CMPXCHG mem64	CMPXCHG8B DATA3	8086	—
		8088	—
		80286	—
		80386	—
		80486	—
		Pentium	10

CPUID CPU identification code to EAX

00001111 10100010

O	D	I	T	S	Z	A	P	C

Example	Clocks	
CPUID	8086	—
	8088	—
	80286	—
	80386	—
	80486	—
	Pentium	14

CWD — Convert word to doubleword

10011001

Example	O D I T S Z A P C	Clocks
CWD	8086	5
	8088	5
	80286	2
	80386	2
	80486	3
	Pentium	2

CWDE — Convert word to extended doubleword

10011000

Example	O D I T S Z A P C	Clocks
CWDE	8086	—
	8088	—
	80286	—
	80386	3
	80486	3
	Pentium	3

DAA — Decimal adjust AL after addition

00100111

O D I T S Z A P C
? * * * * *

Example		Clocks
DAA	8086	4
	8088	4
	80286	3
	80386	4
	80486	2
	Pentium	3

DAS — Decimal adjust AL after subtraction

00101111

O	D	I	T	S	Z	A	P	C
?				*	*	*	*	*

Example		Clocks
DAS	8086	4
	8088	4
	80286	3
	80386	4
	80486	2
	Pentium	3

DEC — Decrement

1111111w oo001mmm disp

O	D	I	T	S	Z	A	P	C
*				*	*	*	*	

Format	Examples		Clocks
DEC reg8	DEC BL DEC BH DEC CL DEC DH DEC AH	8086	3
		8088	3
		80286	2
		80386	2
		80486	1
		Pentium	1
DEC mem	DEC DATAR DEC LIST DEC DATA [SI] DEC BYTE PTR [EAX] DEC WORD PTR [DI]	8086	15 + ea
		8088	23 + ea
		80286	7
		80386	6
		80486	3
		Pentium	3

01001rrr Format	Examples		Clocks
DEC reg16 DEC reg32	DEC AX DEC EAX DEC CX DEC EBX DEC DI	8086	3
		8088	3
		80286	2
		80386	2
		80486	1
		Pentium	1

DIV Unsigned division

1111011w oo110mmm disp Format	 Examples	O D I T ?	S Z A P C ? ? ? ? ? Clocks
DIV reg	DIV BL DIV BH DIV ECX DIV BH DIV CH	8086	162
		8088	162
		80286	22
		80386	38
		80486	40
		Pentium	17–41
DIV mem	DIV DATA2 DIV LIST DIV DATA3 [DI] DIV BYTE PTR [EAX] DIV WORD PTR [DI]	8086	168
		8088	176
		80286	25
		80386	41
		80486	40
		Pentium	17–41

ENTER Create a stack frame

11001000 data

O D I T S Z A P C

Format	Examples		Clocks
ENTER imm,0	ENTER 4,0 ENTER 8,0 ENTER 100,0 ENTER 200,0 ENTER 1024,0	8086	—
		8088	—
		80286	11
		80386	10
		80486	14
		Pentium	11
ENTER imm,1	ENTER 4,1 ENTER 10,1	8086	—
		8088	—
		80286	15
		80386	12
		80486	17
		Pentium	15
ENTER imm,imm	ENTER 3,6 ENTER 100,3	8086	—
		8088	—
		80286	12
		80386	15
		80486	17
		Pentium	15

ESC Escape

11011nnn oonnnmmm — O D I T S Z A P C
nnnnnn = opcode for coprocessor

Format	Examples		Clocks
ESC imm,reg	ESC 5,AL ESC 5,BH ESC 6,CH FADD ST,ST(3)	8086	2
		8088	2
		80286	20
		80386	var
		80486	var
		Pentium	var
ESC imm,mem	ESC 2,DATAS ESC 3,FROG FADD DATAE FMUL FROG	8086	8 + ea
		8088	12 + ea
		80286	20
		80386	var
		80486	var
		Pentium	var

HLT Halt

11110100 — O D I T S Z A P C

Example		Clocks
HLT	8086	2
	8088	2
	80286	2
	80386	5
	80486	4
	Pentium	12

IDIV Signed division

1111011w oo111mmm disp

O	D	I	T	S	Z	A	P	C
?				?	?	?	?	?

Format	Examples	Clocks	
IDIV reg	IDIV BL IDIV BH IDIV ECX IDIV BH IDIV CX	8086	184
		8088	184
		80286	25
		80386	43
		80486	43
		Pentium	22–46
IDIV mem	IDIV DATAF IDIV LIST IDIV DATAG [DI] IDIV BYTE PTR [EAX] IDIV WORD PTR [DI]	8086	190
		8088	194
		80286	28
		80386	46
		80486	44
		Pentium	22–46

IMUL Signed multiplication

1111011w oo101mmm disp

O	D	I	T	S	Z	A	P	C
*				?	?	?	?	*

Format	Examples	Clocks	
IMUL reg	IMUL BL IMUL CL IMUL CX IMUL ECX IMUL EBX	8086	154
		8088	154
		80286	21
		80386	38
		80486	42
		Pentium	10–11

Format	Examples		Clocks
IMUL mem	IMUL DATAN IMUL LIST IMUL DATAK [SI] IMUL BYTE PTR [EAX] IMUL WORD PTR [DI]	8086	160
		8088	164
		80286	24
		80386	41
		80486	42
		Pentium	10–11

011010sl oorrrmmm disp data

Format	Examples		Clocks
IMUL reg,imm	IMUL CX,16 IMUL DX,100 IMUL EAX,20	8086	—
		8088	—
		80286	21
		80386	38
		80486	42
		Pentium	10
IMUL reg,reg,imm	IMUL DX,AX,2 IMUL CX,DX,3 IMUL BX,AX,33	8086	—
		8088	—
		80286	21
		80386	38
		80486	42
		Pentium	10
IMUL reg,mem,imm	IMUL CX,DATA,4	8086	—
		8088	—
		80286	24
		80386	38
		80486	42
		Pentium	10

00001111 10101111 oorrrmmm disp

Format	Examples		Clocks
IMUL reg,reg	IMUL CX,DX IMUL DX,BX IMUL EAX,ECX	8086	—
		8088	—
		80286	—
		80386	38
		80486	42
		Pentium	10
IMUL reg,mem	IMUL DX,DATA IMUL CX,FROG IMUL BX,LISTS	8086	—
		8088	—
		80286	—
		80386	41
		80486	42
		Pentium	10

IN Input data from port

1110010w port number O D I T S Z A P C

Format	Examples		Clocks
IN acc,pt	IN AL,12H IN AX,12H IN AL,0FFH IN AX,0FFH IN EAX,10H	8086	10
		8088	14
		80286	5
		80386	12
		80486	14
		Pentium	7

1110110w Format	Examples		Clocks
IN acc,DX	IN AL,DX IN AX,DX IN EAX,DX	8086	8
		8088	12
		80286	5
		80386	13
		80486	14
		Pentium	7

INC Increment

1111111w oo000mmm disp

O	D	I	T	S	Z	A	P	C
*				*	*	*	*	*

Format	Examples		Clocks
INC reg8	INC BL INC BH INC CL INC DH INC AH	8086	3
		8088	3
		80286	2
		80386	2
		80486	1
		Pentium	1
INC mem	INC DATA2 INC LIST INC DATA3 [BX] INC BYTE PTR [EAX] INC WORD PTR [BX] INC DWORD PTR [ECX]	8086	15 + ea
		8088	23 + ea
		80286	7
		80386	6
		80486	3
		Pentium	3

01000rrr Format	Examples		Clocks
INC reg16 INC reg32	INC AX INC EAX INC CX INC EBX INC DI	8086	3
		8088	3
		80286	2
		80386	2
		80486	1
		Pentium	1

INS Input string from port

0110110w Format	Examples	O D I T	S Z A P C Clocks
INSB INSW INSD	INSB INSW INSD INS DATA REP INSB	8086	—
		8088	—
		80286	5
		80386	15
		80486	17
		Pentium	9

INT Interrupt

11001101 type Format	Examples	O D I T	S Z A P C Clocks
INT type	INT 10H INT 255 INT 21H INT 20H INT 15H	8086	51
		8088	71
		80286	23
		80386	37
		80486	30
		Pentium	16

11001100 Example		Clocks
INT 3	8086	52
	8088	72
	80286	23
	80386	33
	80486	26
	Pentium	13

INTO Interrupt on overflow

11001110 Example	O D I T	S Z A P C Clocks
INTO	8086	53
	8088	73
	80286	24
	80386	35
	80486	28
	Pentium	13

INVD Invalidate data cache

00001111 00001000 Example	O D I T	S Z A P C Clocks
INVD	8086	—
	8088	—
	80286	—
	80386	—
	80486	4
	Pentium	15

INVLPG Invalidate TLB entry

00001111 00000001 oo111mmm — O D I T S Z A P C

Format	Examples		Clocks
INVLPG mem	INVLPG DATA INVLPG LIST	8086	—
		8088	—
		80286	—
		80386	—
		80486	12
		Pentium	25

IRET/IRETD Interrupt return

11001101 data

O	D	I	T	S	Z	A	P	C
*	*	*	*	*	*	*	*	*

Format	Examples		Clocks
IRET IRETD	IRET IRETD IRET 10H	8086	32
		8088	44
		80286	17
		80386	22
		80486	15
		Pentium	8

Jconditional Conditional jump

0111cccc disp — O D I T S Z A P C

Format	Examples		Clocks
Jcc label (8-bit disp)	JA BELOW JB ABOVE JG GREATER JE EQUAL JZ ZERO	8086	16/4
		8088	16/4
		80286	7/3
		80386	7/3
		80486	3/1
		Pentium	1

00001111 1000cccc disp

Format	Examples	Clocks	
Jcc label (16-bit disp)	JNE NOT_MORE JLE LESS_THAN	8086	—
		8088	—
		80286	—
		80386	7/3
		80486	3/1
		Pentium	1

Condition Codes	Mnemonic	Flag	Description
0000	JO	O = 1	Jump if overflow
0001	JNO	O = 0	Jump if no overflow
0010	JB/JNAE	C = 1	Jump if below
0011	JAE/JNB	C = 0	Jump if above or equal
0100	JE/JZ	Z = 1	Jump if equal/zero
0101	JNE/JNZ	Z = 0	Jump if not equal/not zero
0110	JBE/JNA	C = 1 + Z = 1	Jump if below or equal
0111	JA/JNBE	C = 0 • Z = 0	Jump if above
1000	JS	S = 1	Jump if sign
1001	JNS	S = 0	Jump if no sign
1010	JP/JPE	P = 1	Jump if parity even
1011	JNP/JPO	P = 0	Jump if parity odd
1100	JL/JNGE	S • O	Jump if less than
1101	JGE/JNL	S = O	Jump if greater or equal
1110	JLE/JNG	Z = 1 + S • O	Jump if less than or equal
1111	JG/JNLE	Z = 0 + S = O	Jump if greater

JCXZ/JECXZ Jump if CX (ECX) equals zero

11100011 ODIT SZAPC

Format	Examples	Clocks	
JCXZ label JECXZ label	JCXZ LOTSA JCXZ OVER JECXZ UPPER JECXZ UNDER JCXZ NEXT	8086	18/6
		8088	18/6
		80286	8/4
		80386	9/5
		80486	8/5
		Pentium	6/5

JMP	Unconditional jump		
11101011 disp		O D I T	S Z A P C
Format	Examples		Clocks
JMP label (short)	JMP SHORT UP JMP SHORT DOWN JMP SHORT OVER JMP SHORT CIRCUIT JMP SHORT ARM	8086	15
		8088	15
		80286	7
		80386	7
		80486	3
		Pentium	1
11101001 disp			
Format	Examples		Clocks
JMP label (near)	JMP VER JMP FROG JMP UNDER JMP NEAR PTR OVER	8086	15
		8088	15
		80286	7
		80386	7
		80486	3
		Pentium	1
11101010 disp			
Format	Examples		Clocks
JMP label (far)	JMP VER JMP FROG JMP UNDER JMP FAR PTR THERE	8086	15
		8088	15
		80286	11
		80386	12
		80486	17
		Pentium	3

11111111 oo100mmm Format	Examples		Clocks
JMP reg (near)	JMP AX JMP EAX JMP CX JMP DX	8086	11
		8088	11
		80286	7
		80386	7
		80486	3
		Pentium	2
JMP mem (near)	JMP DATA JMP LIST JMP DATA [DI+2]	8086	18 + ea
		8088	18 + ea
		80286	11
		80386	10
		80486	5
		Pentium	4

11111111 oo101mmm Format	Examples		Clocks
JMP mem (far)	JMP WAY_OFF JMP TABLE JMP UP	8086	24 + ea
		8088	24 + ea
		80286	15
		80386	12
		80486	13
		Pentium	4

LAHF Load AH from flags

10011111 Example	O D I T S Z A P C Clocks	
LAHF	8086	4
	8088	4
	80286	2
	80386	2
	80486	3
	Pentium	2

LAR Load access rights byte

00001111 00000010 oorrrmmm disp

O	D	I	T	S	Z	A	P	C
					*			

Format	Examples		Clocks
LAR reg,reg	LAR AX,BX LAR CX,DX LAR EAX,ECX	8086	—
		8088	—
		80286	14
		80386	15
		80486	11
		Pentium	8
LAR reg,mem	LAR CX,DATA LAR AX,LIST LAR ECX,FROG	8086	—
		8088	—
		80286	16
		80386	16
		80486	11
		Pentium	8

LDS Load far pointer

11000101 oorrrmmm

O	D	I	T	S	Z	A	P	C

Format	Examples		Clocks

LDS reg,mem	LDS DI,DATA
	LDS SI,LIST
	LDS BX,ARRAY
	LDS CX,PNTR
		8086	16 + ea

		8088	24 + ea
		80286	7
		80386	7
		80486	6
		Pentium	4

LEA Load effective address

10001101 oorrrmmm disp		O D I T S Z A P C	
Format	Examples		Clocks
LEA reg,mem	LEA DI,DATA1 LEA SI,LIST LEA BX,ARRAY LEA CX,PNTR LEA BP,ADDR	8086	2 + ea
		8088	2 + ea
		80286	3
		80386	2
		80486	2
		Pentium	1

LEAVE Leave high-level procedure

11001001	O D I T S Z A P C	
Example		Clocks
LEAVE	8086	—
	8088	—
	80286	5
	80386	4
	80486	5
	Pentium	3

LES Load far pointer

11000100 oorrrmmm		O D I T S Z A P C	
Format	Examples		Clocks
LES reg,mem	LES DI,DATA LES SI,LIST LES BX,ARRAY LES CX,PNTR	8086	16 + ea
		8088	24 + ea
		80286	7
		80386	7
		80486	6
		Pentium	4

LFS — Load far pointer

00001111 10110100 oorrrmmm disp — O D I T S Z A P C

Format	Examples		Clocks
LFS reg,mem	LFS DI,DATA LFS SI,LIST LFS BX,ARRAY LFS CX,PNTR	8086	—
		8088	—
		80286	—
		80386	7
		80486	6
		Pentium	4

LGDT — Load global descriptor table

00001111 00000001 oo010mmm disp — O D I T S Z A P C

Format	Examples		Clocks
LGDT mem64	LGDT DESCRIP LGDT TABLE	8086	—
		8088	—
		80286	11
		80386	11
		80486	11
		Pentium	6

LGS — Load far pointer

00001111 10110101 oorrrmmm disp — O D I T S Z A P C

Format	Examples		Clocks
LGS reg,mem	LGS DI,DATA LGS SI,LIST LGS BX,ARRAY LGS CX,PNTR	8086	—
		8088	—
		80286	—
		80386	7
		80486	6
		Pentium	4

LIDT	Load interrupt descriptor table		
00001111 00000001 oo011mmm disp		O D I T	S Z A P C
Format	Examples		Clocks
LIDT mem64	LIDT DATA LIDT DESCRIP	8086	—
		8088	—
		80286	12
		80386	11
		80486	11
		Pentium	6

LLDT	Load local descriptor table		
00001111 00000000 oo010mmm disp		O D I T	S Z A P C
Format	Examples		Clocks
LLDT reg	LLDT AX LLDT CX	8086	—
		8088	—
		80286	17
		80386	20
		80486	11
		Pentium	9
LLDT mem	LLDT DATA2 LLDT LIST	8086	—
		8088	—
		80286	19
		80386	24
		80486	11
		Pentium	9

LMSW Load machine status word

00001111 00000001 oo110mmm disp — O D I T S Z A P C

should only be used with the 80286

Format	Examples		Clocks
LMSW reg	LMSW AX LMSW CX	8086	—
		8088	—
		80286	3
		80386	10
		80486	2
		Pentium	8
LMSW mem	LMSW DATA4 LMSW LIST	8086	—
		8088	—
		80286	6
		80386	13
		80486	3
		Pentium	8

LOCK Lock the bus

11110000 — O D I T S Z A P C

Format	Examples		Clocks
LOCK inst	LOCK:XCHG AX,BX LOCK:MOV AL,AH	8086	2
		8088	2
		80286	0
		80386	0
		80486	1
		Pentium	1

LODS Load string operand

1010110w Format	Examples	O D I T S Z A P C	Clocks
LODSB LODSW LODSD	LODSB LODSW LODSD LODS DATA5 LODS ES:DATA6	8086	12
		8088	16
		80286	5
		80386	5
		80486	5
		Pentium	2

LOOP Loop until CX = 0

11100010 disp Format	Examples	O D I T S Z A P C	Clocks
LOOP label	LOOP DATA_FROG LOOP BACK LOOPD LOOPS	8086	17/5
		8088	17/5
		80286	8/4
		80386	11
		80486	7/6
		Pentium	5/6

LOOPE Loop while equal

11100001 disp Format	Examples	O D I T S Z A P C	Clocks
LOOPE label LOOPZ label	LOOPE NEXT LOOPE AGAIN LOOPZ REPEAT LOOPED NOW	8086	18/6
		8088	18/6
		80286	8/4
		80386	11
		80486	9/6
		Pentium	7/8

LOOPNE Loop while not equal

11100000 disp — O D I T S Z A P C

Format	Examples		Clocks
LOOPNE label LOOPNZ label	LOOPNE AGAIN LOOPNE BACK LOOPNZ REPL LOOPED WHEN	8086	19/5
		8088	19/5
		80286	8/4
		80386	11
		80486	9/6
		Pentium	7/8

LSL Load segment limit

00001111 00000011 oorrrmmm disp — O D I T S Z A P C (Z: *)

Format	Examples		Clocks
LSL reg,reg	LSL AX,BX LSL CX,BX LSL DX,AX	8086	—
		8088	—
		80286	14
		80386	25
		80486	10
		Pentium	8
LSL reg,mem	LSL AX,LIMIT LSL EAX,NUMB	8086	—
		8088	—
		80286	16
		80386	26
		80486	10
		Pentium	8

LSS Load far pointer

00001111 10110010 oorrrmmm disp

O D I T S Z A P C

Format	Examples		Clocks
LSS reg,mem	LSS DI,DATA LSS SI,LIST LSS BX,ARRAY LSS CX,PNTR	8086	—
		8088	—
		80286	—
		80386	7
		80486	6
		Pentium	4

LTR Load task register

00001111 00000000 oo001mmm disp

O D I T S Z A P C

Format	Examples		Clocks
LTR reg	LTR AX LTR CX LTR DX	8086	—
		8088	—
		80286	17
		80386	23
		80486	20
		Pentium	10
LTR mem	LTR TASK LTR EDGE	8086	—
		8088	—
		80286	19
		80386	27
		80486	20
		Pentium	10

MOV Move data

100010dw oorrrmmm disp Format	Examples	O D I T S Z A P C Clocks	

MOV reg,reg	MOV CL,CH
	MOV BH,CL
	MOV CX,DX
	MOV EAX,ECX
	MOV EBP,ESI
		8086	2

		8088	2
		80286	2
		80386	2
		80486	1
		Pentium	1

MOV mem,reg	MOV DATA7,DL
	MOV NUMB,CX
	MOV TEMP,EBX
	MOV TEMP1,CH
	MOV DATA2,CL
		8086	9 + ea

		8088	13 + ea
		80286	3
		80386	2
		80486	1
		Pentium	1

MOV reg,mem	MOV DL,DATA42
	MOV DX,NUMB
	MOV EBX,TEMP
	MOV CH,TEMP1
	MOV CL,DATA2
		8086	10 + ea

		8088	12 + ea
		80286	5
		80386	4
		80486	1
		Pentium	1

1100011w oo000mmm disp data Format	Examples	Clocks	

MOV mem,imm	MOV DATAF,23H
	MOV LIST,12H
	MOV BYTE PTR [DI],2
	MOV NUMB,234H
	MOV DWORD PTR[SI],100
		8086	10 + ea

		8088	14 + ea
		80286	3
		80386	2
		80486	1
		Pentium	1

1011wrrr data Format	Examples		Clocks

MOV reg,imm	MOV BX,23H
	MOV CX,12H
	MOV CL,2
	MOV ECX,123423H
	MOV DI,100
		8086	4

		8088	4
		80286	3
		80386	2
		80486	1
		Pentium	1

101000dw disp Format	Examples		Clocks

MOV mem,acc	MOV DATA_FOO,AL
	MOV NUMB,AX
	MOV NUMB1,EAX
		8086	10

		8088	14
		80286	3
		80386	2
		80486	1
		Pentium	1

MOV acc,mem	MOV AL,DATAE
	MOV AX,NUMB
	MOV EAX,TEMP
		8086	10

		8088	14
		80286	5
		80386	4
		80486	1
		Pentium	1

100011d0 oosssmmm disp Format	Examples		Clocks

MOV seg,reg	MOV SS,AX
	MOV DS,DX
	MOV ES,CX
		8086	2

		8088	2
		80286	2
		80386	2
		80486	1
		Pentium	1

MOV seg,mem	MOV SS,DATA9
	MOV DS,NUMB
	MOV ES,TEMP1
		8086	8 + ea

		8088	12 + ea
		80286	2
		80386	2
		80486	1
		Pentium	2/3

MOV reg,seg	MOV AX,DS
	MOV DX,ES
	MOV CX,CS
		8086	2

		8088	2
		80286	2
		80386	2
		80486	1
		Pentium	1

MOV mem,seg	MOV DATAA,SS
	MOV NUMB,ES
	MOV TEMP1,DS
		8086	9 + ea

		8088	13 + ea
		80286	3
		80386	2
		80486	1
		Pentium	1

00001111 001000d0 11rrrmmm

Format	Examples		Clocks

MOV reg,cr	MOV EAX,CR0
	MOV EBX,CR2
	MOV ECX,CR3
		8086	—

		8088	—
		80286	—
		80386	6
		80486	4
		Pentium	4

MOV cr,reg	MOV CR0,EAX
	MOV CR2,EBX
	MOV CR3,ECX
		8086	—

		8088	—
		80286	—
		80386	10
		80486	4
		Pentium	12–46
00001111 001000d1 11rrrmmm Format	Examples		Clocks

MOV reg,dr	MOV EBX,DR6
	MOV EAX,DR6
	MOV EDX,DR1
		8086	—

		8088	—
		80286	—
		80386	22
		80486	10
		Pentium	11

MOV dr,reg	MOV DR1,ECX
	MOV DR2,ESI
	MOV DR6,EBP
		8086	—

		8088	—
		80286	—
		80386	22
		80486	11
		Pentium	11
00001111 001001d0 11rrrmmm Format	Examples		Clocks
MOV reg,tr	MOV EAX,TR6 MOV EDX,TR7	8086	—
		8088	—
		80286	—
		80386	12
		80486	4
		Pentium	11

MOV tr,seg	MOV TR6,EDX MOV TR7,ESI	8086	—
		8088	—
		80286	—
		80386	12
		80486	6
		Pentium	11

MOVS Move string data

1010010w — O D I T S Z A P C

Format	Examples		Clocks
MOVSB MOVSW MOVSD	MOVSB MOVSW MOVSD MOVS DAT1,DAT2 REP MOVSB	8086	18
		8088	26
		80286	5
		80386	7
		80486	7
		Pentium	4

MOVSX Move with sign extend

00001111 1011111w oorrrmmm disp — O D I T S Z A P C

Format	Examples		Clocks
MOVSX reg,reg	MOVSX BX,AL MOVSX EAX,DX	8086	—
		8088	—
		80286	—
		80386	3
		80486	3
		Pentium	3
MOVSX reg,mem	MOVSX AX,DATA34 MOVSX EAX,NUMB	8086	—
		8088	—
		80286	—
		80386	6
		80486	3
		Pentium	3

MOVZX Move with zero extend

00001111 1011011w oorrrmmm disp

O	D	I	T	S	Z	A	P	C

Format	Examples		Clocks
MOVZX reg,reg	MOVZX BX,AL MOVZX EAX,DX	8086	—
		8088	—
		80286	—
		80386	3
		80486	3
		Pentium	3
MOVZX reg,mem	MOVZX AX,DATADE MOVZX EAX,NUMB	8086	—
		8088	—
		80286	—
		80386	6
		80486	3
		Pentium	3

MUL Unsigned multiplication

1111011w oo100mmm disp

O	D	I	T	S	Z	A	P	C
*				?	?	?	?	*

Format	Examples		Clocks
MUL reg	MUL BL MUL CX MUL ECX	8086	118
		8088	143
		80286	21
		80386	38
		80486	42
		Pentium	11

MUL mem	MUL DATA69 MUL BYTE PTR [SI] MUL WORD PTR [SI] MUL DWORD PTR [ECX]	8086	139
		8088	143
		80286	24
		80386	41
		80486	42
		Pentium	11

NEG Negate

1111011w oo011mmm disp

O	D	I	T	S	Z	A	P	C
*				*	*	*	*	*

Format	Examples		Clocks
NEG reg	NEG AX NEG CX NEG EDX	8086	3
		8088	3
		80286	2
		80386	2
		80486	1
		Pentium	1
NEG mem	NEG DATA_UP NEG NUMB NEG WORD PTR [DI]	8086	16 + ea
		8088	24 + ea
		80286	7
		80386	6
		80486	3
		Pentium	1/3

NOP	No operation		
10010000 Example		O D I T S Z A P C	Clocks
NOP		8086	3
		8088	3
		80286	3
		80386	3
		80486	3
		Pentium	1

NOT	One's complement		
1111011w oo010mmm disp Format	Examples	O D I T S Z A P C	Clocks
NOT reg	NOT AX NOT CX NOT EDX	8086	3
		8088	3
		80286	2
		80386	2
		80486	1
		Pentium	1
NOT mem	NOT DATA_TRUE NOT NUMB NOT WORD PTR [DI]	8086	16 + ea
		8088	24 + ea
		80286	7
		80386	6
		80486	3
		Pentium	3

OR Inclusive-OR

000010dw oorrrmmm disp

O	D	I	T	S	Z	A	P	C
0				*	*	?	*	0

Format	Examples		Clocks
OR reg,reg	OR CL,BL OR CX,DX OR ECX,EBX	8086	3
		8088	3
		80286	2
		80386	2
		80486	1
		Pentium	1
OR mem,reg	OR DATAZ,CL OR NUMB,CX OR [DI],CX	8086	16 + ea
		8088	24 + ea
		80286	7
		80386	7
		80486	3
		Pentium	3
OR reg,mem	OR CL,DATA_R_US OR CX,NUMB OR CX,[SI]	8086	9 + ea
		8088	13 + ea
		80286	7
		80386	6
		80486	2
		Pentium	2

100000sw oo001mmm disp data

Format	Examples		Clocks
OR reg,imm	OR CL,3 OR DX,1000H OR EBX,100000H	8086	4
		8088	4
		80286	3
		80386	2
		80486	1
		Pentium	1

Format	Examples		Clocks
OR mem,imm	OR DATA9,33 OR NUMB,4AH OR NUMS,123498H OR BYTE PTR [ECX],2	8086	17 + ea
		8088	25 + ea
		80286	7
		80386	7
		80486	3
		Pentium	3
0000110w data Format	Examples		Clocks
OR acc,imm	OR AL,3 OR AX,1000H OR EAX,100000H	8086	4
		8088	4
		80286	3
		80386	2
		80486	1
		Pentium	1

OUT Output data to port

1110011w port number Format	Examples	O D I T S Z A P C	Clocks
OUT pt,acc	OUT 12H,AL OUT 12H,AX OUT 0FFH,AL OUT 0FEH,AX OUT 10H,EAX	8086	10
		8088	14
		80286	3
		80386	10
		80486	10
		Pentium	12
1110111w Format	Examples		Clocks
OUT DX,acc	OUT DX,AL OUT DX,AX OUT DX,EAX	8086	8
		8088	12
		80286	3
		80386	11
		80486	10
		Pentium	13

OUTS — Output string data to port

1110011w port number		O D I T S Z A P C	
Format	**Examples**		**Clocks**
OUTSB OUTSW OUTSD	OUTSB OUTSW OUTSD OUTS DATAD REP OUTSB	8086	—
		8088	—
		80286	5
		80386	14
		80486	10
		Pentium	13

POP — Pop data from stack

01011rrr		O D I T S Z A P C	
Format	**Examples**		**Clocks**
POP reg	POP CX POP AX POP EBX	8086	8
		8088	12
		80286	5
		80386	4
		80486	1
		Pentium	1

10001111 oo000mmm disp			
Format	**Examples**		**Clocks**
POP mem	POP DATAA POP LISTS POP NUMBS	8086	17 + ea
		8088	25 + ea
		80286	5
		80386	5
		80486	4
		Pentium	3

00sss111 Format	Examples	Clocks	
POP seg	POP DS POP ES POP SS	8086	8
		8088	12
		80286	5
		80386	7
		80486	3
		Pentium	3

00001111 10sss001 Format	Examples	Clocks	
POP seg	POP FS POP GS	8086	—
		8088	—
		80286	—
		80386	7
		80486	3
		Pentium	3

POPA/POPAD Pop all registers from stack

01100001 O D I T S Z A P C

Examples	Clocks	
POPA POPAD	8086	—
	8088	—
	80286	19
	80386	24
	80486	9
	Pentium	5

POPF/POPFD Pop flags from stack

10011101

O	D	I	T	S	Z	A	P	C
*	*	*	*	*	*	*	*	*

Examples		Clocks
POPF POPFD	8086	—
	8088	—
	80286	5
	80386	5
	80486	6
	Pentium	6

PUSH Push data onto stack

01010rrr

O	D	I	T	S	Z	A	P	C

Format	Examples		Clocks
PUSH reg	PUSH CX PUSH AX PUSH ECX	8086	11
		8088	15
		80286	3
		80386	2
		80486	1
		Pentium	1

11111111 oo110mmm disp

Format	Examples		Clocks
PUSH mem	PUSH DATAD PUSH LISTS PUSH NUMB PUSH DWORD PTR [ECX]	8086	16 + ea
		8088	24 + ea
		80286	5
		80386	5
		80486	4
		Pentium	2

00sss110 Format	Examples		Clocks
PUSH seg	PUSH DS PUSH CS PUSH ES	8086	10
		8088	14
		80286	3
		80386	2
		80486	3
		Pentium	1

00001111 10sss000 Format	Examples		Clocks
PUSH seg	PUSH FS PUSH GS	8086	—
		8088	—
		80286	—
		80386	2
		80486	3
		Pentium	1

011010s0 data Format	Examples		Clocks
PUSH imm	PUSH 2000H PUSH 5322H PUSHW 10H PUSHD 100000H	8086	—
		8088	—
		80286	3
		80386	2
		80486	1
		Pentium	1

PUSHA/PUSHAD — Push all registers

01100000 — O D I T S Z A P C

Example		Clocks
PUSHA PUSHAD	8086	—
	8088	—
	80286	17
	80386	18
	80486	11
	Pentium	5

PUSHF/PUSHFD — Push flags onto stack

10011100 — O D I T S Z A P C

Examples		Clocks
PUSHF PUSHFD	8086	10
	8088	14
	80286	3
	80386	4
	80486	3
	Pentium	4

RCL/RCR/ROL/ROR — Rotate

1101000w ooTTTmmm disp

O	D	I	T	S	Z	A	P	C
*								*

TTT = 000 = ROL
TTT = 001 = ROR
TTT = 010 = RCL
TTT = 011 = RCR

Format	Examples		Clocks
ROL reg,1 ROR reg,1	ROL CL,1 ROL DX,1 ROR CH,1 ROL SI,1	8086	2
		8088	2
		80286	2
		80386	3
		80486	3
		Pentium	1

RCL reg,1 RCR reg,1	RCL CL,1 RCL SI,1 RCR AH,1 RCR EBX,1	8086	2
		8088	2
		80286	2
		80386	9
		80486	3
		Pentium	3
ROL mem,1 ROR mem,1	ROL DATA4,1 ROL BYTE PTR [DI],1 ROR NUMB,1 ROR DWORD PTR [ECX],1	8086	15 + ea
		8088	23 + ea
		80286	7
		80386	7
		80486	4
		Pentium	3
RCL mem,1 RCR mem,1	RCL DATA7,1 RCL BYTE PTR [DI],1 RCR NUMB,1 RCR WORD PTR [ECX],1	8086	15 + ea
		8088	23 + ea
		80286	7
		80386	10
		80486	4
		Pentium	3
1101001w ooTTTmmm disp Format	Examples		Clocks
ROL reg,CL ROR reg,CL	ROL CH,CL ROL DX,CL ROR CH,CL ROL SI,CL	8086	8 + 4n
		8088	8 + 4n
		80286	5 + n
		80386	3
		80486	3
		Pentium	7–24
RCL reg,CL RCR reg,CL	RCL DL,CL RCL SI,CL RCR AH,CL RCR BX,CL	8086	8 + 4n
		8088	8 + 4n
		80286	5 + n
		80386	9
		80486	8
		Pentium	7–24

Format	Examples		Clocks
ROL mem,CL ROR mem,CL	ROL DATA3,CL ROL BYTE PTR [DI],CL ROR NUMB,CL ROR WORD PTR [ECX],CL	8086	20 + a + 4n
		8088	28 + a + 4n
		80286	8 + n
		80386	7
		80486	4
		Pentium	9–26
RCL mem,CL RCR mem,CL	RCL DATA0,CL RCL BYTE PTR [DI],CL RCR NUMB,CL RCR WORD PTR [ECX],CL	8086	20 + a + 4n
		8088	28 + a + 4n
		80286	8 + n
		80386	10
		80486	9
		Pentium	9–26

1100000w ooTTTmmm disp data

Format	Examples		Clocks
ROL reg,imm ROR reg,imm	ROL CL,4 ROL DX,5 ROR CH,12 ROL SI,9	8086	—
		8088	—
		80286	5 + n
		80386	3
		80486	2
		Pentium	1
RCL reg,imm RCR reg,imm	RCL CL,2 RCL SI,3 RCR AH,5 RCR BX,13	8086	—
		8088	—
		80286	5 + n
		80386	9
		80486	8
		Pentium	1

ROL mem,imm ROR mem,imm	ROL DATAH,4 ROL BYTE PTR [DI],2 ROR NUMB,2 ROR WORD PTR [ECX],3	8086	—
		8088	—
		80286	8 + n
		80386	7
		80486	4
		Pentium	3
RCL mem,imm RCR mem,imm	RCL DATAK,6 RCL BYTE PTR [DI],7 RCR NUMB,6 RCR WORD PTR [ECX],5	8086	—
		8088	—
		80286	8 + n
		80386	10
		80486	9
		Pentium	3

RDMSR Read from model specific register

00001111 00110010 — O D I T S Z A P C

Example		Clocks
RDMSR	8086	—
	8088	—
	80286	—
	80386	—
	80486	—
	Pentium	20–24

REP Repeat prefix

11110011 1010010w — O D I T S Z A P C

Format	Examples		Clocks
REP MOVS	REP MOVSB REP MOVSW REP MOVSD REP MOVS DATA1,DATA2	8086	9 + 17n
		8088	9 + 25n
		80286	5 + 4n
		80386	8 + 4n
		80486	12 + 3n
		Pentium	13n

11110011 1010101w Format	Examples		Clocks
REP STOS	REP STOSB REP STOSW REP STOSD REP STOS DATA3	8086	9 + 10n
		8088	9 + 14n
		80286	4 + 3n
		80386	5 + 5n
		80486	7 + 4n
		Pentium	9 + 3n

11110011 0110110w Format	Examples		Clocks
REP INS	REP INSB REP INSW REP INSD REP INS DATA4	8086	—
		8088	—
		80286	5 + 4n
		80386	13 + 6n
		80486	16 + 8n
		Pentium	11 + 3n

11110011 0110111w Format	Examples		Clocks
REP OUTS	REP OUTSB REP OUTSW REP OUTSD REP OUTS DATA5	8086	—
		8088	—
		80286	5 + 4n
		80386	12 + 5n
		80486	17 + 5n
		Pentium	13 + 4n

REPE/REPNE Repeat conditional

11110011 1010011w Format	Examples	O D I T S Z A P C *	Clocks
REPE CMPS	REPE CMPSB REPE CMPSW REPE CMPSD REPE CMPS DATA6,DATA7	8086	9 + 22n
		8088	9 + 30n
		80286	5 + 9n
		80386	5 + 9n
		80486	7 + 7n
		Pentium	9 + 4n

11110011 1010111w Format	Examples		Clocks
REPE SCAS	REPE SCASB REPE SCASW REPE SCASD REPE SCAS DATA8	8086	9 + 15n
		8088	9 + 19n
		80286	5 + 8n
		80386	5 + 8n
		80486	7 + 5n
		Pentium	9 + 4n

11110010 1010011w Format	Examples		Clocks
REPNE CMPS	REPNE CMPSB REPNE CMPSW REPNE CMPSD REPNE CMPS DATA9,DATA10	8086	9 + 22n
		8088	9 + 30n
		80286	5 + 9n
		80386	5 + 9n
		80486	7 + 7n
		Pentium	8 + 4n

11110010 1010111w Format	Examples		Clocks
REPNE SCAS	REPNE SCASB REPNE SCASW REPNE SCASD REPNE SCAS DATA11	8086	9 + 15n
		8088	9 + 19n
		80286	5 + 8n
		80386	5 + 8n
		80486	7 + 5n
		Pentium	9 + 4n

RET Return from procedure

11000011 Example	O D I T S Z A P C	
		Clocks
RET (near)	8086	16
	8088	20
	80286	11
	80386	10
	80486	5
	Pentium	2

11000010 data Format	Examples		Clocks
RET imm (near)	RET 4 RET 100H	8086	20
		8088	24
		80286	11
		80386	10
		80486	5
		Pentium	3

11001011 Example		Clocks
RET (far)	8086	26
	8088	34
	80286	15
	80386	18
	80486	13
	Pentium	23

11001010 data Format	Examples		Clocks
RET imm (far)	RET 4 RET 100H	8086	25
		8088	33
		80286	11
		80386	10
		80486	5
		Pentium	23

RSM — Resume from system management mode

00001111 10101010

O	D	I	T	S	Z	A	P	C
*	*	*	*	*	*	*	*	*

Example	CPU	Clocks
RSM	8086	—
	8088	—
	80286	—
	80386	—
	80486	—
	Pentium	83

SAHF — Store AH into flags

10011110

O	D	I	T	S	Z	A	P	C
				*	*	*	*	*

Example	CPU	Clocks
SAHF	8086	4
	8088	4
	80286	2
	80386	3
	80486	2
	Pentium	2

SAL/SAR/SHL/SHR — Shift

1101000w ooTTTmmm disp

O	D	I	T	S	Z	A	P	C
*				*	*	?	*	*

TTT = 100 = SHL/SAL
TTT = 101 = SHR
TTT = 111 = SAR

Format	Examples	CPU	Clocks
SAL reg,1	SAL CL,1	8086	2
SHL reg,1	SHL DX,1	8088	2
SHR reg,1	SHR CH,1	80286	2
SAR reg,1	SAR SI,1	80386	3
		80486	3
		Pentium	1

Format	Examples		Clocks
SAL mem,1 SHL mem,1 SHR mem,1 SAR mem,1	SAL DATAR,1 SHL BYTE PTR [DI],1 SHR NUMB,1 SAR WORD PTR [ECX],1	8086	15 + ea
		8088	23 + ea
		80286	7
		80386	7
		80486	4
		Pentium	3

1101001w ooTTTmmm disp

Format	Examples		Clocks
SAL reg,CL SHL reg,CL SHR reg,CL SAR reg,CL	SAL CH,CL SHL DX,CL SHR CH,CL SAR SI,CL	8086	8 + 4n
		8088	8 + 4n
		80286	5 + n
		80386	3
		80486	3
		Pentium	4
SAL mem,CL SHL mem,CL SHR mem,CL SAR mem,CL	SAL DATAU,CL SHL BYTE PTR [DI],CL SHR NUMB,CL SAR WORD PTR [ECX],CL	8086	20 + a + 4n
		8088	28 + a + 4n
		80286	8 + n
		80386	7
		80486	4
		Pentium	4

1100000w ooTTTmmm disp data

Format	Examples		Clocks
SAL reg,imm SHL reg,imm SHR reg,imm SAR reg,imm	SAL CL,4 SHL DX,5 SHR CH,12 SAR SI,9	8086	—
		8088	—
		80286	5 + n
		80386	3
		80486	2
		Pentium	1

SAL mem,imm SHL mem,imm SHR mem,imm SAR mem,imm	SAL DATAP,6 SHL BYTE PTR [DI],7 SHR NUMB,6 SAR WORD PTR [ECX],5	8086	—
		8088	—
		80286	8 + n
		80386	7
		80486	4
		Pentium	3

SBB Subtract with borrow

000110dw oorrrmmm disp

O	D	I	T	S	Z	A	P	C
*				*	*	*	*	*

Format	Examples		Clocks
SBB reg,reg	SBB CL,DL SBB AX,DX SBB CH,CL SBB EAX,EBX SBB EBX,[ECX+2*EDX]	8086	3
		8088	3
		80286	2
		80386	2
		80486	1
		Pentium	1
SBB mem,reg	SBB DATAJ,CL SBB BYTES,CX SBB NUMBS,ECX SBB [EAX],CX	8086	16 + ea
		8088	24 + ea
		80286	7
		80386	6
		80486	3
		Pentium	3
SBB reg,mem	SBB CL,DATAL SBB CX,BYTES SBB ECX,NUMBS SBB CX,[EDX]	8086	9 + ea
		8088	13 + ea
		80286	7
		80386	7
		80486	2
		Pentium	2

100000sw oo011mmm disp data Format	Examples		Clocks
SBB reg,imm	SBB CL,4 SBB DX,5 SBB CH,12 SBB SI,9	8086	4
		8088	4
		80286	3
		80386	2
		80486	1
		Pentium	1
SBB mem,imm	SBB DATAF,6 SBB BYTE PTR [DI],7 SBB NUMB,6 SBB WORD PTR [ECX],5	8086	17 + ea
		8088	25 + ea
		80286	7
		80386	7
		80486	3
		Pentium	3

0001110w data Format	Examples		Clocks
SBB acc,imm	SBB AL,4 SBB AX,5 SBB AH,12 SBB AX,9	8086	4
		8088	4
		80286	3
		80386	2
		80486	1
		Pentium	1

SCAS Scan string

1010111w

O	D	I	T	S	Z	A	P	C
*				*	*	*	*	*

Format	Examples		Clocks
SCASB SCASW SCASD	SCASB SCASW SCASD SCAS DATAH REP SCASB	8086	15
		8088	19
		80286	7
		80386	7
		80486	6
		Pentium	4

SET Set on condition

00001111 1001cccc oo000mmm — O D I T S Z A P C

Format	Examples	Clocks	
SETcd reg8	SETA BL SETB CH SETG DL SETE BH SETZ AL	8086	—
		8088	—
		80286	—
		80386	4
		80486	3
		Pentium	4
SETcd mem8	SETE DATAK SETLE BYTES	8086	—
		8088	—
		80286	—
		80386	5
		80486	3
		Pentium	4

Condition Codes	Mnemonic	Flag	Description
0000	SETO	O = 1	Set if overflow
0001	SETNO	O = 0	Set if no overflow
0010	SETB/SETNAE	C = 1	Set if below
0011	SETAE/SETNB	C = 0	Set if above or equal
0100	SETE/SETZ	Z = 1	Set if equal/zero
0101	SETNE/SETNZ	Z = 0	Set if not equal/not zero
0110	SETBE/SETNA	C = 1 + Z = 1	Set if below or equal
0111	SETA/SETNBE	C = 0 • Z = 0	Set if above
1000	SETS	S = 1	Set if sign
1001	SETNS	S = 0	Set if no sign
1010	SETP/SETPE	P = 1	Set if parity even
1011	SETNP/SETPO	P = 0	Set if parity odd
1100	SETL/SETNGE	S • O	Set if less than
1101	SETGE/SETNL	S = O	Set if greater or equal
1110	SETLE/SETNG	Z = 1 + S • O	Set if less than or equal
1111	SETG/SETNLE	Z = 0 + S = O	Set if greater

SGDT/SIDT/SLDT Store descriptor table

00001111 00000001 oo000mmm disp — O D I T S Z A P C

Format	Examples		Clocks
SGDT mem	SGDT MEMORY SGDT GLOBAL	8086	—
		8088	—
		80286	11
		80386	9
		80486	10
		Pentium	4

00001111 00000001 oo001mmm disp

Format	Examples		Clocks
SIDT mem	SIDT DATAS SIDT INTERRUPT	8086	—
		8088	—
		80286	12
		80386	9
		80486	10
		Pentium	4

00001111 00000000 oo000mmm disp

Format	Examples		Clocks
SLDT reg	SLDT CX SLDT DX	8086	—
		8088	—
		80286	2
		80386	2
		80486	2
		Pentium	2
SLDT mem	SLDT NUMBS SLDT LOCALS	8086	—
		8088	—
		80286	3
		80386	2
		80486	3
		Pentium	2

SHLD/SHRD Double precision shift

00001111 10100100 oorrrmmm disp data

O	D	I	T	S	Z	A	P	C
?				*	*	?	*	*

Format	Examples		Clocks
SHLD reg,reg,imm	SHLD AX,CX,10 SHLD DX,BX,8 SHLD CX,DX,2	8086	—
		8088	—
		80286	—
		80386	3
		80486	2
		Pentium	4
SHLD mem,reg,imm	DATAQ,CX,8	8086	—
		8088	—
		80286	—
		80386	7
		80486	3
		Pentium	4

00001111 10101100 oorrrmmm disp data

Format	Examples		Clocks
SHRD reg,reg,imm	SHRD CX,DX,2	8086	—
		8088	—
		80286	—
		80386	3
		80486	2
		Pentium	4
SHRD mem,reg,imm	SHRD DATAT,CX,3	8086	—
		8088	—
		80286	—
		80386	7
		80486	3
		Pentium	4

00001111 10100101 oorrrmmm disp Format	Examples		Clocks
SHLD reg,reg,CL	SHLD DX,BX,CL	8086	—
		8088	—
		80286	—
		80386	3
		80486	3
		Pentium	4/5
SHLD mem,reg,CL	SHLD DATA4,AX,CL	8086	—
		8088	—
		80286	—
		80386	7
		80486	3
		Pentium	4/5

00001111 10101101 oorrrmmm disp Format	Examples		Clocks
SHRD mem,reg,CL	SHLD DX,BX,CL	8086	—
		8088	—
		80286	—
		80386	3
		80486	3
		Pentium	4/5
SHRD mem,reg,CL	SHRD DATA3,AX,CL	8086	—
		8088	—
		80286	—
		80386	7
		80486	3
		Pentium	4/5

SMSW Store machine status word

00001111 00000001 oo100mmm disp (should only be used by the 80286)

O D I T S Z A P C

Format	Examples		Clocks
SMSW reg	SMSW AX SMSW DX SMSW CX	8086	—
		8088	—
		80286	2
		80386	10
		80486	2
		Pentium	4
SMSW mem	SMSW DATAZ1	8086	—
		8088	—
		80286	3
		80386	3
		80486	3
		Pentium	4

STC Set carry flag

11111001

O D I T S Z A P C
C = 1

Example		Clocks
STC	8086	2
	8088	2
	80286	2
	80386	2
	80486	2
	Pentium	2

STD — Set direction flag

11111101

O	D	I	T	S	Z	A	P	C
	1							

Example		Clocks
STD	8086	2
	8088	2
	80286	2
	80386	2
	80486	2
	Pentium	2

STI — Set interrupt flag

11111011

O	D	I	T	S	Z	A	P	C
		1						

Example		Clocks
STI	8086	2
	8088	2
	80286	2
	80386	3
	80486	5
	Pentium	7

STOS — Store string data

1010101w

O	D	I	T	S	Z	A	P	C

Format	Examples		Clocks
STOSB STOSW STOSD	STOSB STOSW STOSD STOS DATA4 REP STOSB	8086	11
		8088	15
		80286	3
		80386	4
		80486	5
		Pentium	3

STR Store task register

00001111 00000000 oo001mmm disp

O	D	I	T	S	Z	A	P	C

Format	Examples		Clocks
STR reg	STR DX STR CX STR AX	8086	—
		8088	—
		80286	2
		80386	2
		80486	2
		Pentium	2
STR mem	STR DATA3	8086	—
		8088	—
		80286	3
		80386	2
		80486	3
		Pentium	2

SUB Subtract

001010dw oorrrmmm disp

O	D	I	T	S	Z	A	P	C
*				*	*	*	*	*

Format	Examples		Clocks

SUB reg,reg	SUB CL,DL
	SUB AX,DX
	SUB CH,CL
	SUB EAX,EBX
		8086	3

		8088	3
		80286	2
		80386	2
		80486	1
		Pentium	1

SUB mem,reg	SUB BEAVUS,CL
	SUB BYTES,CX
	SUB NUMBS,ECX
	SUB [EAX],CX
		8086	16 + ea

		8088	24 + ea
		80286	7
		80386	6
		80486	3
		Pentium	3

Format	Examples		Clocks

SUB reg,mem	SUB CL,BUTHEAD
	SUB CX,BYTES
	SUB ECX,NUMBS
	SUB CX,[EDX]
	SUB AX,[EDI+2*ECX]
		8086	9 + ea

		8088	13 + ea
		80286	7
		80386	7
		80486	2
		Pentium	2

100000sw oo101mmm disp data

Format	Examples		Clocks

SUB reg,imm	SUB CL,4
	SUB DX,5
	SUB CH,12
	SUB SI,9
		8086	4

		8088	4
		80286	3
		80386	2
		80486	1
		Pentium	1

SUB mem,imm	SUB MTV,6
	SUB BYTE PTR [DI],7
	SUB NUMB,6
	SUB WORD PTR [ECX],5
	SUB DWORD PTR [EDI],1
		8086	17 + ea

		8088	25 + ea
		80286	7
		80386	7
		80486	3
		Pentium	3

0010110w data

Format	Examples		Clocks

SUB acc,imm	SUB AL,4
	SUB AX,5
	SUB AH,12
	SUB AX,9
		8086	4

		8088	4
		80286	3
		80386	2
		80486	1
		Pentium	1

TEST Test operands (logical compare)

1000011w oorrrmmm disp

O	D	I	T	S	Z	A	P	C
0				*	*	?	*	0

Format	Examples		Clocks
TEST reg,reg	TEST CL,DL TEST CX,DX TEST CL,CH TEST ECX,EBX	8086	5
		8088	5
		80286	2
		80386	2
		80486	1
		Pentium	1
TEST reg,mem mem,reg	TEST DATA1,CL TEST CL,DATA1	8086	9 + ea
		8088	13 + ea
		80286	6
		80386	5
		80486	2
		Pentium	2

1111011w oo000mmm disp data

Format	Examples		Clocks
TEST reg,imm	TEST CL,4 TEST DX,5 TEST CH,12H TEST SI,256	8086	4
		8088	4
		80286	3
		80386	2
		80486	1
		Pentium	1
TEST mem,imm	TESTDATA3,6	8086	11 + ea
		8088	11 + ea
		80286	6
		80386	5
		80486	2
		Pentium	2

1010100w data Format	Examples	Clocks	
TEST acc,imm	TEST AL,4 TEST AX,5 TEST AH,12 TEST AX,9 TEST EAX,2	8086	4
		8088	4
		80286	3
		80386	2
		80486	1
		Pentium	1

VERR/VERW Verify read or write

00001111 00000000 oo100mmm disp

O	D	I	T	S	Z	A	P	C
					*			

Format	Examples	Clocks	
VERR reg	VERR BX VERR CX VERR DX	8086	—
		8088	—
		80286	14
		80386	10
		80486	11
		Pentium	7
VERR mem	VERR DATA_SEVEN	8086	—
		8088	—
		80286	16
		80386	11
		80486	11
		Pentium	7

00001111 00000000 oo101mmm disp Format	Examples	Clocks	
VERW reg	VERW AX VERW CX VERW DX	8086	—
		8088	—
		80286	14
		80386	15
		80486	11
		Pentium	7

VERW mem	VERW DATA_SEG	8086	—
		8088	—
		80286	16
		80386	16
		80486	11
		Pentium	7

WAIT Wait for coprocessor

10011011 O D I T S Z A P C

Examples		Clocks
WAIT FWAIT	8086	4
	8088	4
	80286	3
	80386	6
	80486	6
	Pentium	1

WBINVD Write-back and invalidate data cache

00001111 00001001 O D I T S Z A P C

Example		Clocks
WBINVD	8086	—
	8088	—
	80286	—
	80386	—
	80486	5
	Pentium	2000+

WRMSR Write to model specific register

00001111 00110000 Example		O D I T S Z A P C Clocks
WRMSR	8086	—
	8088	—
	80286	—
	80386	—
	80486	—
	Pentium	30–45

XADD Exchange and add

00001111 1100000w 11rrrrrr

O	D	I	T	S	Z	A	P	C
*				*	*	*	*	*

Format	Examples		Clocks
XADD reg,reg	XADD EBX,ECX XADD EDX,EAX XADD EDI,EBP	8086	—
		8088	—
		80286	—
		80386	—
		80486	3
		Pentium	3

00001111 1100000w oorrrmmm disp

Format	Examples		Clocks
XADD mem,reg	XADD DATA5,EAX XADD [DI],EAX XADD [ECX],EDX	8086	—
		8088	—
		80286	—
		80386	—
		80486	4
		Pentium	4

XCHG Exchange

1000011w 1oorrrmmm — O D I T S Z A P C

Format	Examples		Clocks
XCHG reg,reg	XCHG BL,CL XCHG AX,DX XCHG EDI,EBP	8086	4
		8088	4
		80286	3
		80386	3
		80486	3
		Pentium	3
XCHG reg,mem mem,reg	XCHG CL,DATA4 XCHG DATA7,CL XCHG DX,[DI] XCHG ECX,[EBP]	8086	17 + ea
		8088	25 + ea
		80286	5
		80386	5
		80486	5
		Pentium	3

10010reg

Format	Examples		Clocks
XCHG acc,reg XCHG reg,acc	XCHG AL,CL XCHG DX,AX	8086	3
		8088	3
		80286	3
		80386	3
		80486	3
		Pentium	2

XLAT Translate

11010111 — O D I T S Z A P C

Example		Clocks
XLAT	8086	11
	8088	11
	80286	5
	80386	3
	80486	4
	Pentium	4

XOR Exclusive-OR

001100dw oorrrmmm disp

O	D	I	T	S	Z	A	P	C
0				*	*	?	*	0

Format	Examples		Clocks
XOR reg,reg	XOR BL,CL XOR CX,DX XOR CH,CL XOR EAX,EBX	8086	3
		8088	3
		80286	2
		80386	2
		80486	1
		Pentium	1
XOR mem,reg	XOR DATAD,CL XOR BYTES,CX XOR NUMBS,ECX XOR [EAX],CX	8086	16 + ea
		8088	24 + ea
		80286	7
		80386	6
		80486	3
		Pentium	2
XOR reg,mem	XOR CL,DATAB XOR CX,BYTES XOR ECX,NUMBS XOR CX,[EDX]	8086	9 + ea
		8088	13 + ea
		80286	7
		80386	7
		80486	2
		Pentium	2

100000sw oo110mmm disp data

Format	Examples		Clocks
XOR reg,imm	XOR BL,33 XOR CX,234H XOR CH,'A' XOR EAX,123445	8086	4
		8088	4
		80286	3
		80386	2
		80486	1
		Pentium	1

XOR mem,imm	XOR DATA_NEW,34 XOR BYTES,1234 XOR NUMBS,123 XOR [EAX],11	8086	17 + ea
		8088	25 + ea
		80286	7
		80386	7
		80486	3
		Pentium	3
0011010w data Format	Examples		Clocks
XOR acc,imm	XOR AL,33 XOR AX,234H XOR AL,'A' XOR EAX,123445	8086	4
		8088	4
		80286	3
		80386	2
		80486	1
		Pentium	1

APPENDIX C

Flag-Bit Changes

This appendix shows only instructions that actually change the flag bits. Any instruction not listed does not affect any of the flag bits.

Instruction	*Flags*								
	O	D	I	T	S	Z	A	P	C
AAA	?				?	?	*	?	*
AAD	?				*	*	?	*	?
AAM	?				*	*	?	*	?
AAS	?				?	?	*	?	*
ADC	*				*	*	*	*	*
ADD	*				*	*	*	*	*
AND	0				*	*	?	*	0
ARPL						*			
BSF						*			
BSR						*			
BT									*
BTC									*
BTR									*
BTS									*
CLC									0
CLD		0							
CLI			0						
CMC									*
CMP	*				*	*	*	*	*
CMPS	*				*	*	*	*	*

Instruction	*Flags*								
	O	D	I	T	S	Z	A	P	C
CMPXCHG	*				*	*	*	*	*
CMPXCHG8B						*			
DAA	?				*	*	*	*	*
DAS	?				*	*	*	*	*
DEC	*				*	*	*	*	
DIV	?				?	?	?	?	?
IDIV	?				?	?	?	?	?
IMUL	*				?	?	?	?	*
INC	*				*	*	*	*	
IRET	*	*	*	*	*	*	*	*	*
LAR						*			
LSL						*			
MUL	*				?	?	?	?	*
NEG	*				*	*	*	*	*
OR	0				*	*	?	*	0
POPF/POPFD	*	*	*	*	*	*	*	*	*
RCL/RCR	*								*
REPE/REPNE						*			
ROL/ROR	*								*
SAHF					*	*	*	*	*
SAL/SAR	*				*	*	?	*	*
SHL/SHR	*				*	*	?	*	*
SBB	*				*	*	*	*	*
SCAS	*				*	*	*	*	*
SHLD/SHRD	?				*	*	?	*	*
STC									1
STD		1							
STI			1						
SUB	*				*	*	*	*	*
TEST	0				*	*	?	*	0
VERR/VERW						*			
XADD	*				*	*	*	*	*
XOR	0				*	*	?	*	0

APPENDIX D

Answers to Selected Even-Numbered Questions and Problems

CHAPTER 1

2. Herman Hollerith
4. Konrad Zuse
6. ENIAC
8. Augusta Ada Byron
10. A von Nuemann machine stores a program in the memory system for execution by the computer.
12. 200,000,000
14. 32M
16. 1993
18. Complex instruction set computer
20. 1,024
22. 1,024
24. TPA and systems
26. 384K bytes
28. 16M
30. Extended memory or the extended memory system (XMS)
32. The DOS, or disk operating system, is a program that controls the microprocessor and supervises the organization and storage of files and programs on the disk.
34. The VESA local bus is a bus system found in newer systems that allows the microprocessor to communicate with I/O at the microprocessor clock speed.
36. The XMS is the extended memory system, which begins above the first 1M byte of memory.
38. TPA or upper memory area.
40. A TSR (terminate and stay resident program) is accessed through an interrupt or hot key.
42. The AUTOEXEC.BAT file executes DOS commands without their being entered each time the system is started.
44. 64K

46. CONFIG.SYS file
48. The video BIOS is located on a ROM on the video card, at addresses C0000H through C7FFFH.
50. The microprocessor is the controlling element in the personal computer. It executes instructions that transfer data, perform simple arithmetic and logic operations, and make simple decisions.
52. Address bus
54. The $\overline{\text{IORC}}$ (I/O read control) signal activates an I/O device for a read operation. This inputs data from the I/O device to the microprocessor.
56. (a) 13.25, (b) 57.1875, (c) 43.3125, (d) 7.0625
58. (a) 163.1875, (b) 297.75, (c) 172.859375, (d) 4011.1875, (e) 3000.05078125
60. (a) 0.101_2 0.5_8 $0.A_{16}$
    (b) $.00000001_2$ $.002_8$ $.01_{16}$
    (c) $.10100001_2$ $.502_8$ $.A1_{16}$
    (d) $.11_2$ 0.6_8 $0.C_{16}$
    (e) $.1111_2$ $.74_8$ $0.F_{16}$
62. (a) C2H, (b) 10FDH, (c) B.CH, (d) 10H, (e) 8BAH
64. (a) 0111 1111, (b) 0101 0100, (c) 0101 0001, (d) 1000 0000
66. (a) FROG = 46 52 4F 47
    (b) Arc = 41 72 63
    (c) Water. = 57 61 74 65 72 2E
    (d) Well = 57 65 6C 6C
68. DB 'What time is it?'
70. (a) 0000 0011 1110 1000
    (b) 1111 1111 1000 1000
    (c) 0000 0011 0010 0000
    (d) 1111 0011 0111 0100
72. (a) 34 12, (b) 22 A1, (c) 00 B1
74. DW 123AH
76. (a) –128, (b) +51, (c) –110, (d) –119
78. (a) 0 01111111 10000000000000000000000
    (b) 1 10000010 01010100000000000000000
    (c) 0 10000101 10010001000000000000000
    (d) 1 10001001 00101100000000000000000

CHAPTER 2

2. 3
4. The main directory on the disk drive, which contains a limited number of entries
6. Fixed disk drive
8. Sector
10. 1.44M
12. The boot sector contains a program (bootstrap loader) that loads DOS into memory.
14. A cluster is a group of sectors.
16. 112

18. File name, file extension, attributes, creation time and date, file length, and starting cluster
20. A compressed disk partition is one where the files have been compressed. A compressed file is one where the data has been encoded to save space.
22. The data are stored in a hidden file located in the root directory.
24. An external DOS command is one that is stored as a program in the DOS directory.
26. (a) displays a directory of the files, programs, and other directories and subdirectories located in the current directory, (b) displays a wide version of (a), (c) displays a directory as in (a), but it pauses when the screen fills, (d) provides help about the DIR command
28. The DOS wild card * represents any file name or extension.
30.
```
CD C:\
RD TEST1.3
```
32. TREE displays a tree or graphical list of all subdirectories in the current directory.
34. (a) DEL *.TXT
    (b) DEL C:\TEST\FROG.ANY
    (c) DEL *.*
    (d) DEL D???????.*
36. PROMPT
38. The DOS internal command executes faster, because no disk access is required.
40. FORMAT A: /F:720
42. The CHKDSK program checks the directory and file structure of a disk. If it finds an error, it reports it to the user.
44. ATTRIB +R+H C:\TEMP\TEST.TXT
46. DOSKEY DELL=DEL *.BAK
48.
```
DEVICE=C:\DOS\HIMEM.SYS
DEVICE=C;\DOS\EMM386.EXE 1024
DOS=UMB,HIGH
```
50.
```
ECHO FORMAT MENU
ECHO.
ECHO 1 Format Drive A as a 360K DSDD 5-1/4" floppy disk
ECHO 2 Format Drive A as a 1.2M HD 5-1/4" floppy disk
ECHO.
CHOICE /C:12 Enter choice
IF ERRORLEVEL 2 GOTO HD
IF ERRORLEVEL 1 GOTO DSDD
:HD
ECHO.
FORMAT A: /F:720 /U
GOTO END
:DSDD
ECHO.
FORMAT A: /F1200 /U
:END
```
52. The [MENU] command allows the user to develop a CONFIG.SYS program that allows choices on configurations.
54. The MENUDEFAULT command allows the default menu to be chosen after a given amount of time has elapsed.

CHAPTER 3

2. An icon is a small picture that represents a program or application in the Windows environment.
4. To execute a program is to launch it.
6. The mouse pointer is moved to the application icon and then the left mouse button is double-clicked.
8. The Windows program requires at least 4M bytes of memory to operate efficiently, much more than that required for DOS. This huge amount of memory is required because of the way Windows presents data on the video display.
10. The commands are the same except that, in Windows, the file manager displays them and in DOS commands must be memorized or learned.
12. The WIN.INI and SYSTEM.INI files are modified with the Windows notepad or the DOS EDIT program.
14. The 32BitDiskAccess enables faster hard disk access in most systems.
16. Standard-mode operation assumes that the user has an 80286 microprocessor. Enhanced mode assumes a 80386 or better microprocessor. Also, enhanced mode allows the user to select more features, such as permanent swap files and a virtual memory system.
18. The SMARTDRV program speeds the operation of Windows and is, therefore, an important feature of the system.
20. DELTREE, RAMDRIVE.SYS, GRAPHICS, PRINTER.SYS, and FASTOPEN, to name a few
22. CGA??????.FON, EGA?????.FON, CPWIN386.CPL, DOSAPP.FON, *3GR, *.386, WIN386.EXE, WIN386.PS2, and WINOA386.MOD.
24. A PIF is modified in the MAIN icon of the program manager by selecting the PIF editor.
26. Enhanced mode

CHAPTER 4

2. 16
4. EBX
6. The instruction pointer is used by the microprocessor in conjunction with the CS register to locate the next step or instruction in the program.
8. No, because an FFH (–1) plus a 01H (+1) results in the sum of 00H, not an overflow.
10. The interrupt flag bit.
12. The segment register addresses the start of a 64K byte memory segment in the real mode.
14. (a) 12000H, (b) 21000H, (c) 24A00H, (d) 25000H, (e) 3F12DH
16. DI
18. SS plus SP
20. (a) 12000H, (b) 21002H, (c) 26200H, (d) A1000H, (e) 2CA00H
22. Memory locations 000000H–FFFFFFH
24. The segment register contains a descriptor number, table selector, and requested privilege bits in the protected mode. The descriptor describes the start of the memory segment, its length, and its access rights.

26. A00000H–A01000H
28. 00280000H–00290000H
30. 3
32. 64K bytes
34. See Figure D–1

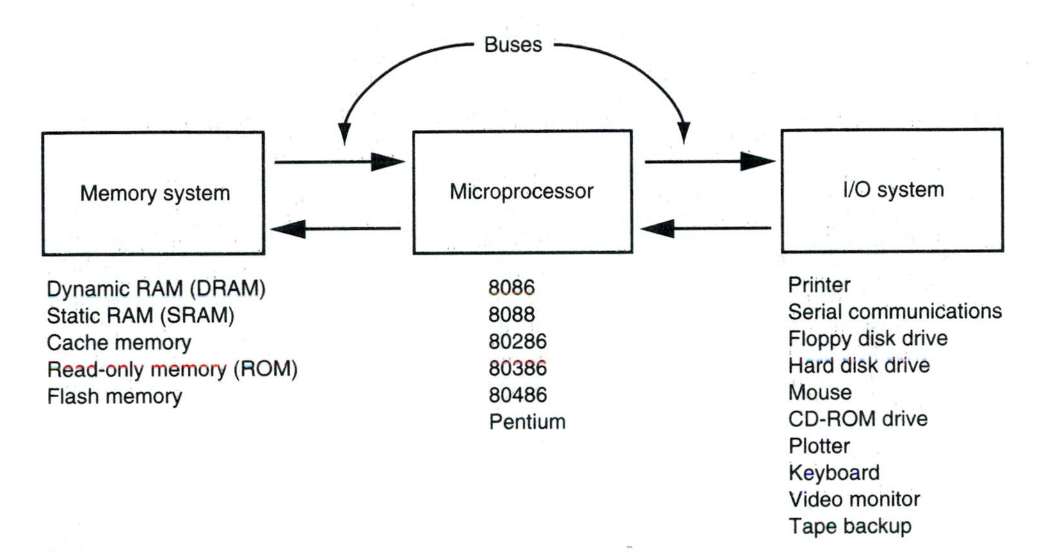

FIGURE D–1

36. The local descriptor table is accessed when the TI bit in the segment register is set to a logic 1.
38. The program-invisible registers are cache portions of the segment registers which contain the protected-mode base address, limit, and access rights. The base address of each descriptor table is stored in a program-invisible register, as are the task and local selectors.
40. 4K bytes
42. 1,024
44. Page directory entry 0 and page table entry 200H.
46. The TLB stores the last 32 page-translation addresses, so if an address that has been looked up recently is looked up in the paging mechanism, no reference is required to the memory.

CHAPTER 5

2. AH, AL, BH, BL, CH, CL, DH, and DL
4. EAX, EBX, ECX, EDX, ESP, EBP, ESI, and EDI
6. Mixed register sizes are not allowed.
8. (a) MOV EDX,EBX
   (b) MOV CL,BL
   (c) MOV BX,SI
   (d) MOV AX,DS
   (e) MOV AH,AL

10. #
12. .CODE
14. The opcode field
16. The .EXIT directive returns control to the DOS operating system.
18. The .STARTUP directive indicates the start of the program and loads DS with the segment address of the data segment.
20. The [ ] symbols denote indirect addressing.
22. Memory-to-memory transfers are not allowed, except for the MOVS instructions.
24. MOV WORD PTR [DI],3
26. The MOV BX,DATA instruction copies the contents of memory location DATA into BX, while the MOV BX,OFFSET DATA instruction copies the offset address of DATA into BX.
28. Nothing; this is an alternate way to form the MOV AL,[BX+SI] instruction.
30. (a) 11750H, (b) 11950H, (c) 11700H
32. The BP or base pointer register address data in the stack segment by default.
34.
```
FIELDS    STRUC

F1        DW    ?
F2        DW    ?
F3        DW    ?
F4        DW    ?
F5        DW    ?

FIELDS    ENDS
```
36. Direct, relative, and indirect
38. The intersegment jump is considered a far jump that allows access to any location within the real memory system. The intrasegment jump is considered a near jump that allows access to any location within the current code segment.
40. 32-bit
42. Short
44. JMP BX
46. The PUSH instruction stores 2 bytes if it is a 16-bit push or 4 bytes if it is a 32-bit push. Exceptions are the PUSHA and PUSHAD instructions, which push the entire register set onto the stack.
48. AX, CX, DX, BX, SP, BP, DI, and SI
50. PUSHFD

CHAPTER 6

2. The D bit selects the direction of flow (reg to r/m or r/m to reg) and the W bit selects the size of the register (byte or word/doubleword).
4. DL
6. DS:[BX+DI]
8. MOV AX,[BX]
10. 8B7702
12. The CS register cannot be changed with a MOV instruction.
14. 16

16. AX, CX, DX, BX, SP, BP, DI, and SI
18. (a) pushes the contents of AX onto the stack
    (b) removes a 32-bit number from the stack and places it into ESI
    (c) pushes a word from the data segment memory location addressed by BX onto the stack
    (d) pushes the EFLAG register onto the stack
    (e) removes a 16-bit number from the stack and places it into DS
    (f) pushes a 00000004H onto the stack
20. The stack memory locations are changed as illustrated in Figure D-2.

FIGURE D–2

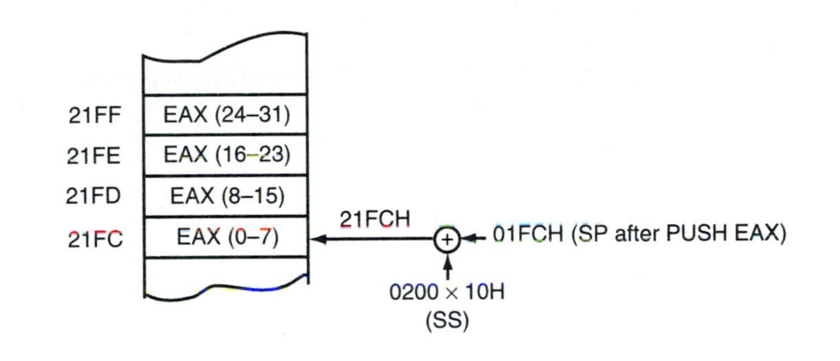

22. A possible combination for SP and SS is: SS = 0200H and SP = 0200H.
24. Both instructions load the offset address of NUMB to SI, but only the MOV SI,OFFSET NUMB instruction is stored as a move immediate.
26. The LDS BX,NUMB instruction loads BX with the contents of NUMB and NUMB+1 and DS with the contents of NUMB+2 and NUMB+3.
28.
```
MOV BX,NUMB
MOV DX,BX
MOV SI,DX
```
30. The CLD instruction clears the direction flag and the STD instruction sets it.
32. The LODSB instruction loads AL from the byte-sized memory location addressed by DS:SI and then either increments or decrements SI by one, depending on the state of the direction flag.
34. The OUTSB instruction outputs the contents of the data segment memory location addressed by SI to the I/O port addressed by DX. After the byte is transferred, the contents of SI either increment or decrement by one, depending on the state of the direction flag.
36.
```
MOV CX,12
CLD
MOV SI,OFFSET SOURCE
MOV DI,OFFSET DEST
REP MOVSB
```
38. XCHG EBX,ESI
40. The XLAT instruction addresses the data segment memory location indexed by BX+AL. The contents of this location replaces the value in AL.
42. The IN AL,12H instruction copies a byte from I/O port 0012H and places it into AL.

44. The segment override prefix is used to change from the default segment to any other segment. For example, the MOV AL,ES:[BX] instruction addresses the extra segment.
46.
```
XCHG AX,BX
XCHG ECX,EDX
XCHG SI,DI
```
48. These directives define bytes, words, and doublewords.
50. The EQU directive equates a label with a value or another label.
52. The .MODEL directive is used to select the memory model for the program under development.
54. Full-segment definitions
56. The PROC directive indicates the start of a procedure and the ENDP directive indicates its end.
58. The segment statement contains the USE16 directive or switch.
60.
```
COP     PROC    FAR

        MOV     AX,CS:DATA1
        MOV     BX,AX
        MOV     CX,AX
        MOV     DX,AX
        MOV     SI,AX
        RET

COP     ENDP
```

CHAPTER 7

2. You may never use mixed register sizes.
4. Sum = 3100H, C = 0, A = 1, S = 0, Z = 0, and = 0
6.
```
ADD AX,BX
ADD AX,CX
ADD AX,DX
ADD AX,SP
MOV DI,AX
```
8. ADC DX,BX
10. The assembler cannot determine whether the memory location is a byte, word, or doubleword. The BYTE PTR, WORD PTR, or DWORD PTR must appear before [BX] to specify the size of the memory location.
12. Difference = 81H, C = 0, A = 0, S = 1, P = 1, Z = 0, O = 0
14. DEC EBX
16. Both instructions subtract, but only the CMP instruction changes the flags.
18. DX holds the most significant and AX the least.
20. EDX and EAX
22.
```
MOV DL,5
MOV AX,0
MOV AL,DL
MUL DL
MUL DL
```
24. AL
26. Divide overflow and divide by zero

28. AH
30. DAA and DAS
32. The AAM instruction divides AX by 10. Unlike a normal division, the remainder is found in AL and the quotient is found in AH.
34.
```
PUSH AX
MOV  AL,BL
ADD  AL,DL
DAA
MOV  DL,AL
MOV  AL,BH
ADD  AL,DH
DAA
MOV  DH,AL
POP  BX
MOV  AL,BL
ADD  AL,CL
DAA
MOV  CL,AL
MOV  AL,BH
ADD  AL
DAA
MOV  CH,AL
```
36.
```
AND  DH,1FH
MOV  BH,DH
```
38.
```
OR   DI,1FH
MOV  SI,DI
```
40. A TEST instruction generally tests a bit of a number, while the CMP instruction tests all of the bits of a number.
42. TEST CH,4
44. (a) SHR DI,3
    (b) SHL AL,1
    (c) ROL AL,3
    (d) RCR EDX,1
    (e) SAR DH,1
46. Extra
48. The REPE instruction repeats the SCASB instruction until either CX counts down to zero or the comparison performed by SCASB yields a not-equal condition.
50. The CMPSB instruction compares the memory bytes at the locations addressed by DS:SI with ES:DI and then increments or decrements the SI and DI register by one, depending on the state of the direction flag.
52. When AH = 02H, DOS INT 21H selects the video display function. The video display function takes the ASCII-coded character from register DL and displays it at the current cursor position on the video display. When DL contains a 43H, the letter C is displayed.

CHAPTER 8

2. A near jump
4. The far jump
6. (a) near, (b) short, (c) far

8. IP
10. The JMP AX instruction jumps to the offset address located in register AX. This is a near jump.
12. The JMP [DI] instruction performs a near jump, while the JMP FAR PTR [DI] performs a far jump.
14. The JA instruction jumps if the result is above. For example, suppose that AL is compared with a 6. If AL is above a 6 (07H–FFH), the jump occurs.
16. JE, JNE, JG, JGE, JL, and JLE
18. JA, JBE, JG, and JLE
20. SETZ
22. ECX
24.
```
        MOV  DI,OFFSET DATA
        CLD
        MOV  CX,150H
        XOR  AL,AL
D1:
        STOSB
        LOOP D1
```
26. A procedure is a group of instructions that perform one task. The procedure is invoked with the CALL instruction and ends with the RET instruction. A procedure is stored in the memory once, but can be used as many times as necessary.
28. The near RET instruction removes a word from the stack and places it into the instruction pointer to return from a procedure.
30. PROC
32. The RET 6 instruction adds a 6 to the stack pointer after retrieving the return address from the stack. This form of the return instruction is used to delete data placed on the stack in procedures used with C or PASCAL.
34.
```
MUL  PROC NEAR

     PUSH DX
     PUSH CX
     MOV  AX,DI
     MUL  SI
     MOV  CL,8
     SHR  AX,CL
     POP  CX
     POP  DX
     RET

MUL  ENDP
```
36. An interrupt is an event that interrupts the currently executing program to call an interrupt service procedure. An interrupt is generated by an external hardware event or by a software event.
38. 256
40. The interrupt vector contains four bytes. The first two bytes hold the offset address of the interrupt procedure, and the last two bytes contain the segment address of the interrupt procedure. Both addresses are stored in the little endian format.
42. The IRETD instruction returns from an interrupt service procedure, but rather than removing the FLAGS from the stack, it removes the EFLAGS.

44. 100H–103H
46. INT 17H
48. WAIT or FWAIT
50. 16
52. ESC or escape
54. The program displays two character strings, one is stored at data segment address MES1 and the other at MES2. To display the strings, the offset address of the string is loaded to SI and procedure STRING is called. The STRING procedure displays the ASCII-null string or ASCII-Z string addressed by SI. The STRING procedure loads AL with a character from the string and then tests it for a null. If the null is found, the procedure returns. If the null is not found, DOS function 02H is used to display the character from the string.

CHAPTER 9

2. TEST.OBJ, TEST.LST, and TEST.CRF
4. The EXE (execute) file contains a program and data of any length, while the COM (command) file is limited to a program and data of a length of no more than 64K bytes.
6. The EXTRN directive indicates that a label is external to the current file or programming module.
8. The library file contains a collection of procedures. When a library file is linked to a program, only the procedures needed by the program are extracted and stored in the execution file.
10. A macro sequence is a group of instructions, stored in the program, that execute when invoked by the name of the macro.
12.
```
ADD32   MACRO

        ADD AX,CX
        ADC BX,DX

        ENDM
```
14. The LOCAL directive identifies all local labels in the macro. The LOCAL directive must immediately follow the MACRO statement.
16.
```
ADDM      MACRO LIST,LENGTH
          LOCAL A1,A2

          XOR AX,AX
          MOV SI,OFFSET LIST
          MOV CX,LENGTH
     A1:
          ADD AL,[SI]
          JNC A2
          INC AH
     A2:
          INC SI
          LOOP A1

          ENDM
```

18.
```
RANDOM   PROC NEAR

         INC  CL
         MOV  AH,6
         MOV  DL,-1
         INT  21H
         JZ   RANDOM
         RET
RANDOM   ENDP
```
20.
```
STR    PROC NEAR

       PUSH DX
       PUSH SI
       MOV SI,DX
S1:
       LODSB
       OR   AL,AL
       JZ   S2
       MOV DL,AL
       MOV AH,6
       INT 21H
       JMP S1
S2:
       POP SI
       POP DX
       RET

STR    ENDP
```
22.
```
NEW    PROC NEAR

       MOV AH,2
       MOV BH,0
       MOV DX,0306H
       INT 10H

NEW    ENDP
```
24. The number is divided by 10. After each division, the remainder is saved as a significant BCD digit. The process continues until the quotient is zero.
26. 30H
28. The number is first converted to BCD by subtracting a 30H from each digit. Next, the binary result is cleared to zero. Finally, the result is multiplied by 10 and a BCD digit is added to the product. The multiplication and addition process continues for each BCD digit.
30.
```
LOOK PROC  NEAR

     MOV  BX,OFFSET TAB
     XLAT CS:TAB
     RET

LOOK ENDP

TAB  DB 30H,31H,32H,33H
     DB 34H,35H,36H,37H
```

```
     DB 38H,39H,41H,42H
     DB 43H,44H,45H,46H
```

32. XLAT SS:LOOK

34.
```
DATA SEGMENT

M1   DB 13,10,'Two raised to the power of $'
M2  DB ' = $'

DATA ENDS

CODE    SEGMENT 'code'
        ASSUME CS:CODE

MAIN    PROC FAR
        MOV  BL,8
MAIN1:
        MOV  AH,9
        MOV  DX,OFFSET M1
        INT  21H
        MOV  DL,8
        SUB  DL,BL
        ADD  DL,30H
        MOV  AH,6
        INT  21H
        MOV  AH,9
        MOV  DX,OFFSET M2
        INT  21H
        MOV  AL,1
        MOV  CL,8
        SUB  CL,BL
        SHL  AL,CL
        CALL DIS
        DEC  BL
        JNZ  MAIN1
        MOV  AX,4C00H
        INT  21H

MAIN    ENDP

DIS     PROC NEAR

        MOV  CL,10
        MOV  DX,-1
        PUSH DX
DIS1:
        XOR  AH,AH
        DIV  CL
        PUSH AX
        CMP  AL,0
        JNZ  DIS1
DIS2:
        POP  DX
        CMP  DX,-1
        JE   DIS3
        MOV  DL,DH
```

```
        ADD  DL,30H
        MOV  AH,6
        INT  21H
        JMP  DIS2
DIS3:
        RET

DIS     ENDP

CODE    ENDS
        END  MAIN
```

36.
```
RAN     PROC NEAR

        MOV  CL,1
RAN1:
        INC  CL
        CMP  CL,48
        JNE  RAN2
        MOV  CL,1
RAN2:
        MOV AH,6
        MOV DL,-1
        INT 21H
        JZ  RAN1
        RET

RAN     ENDP
```

38.
```
        .MODEL TINY
        .CODE
        .STARTUP
        MOV  AX,123AH         ;load test data
        MOV  BX,8             ;load BX with 8 for octal
        PUSH BX               ;save end indicator
LABEL1:
        MOV  DX,0
        DIV  BX               ;find remainder 0-7
        PUSH DX               ;save remainder
        CMP  AX,0
        JNZ  LABEL1           ;repeat until quotient = 0
LABEL2:
        POP  DX               ;get remainder
        CMP  DX,BX            ;test for end
        JZ   LABEL3           ;if remainder = 8
        ADD  DL,30H           ;convert to ASCII
        MOV  AH,6             ;select display function
        INT  21H              ;display digit
        JMP  LABEL2           ;repeat until remainder = 8
LABEL3:
        .EXIT                 ;exit to DOS
        END
```

CHAPTER 10

2. The purpose of the .ENDIF statement is to terminate or end the .IF sequence.
4.
```
SH   PROC NEAR
     MOV  AH,0
```

```
        AAM
        .IF AH != 0
               MOV  DL,AH
               ADD  DL,30H
               PUSH AX
               MOV  AH,2
               INT  21H
               POP  AX
        .ENDIF
        MOV    DL,AL
        ADD    DL,30H
        MOV    AH,2
        INT    21H
        RET
SH      ENDP
```

6. The .BREAK statement is used to break out of an infinite loop.
8. On the same line as the .UNTIL statement, following .UNTIL
10.

```
SH      PROC NEAR
        MOV  AL,0
        .REPEAT
             MOV  [SI],AL
             INC   SI
        .UNTILCXZ
        RET
SH      ENDP
```

12. Yes
14. The instructions that follow the .IF AL < 3 && AL > 5 statement up through the .ENDIF statement execute if AL is less than 3 and AL is greater than 5. In other words, the statements don't execute for values of 3, 4, and 5.
16. Usually at locations B8000H–BFFFFH
18. To change to a new video mode, the new mode is loaded into AX before the INT 10H instruction is executed.
20.

```
.MODEL TINY
.CODE
.STARTUP
MOV   AX,13H       ;select mode 13H
INT   10H
MOV   AX,0A000H    ;address video display memory
MOV   ES,AX
MOV   DI,320*20    ;find 20 PELs from the top
CLD                ;select increment
MOV   CX,50        ;load count
.REPEAT
      PUSH  CX
      PUSH  DI
      MOV   CX,30
      .REPEAT
      MOV   AL,4
      STOSB
      .UNTILCXZ
      POP   DI
      ADD   DI,320
      POP   CX
.UNTILCXZ
```

```
MOV   AH,1        ;wait for any key
INT   21H
MOV   AX,3        ;back to video mode 3
INT   10H
.EXIT
END
```

22. The bit-mask register selects which bit or bits will change in each byte of the video display. Recall that each byte holds 8 picture elements.

24.
```
.MODEL TINY
.CODE
.STARTUP
MOV   AX,12H              ;select mode 12H
INT   10H
MOV   AX,0A000H           ;display video memory
MOV   DS,AX
MOV   DI,40               ;address top center
MOV   DX,3CEH             ;select middle bit in MMR
MOV   AX,1008H
OUT   DX,AX
MOV   CX,480
MOV   DX,3C4H
MOV   AL,2
OUT   DX,AL
INC   DX
.REPEAT
      MOV   AL,0FFH       ;enable all planes
      OUT   DX,AL
      MOV   AL,[DI]       ;read color
      MOV   BYTE PTR [DI],0
      MOV   AL,9          ;bright blue
      OUT   DX,AL
      MOV   BYTE PTR [DI],0FFH
      ADD   DI,80
.UNTILCXZ
MOV   AH,1                ;wait for any key
INT   21H
MOV   AX,3
INT   10H                 ;back to video mode 3
.EXIT
END
```

26.
```
.MODEL TINY
.CODE
.STARTUP
MOV   AX,12H      ;select video mode 12H
INT   10H
MOV   AX,0A000H   ;address video memory
MOV   DS,AX
CLD               ;select increment
MOV   DI,0
MOV   DX,3CEH     ;select all bits in MMR
MOV   AX,0FF08H
OUT   DX,AX
MOV   DX,3C4H
MOV   AL,2
OUT   DX,AL
```

```
INC   DX
MOV   CX,480
.REPEAT
      PUSH  CX
      MOV   CX,16
      MOV   AH,0
      .REPEAT
            PUSH  CX
            MOV   CX,5
            .REPEAT
            MOV   AL,0FFH      ;enable all planes
            OUT   DX,AL
            MOV   AL,[DI]      ;read color
            MOV   BYTE PTR [DI],0
            MOV   AL,AH        ;get color
            OUT   DX,AL
            MOV   BYTE PTR [DI],0FFH
            INC   DI
            .UNTILCXZ
            POP   CX
            INC   AH
      .UNTILCXZ
      POP   CX
.UNTILCXZ
MOV  AH,1           ;wait for any key
INT  21H
MOV  AX,3           ;back to video mode 3
INT  10H
END
```

CHAPTER 11

2. Eight bytes store the name and three bytes store the extension to the name.
4.
```
MOV AH,3CH
MOV CX,2
MOV DX,OFFSET FILES
INT 21H
```
6.
```
READ MACRO BUFFER,COUNT,HANDLE

     MOV AH,3FH
     MOV BX,HANDLE
     MOV CX,COUNT
     MOV DX,OFFSET BUFFER
     INT 21H

     ENDM
```
8.
```
     .MODEL SMALL
     .DATA
NAM  DB 'RAN.FIL',0
BUF  DB 100 DUP (0)
     .CODE
     .STARTUP
     MOV BP,200
     MOV AX,3D00H
     MOV DX,OFFSET NAM
```

```
     INT 21H
     JC  ERROR
     MOV BX,AX
MAIN1:
     MOV AH,40H
     MOV CX,100
     MOV DX,OFFSET BUF
     INT 21H
     JC  ERROR
     DEC BP
     JNZ MAIN1
     MOV AH,3FH
     INT 21H
     MOV AL,0     ;AL = 0 for no error
     JNC MAIN2
ERROR:
     MOV AL,1     ;AL = 1 for error
MAIN2:
     .EXIT
     END
```

10.
```
ACCESS   MACRO T,BUFFER,RECORD,HANDLE

         MOV  BX,HANDLE
         MOV  AX,100
         MOV  CX,RECORD
         MUL  CX
         XCHG DX,AX
         MOV  CX,AX
         MOV  AX,4200H
         INT  21H
    IF <T> = 'R'
         MOV  AH,3FH
    ELSE
         MOV  AH,40H
    ENDIF
         MOV  CX,100
         MOV  DX,OFFSET BUFFER
         INT  21H
         ENDM
```

16. The first file name is found in the data segment stored at the location addressed by DX as an ASCII-Z string.
18.
```
MOV AH,0EH
MOV DL,2
INT 21H
MOV AH,5AH
MOV CX,0
MOV DX,OFFSET NAME
INT 21H
```

20. The mouse is tested by first checking the INT 33H vector for an address other than 0000:0000. If another address exits, its contents are tested for an IRET instruction. If the IRET instruction is missing, function AX = 0000H is invoked using INT 33H, and the AX register is then tested for a 0000H. If 0000H is returned, the mouse is not present; otherwise, a mouse is present.
22. The right mouse button is pressed when the status of the mouse is checked (function 05H) and bit position 1 contains a 1.

CHAPTER 12

2. 35H
4. 25H
6. The COM (command) file must be organized so the first executable instruction is at offset 100H. The file must be organized as a single segment (tiny model).
8.

```
                .MODEL TINY
                .CODE
                .STARTUP
                MOV AL,0B6H     ;condition timer 2
                OUT 43H,AL
                MOV DX,12H      ;find count for timer 2
                MOV AX,34DCH
                MOV CX,3000
                DIV CX
                OUT 42H,AL      ;program timer 2 count
                MOV AL,AH
                OUT 42H,AL
                IN  AL,61H      ;enable speaker
                OR  AL,3
                OUT 61H,AL
                XOR CX,CX
                MOV ES,CX
                MOV DX,36       ;get clock tick count and add 36
                ADD DX,ES:[46CH]
                ADC CX,ES:[46EH]
MAIN1:
                MOV BX,ES:[46CH]
                MOV AX,ES:[46EH]
                SUB BX,DX
                SBB AX,CX
                JC  MAIN1       ;wait for 2 seconds
                IN  AL,61H      ;turn speaker off
                AND AL,0FCH
                OUT 61H,AL
                .EXIT
                END
```

10. The number stored at the TONE EQU 1000 is changed to change the frequency audio tone. When this is changed, the program must be assembled again.
12. The number of paragraphs (16-byte increments) indicates the length of the TSR. This value is passed to INT 21H function AH = 31H (terminate and stay resident) in the DX register.
14. A hot key is a multiple-key combination that activates a TSR.
16. By using INT 15H
18. Bit position 0
20. The video BIOS is very slow when compared to direct access.

CHAPTER 13

2. The 16-bit word integer has a value of ± 32K, the 32-bit short integer has a value of $\pm 2 \times 10^{+9}$, and the 64-bit long integer has a value of $\pm 9 \times 10^{+18}$.
4. The 32-bit single precision, the 64-bit double precision, and the 80-bit extended precision
6. (a) –3.75, (b) +0.28125, (c) +308, (d) +2.0, (e) +10, (f) +0.0

8. The microprocessor is idle or executing an integer instruction.
10. The coprocessor status word is copied into AX.
12. The SAHF instruction copies the contents of AH into the flag register after the coprocessor status register is loaded to AX with the FSTSW AX instruction. The flags can then be tested after the FCOM ST,ST(2) instruction executes.
14. FSTSW AX
16. In 80-bit extended precision form
18. Zero
20. The affine infinity control selects signed infinity and the projective infinity control selects unsigned infinity.
22. Double precision
24. The FST DATA instruction copies the top of the stack to memory location DATA.
26. FADD ST,ST(3)
28. FSUB ST(2),ST
30. The FDIV instruction divides ST(1) by ST, while the FDIVR divides ST by ST(1). This answer remains on the stack top with ST(1) popped off the stack.
32. The FTST instruction compares the stack top with zero, while the FXAM examines the stack top.
34. The IE (invalid error) bit
36. FLD1
38. FSTENV
40.

```
AREA PROC NEAR

     FLD L
     FMUL W
     FSTP A

AREA ENDP
```

42.

```
ROOT PROC NEAR

     MOV DI,OFFSET ROOTS
     MOV CX,9
R1:
     MOV AX,11
     SUB AX,CX
     MOV TEMP,AX
     FILD TEMP
     FSQRT
     FSTP WORD PTR [DI]
     ADD DI,4
     LOOP R1
     RET

ROOT ENDP

TEMP DW ?
```

44. The FSTSW causes a FWAIT. The FNSTSW does not.
46.

```
COS  PROC NEAR

     MOV TEMP,EAX
     FLDPI
```

```
        FDIV N180
        FLD TEMP
        FDIVR
        FCOS
        FSTP TEMP
        MOV EAX,TEMP
        RET

COS     ENDP

TEMP DD ?
N180 REAL4 180.0
```

48.
```
DO      PROC NEAR

        MOV  TEMP,EBX
        FLDPI
        FMUL TEMP
        FSTP TEMP
        MOV  EBX,TEMP
        RET

DO      ENDP

TEMP DD  ?
```

50.
```
        LOG10 PROC NEAR

             FLD1
             FXCH
             FYL2X
             FLD1
             FLD TEN
             FYL2X
             FLD1
             FDIVR
             FMUL
             RET

TEN        DD   10.0

        LOG10 ENDP
```

CHAPTER 14

2. The child is the program executed with the EXEC function. It is stored in the data segment at the location addressed by DX as an ASCII-Z string.
4. The parameter block contains the segment address of the environment block, the address of the command line, and two file-control blocks that are today obsolete.
6. HIMEM.SYS, EMM386.EXE, VCPI, and DPMI
8. EMM386.EXE
10. INT 67H
12. HIMEM.SYS
14. The DOS multiplex interrupt, function AX = 4300H, is used to determine whether XMS is installed.

16. VCPI allows access to both extended and expanded memory.
18. The DOS protected-mode interface, or DPMI, is provided by Windows.
20. The client is the program that uses the services provided by DPMI.
22. Character
24. (a) the console (CON) is the system keyboard and video display.
    (b) LPT1 is the parallel line printer port number 1.
    (c) COM2 is the serial interface number 2.
    (d) PRN is the standard printer port, usually LPT1.
26. Header, strategy, and interrupt
28. The address of the request header
30. Display the initialization message, initialize hardware and/or software used by the driver, and DOS expects the size of the driver in return.
32. The read command reads the character and signals the interface to clear it from its buffer. The nondestructive read command reads the character, but the interface buffer is not cleared.
34. The DPMI, provided by Windows, is installed if the DOS multiplex interrupt, called with AX = 1687H, returns a zero in AX.
36. The switch to protected mode is made by calling the DPMI entry address obtained by the DOS multiplex interrupt, function AX = 1687H.
38. The DOS INT 21H function 4CH terminates the application from protected mode or from real mode.

GLOSSARY

Abacus A mechanical calculator that uses beads for calculations. Invented by the ancient Babylonians before 500 B.C.

Absolute memory location An actual memory location, as in the instruction MOV AX,DS:[1234H], where 1234H is the offset address of the memory operand.

ADC (add-with-carry) A microprocessor instruction that adds the source and the carry flag to the destination.

Arithmetic coprocessor A special-purpose microprocessor designed to operate concurrently with the microprocessor. The arithmetic coprocessor processes arithmetic operations using signed integers, BCD data, or real numbers. Functions provided by the arithmetic coprocessor include addition, subtraction, multiplication, division, sine, cosine, tangent, and various other logarithmic and other functions.

ASCII (American Standard Code for Information Interchange) A 7-bit code, most commonly used for storing alphabetic and numeric data in a computer system. Variations include extended ASCII code.

ASCII-Z string A character string that is stored in the memory and ends with a null (OOH). The conventional character string ends with a carriage return (ODH) and line feed (OAH).

Assembler program A program that converts symbolic assembly language into the native binary machine language of the microprocessor. The symbolic input to the assembler is called the source program and the output from the assembler is called the object program.

ATTRIB A DOS external command that displays or changes the attributes associated with a file or directory. The attributes are R (read-only), H (hidden), S (system), and A (archive).

AUTOEXEC.BAT A batch program that automates the tedious task of loading programs into the memory system at boot time.

Auxilary carry flag (A) A flag that indicates conditions after a BCD addition. Used by DAA and DAS to adjust the contents of AL.

Base-plus-index addressing An addressing mode that uses a combination of two registers to indirectly address a memory operand. In the 80386–Pentium, any 32-bit register may be combined with any other 32-bit register (except ESP) to indirectly address the memory operand. Also available are register combinations [BX+SI], [BX+DI], [BP+SI], and [BP+DI].

Base relative-plus-index addressing An addressing mode that uses two registers plus a constant to indirectly address the memory operand. An example is the MOV AX,[BX+DI+3] instruction.

BCD (binary-coded decimal) A decimal number coded in 4-bit binary nibbles that are assembled in unpacked form as one digit per byte or in packed form as two digits per byte.

Big endian A method of storing a multiple-byte number in the memory so that the most significant byte is stored in the lowest-numbered memory location.

BIOS (basic I/O system) A collection of programs, stored in a read-only memory, that operate many of the I/O devices connected to the computer system. The system BIOS controls the keyboard and disk drives connected to the computer system. It provides service to the system clock, serial COM ports, and parallel LPT ports.

Bit A binary digit with a value of 1 or 0.

Block device A device that processes one block of data at a time, such as a disk driver.

Boot The act of starting a computer system, which loads the disk operating system into the memory for execution.

Boot sector The boot sector contains a program called the bootstrap loader and occupies the first sector of the outer track on the disk.

Bootstrap loader A program that loads DOS or any other operating system into the memory whenever power is applied to the system or whenever the control+alternate+delete key sequence is activated.

Bus A common set of connections or wires that thread the components in a computer system and carry the same type of information between these components.

Byte Generally, an 8-bit wide number. If it is in memory with parity, then it is nine bits wide: eight bits for data and the ninth bit for parity or error control. Note that the byte still contains eight bits of data.

CAD Computer aided drafting/design.

CALL A microprocessor instruction used to link to a procedure by pushing the return address on the stack and jumping to the procedure.

Carriage return An ASCII-coded character (0DH) that returns the print-head or cursor to the left margin of the paper or video display. The carriage return is entered by using the enter key or a control+M keyboard combination.

Carry flag (C) A bit in the flag register that holds the carry after an addition or the borrow after a subtraction. The carry flag is also used by DOS INT 21H to signal an error.

CD A DOS internal command that displays the current directory or allows a switch to be made to any other directory or subdirectory.

CDROM (compact disk read-only memory) A form of optical memory. The CDROM holds up to 660M bytes of digital data, 70 minutes of audio data, or any combination of the two.

CGA (color graphics adapter) A feature that allows color information to be displayed on the video monitor with a resolution of 320 pixels per raster line with 200 lines, or a resolution of 320 × 200. At this resolution, four colors are allowed: pink green, white, and black.

Character device A device that processes a single byte at a time in either binary or ASCII format. Typical character devices are keyboard, mouse, and printer.

Child directory A subdirectory displayed as a .. in a directory listing. When used with the EXEC function, *child* refers to a child process loaded into memory by the parent and executed.

CHKDSK A DOS external command that checks a disk for any defects. Never use this command from within Windows, even if at the DOS prompt from the DOS shell in Windows.

CISC (complex instruction set computer) A computer that uses an extensive set of instructions to accomplish programming tasks. The 8086–Pentium are examples of CISC machines.

Classic stack addressing An addressing mode used with the arithmetic coprocessor. The source operand is at ST and the destination operand is at ST(1). After the classic stack instruction is executed, the result remains at ST, but neither the source nor the destination remain in the coprocessor. The FSUB instruction is an example that subtracts ST(1) from ST and leaves the difference at ST.

Client The program or system that calls upon the host to provide service to the computer system. Typical clients are DOS and Windows; a typical host is DPMI.

Cluster A unit of size on a disk that refers to the number of sectors in each file allocation unit.

CMP (comparison) A microprocessor instruction that compares the source and destination operands. The comparison subtracts the source from the destination: the destination does not change, but the flags do.

Code segment (CS) A section of memory that holds the code or program. The code segment is addressed by the CS register in the real mode or by a descriptor selected by the CS register in the protected mode.

Command file A file, ending with the extension .COM, that contains a program which must be contained in one 64K-byte segment originated at location 100H.

CONFIG.SYS A file that configures the DOS-based personal computer system.

CPU The central processing unit, sometimes referred to as a microprocessor.

CS See *Code segment*

Cylinder A group of tracks. On a floppy disk, a cylinder consists of an upper and lower track. On a hard disk, a cylinder can be any number of tracks up to 16.

Data segment The section of memory that holds most of the data. The start of the data segment is addressed with the DS register in the real mode and with the descriptor selected by the DS register in the protected mode. Data may also be addressed by any other segment register, using segment override prefixes.

DB (define byte) An assembler directive that allows bytes of data to be stored in the computer system memory with the assembler program.

DD (define doubleword) An assembler directive that allows 32-bit doublewords to be stored in the computer system memory with the assembler program.

Decrement subtraction The act of subtracting one from the contents of any register or memory location.

Descriptor An 8-byte entry in the descriptor table that describes the memory segment base or starting address, its length, and access rights for the segment in the protected mode.

Device driver A special program installed in the CONFIG.SYS file with the DEVICE or DEVICEHIGH directive that provides a means of controlling installable devices. The device driver is a special COM file that contains the extension SYS or EXE. A device driver contains three parts: the header, strategy, and interrupt.

Direct addressing An addressing mode that directly addresses a memory location for the source or destination operand, using the address of the memory location or a label that represents the address.

Direction flag (D) A flag bit used to select whether the source/destination index registers increment or decrement with string instructions. The CLD (clear direction) instruction selects increment and the STD (set direction) instruction selects decrement.

DOS (disk operating system) The system that controls the operation of the personal computer and manages the information and programs stored on the disk.

Double-precision A form of floating-point storage that requires eight bytes of memory.

Doubleword A 32-bit binary number in the personal computer system.

DPMI The DOS protected-mode interface that allows DOS programs to switch to protected mode to access extended memory and other features provided by protected-mode operation. DPMI services are provided by Windows in a typical personal computer system.

DQ (define quadword) An assembler directive that allows 64-bit binary numbers to be stored in memory with the assembler program.

Driver A program designed to drive or control an I/O device.

DS See *Data segment*

DUP An assembly language directive that informs the assembler to store duplicates of bytes, words, doublewords, quadwords, or ten-bytes.

DW (define word) An assembler directive that allows 16-bit words of data to be stored in the computer system memory with the assembler program.

EAX A multipurpose 32-bit wide register, also known as the accumulator register. The EAX register is addressed as EAX, AX for 16-bit data, and as AH or AL for 8-bit data. The EAX register holds the product after multiplication and the dividend, quotient, and remainder for division. It is also used to address memory data.

EBP A multipurpose 32-bit wide register, also known as the base pointer register. The EBP register is addressed as EBP for 32-bit data and BP for 16-bit data. The EBP or BP register also address memory data.

EBX A multipurpose 32-bit wide register, also known as the base index register. The EBX register is addressed as EBX for 32-bit data, BX for 16-bit data, and as either BH or BL for 8-bit data. The BX or EBX register also addresses memory data.

Echo Send a character to the video display from the keyboard.

ECX A multipurpose 32-bit wide register, also known as the count register. The ECX register is addressed as ECX for 32-bit data, CX for 16-bit data, and as either CH or CL for 8-bit data. The ECX register holds a count for repeated string instructions, shift, rotate, and the LOOP/LOOPD instruction. The ECX register is also used to address memory.

EDI A 32-bit wide multipurpose register, also known as the destination index register. The EDI register is addressed as EDI for 32-bit data and as DI for 16-bit data. The EDI or DI register is used to address memory data.

EDX A multipurpose register, also known as the data register. The EDX register is addressed as EDX for 32-bit data, DX for 16-bit data, and as either DH or DL for 8-bit data. The EDX register holds part of the product after a multiplication and is also used in division. The EDX register also addresses memory.

EFLAGS A special-purpose extended flag register that indicates the condition of the microprocessor and is used to select certain operating conditions for the microprocessor.

EGA (enhanced graphics adapter) A short-lived video standard that displayed information with a resolution of 640 × 300. Most EGA display adapters allow either 16 or 256 colors.

EIP A 32-bit wide special-purpose register, also known as the instruction pointer register. The EIP register functions as a 16-bit register (IP) in the real mode and as a 32-bit register in the protected mode. The EIP register is used in conjunction with the CS register to address the next instruction to be executed in a program.

EISA (extended industry standard architecture) A feature designed to interface 8-, 16-, and 32-bit peripherals to the personal computer system.

EQU An assembly language directive that equates a label to data or another label.

ES See *Extra segment*

ESI A 32-bit wide multipurpose register, also known as the source index register. The ESI register is addressed as ESI for 32-bit data and as SI for 16-bit data. The ESI register is also used to address memory data.

ESP A 32-bit wide special-purpose register, also known as the stack pointer register. The stack pointer register is used in conjunction with the SS register to address the stack memory.

Execution file A file, which contains the extension .EXE, that is run from the DOS command line by typing the name of the program.

Expanded memory system (EMS) An older memory standard that allowed additional memory to be added to the systems area in 64K-byte banks located in an area called a page frame.

Exponent A part of a floating-point number that represents the position of the binary point.

Extended memory manager A program designed to allow DOS programs access to extended memory. The HIMEM.SYS program often provides a system with a procedure to access extended memory.

Extended memory system (XMS) The memory that exists above the first 1M byte in a personal computer system.

Extra segment A section of the memory that holds string destination data for string instructions or other data. The start of the extra segment is addressed by the ES register in the real mode and by a descriptor selected by the ES register in the protected mode.

FAT (file allocation table) A table that contains codes to indicate whether file allocation units are free, in use, or bad.

File handle A number that refers to a file once the file is opened or created. A file handle can also refer to the keyboard, video display, COM port, LPT port, or any driver program.

File manager A Windows program that replaces most of the functions selected from the DOS command line, such as COPY, DEL, and so forth.

File name extension An optional 3-character name, preceded by a period and the file name, that usually defines the type of file or program.

File pointer A 32-bit number that addresses any byte in a file. When a file is opened or created, the file pointer addresses the first byte in the file.

G bit See *Granularity bit*

Global descriptor table (GDT) A table containing descriptors used by all applications.

Global descriptor/variable A variable or descriptor that is available to all applications.

Granularity bit A bit located in the 80386, 80486 or Pentium descriptor that selects a multiplier of 1 or 4K. The multiplier is applied to the limit portion of the descriptor in the protected mode to select up to a 1M or 4G segment.

Graphical user interface (GUI) A computer interface that uses graphic bit-mapped data to display information instead of ASCII data, which is used in text-mode displays.

Head Refers to either the upper or lower side of a floppy disk and corresponds to the read/write head in the disk drive.

Header A portion of the device driver that contains the name of the driver and pointers used by the driver.

Hexadecimal code A code written in base 16 that contains the digits 0 through 9 and A through F.

High memory An area of real memory that extends from memory location 0FFFF0H to 10FFEFH when the HIMEM.SYS driver is installed and often holds much of the DOS program.

Home position The upper left corner of the video display screen.

Hot key A special key combination that accesses a software application.

Host program A program such as DPMI that provides an interface to the microprocessor and its resources.

Icon A picture that represents an application in the Windows environment.

Immediate Addressing An addressing mode that allows immediate constants, 8-, 16-, or 32-bits long, to be used as the source operand.

Increment addition The act of adding one to the contents of a register or memory location.

Interrupt A hardware- or software-generated call to a special procedure called an interrupt handler or interrupt service procedure.

Interrupt flag (I) A flag that controls the state of the INTR input pin used to request interrupts. If I = 0, interrupts are turned off. If I = 1, interrupts are turned on.

Interrupt handler A program that responds and services interrupts, sometimes called an interrupt service procedure.

Interrupt hook A way to intercept the normal interrupt to add features for the keyboard, clock-tick, and other interrupt driven I/O devices.

Interrupt procedure A part of a device driver that handles requests from DOS and controls the installable device.

Interrupt vector A 4-byte number that contains the address of an interrupt service procedure located in the first 1K byte of the memory system.

Intersegment A term that indicates movement between segments. In an intersegment jump or call, we jump to or call an address located in any segment.

Intrasegment A term that indicates movement within a segment.

ISA (industry standard architecture) A feature designed to describe the input/output bus in computer systems designed to interface either 8- or 16-bit peripherals.

K (kilo or kilobyte) 1,024 bytes in a computer system.

KIP Kilo-instructions per second.

Launch A term used with Windows that is equivalent to *execute* in DOS.

LDTR See *Local descriptor table*

Linear address An address generated by a program that is often converted to a physical address by the microprocessor.

Linear program A program that flows from the top to the bottom without any calls to procedures.

Linker program A program that links object modules and libraries to create an execution file or run-time program.

Little endian A method of storing multiple-byte numbers in memory where the least-significant byte is stored in the lowest-numbered memory location.

Local data/descriptor Data or a descriptor that is unique or only available to the current application.

Local descriptor table A table of descriptors used by a specific application.

Local bus An interface standard that allows 32- and 64-bit peripherals to be interfaced to the personal computer system.

LODS A microprocessor instruction that loads the accumulator from the data segment memory location addressed by the SI register. After the load operation, the contents of SI are either incremented or decremented, depending on the state of the direction flag.

M (megabyte) 1,024K bytes or 1,048,576 bytes in a computer system.

Machine language The native binary code that a microprocessor understands and uses to control its operation.

Mantissa The part of a floating-point number that contains the fraction or significand of the number. The mantissa is normalized and its value is between 0.0 and 1.1111.

MD (or MKDIR) A DOS internal command that makes a new directory or subdirectory.

Memory address A number, usually written in hexadecimal code, that describes the location of data in the memory system.

Micro-floppy disk A $3^1/_2''$ disk that records digital data as 720K bytes, 1.44M bytes, or 2.88M bytes per disk.

Microprocessor The controlling element in a microprocessor-based personal computer system.

Mini-floppy disk A $5^1/_4''$ disk that stores digital data as either 360K bytes or 1.2M bytes per disk.

MOV (move data) A microprocessor instruction that copies the contents of the source operand into the destination operand. The source never changes with the MOV instruction.

MSDOS Microsoft disk operating system.

Multipurpose register A register that has many purposes and can hold data of varying sizes and values. In the 80386, 80486, and Pentium, the multipurpose registers are EAX, EBX, ECX, EDX, EBP, EDI, and ESI.

NAN (not a number) An invalid floating-point number that has all ones in the exponent with a significand that is not all zeros.

Nibble A 4-bit wide binary number.

OFFSET An assembler directive that informs the assembler to use the offset address of a label in place of its contents. An example is MOV DI,OFFSET TOP, where the offset address TOP is loaded to DI in place of its contents.

Offset address Defines the distance above the start of the segment where the data or next instruction is located. The linear memory address is generated by a segment-plus-offset address combination.

Opcode (operation code) A binary or symbolic instruction that tells the microprocessor what operation to perform.

Overflow flag (O) The flag that indicates that the most recent addition or subtraction has resulted in an overflow condition.

Paragraph A 16-byte address boundary.

Parent When associated with directories, the *parent* is the directory and the child is a subdirectory. The parent directory is listed as a . in the directory display. When associated with the EXEC function, *parent* refers to the program that loads memory and executes another program.

Parity A count of ones, which indicates whether a number contains an even or an odd number of one-bits.

Parity flag (P) A bit in the flag register that indicates even or odd parity.

Partition A section of the surface of a hard disk drive that contains file storage space for an operating system or other logical disk drive.

PCDOS IBM's personal computer disk operating system.

Pixel (pel) A picture element; the smallest displayable part of a video image.

Pop-up program Another name for a program accessed by a hot key.

Procedure A group of instructions that usually perform one task. The instructions in a procedure are stored once in the memory, but can be used many times, from several points in a program. The CALL instruction links to the procedure and the RET instruction returns from the procedure.

Program A collection of instructions organized in such a manner as to perform and direct the computer to perform a useful task.

Program-invisible registers Registers that are not normally used by applications, but may be used during system programming to control the management of the memory and other hardware features of the microprocessor.

Program manager The main menu, replete with icons that represent applications, in the Windows program. The program manager is used to launch application windows or applications.

Program-visible registers Registers used during normal programming of the microprocessor, as specified by its instructions.

RAM (random access memory) Read/write memory that can store information and be changed by a program.

RD An internal DOS command that allows a directory or subdirectory to be erased, provided it contains no files or subdirectories.

Real memory The first megabyte of system memory.

Register Addressing An addressing mode that uses an 8-bit (AH, AL, BH, BL, CH, CL, DH, or DL), 16-bit (AX, BX, CX, DX, SP, BP, SI, or DI), or 32-bit (EAX, EBX, ECX, EDX, ESP, EBP, ESI, or EDI) register as an operand. Some instructions (MOV, PUSH, and POP) also allow the use of a segment (CS, DS, ES, or SS) register as an operand.

Register-Indirect Addressing An addressing mode that uses the contents of a register to hold the offset address of the memory operand. In the 8086–80286, registers BP, BX, SI, and DI address memory indirectly. In the 80386–Pentium, the EAX, EBX, ECX, EDX, EBP, ESI, and EDI registers are also used, to indirectly address the memory operand.

Register-Relative Addressing An addressing mode that uses one register to indirectly address memory plus a constant. An example is the MOV AX,[BX+3] instruction.

Relocatable data/program Data or a program that can be placed anywhere in the memory system and accessed or executed without modification to the data or program.

REPE/REPNE Special microprocessor instructions used with string comparisons to repeat the comparison while the compared condition is equal (REPE) or not equal (REPNE). These instructions, like REP, use CX as a counter.

Request header A variable-length memory buffer that contains commands and other information used by DOS to communicate between itself and a device driver.

RET A microprocessor instruction that returns from a procedure to the point just following the CALL instruction. This is accomplished by retrieving the return address from the stack.

RISC Reduced instruction set computer.

Root directory The main directory on a disk drive, which contains files and directory names. The root directory contains a limited number of available spaces for file and directory names.

Scaled-Index Addressing An addressing mode, available to only the 80386–Pentium, that uses two 32-bit registers to indirectly address memory. The second (index) register is multiplied by a scaling factor or 1, 2, 4, or 8 to indirectly address bytes, words, doublewords, or quadwords of memory data.

Scaled-index byte A part of a binary machine language instruction that defines the scaling factor, base, and index registers for the scaled-index addressing mode.

Sector A data subdivision on a disk memory that normally stores 512 bytes of data.

Segment address The beginning address of a 64K byte segment of the memory when the microprocessor operates in the real mode. In the protected mode, the segment starting address is selected by the base address in a descriptor and, depending on the microprocessor, can be up to 4G bytes in length.

Selector A 13-bit part of the segment register in the protected mode that addresses a descriptor from one of two descriptor tables (local and global).

Sign flag (S) A flag that indicates the sign of the result of an arithmetic or logic operation. If S = 0, the result is positive. If S = 1, the result is negative.

Single-precision data A form of floating-point data that requires four bytes of memory for storage.

Software A term that refers to a computer's programs.

Spawning The act of loading a child process into the memory system by the EXEC function.

Special-purpose computer A computer system designed to perform one task. An example is the ingition-control computer in an automobile.

Special-purpose registers Registers designed to perform a special task in a microprocessor.

SS See *Stack segment*

Stack segment A section of the memory that holds the stack data used by PUSH, POP, CALL, and RET. The start of the stack segment is addressed by the SS register in the real mode, and by the descriptor selected by the SS register in the protected mode.

STOS A microprocessor instruction that stores the contents of the accumulator into the extra segment memory location addressed by the DI register. After the data are stored, the contents of DI increment or decrement depending on the state of the direction flag.

Subroutine See *Procedure*

Switch A term that applies to the / when it follows a DOS command. The switch applies options that are available to the DOS command such as /? for help.

Symbolic memory location A location addressed by a symbolic label in place of the actual memory location. An example is JMP UP, where UP is a symbolic memory location.

System area A section of memory that contains system programs stored on read-only memory, video memory, and other system components.

Terminate and stay resident A term that applies to a program left in memory and accessed via an interrupt or hot key.

TPA (transient program area) The place programs are stored for execution in a computer system. In the personal computer, this area is located in the first 640K bytes of the memory system.

Track A concentric, nonspiraling ring that stores data on a floppy or hard disk memory.

TSR See *Terminate and stay resident*

Two's complement form A method of storing a negative-signed binary number in the memory system.

Upper memory A section of the memory, located in the system area, that is added by the EMM386.EXE program.

Utility program See *External DOS command*

Variable-port address An I/O address, located in register DX, that can be varied and ranges in size from 0000H to FFFFH.

VCPI (virtual control program interface) An interface that allows access to both extended and expanded memory if installed with EMM386.EXE (included with DOS version 6.X). The extended memory resources are allocated as needed to either extended or expanded memory by this interface.

VESA local bus Video electronics standards association. See also *Local bus.*

Video BIOS Programs that control the video display stored on a read-only memory at system memory locations C0000H–C7FFFH.

VGA (variable graphics array) A video display standard, used with computer systems, that has a resolution of 640 pixels across a line with 480 raster lines, written as 640 × 480.

Wild-card A character that acts like another character or characters at the DOS command line. The ? is a wild card used to replace a single character and the * is a wild card that replaces an entire file name or file name extension.

WIN.INI The file used to initialize Windows, much the same way CONFIG.SYS initializes DOS.

Windows Microsoft's graphical user interface, which allows easier access and use of personal computers than DOS.

Word A 16-bit binary number in the personal computer system.

WYSIWYG What you see is what you get.

XMS (extended memory system) The memory that exists above the first IM byte in a personal computer system.

Zero flag (Z) A flag that indicates whether the result from an arithmetic or logic operation is zero. If Z = 0, the result is *not* zero. If Z = 1, the result *is* zero.

INDEX

FRAMINGHAM STATE COLLEGE
3 3014 00100 2247